# Extractive Industries

UNU World Institute for Development Economics Research (UNU-WIDER) was established by the United Nations University as its first research and training centre and started work in Helsinki, Finland, in 1985. The mandate of the institute is to undertake applied research and policy analysis on structural changes affecting developing and transitional economies, to provide a forum for the advocacy of policies leading to robust, equitable, and environmentally sustainable growth, and to promote capacity strengthening and training in the field of economic and social policymaking. Its work is carried out by staff researchers and visiting scholars in Helsinki and via networks of collaborating scholars and institutions around the world.

*United Nations University World Institute for*
*Development Economics Research—UNU-WIDER*
*Katajanokanlaituri 6B, 00160 Helsinki, Finland*
*www.wider.unu.edu*

# Extractive Industries

## The Management of Resources as a Driver of Sustainable Development

Edited by
Tony Addison and Alan Roe

A study prepared by the United Nations University World Institute for Development Economics Research (UNU-WIDER)

OXFORD
UNIVERSITY PRESS

OXFORD
UNIVERSITY PRESS

Great Clarendon Street, Oxford, OX2 6DP,
United Kingdom

Oxford University Press is a department of the University of Oxford.
It furthers the University's objective of excellence in research, scholarship,
and education by publishing worldwide. Oxford is a registered trade mark of
Oxford University Press in the UK and in certain other countries



First Edition published in 2018



Published in the United States of America by Oxford University Press
198 Madison Avenue, New York, NY 10016, United States of America

British Library Cataloguing in Publication Data
Data available

Library of Congress Control Number: 2018944297

ISBN 978–0–19–881736–9

To Lynda and Susan, for all their encouragement and support.

# Foreword

For some developing countries natural resource wealth has turned out to be a curse rather than a blessing. Economic growth based exclusively on natural resources—such as oil, gas, and minerals—is often of a very narrow kind, failing to provide opportunities for much of the population. The wealth from a nation's extractive industries needs to be carefully managed if inclusive and sustainable growth is to be achieved. Resource wealth has often been associated with political instability, corruption and the non-transparent use of the revenues for private gain instead of national development. Many low- and middle-income countries continue to struggle with their resource wealth, with conflict and civil war being the unfortunate outcome for some.

This natural resource curse, and all that it brings, is a motivation for UNU-WIDER's Extractives for Development project—launched in early 2016 and led by Tony Addison and Alan Roe—bringing together a network of researchers, practitioners, and policy makers, international and local, government and corporate, to share their accumulated learning. By comparing experiences of the extractive industries across countries—opportunities and challenges, successes and failures—the project has built a comprehensive body of knowledge, potentially transferable between countries when suitably adapted to local circumstances. This book will help national policy makers and their international partners in the task of creating development strategies that use resource revenues for inclusive and sustainable development, better manage the macroeconomic risks, and prepare their societies for the impact on the extractive sectors of the transition from fossil fuels to renewable energy and other low-carbon technologies.

I sincerely thank the contributors for their studies which share with us their individual expertise within the field, and the editors, Tony Addison and Alan Roe, for bringing this academic work to full fruition and particularly for their analytical and authorship skills, evidenced so clearly in this fascinating book.

UNU-WIDER gratefully acknowledges the support and financial contributions to its research programme by the governments of Finland, Sweden, and the United Kingdom. Without this vital funding our research and policy advisory work would be impossible.

*Finn Tarp*
*Director, UNU-WIDER*
*Helsinki, June 2018*

# Preface

This book is an output of the UNU-WIDER 2014–18 work programme, entitled 'Transformation, Inclusion and Sustainability: Creating and Sharing Knowledge for Development'. The topic of extractives and development touches almost every aspect of development, and is relevant to each of the three themes of the UNU-WIDER programme. It is also highly relevant to the United Nations' Sustainable Development Goals (SDGs), and especially to UN member states which are looking to strengthen the contribution that the extractive sectors can make to national development.

Transforming the structure of economies to achieve higher rates of economic growth and greater resilience to shocks has proven to be more difficult in resource-rich countries than was expected in the early years of development thinking and practice from the 1950s onwards. Many such countries are today more dependent on the extractive industries despite, in many cases, having achieved higher levels of per capita income. Other countries have seen their living standards slide as political instability and conflict have taken hold. While some countries have managed to use the revenues from extractive resources to advance social inclusion by investing in education, healthcare, and development infrastructure, many resource-rich countries have seen social exclusion rather than greater inclusion.

Today there is an additional challenge. Extractives sectors in all countries need to reduce both their own environmental footprints and especially their greenhouse gas emissions. Countries and extractive companies (both minerals and oil and gas) need to prepare for the accelerating shift now underway from fossil fuels to renewable energy, and for the multiple revenue and other economic consequences of a global transition to low-carbon pathways for economies and societies. In sum, the subject of extractives and development has an undoubted and increasing relevance for the goals of transformation, inclusion, and sustainability—the three pillars of UNU-WIDER's work programme.

In addition to this book, the extractives for development project at UNU-WIDER has many other outputs, and we plan to add further to these in the coming years. Over more than thirty years, UNU-WIDER's research projects and conferences have generated a very large amount of knowledge on a wide range of development policy issues. This knowledge is shared via our website www.wider.unu.edu. There you can also find links to many other research papers, policy briefs, and blogs, as well as to the videos that UNU-WIDER increasingly produces. These include interviews with many of the contributors to this book, giving readers the opportunity to learn more about the topic. The website also provides users with ample opportunities to engage with UNU-WIDER through our social media outlets and to keep up-to-date with our work, including that on extractives and development.

In its more than thirty years of existence, UNU-WIDER has engaged with a very large number of researchers and practitioners. This book has enlarged and deepened our network further. It has been excellent to work with Alan Roe as the project's co-leader, and the book's co-editor. He brings long-standing expertise on the subject, very clear insight into the ways that development policy works, and has been central to the design, construction, and process of preparing this UNU-WIDER book.

There are few subjects as important as development policy—getting it right can lift millions from poverty. At the same time, development in all its dimensions is the subject of a continual conversation with many actors across our world, offering many different perspectives and experiences. We at UNU-WIDER believe that this book offers a useful contribution to that debate, drawing as it does on a large amount of expertise and experience.

*Tony Addison*
*Chief Economist-Deputy Director, UNU-WIDER*
*Helsinki, June 2018*

# Acknowledgements

Many people, the UNU-WIDER team and beyond, have been engaged in designing, developing and bringing this book to its successful conclusion. Finn Tarp, UNU-WIDER's director, provided encouragement as well as valuable guidance throughout our endeavour. Support from UNU-WIDER staff was outstanding. Lorraine Telfer-Taivainen, UNU-WIDER's publications 'maestro', drew on years of publications experience to guide this large book through to its successful completion, especially in the final pre-publication stage. Thorunn Sigurdardottir and Ann-Mari Sundsten ably handled the practical management of the project in its earlier and later stages. Anna-Mari Vesterinen and later Ans Vehmaanperä did great work in helping to prepare many of the studies for early exposure in WIDER Working Papers. Our thanks are also due to the UNU-WIDER support teams, including: Lay Poh Allonen, Mayra Da Silva De Gouveia, Dominik Etienne, Amanda Hajnal, Tuuli Levit, Anne Ruohonen, Sherry Ruuskanen, Paul Silfvenius, and Marian Vo. The project website, developed soon after a first project meeting, has benefitted from excellent editing and production from the UNU-WIDER communications and IT teams, including Kennedy Ambang, Anu Laakso, Bruck Tadesse, and Annett Victorero. The website will continue to be a source of much valuable supplementary information.

A project meeting was held in Helsinki on 11–12 April 2016, to share and discuss initial ideas and early drafts of papers. Roger Williamson chaired that meeting with great skill combined with his usual aplomb. Participants at the meeting (many of whom later contributed chapters to this book) included: Dede Adzovi, Wisdom Akpalu, Sheila Agyemang, Benjamin Anang, Hernan Araneda, Toni Aubynn, Millicent Awuku, Mahamudu Bawumia, Mark Beare, Ruth Greenspan Bell, Joanna Buckley, Jim Cust, Evelyn Dietsche, Lasse Djahlin, Samantha Dodd, Mark Essex, Liesel Mack Filgueiras, Holger Grundel, Patrick Heller, Tony Hodge, Jussi Huopaniemi, Malik Iddrisu, Maja Jakobsen, Anton Löf, Neil McCulloch, Catherine Macdonald, Kathryn McPhail, Angel Mondoloka, Senia Nhamo, Robert Osei, Richard Osei, Olle Östensson, James Otto, Nadia Ouedraogo, Caroline Schimanski, Jeffery Round, Paul Stevens, Adam Swallow, Kathryn Tomlinson, Nick Travis, Rick van der Ploeg, and Sophie Witter. Bob Denham and his team from Econ Films conducted video interviews with many of the meeting participants for the UNU-WIDER video series on Extractives and Development.

Adam Swallow and Katie Bishop from Oxford University Press gave us very useful advice in planning and organizing the book for publication. The constructive suggestions of OUP's anonymous peer reviewers also helped to improve the book in several ways. We are delighted that the book is included in the UNU-WIDER series with OUP.

Our contributing authors dealt with our numerous editorial—and no doubt occasionally irritating—interventions with great patience and have provided us with excellent co-operation throughout the process. Together they constitute an immense and very diverse source of wisdom and insight. We have been very fortunate to have been able to draw on such a team and we thank them most sincerely for all for their hard work and assistance in bringing this study to publication.

Finally, we would like to thank Lynda Addison and Susan Roe for their great encouragement and support, and of course for their patience over the many hours when they lost us to the labours involved. This book is dedicated to them.

*Tony Addison and Alan Roe*
*June 2018*

# Contents

# List of Figures

# List of Tables

# List of Abbreviations

| | |
|---|---|
| ACEP | African Centre for Energy Policy |
| AfDB | African Development Bank |
| AMV | African Mining Vision |
| API | American Petroleum Institute |
| ARM | Alliance for Responsible Mining |
| ASM | artisanal and small-scale mining |
| AU | African Union |
| BBOP | Business and Biodiversity Offsets Programme |
| BEITI | Bayelsa Expenditure and Income Transparency Initiative |
| BL | backward linkage |
| BTU | British thermal unit |
| CAP | Community Action Plan |
| CCCMC | China Chamber of Commerce of Metals Minerals and Chemicals Importers and Exporters |
| CCD | corporate community development |
| CCS | carbon capture and storage |
| CCSI | Columbia Center on Sustainable Investment |
| CDA | Community Development Agreement |
| CEIM | Centro de Entrenamiento Industrial y Minero |
| CFSI | Conflict-Free Sourcing Initiative |
| CFSP | Conflict-Free Smelter Program |
| CGE | computable general equilibrium |
| CLM | China Luanshya Mine |
| CMV | Country Mining Visions |
| CNMC | China Nonferrous Metal Mining Corporation |
| CRCF | Copper Revenue Compensation Fund |
| CSER | corporate, social, and environmental responsibility |
| CSIRO | Commonwealth Scientific and Industrial Research Organisation |
| CSF | Copper Stabilization Fund |

| | |
|---|---|
| CSF | Critical Success Factors |
| CSO | civil society organization |
| CSP | concentrated solar power system |
| CSR | corporate social responsibility |
| CT | cash transfer |
| DA | district assembly |
| DAC | Development Assistance Committee |
| DDF | District Development Funds |
| DDI | Diamond Development Initiative |
| DFID | Department for International Development |
| DJSI | Dow Jones Sustainability Index |
| DRC | Democratic Republic of the Congo |
| ECZ | Environmental Council of Zambia |
| EEZ | Exclusive Economic Zone |
| EIA | environmental impact assessment |
| EICC | Electronic Industry Citizenship Coalition |
| EIR | Extractive Industries Review |
| EITI | Extractive Industries Transparency Initiative |
| ELAW | Environmental Law Alliance Worldwide |
| EMP | environmental management plan |
| EPA | Environmental Protection Agency |
| ERP | Economic Recovery Program |
| ESG | environmental, social, and governance |
| ESIA | Environmental and Social Impact Assessment |
| ESSF | Economic and Social Stabilization Fund |
| ETS | emissions trading scheme |
| ETP | Energy Technology Perspective |
| EWURA | Energy and Water Utility Regulatory Authority |
| FDI | foreign direct investment |
| FL | forward linkage |
| FOSTER | Facility for Oil Sector Transparency and Reform |
| FPIC | free, prior, and informed consent |
| FQM | First Quantum Minerals Ltd |
| GBV | gender-based violence |
| GCC | Gulf Cooperation Council |
| GDP | gross domestic product |

| | |
|---|---|
| GFS | Government Finance Statistics |
| GHG | greenhouse gas |
| GMI | Global Mining Initiative |
| GMoU | Global Memorandum of Understanding |
| GNI | gross national income |
| GOGIG | Ghana Oil and Gas for Inclusive Growth |
| HDPE | high-density pipe |
| HIC | high-income country |
| HIE | high-income economies |
| HIPC | heavily indebted poor countries |
| HSE | Health, Safety, and Environment |
| ICMC | International Cyanide Management Code |
| ICMM | International Council on Mining and Metals |
| ICRG | International Country Risk Guide |
| ICTD | International Centre for Tax and Development |
| IEA | International Energy Agency |
| IFC | International Finance Corporation |
| IFI | international finance institution |
| IIED | International Institute for Environment and Development |
| IMF | International Monetary Fund |
| INDC | Intended Nationally Determined Contributions |
| I–O | input–output |
| IOC | international oil company |
| IP | indigenous peoples |
| IPCC | Intergovernmental Panel on Climate Change |
| IPIECA | International Petroleum Industry Environmental Conservation Association |
| IRMA | Initiative for Responsible Mining Assurance |
| ISI | import substitution industrialization |
| KMC | Kalulushi Municipal Council |
| KPCS | Kimberley Process Certification Scheme |
| LCA | Local Content Act |
| LCD | Local Contractor Development |
| LCDF | Lumwana Community Development Forum |
| LGS | Licensed Gold Exporter |
| LIC | low-income countries |

| | |
|---|---|
| LIE | low-income economies |
| LMIE | lower-middle-income economies |
| LNG | liquified natural gas |
| LSM | large-scale mining companies |
| MCI | Mining Contribution Index |
| MCI-W | Mining Contribution Index WIDER |
| MCM | Mopani Copper Mines |
| MCRB | Myanmar Centre for Responsible Business |
| MDC | Maputo Development Corridor |
| MDRI | Multilateral Debt Relief Initiative |
| MDTF | Multi-Donor Trust Fund |
| MEM | Ministry of Energy and Minerals |
| MIC | middle-income countries |
| MMSD | Mining, Minerals and Sustainable Development Project |
| MOF | Ministry of Finance |
| MOM | self-mobilization, self-organization, and self-management |
| MPD | Mining: Partnerships for Development |
| MVM | Mineral Value Management |
| NBER | National Bureau of Economic Research |
| NDC | National Democratic Congress |
| NDRC | National Development and Reform Commission |
| NEPA | National Environmental Policy Act |
| NFCA | Non-Ferrous China Africa |
| NGO | non-governmental organization |
| NNPC | Nigerian National Petroleum Corporation |
| NOC | national oil company |
| NPP | New Patriotic Party |
| NRC | Natural Resource Charter |
| NRGI | Natural Resources Governance Institute |
| NRTEE | National Round Table on the Environment and the Economy |
| O&M | operations and maintenance |
| O&OD | Obstacles and Opportunities for Development |
| OAPEC | Organization of Arab Petroleum Exporting Countries |
| OCI | Oil Change International |
| ODA | overseas development assistance |
| OECD | Organisation for Economic Co-operation and Development |

| | |
|---|---|
| OEM | original equipment manufacturer |
| ODI | Overseas Development Institute |
| OPEC | Organization of Petroleum Exporting Countries |
| OPM | Oxford Policy Management |
| OSI | Open Society Institute |
| PACI | Partnering Against Corruption Initiative |
| PDAC | Prospectors and Developers Association of Canada |
| PDVSA | Petróleos de Venezuela |
| PEA | political economy analysis |
| PFM | public finance management |
| PIAC | Public Interest Accountability Committee |
| PIH | permanent income hypothesis |
| PNG | Papua New Guinea |
| PPP | public–private partnership |
| PPP | purchasing power parity |
| PPSP | public–private social partnership |
| PRA | participatory rural appraisal |
| PRF | Pension Reserve Fund |
| PRMA | Petroleum Revenue Management Act |
| PSA | production-sharing agreement |
| PV | photovoltaic |
| PWYP | Publish What You Pay |
| REi | Resource Endowment initiative |
| RER | real exchange rate |
| RMDI | Responsible Mineral Development Initiative |
| R-SNDP | Revised Sixth National Development Plan |
| RWG | Resettlement Working Group |
| RWI | Revenue Watch Institute |
| SAP | Structural Adjustment Programme |
| SB | structural balance |
| SDC | Swiss Agency for Development Co-operation |
| SEC | Securities and Exchange Commission |
| SI | social investment |
| SIA | social impact assessment |
| SITC | Standard International Trade Classification |
| SLO | social licence to operate |

| | |
|---|---|
| SMT | specific mining tax |
| SNPC | Société Nationale des Pétroles du Congo |
| SOE | state-owned enterprise |
| SSM | small-scale miner |
| SWF | sovereign wealth fund |
| SWIA | sector-wide impact assessment |
| TL | total linkage |
| TPDC | Tanzania Petroleum Development Corporation |
| UAE | United Arab Emirates |
| UMIE | upper-middle-income economies |
| UNCTAD | UN Conference on Trade and Development |
| UNDP | UN Development Programme |
| UNEP | United Nations Environment Programme |
| UNFCCC | United Nations Framework Convention on Climate Change |
| UN-SDGs | UN Sustainable Development Goals |
| VPs | Voluntary Principles on Security and Human Rights |
| WEF | World Economic Forum |
| WIOML | World Initiative of Mining Lawyers |
| WTO | World Trade Organization |
| WWF | World Wildlife Fund for Nature |
| ZCCM | Zambia Consolidated Copper Mines |

# Notes on Contributors

**Isabel Aché Pillar** has been head of Instituto Credit Suisse Hedging-Griffo in Brazil since March 2017, after three years as social and political analyst at Vale SA, and six years as CSR specialist at Vale Foundation. She holds a Master's in economics from IBMEC/Rio de Janeiro and a BA in economics from Universidade Federal do Rio de Janeiro (UFRJ). Since 2015, she has been a member of the Network of Political Action for Sustainability.

**Tony Addison** is Chief Economist-Deputy Director of UNU-WIDER in Helsinki. He was previously: professor of development studies, University of Manchester; executive director of the Brooks World Poverty Institute (BWPI) University of Manchester; associate director of the Chronic Poverty Research Centre (CPRC). He has also held positions at the University of Warwick, the School of Oriental and African Studies (London University), and the Overseas Development Institute (London). He has published widely in books and journals on the issues of post-conflict reconstruction, debt relief, fiscal policy, poverty dynamics, development finance, and foreign aid.

**Toni Aubynn** has served as the CEO of the Ghana Minerals Commission, 2014–17. Prior, he was the CEO of the Ghana Chamber of Mines, the main advocacy body of Ghana's mining industry. He has held various senior executive positions at Tullow Oil Ghana Limited, and the South Africa-headquartered Gold Fields, the leading gold producing company in Ghana. He is a former post-graduate student of the Universities of Oslo and Tampere and a fellow of the United Nations University, Institute of Development Studies, University of Helsinki and Tokyo University. His many academic publications have been in the areas of mining and policy, mining and sustainable development, small-scale mining, and elections and democracy.

**Mahamudu Bawumia** is Vice President of the Republic of Ghana. Earlier he was a lecturer in monetary economics and international finance at the Emile Woolf College of Accountancy in London, economist at the Research Department of the IMF in Washington, DC, and assistant professor of economics at Hankamer School of Business, Baylor University, Texas, where he also received the Young Researcher Award in 1998 and made the list of *Who is Who Among America's Teachers* in 1999. He later joined the Bank of Ghana, rising through the ranks from senior economist to deputy governor. He holds a Master's degree in economics from Oxford University, and a PhD in economics from Simon Fraser University, Vancouver. His areas of specialization include macroeconomics, international economics, development economics, and monetary policy.

**Ruth Greenspan Bell** is Public Policy Scholar at Woodrow Wilson International Center for Scholars in Washington, DC with a long-standing focus on the governance issues inherent in managing pollution and greenhouse gas emissions, domestically and internationally. She worked extensively on these issues in the former Soviet bloc and later in Asia, first at the US Environmental Protection Agency, and later at Washington, DC think tanks. Bell publishes extensively, addressing policy and environmental audiences (*Foreign Affairs and Issues in Science and Technology* among the periodicals); she was co-author of *Building International Climate Cooperation: Lessons from the Weapons and Trade Regimes for Achieving International Climate Goals*). Bell is a long-standing member of the Council on Foreign Relations.

**Joanna Buckley** is a consultant at Oxford Policy Management with expertise in economic and private sector development. Over the last decade, she has conducted a wide range of research and consultancy assignments in the areas of conflict sensitivity, job creation, stabilization, and social exclusion. She served as project manager of FOSTER during 2015 and 2016.

**James Cust** was previously the director of research and data at the Natural Resource Governance Institute and the founding staff member of the Natural Resource Charter. He now is based in the Office of the Chief Economist, Africa Region at the World Bank, and an external research associate of the Oxford Centre for the Analysis of Resource Rich Economies (OxCarre) at the Department of Economics in Oxford. His research examines the role of government and governance in harnessing resources for growth and poverty reduction. He holds a DPhil (PhD) in economics from the University of Oxford, an MSc from Oxford, and a BA from Cambridge.

**Evelyn Dietsche** is a non-resident senior research fellow with UNU-WIDER and an associate fellow of Chatham House. She has twenty years of practical work experience in development and the extractive resources sector, which she has combined with thought-leading contributions to international research projects on these two sectors. Her interests and strengths lie with the political-economic and structural challenges of natural resource exploitation.

**Samantha Dodd** was formerly a consultant at Oxford Policy Management and is now an adviser with the Behavioural Insights team at the London School of Economics. A South African national, her academic qualifications are multi-disciplinary, spanning economics, psychology, and international development. She has worked extensively on mining-related projects for a number of organizations, including DFID and ICMM.

**Magnus Ericsson** is a veteran observer of the global mining sector, co-founder of the Raw Materials Group in Stockholm, and Consulting professor at the Luleå University of Technology.

**Liesel Mack Filgueiras** has been working at Vale for twelve years. She is responsible for Vale Human Rights Agenda and Indigenous Community Relations. She joined Vale Foundation in 2005 and in 2007 joined the sustainability area at Vale. Liesel has completed executive training on extractives and sustainable development at Columbia University, leadership training at MIT, and has a Master's degree in business administration and a graduation degree in psychology, both from PUC-RIO.

**Diego Calderón Guajardo** is a Chilean economist. He holds a BA and MA in economics from the University of Chile. Currently, he is pursuing a PhD in economics at the University of Warwick.

**Håvard Halland** is a visiting scholar at the Stanford Global Projects Center, Stanford University, where his research focuses on strategic investment funds and sovereign wealth funds. From 2012 to 2018 he was a senior economist at the World Bank, where he provided advice on fiscal management to governments of resource-rich countries. He has published widely on the public economics and management of oil and mineral revenues. He earned a PhD in economics from the University of Cambridge.

**Patrick R. P. Heller** is an advisor at the Natural Resource Governance Institute and a senior visiting fellow at the Center on Law, Energy, and Environment at the University of California, Berkeley. He advises governments and civil society groups in Africa, Asia, and Latin America on good governance and legal frameworks for extractive industries. He has published and lectured extensively on the roles of state-owned enterprises, legislative reform, and the management of oil and mineral contracts.

**Mark Henstridge** is Chief Economist at Oxford Policy Management. He is programme director of two major academic research programmes—on 'Economic Development and Institutions', and on 'Research into Improving Systems of Education' (RISE). He also works on a range of issues in development economics and policy diagnostics. He has previously been macroeconomic policy advisor in the Ministry of Finance and Economic Planning in Uganda, and an economist at the IMF and at BP Plc; he was also an ODI fellow in Uganda.

**R. Anthony Hodge** is currently a senior scholar in residence at the Robert M. Buchan Department of Mining Engineering, Queen's University in Kingston, Ontario. During 2008–15 he served as president of the International Council on Mining and Metals. His interests and skills lie in grasping both technical and social dimensions of achieving human and ecosystem well-being. He has continuously sought common ground between actors, integrating multiple values into the solutions to some of the most difficult socio-technical-environmental-financial challenges facing society.

**Maja Jakobsen** is the manager of Oxford Policy Management's Natural Resources and Energy portfolio. Since joining Oxford Policy Management, Maja has been engaged in a variety of projects with a focus on natural resources and social sectors. She has carried out an analysis of contributions from mining to economic and social outcomes in the Democratic Republic of the Congo and Zambia; undertaken a study of the sustainability of Indonesia's recent growth in the light of the latest commodity boom; contributed to an analysis of vulnerabilities to corruption in NORAD's Oil for Development (OfD) programme in Uganda, Mozambique, and Bolivia; carried out impact evaluations of social protection programmes; and captured community perceptions using qualitative interviewing techniques and tools. She holds an MSc in economics from the University of Copenhagen and an MSc in development economics from the School of Oriental and African Studies in London.

**Glada Lahn** is a senior research fellow with the Energy, Environment & Resources Department, Chatham House, London. Her research is in economic and energy transitions in hydrocarbons-export-dependent countries, energy access, and energy and water resilience in displacement-affected areas. She has published widely in the fields of extractives governance, investment, and sustainable energy policy, most recently focusing on the implications of climate policy and technology shifts for hydrocarbons growth strategies.

**Anton Löf** works as a consultant for both the private and public sector, including various international organizations. His research work focuses on the iron ore market and downstream processing.

**Olof Löf** is a researcher and has previously worked for the Raw Materials Group in Stockholm, Sweden.

**Catherine Macdonald's** longstanding sensitivity to the importance of gender in human development was demonstrated by her doctorate on social change among female-headed households in Tanzania, completed in 1996. In the following twenty years, she has worked in the extractives sector community development field in many countries, particularly in Asia and Africa. She specializes in encouraging community participation, especially of women, in sustainable local development planning.

**Neil McCulloch** is a development economist with focus on the political economy of energy in developing countries. He was the project director of the Facility for Oil Sector Transparency (FOSTER) in Nigeria and has also done extensive work on the political economy of fuel subsidies, in both Indonesia and Nigeria, and the political economy of aid for power sector reform in sub-Saharan Africa and South Asia. Previously, he was the director of the Economic Policy Programme at Oxford Policy Management and, before that, the lead economist of the Australian Aid programme in Indonesia.

**Kathryn McPhail**, as the CEO of EnergyCC, is a recognized industry leader on matters relating to sustainability. Formerly a staff member of the World Bank Group, she has also worked as an executive with various oil and gas and mining industry FTSE100 companies. She is now a governor of the University of Aberdeen and serves on its Remunerations Committee; and a member of the Technical Advisory Group of the Natural Resources Governance Institute, the United Nations Sustainable Development Solutions Network on extractives and land resources, and the Singapore Institute of Directors where she also serves on its Sustainability Committee. She is also a former member of the World Economic Forum's Global Agenda Councils.

**Angel Mondoloka** is a development consultant with a thirty-year career spanning several jurisdictions. His passion is to help Zambia—and other countries in Southern Africa—leverage the political, economic, and regulatory changes in the global marketplace to, hence, reposition themselves on a critical path towards inclusive growth and sustainable development. Angel has consulted to governments, development agencies, international and local NGOs, and private corporations. He has written authoritatively on many development topics.

**Olle Östensson** is an independent consultant and writer specializing in development and governance issues around extractive industries. Both his consultancy work and his main research interests concern mining taxation and legislation, employment in mining, the relationship between extractive industries and the rest of the economy, and commodities trade issues.

**James M. Otto** is an independent mining lawyer and mineral economist and has worked in over sixty nations for governments, the private sector, multi-lateral institutions, and universities. He has undertaken a wide variety of natural resources assignments related to: the development of national mining policies, laws, model agreements and fiscal systems; sustainable development; environmental impact mitigation; mineral-sector-driven poverty alleviation, and country risk analysis.

**Andreia Rabetim** is experienced in public–private social partnerships and fundraising for social investments, acting for more than twenty years in the third sector. She has published articles in the social area in Brazil, especially on issues related to education and voluntary private social investment.

**Alan Roe** has worked for more than forty-five years as an academic economist and as a policy adviser. Now retired, he remains as a senior associate at Oxford Policy Management and a senior non-resident research fellow at UNU-WIDER. He has undertaken a number of case studies on the impacts of mining and oil and gas for the World Bank, DFID UK, the ICMM, and a number of extractives companies.

**Jeffery Round** is an emeritus reader in economics at the University of Warwick. His main research interests include: the design and use of social accounting matrices (SAMs) for development policy analysis; poverty and inequality; as well as CGE modelling and regional modelling, especially multiplier analysis. He has contributed to OPM studies on mining and extractive industries for Romania, Brazil, Zambia, and Chad.

**Keith Slack** is director of the Extractive Industries Global Program in the Washington, DC office of Oxfam America. He directs the organization's extractive industries reform program, which promotes greater respect by corporations, international financial institutions, and governments for the basic rights of communities in developing countries affected by oil and mining operations. He has published widely on issues relating to extractive industries and development and serves on the editorial board of the journal *Extractive Industries and Society*.

**Andres Solimano** received his PhD in economics from MIT and is founder and chairman of the International Center for Globalization and Development in Santiago, Chile. He has published several books on economic elites, crisis, global capitalism, inequality, international migration, and talent mobility. He has held senior positions at the World Bank, United Nations, and the Inter-American Development Bank.

**Paul Stevens** is a distinguished fellow at Chatham House, London, a professor emeritus at the University of Dundee, a distinguished fellow at the Institute of Energy Economics Japan (IEEJ) in Tokyo and, until recently, a visiting professor at University College London (Australia). His career in academia and consultancy spans from the oil shocks of the early 1970s to the current energy transition from molecules to electrons. He has published extensively on energy economics, the international petroleum industry, economic development issues, and the political economy of the Gulf.

**Kathryn Tomlinson** currently works as an independent consultant, and has over ten years' international experience working on best-practice approaches to dealing with social and human rights impacts surrounding private sector projects, with a particular focus on the extractive industries. She holds a PhD in anthropology from the University of Sussex, completed in 2005, which explored indigenous peoples' land rights in

Venezuela during the development and aftermath of a conflict over the building of an electricity power line through indigenous territories.

**Nicholas Travis** is a consultant at Oxford Policy Management with expertise in public financial management, currently based in Myanmar. Over the last decade he has conducted a wide range of research and consultancy assignments in the areas of macro-fiscal policy, natural resource revenue management and public expenditure. He served as the project manager of FOSTER between 2012 and 2015.

**Frederick van der Ploeg** is a professor of economics and research director of the Oxford Centre for the Analysis of Resource Rich Economies at the University of Oxford. His main current research is in the macroeconomic analysis of resource-rich developing economies, the empirical drivers of carbon and mineral reserves, and the economics of climate change. He has published widely in the fields of international macroeconomics, international finance, public economics, and environmental and resource economics.

**Anthony J. Venables** is the BP Professor of Economics and director of the Oxford Centre for the Analysis of Resource Rich Economies at the University of Oxford. His main current research is in the analysis of resource-rich developing economies and the development of cities as drivers of development. He has published widely in the areas of international trade and spatial economics, including work on trade and imperfect competition, economic integration, multinational firms, and economic geography.

**Sophie Witter** is professor of international health financing and health systems at Queen Margaret University, Edinburgh. She is a health economist who has worked in international health for the past twenty-eight years in Asia, Africa, and transitional economies, publishing more than eighty peer-reviewed articles in leading journals. Her research focuses on health financing in low- and middle-income countries, including fragile and post-conflict states, on ensuring financial access and appropriate incentives for health workers, and more broadly on the political economy of evidence-based policymaking.

# Part I

# Overview

# 1

# Extractives for Development

## Introduction and Ten Main Messages

*Tony Addison and Alan Roe*

### 1.1 Introduction

This book is about the challenges and opportunities that countries face in using their extractive industries to achieve inclusive development. Its focus is on the developing world, both low-income countries (LICs) and middle-income countries (MICs), drawing upon the experiences of high-income countries (HICs) when relevant. Extractive industries have shaped the economies, societies, and politics of nations—for good and bad. Today's richest nations owe at least part of their high living standards to the extractive industries. Yet while a large national income can result from resource wealth, it can also be associated with acute social inequality and deep poverty—the very opposites of inclusive development. Many LICs and MICs struggle to diversify their economies, and create redistributive fiscal systems, in ways that reduce poverty, inequality, and social division. The very worst cases see violent conflict and civil war.

The phrase 'resource curse' became common coin by the turn of the millennium. Crises and resource-wars were important catalysts in a new determination to improve the sector's governance. Global civil society, notably Global Witness and Oxfam America, together with the Natural Resources Governance Institute (NRGI) and industry bodies such as the International Council on Mining and Metals (ICMM) have led efforts to achieve improved outcomes for the extractives sector. One of the most notable manifestations of this was the Extractive Industries Transparency Initiative (EITI) launched in 2002. These all recognize, in different ways, that natural resources can provide a means, when properly used, for poorer nations to decisively break with poverty.

National ambitions for the extractives sector were given a major boost in the years after the millennium by an upswing in the prices of metals and fuels, following low prices in the 1980s and 1990s. China's economic boom resulted in a seemingly insatiable demand for commodities of all kinds and this, together with limited supplies after years of low investment, created a 'commodities super-cycle'. Growth elsewhere in the global economy added to demand. Euphoria returned during the super-cycle years from 2002 with buoyant export earnings and public revenues, as well as higher economic growth (though often of a narrow and undiversified kind). Very large investments were made in mining as well as in oil and gas after many years of moderating capacity; much of this was in the LICs and MICs.[1] Producers rode the super-cycle, including a sharp dip during the 2008–9 financial crisis, for more than a decade until it finally stalled in 2011–12.

However, the price slump did remind companies and governments of the commercial and economic risks associated with the extractives sector. Companies cut production and scaled back many of the investment plans made during the super-cycle years. Host countries initiated macroeconomic adjustments in response to lower revenues, their options limited in many cases by a failure to build fiscal buffers and diversify their economies during the good times. Painful adjustments are still ongoing in countries that over-accumulated debt.

Prices are now (January 2018) above their lowest levels, though for oil and gas, as well as many metals, they are still far from their super-cycle peak. Nevertheless, prices in real terms remain above their long-term trend levels in the case of most metals as well as oil (see Stevens, Chapter 4, this volume). Some metals required in the manufacture of batteries for electric vehicles and in renewable energy technologies, notably cobalt and lithium, experienced spectacular price increases in 2017/18 because of significant supply–demand imbalances.

The size and direction of future price changes are inherently uncertain. A myriad of forces will determine the future of each extractives sector, favouring some fuels, minerals, and metals at the expense of others. The forces driving the future include: the pace and pattern of global economic growth (with India perhaps nudging China's position as the leading driver of Asian commodities demand); the speed of the shift from fossil fuels to renewables in the overall energy mix (in turn a product of technical change as well as national and international climate action); and changes in the industry itself,

[1] Oil and gas exploration and production spending globally rose over fourfold between 2000 and 2012 and the global exploration spend of mining companies was almost US$30 billion in 2012 as compared to less than US$3 billion in 2000 (World Bank 2015).

including the pace of technical change that makes it easier to develop new deposits in more difficult locations, as well as the adoption of 'green mining' in order to minimize the sector's own environmental footprint. Developing countries need a strong understanding of these trends as they play out. LICs and MICs have become significantly more important as producers of extractives in recent years (see Roe and Dodd, and Ericsson and Löf—Chapters 2 and 3, respectively, in this volume). Since this tendency will almost certainly continue, their prospects for growth and poverty reduction significantly depend upon their extractives sectors and how these are managed. Although 'keep it in the ground' is a theoretical policy option, it is not an option that is likely to be widely adopted in practice (see Lahn and Stevens, Chapter 5).

### 1.1.1 *This Book*

This book aims to provide a comprehensive contribution to a lively and ongoing debate, in which many stakeholders now participate: governments and their international partners (bilateral and multilateral development agencies); the industry itself (the companies together with industrial associations such as ICMM and IPIECA);[2] community-based organizations (and their NGO and INGO partners); the national and international media; and the research community in universities and think tanks. This debate centres on achieving *practical action* to deliver inclusive development using resource wealth, protect often fragile environments from damage, enhance the rights of affected communities (and the benefits to them), and support climate change action. Central to these tasks is the creation of a set of effective and accountable institutions to manage the extractives sector and maximize its potential for development impact. In addition to capturing the flavour of current debate on extractives and development, this book offers ideas and some recommendations in most of the main policy areas.

Since no single person has expertise on every facet of extractives and development, this book brings together a range of international experts from many disciplines and organizations; it therefore represents a large amount of collective insight and experience. The book is a major output of UNU-WIDER's latest research project on extractives and development which began in 2015. The book is accompanied by a website which makes available additional materials.[3] The book does not seek to define a single formula for

[2] IPIECA is the global oil and gas industry association for environmental and social issues.

[3] http://www.wider.unu.edu/project/extractives-development-e4d. The website makes available further papers, including more country case studies and papers that discuss topics not covered in this book. It also links to video interviews with the authors of the chapters in this volume, as well as other experts, which are located on UNU-WIDER's YouTube site.

'success'. Instead it offers a comprehensive but integrated account of the multiple ingredients that are needed to turn the undoubted potential of extractives wealth into the reality of sustained improvements in living standards and social well-being. Section 1.2 presents *ten* of the most important messages of this book. The chapter concludes with a restatement of what we believe this book offers to ongoing debate and action on this most vital of development issues.

## 1.2 The Main Messages of This Book

The book covers a large amount of ground and many messages emerge from the individual chapters and their authors. Here we focus on ten key overarching, but inter-connected messages, emerging from the detailed analysis of the individual chapters.

### 1.2.1 *Message 1: Extractive Industries Are Important in Developing Economies and Will Remain Important*

Extractive industries have become more important to export revenues and government revenues in many LICs and MICs, over the last twenty years. Of the seventy-two LICs and MICs that we identify as *most* dependent at least in terms of extractives *exports*, sixty-three have increased their dependence on extractives resources over the past fifteen to twenty years (see Roe and Dodd, Chapter 2). While development economics emphasizes the desirability of structural transformation to diversify economies away from dependence on primary products including extractives, dependence is increasing in many developing countries.

Why is this? The main reason is that many of the newer extractive investments have been so large relative to the size of national economies. A second reason is of course the highly favourable prices for extractive commodities over an extended period. However, a failure to prioritize diversification is often an important further reason. Resource wealth is notorious for stymieing policy action, when non-renewable resources are viewed as perpetual generators of wealth, and not as finite (time-limited) opportunities. Sometimes diversification is attempted, but when the new investments to achieve this are badly selected, the economy's vigour is reduced rather than enhanced—leading to more, not less, dependence on extractive exports and revenues. Although it is the case that the shares of extractives in the national economy, export earnings, and public revenues have risen partly because prices have risen, levels of extractives dependence in LICs and MICs remain at historically high levels even after the adjustments provoked by the price falls of 2011 and 2012 (Roe and Dodd, Chapter 2).

Box 1.1 provides a summary of some of the main factors that lie behind the recent observed trends.

**Box 1.1** GLOBAL MARKET TRENDS: SELECTED INFLUENCES

China has been, and continues to be, easily the biggest driver in this story. Chinese demand for commodities grew through the 1990s, but really took off after 2000, with its WTO accession in December 2001 super-charging its exports, and consequently its imports of essential inputs (Erten and Ocampo 2013). China is a net importer of most metals as well as oil and gas, with an especially strong appetite for bauxite, copper, iron ore, nickel, and uranium (Pigato and Tang 2015). In the period 2001–6, when world prices of metals almost tripled, China accounted for more than 50 per cent of the rise in demand (Francis 2007: 20). China's sustained demand was also one reason why prices rebounded following their dip during the 2008–9 financial crisis.

Commodity prices move in cycles along with the global business cycle. The fall in the oil price from 2011, and especially from 2014–16, was dramatic and many metals prices—especially iron ore—saw similar sharp falls. Market analysts have been calling the end of the 'commodity super-cycle' for the past two to three years, though forecasting is notoriously difficult—some would say impossible—in this area. Although these falls have disturbed the markets, it is also significant that in real terms, commodity prices are mostly higher than was the case twenty years ago, and some of the forces driving up prices over the longer term are still in play.[4] These include China's continued economic growth. Although this has slowed to around 6.5 per cent annually, China's economy is now some four times its size at the turn of the millennium when growth was much higher (at around 9 per cent). More generally, the past twenty years have seen the physical demand for metals growing significantly faster than global GDP and this high elasticity seems well established (ICMM 2016). The demand for energy and metals continues to be robust as MICs that currently consume fewer extractives products (per capita) than richer countries accelerate their economic growth.

Whereas the market outlook for commodities remains positive overall, there will be plenty of change in *relative* prices within the commodities category, as well as winners and losers amongst nations (and companies). The fossil fuels (oil, gas, and coal) illustrate both points; they will see sustained demand assuming continued global economic growth, but their market share in global energy production will decline as climate-change actions accelerate (see Message 10) and OPEC remains under pressure from US shale production (with the United States likely to become the world's largest exporter of liquified natural gas (LNG) in coming years).[5] The prospects for metals are generally good, as renewables gain market share in energy production (see Message 10).

[4] ICMM (2016) provides some analysis of these forces as they apply to metals.

[5] There is an intense sparring match between Saudi Arabia and US shale producers. To date the lower oil price has failed to drive US shale oil producers out of business. Every time the oil price looks set to move out of its present US$40–70 a barrel range, shale production comes back into the market.

### 1.2.2 *Message 2: Developing Economies Have Great Potential to Develop their Extractive Resources Further*

Extractive industries have increased, rather than decreased, in importance for many LICs and MICs (Message 1). Additionally, it is anticipated that many, though not all, extractives sectors will see considerable further investment in the future; this is certainly so for most sectors and especially for metals (see analysis by the McKinsey Global Institute (MGI 2013)), notwithstanding the adjustments associated with climate change policies (see also Addison, Chapter 22). The second main message is that the LICs and MICs are in many cases very well positioned to take advantage of this expected growth in commodities demand—*should they choose to do so.*

Many of these countries have huge unexploited reserves of resources. Data from MGI (2013) reproduced as Figure 1.1 indicates the scale of that potential. In many of the LICs and MICs shown in the figure, reserves are a huge multiple of prevailing per capita income levels (remembering that the vertical scale is logarithmic). In some country cases the calculus of future potential is dramatic. Guinea, for example (per capita income of US$508 in 2016), is

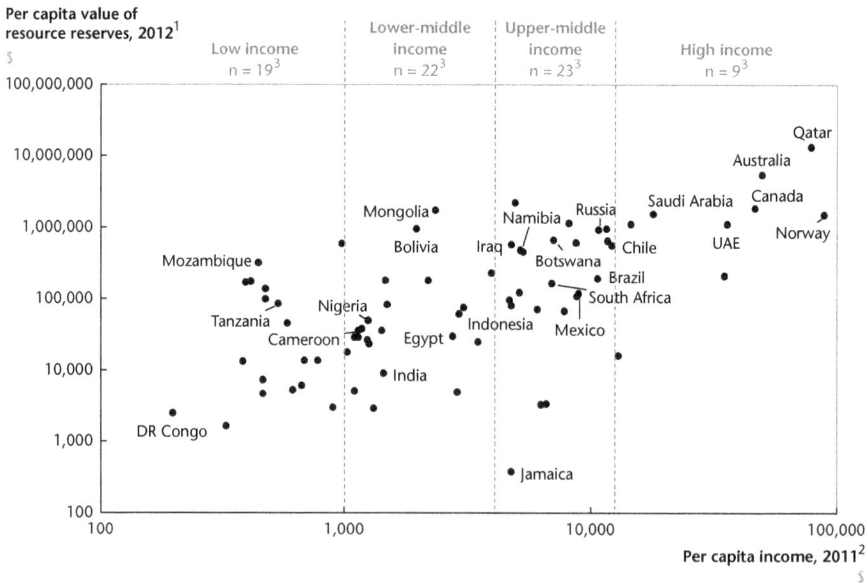


1 Includes reserves of oil, gas, iron ore, coal, copper, gold, nickel, silver, potash, and phosphate rocks (valued in current prices).

2 Per capita GNI in current prices; 2011 World Bank thresholds for categorization are $1,026 for lower-middle income, $4,036 for upper-middle income, and $12,476 for high income.

3 The sample size includes future resource-driven countries identified by the IMF (Afghanistan, Guatemala, Madagascar, São Tomé and Príncipe, Togo, and Uganda); 14 countries were excluded due to lack of data.

**Figure 1.1.** Known reserves (2012) and per capita incomes (2011)
*Source*: MGI (2013: 30: Exhibit 3); reproduced here with permission.

thought to have one quarter of the world's total reserves of bauxite—most high grade and unexploited. But it produces only at the level of production in China and India, which have only one tenth Guinea's reserves. Furthermore, these data significantly underestimate the developing world's full potential because in relative terms so little geological exploration has taken place in LICs (as well as many MICs), and this is only now changing.[6]

The reason why we enter the caveat above, 'should they choose to do so', is clear. In the past many LICs and MICs endowed with large extractives resources have not seen significant growth or broader development benefits from exploiting these. Data from the MGI (2013) study suggests that in the period 1995 through 2011, there were slightly more resource-dependent countries with below average per capita incomes that fell further behind (in terms of per capita income) than the number that caught up or surpassed the average growth of per capita income (MGI 2013: 34).

This relationship between per capita income, resource wealth, and subsequent economic growth can also be interpreted as highlighting the dangers that resource wealth poses to a country's chances of raising its average living standard. A very risk-averse development strategy might conclude: 'leave the resources in the ground'. But that carries a high opportunity cost as the revenues (which can far exceed any alternative source, including foreign aid) can fund major pro-poor spending and development infrastructure (including that needed to diversify the economy itself). Moreover, while the risks are indeed high, development disaster is not a foregone conclusion: there are countries like Botswana that have achieved upper-middle-income status (Leith 2005). More generally, the MGI study also shows that *if* the record of the more successful resource-driven countries could be replicated by LICs and lower-middle-income countries then there is the potential to lift more than 500 million more people out of poverty by 2030 (MGI 2013: 23). This is a tantalizing prize and it is easy to see why developing countries are unlikely to be persuaded by arguments that they should forgo the use of their resource endowments.

A key task of development advisers and analysts (as well as this book) is to explore all possible ways to deliver the best possible outcomes for sustainable and inclusive development. This includes efforts to improve governance of the extractives sector, as poor governance—implying high risks for potential investors, both domestic and foreign—is a major reason why many LICs have huge unexploited reserves (see Message 6). That said, the market for fossil fuels faces strong headwinds from climate change action and the rapidly declining

[6] For example, non-ferrous metal exploration in Africa increased from less than US$17 per sq km in 2000 to US$189 per sq km in 2012. Similar statistics can be cited in relation to oil and gas (MGI 2013: 32).

cost of renewable energy. Reserves of fossil fuels will eventually constitute 'stranded assets' (see Message 10).

### 1.2.3 *Message 3: Strategies Should Be Guided by Realism: Neither Euphoria Nor Despair Is Helpful*

Many have dreamt, and many still do, of striking it rich by discovering a valuable mineral or, better still, oil. Thriving mine towns had a powerful hold on the imagination of nineteenth-century America. The soaring architecture of the Gulf states, built on oil wealth, captivates the modern imagination. The scale of the potential rewards inspires explorers, miners, and investors, while politicians dream of fast-tracks to national prosperity. But alongside such euphoria sit dismal images of the ghost towns left behind once mines closed, worthless infrastructure and industries erected during the good times, and environments and communities devastated by mine tailings and oil spills. For many, the phrase 'boom and bust' sums up the extractives sector.

The watchword of this volume is 'realism'. In crafting strategies to deliver prosperity from resource wealth, countries need to thoroughly understand the risks—drawing on the many available lessons—while putting in place policies and investments that realize the development rewards. Extractive industries, when reasonably well managed for the good of the nation, are one of the few means available to poor countries to escape poverty: above all they avoid many of the economic preconditions required to attract large-scale investment in other productive areas such as manufacturing. However, their good management requires realism about the likely level and duration of the associated government revenues, close attention to price volatility (and the consequent revenue risks), and the creation of fiscal buffers to accumulate savings in good times in order to protect essential development spending in bad times (see van der Ploeg and Venables, Chapter 9). Good management also entails avoiding ill-judged investments that carry high risks of failure (see Östensson and Löf, Chapter 25) while investing in human capital and sectors, existing and new, with a good potential.

Discussion and practice on extractives and development in the past few decades has been marked by both optimism and pessimism, with one or the other being in the ascendant at any particular time, but neither being dominant across time. The thesis of the resource curse (and 'Dutch Disease') prevailed for many years from the mid-1980s onwards (see Lahn and Stevens, Chapter 5). However, something of a turning point on the need for change was reached around the start of the new millennium as pressures for improvement converged from different directions. One pressure came from the pain of macroeconomic adjustment, in response to low commodity prices in much of the 1980s and 1990s. Wars in which combatants fought for control of valuable resources had left catastrophic human and economic damage, not least in West Africa, in the last years of the twentieth century. The media, community-based organizations, and NGOs had exposed abuses of communities around mining sites (especially of indigenous communities), environmental damage (sometimes catastrophic), and extensive corruption and the appropriation of revenues for private gain. High-profile reports and campaigns by Global Witness, Human Rights Watch, and other NGOs put pressure on the industry to start paying greater attention to its social and environmental impacts.

Another catalyst for change arose in the World Bank which, given its prominence, is often a lightning rod when development crises erupt. Amid mounting and well-organized criticism by the NGOs of the mining industry, World Bank President James Wolfensohn commissioned a review of World Bank operations in the sector, namely the *Extractive Industries Review* (EIR) (World Bank 2003). Led by Emil Salim, a former Indonesian environmental minister, the EIR process listened hard to all stakeholders. By asking whether extractives projects were compatible with the Bank's goals of poverty reduction and sustainable development, the EIR's initial recommendations placed a serious question mark over the Bank's future work in the sector. The review concluded that the Bank should only continue to support investments in the extractives sector if its interventions could be shown to contribute to poverty alleviation through sustainable development. And this was only possible if three enabling conditions were met: (i) pro-poor public and corporate governance; (ii) much more effective social and environmental policies; and (iii) greater respect for human rights (World Bank 2003). The review concluded that the Bank should stay out of further investments in the industry until the sector's 'governance' was significantly improved.[7]

The EIR process together with the NGO campaigns also strengthened the forces for change within the mining industry. These were in any case gathering momentum by the new millennium. The Global Mining Initiative (GMI) was created by a group of mining company CEOs and building on this nine of the world's largest mining companies commissioned a major new initiative, the Mining, Minerals and Sustainable Development Project (MMSD), to examine how minerals and mining could best contribute to global sustainable development (MMSD 2002). As it proceeded more companies joined the MMSD process as did a number of donor agencies. The mining companies driving the MMSD process eventually constituted themselves as the International Council on Mining and Metals (ICMM), and the MMSD germinated ICMM's ten existing sustainability principles. In contrast to the mining

[7] Given that the term 'governance' was at that stage only loosely defined, this recommendation encouraged significant work to try to pin down exactly what it meant: an issue summarized in Dietsche (Chapter 6).

industry, the oil and gas industry did not undertake a comprehensive sustainable development review, let alone one led by company CEOs, and responses to campaigns were not coordinated at the industry level; any initiatives tended to come from individual companies, notably Shell and BP (during John Brown's tenure as BP's CEO over 1995–2007) (Tomlinson, Chapter 20).[8]

Several new international initiatives also emerged which aimed to encourage more transparency in natural resources management. George Soros established the 'Revenue Watch' programme under his Open Society Initiative to investigate the flow of funds from oil companies to governments in the Caspian region. International initiatives culminated in the Extractive Industries Transparency Initiative (EITI). At an international conference convened in London by the Department for International Development (DFID) in June 2003, a Statement of 12 Principles to increase transparency of payments and revenues in the extractives sector was agreed; EITI was then founded to give effect to these (Cust, Chapter 19).

In summary, while some observers remain deeply pessimistic about the developmental benefits of the extractives sector, in the past two decades a great deal of national and international effort has been devoted to trying to avoid this becoming a foregone conclusion. (Hodge, Chapter 18, provides a fuller listing of the very many post-2000 initiatives.) During the super-cycle period an 'extractives for growth' agenda had become accepted by some donor agencies in an era of much greater optimism. But as explained in detail in Lahn and Stevens (Chapter 5), excessive optimism is scarcely justified. Instead what is needed is a well-grounded realism that involves, above all, trying to learn from, and then avoid, the mistakes of the past, being cautious and avoiding over-optimism, while recognizing that extractive industries (EIs) if well managed can certainly support sustainable development.

### 1.2.4 *Message 4: Diversifying Economies Is Critical, But Hard to Achieve*

The academic literature has become somewhat fixated on the phenomenon of 'Dutch Disease' and the 'resource curse' more broadly. Lahn and Stevens (Chapter 5) argue persuasively that the resource curse propositions offer no real guide to practical policy. They suggest that these propositions should be stood on their head to ask the following question: why, in many countries, has the (often large) extractives sector failed to become the leading sector for the rest of the economy?

Such an approach suggests that much greater focus and attention than is common should be given to the role of the sector over time in a country's

[8] A global oil and gas industry association for environmental and social issues, IPIECA, was established in 1974.

overall and long-term development strategy. Bearing in mind that extractives are always depletable resources (even though the time horizon for depletion can be very long in some cases), other productive activities will in time need to replace them if any initial growth and development is to be sustained. This approach leads naturally to a fuller assessment than is normal of the manner in which an extractives activity might function over a long time-period and during that time might catalyse other non-extractive activities: what actions are needed to promote and maximize such effects? Similarly, it draws attention to pitfalls that could undermine the catalysing influence coming from extractives: are these impediments purely macroeconomic in nature (see chapters in this volume by van der Ploeg and Venables (Chapter 9), Solimano and Guajardo (Chapter 10), and Bawumia and Halland (Chapter 11)) or might sectoral policies and issues be equally or more significant? What services, jobs, and business opportunities can realistically be generated from the extractives operations, including opportunities to provide more and cheaper domestic energy linked to possible oil and gas resources? These and some related questions are addressed using specific country examples in Chapter 24 by Östensson, and in Chapter 25 by Östensson and Löf.

The literature review by Stevens and Lahn and the focus it brings to bear on *economic diversification* provides a natural lead in to a sequence of other chapters in the book—by respectively Roe and Round (Chapter 23); Östensson (Chapter 24); Östensson and Löf (Chapter 25); and Witter and Jakobsen (Chapter 26)—that analyse several distinct aspects of the linkage effects of extractives and the policies needed to achieve these in much greater detail, and to the further chapter by McPhail (Chapter 17) that looks at an 'all of government' approach to policies for extractives sectors. Because of its attention to broader development strategies it also brings into the limelight the controversial suggestion that some countries might be well advised to hold back from developing certain extractives reserves or to develop them at an appropriate pace against the inevitable pressures from popular, political, investor (shareholder) interests to ramp up production as quickly as possible.

### 1.2.5 *Message 5: Better Institutions Are Vital to Success But Technocratic Institution-building Has its Limitations*

Creating effective institutions is vital to success both in a technocratic sense (personnel, processes, etc.) and in the politics around these processes (whether they are open to capture by personal political interests etc.). This is hardly a new message—development debate has been replete with references to institutional development for the last two decades at least—but it remains especially relevant to resource-rich countries where progress has often been especially difficult. Many of the chapters in this book refer to the role and importance of institutions. Those chapters address many different aspects of the institutional and governance challenge including: government management of macroeconomic policy (all of Part IV of this volume); taxation (Otto, Chapter 14) and public expenditure policies; the use of quasi-government agencies such as state oil companies (Heller, Chapter 15) and sovereign wealth funds (van der Ploeg and Venables, Chapter 9); gender inequality (Macdonald, Chapter 21); the sectoral management issues around local content (Östensson, Chapter 24) and the stimulation of downstream activity (Östensson and Löf, Chapter 25); environmental regulation (Greenspan Bell, Chapter 16) and the management of community relations (all of Part VIII of this volume). Across the chapters, the political economy realities and difficulties are taken full account of, and so too is the pre-eminent desirability of a high level of coordination across these many different aspects of governance: a requirement that is often lacking in practice.

Using a framework originally advanced by Oliver Williamson (2000), Dietsche (Chapter 6) relates extractives issues to four levels of institutions:

- Level 1: *social embeddedness* (customs, tradition, norms, religion). These affect the way that people, through their social networks, see their rights with respect to resource revenues, jobs, and other benefits.[9] These are often subject to considerable inertia.
- Level 2: the *institutional environment* (formal rules of the game)—that is, constitutional rules regarding ownership of primary resources (and how the different levels of government engage with extractives). This includes the state's capacity (at all levels, from central to state/provincial to local) to provide public goods and services that can maximize the benefit of the extractives sector.
- Level 3: *governance* or 'play of the game' (how specific rules are set up between companies and governments) including written contracts, safety rules (balancing costs versus safety), and whether countries have mandatory obligations to communities, or only voluntary codes of conduct.
- Level 4: *resource allocation and employment*: the set of commercial decisions for selecting and implementing projects, how much labour to use versus labour-saving technologies and so forth.

Policies to enhance the contribution of extractive industries to inclusive and sustainable development most commonly focus on Levels 2 and 3—through such measures as revising legal frameworks, the drafting of fairer contracts etc. These are typically the levels at which international action seeks improvements, and at which most aid donors concentrate their support to governments.

[9] Economic sociology, notably the work of Karl Polanyi (2001 [1944]), emphasizes how economic systems are embedded in social networks and their relationships.

However, it can be difficult to make progress when social networks (Level 1) are resistant and well organized. This is especially so when they exercise control of the state, or have a high degree of influence upon it. The Chad–Cameroon Petroleum Development and Pipeline Project makes this point painfully well. It was a model of its kind when approved (i.e. a good outcome at Level 2), but was unable to deal adequately with a governing elite that grabbed most of the resource rents to fund the military and defend itself (Level 1). Powerful family networks and their allies may constitute 'deep states', enabling them to control public money, including resource taxes, and to allocate mining rights, as well as oil and gas concessions, in favour of businesses controlled by the network itself and to the detriment of Levels 2 and 3. Such networks are especially prevalent in nations with histories of extended violent conflict (Addison 2003). In a worst-case scenario, such as South Sudan, competing networks will disregard any formal rules designed to peacefully manage competing pressures, and will instead go to war.

The democratic electoral cycle, which provides an essential means for voters to express their preferences over the use of natural resource wealth, also interacts with Level 1. Long-standing networks provide an instrument for political incumbents and challengers to mobilize support, and build winning coalitions. In doing so, the temptation is to spend and borrow to win elections, perhaps overriding Level 2 institutional limitations that are designed to preserve macroeconomic stability.

Ghana's recent experience makes this point. A decade ago, as Ghana looked to its future as an oil economy, policy makers examined carefully the lessons of oil-rich Nigeria. They were determined to avoid Nigeria's mistakes, and created what were thought to be effective institutional checks and balances on over-spending and over-borrowing. Nevertheless, the government over-spent and over-borrowed in the lead-up to the 2012 elections. In terms of Oliver Williamson's fourfold institutional typology, the second level was organized to a high standard, notably through the 2015 Petroleum Revenue Management Act (PRMA) which created a sovereign wealth fund (SWF). But its effectiveness was then overwhelmed by Level 1 pressures, in part expressed through electoral politics: a point strongly endorsed by Bawumia and Halland in Chapter 11.

The widespread transition from authoritarianism to multiparty democracy over the last thirty years must be celebrated. But both old and new democracies face the challenge of building and maintaining formal institutional systems, including the checks and balances (as well as transparency) necessary to manage competing demands on the public purse. This frequently has a regional dimension as well, where systems for the distribution of resource rents are often associated with tensions between the centre and local political levels (Nigeria being one example).

In sum, 'institutions' are vital and many of the contributions to this book drill down into their various meanings in the context of many specific aspects of extractive industry management. In this manner, the book gives substance to the proposition that it is not enough merely to assert that 'good governance' and 'good institutions' are important. These complex topics need to be unpicked and assessed in all their multiple aspects. This book seeks to do this based on the underlying view from most of its contributions that positive change based on extractive industries is possible, but also that the institutional ingredients of such success are by no means assured.

### 1.2.6 *Message 6: Effective and Inclusive Government Working with Enlightened Companies Is the Ideal Combination. Ineffective and Divisive Government Combined with Rogue Companies Is the Worst*

Figure 1.2 below sets out a simplified taxonomy of the different situations that we find in extractives-dependent economies around the world. It illustrates a spectrum of governance and institutions from *effective and inclusive* through to *ineffective and divisive* (on the vertical axis). By effective, we mean a state capable of formulating and delivering a development strategy, and by inclusive we mean a strategy that aims to achieve development for everyone in

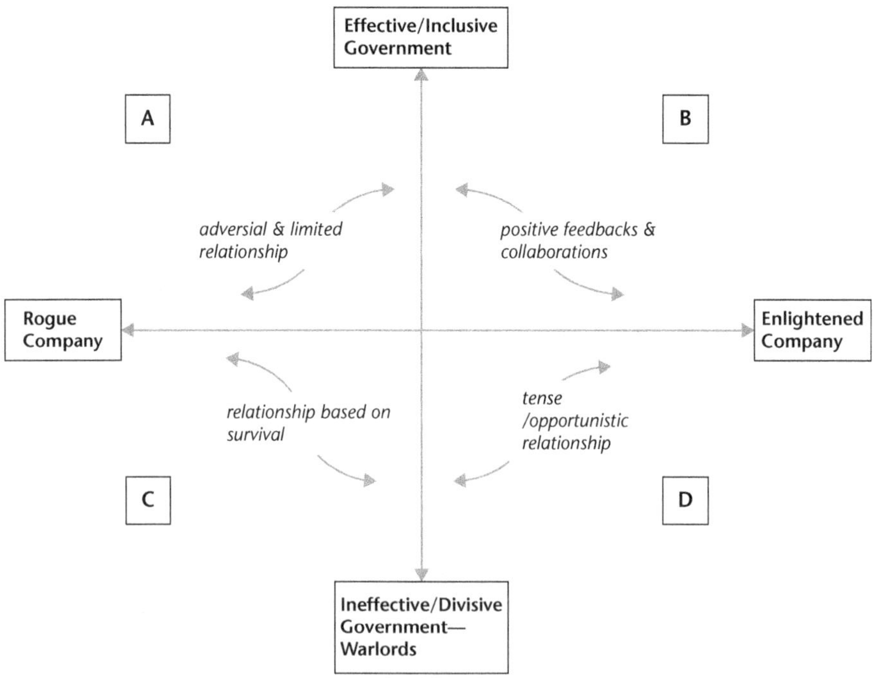


**Figure 1.2.** A simple taxonomy of governance and company behaviours
*Source*: authors' illustration.

society, but especially for the poor (and one that protects and sustains the environment on which inclusive development is ultimately based). Figure 1.2 also shows a spectrum of company behaviours from *highly enlightened* (working with communities and states to deliver inclusive and sustainable development) to 'rogue' (no concern whatsoever for the company's social, political, and environmental impact) on the horizontal axis.

Zone B is the one in which there is likely to be the greatest prospect of outcomes that are inclusive and sustainable as an enlightened company works collaboratively with a relatively well-organized and inclusive government. Both parties may have technical limitations but the necessary conditions are in place for cooperation to evolve in ways that deliver successful outcomes for both parties. The numerous post-2000 initiatives discussed earlier (and in more detail in Hodge, Chapter 18) can be thought of as actions that together aim to help more countries and companies move into the Zone B space.

Zone C is the exact opposite of Zone B. Governments in Zone C are neither interested nor capable of delivering inclusive sustainable development: in the worst cases the state (at either the central and/or local levels) is controlled by self-serving warlords.[10] In this zone mining or oil companies that are 'rogues' focus narrowly on extracting a resource for the greatest possible profit, without concern for community, development, or environmental impacts. Such governments and companies both work to survive and to defend their own self-serving interests, and civil society is likely to be both weak and repressed. Accordingly, we should not expect external interventions such as the EITI or those proposed by the NRGI to achieve any significant early improvement in Zone C situations. Nevertheless, helping strengthen progressive civil society organizations together with robust international action on transparency, corruption, and money laundering—to begin to nudge the behaviour of companies and state actors towards improvement—might over time increase the prospects of a more inclusive development path.

The other two zones A and D are inferior to Zone B but offer more hope of progress than Zone C, since at least one of the two sets of actors favours reform and can work, with international assistance and initiatives, to try and alter the behaviour of the other. In Zone A an effective and development-orientated government can build institutions (at Levels 2 and 3) to rein in any rogue company, at the limit, withdrawing its licence to operate (and if there are competing extractives companies, this should be at low cost). Support from EITI, NRGI, and other international initiatives, as well as civil society and the

[10] At its worst, warlords act as 'roving bandits'—using the terminology of Olson (2000)—treating natural resources as only so much loot, and possessing little if any incentive to develop the territory which they (often temporarily) control. Predation rules, discouraging most investment, driving the economy downward, and thereby further tightening the resource constraint—with the excluded populace bearing the cost (Addison and Murshed 2005). This is inherently unstable politically.

media, should be helpful here. International companies increasingly take heed of reputation and most are cognizant of their need to obtain and retain a social licence to operate; as well as the need to respect ever more demanding international standards. We now discuss issues arising especially in Zones A and D in more detail in our Messages 7 and 8.

### 1.2.7 *Message 7: Delivering Effective and Inclusive Governance Is Vital to Improving Outcomes in the Extractives Sector*

Looking again along the vertical axis of our framework diagram (Figure 1.2), countries will lie along a spectrum. At the worst end, exclusion is imposed by mass violence (warlordism) and, at the best end, inclusion is supported and achieved by good policy.

In between are a variety of positions, including: (i) countries with systems involving *exclusion* (perhaps of indigenous communities) but tempered by an otherwise politically stable and prosperous society, and (ii) countries with governments that seek to improve inclusion but do so by driving public spending and debt up to unsustainable levels (an eventually self-defeating form of populism).

Nations and their governments can move up and down this spectrum: some may, via bad policy and bad politics, find themselves in the C or D zones and at risk of civil conflict (Venezuela today). Some may nevertheless have enough institutional robustness to pull back and recover (Zambia from the 1990s on; Zimbabwe, perhaps in the future). Others descend into civil war, which is difficult both to halt, and to recover from (presently: Libya, Somalia, South Sudan, and Syria). Some can recover but remain vulnerable and fail to reach the excluded, with the politics around their resource wealth being a key obstacle (Algeria and Angola). Others pull themselves out of deep crisis (sometimes war), with help from the international community: Ghana starting in 1983, Mozambique in 1992, and Indonesia after 1998 (but may then run into economic difficulty years later: both Ghana and Mozambique today). Some make remarkable political transitions, but then stall (South Africa today). Some are written off as hopeless cases but then make surprising transitions back towards democracy (Nigeria and perhaps Myanmar).

This movement up and down the scale from exclusion to inclusion can be driven by exogenous economic shocks (price shocks especially). Resource booms, if handled well, provide new revenues and economic opportunities for strategies of inclusive development. Price collapses threaten inclusion if the resulting macroeconomic adjustment is unduly borne by the poor; this risk is reduced if the country has saved some of the windfall in good years, giving it more fiscal room to manoeuvre (as Chile has done; see Solimano and Guajardo, Chapter 10).

In discussing economic policy, states are often described as monoliths: acting with a single set of objectives, a single set of views, and a homogeneous voice. However, real-world states combine a wide variety of actors, with different interests: those benefitting from the *status quo* (from exclusion, for example) may be dominant for a while, but reformist elements can come to the fore within a government and the public administration. Anti-reformers can change their stripes when they recognize that their own political base is under threat from a sagging economy, rising grievances among the excluded, or a threat from another state that requires the government to mobilize the nation around a more inclusive agenda.

Smart reformers will craft broad coalitions, working with the 'grain of society' (especially its Level 1 social norms and culture) in ways that may not be fully understandable to outsiders (hence donors should be wary of imposing themselves: there are plenty of examples of externally induced reforms that have stalled). Community-based organizations and NGOs can successfully widen the space of the political debate, leading to political openings that transform the possibilities for public action—around the rights of indigenous communities in mining areas, for example. In these ways, actions at Levels 2 to 4 to create institutions for more inclusive development can very occasionally make headway, sometimes overcoming the powerful inertia at Level 1. If successful, countries will move up the vertical plane in our Figure 1.2.

For analytical purposes, some helpful positioning of individual countries in the vertical plane of our framework diagram is possible using the metrics produced by the NRGI and its NRGI Index. For example, in the 2017 version of this Index the NRGI has ranked eighty-nine 'countries' (including double counting of countries with both minerals and oil and gas) that have a significant engagement with extractives in terms of the quality of their governance systems. It employs a scoring system that is used to classify these countries into five categories according to the assessed quality of their governance.[11] These categories are good, fair, satisfactory, poor, and failing. The 'good' performing country is defined as one that 'has established laws and practices that are likely to result in extractive resource wealth benefitting citizens, although there may be some costs to society'. A 'failing' country is described as one that 'has almost no governance framework to ensure resource extraction benefits society. It is highly likely that benefits flow only to some companies and elites.' These two extremes could reasonably be thought of as indicating the highest and lowest points respectively on the vertical axis of our stylized diagram (Figure 1.2).

[11] The methodology is based around 149 questions and scores and subsequent expert review of three main areas of governance, namely the enabling environment, revenue management, and value realization. Within each of these three areas there are a total of fourteen sub-categories. See NRGI (2017: 7).

In the 2017 results only one MIC (but no LICs) are classified in the 'good' categories, namely Chile. But a number of mostly MICs are classified as satisfactory including (in order of their index scores) Brazil, Columbia, India, Indonesia, and Ghana. Overall thirty-nine of the eighty-nine country cases are rated as 'poor' or 'failing' and these include a large number of other MICs and LICs. The ten failing countries include, not surprisingly, Zimbabwe, Mauritania, Myanmar, the Democratic Republic of the Congo (DRC), Equatorial Guinea, Sudan, Libya, and Eritrea—mostly poor African economies. So there is a long way to go with so many countries closer to Zones C and D than to Zone B before the aspiration of Message 7 can be said to be achieved. However, even in difficult environments, politically sensitized interventions have been able to achieve some progress as explained by Buckley, McCulloch and Travis in Chapter 27.

### 1.2.8 *Message 8: Improving the Practices of Companies Can Now Draw upon a Great Deal of Accumulated Experience*

In the horizontal plane of our framework diagram (Figure 1.2), companies range along a continuum from those that care nothing for social and environmental obligations (and may even deny that any such obligations exist: the only purpose being to turn a profit) to those that fully incorporate social and environmental objectives into their strategies and day-to-day operations. We term the former 'rogues' and the latter 'enlightened'.

ICMM would claim, with some justification, that its twenty-five large multinational company members are in the socially responsible ('enlightened') category (as ICMM applies stringent membership criteria for corporate members, involving a rigorous admissions process followed by regular monitoring after admission).[12] Companies with a long history tend to improve their practices (environmental impact, workers' rights, etc.) relative to the norms of some thirty to forty years ago; in part because they learn from past mistakes. Newmont, a founding member of ICMM, is one example.[13]

Many companies have responded to NGO campaigns, international public opinion, international initiatives—and their own self-analysis of the benefits of a 'social licence to operate'—by creating in-house sustainability departments, responsible for advising on the social and environmental impact of

[12] Additionally, ICMM has membership from thirty-four national and regional mining associations and global commodity associations. Through these associations it also connects to another 1,500 companies in the sector. http://www.icmm.com/en-gb/members.

[13] Newmont had a relatively poor record of corporate social and environmental achievement in its mines in Papua New Guinea. However, when sustainability managers familiar with that record moved to work at a new Newmont mine in Ghana they resolved to establish very high standards. They achieved this, as evidenced by the mine's early receipt of IFC capital funding, implying its compliance with the IFC's rigorous standards.

operations. The formation of the extractive industry associations, ICMM and IPIECA, is merely one aspect of this tendency. Today, the major mining and oil and gas companies that are ICMM or IPIECA members would claim to be socially and environmentally responsible. Since the turn of the millennium there has been a clear trend for more enlightened corporate behaviours to emerge and this is well documented for the oil and gas industry by Kathryn Tomlinson in Chapter 20. Some of the large Chinese mining companies operating outside China are also seeking to adopt very similar practices to those recommended by Western industry associations.[14]

In short, it has become harder to be a 'rogue' company in the extractives business. Over the last two decades, INGOs and civil society have pushed to increase awareness of irresponsible company behaviour (details of some of these pressures are discussed in Chapter 16 by Greenspan Bell). Before the Internet age, it was much easier for companies to get away with environmental and social damage, especially in the world's remoter areas (including conflicts between miners and indigenous communities). Moreover, the goals of environmental sustainability and social responsibility are increasingly central to a growing class of financial institutions which provide ethical investment instruments to an increasing number of pension funds and private individuals. These investors are now highly influential in reshaping the extractives industry, notably via disinvestments in fossil fuels (especially coal), and in requiring manufacturers to source materials and metals from ethical supply chains that avoid environmental damage and human rights abuses such as child labour (Addison, Chapter 22).

The sector has come a long way in recent years and many companies are probably now positioned somewhere to the right of the vertical dividing line in Figure 1.2. Nevertheless, even as a reputation for environmental and social responsibility becomes an increasingly important determinant of company valuations over the long run, some individuals down the chain of corporate management may still face pressures to downplay the environmental and social footprint of their companies. At the highest levels of management, the drive to maximize shareholder value on an annual basis (exacerbated by linking CEO remuneration to the company's share price, combined with a rapid turnover in top management) can lead some companies into behaviours that dilute their standards (thereby risking both minor and major catastrophic events), neglect the quality of their supply chains, and even use corrupt methods to secure access to mining and drilling rights on favourable terms.

[14] In October 2014, the China Chamber of Commerce of Metals, Minerals and Chemicals Importers and Exporters (CCCMC) launched the Chinese Guidelines for Social Responsibility in Outbound Mining Investment at the China Exploration Exchange in Beijing. This has drawn some of its inspiration from the earlier ICMM arrangements.

Periods of price weakness, such as those after 2011, when companies looked to cut costs, can also weaken corporate commitments to their sustainability and social impact agendas (especially when the responsible departments have been downsized). In such times, marginal mines are likely to be mothballed, shut down or sold (perhaps to companies having weaker environmental and social commitments than the previous owners) and tax and royalty payments are likely to be reduced.[15] A big question therefore remains on the table: how to build robust mechanisms and company cultures that keep sustainability and social impact agendas to the fore—and irrespective of management turnover and market fluctuations.

The toughest challenges face companies working in Zones C and D of Figure 1.2 (countries characterized by ineffective and divisive government). Autocracy and national resource wealth often go together, given the large personal fortunes to be made via control of the state. Some autocracies have capable states, but many have states incapable of delivering even the most basic of services (note the NRGI definition of 'failing' given earlier). This poses a dilemma for any socially responsible company. It may decide to live with the situation, perhaps providing some benefits to local communities when the central and/or local governments are disinterested, or incapable of, doing so. There are now many examples in which communities see extractives companies as quasi-government agencies, expecting them to provide public goods that are otherwise unavailable from their own governments. This can have significant financial implications for the companies. When autocratic governments commit human rights abuses, extractives companies risk their reputations by continuing to operate in the country; examples have included Indonesia's Aceh province and Nigeria's Niger Delta (Coll 2013).

An enlightened company might nonetheless achieve localized benefits (increased local employment and infrastructure provision for communities around the mine site, for example). If an otherwise self-interested government and public administration contains at least a few progressive elements, then there is scope for an enlightened company to work with those progressive elements and also with community groups, with NGOs, and perhaps with like-minded donors to achieve at least a modicum of localized improvement. This can increase the country's chances of reaching a better future (Zone B).

The worst dilemmas occur when the country is in civil conflict and war. Access to rights for the exploration and export of valuable minerals and oil can then be highly personalized. Some international companies may decide to avoid the country entirely (even if international sanctions do not actually

[15] For some metals, the statistical data suggest that lower royalty payments as well as lower wage bills were indeed one important element in the cost-cutting that companies managed to achieve: see ICMM (2016).

prohibit their engagement). Others may face the moral dilemma of either abandoning well-established mines and loyal workers or staying the course despite the reputational and other problems this may pose for them. Some less enlightened companies (i.e. in Zone C) may find the riches too tempting, and so cut deals with the government, its officials, or warlords who control the relevant territory. State officials and warlords often create their own domestic businesses to profit in this way. Still others profit from offering protection to miners. Some of Africa's alluvial mining is linked to networks of organized crime.

Statements about the potential for progress from Zones C and D to Zone B are necessarily speculative and highly contingent, but they should be a crucial part of the overall improvements for extractives and development that most people seek. Such progress in the resource sectors depends on shifts in the broader political and economic climate of the country. Game-changers include: a move from authoritarianism to democracy (and whether stability is maintained); the nature of the political settlement that ends a violent conflict; the creation of multiparty democracies that are not simply covers for old (or new) authoritarian elites; and whether new governments take firm action to reduce the 'horizontal' and regional inequalities that feed grievances, including conflict over the distribution of resource wealth (Addison 2009, 2012; Stewart 2009). The hand of both enlightened companies and donors will be strengthened if the self-interested elite, or elements within it, become convinced that a more inclusive development path can secure their futures. But there is no one lever that can be pulled to yield such success. And promising transitions can stall (Myanmar today).

### 1.2.9 *Message 9: There Are Now Many More Ways in Which Interventions by External Stakeholders, Especially Aid Donors, Can Improve Outcomes*

With the increase in attention by the international community (donors and NGOs) to the extractive industries since the millennium, there has been a veritable sea change in the quality and breadth of understanding of the challenge of turning resource wealth into inclusive and sustainable development. This has encouraged a vast array of new initiatives—many of them international but others purely local. As already noted, Hodge (Chapter 18) identifies at least forty-six specific initiatives—nearly all initiated since the millennium—and additional to the many reforms in the formal legal and regulatory systems of individual national governments. This remarkable change in the situation warrants far more comment than it has received thus far.

Most of the new international and local initiatives relating to mandatory or voluntary improvements in the industry's regulation involve donors as significant players, and have potential traction over one or both dimensions—government and corporate—of our framework diagram (Figure 1.2). The Natural Resource Charter, for example, provides a coherent set of protocols to help governments understand better the various component tasks that they face. The benchmarking framework now managed by the NRGI is an especially powerful diagnostics tool that helps assess where institutional and other improvements in public policy are most needed. Companies can also draw on a substantial body of external guidance, provided by the International Finance Corporation (IFC) (with its detailed Environmental and Social Performance Standards), together with the ICMM and China's CCCMC. Some of these standards have serious teeth: they can grant or deny access to loan and equity finance in the case of the IFC standards. The EITI for its part gives guidance to both companies and governments about the better custodianship of the fiscal revenues and royalties from extractives activity. Although these initiatives have come in for criticism, there seems little doubt that their collective influence has been considerable, at least in the most conducive country situations (those represented as Zone B of our framework diagram). Finally, the powerful voices of some NGOs are also exerting a significant influence on some policy decisions (see Slack, Chapter 31).

At the level of *national governments* there is often a need for greater recognition of the very broad range of mainly new international initiatives that these host governments can draw on to boost their own regulatory efforts should they choose to do so: from aid donors, companies, and NGOs. In combining their sovereign right to regulate their own extractives sector with some of these external supports, there is also a case for adopting a coordinated 'all of government approach'. The objective of using extractives to stimulate eventual diversification of the economy must recognize the broad range of government ministries and agencies that are needed to attain that long-term objective. So, the interfaces with the external players that are of relevance to the extractive industries should not be narrowly construed to involve only the ministries of mines and petroleum, but should also include those of labour, health, education, etc. (Witter and Jakobsen, Chapter 26).

The initiatives and standards originated by *extractives companies* are increasingly being refined and applied more broadly and these can also be of great assistance to a receptive host country government. However, there are still some important further steps to be taken by most companies to embed their work into a comprehensive framework of 'contribution to host societies' that embraces broad concepts of both *sustainability* and *sustainable development* and that can also be challenged by becoming more amenable to rigorous comparison (across companies and countries), as well as to monitoring and evaluation (see, e.g., Chapter 29 by Mondoloka, Chapter 28 by Macdonald, and Chapter 18 by Hodge).

Many extractives companies now address their *responsibilities to local communities* in which they operate by reference to coherent frameworks of 'good practice'. However, there is room for major improvements in the specifics and consistency of actual practice on the ground. There is also a rising call for a greater reliance on formal legislation, regarding the duties of extractive companies to local communities, rather than the voluntary approaches that have so far been more common (Otto, Chapter 32).

*Donor agencies* were slow off the mark but are now beginning to support innovative new ways to address governance weaknesses of relevance to extractives, including in the less propitious 'failing' environments identified in our framework diagram (e.g. those in areas lower than Zone B. See also Chapter 27 by Buckley, McCulloch and Travis). Donors have collectively come up with many innovative ideas. These have increasingly been supported in various ways by the corporate players and by various NGOs. As one example: for some years mining companies led by ICMM have suggested that donor agencies can have a significant role to play in a variety of partnerships involving host governments, local communities, NGOs, and also the extractive companies themselves. ICMM (2010) identified and discussed five[16] main sets of partnership arrangements, all involving some combination of these various partners, and documented the success factors and other elements in some thirty specific cases.[17] These five sets were:

- Mining and Poverty Reduction: with six specific documented examples for countries such as Indonesia, Peru, and South Africa. The Indonesian example involved a partnership between Freeport Indonesia, USAID, and PADA,[18] with the Catholic Church also playing a role.
- Mining and Revenue Management: with examples from three countries, one of which is Ghana: a good example of the multiple partners involved in an EITI implementation.
- Mining and Regional Development: with examples from three countries including Madagascar where Rio Tinto partnered in projects with the World Bank.
- Mining and Local Content: with examples from seven countries including Mozambique where BHP Billiton partnered in various ways with the IFC.

[16] A sixth set described in that publication, namely *dispute resolution*, is not discussed here because the identified examples did not involve specific roles for donors.

[17] Subsequent to the 2010 publication ICMM also refined and re-issued its 'Mining Partnership for Development Toolkit' which offers many general principles as well as further country-specific examples that continue to be added to through new in-depth country case studies (ICMM 2011).

[18] Papuan Agribusiness Development Alliance.

- Mining and Social Investment: with examples from six countries including South Africa where Anglo American partnered with PEPFAR[19] and a number of local NGOs.

In addition to a role in some or all of these areas, a few donors have started to engage more explicitly with the difficult *political economy* issues that are involved with policies for extractive industries, and in the process have experimented with novel tools and modes of engagement. The United Kingdom's DFID can claim a leadership role in this area with its innovative projects in Nigeria (FOSTER)[20] and Ghana (GOGIG)[21] and more recently in Kenya. The assessment of the FOSTER example in Buckley et al. (Chapter 27) demonstrates: (i) how a regularly updated political assessment is essential to understanding the underlying incentives of key actors and whether possible new interventions may work; (ii) how interventions need to be 'locally led'—not necessarily led by government but instead by relevant parts of civil society—in order to provide sufficient legitimacy to the reform effort; and (iii) how any interventions need to be flexible and adaptive in order to deal with a context that can change quickly. Additionally, their analysis illustrates that to take on any explicit role involving political economy, donors must be prepared to accept a certain degree of risk since the connection between inputs of effort and outcomes cannot be assured.

Finally, in reviewing the underpinnings and some of the main messages of this book we need to confront arguably the most challenging issue that today confronts the extractive industries and the countries that host them: climate change.

### 1.2.10 *Message 10: Climate Action Will Create New Winners and New Losers among the Extractives Sectors*

Atmospheric concentrations of carbon dioxide are now at levels last experienced on earth some 800,000 years ago when the temperature was 2°–3°C warmer, and the sea level was 10–20 metres higher, than it is today (WMO 2017). Meeting the international target of avoiding a more than 2°C global temperature rise requires emissions of carbon dioxide, methane, and other greenhouse gases to peak within the next few decades, and the world must then achieve zero net emissions (IPCC 2015). Whether the world can meet this challenge while simultaneously ensuring the continued growth of global prosperity, and the reduction of poverty, will define much of the character

[19] The US President's Emergency Plan for AIDS Relief.
[20] The Facility for Oil Sector Transparency and Reform.
[21] Ghana Oil and Gas for Inclusive Growth.

of the rest of this century. Failure puts the planet at risk of catastrophic climate change and rising levels of poverty.

The extractives sector is central to this unprecedented challenge. In the history of humanity's search for new sources of energy, the long era of fossil fuels is entering its last stages, and a new era—that of renewable energy—is beginning (IPCC 2015; Stern 2015; Addison, Chapter 22). This new era is already seeing increasing demands for those metals and materials needed for renewable energy systems, as well as zero-carbon buildings and new forms of transportation. Some of this increase can be met by increased rates of recycling, but much will have to come from new supplies. The challenge for the mining sector, and for nations endowed with the resources in question, is to increase mining while ensuring that the sector's own emissions are contained (including switching to renewable energy for its power needs), protecting the local environment (especially the renewable resources of waters, forests, soils, and other components of the ecosystems that underlie sustainable development), and managing the other social impacts. Achieving all of this will be an important determinant of whether commitments to reduce emissions (the 2015 UN Paris Climate Agreement), together with commitments to the United Nations Sustainable Development Goals (UN-SDG), are fully realized.

Countries with reserves of metals have a bright future. Thus, while copper and nickel experienced significant price falls after 2012 (and a partial price recovery since then), the need for such metals is set to grow, largely irrespective of the exact mix of renewable energy and transportation technologies that evolves. Recent price forecasts indicate a 20 per cent increase in the copper price in real terms over the next ten years (World Bank 2016). Battery storage, to overcome the intermittency of supplies of electricity from renewable energy sources and to replace the internal combustion engine in vehicles, will be important to the demand for nickel, but especially important for the rarer metals of cobalt and lithium (the prices of which rose strongly over 2017/18 as supply lagged demand, especially from manufacturers of electric vehicles). The most thorough study to date of the implications of the low-carbon future for minerals and metals concludes that: 'the technologies assumed to populate the clean energy shift—wind, solar, hydrogen, and electricity systems—are in fact significantly *more* material intensive in their composition than current traditional fossil-fuel-based energy supply systems' (World Bank 2017: xii). Africa is especially mineral-rich, and can expect high and rising demand as the technologies of the low-carbon future are highly materials-intensive.

So, making the link with our Message 2, there is clearly a major opportunity for some LICs and MICs to use extractives more fully than in the past, and thereby to achieve the accelerated development in the ways discussed in this book. But there are also serious concerns, as discussed more fully in Chapter 22 by Addison. Countries and companies expecting to meet increasing demands for both base and rarer metals by raising their extraction and production levels may find it difficult to do so while at the same time reducing their own emissions (especially when ore grades are low). Large investments in mining together with supporting infrastructure are required to achieve this balancing act. However, many countries with large reserves of minerals also have very weak governance which holds back investment; thus constraining their ability to participate in any future mineral boom associated with the accelerating uptake of low-carbon technologies. Consequently, it is an open question whether most citizens of such countries will benefit from any boom; the history of extractives sectors in fragile states (e.g. Zone C and D countries) has largely been one in which the ruling elites are able to capture a disproportionate share of the gains. The supply chains for some of the minerals are characterized by human rights abuses (such as the use of child labour), notably that of unregulated artisanal mining of cobalt in the DRC (which has the world's largest reserves). Those supply chains need to be embedded in much better frameworks of governance and transparency.

While the cost of renewables is falling at a faster rate than expected, fossil fuels will still be necessary to energy production for many years to come (and overall fossil fuels still enjoy subsidies that exceed the level of support to renewables). Natural gas is taking over from coal as efficient gas-fired power stations have less than half the carbon emissions of coal (and coal is responsible for much of the drastic deterioration of air quality, especially in China and India's big cities). At the same time, there are still big environmental concerns around natural gas (notably shale gas). Yet many energy decision-makers do see it as an energy 'bridge' to a renewables future, albeit one that must end in less than two decades if international emissions targets are to be met. Many countries with large oil and gas reserves still cling to the hope that prices will recover their all-time highs of recent years, but the headwinds against oil and gas are strong, not least from international action on climate change. With the notable exception of the Trump administration, signatories to the Paris Agreement remain committed (indeed China sees a vast commercial opportunity in the technologies of the low-carbon future). However, countries with reserves of fossil fuels run the risk that they will be unusable, and they must surely reduce the vulnerability of their public finances by expanding their tax bases, and doing more to develop new economic sectors.

## 1.3 Conclusions

This volume has emerged from a large body of research on the extractive industries conducted at UNU-WIDER over the last two to three years. It is part of the 'transformation' theme which has been a major pillar of UNU-WIDER's overall research programme, and is also of great relevance to the other two pillars of UNU-WIDER's work, namely 'inclusion' and 'sustainability'. A major concern of the Institute is to focus on the questions of most relevance to policy makers in the developing world, and to their development partners. How to manage a nation's resource wealth is without question one of the biggest issues that policymaking in the LICs and MICs faces. Getting it right can yield enormous benefits, both for the public finances—and therefore for the expansion of public services and infrastructure provision—but also in enabling structural transformation of economies in ways that increase employment and livelihoods, thereby reducing poverty. But getting it wrong can lead to serious economic, social, and environmental impacts that can be difficult to reverse. The contributors to this volume discuss the many lessons that have been learned in the past few decades, the traps to be avoided, the ways forward, and the new opportunities. It is to be hoped that their collective insights will together make a valuable contribution to improved outcomes in the future.

There are many challenges for resource-rich countries. It is not just governments and their international partners that drive the process of change, but also the extractive industry itself, communities, and their NGO and INGO partners. This book recognizes the severe difficulties of achieving positive change that are often encountered in resource-wealthy nations; we are 'realistic optimists'. The book focuses on recommendations that might (but are certainly not guaranteed to) work in delivering inclusive and sustainable development. We are keen to identify actions, policies, and institutions that are potentially transferable across countries, with suitable adaption to local circumstances. Overall, our expectation is that strategies and practices around the extractives sector can improve—but not without a struggle.

## References

Addison, T. (ed.) (2003). *From Conflict to Recovery in Africa.* Oxford: Oxford University Press for UNU-WIDER.

Addison, T. (2009). 'The Political Economy of the Transition from Authoritarianism', in P. de Greiff and R. Duthie (eds), *Transitional Justice and Development: Making Connections*, 110–40. New York: Social Science Research Council for the International Center for Transitional Justice.

Addison, T. (2012). 'The Political Economy of Fragile States', in G. K. Brown and A. Langer (eds), *Elgar Handbook of Civil War and Fragile States*, 363–78. Cheltenham: Edward Elgar.

Addison, T. and S. M. Murshed (2005). 'Post-Conflict Reconstruction in Africa: Some Analytical Issues', in A. K. Fosu and P. Collier (eds), *Post-Conflict Economies in Africa*, 3–17. Basingstoke: Palgrave Macmillan.

Coll, S. (2013). *Private Empire: ExxonMobil and American Power*, London: Penguin.

Erten, B. and J. A. Ocampo (2013). 'Super Cycles of Commodity Prices since the Mid-Nineteenth Century', *World Development*, 44: 14–30.

Francis, M. (2007). 'The Effect of China on Global Prices', *Bank of Canada Review*, (Autumn): 13–25.

ICMM (2010). 'Mapping In-country Partnerships: Mining Partnerships for Development, London: ICMM.

ICMM (2011). 'Mining: Partnerships for Development: Toolkit', London: ICCM.

ICMM (2016). *Role of Mining in National Economies*, 3rd edition. London: ICMM.

IPCC (2015). *Fifth Assessment Report 2015*. Geneva: Intergovernmental Panel on Climate Change. https://www.ipcc.ch/report/ar5/.

Leith, C. (2005). *Why Botswana Prospered*. Montreal: McGill-Queens University Press.

MGI (2013). 'Reverse the Curse: Maximizing the Potential of Resource-driven Economies', McKinsey Global Institute, December.

MMSD (2002). *Breaking New Ground: The Report of the Mining, Minerals and Sustainable Development Project*. London: Mining, Minerals and Sustainable Development Project.

NRGI (2017). *2017 Resource Governance Index*. New York: Natural Resource Governance Institute.

Olson, M. (2000). *Power and Prosperity: Outgrowing Communist and Capitalist Dictatorships*. New York: Basic Books.

Pigato, M. A. and W. Tang (2015). 'China and Africa: Expanding Economic Ties in an Evolving Global Context', Washington, DC: World Bank.

Polanyi, K. (2001 [1944]). *The Great Transformation: The Political and Economic Origins of our Time*. Boston, MA: Beacon Press.

Stern, N. (2015). *Why Are We Waiting: The Logic, Urgency, and Promise of Tackling Climate Change*. Cambridge, MA: MIT Press.

Stewart, F. (2009). 'Policies towards Horizontal Inequalities in Post-conflict Reconstruction', in T. Addison and T. Bruck (eds), *Making Peace Work: The Challenges of Social and Economic Reconstruction*, 136–74. Basingstoke: Palgrave Macmillan for UNU-WIDER.

Williamson, O. (2000). 'The New Institutional Economics: Taking Stock, Looking Ahead', *Journal of Economics Literature*, 38(3): 595–613.

WMO (2017). 'The State of Greenhouse Gases in the Atmosphere Based on Global Observations through 2016', in *WMO Greenhouse Bulletin* No.13. Geneva: World Meteorological Organization.

World Bank (2003). *The Final Report of the Extractive Industries Review*. Washington, DC: World Bank.

World Bank (2015). 'After the Commodities Boom: What Next for Low-Income Countries?', *Global Economic Prospects*, Special Feature 2, June: 91–106.

World Bank (2016). *Commodity Price Forecast: April 19, 2016 Release*. Washington, DC: World Bank.

World Bank (2017). *The Growing Role of Minerals and Metals for a Low Carbon Future*. Washington, DC: World Bank.

# Part II

# Minerals and Oil and Gas in the Global Context

# 2

# Dependence on Extractive Industries in Lower-income Countries

## The Statistical Tendencies

*Alan Roe and Samantha Dodd*

## 2.1 Introduction

The central proposition of this chapter is that extractive industries—minerals (metals and other minerals) plus oil and gas[1]—today play a highly significant role in the economic situations of many low- and middle-income developing economies. This being the case, it is important to understand better the various dimensions of that role; to design approaches and policies that can enhance the positive contributions of extractives; to improve the approaches that can mitigate the potential negative impacts; and generally to embrace the extractives sector as a key contributor to long-run economic and social development. In adopting such an approach, this chapter does not seek to deny the multiple arguments that together constitute the so-called resource curse paradigm. Nor does it take an excessively rose-tinted view of the potential developmental benefits of extractives. Rather, it asserts the core fact that extractives are important and therefore need to be embraced more fully (than is commonly the case) in debates about broader economic and social development.

The purpose of this chapter is to synthesize some statistical and other relevant facts to evidence the central proposition that extractive industries

[1] It is readily accepted that some of the analysis of the book might also be applied to other natural resource industries such as forestry.

are indeed of great significance in many low- and middle-income developing economies. There are four sub-components of this central proposition that are addressed in the subsequent sections of the chapter, namely:

- Section 2.2: How great today is the statistical dependence on extractive industries in low- and middle-income developing countries?
- Section 2.3: Has that level of statistical dependence changed over time—in fact, over the past two decades, since 1996?
- Section 2.4: Has the level of dependence changed as a result of the sharp drop in the prices of extracted commodities in the past three to four years—that is, since the end of the so-called super-cycle?
- Section 2.5: What can we say about the likely future implications of extractives dependence given both the recent softness of prices and some obvious global structural changes such as the phasing down of global dependence on fossil fuels?[2]

## 2.2 Current Levels of Dependence on Extractives

### 2.2.1 *Minerals*

The ICMM publication *The Role of Mining in National Economies* (ICMM 2014) provides data on various measurable aspects of the contribution of mining (but not oil and gas) for every economy in the world (214 countries in total) for the year 2012.[3] By combining the data for three key indicators, that publication provided an updated version of what it terms the Mining Contribution Index (MCI).[4] The ranking of the MCI results for all 214 countries shows that among the top fifty countries, no fewer than thirty-one are either low-income (fourteen countries) or lower-middle-income (seventeen countries) as

[2] The answers to these questions that we are able to provide at this stage rely a great deal on work that has previously been done by ourselves and others under the auspices of the International Council of Mining and Metals (ICMM)—see ICMM (2012, 2014). For this present chapter we have produced only partial updates to that published work.

[3] ICMM has published the third edition of this publication (ICMM 2016), which updates the statistical results on the MCI to the year 2014. These results would marginally alter some of the findings in this section of this chapter (which are based on the MCI data for 2012) but do not change the basic propositions presented here, or the analysis of the later sections, which, consistently with the later ICMM publication, uses export data up to 2014.

[4] The three indicators were: (i) exports of minerals including coal as a share of total merchandise exports in 2012; (ii) the percentage point change in that same indicator measured between 2007 and 2012; and (iii) the total production of metallic minerals including coal expressed as a percentage of GDP. The revised version published in 2016 now uses four indicators rather than three.

classified by the World Bank.[5] There are also four higher-middle-income countries (Botswana, Suriname, Mongolia, and Namibia) among the fifty countries recording the highest levels of overall mineral dependence. So, even if we were to focus only on minerals and ignore oil and gas, there are some thirty-five relatively low-income countries where extractive activity is of very great significance.

The decomposition of the MCI data into two of its three component parts (exports and production) reveals a further important point: that *the most important countries in terms of production* (both US$ value and shares of total world production) *are almost all high-income countries* (notably Australia, Chile, Russia, the United States, and Canada) or higher-middle-income countries (notably Brazil, China, and South Africa). Although a few lower-income countries, such as Mauritania, Guyana, Guinea, and Zimbabwe, have production levels that are high relative to their levels of GDP, their *absolute* levels of production are small relative to those of the richer economies. By contrast, the list of the *most significant countries in terms of export contribution* (mineral exports as a percentage of total exports) *is dominated by low- or lower-middle-income countries*. For example, in 2012, Botswana, Zambia, Eritrea, and Guinea had particularly high mineral export shares of 91.6 per cent, 69.2 per cent, 60.5 per cent, and 60.1 per cent, respectively. Only three of the top twenty countries on this indicator were classified as high-income.[6] Table 2.1 lists all thirty-seven countries that had a mineral export share greater than 30 per cent. All but five of these countries (shaded rows)—so thirty-two countries in total—were in the low- or middle-income categories.

### 2.2.2 *Oil and Gas*

For oil and gas dependence we can assess the situation (also for 2012) by using the detailed UNCTAD data on exports that are available for 215 countries. Those data show that no fewer than forty-eight countries had oil and gas exports in that year greater than 30 per cent of their total merchandise exports.[7] These countries are listed in Table 2.2, which again also shows their World Bank classification (low-income, lower- or upper-middle-income, or high-income).

It should be noted that nineteen of the countries listed are classified as high-income (shaded rows), which means that the remaining twenty-nine[8] countries are either low- or middle-income.

[5] We here use the Bank's most recent classification, which differs slightly from the one that would have applied in 2012—the year to which the MCI data relate.

[6] ICMM (2014: Table 2).

[7] This compares with forty-seven countries in 2011.

[8] This compares with thirty in 2011.

**Table 2.1.** Mineral export dependence in 2012

| | | Country classification | Minerals incl. coal as % of total |
|---|---|---|---|
| 1 | Nauru | Small Island State | 95.9 |
| 2 | Botswana | Upper MY | 91.7 |
| 3 | Mongolia | Upper MY | 83.1 |
| 4 | DRC | Low | 81.5 |
| 5 | Suriname | Upper MY | 75.7 |
| 6 | Zambia | Low MY | 69.2 |
| 7 | French Polynesia | High | 64.6 |
| 8 | Mauritania | Low MY | 62.9 |
| 9 | Chile | High | 61.6 |
| 10 | Eritrea | Low | 60.5 |
| 11 | Peru | Upper MY | 60.1 |
| 12 | Guinea | Low | 60.1 |
| 13 | Guyana | Low MY | 58.8 |
| 14 | Tajikistan | Low MY | 58.5 |
| 15 | Australia | High | 57.3 |
| 16 | Korea, Dem. People's Rep. | Low | 54.4 |
| 17 | Namibia | Upper MY | 53.4 |
| 18 | Papua New Guinea | Low MY | 51.3 |
| 19 | Sierra Leone | Low | 50.6 |
| 20 | Mozambique | Low | 47.5 |
| 21 | Burkina Faso | Low | 46.3 |
| 22 | Sudan | Low | 45.8 |
| 23 | Montserrat | ? | 45.7 |
| 24 | Montenegro | Upper MY | 44.7 |
| 25 | Armenia | Low MY | 44.5 |
| 26 | Lesotho | Low MY | 44.5 |
| 27 | Central African Republic | Low | 44.4 |
| 28 | Mali | Low | 42.3 |
| 29 | Lao People's Dem. Rep. | Low MY | 39.6 |
| 30 | Rwanda | Low | 39.1 |
| 31 | Jamaica | Upper MY | 39.1 |
| 32 | South Africa | Upper MY | 38.8 |
| 33 | New Caledonia | High | 37.8 |
| 34 | Zimbabwe | Low | 37.8 |
| 35 | Iceland | High | 37.7 |
| 36 | United Republic of Tanzania | Low | 35.3 |
| 37 | Lebanon | Upper MY | 32.5 |

*Source*: authors' calculations based on data from UN (available at: https://comtrade.un.org/data/).

It is noteworthy also that only three countries appear in *both* Table 2.1 (minerals) and Table 2.2 (oil and gas). These are the Democratic Republic of Korea, Mozambique, and Sudan.[9] Thus the combining of the two tables would show that in 2012 there were no fewer than fifty-eight low- and middle-income countries (thirty-two plus twenty-nine minus three) that could be

[9] Interestingly, there were also three (different) overlap countries in 2011: Bahrain, Bolivia, and Guinea.

**Table 2.2.** Oil and gas export dependence in 2012

| | | Country classification | Oil and gas % of total |
|---|---|---|---|
| 1 | Iraq | Upper MY | 98.8 |
| 2 | Algeria | Upper MY | 98.4 |
| 3 | Angola | Upper MY | 98.3 |
| 4 | Libya | Upper MY | 97.3 |
| 5 | Timor-Leste | Low MY | 97.1 |
| 6 | Brunei Darussalam | High Y | 96.2 |
| 7 | Equatorial Guinea | High Y | 95.0 |
| 8 | Chad | Low | 93.7 |
| 9 | Nigeria | Low MY | 93.5 |
| 10 | Azerbaijan | Upper MY | 93.4 |
| 11 | Qatar | High Y | 91.4 |
| 12 | Kuwait | High Y | 90.9 |
| 13 | Congo | Low | 88.2 |
| 14 | Yemen | Low MY | 85.9 |
| 15 | Saudi Arabia | High Y | 85.2 |
| 16 | Aruba | High Y | 81.9 |
| 17 | Venezuela | High Y | 81.2 |
| 18 | Gabon | Upper MY | 78.7 |
| 19 | Gibraltar | High Y | 73.7 |
| 20 | Oman | High Y | 71.2 |
| 21 | Iran (Islamic Republic of) | Upper MY | 70.4 |
| 22 | Russian Federation | High Y | 70.3 |
| 23 | Kazakhstan | Upper MY | 69.9 |
| 24 | Norway | High Y | 69.8 |
| 25 | Bahamas | High Y | 66.0 |
| 26 | Colombia | Upper MY | 65.7 |
| 27 | Trinidad and Tobago | High Y | 64.9 |
| 28 | Turkmenistan | Upper MY | 62.6 |
| 29 | Ecuador | Upper MY | 57.8 |
| 30 | United Arab Emirates | High Y | 57.4 |
| 31 | Bolivia (Plurinational State of) | Low MY | 50.9 |
| 32 | Cameroon | Low MY | 50.4 |
| 33 | Bahrain | High Y | 41.5 |
| 34 | Malta | High Y | 41.2 |
| 35 | Korea, Dem. People's Rep. of | Low | 40.1 |
| 36 | Myanmar | Low MY | 39.3 |
| 37 | Cyprus | High Y | 38.1 |
| 38 | Greece | High Y | 37.0 |
| 39 | American Samoa | Upper MY | 36.7 |
| 40 | Egypt | Low MY | 36.0 |
| 41 | Mozambique | Low | 35.8 |
| 42 | Niger | Low | 35.6 |
| 43 | Belarus | Upper MY | 35.6 |
| 44 | Sudan | Low | 34.2 |
| 45 | Indonesia | Low MY | 33.3 |
| 46 | Guam | High Y | 32.9 |
| 47 | Saint Lucia | Upper MY | 31.8 |
| 48 | Ghana | Low MY | 30.5 |

*Source*: authors' calculations based on data from UN (available at: http://unctadstat.unctad.org/EN/).

**Table 2.3.** Changes in extractives export dependence since 1996

| | Income | Country | Minerals as % of total | | | Minerals incl. coal and oil & gas as % of total | | | Change 1996 to 2012 (percentage points) | Change 1996 to 2014 (percentage points) |
|---|---|---|---|---|---|---|---|---|---|---|
| | | | 1996 | 2012 | 2014 | 1996 | 2012 | 2014 | | |
| 1 | Low | Benin | 1 | 21 | 12 | 5 | 37 | 26 | 31 | 21 |
| 2 | Low | Burkina Faso | 8 | 46 | 50 | 23 | 46 | 57 | 23 | 33 |
| 3 | Low | Central African Republic | 56 | 44 | 45 | 56 | 45 | 46 | −12 | −10 |
| 4 | Low | Chad | 0 | 0 | 0 | 0 | 94 | 94 | 94 | 94 |
| 5 | Low | DRC | 72 | 81 | 78 | 83 | 92 | 93 | 9 | 10 |
| 6 | Low | Eritrea | 62 | 61 | 36 | 63 | 61 | 36 | −2 | −27 |
| 7 | Low | Guinea | 76 | 60 | 53 | 81 | 87 | 92 | 7 | 11 |
| 8 | Low | Korea, Dem. People's Rep. | 9 | 16 | 15 | 11 | 56 | 50 | 44 | 39 |
| 9 | Low | Liberia | 49 | 24 | 43 | 50 | 41 | 44 | −9 | −7 |
| 10 | Low | Madagascar | 8 | 18 | 34 | 11 | 20 | 36 | 9 | 25 |
| 11 | Low | Mali | 8 | 42 | 47 | 10 | 43 | 50 | 33 | 40 |
| 12 | Low | Mozambique | 6 | 36 | 42 | 8 | 72 | 68 | 64 | 61 |
| 13 | Low | Niger | 21 | 22 | 21 | 40 | 57 | 57 | 17 | 17 |
| 14 | Low | Rwanda | 3 | 39 | 45 | 3 | 47 | 55 | 44 | 52 |
| 15 | Low | Sierra Leone | 28 | 51 | 46 | 29 | 51 | 46 | 22 | 17 |
| 16 | Low | Togo | 33 | 28 | 18 | 40 | 43 | 34 | 3 | −6 |
| 17 | Low | United Republic Tanzania | 4 | 35 | 33 | 4 | 37 | 34 | 32 | 30 |
| 18 | Low | Zimbabwe | 15 | 27 | 19 | 17 | 38 | 31 | 22 | 14 |
| 19 | Low MY | Armenia | 25 | 45 | 47 | 27 | 50 | 52 | 23 | 25 |
| 20 | Low MY | Bhutan | 3 | 15 | 16 | 23 | 36 | 27 | 12 | 4 |
| 21 | Low MY | Bolivia | 30 | 30 | 27 | 43 | 81 | 81 | 38 | 38 |
| 22 | Low MY | Cameroon | 5 | 5 | 3 | 40 | 5 | 57 | −36 | 16 |
| 23 | Low MY | Congo | 2 | 4 | 8 | 87 | 92 | 92 | 5 | 5 |
| 24 | Low MY | Côte d'Ivoire | 1 | 6 | 6 | 16 | 35 | 26 | 19 | 10 |
| 25 | Low MY | Djibouti | 6 | 17 | 17 | 30 | 23 | 25 | −6 | −4 |
| 26 | Low MY | Egypt | 5 | 9 | 7 | 56 | 45 | 35 | −11 | −22 |
| 27 | Low MY | Ghana | 28 | 18 | 21 | 33 | 48 | 55 | 15 | 22 |
| 28 | Low MY | Guyana | 37 | 58 | 52 | 37 | 59 | 52 | 21 | 14 |
| 29 | Low MY | India | 16 | 11 | 12 | 18 | 30 | 31 | 12 | 14 |
| 30 | Low MY | Indonesia | 6 | 7 | 6 | 32 | 41 | 35 | 9 | 3 |

|  |  |  |  |  |  |  |  |  |  |  |
|---|---|---|---|---|---|---|---|---|---|---|
| 31 | Low MY | Kyrgyzstan | 7 | 17 | 26 | 22 | 31 | 37 | 9 | 15 |
| 32 | Low MY | Lao People's Dem. Rep. | 1 | 39 | 30 | 2 | 55 | 45 | 54 | 43 |
| 33 | Low MY | Lesotho | 4 | 44 | 38 | 4 | 44 | 38 | 41 | 34 |
| 34 | Low MY | Mauritania | 36 | 63 | 59 | 36 | 72 | 68 | 36 | 31 |
| 35 | Low MY | Myanmar | 8 | 18 | 19 | 8 | 58 | 35 | 49 | 27 |
| 36 | Low MY | Nigeria | 0 | 1 | 1 | 94 | 94 | 95 | 0 | 1 |
| 37 | Low MY | Papua New Guinea | 24 | 51 | 39 | 55 | 69 | 70 | 14 | 15 |
| 38 | Low MY | Senegal | 10 | 13 | 16 | 29 | 34 | 36 | 5 | 7 |
| 39 | Low MY | Sudan | 4 | 46 | 25 | 4 | 80 | 88 | 76 | 84 |
| 40 | Low MY | Tajikistan | 30 | 59 | 49 | 33 | 60 | 51 | 27 | 18 |
| 41 | Low MY | Uzbekistan | 10 | 19 | 35 | 16 | 26 | 52 | 10 | 36 |
| 42 | Low MY | Yemen | 1 | 3 | 3 | 94 | 89 | 91 | –5 | –3 |
| 43 | Low MY | Zambia | 76 | 69 | 69 | 80 | 70 | 71 | –10 | –9 |
| 44 | Upper MY | Albania | 9 | 12 | 8 | 15 | 38 | 27 | 23 | 12 |
| 45 | Upper MY | Algeria | 1 | 0 | 0 | 78 | 99 | 98 | 20 | 20 |
| 46 | Upper MY | American Samoa | 0 | 3 | 5 | 0 | 40 | 5 | 40 | 5 |
| 47 | Upper MY | Angola | 5 | 1 | 2 | 99 | 100 | 100 | 1 | 1 |
| 48 | Upper MY | Azerbaijan | 2 | 1 | 1 | 64 | 94 | 94 | 30 | 30 |
| 49 | Upper MY | Belarus | 1 | 1 | 1 | 9 | 36 | 34 | 27 | 25 |
| 50 | Upper MY | Botswana | 81 | 92 | 92 | 81 | 92 | 92 | 11 | 11 |
| 51 | Upper MY | Brazil | 12 | 17 | 16 | 12 | 28 | 26 | 16 | 14 |
| 52 | Upper MY | Bulgaria | 10 | 17 | 15 | 16 | 34 | 27 | 17 | 11 |
| 53 | Upper MY | Colombia | 4 | 7 | 4 | 40 | 73 | 70 | 33 | 30 |
| 54 | Upper MY | Cuba | 16 | 22 | 20 | 17 | 33 | 28 | 16 | 11 |
| 55 | Upper MY | Ecuador | 3 | 2 | 5 | 39 | 60 | 56 | 21 | 17 |
| 56 | Upper MY | Fiji | 9 | 9 | 6 | 10 | 34 | 31 | 25 | 21 |
| 57 | Upper MY | Gabon | 4 | 6 | 7 | 83 | 85 | 72 | 2 | –11 |
| 58 | Upper MY | Iran | 1 | 5 | 6 | 81 | 75 | 71 | –6 | –10 |
| 59 | Upper MY | Iraq | 0 | 0 | 0 | 85 | 99 | 98 | 15 | 13 |
| 60 | Upper MY | Jamaica | 50 | 39 | 48 | 50 | 62 | 69 | 12 | 19 |
| 61 | Upper MY | Kazakhstan | 20 | 14 | 9 | 53 | 84 | 87 | 31 | 34 |
| 62 | Upper MY | Lebanon | 11 | 32 | 26 | 11 | 35 | 30 | 24 | 19 |
| 63 | Upper MY | Libya | 0 | 1 | 2 | 94 | 98 | 97 | 5 | 3 |
| 64 | Upper MY | Malaysia | 1 | 3 | 3 | 9 | 23 | 25 | 14 | 16 |
| 65 | Upper MY | Mongolia | 57 | 75 | 64 | 58 | 86 | 93 | 29 | 35 |
| 66 | Upper MY | Namibia | 38 | 53 | 38 | 40 | 54 | 40 | 15 | 0 |

*(continued)*

**Table 2.3.** Continued

| | Income | Country | Minerals as % of total | | | Minerals incl. coal and oil & gas as % of total | | | Change 1996 to 2012 (percentage points) | Change 1996 to 2014 (percentage points) |
|---|---|---|---|---|---|---|---|---|---|---|
| | | | 1996 | 2012 | 2014 | 1996 | 2012 | 2014 | | |
| 67 | Upper MY | Panama | 3 | 3 | 5 | 9 | 15 | 32 | 7 | 23 |
| 68 | Upper MY | Peru | 48 | 60 | 54 | 55 | 72 | 66 | 17 | 11 |
| 69 | Upper MY | Saint Lucia | 0 | 2 | 2 | 0 | 34 | 33 | 34 | 33 |
| 70 | Upper MY | South Africa | 30 | 33 | 33 | 41 | 42 | 43 | 1 | 2 |
| 71 | Upper MY | Suriname | 69 | 76 | 23 | 73 | 85 | 37 | 12 | –36 |
| 72 | Upper MY | Turkmenistan | 1 | 0 | 1 | 75 | 63 | 89 | –12 | 15 |

*Source*: authors' calculations based on data from UN (available at: http://unctadstat.unctad.org/EN/).

said to be highly dependent on extractive industries—at least in the dimension of their export trade.

## 2.3 Changes in Extractives Dependence since 1996

Unfortunately, the full MCI data have been produced only for the years back to 2012. So, in order to examine changes in dependence over an extended period, we once again need to rely mainly on the UNCTAD export trade data. Changes over time were assessed by looking first at just the 58 low- and middle-income countries identified in Tables 2.1 and 2.2[10] and comparing their levels of export dependence on extractives in 1996, 2012, and 2014 (the latest year for which UNCTAD data were available at the time of writing). The rows of Table 2.3 show the results of this over-time comparison, with the percentage point changes (1996–2012 and 1996–2014) given in the last two columns. We have supplemented the list of countries identified above with data for a further nine countries, where the joint contribution of minerals and oil and gas exports brings the 2014 level of export dependence above the 30 per cent cut-off that was used for both Table 2.1 and Table 2.2. These additional countries are Liberia, Madagascar, Togo, India, Kyrgyzstan, Senegal, Uzbekistan, Fiji, and Panama. Finally, we have included a few countries that have large exports of extractives but do not quite attain the 30 per cent cut-off point. The results are shown both for metals and for the total of metals and oil and gas.

The results of the comparisons are clear and unambiguous. The final two columns of Table 2.3 show that for most of the countries listed there was a strong and positive increase in their levels of dependence on the export of extractives from 1996 onwards. Between 1996 and 2012, no fewer than sixty-two of the seventy-two countries listed saw increases in their export dependence ratios,[11] the percentage point increases being as high as 94 per cent (Chad), 76 per cent (Sudan), 64 per cent (Mozambique), and 54 per cent (Lao). Only ten countries saw a decline in the ratio in that period (marked by shading in the penultimate column). The simple average increase over that sixteen-year period was 18 percentage points.

For the longer period, from 1996 to 2014, the changes were still predominantly upwards but slightly smaller in magnitude: the simple average increase across all the countries was 17 percentage points. In that period, eleven of the

[10] For some of these countries the over-time comparison was not possible because of missing data in the earlier year.

[11] When the comparison is run from 1996 to 2011 (rather than 2012), fifty-eight of the total of sixty-eight countries saw increases.

seventy-two countries listed saw falls in their levels of extractives dependence (marked by shading in the final column), but sixty-one countries saw increases.[12]

Among the eighteen *low-income* countries that are listed, fifteen saw an increase in the dependence ratio up to 2012 and fourteen saw an increase up to 2014. Of the twenty-five *lower-middle-income* countries, twenty saw an increase and one no change (Nigeria) to 2012, and the same twenty-one countries saw an increase to 2014.

The conclusion from this set of simple over-time comparisons is quite clear: dependence on the extractive industries has increased in the eighteen-year period between 1996 and 2014 in low- and middle-income countries, which are the main concern of this chapter. The high level of dependence that we identified in Section 2.2 for the year 2012 was largely sustained until 2014 and has been the result of an extended period of change from 1996 onwards, during which the general tendency has been for greater, rather than reduced, dependence.

However, we need to be aware of certain limitations of this analysis: in particular, it is for the moment based predominantly on export trade data and it extends only to 2014. Correction for these limitations would be likely to strengthen the main conclusions for several reasons. First, there are a number of newer extractive countries, such as Afghanistan, Kenya, and Uganda, for which the 2014 UNCTAD data do not yet capture the increases in dependence ratios that seem likely to occur in the next few years. By 2014, Afghanistan and Kenya, for example, had export dependence ratios of 19 per cent and 12 per cent, respectively, and these seem certain to rise in future years. Second, there are some very important extractives producers that are not statistically important in the export markets. These include principally China (15.8 per cent of world metal production in 2012), Brazil (8.5 per cent), and India (3.4 per cent). The inclusion of these large producers in the datasets would obviously strengthen our main conclusion.

## 2.4 The Impact of the End of the Commodity Price Cycle

It is difficult to date precisely the point at which the so-called super-cycle of commodity prices ended. As shown in Figure 2.1, the prices of different metal commodities headed south around 2011, but did so at various dates and with various degrees of associated volatility. But it is clear, in the case of metals at least, that the downward tendency of prices was well established by 2012. The same is true of the crude oil price (Brent oil prices hit a high of US$127 a barrel

[12] For the comparisons from 1996 to 2011, these numbers change to eleven countries seeing falls and fifty-seven countries seeing increases.

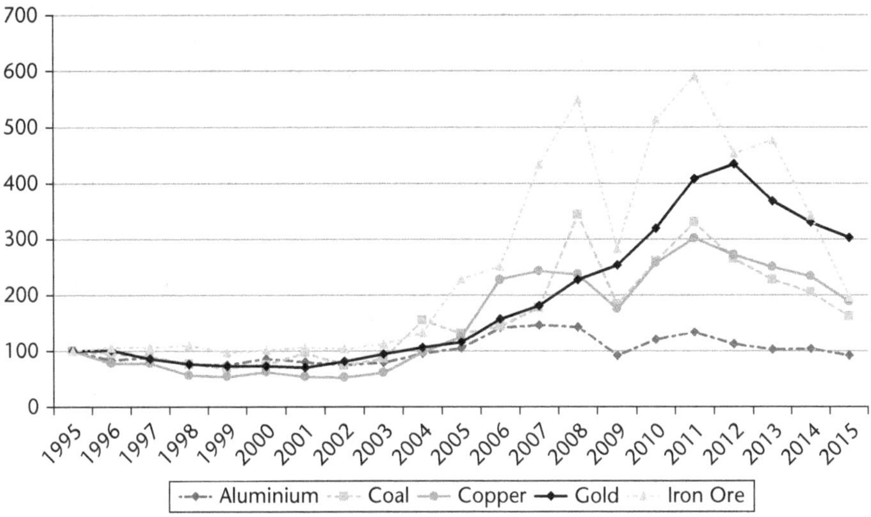


**Figure 2.1.** Prices for selected metals, 1995–2015 (index 1990 = 100)
*Source*: World Bank Commodity Price Data (The Pink Sheet), 2016.

in April 2011 as the conflict in Libya shut down its supplies). It is too early to say what longer-term effect the price falls of the past four years will have on the levels of extractives dependence in the countries we have identified. Although significant extractives capacity has been mothballed, it is not clear how much may eventually be closed or how much planned new investment will eventually have to be cancelled. For some metals, including copper, the price falls to date have arguably been offset by lower production costs and have affected only the very highest-cost producers.[13]

Data such as those shown in Table 2.3 enable us to tell merely an initial and partial story about the impact of the price collapse on the levels of extractive industry dependence. We have used 2011 as the assumed start date of the commodity price collapse, although in reality the actual start date was somewhere between 2011 and 2012. Data comparable to those shown in Table 2.3 were compiled for sixty-eight countries[14] for both 2011 and 2014 and then compared. The results of this for these sixty-eight countries are shown graphically in Figure 2 in Roe and Dodd (2017).

[13] For example, a March 2015 press release by SNL Metals noted that the copper price had fallen to a level just above the 9th decile of high-cost producers, which indicated, they argued, that the copper price would need to fall further before significant capacity became vulnerable to closure. Also 'the... spot price of copper averaged only US$5,700/t in February, but remains comfortably above SNL's prediction of this year's average mined cost of 168c/b, calculated on a co-product basis'.

[14] These sixty-eight countries emerged from the 2011 data in the same way that the seventy-two countries in Table 2.3 emerged from the 2012 data (see the explanations in Section 2.3).

The results of this limited experiment are ambiguous. Using the new data, forty-two of the sixty-eight countries did experience some decline in measured export dependence between 2011 and 2014. In nine country cases, this decline was large, at more than 10 percentage points.[15] In the other thirty-three cases of decline, the decline was quite small. By contrast, in twenty-six country cases there was no decline—rather an increase in measured dependence with quite substantial percentage point increases in many cases. These results suggest that the commodity price collapse did indeed disrupt the previously strong upward tendency in extractives dependence evidenced in the earlier part of this chapter. However, that disruption was not general across all countries and a significant number among the sixty-eight saw an ongoing and often strong increase in their levels of dependence. We will need to await the similar data up to 2017 to see whether this somewhat ambiguous pattern has persisted.

## 2.5 Implications of High Extractives Dependence

Section 2.5 is intended to provide an assessment of some of the implications for low- and middle-income countries of the high levels of extractives dependence that the chapter has identified. The material presented here is highly derivative from the published work of others and is intended mainly to support a broader debate on the topic. The discussion is sub-divided between the short-term consequences and the longer-term perspective.

### 2.5.1 *Short-term Consequences*

Several papers have been published in the past three years or so assessing the implications for developing economies of the sharp drop in the prices of commodities in general but extractive products in particular.[16] Thomas Lassourd and David Manley of the NRGI have identified ten significant economic and political consequences.[17] Foremost among these are the significant losses of *fiscal revenue* and the associated painful fiscal adjustment that many extractives-dependent economies now face. They note that these pressures apply also to some richer economies, such as Saudi Arabia and Norway, which

[15] These nine countries were Eritrea (25%), Myanmar (21%), Benin (19%), Syrian Arab Republic (17%), Gabon (13%), Mali (13%), Tanzania (12%), Egypt (11%), and Lao People's Dem. Rep. (11%).

[16] For example, Zhenbo Hou et al. (2015). However, as an indication of how quickly things have changed, another ODI paper, published as relatively recently as August 2012, was still mainly concerned with the very high prices (for oil) and predicting that 'in terms of real GDP, African countries may suffer up to a 3% loss from a doubling of oil prices' (Cantore et al. 2012).

[17] Lassourd and Manley (2015).

have large, established reserve funds that can absorb some of the pressures. This mitigating factor, however, is not applicable to all extractives-dependent countries or indeed to most of the low- and middle-income economies that are our main concern: the authors mention the cases of Yemen (a low-income country with a projected fiscal deficit of 10 per cent of GDP) and Venezuela (a high-income country with inflation of over 200 per cent and a projected fiscal deficit of almost 20 per cent of GDP).[18] Further, anticipating a later chapter in this volume (Solimano and Guajardo, Chapter 10, and Solimano 2017), it is well known that some countries (notably Chile) have strong counter-cyclical fiscal arrangements in place that can mitigate some of the consequences of lower prices, but most countries do not.

As a consequence of lower prices for their extractive exports, many dependent economies can also expect depreciating *exchange rates* and higher rates of *inflation* in addition to the inevitable impact on real incomes that lower prices will cause. The lower prices are also likely to affect prospective income, as many of the early-stage exploratory investments in certain lower-income countries are delayed or even abandoned. Lassourd and Manley (2015) mention important prospective investments in the cases of Guinea (iron ore), Mozambique (oil and gas), Uganda (oil), and Tanzania (natural gas), but there are many others. ICMM (2016) provides evidence of an 80 per cent decline in exploration spending by major mining companies between 2012 and 2015.

However, not all of the consequences are necessarily negative and some opportunities are also referred to. Foremost among these are the opportunities that are presented by much lower fuel prices to reduce or eliminate longstanding but fiscally costly fuel subsidies: an example of a country where this opportunity is being seized is Indonesia. It might be added that Tanzania has been fortunate in timing in that its new near-shore gas came on stream in late 2015 and is already fuelling new gas-fired power generation near Dar es Salaam that is enabling the government to reduce its dependence on high-cost imported feedstocks and so reduce associated large fiscal subsidies; see Chapter 25 by Östensson and Löf for more detail.[19] Lassourd and Manley (2015) also suggest that a period of low fuel prices provides an opportunity for governments to introduce or raise carbon taxes and take other measures to discourage fossil-fuel consumption and encourage the greater uptake of renewables.

Finally, they draw attention to a range of *political* consequences that could arise from lower prices. For example, lower prices could either mitigate or exacerbate political repression in countries like Azerbaijan, and ongoing resource-related conflicts in Libya, South Sudan, and Iraq. Potentially, lower

[18] These are the figures quoted by the authors and are not necessarily the most up-to-date figures. In the case of Venezuela, there has been no formal IMF Article IV review since 2004.

[19] See also Roe (2016).

resource revenues available for governments 'to buy social peace' might spark conflict in countries such as Bahrain, where they argue that the leaders have used resource revenues to keep a lid on growing discontent. More generally, 'a decline in natural resource prices will reduce the incumbency advantages held by the leaders controlling the tap' (Lassourd and Manley 2015).

The manner in which these potential impacts will apply in practice is obviously country- and case-specific. For example, countries dependent on gold mining have faced smaller fiscal and other economic problems because of the relative strength of the gold price. But few if any of the seventy-two most extractives-dependent economies that this chapter has identified will be able to escape all of the consequences. So there is undoubtedly a very large agenda of work needed in most of those countries to address the situation that they are now experiencing—and will continue to experience for as long as commodity prices remain relatively soft.

### 2.5.2 *The Longer-term Perspective*

The purpose here is to identify some of the (positive) longer-term aspects of resource dependence for the type of country we have identified. The ideas are based squarely on the detailed analysis published in December 2013 by the McKinsey Global Institute. Although this study was published before the commodity price slump intensified, many of its broad findings remain relevant to the majority of the seventy-two countries we have identified. McKinsey (2013b) identified four main drivers of what it terms 'resource market dynamics'. These are:

- the growing levels of global market demand for resources, including minerals and oil and gas (evidenced also by an earlier study, Mckinsey 2013a)[20]
- the gradual shift of the sources of supply to more challenging locations—many in less-developed countries
- the environmental pressures that will both raise the costs of mining and constrain outputs (especially of carbon-intensive fuels)
- the technology improvements that will allow more efficient and lower-cost extraction and extraction in more difficult sites.

[20] As one aspect of this, McKinsey draws on OECD forecasts that the numbers of global middle-class consumers will increase by 3 billion over the next twenty years. The research defines 'middle-class' as having daily per capita spending of US$10 to US$100 in PPP terms and is based on a 2010 paper by Homi Kharas (2010). Increased income inequality would likely add to the estimated number of middle-class consumers. A faster rate of change in energy efficiency would somewhat dilute the estimated impact on resource needs.

The influences of these various component drivers on future levels of demand for extractive products, the outputs of such products, and the costs of producing them are to an extent offsetting. Overall, however, the perspective that McKinsey presents is one that will require investment in mineral and oil and gas extraction of between US$11 trillion and US$17 trillion cumulatively by 2030. This is 65 per cent higher than the historical annual rate of investment in these sectors, even when it allows for the reducing effects of climate change initiatives. Figure 2.2 summarizes McKinsey's various historical and projected data on the investments needed in these sectors both to replace existing supplies and to provide net new capacity. This is done using two alternative scenarios. The first assumes a supply change that allows for 'business-as-usual' improvements in resource productivity (e.g. more efficient vehicles). The second builds in larger supply reductions due not only to productivity improvements but also to changes in the energy supply mix towards the greater use of renewables.[21]

It needs to be recognized that these McKinsey scenarios were constructed at a time when the decline in commodity prices was only in its early stages (2012/13). The large declines in some prices since then will have affected the realism of those scenarios, but to what extent we do not know. However, as we

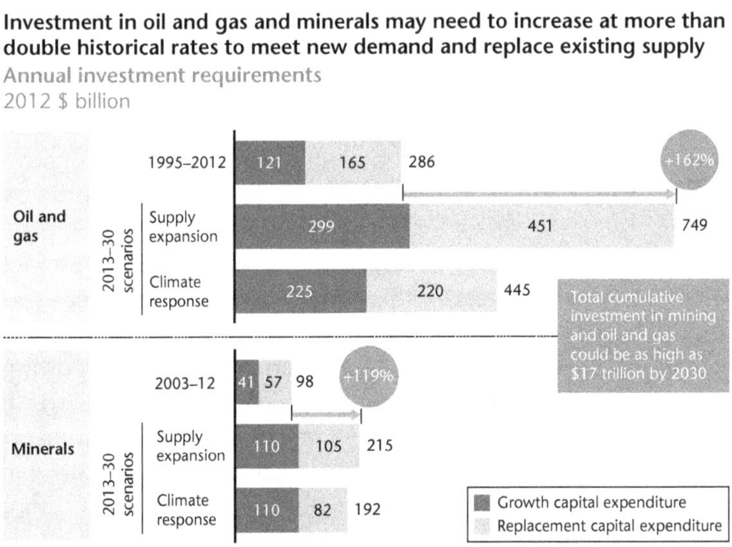


**Figure 2.2.** Investments in oil and gas, and minerals, 1995–2030

*Source*: Exhibit 5 in McKinsey Global Institute (2013b: 31); reproduced here with permission from McKinsey Global Institute.

[21] McKinsey (2013b: appendix 1).

can see from Figure 2.2, we would need to reduce the smaller of the two projected investment amounts (the scenario that factors in climate change responses) by almost 40 per cent in the case of oil and gas before it would become equal to the 1995–2012 annual average investment (i.e. from US$445 billion to US$286 billion) and by almost 50 per cent in the case of minerals before it would become equal to the 2003–2012 annual average (i.e. from US $192 billion to US$98 billion). Any re-running of the McKinsey scenarios would be unlikely to reduce the projected figures by such large percentage amounts.[22] So it seems safe to assume that a feasible scenario to 2030 would still show the necessary levels of future new investment to be high relative to the historical averages. Since the historical period used in the McKinsey analysis (1995–2012) coincides with the period for which we have shown extractives dependence in many low- and middle-income countries to have risen significantly (Table 2.3), it is a further reasonable assumption that that level of dependence will be sustained until 2030 and may even increase further.

This being the case, there are substantial potential opportunities for the developing countries that produce minerals and/or oil and gas. In terms of investment, McKinsey (2013b) suggests that in its 'potential upside' case, up to US$3 trillion of the total global investment in scenario 1 could be expected to be invested in low- and low-middle income countries (cumulatively by 2030). This would represent a tripling of the levels of investment in the extractives sectors seen in those countries since 1995. Even in its lower, 'base case', scenario the cumulative investment in these countries could be at an annual rate 50 per cent higher than in the past twenty years. McKinsey further estimates that on some assumptions this investment and the growth that it could sustain could lift some 540 million more people out of poverty—a figure higher than the poverty reduction achieved in China in the previous twenty years.[23]

If one accepts the McKinsey investment scenarios (or something close to them), the *necessary* conditions for this dramatic gain in investment and living standards can be expected to be in place over the next few years. However, the *sufficient* conditions (in terms of governance arrangements, supporting infrastructure, etc.) are certainly not guaranteed to be available to deliver the full pay-off that seems to be possible. Indeed, one of the central purposes of this and several other chapters in this book is to explore in some detail what these further conditions might be and how they might be enhanced through the work of a range of actors—government, companies, donors, and others.

[22] Further, we can expect that lower prices and their consequences will in some cases erode the earlier enthusiasm for climate change adjustments. A current example is very high energy prices in the United Kingdom, which threaten the survival of domestic steel production (an intensive user of fossil fuels), that these have now been shown to create.

[23] McKinsey (2013b: 33, exhibit 7).

In the partial absence of these further conditions, one needs to take on board the warning notes also voiced by the McKinsey study. This showed that between 1995 and 2011 (broadly the period over which we have reviewed the change in extractives dependence), some 52 per cent of the lower-income resource-driven countries of the world had failed to make significant progress in catching up with the incomes of higher-income countries.[24] The slightly more positive insight was that, of the 77 per cent of resource-driven countries with below-average per capita income (of US$10,000) in 1995, 48 per cent did achieve some catch-up.

## 2.6 Conclusions

The statistics outlined in this chapter highlight several important facts that have significant implications for many low- and middle-income countries. Three points are particularly salient.

First, although high-income countries dominate extractives production figures, countries with the highest levels of export dependence on extractives are predominantly low- and middle-income countries. Second, export dependence has shown a clear upward trend, with sustained increases in most countries over the last two decades. Finally, although the overall trend since the end of the commodity super-cycle is somewhat ambiguous, this upward trend has nevertheless continued in many countries despite the recent commodity price reversals.

High levels of export dependence in many low- and middle-income countries, coupled with lower commodity prices, have serious economic and political implications for these countries. The primary economic concerns over the short and medium term are loss of fiscal revenue, exchange rate depreciation, higher inflation, and a reduction in real income. While the principal economic impacts are negative, lower prices do open up the possibility of reform of costly fuel subsidies and the introduction of carbon taxes, both of which would be positive developments. Political implications are more ambiguous; it is unclear whether the reduced prices will exacerbate or mitigate political repression and (current and potential future) resource-related conflicts.

Projected trends in investment suggest that, despite reduced commodity prices, investment in extractive industries over the next fifteen years is likely to be high relative to historical averages. This suggests that, over the longer term, dependence on extractives in low- and middle-income countries may increase further. Increased investment could lead to improved living standards. However,

[24] McKinsey (2013b: 34, exhibit 8).

this outcome is contingent upon the decisions and actions of a range of actors, including government, companies, and donors.

## References


Cantore, N., A. Antimiani, and P. R. Anciaes (2012). 'Energy Price Shocks: Sweet and Sour Consequences for Developing Countries', ODI Working Paper 355. London: ODI.

Hou, Z., J. Keane, J. Kennan, and D. W. te Velde (2015). 'The Oil Price Shock of 2014: Drivers, Impacts and Policy Implications', ODI Working Paper 415. London: ODI.

ICMM (2012). *The Role of Mining in National Economies*. London: ICMM.

ICMM (2014). *The Role of Mining in National Economies*, 2nd edition. London: ICMM.

ICMM (2016). *The Role of Mining in National Economies*, 3rd edition. London: ICMM.

Kharas, H. (2010). 'The Emerging Middle Class in Developing Countries', OECD Development Centre Working Paper 285. Paris: OECD.

Lassourd, T. and D. Manley (2015). '10 Consequences of the Commodity Crash', Guest Post, NRGI, February.

McKinsey Global Institute (2013a). 'Resource Revolution: Tracking Global Commodity Markets', September 2013. London: MGI.

McKinsey Global Institute (2013b). 'Reverse the Curse: Maximizing the Potential of Resource-driven Economies', December 2013. London: MGI.

Roe, A. R. (2016). 'Tanzania: From Mining to Oil and Gas', WIDER Working Paper 2016/79. Helsinki: UNU-WIDER.

Roe, A. R. and S. Dodd (2017). 'Dependence on Extractive Industries in Lower-income Countries: The Statistical Tendencies', WIDER Working Paper 2017/98. Helsinki: UNU-WIDER.

Solimano, A. (2017). 'Sector, Fiscal Rules, and Stabilization Funds in Chile: Scope and Limits', WIDER Working Paper 2017/53. Helsinki: UNU-WIDER.

# 3

# Mining's Contribution to Low- and Middle-income Economies

*Magnus Ericsson and Olof Löf*

## 3.1 Introduction

This chapter is designed to provide an up-to-date statistical analysis of the scale of the current dependency of low- and middle-income economies on various extractive resources in dimensions such as production, income (GDP), exports, government revenues, exploration, and employment. The study also attempts to explain and document how country levels of minerals dependency have changed in the past twenty years.

Drawing on the detailed data available for the minerals sector, an analysis is carried out of the situation in 2014, and of recent trends in mining's contribution to the economic development of low- and middle-income countries for the years 1996–2014. By using data on variables such as production, prices, mineral rents, exploration expenditure, government revenues, and employment, this chapter offers answers to questions such as:[1]

- What is the magnitude of the statistical dependency on mining industries in low- and middle-income developing countries today?
- Has that level of statistical dependency changed over the past twenty years, from 1996 to 2015?
- Has the level of dependency changed as a result of the sharp drop in prices of most extracted commodities since about 2011, after the end of the so-called super-cycle?

[1] This chapter complements an earlier paper examining similar questions for both mining and oil and gas (see Roe and Dodd 2016; Roe and Round 2017).

The methodology is based on earlier work coordinated by the ICMM, in which the authors participated in 2010 and 2014 (ICMM 2010, 2014).

## 3.2 Methodology

### 3.2.1 *Mining Contribution Index WIDER*

One existing approach to assessing the magnitude of the dependency of countries on extractive resources is the MCI developed by the ICMM (2010, 2014, 2016).

In this chapter MCI is updated and also further developed. Our revised version is called the Mining Contribution Index WIDER (MCI-W), and is based on four indicators:

1. exports of minerals including coal as a share of total merchandise exports
2. the total production value at mine stage of metallic minerals, industrial minerals, and coal, expressed as a percentage of GDP
3. mineral rents as a percentage of GDP
4. exploration expenditure.

MCI and MCI-W are similar, but use two different ways of combining some measurable indicators. MCI-W uses GDP purchasing power parity (PPP, real US$ with 2011 as the base year) from the World Bank.

### 3.2.2 *Indicators*

The rationale for including each of our four indicators is as follows.

#### 3.2.2.1 EXPORTS

International trade in metals reflects regional and national advantages and specializations along the value chain (Tercero Espinoza and Soulier 2016). Mineral and metal export contribution in 2014 provides a measure for the scale of mining in relation to other productive activities, in particular for small low- to middle-income countries. UNCTAD validates and compiles a wide range of data collected from national and international sources to provide reliable statistics to facilitate analyses of the most urgent and emerging issues. UNCTAD covers international trade and exports of metals and minerals. The specific trade groups used are: non-ferrous metals (Standard International Trade Classification (SITC) 68); other ores and metals (SITC 27 and 28); pearls, precious stones, and non-monetary gold (SITC 667 and 971); coal, whether or not pulverized, not agglomerated (SITC 321);

coke and semi-cokes of coal, lignite, or peat, and retort carbon (SITC 325) (UN Comtrade).

#### 3.2.2.2 VALUE OF MINE PRODUCTION

This is non-fuel mineral production value expressed as a percentage of GDP (1996–2014). It provides a sense of the scale of value of production relative to the size of the economy. Note that it does not represent the contribution of mining to GDP—on average perhaps only a third of production value represents value addition to the national economy.

The value of mine production is based on figures obtained from Raw Materials Group data until 2013. Figures for 2014 were collected and computed by the authors using the same methodology (Raw Materials Group 1997: 497). A list of minerals and metals included is given in Figure 3.1. Uranium, aggregates, and limestone are not included.

#### 3.2.2.3 MINERAL RENTS

Mineral rents are the difference between the value of production for a stock of minerals at world prices and their total costs of production including 'normal' profit. Minerals included in the calculation are tin, gold, lead, zinc, iron, copper, nickel, silver, bauxite, and phosphate. Mineral rent statistics are derived from the World Development Indicators created by the World Bank.

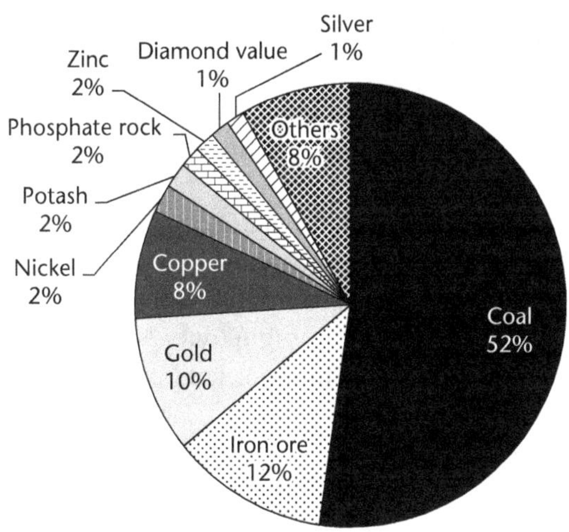


**Figure 3.1.** Value of mine production by commodity (%), 2014

*Note*: Others include: salt, lead, chromite, manganese ore, molybdenum, bauxite, tin, palladium, graphite, rare earth elements, kaolin, boron, fluorspar and feldspar.

*Source*: authors' illustration based on data from British Geological Survey, US Geological Survey, World Mineral Statistics, and Raw Materials Data.

3.2.2.4 EXPLORATION

The exploration expenditure data produced by SNL Mining & Metals (2016)[2] provides a forward-looking indication of the likelihood of continued mining activity in a country.

### 3.2.3 *Calculation*

MCI-W is calculated as follows: countries are ranked in descending order for each of the four MCI indicators. Countries for which data do not exist are omitted from the ranking. As a result, indicator 1 is ranked out of 216 countries, indicator 2 is ranked out of 127 countries, indicator 3 is ranked out of 125, and indicator 4 is ranked out of 122 countries. For each country percentile ranks are calculated based on the four indicators, by dividing the country rank by the maximum rank within that indicator to generate a ranking between 0 and 1. Finally, the four MCI indicators are weighted equally at 1/4, summed up, and multiplied by 100 (ICMM 2014).

In this chapter the focus is on the low- and middle-income economies for the years 1996–2014.[3]

## 3.3 Current Levels of Mining Contribution to National Economies

Our MCI-W results confirm that mining is indeed the backbone of several nations' economies. In some nations, mining accounts for a dominant share of the national wealth, with more than 50 per cent of exports and around 10–20 per cent of GDP: many of these countries are low- and middle-income economies. The distinction between different regions is shown graphically in Figure 3.2, the black areas showing the highest levels of dependency. Regions where mining makes a particularly high contribution are Western, Southern, and Central Africa, Oceania, Central Asia, and Latin America. Almost all countries have some, often small-scale, mining activity producing, for example, coal and aggregates for domestic use. These mineral products are most often not

[2] SNL Metals & Mining (2016) focuses on corporate spending. In reality, if one adds metals and minerals not included by SNL Mining & Metals, and if one counts exploration undertaken by entities not surveyed, total exploration on either a national or a global basis is definitely higher than indicated by SNL for each country. In this chapter this difference is considered to be of minor importance.

[3] Low-income economies are defined by the World Bank as those with a gross national income (GNI) per capita of US$1,025 or less in 2015; lower-middle-income economies are those with a GNI per capita between US$1,026 and US$4,035; upper-middle-income economies are those with a GNI per capita between US$4,036 and US$12,475; high-income economies are those with a GNI per capita of US$12,476 or more.

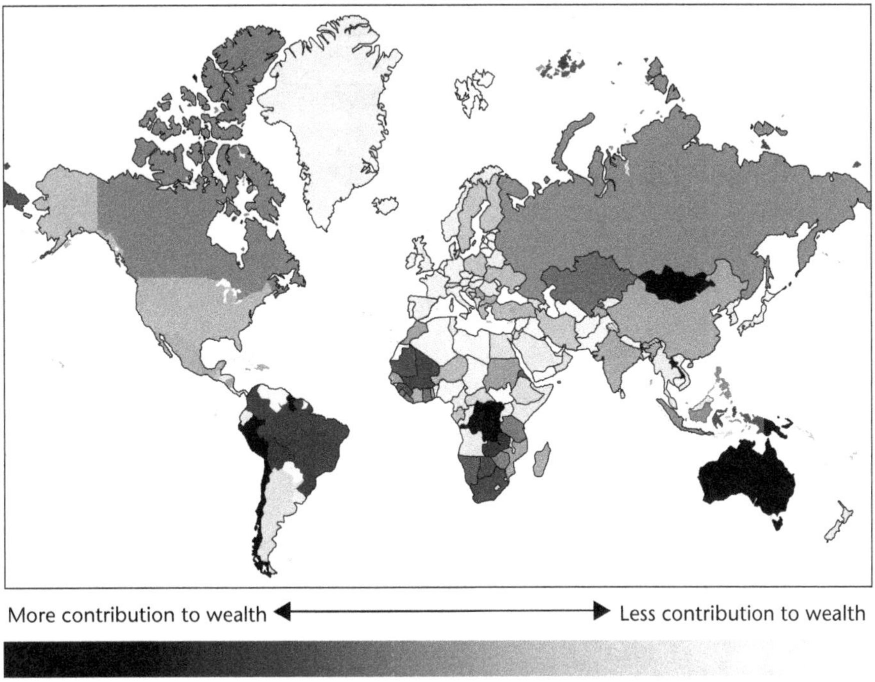


**Figure 3.2.** MCI-W score by country, 2014
*Source*: authors' calculations.

exported, as their low value does not allow transport over longer distances, and hence the combined contribution by production and exports is small. There are some regions or countries where mining contributes less to national wealth: Western Europe, the Middle East and North Africa, Japan, and South-East Asia (lighter areas in Figure 3.2).

### 3.3.1 *Country Rankings*

In MCI-W based on the latest available data for 2014, the DRC is ranked as the country with the largest contribution of mining to its economy (Table 3.1). Mineral exports constitute 81 per cent of total exports there, and DRC is ranked the fourth most important country in relation to mineral export contribution. Mineral production value at the mine stage was US$8 billion in 2014, and the mineral production value as a percentage of GDP was 15 per cent: on this indicator, DRC is ranked number three. Exploration expenditure was US$300 million in 2014, placing DRC in tenth place globally. Mineral rents constituted 20 per cent of total GDP, and DRC is ranked number two in 2014. These four variables give the composite score of 97.6 out of 100 in

**Table 3.1.** MCI-W top 20, 2014

| Country | Ranking | MCI-W score |
|---|---|---|
| DRC | 1 | 97.6 |
| Chile | 2 | 95.2 |
| Australia | 3 | 95.0 |
| Mongolia | 4 | 93.9 |
| Papua New Guinea | 5 | 93.4 |
| Zambia | 6 | 92.6 |
| Peru | 7 | 91.4 |
| Burkina Faso | 8 | 90.5 |
| Mali | 9 | 89.9 |
| Guyana | 10 | 89.9 |
| South Africa | 11 | 89.2 |
| Botswana | 12 | 89.0 |
| Guinea | 13 | 88.6 |
| Mauritania | 14 | 88.5 |
| Eritrea | 15 | 86.4 |
| Namibia | 16 | 86.2 |
| Ghana | 17 | 84.5 |
| Lao PDR | 18 | 83.5 |
| Sierra Leone | 19 | 82.5 |
| Uzbekistan | 20 | 81.2 |

*Source*: authors' calculations.

the index for DRC. The top ten countries in the 2014 MCI-W ranking in descending order are DRC, Chile, Australia, Mongolia, Papua New Guinea, Zambia, Peru, Burkina Faso, Mali, and Guyana.

Of the top fifty countries in MCI-W 2014, there are only four high-income economies (HIE), but sixteen upper-middle-income economies (UMIE), eighteen lower-middle-income economies (LMIE), and twelve low-income economies (LIE).

While there are two high-income countries, Chile and Australia, among the five countries with the highest MCI-W scores, there are only two additional high-income countries among the top fifty (Canada and the Russian Federation). It should also be noted that all five of the BRICS countries (Brazil, the Russian Federation, India, China, and South Africa) are among the MCI-W top forty-five.

In Figure 3.3 we present a four-dimensional chart with the export contribution shown on the x-axis and mineral value as percentage of GDP on the y-axis. The size of the circles is proportional to the value of mine production in absolute terms (US dollars). The fourth dimension is time, the data being presented only for 2014 in Figure 3.3. The figure shows the top twenty MCI-W countries. Australia has by far the largest mining industry by value of production, and the high value is represented by the size of the circle. The export contribution ranking is topped by Mongolia, DRC, and Botswana at levels of 80–90 per cent of total exports, followed by Zambia, Mauritania, and Mali

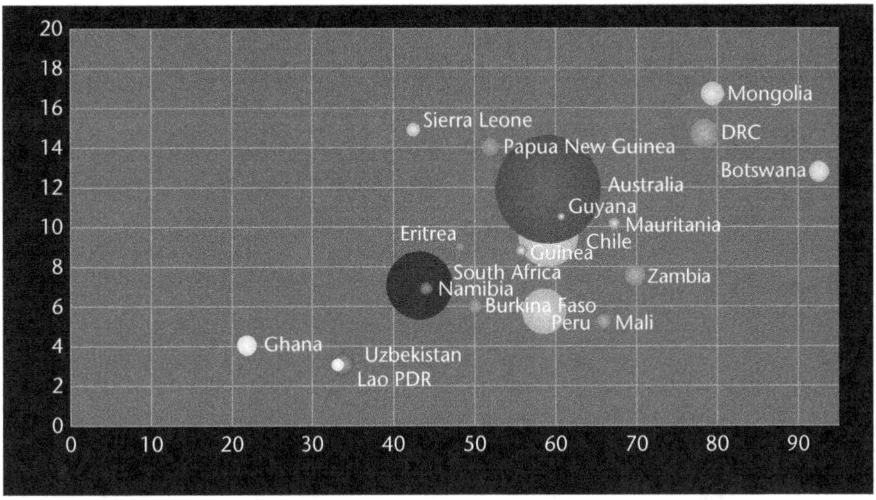


**Figure 3.3.** MCI-W top 20, 2014
*Source*: authors' calculations.

with export contribution levels at around 60–70 per cent. The graphic confirms that the countries with the highest levels of export contribution are mainly LIE or LMIE. Eritrea, with only one mine of industrial scale in operation in 2014, is represented by the small circle at production 9 per cent and export 48 per cent.

## 3.3.2 *Value of Mine Production*

While there are thirty LIE and LMIE among the top fifty MCI-W countries, the HIE and UMIE are substantially more important in terms of production value—for example, China, Australia, the United States, Canada, Chile, the Russian Federation, South Africa, and Brazil (Figure 3.4). It should be noted that the main engine of metal demand, China, is also by far the most important mining country when coal is included in the production total. If coal is not considered, but only metals and industrial minerals, Australia and China are roughly the same size. The absolute levels of production are relatively small for several of the states in the MCI-W top fifty—such as Guyana, Eritrea, and Guinea—but for the economy in the broader sense, mining is an important contributor to all the MCI-W top fifty states.

Figure 3.4 clearly shows that the total value of mineral production at the mine stage is dominated by coal. Coal constitutes roughly half of the total value of industry production globally. Iron ore (Fe), copper (Cu), and gold (Au) follow next. The industrially important metals nickel (Ni) and zinc (Zn) are each roughly an order of magnitude smaller. These metals are of the same

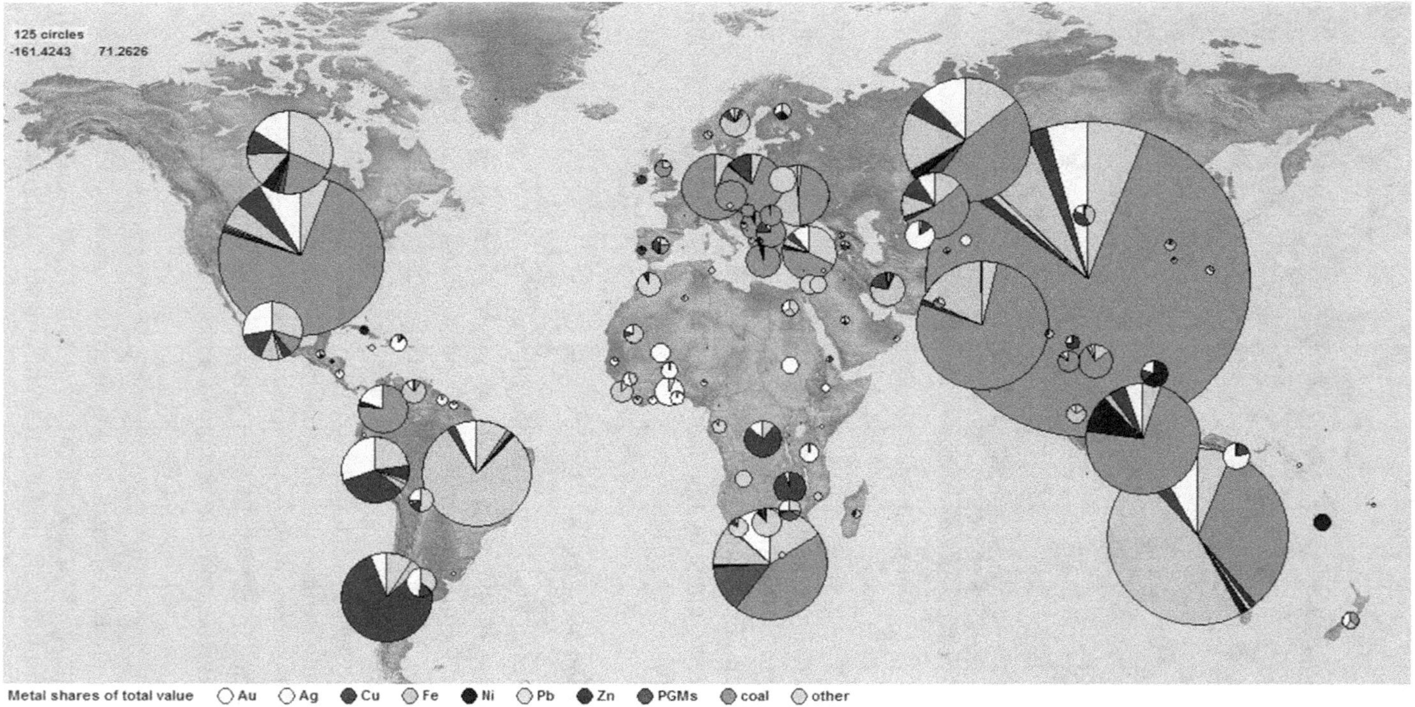


**Figure 3.4.** Value of mine production by country, 2014

*Notes*: Circles are proportional to value of mine production. Coal accounts for 3/4 of Indian production value. Iron ore accounts for >3/4 of Brazilian production value. Gold accounts for approximately 1/3 of Peruvian production value.

*Source*: Raw Materials Data.

value in total global production as the fertilizer minerals—i.e. phosphate and potash—at two to three per cent of the total value of production. Thereafter there are a number of metals and industrial minerals that each contribute less than one per cent of total global value. (See Figure 3.1 for a complete list of the minerals included in total mine production value.) China is by far the most important country in terms of total production value, followed by Australia and the United States. The top ten countries in terms of the value of their mine production contribute almost 80 per cent of the total value of non-fuel mineral production at the mine stage globally.

For each of the MCI-W top twenty LIE and middle-income economies (MIE), Figure 3.5 shows how metals and minerals contributed to the total value of their mine production in 2014. Gold mining is the major mineral contributor in no fewer than nine countries in this top twenty. In Mali, gold is the only mineral mined and hence contributes 100 per cent of the total value; in Burkina Faso, Guyana, Ghana, Uzbekistan, Suriname, and Tanzania, gold mining contributes between 75 and 94 per cent. Copper is the most important commodity in Zambia, DRC, and Lao PDR. In Namibia and Botswana, diamonds are the main contributor.

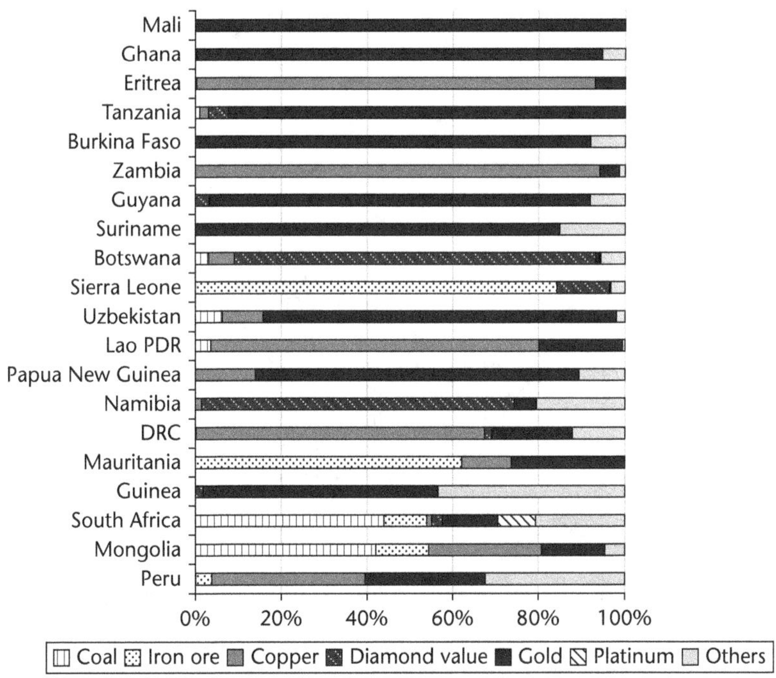


**Figure 3.5.** Contribution by commodity to MCI-W top 20 LIE and MIE
*Source*: authors' illustration based on Raw Materials Data.

In 2014, the total global value of mine production at the mine stage including coal was around US$1,200 billion. Coal contributed US$650 billion, and iron ore is estimated at US$145 billion. The change over time in the total global value of mineral production follows the general metal/mineral prices, as seen in Figure 3.6. However, for some individual countries, the changes in the level of production have also been very important.[4] For example, copper production in DRC has increased tenfold over the last ten years and is now twice as large as during the previous peak in the 1980s.

3.3.2.1 CHANGE OF MINING CONTRIBUTION OVER TIME, 1996–2014

Metal and mineral prices reached a peak in 2011, but have since been in a five-year downturn that was showing some signs of correcting in 2016–17. It should be noted, however, that most metal prices in nominal terms are still higher than they were in the early 2000s. Our price index is made up by a variety of metals/minerals (coal, copper, gold, iron ore, nickel, and zinc). The weighting on the price index was calculated as an average based on the total

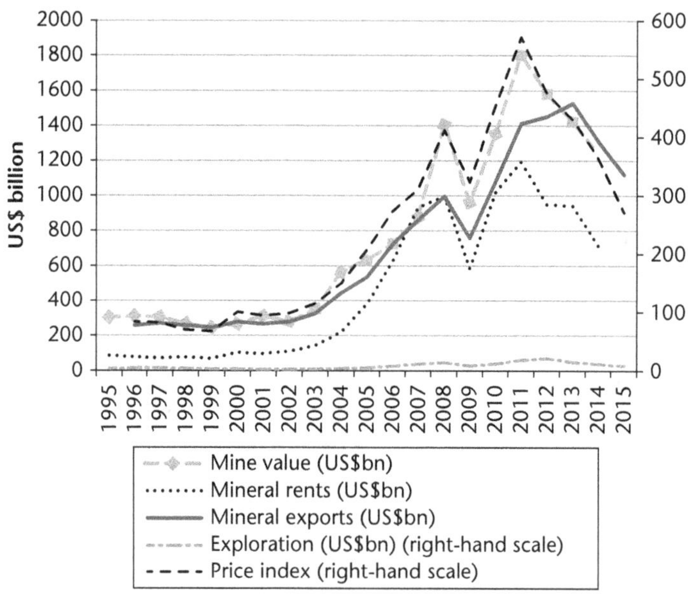


**Figure 3.6.** Mining development trends, 1995–2015: prices, exports, exploration, value of mine production, mineral rents

*Sources*: authors' compilation based on data from Raw Materials Group, World Bank, SNL Metals & Mining, and UNCTAD.

[4] See e.g. Eritrea and some other high-ranking MCI-W countries. Annual production data by country for all of the countries covered are not yet available for 2015.

value of products of the mining industry. The weighting was used to combine the price development of different products into one index.

As Figure 3.6 shows, the price index has been on a downward trend since 2011, with a flattening-out beginning in early 2016. It is certain that the global production value will also have dropped for 2015, but we see several important indicators making us believe that the bottom in terms of production value was reached in late 2016 or early 2017. As can also be seen from Figure 3.6, mineral prices are an important but not the sole determinant of the changing levels of exports, value of mine production, mineral rents, and exploration expenditures.

### 3.3.3 *Export Contribution*

Non-fuel minerals and metals are the major contributor to many nations' exports. Among the top fifty countries with the highest mineral exports relative to total exports in 2014, there were seventeen nations with a total mineral export of more than 50 per cent of the total. Among the top fifty ranked by export contribution, no fewer than 34 per cent are LIE and 28 per cent are LMIE. Only eight countries or 16 per cent are HIE. The export contribution to the MCI-W score in LIE and MIE is the most important factor explaining their high ranks. Sierra Leone is number one with a mineral export contribution of no less than 94 per cent of total exports. Botswana, DRC, Mongolia, and Zambia are all countries where mineral exports contribute more than 70 per cent.

### 3.3.4 *Exploration*

Exploration activity and spending is mainly driven by expectations of future, mostly short-term mineral demand and prices. In reality, exploration expenditure in a given year is closely related to metal prices in the preceding year (Canadian Intergovernmental Working Group on the Mineral Industry 2001: 20–1). This means that future metal demand, which should logically determine levels of exploration, is not a prime driver. This is a failure of the market for this specific service. Some attempts to stimulate exploration have been made in certain countries, with varying success. Examples are financial support to risk-willing investors in Canada and Australia (flow-through shares), and government-funded exploration work in China, India, and Finland.

### 3.3.5 *Mineral Rents*

It is important to note that diamonds are not included in the list of minerals for which the World Bank calculates mineral rent. Thus, countries such as Botswana and Namibia, where diamonds are the main mineral contributor to the economy, will get a lower MCI-W score than if diamond rents were also included. Mineral rent is a theoretical approach to calculate some concept of the surplus from the mineral sector.

### 3.3.6 *Other Factors*

#### 3.3.6.1 GOVERNMENT REVENUES FROM MINING

The capturing by government of some part of total resource revenues as government revenues (mainly taxes and royalties) is crucial to generate development for many reasons, not least that mineral resources are considered non-renewable.

#### 3.3.6.2 EMPLOYMENT

The *direct* contribution of mining to the total formal employment of a country is seldom more than 1–4 per cent in countries with large mining sectors. The number of direct jobs created is normally relatively small, as mining is capital-intensive; but mining also generates *indirect* jobs, which are more difficult to measure. Furthermore, mines are often located in remote areas with limited other opportunities. However, the jobs created by large mining companies are normally well paid compared with other similar jobs in the same country. This means that the mining contribution to the total wage bill of a country is often proportionately larger than its contribution to job numbers.

To sum up, direct employment in the mining sector most often varies between 1 and 3 per cent, but there are examples of much higher levels. This is invariably the case, in particular, if informal/artisanal-sector employment is also included. Employment is an important stabilizing factor in the contribution of mining in many mineral-rich countries. Employment has also been generally rising in the past ten years, and has not declined as much recently as the value of mine production, exports, and other factors directly related to commodity prices. Employment is also somewhat less volatile than the other factors under study, and there was for example only a marginal dip during the global financial crisis in 2008–9.

## 3.4 Changes in MCI-W since 1996

The 1996 value of mineral production at the mine stage was US$300 billion (in nominal terms), equivalent to 0.6 per cent of total world GDP PPP (World Bank 2016). In 2011 mine value peaked at US$1,800 billion (1.9 per cent of global GDP); it has since fallen back to US$1,200 billion and 1.2 per cent of world total GDP. The super-cycle—the long boom in metal and mineral markets and prices beginning in 2003—made mining a more important part of GDP in almost all mining countries. The share of mining in global GDP doubled in four years, and peaked at three times higher in 2011 than in 1996. These dramatic changes in the preconditions for mining's contribution to national economies also had strong effects on MCI-W. In 1996 Chile was number one in the MCI-W ranking while DRC, which is number one in 2014, was ranked only at number 24.

Among the twenty LIE and MIE which had the highest MCI-W ranking in 1996, no fewer than thirteen economies have climbed up one step in the World Bank's income group classification by 2014. In 1996 the MCI-W top fifty included six HIE, five UMIE, twenty-one LMIE, and eighteen LIE. By contrast, in 2014 the numbers are: four HIE, sixteen UMIE, eighteen LMIE, and twelve LIE. Zambia, Ghana, Guyana, Mauretania, Mongolia, and Tajikistan were classified as LIE in 1996 but LMIE in 2014. Countries classified as LMIE in 1996 but UMIE in 2014 are: Peru, Kazakhstan, Suriname, Botswana, Namibia, Fiji, Cuba, and Venezuela. Chile and the Russian Federation became HIE between 1996 and 2014.[5] There are of course many factors influencing these gradual economic developments, but it seems likely that the contribution of mining and minerals is one important factor.

When comparing the mining contribution to national economies between 1996 and 2014 at the global level, we see a broadly similar picture. There are, however, regions and specific countries that have climbed up the rankings very significantly. West Africa, for example, is a region that has now moved to the top of the MCI-W rankings.

Individual countries which have climbed most in the MCI-W rankings are in following order: Lao PDR, Eritrea, Côte d'Ivoire, Burkina Faso, Sudan, Mozambique, Serbia, Togo, Mali, and DRC. Lao PDR and Eritrea did not have any industrial-scale mining in 1996, so when mining started they went from almost zero to a point today where mining is contributing considerably to their economies. African mining countries in particular have gained an increase in MCI-W score. Among the sixteen countries whose MCI-W score increased more than 25 per cent between 1996 and 2014, no fewer than thirteen are in Africa.

In summary, mining quite clearly increased its contribution to economic activity in the low- and middle-income countries between 1996 and 2014. The increase in contribution is higher in LIE than in MIE. Mining's share of GDP tripled during these years for these two categories of country. The share was 3.1 per cent in 2014, compared with 1.1 per cent in 1996. Mineral exports' share of total exports in those countries increased by 50 per cent in the same period. Mineral rents followed the general price developments and reached a

[5] The Russian Federation is among the UMIE again in 2015.

peak in 2011, but have declined since, although they were still higher in 2014 than they were in the 1990s. Exploration spending in the countries studied increased over the period as a whole, but has been declining steeply since 2013. Several LIE and MIE with high MCI-W scores in 1996 have developed successfully and risen in the World Bank GNI classification from LIE to MIE and from LMIE to UMIE. The MCI-W index for individual countries has moved up and down depending on the performance of their mining sector relative to other sectors of the economy. It is difficult to draw any general conclusions from this relative index. There is a need to further develop the contribution index with this in mind.

## 3.5 The Impact of the End of the Super-cycle

Over the first decade of the new millennium, the global mining industry moved from a long period of low prices, unacceptable levels of return, and limited investments to a boom with record high metal prices, improved profitability, and a flurry of new projects. The main driving force behind this change back in 2003–4 was strong demand for metals and minerals, especially from China. This spurred high levels of investment into the extractive industry in order to increase supply to meet growing demand. Since 2011–12 metal prices have dropped, but, excluding nickel, not to pre-boom price levels.

Among the most important metals, gold stands out in that its price has not fallen as precipitously as those of the other minerals, and indeed has already started to move upwards again.

As shown in Figure 3.5, gold is the single most important metal for the LIE and MIE with the highest MCI-W rankings. Forty-five per cent of their total mine value is from gold mining, and it is the main contributor in nine of these twenty individual countries. In seventeen countries in the MCI-W top fifty ranking, gold mining contributed more than 50 per cent of the total value of all mineral production. In Côte d'Ivoire, Mali, Nicaragua, and Sudan, gold contributed 100 per cent of total value. Among all the LIE and MIE together, there are a total of thirty-one nations where gold mining is the main contributor. When small-scale/artisanal gold mining is also considered (such production is not always fully accounted for in the national statistics used), the importance of gold production and the significance of the relative stability of the gold price are even greater. This is also valid for a number of LIE such as Sudan, Burundi, and Cameroon, where small-scale/artisanal gold production is considerable.

One conclusion is that LIE and MIE dependent on gold mining have not been affected as severely by the end of the super-cycle as countries producing certain other metals, such as nickel and iron ore. An example is visualized in

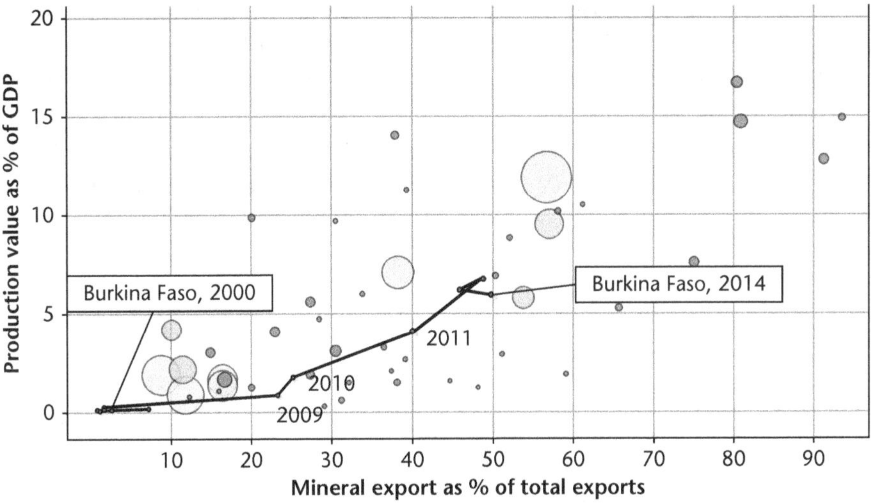


**Figure 3.7.** Burkina Faso, development in export and production values, 2000–14
*Note*: Circles are proportional to value of mine production.
*Source*: authors' calculations.

Figure 3.7. The figure shows a circle for each year between 2000 and 2014 for Burkina Faso's position on the x-axis (mineral export as a percentage of total exports) and y-axis (production value as a percentage of GDP). The line joins these together in chronological order. Other circles in Figure 3.7 represent other countries and their position in 2014. The size of the circles represents the magnitude of mining. In 2000 Burkina Faso had limited mining, the production value as a percentage of GDP was close to zero, and exports were just a few per cent. By 2014 production value as a percentage of GDP was around 6 per cent, and exports as a percentage of total exports were 50 per cent. Gold output in Burkina Faso was fairly constant between 2011 and 2014 at around 30–35 tonnes, while the gold price decreased 24 per cent between 2012 and 2014. However, the levels of mine value as a percentage of GDP and mineral exports were roughly the same in 2012 as in 2014. The example confirms that the impact of the end of the super-cycle has been smaller for Burkina Faso and other LIE and MIE where gold mining is important.

To sum up, the end of the super-cycle has hit countries in different ways, depending on the composition of their mineral production and many other factors. Gold mining countries are experiencing slower but still continuing growth. The level of export dependency and mining's share of GDP reached a maximum at the peak of the mining boom in 2011, when the GDP contribution reached as high as 25 per cent for some countries and export dependency went over 85 per cent. Naturally these figures had declined for some countries by 2014, but the situation for most countries was still a significantly larger contribution of/dependency on mining than in 1996. For some countries, production value as a percentage of GDP and mineral exports was even higher in 2014 after the price peak in 2011, because of a strong growth in production: this is the case for DRC, Sierra Leone, and Eritrea. Countries with a higher share of mineral exports in 2014 compared with 2011 are Burkina Faso, Mali, Guyana, Ghana, Namibia, Mauritania, Guinea, and Botswana.

## 3.6 Future Implications of Extractives Dependency

Metal and mineral prices are at present low relative to the peaks of 2011, but still well above the low levels of the early 2000s. Exploration expenditure is also low, and investments into new mines are also at a relatively low level. At the same time, it is clear that demand for metals and minerals in general has not dropped as much as prices have. There are clear indications that the price trough is generated more by an oversupply situation than by a fall in demand (see e.g. Worstall 2015). With the gradual improvement in standards of living, increased life expectancy, and continuing urbanization, which constitute the three major long-term drivers of metal and mineral use, it seems as if there will be a continuing, slow, and gradual increase in metal demand (McKinsey Global Institute 2013). Increased recycling and alternative energy sources might change this situation in the long-term future, but will not affect mid-term scenarios. One of the major reasons for the 2003–11 super-cycle was the slow response of the mining industry to increased demand. It takes a minimum of three to five years to increase mine capacity, and this time lag is increasing all the time due to the increasing advantages of scale economies, i.e. bigger mines with larger investments and longer and more difficult permitting processes. In short there are no signs of the lag time decreasing—rather the opposite. In principle, the global mining industry faces a similar situation during the next few years as it did in the early 2000s: slowly increasing demand, but some hesitancy about investing, and hence a low elasticity in mine production in response to demand. There is today less indication of such a strong growth in demand as was seen in the early 2000s. Nevertheless, metal prices might shoot up when supply gets short. The situation might also be exacerbated by the fact that investments into exploration have dropped dramatically in the recent past, and this might be a factor slowing the opening of new mines when new capacity is needed.

In the second half of 2016 (the time of writing) there are some indications that the bottom of the present cycle has been reached. However, the question remains as to how long prices will remain at their present relatively low levels.

The possibility of a steeper upturn than expected is not completely unrealistic (see e.g. Keen 2016). Given the long lead times for a mining project to get into production, it is important for mineral-rich countries not to focus too much on present metal prices, but to maintain a long-term approach to their national mineral resources.

As noted earlier, of the twenty LIE and MIE economies with the highest MCI-W scores in 1996, no fewer than sixteen have climbed one step on the Word Bank economic development classification. At the other end of the MCI-W rankings, when we compare the World Bank classification of the bottom twenty LIE and MIE in 1996 and 2014, there are only nine countries that have moved up one step. There are certainly many reasons why countries have *not* developed in this period, and naturally not only because of a lack of mining activity. Nevertheless, a statistical conclusion from this chapter is that mining can and has triggered development in several countries. When the analysis is expanded to include how the Gini coefficient has developed in the mineral-rich countries, it further seems as if inequalities have decreased. In this sample of the twenty LIE and MIE with the highest MCI-W scores in 1996, the Gini coefficient has remained constant or decreased, i.e. inequalities have diminished in fourteen countries and increased in six countries. Further, in one of the countries exhibiting a higher Gini coefficient in 2014, the increase was marginal.[6]

## 3.7 Conclusions

'Contribution' or 'dependency': even the choice of words to describe the relationship between national economies and the extractive sector poses a fundamental choice between good and bad. The traditional perspective in many historically resource-rich countries—such as our own country, Sweden—has been to view mineral resources as fountains from which wealth flows and development grows. To express it poetically, 'Through Swedish history sounds a mighty ringing of iron and copper from medieval times until today' (Furuskog 1935: 65)—clearly an analysis of the *contribution* of minerals to Swedish development.

From the 1990s until just a few years ago, however, the *dependency* approach was the dominant norm. The resource curse paradigm was the starting point for critical analyses in a host of works on mining during the past twenty years.

[6] The Gini coefficient is not updated every year for all countries by the World Bank or UNU-WIDER World Income Inequality Database. In cases where the years 1996 and 2014 were not available, the closest year was selected.

During the super-cycle of high metal prices and high oil prices, this a priori negative starting point was sometimes abandoned. There was an increasingly important view based on the hypothesis that the problem might not be the minerals as such, but rather the way the economic results they created were handled. McKinsey Global Institute's (2013) report entitled 'Reverse the Curse' is but one example of this recent turnaround in thinking. Another example is the discussion about mining's potential role as a catalyst for the diversification of national economies (Bastida 2014), the World Bank report on 'The Contribution of the Mining Sector to Socioeconomic and Human Development' (McMahon and Moreira 2014), and the study 'Local Industrial Shocks, Female Empowerment and Infant Health: Evidence from Africa's Gold Mining Industry' (Tolonen 2014).

This chapter provides backing for this reversal and reorientation by presenting a thorough statistical analysis of almost all countries in the world, including in particular all metal- and industrial mineral-producing countries. We therefore prefer the word 'contribution', as we cannot imagine a world without metals and minerals, and hence mineral resources need not be viewed as a curse if managed carefully.

### 3.7.1 *Contribution of Mining Industries in Low- and Middle-income Countries*

Among the fifty countries with the highest MCI-W scores, thirty-four are middle-income countries, twelve low-income countries, and only four high-income countries. Clearly mining plays a particularly important role in many low- and middle-income countries. Among the top twenty countries, DRC has the highest score, followed by Mongolia, Papua New Guinea, Zambia, Peru, and Guyana among the middle-income countries, and by Burkina Faso and Mali among the low-income countries (rankings eight and nine). The high-income countries Chile and Australia are ranked two and three respectively, demonstrating that in high-income countries too, mining can and does remain an important contributor to the national economy. Among the twenty highest-ranking countries, Africa dominates with twelve countries. The vision of minerals as an important part of African economic development is clearly well founded. There are only three countries each from Asia and Latin America, and two from Oceania, in the top twenty.

Of the world's ten largest mineral producers, in order of production value, China ranks 45th in MCI-W, Australia ranks at number 3, the United States is not even in the top fifty, the Russian Federation ranks 30, India ranks 42, South Africa ranks 11, Indonesia ranks 31, Brazil ranks 29, Chile ranks 2, and Canada ranks 27. This confirms that a high *absolute* value of mine production does not automatically translate into an important contribution to GDP and exports.

### 3.7.2 *Change in Contribution over the Past Twenty Years*

Among the twenty low- and middle-income countries with the highest MCI-W score in 1996, no fewer than thirteen have climbed up one step on the GNI development classification to the lower-middle-, upper-middle- or high-income category. There are of course many factors contributing to this development, but it seems likely that mining and minerals are one important factor. Geographically, Africa has benefitted most, and in particular West Africa—a region of growing mineral importance—is the prime example of this. Among the sixteen countries where the MCI-W score increased by more than 25 per cent between 1996 and 2014, no fewer than thirteen are in Africa.

The value of mineral production measured as a percentage of GDP grew from 1.1 per cent in 1996 to 3.1 per cent in 2014: on average, a growth of 200 per cent. In 1996 mineral exports as a percentage of total exports of the LIE and MIE taken together were 12.1 per cent. By 2014 that figure had increased to 17.4 per cent. Furthermore, the figures for both GDP and export share of minerals and mining are considerably higher on average for LIE than for MIE. The levels of GDP and export contribution in 2014 were still at a higher level than in 1996, in spite of the drop in metal prices since the end of the super-cycle.

It has not been possible to include employment in the mineral sector as one of the contributing factors to our mining contribution index, because of a lack of data. Nevertheless, the countries for which statistics are available clearly demonstrate that employment is a stabilizing factor, as it does not vary as rapidly as the other factors studied. Further, employment levels in general increased over the period 1996–2014.

### 3.7.3 *Impact of the End of the Super-cycle*

The contribution of minerals and mining to GDP and exports reached a maximum at the peak of the mining boom in 2011. Naturally, the figures for mining's contribution had declined for most countries by 2014, but importantly the levels were still considerably higher than in 1996.

The results of this survey do not support the widespread view that mineral resources create a difficult dependency which might not be conducive to economic and social development—rather the opposite. Certainly, the indicators on which we base our chapter only shed light on some aspects of economic and social development. But we think we have enough substance to claim that if additional low- and middle-income countries could locate additional mineral resources, their chances of economic development would be better than they are at present, when only limited mineral resources are known.

## References


Bastida, A. E. (ed.) (2014). 'From *Extractive* to *Transformative* Industries: Paths for Linkages and Diversification for Resource-driven Development', Special Issue, *Mineral Economics*, 27(2–3): 73–87.

Canadian Intergovernmental Working Group on the Mineral Industry (2001). 'Overview of Trends in Canadian Mineral Exploration 2000'. Ottawa: Natural Resources Canada.

Furuskog, J. (1935). 'Berget', in *Svenska Tursitföreningens Årsskrift 1935*. Stockholm: Svenska Turistföreningens Förlag.

ICMM (2010). *The Role of Mining in National Economies*, 1st edition. London: ICMM.

ICMM (2014). *The Role of Mining in National Economies*, 2nd edition. London: ICMM.

ICMM (2016). *The Role of Mining in National Economies*, 3rd edition. London: ICMM.

Keen, K. A. (2016). 'Next Upswing in Metal Prices Inevitable with Few Quality Mines Coming Online', *S&P Global Market Intelligence*, 7 September.

McKinsey Global Institute (2013). 'Reverse the Curse: Maximizing the Potential of Resource-Driven Economies'. New York: McKinsey & Company.

McMahon, G. and S. Moreira (2014). 'The Contribution of the Mining Sector to Socio-economic and Human Development'. Extractive Industries for Development Series 30. Washington, DC: World Bank.

Raw Materials Group (1997). *Who Owns Who in Mining*. London: Roskill Information Services.

Roe, A. and S. Dodd (2016). 'The Statistical Tendencies', paper presented to UNU-WIDER project meeting 'Extractive Industries and Development', Helsinki, April.

Roe, A. R. and J. I. Round (2017). 'Framework: The Channels for Indirect Impacts', WIDER Working Paper 2017/79. Helsinki: UNU-WIDER.

SNL Metals & Mining (2016). 'Corporate Exploration Strategies 2016'. Halifax, NS: SNL Metals & Mining.

Tercero Espinoza, L. A. and M. Soulier (2016). 'An Examination of Copper Contained in International Trade', *Mineral Economics*, 29: 47–56.

Tolonen, A. (2014). *Local Industrial Shocks, Female Empowerment, and Infant Health: Evidence from Africa's Gold Mining Industry*. Gothenburg: University of Gothenburg.

UN Comtrade database.

World Bank (2016). 'Zambia Mining Investment and Governance Review'. Washington, DC: World Bank.

Worstall, T. (2015). 'Rio Tinto and Vale Killed the Commodities "Super Cycle" Not China or the Fed', *Forbes*, 29 November.

# 4

# The Role of Oil and Gas in the Economic Development of the Global Economy

*Paul Stevens*

## 4.1 Introduction

This chapter is concerned with the role of oil and gas in the economic development of the global economy. Its focus is very much on the context in which established and newer developing-country oil and gas producers must frame their policies in order to optimize the benefits of producing such resources. It begins by outlining a brief history of the issue over the last twenty-five years. It considers oil and gas as factor inputs, their role in global trade, the role of oil prices in the macroeconomy and the impact of the geopolitics of oil and gas over the same period. The chapter then considers various conventional views of the future of oil and gas in the primary energy mix, trying to explain why there is such a tendency to consensus in the different forecasts. Finally, it seeks to challenge the various drivers behind these conventional views of the future with an emphasis on why they may prove to be very different from what is expected and how this may change the context in which producers must frame their policy responses. The principal purpose is to explain recent developments in energy, particularly those that have led to lower oil prices since 2014, and to argue that lower prices may well be the norm going forward. For low- and middle-income countries that were anticipating significant revenues from oil and gas when expected prices were much higher, this will require a number of adjustments. In particular, they will have to lower their spending levels and seek to raise alternative sources of revenue. They must also make plans to manage the macroeconomic consequences of lower revenues. Above all they must give (even more) serious consideration to policies to try and diversify their economies away from dependence on hydrocarbon revenues and away from expectations of dependence.

### 4.1.1 *Recent History*

This section considers some of the background context for established and newer developing-country producers, leading to a picture which can then be developed as they face a more uncertain future in terms of energy prices.

#### 4.1.1.1 AS A FACTOR INPUT

Ever since the Industrial Revolution, energy has always been a key factor of production (Landes 1969). Recent history has proved no exception. The pattern of primary commercial energy consumption[1] since 1965 is presented in Figure 4.1.

Since the start of this century, what is clear is that energy consumption outside of the OECD has grown very strongly while in the OECD it effectively peaked in 2004. As explained in Appendix 1 (Stevens, 2016a), this is because of the lagged relationship between prices and energy consumption following the oil shocks of the 1970s and represents an important part of the story, especially when the future is considered in Section 4.4.

As to fuel mix, Figure 4.2 illustrates the patterns since 1965. As can be seen, oil has dominated the mix in this period as a result of its innate advantages. Thus, because it is liquid and flows in three-dimensional space it attracts very large economies of scale.[2] It also has much higher energy content than other fuels.[3]

The effect of the oil price shocks of the 1970s was to push oil out from the static sectors (such as power generation), replaced initially by coal and gas. However, it retained its favoured position in the transport sector. Until recently, gas remained a constrained fuel limited by its very high transportation costs.[4] Also its use in the power sector was prescribed in the United States and the European Union by regulation between 1975 and 1990.[5] In emerging-market

[1] Commercial energy is distinct from traditional energy. Commercial energy can be defined as energy that moves in corporately controlled markets and involves some level of foreign exchange input. This includes coal, oil, gas, nuclear, and modern renewables including hydro. This is as opposed to traditional energy that either moves in small local markets or is non-commercial and involves no foreign exchange input. This includes wood fuel, charcoal, animal and vegetable residues (Stevens and Pearson 1984). Because of the lack of accurate statistics, traditional energy is not included in the data, although in many parts of the developing world, those fuels constitute the majority of primary energy consumption.

[2] The capital cost of a storage tank is a function of surface area and its output a function of capacity. There exists an exponential relationship between the two. Doubling the dimensions of a tank halves the average cost of storage. 'Tanks' in various shapes and forms (pipelines, oil tankers, refineries, etc.) dominate the oil industry.

[3] A tonne of oil contains 1.5 times more energy than a tonne of hard coal and a cubic metre of oil contains vastly more energy than a cubic metre of gas.

[4] Because of its low energy content, gas is a high-volume low-value commodity which severely limits its effective geographic market.

[5] Gas was seen as a premium fuel and the view was taken it should be preserved for premium markets. It was viewed as too valuable to burn in power stations.

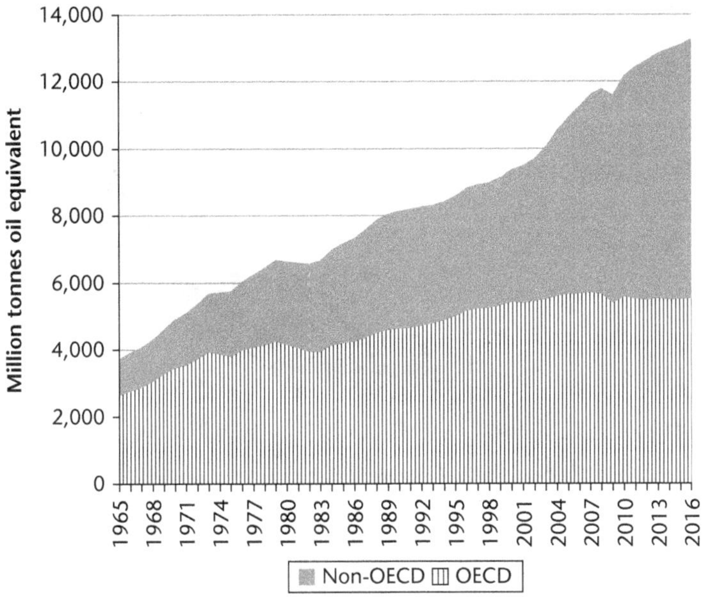


**Figure 4.1.** World primary energy consumption by region, 1965–2016
*Source*: author's illustration; unless otherwise stated, all the data in this chapter are taken from BP (2016).

economies, its domestic development by the (foreign) companies that had discovered the gas was also constrained in the face of non-convertible currencies. Thus the gas could only earn revenue that the foreign company could not remit back to its shareholders.

However, despite the growing use of commercial energy, the world faced very considerable fuel poverty. Many remain without access to electricity and to modern cooking fuels. It is interesting to observe that the Millennium Development Goals did not relate at all to energy. However, the more recent Sustainable Development Goals explicitly refer to energy.[6] As will be discussed, this will carry important implications for the future of oil and gas in the development process not least because the price of solar electricity is falling far more rapidly than many expected.

#### 4.1.1.2 IN GLOBAL TRADE

Oil and gas are major internationally traded commodities. Because of economies of scale, transporting oil and oil products is extremely easy and extremely cheap. This goes a long way to explain why oil is a truly internationally traded

[6] The seventh development goal is to ensure 'access to affordable, reliable, sustainable and modern energy for all' (UN 2015).

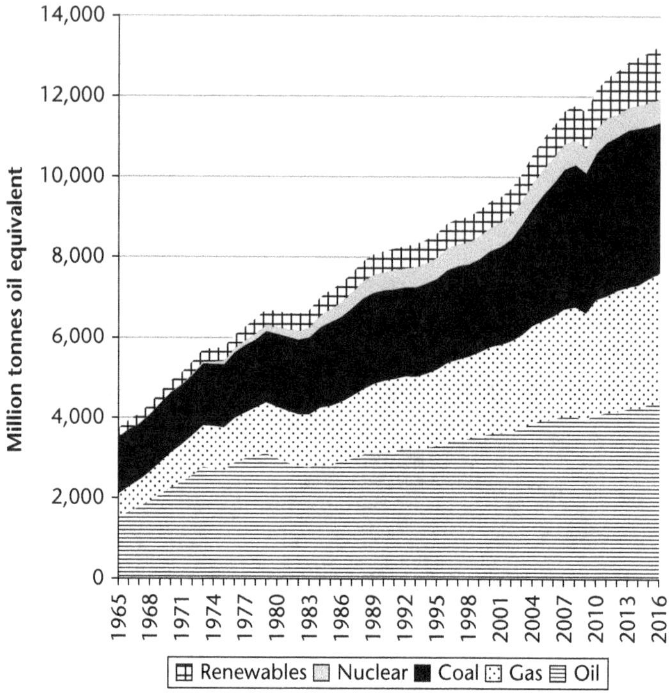


**Figure 4.2.** World primary energy consumption by fuel, 1965–2016

*Source*: author's illustration; unless otherwise stated, all the data in this chapter are taken from BP (2016).

commodity with a unified market while gas remains traded in regional markets.[7] Gas suffers from what has become known as 'the tyranny of distance'. Thus a much smaller proportion of gas consumption is traded internationally. For many countries their oil and gas trade makes a significant contribution (both positive and negative) to the current account in the balance of payments. In particular, a number of oil exporters are highly dependent upon oil exports for their foreign exchange.

#### 4.1.1.3 PRICES AND THE MACRO ECONOMY

Ever since the oil price shocks of the 1970s, attention has been paid to the relationship between oil prices and global GDP growth. Casual observation suggests that the changes in oil prices in the 1970s and 1980s did affect economic growth in the OECD. There are good economic grounds to expect this negative relationship. Higher oil prices shifted income from the OECD to the OPEC countries. The OECD countries had a much higher propensity to

[7] See Stevens (2016a: Appendix 2).

consume compared to the OPEC countries, which in this period lacked the institutional capacity to spend fully their windfall revenues. A large proportion of this windfall was simply left in Western banks. Thus global aggregate demand fell, generating economic recession.[8]

The higher prices seen from 2004 to 2014 in all natural resources have brought the issue of 'resource curse' back into the policy debate (Stevens et al. 2015).[9] Linked to the debate on resource curse has also been a revived interest in economic diversification in those countries highly dependent upon oil and gas revenues. The key point here is that oil revenue is not income. Rather, it simply represents the re-shuffling of the nation's portfolio of assets. Figure 4.3[10] typifies the production profile of an oil producer.

In the initial stages, production provides financial resources, which should be used to promote economic development. Production then reaches a plateau.[11] In this transition phase, it is crucial that the financial resources released by producing the oil are used to create an alternative income-producing asset. Eventually, production will decline as a result of natural depletion or

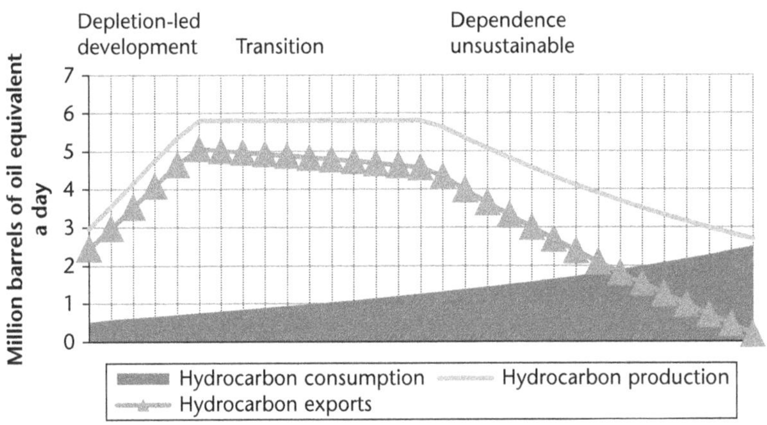


**Figure 4.3.** Depletion and development

*Source*: Author's illustration redone from Mitchell and Stevens (2008). Figure reproduced with the permission of Chatham House, the Royal Institute of International Affairs.

[8] This is a very simplistic explanation. There were many other factors in play and the precise relationship between oil price and global GDP has been much debated in the economics literature.

[9] This is discussed in great detail in Chapter 5 of this book.

[10] The author is grateful to Chatham House, the Royal Institute of International Affairs, for permission to reproduce the figure entitled 'Example: Turning points in the hydrocarbon balance of Nigeria' from the Chatham House Report published in 2008, *Ending Dependence: Hard Choices for Oil-Exporting States* by John V. Mitchell and Paul Stevens.

[11] The idea that production would somehow peak, rather than reach a plateau, is totally unrealistic. The infrastructure to produce the oil is expensive and the economics only work if it is at full capacity. Therefore, the idea that infrastructure would be built to accommodate a short-lived peak is not viable. Thus, the production profile would be a plateau.

because of a lack of demand. For export earnings, this process is accelerated as domestic consumption rises. In this stage—'dependence unsustainable' in Figure 4.3—it is no longer feasible to rely on oil revenues to support the rest of the economy. Therefore, oil-dominated economies should aim, from the very early stage of production, to diversify away from dependence on oil revenues. The success or otherwise of this strategy can be measured by the non-hydrocarbon fiscal deficit.[12]

#### 4.1.1.4 GEOPOLITICS AND THEIR ECONOMIC IMPACT

Geopolitics and oil are inextricably linked. There are two oil markets—the wet barrel market where real barrels of crude oil are bought and sold, and the paper barrel market where promises (written on paper) to deliver or take delivery of oil are exchanged. To understand the wet barrel needs 'Economics 101'. To understand the paper barrel market needs 'Psychology 101'. The relationship between the two markets is complex and controversial but comes down to issues of perception. Those negotiating contracts in the wet barrel market will look to the paper barrel market to give an indication of what prices might be. Those in the paper barrel market deciding where to invest will look to the wet barrel market for signs of surplus or shortage.

There are two problems here that aggravate oil price volatility. First, many of those playing in the paper markets do not really understand the oil industry and frequently misread the state of the wet barrel market, often assuming shortages when there are none. While this may sound improbable, a similar situation exists in foreign exchange markets, known as 'scapegoat theory' (Bacchetta and Van Wincoop 2004). Thus, economists trying to predict exchange rates using the usual economic metrics will often get it wrong because those setting the rate—that is, the traders—look at a totally different set of metrics. The second source of price volatility is that perceptions can change in the blink of an eye and with them so can prices!

In this context, it is easy to see the relevance of geopolitics in terms of recent price history. Any loss of oil supplies as a result of geopolitical events, such as a war, has affected physical supply in the wet barrel market. Wars and rumours of war have also affected expectations in the paper barrel market. Politics have clearly affected government policy (Stevens et al. 2013). Security-of-supply concerns drive the energy policies of energy-importing nations, which in turn will impact energy consumption levels and the energy mix. Equally, security-of-demand concerns drive the depletion policies of producer governments and the rise (or fall) of resource nationalism. This century there has been a

[12] This measures the extent to which a fiscal deficit in the non-oil sector can be funded by the oil revenues. A similar measure for foreign exchange is the non-hydrocarbon current account deficit.

significant rise in resource nationalism that has coincided with the upturn of the commodity super-cycle.

## 4.2 Current Conventional Views of the Future for Oil and Gas

As indicated at the start of the chapter, the future of energy markets will frame the need for the established and newer developing-country producers to develop their own policies against a background of highly uncertain energy prices. Thus it is important to begin by considering what the conventional views of this future suggest as the most likely path. Figures 4.4 and 4.5 give a flavour of some of the views of future energy demand from a variety of sources.

The first point to note is that there is a fairly strong consensus on the direction of travel. Virtually all of the growth in primary energy demand is

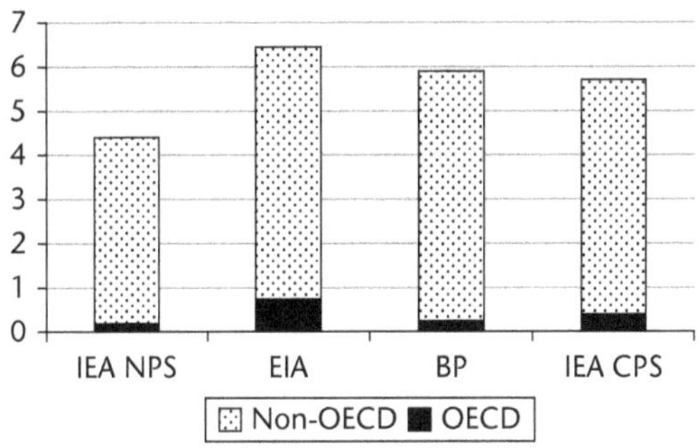


**Figure 4.4.** Increases in energy consumption by region, 2010–35 (billion toe)

*Note*: IEA NPS is the International Energy Agency's New Policies Scenario; EIA is the United States Energy Administration's Energy Outlook 2014. BP is BP's Energy Outlook 2014; IEA CPS is the International Energy Agency's Current Policies Scenario.

*Source*: author's illustration; data from BP Energy Outlook 2014.

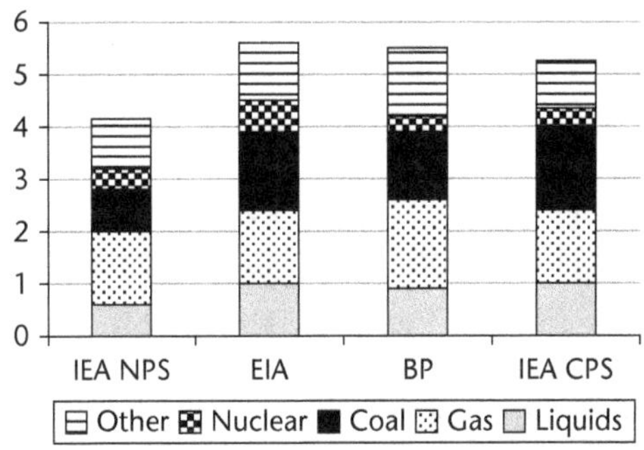


**Figure 4.5.** Increases in energy consumption by fuel, 2010–35 (billion toe)

*Source*: author's illustration; data from BP Energy Outlook 2014.

thought likely to occur in the non-OECD countries and fossil fuels are thought likely to continue to dominate supply.

It is interesting to speculate why there is such consensus. One explanation is that the forecasters have models (quantitative and qualitative) based upon similar drivers. How much certainty there is over these drivers will be discussed in Sections 4.3 and 4.4. There are, however, other more controversial explanations. First there is safety in consensus. Thus, if your forecast is wrong, so is everyone else's forecast. Second, there is also a real problem of forecasting discontinuities. Forecasting trends based upon business as usual is fairly straightforward. Forecasting bends in trends can also be done with some imaginative thinking. However, forecasting major discontinuities is next to impossible. Even a brief acquaintance with the history of energy markets since the 1970s shows it has been littered with major discontinuities driven by economics, technology, and pure accident. There were the three oil price shocks of 1973, 1979, and 1986. There have been various accidents ranging from Three Mile Island to Chernobyl to Fukushima for nuclear and the Macondo spill for oil. There has been the shale technology revolution that has had such a dramatic impact on oil supplies. The list could go on.

Finally, there is the problem of political and vested interests behind the mutually supporting consensus of most forecasts. The IEA, for example, was created to persuade oil-importing OECD countries to reduce dependence on imported oil.[13] Projecting shortages in the future might be seen as a good method to encourage consumer governments to take steps to reduce dependence. Equally, the oil companies have a vested interest in persuading their shareholders that oil demand will continue to increase and that better times in the market lie just ahead (Stevens 2016b).

Such consensus views of a business-as-usual future are in danger of giving false expectations to oil and gas producers. Sections 4.3 and 4.4 consider why these conventional views of the future might be wrong, thereby forcing the established and newer developing-country producers to consider carefully their future policy options. Section 4.3 looks at economic and technological concerns and Section 4.4 considers geopolitical threats to such futures.

## 4.3 Drivers of Future Trends: What Economic and Technical Factors Might Derail Conventional Views of the Future?

As implied in Section 4.2, the consensus associated with forecasts of energy supply and demand in part arises because the forecasters tend to take similar

[13] It is worth remembering that the IEA was created in 1974 as Henry Kissinger's response to the perceived growing power of OPEC.

views of the drivers behind the forecasts. However, as suggested, the history of energy markets has been characterized by frequent deviations of the drivers from the expected path. This section considers what factors might cause the future drivers underlying the forecasts to be different from today's expectations.

### 4.3.1 *Technology*

Invention and innovation are embedded in the forecasts.[14] There are three particular areas where change could impact how the future unfolds: carbon capture and storage (CCS); the development of batteries; and the spread of shale technology.

*CCS* is a well-established technology. However, the costs and problems with its adoption are considerable. First capturing the carbon adds significantly to the cost of power generation. Also there are serious problems in capturing from small-scale emitters. Storage faces many problems, not least ensuring that once the carbon is stored there are no leakages. For example, a major problem in the United States is that property rights to the sub-soil make it unclear where responsibility would lie for maintaining the storage once the $CO_2$ has been injected. Solving these problems would mean that the very large reserves of hydrocarbons could be burnt without aggravating climate change.

Improvement in *battery technology* would greatly assist in the key problem with renewables: namely that of intermittency. Technological developments could make batteries smaller, easier to charge, and cheaper. A good example of possible change relates to electric vehicles. A large stock of electric vehicles could provide considerable storage. Modern metering technology means that electric vehicles could supply power to the grid and be paid for it.[15] Furthermore, if roads could be built with induction strips, electric vehicles could charge or discharge while travelling.[16]

Offering a 'longitude prize' might be one way to generate changes to the technology for CCS and batteries. At the start of the eighteenth century, the great scientific question of the age was how to determine where you were at sea. Latitude could be measured by assessing the angle of the sun but

[14] The difference between invention and innovation is that invention has strong elements of being a random process. Innovation, which is the spread of an invention, on the other hand has clear socio-economic and political drivers. That said, the distinction is not always so clear-cut.

[15] For example, in Western Europe the average daily car journey is less than 20 kilometres. Assuming a battery capacity of 100 kilometres (which is currently a conservative estimate), that leaves considerable flexibility to use the car battery as storage.

[16] It is worth pointing out that most of the conventional energy forecasts referred to in Section 4.2, in this author's opinion, are extremely conservative on the penetration of electric vehicles into the car park.

estimating longitude required an accurate timepiece to determine the time at the port of origin. In 1714, the British government offered a very significant prize for whoever came up with such a clock. It was eventually awarded in 1793 (Sobel 2011). The COP21 meeting in Paris in December 2015 actually created the institutions that could well make a similar prize feasible in the two areas of CCS and batteries. Thus there was the Mission Innovation group of twenty governments, which was a global initiative to accelerate clean-energy innovation, and the Breakthrough Energy Coalition, which was a global group of twenty-eight high-net-worth investors from ten countries committed to the initiatives of Mission Innovation.

The *shale technology revolution* has the capability to transform the prospects for oil and gas supply. In the United States it has led already to a dramatic increase in oil supply and an equally impressive increase in domestic supply of gas.[17] The revolution consisted of the development of horizontal drilling and hydraulic fracturing.[18] A key question is how far this technology can be used outside of the United States to produce similarly dramatic increases in supply. Certainly, there are ample technically recoverable resources of shale hydrocarbons globally. However, there are many barriers to converting these resources into producing capacity. Table 4.1 outlines the conditions present in the United States that effectively led to the shale technology revolution. By contrast, it also shows that these conditions are for the most part not present in the United Kingdom and a similar story is true for most other parts of the world.

Of course, the shale technology is constantly changing and improving. Many of the characteristics that are not currently present outside of the United States could be made available as a result of improved technology and institutional changes.[19] For example, a major barrier to using hydraulic fracturing in Europe is concern over the impact on water supplies and the danger of contamination of aquifers. Currently, experiments are underway to use sonic waves rather than water and chemicals to frack the rocks and if successful this alternative technology might allay these fears.

Changes in any of these and other technologies could obviously seriously impact the consensus views of the future embodied in the forecast described in Section 4.2.

[17] Thus between 2008 and 2015 oil production increased from 6.8m b/d to 12.7m b/d and gas production between 2005 to 2015 increased from 511bcm to 767bcm (BP 2016).

[18] There were other technical aspects of the revolution such as the development of three- and four-dimensional seismic modelling and coil tube drilling.

[19] The recent collapse in oil prices after June 2014 has generated large cost savings and improved efficiency in the US industry.

**Table 4.1.** Factors creating the shale gas revolution in the United States

| United States | United Kingdom |
|---|---|
| **Geology** | |
| Large shallow, material plays, implying large technically recoverable resources. Also much of the shale had low clay content, making it easier to fracture. | Reported to have a higher clay content |
| After many years of oil and gas drilling, there were plenty of drill core data publicly available to allow explorers to find the 'sweet spots' on the plays. | No |
| The shale gas had a high liquids content, which greatly enhanced the economics of the operations, especially at a time when gas prices were low. | Not known at this stage |
| **Research** | |
| In 1982 the US government began extensive funding of R&D by the Gas Technology Institute into low permeability hydrocarbon-bearing formations. The results were widely disseminated to the industry. | No |
| **Regulation** | |
| The 2005 Energy Act explicitly excluded hydraulic fracturing from the Environmental Protection Agency's Clean Water Act, the so-called Cheney–Halliburton Loophole. Many shale gas operations were done with little environmental impact assessments. | Strong environmental legislation |
| The 1980 Energy Act gave tax credits amounting to 50 cents per million BTUs. It also introduced the Intangible Drilling Cost Expensing Rule, which covered (typically) more than 70% of the well development costs, crucial for small firms with a limited cash flow. These economic incentives were very important in the early stages of the industry, based upon small, relatively cash-strapped, entrepreneurial companies. | The government is considering introducing some tax breaks. |
| Property rights in the United States make the shale gas the property of the landowner, creating a strong financial incentive for private owners to permit the disruptions associated with shale operations. Also, the population is used to being in proximity to oil and gas operations. | No |
| The system is used to licensing large areas for exploration with fairly vague work programme commitments, which is what is needed when dealing with shale plays. | No |
| **The nature of the gas market** | |
| Pipeline access is based upon 'common carriage', so gas producers have at least some access to pipelines, transforming the economics of shale gas production. The United States also has a very large and extensive gas pipeline grid. | No<br>Access is by Third Party Access. |
| The United States is a 'commodity supply gas market', i.e. it has a lot of buyers and sellers and good price transparency. Gas is easy to sell. | Not as easy as the United States |
| The US domestic gas market experienced strong rising prices in the period after 2002, culminating in a price over US$10 per thousand cubic feet (mcf) in May 2008. | No |
| **Industry** | |
| The industry was dominated by small, entrepreneurial companies, the so-called momma and poppa companies. | No |
| The majority of the work was done by a dynamic, highly competitive service industry. At the height of the Barnet Play in 2008, 199 rigs were operating. | No |
| The capital markets are more willing to provide risk finance for oil and gas activities. | No |

*Source*: based on Stevens (2013b).

### 4.3.2 *Incomes*

Another driver of future energy demand is income. The GDP growth figures underlying the energy demand growth in the forecasts are fairly consistent and generally quite conservative.

It is perfectly possible to imagine a GDP outcome that is very different. Three discontinuities loom large. The first would be another global financial crisis along the lines that followed the collapse of Lehman Brothers in 2008. The second could be a dramatic implosion of the Chinese economy as the political system collapses, faced with serious internal conflict as the economy stagnates and political reform fails to keep pace with economic liberalization. The third might involve 'Trumpian Uncertainty'. This refers to the fact that President Trump, to put it kindly, is extremely unpredictable. It is not clear when a 'tweet' is just a 'tweet' or when it is policy. It is perfectly plausible to argue his economic policies of 'America First' potentially could destroy the basis of globalization, creating trade wars that could plunge the world into economic recession.

Less traumatic but equally impacting on energy demand growth is the relationship between GDP growth and energy demand. In the OECD, rising GDP had been associated with declining total primary energy demand: in other words, improving energy efficiency. However, in the non-OECD rapidly rising GDP has also been accompanied by rapidly rising energy demand. All the forecasts have views of future potential for improved energy efficiency and this is a further factor that could disturb any historical relationship between energy demands and incomes. Greater efficiency will also be driven by improvements in technology relating to energy consumption.[20] The scope for improving energy efficiency is huge. One only has to compare Japanese energy consumption to other countries to imagine the scope for improvement.

### 4.3.3 *Energy Pricing*

Domestic prices are the result of the combined influences of international prices and domestic energy pricing policy. In the forecasts, generally only the international crude oil price is forecast. Forecasting crude oil prices has always been notoriously difficult (Huntington 1994). However, in any case, both for demand and supply, this price is misleading. For demand forecasts, crude prices are misleading because what matters are the final prices paid by the

[20] A good example here would be improvements in metering of electricity that allows consumers to see precisely what each appliance costs to run. A basic assumption in economics is that the quantity demanded is a function of price. However, most consumers have little or no idea of what it costs per hour to run even basic appliances. If that information were to become widely available, it could have a very significant impact on consumer behaviour.

consumer. This requires a view of what sales and other taxes will be imposed by the relevant government on product prices. Taxing oil products is a tax collector's dream. It involves a very large tax base since oil products are involved in all sections of the economy. In the short term demand is very price inelastic, which means high tax rates can be imposed without having much impact on the quantity sold. Finally, collecting the tax is extremely low cost. For many years many OECD governments have taken advantage of all these factors and as a result have imposed very high levels of sales and other taxes to raise revenue. Over the last few years, many other governments have also come to realize this potential source of revenue. One consequence has been that when the crude oil price collapsed after June 2014, many governments took the opportunity to use sales taxes to capture the revenue released by the lower price. At the same time, countries where oil products were subsidized took the opportunity also to reduce the level of subsidies. The result is that oil product prices to the final consumer are now and in future likely to be much higher than those assumed in many of the forecasts.

For supply forecasts as well, the crude price can also be misleading. For producers what matters is the price received after the fiscal system has taken its share of the economic rent. Thus lower prices can be offset by an easing of the fiscal terms.

There is another supply-side complication that appears to be neglected by most forecasters. A consequence of the collapse of oil prices since June 2014 has been a reduction in investment in upstream activities. The international oil companies, faced with a reduction in their cash flows, have been responsible for this (Stevens 2016b). On 14 January 2016, *The Financial Times* reported that Wood Mackenzie had estimated that private oil companies have shelved US$400 billions worth of upstream projects. This involved sixty-eight significant projects accounting for 27 billion barrels of reserves. This was equivalent to the reserves of Norway, Brazil, and Oman combined. A similar story of reduced investment has been true for the national oil companies that account for the bulk of global crude oil supply. This is simply because their governments are desperate to secure increased revenues to buy off political discontent amongst their populations of the sort seen in the Arab Uprisings which began in Tunisia at the start of 2011. In effect, the national oil companies (NOC) are being increasingly starved of funds. Upstream oil is a business where it is necessary to run in order to stand still. Thus, all fields have a natural depletion rate. Globally this averages around 5–7 per cent per annum. Thus, lower investment in upstream producing capacity now gives rise to the possibility in the future that, if oil demand continues to increase, as all the forecasts imply, the oil market could face a sudden supply shortage leading to possible price spikes within five to ten years. Even without actual physical shortages, paper barrel markets traditionally get rather nervous when spare capacity falls, which means they could impose a fear premium on oil prices.

Obviously, the future of crude prices does matter and could well disrupt the consensus forecast discussed in Section 4.2. Since OPEC took a decision not to cut production in November 2014, the price in effect has been launched onto a competitive market for the first time since 1928. The result has been a collapse in price. This in turn was the result of serious oversupply leading to a record increase in oil inventories. A key question concerns the level of ceiling and floor prices. The floor price is the short-run marginal cost (the variable cost) of the highest-cost producer. Leaving aside Canadian oil sands that are too small to make a difference globally, that floor price is defined by reference to the costs of US tight oil production. That figure is not known for sure if only because the technology has constantly been improving, allowing for cost reduction as a result of improved efficiency. However, putting it around US$25–30 per barrel is not unreasonable. The ceiling price currently is set by the 'fracklog'. This refers to the large number of wells in the United States that have been drilled but not completed. They were drilled because the terms of their leases required drilling but they were not fracked because prices were too low. Since the end of May 2016, oil prices have shown signs of recovery. If prices begin to rise, at some level probably between US$50 and US$60, these wells will be completed and new supply will quickly enter the market, relieving any signs of shortage.

All this suggests that absent any major geopolitical event, oil prices will remain below US$60–$70 for a number of years.

### 4.3.4 *Climate Change Policies*

Another area where the consensus energy forecasts may be misguided lies in the aftermath of COP21.[21] The Paris Agreement created a legally binding framework coming into force in 2020, once fifty-five countries accounting for over 55 per cent of global emissions have acceded to it. The agreement includes the long-term goal of holding 'the increase in the global average temperature to well below 2°C above pre-industrial levels, and to pursue efforts to limit the temperature increase to 1.5°C above pre-industrial levels', and of achieving 'a balance between anthropogenic emissions by sources and removals by sinks of greenhouse gases in the second half of this century'. In pursuit of this goal, parties are obliged to submit 'intended nationally determined contributions' (INDCs) setting out their pledges for climate action and to review them every five years.[22]

[21] This section relies heavily on Stevens (2016b).

[22] Now referred to as 'nationally determined contributions' (NDCs).

Current estimates are that the collective impact of the NDCs covering 146 countries, submitted as of 1 October 2015, would have a 50 per cent probability of limiting the global forecast temperature rise to 2.7°C by 2100, with a range of 2.2°C to 3.4°C (subsequent submissions would take the total number of NDCs to 160, covering 187 countries). They would slow global emissions growth by approximately one-third for 2010–30 compared to 1990–2010. To increase the likelihood of maintaining global warming below 2°C over the twenty-first century (with $CO_2$-equivalent concentrations in the atmosphere in 2100 of about 450 parts-per-million or less) would require a global reduction of 40 to 70 per cent of greenhouse-gas emissions by 2050, which would require much deeper emissions cuts than most countries are planning. To address the gap between current action and what is necessary to stay below 2°C, a 'facilitative dialogue' will be held in 2018 to give parties the opportunity to confirm or update their NDCs when the agreement becomes effective in 2020. The first formal review will then begin in 2023, leading to new NDCs.

It is entirely understandable that the implications of this new and major global agenda for the energy forecasts described in Section 4.2 are far from clear. Much will depend upon how far the NDCs will compete with each other. This will affect total energy consumption but, above all, it will determine the likely future energy mix. Currently the debate about unburnable carbon is well underway (McGlade and Ekins 2015; Stevens 2016b), but its outcome is uncertain. Pending the conclusion of this debate, the implications of this for future energy consumption remain uncertain.

## 4.4 Drivers of Future Trends: How Geopolitics May Derail the Conventional Views of the Future

This chapter has so far reviewed the extremely challenging global economic context in which policy decisions by the established and newer developing-country oil and gas producers need to be made. Later chapters of the book examine various components of those policy choices in greater detail. This present chapter is a clear reminder of the inherent economic challenges of engaging with the global oil and gas markets. However, those challenges do not end with the various economic and technological arguments outlined above. In addition, all oil and gas producers need to frame their decisions in the light of a volatile and changing global political economy. Hence this chapter ends with an assessment of some of the main features of the current geopolitics of oil and gas, especially in the Middle East. Obviously, there is a long list of possible events that could derail the forecasts in terms of their impact on supply and expectations. However, two areas of uncertainty dominate. These are the impact of the current lower prices and the parlous state of the Middle East.

Figure 4.6 illustrates estimates of the budgetary break-even prices for oil for the OPEC members in the summer of 2014: that is, shortly before the start of the oil price collapse.[23] The weighted average for OPEC was US$102 per barrel. Given the much lower prices experienced since the end of 2014, many countries face serious budgetary constraints. Some countries had accumulated financial cushions; largely Saudi Arabia, Kuwait, UAE, and Qatar. However, for the rest there was little protection, which means the governments are now struggling to find the revenues necessary to buy off the sort of domestic political dissent that triggered the Arab Uprisings from the start of 2011. Many producers face serious political challenges that in some cases, such as Venezuela and Nigeria, threaten the very existence of the state. This could easily affect their future ability to supply global markets. Such geopolitical disruptions in turn could well frighten the paper barrel markets and lead to higher prices. In this context, a key question is how far Saudi Arabia, as the dominant global producer, will be willing to maintain spare crude oil producing capacity.

As Figure 4.7 illustrates, the level of spare capacity to produce crude oil has been a crucial part of the oil market story and goes a long way to explain the historical pattern of prices. When spare capacity gets close to zero, the market becomes very vulnerable to price spikes. Since the summer of 1985, a central pillar of Saudi Arabia's oil policy has been to maintain spare capacity (of the order of 2 to 2.5 million barrels per day) in order to calm markets in the event of geopolitical supply disruptions.[24] In the summer of 2014, Saudi policy

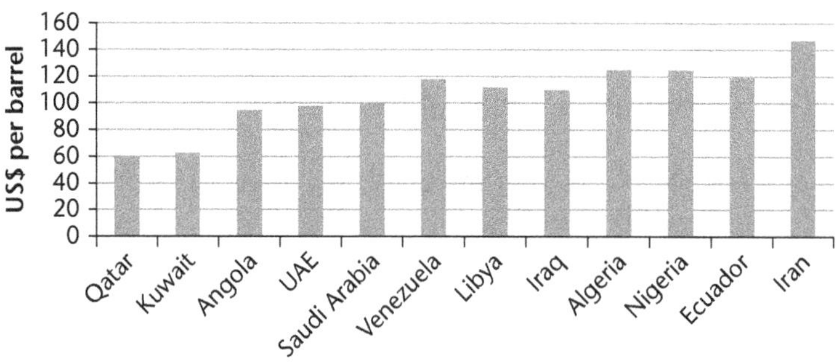


**Figure 4.6.** OPEC median budgetary break-even price, 2014

*Source*: author's illustration based on data in Apicorp Economic Commentary (2014).

[23] The budgetary break-even price is the oil price that would provide sufficient revenue to balance the national budget given expenditure plans and other sources of revenue.

[24] As can be seen from Figure 4.7, when the Iraqi invasion of Kuwait caused a loss of almost 6m b/d as a result of UN-imposed sanctions, the increase in price was very much muted as a result of Saudi Arabia's ability to replace the lost crude from its own spare capacity.

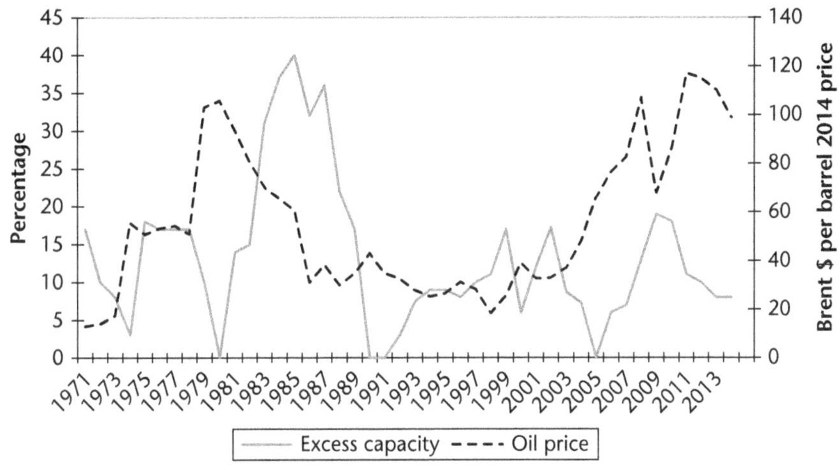


**Figure 4.7.** OPEC spare capacity and oil prices, 1971–2014

*Note*: Excess capacity refers to capacity that can be produce very quickly.

*Source*: author's illustration; data from oil price BP (2015). Excess capacity is author's estimate.

changed its approach, moving away from administering and controlling the market to a policy geared to maintaining market share and leaving price to find its equilibrium level in a competitive market. This to force the supply curve to go the right way for the first time since 1928: the low-cost producers supplying first, and increasingly higher-cost producers supplying the residual.[25] As 2016 developed, the Saudi policy reverted back to trying to defend price by trying to get agreement within OPEC and with some non-OPEC producers to restrict production. It remains to be seen how successful this policy may be. However, as already discussed, the shale technology revolution has significantly increased the price elasticity of supply of crude oil. Thus, any tendency to see increased crude prices is likely to produce a rapid supply response to choke off further increases. Oil prices are therefore likely to be lower for longer than many expect.[26]

The second dominant geopolitical issue that could impact the energy forecasts concerns the state of the Middle East and North Africa region (MENA). For reasons too complex to explain in this short chapter, the MENA region is in turmoil. It is necessary to go back to 1918—the end of the First World War

[25] It was in 1928 that the major oil companies met at Achnacarry Castle in Scotland and formulated the 'As-Is' Agreement, which began the process of administering oil markets and controlling prices. This control of the market stayed in place until the 1970s when OPEC took over the control of the market (Stevens 2013a).

[26] It is important to note that prices since 2014 cannot be regarded as historically low. Between 1986 and 2004 the average price of Brent was US$33 per barrel in real terms.

and the collapse of the Ottoman Empire—to find a period of similar uncertainty in that region compared to today.

One dimension of this uncertainty concerns the deterioration of relations between Saudi Arabia and Iran. Iran, coming out of a sanctions regime following the Joint Comprehensive Plan of Action (JCPOA) agreement on its nuclear programme in 2015, is now determined to restore its crude exports to pre-sanctions levels and beyond. This directly conflicts with Saudi Arabia's determination post 2014 not to cede market share. It was this determination that caused Saudi Arabia to cease its swing role and begin instead to protect its own market share. This in turn was linked to a general perception of 'Shi'a encirclement' within the kingdom in 2015 with President Assad in Syria appearing to secure his position, Iran developing a détente with the United States, and the growing Houthi involvement in Yemen. This sense of threat, real or otherwise, encouraged the new regime of King Salman, fronted by his son Mohammad, to adopt a more aggressive foreign policy. It remains to be seen whether Saudi Arabia's reversion to a policy of controlled supply in 2016 will change this situation.

This Saudi–Iranian rivalry could be dangerous politically for the MENA region. Iran is already deeply frustrated by what it sees (with some justification) as President Trump trying to renege on the JCPOA nuclear deal. Thus, there are threats of heavy fines imposed by the US Office of Foreign Assets Control on banks operating in the United States (which covers virtually all major banks, American or otherwise) and continued sanctions for dealing with the Iranian Revolutionary Guard Corps (IRGC). This is especially significant since the IRGC could be covertly involved in almost any Iranian business operation. As of August 2017, it appears that President Trump intends to take an increasingly hard line against Iran. If as a result Iran's frustration is not assuaged, might it jump before it is pushed and abrogate the JCPOA nuclear deal? Given Iran's ability to cause significant mischief in the region, this could well prove dangerous and could well be another factor de-railing many of the energy forecasts that have determined the current consensus view on the future of oil and gas markets.

## 4.5 Conclusions

The future of oil and gas markets looks very different from the experiences of the last fifteen years—many oil and gas producers, especially in the developing world, thus need to adjust their expectations and adapt their policy options. Expectations of great wealth at the top of the commodity super-cycle are no longer realistic. There are many reasons to think that oil and gas may not necessarily be produced at the levels expected by many low- and middle-income countries and that, even if production does not disappoint, the consequent revenues to their governments may well prove illusory. At the very least this will require them to re-think how to manage the expectations of their peoples, which had been inflated to unrealistic levels.

## References

Bacchetta, P. and E. Van Wincoop (2004). 'A Scapegoat Model of Exchange Rate Fluctuations', NBER Working Paper 10245. Washington, DC: National Bureau of Economic Research.

BP (2016). *The Statistical Review of World Energy 2016*. London: BP.

Huntington, H. G. (1994). 'Oil Price Forecasting in the 1980s: What Went Wrong?' *The Energy Journal*, 15(2): 1–22.

Landes, D. S. (1969). *The Unbound Prometheus: Technological Change and Industrial Development in Western Europe from 1750 to the Present*. Cambridge: Press Syndicate of the University of Cambridge.

McGlade, C. and P. Ekins (2015). 'The Geographical Distribution of Fossil Fuels Unused When Limiting Global Warming to 2°C', *Nature*, 517(7533): 187–90.

Mitchell, J. V. and P. Stevens (2008). *Ending Dependence: Hard Choices for Oil-Exporting States*. A Chatham House Report. London: Chatham House.

Sobel, D. (2011). *Longitude*. London: Harper Perennia.

Stevens, P. (1982). 'Oil Prices: The End of an Era?' *ODI Review*, 2: 1–19.

Stevens, P. (2013a). 'A History of the International Oil Industry', in R. Dannreuther and W. Ostrowski (eds.), *Global Resources: Conflict and Cooperation*, 13–32. Basingstoke: Palgrave Macmillan.

Stevens, P. (2013b). 'Shale Gas in the United Kingdom', Chatham House Briefing Paper, December. London: Chatham House.

Stevens P. (2016a). 'The Role of Oil and Gas in the Development of the Global Economy', WIDER Working Paper 2016/175. https://www.wider.unu.edu/sites/default/files/wp2016-175.pdf.

Stevens, P. (2016b). 'The International Oil Companies: The Death of the Old Business Model', Research Paper. London: Chatham House.

Stevens, P., J. Kooroshy, G. Lahn, and B. Lee (2013). 'Conflict and Coexistence in the Extractive Industries', Chatham House Report, November. London: Chatham House.

Stevens, P., G. Lahn, and J. Kooroshy (2015). 'Resource Curse Revisited', Chatham House Report, August. London: Chatham House.

Stevens, P. and P. J. G. Pearson (1984). 'Integrated Policies for Traditional and Commercial Energy in Developing Countries', *Development Policy Review* 2(2): 131–256.

UN (2015). 'Sustainable Development Knowledge Platform'. Available at: https://sustainabledevelopment.un.org/?menu=1300.

# Part III

# The Academic Literature and the Resources Curse

# 5

# The Curse of the One-size-fits-all Fix

## Re-evaluating What We Know about Extractives and Economic Development

*Glada Lahn and Paul Stevens*

### 5.1 Introduction

The year 2015 marked a notable deterioration in the fortunes of most new and prospective producers of oil, gas, and minerals. Many economies that had been enjoying 7+ per cent annual growth over the previous decade suffered shocks. Under the oil price decline, Ghana hit debt levels of above 70 per cent of GDP; Liberia (where oil has not yet been discovered) dismissed 80 per cent of the staff of its nascent national oil company in an effort to rein in profligate spending; the fall in the price of copper caused Zambia's currency to depreciate by 80 per cent; and Mongolia's debts jumped tenfold in just six years. By 2016, countries hoping for a development boost from their natural resources were having to readjust their expectations. Meanwhile, more experienced exporters, starved of foreign currency, set in motion a combination of spending cutbacks and foreign currency restrictions, which had by 2018 led to a humanitarian crisis in Venezuela. In many resource-exporting or *resource-expecting* countries, the signs of unsustainable spending patterns were evident well before the price crash and accusations of mismanagement of wealth, lost opportunities, and wasted resources will dominate their politics for years to come.[1]

[1] This chapter draws on an earlier Chatham House publication entitled *The Resource Curse Revisited*, and the authors would like to thank everyone who assisted with that study. We would also like to thank Alan Roe for his expert comments and suggestions, and Joseph Laredo for his editing.

The impacts of the debt and austerity measures, meanwhile, will influence the stories we tell about extractives commodities and economic growth.

In the heady days of a rising oil price and seemingly unstoppable growth in raw materials demand from China post 2003, positive stories about the prospects for basing economic growth on extractives were easy to tell. Three decades of economic literature drawing attention to a 'curse of natural resources' became less fashionable in political debates. An industry in how to turn a curse into a blessing in terms of governance advice and consultancy bloomed. However, in reality, the learnings of the past half century and this advice industry seem not to have enabled many exporting countries to avoid the economic ills associated with the resource curse. So perhaps it is time to revisit what might have been missed and also what is new in the current global context for extractives and development.

In this context, this chapter revisits the historical evolution of thinking about the interaction between extractives production and export and economic performance.[2] In particular, it revisits the large body of analysis known as 'resource curse' literature, which seeks to identify a link between resource wealth and poor economic performance. It considers how some of this, particularly the identification of bad decision-making and weak institutions as transmission mechanisms, fed into a political fashion for 'extractives-led growth' that has reigned politically for the last thirteen years or so. The chapter questions this newer agenda, which proposes 'good governance' as the magic formula for turning resource wealth into a blessing. Given that it has not succeeded in saving most resource exporters from some of the resource curse effects—in some cases, economic calamity—where did it go wrong? The chapter identifies three aspects that it contends have been marginalized in recent debates: the very different resource bases and country contexts; the problems posed by too rapid a pace of development; and the influences that resource discovery and development have on the political economy—influences that militate against the effectiveness or even implementation of the recommended 'good governance' practices.

There is merit in returning to the resource curse literature to look at the problem afresh. But instead of asking the familiar question of why extractive resources have caused poor performance, it might be more fruitful to ask why they have not helped the rest of the economy to grow. If, as much of the resource literature contends, dependence on rents is the problem, then economic diversification should be the aim. For states dependent on petrodollars, achieving diversification is currently more urgent than ever. Hirschman's (1977) work on linkages between the extractives sector and the

[2] This literature survey is based upon Stevens (2015). In the interest of brevity, there are several references back to the original survey, which can be found online.

rest of the economy can shed light on the complexity of managing this feat, especially when there is pressure to develop the resources as fast as possible. The chapter does not reject the idea that extractives-led growth is possible, but argues that this possibility must be seen within a transition strategy based on 'worst case' depletion timelines and commodity price scenarios. Not all countries will be equipped to manage the planning for that transition. Looking ahead, the chapter identifies several aspects in a changing global political and environmental context as important considerations for transition countries—and indeed for anyone studying the interactions between extractives and the economy.

## 5.2 An Evolution in Thinking about the Use of Extractive Resources for Economic Growth

Simple logic would suggest that nations presiding over large reserves of a below-ground resource valuable to foreign markets are economically privileged in comparison with less well-endowed territories. Since at least the 1940s, various arguments in economic development theory have supported this assumption. For example, Rosenstein-Rodan (1943) argued that countries must experience a 'big push' in public investment to break out of a self-feeding circle of poverty. The revenue provided by resource extraction can enable this by providing windfall income to finance large-scale capital spending—and, more specifically, allow spending on foreign goods, services, and investment through the generation of foreign exchange. The Harrod–Domar growth model, which dominated economic thinking in the 1960s, supported this idea. The sector should become a 'leading development sector' or 'growth pole', whose success encourages growth in other sectors as wealth and benefits trickle down. Furthermore, historical evidence from industrialized economies such as the United Kingdom tends to reinforce the belief that countries should focus on their *comparative advantage* over others. That view underpins the economic argument that a focus on producing primary products should promote growth (see Stevens 2015: 3).

However, there is an even longer-standing observation that abundance (not necessarily in minerals) leads to decadence and negative economic consequences. This goes back at least as far as the fourteenth century, when the Arab philosopher Ibn Khaldun identified the fifth stage in the evolution of the 'state' as one of waste and squandering (Ibn Khaldun 1967 [1377]). And in the seventeenth century, Spain's wealth acquired from the New World appeared to trigger the decline of that country's fortunes, not to forget, of course, the fate of societies in the lands that had been plundered. In the last century, the oil price shocks of the 1970s led to greater scrutiny of the impacts of extractives exploitation and export growth on national economies (Stevens 2015: 5).

Economists studying development in the 1950s and 1960s had already begun to express concern over countries at an early stage of economic development that depended on the export of primary products. Singer (1950) and Prebisch (1964) argued that such countries would find themselves at a disadvantage in trading with industrialized countries because of deteriorating terms of trade. Others reinforced the argument of the negative consequences of producing and exporting raw materials by emphasizing the limited economic linkages generated from primary sectors compared with manufacturing (Stevens 2015: 5).

The observation of declining economic fortunes in countries with an apparent comparative advantage in extractives gave rise to the thesis of a curse of natural resources (Auty 1993) or that of the 'paradox of plenty' (Karl 1997). A large body of work seeking to establish a negative statistical correlation between abundance of natural resources and GDP performance emerged, beginning in the late 1980s. The evidence at first appeared to support the theory of a negative link; several OPEC countries were shown to have declining GNP per capita over time compared with non-oil-producing countries of similar initial GNP and appeared also to suffer a decline in agriculture and other sectors predating hydrocarbons development. Auty (1994: 22), for example, points out that in Mexico 'by 1982 virtually the entire non-oil economy became non-tradable, that is, in need of total protection or subsidies'.

The Netherlands offers one of the most famous cases of negative impacts on a national economy resulting from hydrocarbons expansion. Here economists drew links between an upsurge in gas export revenue in the 1970s and the declining competitiveness of Dutch manufacturing. As a result of spending natural resource revenues, the economy overheated and the real exchange rate appreciated, thus making national manufactured products far more expensive for other countries to buy. The result was a contraction in the non-oil/gas/mineral-traded sector (Stevens 2015: 14–15).

Over time, however, economists refined their analysis of the mechanisms that turn an apparent blessing into a curse. These mechanisms can be roughly put into two groups. The first relates to the impacts of economic dependence on a primary sector with volatile market prices: long-term decline in the terms of trade, revenue volatility, Dutch disease, crowding-out effects. The second relates to how resource wealth is governed: the increased role of the state, and socio-economic and political impacts (Stevens 2015: 12–27). Naturally, these are not mutually exclusive; in theory, the former could be ameliorated with better policy and spending decisions. The latter largely draw attention to why this does not happen, showing that the presence of resources shapes certain socio-political conditions, which in turn reinforce negative economic impacts.

Interest in the effects of extractives growth on the political economy grew from the 1990s onwards. This literature most frequently examined how the phenomenon of large flows of revenues from a single source (requiring little labour relative to the revenues generated) that generally accrue to government (as owner of the resource) affects power relations, institution-building, and government decision-making. This literature tends to conclude that the special features of extractives-sector development are likely to encourage poor governance and thus poor spending decisions and little fiscal prudence (Stevens 2015: 18).

There are several variations on this theme, national dependence on 'rent' that accrues to a small group of government–business elites being a dominant theme. This is alleged to discourage institution-building and encourage 'rent-seeking' (positioning by groups or individuals to elicit transfers of the resource wealth) rather than entrepreneurship and productivity in the economy. A strand of the literature examines the 'rentier state' and its tendency to crowd out productive private-sector activity and to cement legitimacy through allocation of resources rather than through democratic competition (see Stevens 2015: 20). Auty (2010) attributes many of the ills noted by the resource curse literature not to resources per se but to the dependence on rent, which can also come from other sources (e.g. aid, canal tolls, security rents).

Closely linked to rentier theories is the notion that disproportionate fiscal dependence on petro-dollars in particular affects the capacity of the government to make decisions (e.g. Karl 1997). Having more money to spend tends to weaken prudence and 'due diligence'. Of notable importance is the tendency of governments not to prioritize investments that enable long-term income generation. This can lead to unrestrained spending on 'prestige projects' using expensive foreign expertise and materials (Stevens 2015: 19). In turn, the lack of need for tax revenues from diverse sources may weaken the prospects for democracy and the development of institutions that provide checks and balances on, for example, government power and spending.

Developing this theme, a branch of the literature considers industrial policy, examining how centralized wealth or 'rent' flows can lead to top-heavy government, prone to market control or intervention. A key 'distorting' aspect is the impetus for government to subsidize chosen sectors and key commodities. In the case of domestically produced energy resources, for example, there is a tendency to provide them cheaply to stimulate supposedly job-creating industries and to prop up the agricultural sector. In the 1970s and 1980s, many resource export-dependent governments adopted an industrial policy based on 'import substitution'. This invariably had two components: (a) the introduction of subsidies and growing protectionism in the form of import controls and (b) favourable taxation of the protected activities. Instead of nurturing industry until it was competitive without subsidy, this approach tended to lead to ever increasing dependence on government support, putting an additional drain on government finances as well as creating powerful lobbies that would resist moves to reform policy (Stevens 2015: 21–2).

Many of the dynamics described above are interlinked and can be seen as self-reinforcing. Dirigiste wealth allocation, state capital spending, and rising subsidies in place of policies to enhance productivity, industrial strategy, and institution-building increase a country's dependence on resource income, thus making it more difficult to manage price volatility. They can also lead to a spiral of indebtedness as countries are forced to borrow, not least to maintain political stability, which may be tenuous if government legitimacy is largely based on the distribution of resource wealth.

In this respect, international finance institutions (IFIs) and aid-giving bodies are not necessarily considered neutral parties, and several studies draw attention to their role in worsening resource curse effects (Extractive Industries Review 2003; Hilson and Maconachie 2008; Shaxson 2007).

The resource curse debate peaked in the mid-1990s as analysts looked back at a period of some twenty years of falling commodity prices, and criticisms of the theory began to emerge in the early to mid-2000s as the international prices of oil and other commodities again began to rise. Critiques included claims of selective bias associated with previous studies (choosing only resource-rich countries whose prospects for developing other areas of the economy were poor), a time frame too short to detect real economic impact (the volatile 1970–90 period in particular), and the inadequacy of the technical approaches used for measuring economic progress and the links to extractives (see Stevens 2015: 7–8).

Of course, adherents to resource curse theory as summarized earlier usually note that this 'is not an iron law, rather it is a strong recurrent tendency' (Auty 1994: 12). In this regard, and in response to criticism of the IFIs, interest grew from the early 2000s in those countries that had proved to be exceptions in avoiding, or apparently 'overcoming', the curse and how they had done so. Academics and IFI analysts, for example, often cite the cases of Botswana, Chile, and Malaysia as resource development successes. Norway, Australia, and Canada, as well as the United States and the United Kingdom, could also be seen as having avoided the curse. So why did these countries fare better than others with similar resources? The literature tends to cite as the main reason the strength and quality of their institutions prior to resource development.

The political economy focus that began in the 1990s blossomed into a vibrant debate about governance in extractives-producing countries in the 2000s. Poor governance and institutions became the leading explanations for poor economic outcomes. And this thinking led to the conclusion that good institutions may be able to prevent negative outcomes from resource wealth because they can constrain the predatory behaviour of those who hold political power. Thus, they are a mechanism with which to sever the otherwise negative link between natural resource abundance and poor outcomes (Stevens 2015: 26).

The growth of these ideas was accompanied by the rise of commodity prices—particularly oil from around 2003. Resource exporters' GDP soared. The Gulf Cooperation Council (GCC) members in particular stood out as a sharp retort to those advocating resource curse theories. Here were six petroleum-exporting states that remained highly dependent on their extractives sectors yet could not be judged to be worse off economically than they would have been if they had not developed their resources.[3] As the price of commodities rose, extractives companies made new commercial discoveries in low-income countries and foreign investment flowed in (Stevens et al. 2015). The new questions on the table were: if some countries had done well on the back of resource development, then what was the magic policy formula? And (how) could this be prescribed and transferred successfully to others in the early stages of resource development?

## 5.3 The Extractives-led Growth Agenda Emerges

These questions, in addition to the desire of many interested parties—including aid-giving countries, multilateral financiers, and extractives companies—for a positive development story, paved the way for what we call the *extractives-led growth agenda*.[4] This is built on a consensus between an influential set of international actors around the idea that, if managed properly, extractive industries can help to drive broad-based socio-economic development in developing countries with extractive resources (see, for example, Africa Progress Panel 2013: 6).

A 2013 McKinsey Global Institute report captures the optimism of commodities prices at their height:

> If resource-driven countries, particularly those with low average incomes, use their resources sectors as a platform for broader economic development, this could transform their prospects. We estimate that they could lift almost half the world's poor out of poverty. (Dobbs et al. 2013)

[3] These are Bahrain, Kuwait, Oman, Qatar, Saudi Arabia, and the United Arab Emirates (UAE).

[4] A diverse set of actors promote this narrative—including donors like the World Bank, the UN Development Programme (UNDP), the development agencies of major donor countries including the United Kingdom, Norway, and the United States, (mainly Western) extractives companies, consultancies such as McKinsey and Tony Blair Associates, Adam Smith International, and civil society organizations including the Natural Resource Governance Institute (NRGI) and the Africa Progress Panel.

Although the extractives-led growth agenda did draw on academic critiques of the inevitability of resource curse effects, it was arguably more of a policy fashion than a movement based on theory. It was partly born out of a surge of investment in extractives in more challenging regions amid the high (and rising) commodity prices of the 2000s, coupled with an increase in Asian investment strategies focused on foreign resources. From a Western development agency and IFI perspective, the feeling was that countries benefitting from these benign circumstances would develop resources with or without their assistance. Hence their best bet to remain relevant would be to promote better practices. As the Millennium Development Goals and their only partial achievement made clear, the problem of poverty had not gone away and a booming market for commodities appeared to offer a golden opportunity for some countries to address this problem, as well as to 'graduate from aid'.

In developing countries with resources, technocratic and civil society proponents of greater accountability and the improved handling of resource revenues saw global governance initiatives as an opportunity for much-needed reforms. Leaders in countries with a poor governance record were looking for a way to improve their image internationally and earn assistance (including aid and debt cancellation). And companies were naturally eager to promote their advantages in good governance over new competitors and to avoid penalties in an environment of increasing anti-corruption legislation in their own countries.

The remedy or preventative for resource curse effects prescribed across the board was good governance, institution-building, and best practices in various dimensions of managing the resource. Prescriptions generally involved optimum contractual terms, revenue transparency, institution-building, use of stabilization funds, and local (skills and market) capacity-building to service, and benefit from, the sector. The Extractive Industries Transparency Initiative (EITI), launched in 2003, and the Natural Resource Charter have been particularly successful in signing up both new and more established producers including Azerbaijan, Equatorial Guinea, Ghana, Kazakhstan, Nigeria, and Timor Leste to processes of governance improvement. Many international extractives companies have also signed up to or added their support to these and other voluntary initiatives. Alongside this, several detailed advisory forums and source materials have been developed to guide producers, including the World Bank-funded *Extractive Industries Source Book*, the Canadian-led Intergovernmental Forum on Mining, Minerals, Metals and Sustainable Development, and the NRGI.

### 5.3.1 *Has the Remedy Worked?*

In the past fifteen years, a wealth of global expertise has contributed to building a body of advice on better resource-sector management. However, the record of the new or economically troubled producers at whom much of the advice of the last decade and a half has been aimed, is mixed.

Critiques of the new approaches range from their enabling corrupt governments to gain international legitimacy by cynically joining voluntary governance initiatives (David-Barrett and Okamura 2016; Keblusek 2010) to the problem of advocating 'best practice' standards to countries with little capacity to implement them.[5] The World Bank/Exxon-led Chad–Cameroon pipeline agreement, billed as 'a pioneering effort... to demonstrate that large-scale crude oil projects, when designed to ensure transparency and effective environmental and social mitigation [sic], can significantly improve prospects for sustainable long-term development' (IFC 2003), is a clear example of how such initiatives can fail (World Bank 2008).[6]

The more globally applied EITI has come in for particular criticism for promising more than it can deliver (Darby 2010; Hilson and Maconachie 2008; Keblusek 2010), and a growing body of analysis is now beginning to look back and assess the impact of the EITI in countries that have adopted it (Alstine 2014; Mejía Acosta 2013; Sovacool et al. 2016). Sovacool et al. (2016: 185) find that 'there was not a single governance [or] economic development metric in which EITI countries performed better during EITI candidacy or EITI compliance than pre-EITI as well as better than other country classes.' Furthermore, on most of the standard governance metrics (the exceptions being 'voice and accountability', 'rule of law', and GDP—the last of which can be more readily attributed to the rising oil price during this phase), most member countries registered declines during the candidacy period. This phenomenon is also described by Sovacool and Andrews (2015), who posit that governments may have little incentive to continue improvements once candidacy has been achieved.

The strong belief at the outset of the last commodities boom was that the 'lessons learned' from the oil boom–bust cycles of the 1970s and 1980s, combined with improved fiscal policy to tame resource rent spending, would usher in an era of more even budgeting, thus avoiding the pitfalls of the past. Yet, in a sample of forty-eight countries dependent on exports of oil, gas, and metals,[7] Bova et al. (2016: 17) found that 'adoption of fiscal rules or resource funds [does] not have a significant impact on fiscal cyclicality'. In other words, countries tended to dramatically increase government spending

[5] For example, is it appropriate for a relatively poor country to focus on establishing separate institutions and capacities for the sector in policymaking, regulation, and operations—as Norway does—when it has neither a sufficient number of qualified staff nor enough accumulated knowledge about the sector or the geology? See, for example, Marcel (2013).

[6] Rarely can such a high technical assistance effort to ensure the effective management of oil resources have been derailed so quickly and comprehensively.

[7] Countries where these commodities represent 20 per cent or more of total exports or 15 per cent or more of fiscal revenues for most of the 1970–2013 period.

during the 2003–13 boom in spite of putting in place the prescribed measures to prevent this. Ghana, as discussed elsewhere in this series, is a striking example.

Studies on the effectiveness of stabilization funds tend to agree that they 'must be part of a broader package of institutional reforms designed to improve the country's capacity for resource revenue management' (Dixon and Monk 2011: 5) and that their usefulness depends on the quality of public financial management systems (IMF 2007). As the examples of Chad, Nigeria, Iran, and Cameroon demonstrate, where these are lacking, the original aims and rules of the funds are simply not followed (Collier and Venables 2011: 11–17). Even the existence of well-managed funds does not preclude unsustainable fiscal policies or the use of fund resources as collateral for reckless borrowing (e.g. Kazakhstan in the 2000s) (Stevens and Mitchell 2008: Appendix 1). As might be expected, Bova et al. (2016) find that fiscal measures were effective only where there were also strong political institutions, as in the typically cited success cases: Botswana, Chile, and Norway.

## 5.4 What Are the Extractives-led Development Narrative's Flaws?

### 5.4.1 *There Is a Tendency to Prescribe a One-size-fits-all Solution*

Political history, power relations, geography and geology, demographic structures, culture, economic structure, and level of education all influence how extractives will affect a country's development. And the countries that became the focus of attention on extractives-led growth are vastly different.

Focusing on sub-Saharan Africa, Hilson and Maconachie (2008) divide countries into three groups, drawing attention to inherent political or legal-economic features that subvert attempts at good governance. First, there are the 'petro-economies' Chad, Gabon, Cameroon, and Equatorial Guinea, whose dictatorial policy environments, it is argued, will simply negate the good intentions of voluntary governance regimes. Second, there are the 'lootable' economies (Angola, Sierra Leone, Liberia, Madagascar, and DRC), which contain 'lucrative pockets' of extractives wealth, including diamonds and gemstones. These countries have been subject to frequent violent conflict, the groups having control over these resources being firmly entrenched in the power structure. Third, there are 'conventional mineral producers' such as Ghana, Guinea, Zambia, and Tanzania, where unfavourable mining codes and contracts are argued to have been a greater cause of economic woes than poor governance.

Norway, often highlighted as the success story par excellence of extractives-led growth, had a developed democratic system of governance and a relatively small population at the outset of its discoveries. These discoveries were also all offshore, and therefore less disruptive to local societies.

The rhetoric around extractives-led growth has tended to group even such diverse countries together, considering them all to be 'resource rich'. But this makes little sense. The fact that two countries are sitting on the same volume of minerals does not mean that both are 'rich', as their resulting wealth will depend partly on the cost of extraction and partly on how that wealth is likely to be shared and how many people will enjoy it. For example, in terms of production per capita, the United Kingdom is richer in resources than Nigeria (see Figure 5.1). This may matter in terms of deriving economic and social dividends from resources. Myers (2005), for example, has argued that when their production is below a certain number of barrels of oil equivalent per head per year (b/h/y), non-OECD producing countries tend to underperform on conventional development indicators in comparison with their neighbours.

In this diverse context, policies that contributed to success in Norway or Botswana will not necessarily work in other countries. One glaring problem is that countries and companies with longer periods of economic stability, well-functioning institutions, and experience have evolved high standards of practice. By contrast, low-income countries have lower institutional capacities, fewer skilled and experienced professionals, and different developmental priorities (Marcel 2013). Parachuting the institutions and practices of the former group into the latter countries is likely to be impossible to achieve,

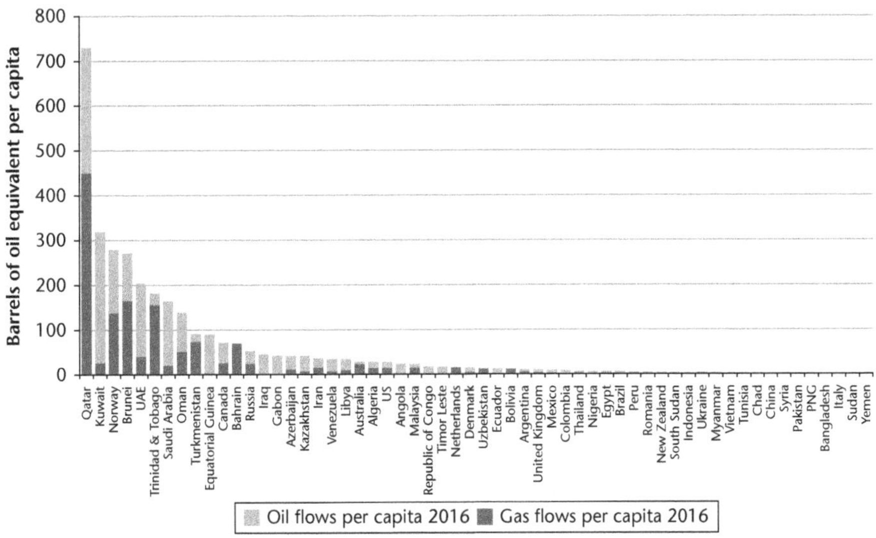


**Figure 5.1.** The world's top 60 per capita oil and gas producers in 2016

*Source*: author's illustration based on Myers (2015), based on BP statistical review and World Bank population data.

whilst consuming a disproportionate amount of resources and efforts relative to the rest of the economy.

### 5.4.2 *Political and Psychological Impacts Can Overpower Attempts at Good Governance*

Several scholars draw attention to the complacency that 'rent'-based growth may induce in government decision-makers, as well as in society more generally, and the natural psychological response to seek opportunities to gather more rent rather than pursue 'productive' enterprise. The tendency to use extractives rents to support other industries, and domestic energy, water, and food consumption in particular, is commonly identified as a 'distortion' that disincentivizes and reduces efficiency and productivity (Stevens et al. 2015: 10).

Achieving institutional good governance in countries with a relatively low capacity to manage the extractive sector at the outset will be a long, hard slog right from the very beginning. It will require both sustained political will and a measure of societal stability. At the same time, while it is known what needs to be done for extractives-led development to have a positive outcome, the reality of the political economy that develops around an extractive industry or other, similar forms of rent-generating activity may make it impossible to achieve that outcome.

### 5.4.3 *Development Is Too Rapid*

There is a strong tendency among countries that discover resources to develop projects as quickly as possible and aim for rapid depletion of those resources. Pressure to do so comes from two sources: the host government and the operating companies. For their part, multilateral financiers and development agencies are eager to apply their expertise to the extractives-led growth agenda, which, even if this is not their intention, in terms of practical policy and economic advice tends to reinforce domestic, government, and investor pressures to 'develop fast' and, as a result, miss out on (the slower gestation) opportunities to develop linkages appropriate to economic diversification and management capacity (Stevens et al. 2015: 32–3).

## 5.5 Considerations for Revising the Advice to Producers

As explained in the literature review, there is economic logic in the argument that resource revenues can improve a country's growth prospects, potentially raising levels of income, savings, and investment. Resource curse studies have tended to focus on negative outcomes and to pose the question of how extractive-sector development has contributed to those outcomes. But a more useful approach may be to ask why the extractive sector has in many cases failed to become the leading sector, thereby serving as the engine to improve opportunities and growth in other parts of the economy. Dissecting the answers to that question might be a better way for countries with extractives reserves to understand the requirements and risks they face in developing their resources. This leads to four interlinked considerations that arguably should guide policy:

1. Diversification of the economy from the outset of decision-making over the extractives sector
2. The capacity for the sector to link with and contribute to the rest of the economy
3. Options for the pace of development of extractives projects
4. Measures of performance that should guide the economy, and the possible models of development these might favour.

We will now examine each of these considerations in turn.

### 5.5.1 *Diversification as Key*

For those countries with large reserves of natural resources per capita, depletion-led diversification—that is, into energy-intensive industries with strong linkages to the sector—has proved a beneficial economic pathway for at least a few countries in past decades (Australia, Canada, GCC states, Trinidad and Tobago). Those economies remain dependent on the extractives sector and their trajectory is not sustainable indefinitely, but we can say that many low-income countries would not mind following in their footsteps in terms of the living standards that their citizens enjoy. Even this development model, however, has been challenged by the fall in the oil price. The governments of Saudi Arabia and the UAE, for example, have now adopted economic diversification as a domestic priority, tightly connected to job provision and social stability.

For less well-endowed countries (on a per capita basis), crunch points may occur much earlier, before even basic infrastructure such as power and clean water has reached the mass of their populations. As the historical problems of harnessing gas for power in sub-Saharan Africa and parts of South Asia demonstrate, the mere existence of a fuel resource does not mean that it can easily be exploited for domestic use. A country will need to at least demonstrate resource potential—and the guarantee of returns that can be remitted to shareholders—to gain investor attention and realize the type of large-scale project that requires, for example, transportation and power infrastructure.

Mitchell and Stevens (2008) and Stevens et al. (2015) have explained the general depletion profile, whereby a country must have begun to diversify its revenues by the time production of the commodity in question reaches a plateau. Failure to do so will likely mean that its spending levels become unsustainable. Naturally, commodity price volatility will also affect the length of that transition period. If there is not sufficient growth in other areas of the economy when the commodity's export volumes go into terminal decline, the country will rapidly fall into deficits—both in its current fiscal account (for public spending) and in its current external account (foreign exchange for purchasing foreign goods).

In promoting governance fixes to enable extractives-led growth, far too little attention has been devoted to whether a country has the characteristics and capacity to make this transition.

More generally if resource-rich countries decide to develop their natural resources, they need to consider how to use the sector as a 'one-off opportunity' for development, bearing in mind the risks of a volatile market. It is doubtful that this is a realistic option for new, small-scale producers that have limited capacity for spending and implementation. A fund to manage revenues over time would be a necessary but not a sufficient condition. But the prospects of such a fund being effective in countries with low capacity to manage it are poor (Stevens et al. 2015: 21). If a slower, 'development first' route is taken, then countries must focus on optimizing economic linkages with the sector.

### 5.5.2 *Understanding Linkages—and their Limitations*

Albert Hirschman's (1977) idea of linkages from extractive projects to the rest of the economy offers some indication of why, in many cases, the extractives sector has failed to lead the rest of the economy upward.

Any extractive project generates linkages to the rest of the economy, which can be categorized as fiscal, forward, and backward. *Fiscal linkages* refers to the revenue generated for the owner of the resource; in most cases outside the United States, this is the state as represented by the government. *Forward linkages* refers to the supply of the sector output to the rest of the economy, which, in the case of oil and gas projects, implies the supply of oil and natural gas products; but this definition can be expanded to include the supply of modern management techniques and managerial capacity to the rest of the economy. *Backward linkages* refers to the inputs into the project from the domestic economy in terms of employment, capital, and material inputs into the value chain.

Where new projects are being developed, it is usually hoped that these linkages will lead to increased productivity and development in other areas. However, Hirschman's analysis pointed to the limitations of the linkages of oil, gas, and mineral projects, which explains why this often does not happen.[8] The 'enclave', high-tech nature of extractive projects makes them more likely to be isolated from the local economy in developing countries lacking the appropriate skills base.

In many cases, the majority of revenues from exporting the raw material accrue directly to the government. This means that government entities, rather than the industry, become by default the critical drivers of economic growth and development. A government's ability to spend revenues and allocate resources effectively then becomes the focus for economic performance. This ability is affected not only by the level of institutional development prior to the arrival of extractives production, but also by the strong political and psychological factors that come into play once (a) public expectations of a new flow of extractives revenue are raised and (b) a state–business elite has developed on the basis of rent capture and positions itself to defend the interests that will arise once that rent begins to flow.

### 5.5.3 *The Pace of Development Matters*

In theory, the rapid-development model offers advantages in terms of early stage cash with which to initiate solutions to immediate and urgent problems in an economy—such as poverty reduction, debt financing, and energy and transport infrastructure development. At the same time, it is in harmony with the idea of a 'one-off' opportunity for development in a volatile market, especially when there are concerns that the resources may not be worth as much in the future. There are several pressures that tend to accelerate development. These include popular demands to deploy the new revenues for better standards of living; manoeuvring among elites aiming to capture the new wealth; and the appetite of foreign companies for economies of scale and rapid returns on investment.

However, for a country, accelerated development may mean depleting its reserves faster than is desirable. It also poses several problems, especially for countries that have limited institutional and regulatory capacity. These include the inability to develop backward linkages in time to benefit from potential new investments and opportunities to serve the sector; the lack of time to introduce regulation of the new sector and to increase capacity for handling and deploying new revenue flows; and vulnerability to the

[8] Hirschman's original framework pointed to negative impacts on non-resource sectors and implicitly pointed to potential alternative sources of development.

boom–bust cycles that are at the heart of resource curse ills (Stevens et al. 2015: 34). The result is that enhanced levels of indebtedness once resource flows wane are inevitable.

### 5.5.4 *Alternative Measures of Performance and Models of Development Overlooked*

The extractives-led development rhetoric may not only overrate the ability of poorer countries to rise to the considerable challenges they face. Perhaps more important, it may also obscure the possible comparative benefits and savings involved in developing other sectors rather than extractives or developing the extractives sector at a much slower pace than that advocated instinctively by industry and investors. That may require a change in the way that companies and investment opportunities are judged, with less emphasis on short-term growth and more on sustainable value creation.

GDP growth is a poor indicator of the contribution of the sector to the rest of the economy. Alternatives to GDP are being pursued by several international institutions (see, for example, Stiglitz et al. 2009). Which measures countries, finance institutions, and banks eventually adopt will affect decision-making regarding extractive resources. As the economist John Talbert (2010) puts it:

> Obsession with GDP growth has spurred policies to liquidate natural capital as quickly as possible... correctly valuing changes in our stocks of natural capital and the ecosystem services that they provide will help advance a science of new metrics capable of inspiring more sustainable policy choices.

Whist new producers may not have the power to address this broader systemic market problem, a better calculation of resource value and depletion and domestic environmental costs, factored into the extractives and non-extractives GDP calculation, would result in a better measure of sustainable diversification. The key point here is that revenue from extractives is not income but an exchange of below-ground resource assets for above-ground cash assets. Not only this, but the process of reshuffling has costs—environmental (variously including land use change, pollution, water demand increase) and often also social, especially where extractives activities displace people whilst others migrate to the area of resource production in search of work. For extraction to be worthwhile on a national basis, its long-term value must exceed these costs. The most logical way of achieving this situation is to invest earnings in income-generating activities or other assets of societal benefit that can replace their value to the national economy over time.

A new measure of declining dependence on the sector—that is, a *measure of sustainable economic diversification*—would be useful here. Tracking the non-hydrocarbon fiscal and current account balances over time can indicate whether a country is moving in the wrong direction (Mitchell and Stevens 2008; Stevens et al. 2015). But reducing non-extractives or non-hydrocarbon fiscal deficits does not necessarily indicate *sustainable* diversification, as non-extractives activities often include industries that are dependent on subsidized extractives input or other government subsidies. Not only can this put an increasing burden on state finances; it can also impose environmental asset losses and inhibit the implementation of efficient energy access and the growth of low-carbon sectors, which most new and prospective producers also aim at (Lahn and Bradley 2016; Bradley et al. 2018).

## 5.6 Conclusions

The resource curse literature provides insights that have been partially sidelined during a period of rising and high commodity prices: a period of heightened international interest in increasing oil, gas, and mineral resource flows. The evolution in thinking about the impacts of extractives on economic and wider patterns of development reveals the folly of our predilection for a single cause or reason for particular outcomes and 'one big solution' to our problems. The shortcomings of the newer extractives-led growth agenda underscore this folly. Country histories, populations, capacities, power dynamics, and reserves bases, in all their diversity, are a critically under-explored set of factors that need to be taken into account when prescribing 'governance' for the sector. While good institutions and good governance are undoubtedly the reason certain countries have been able to benefit from their extractives endowment, this does not mean that the same policies can be applied to other countries easily or with the same effects.

This chapter has argued that the governance challenges for new, low-capacity producers remain immense and that the global context for extraction calls for expectations to be revised. At the time of writing, the downturn in commodity prices is proving challenging to exporters, many of whom benefitted financially during the previous decade of high prices. Indebtedness is likely to frame the next decade for many developing countries that have become dependent on exports. At the same time, reliance on the sale of high-carbon fuels is being challenged by the prospective global shift to lower-carbon technologies and greater energy efficiency. Greater pressures on the environment are increasing the value of assets that can be damaged in the process of extractives development, production, and use in industry.

Turning around the resource curse question to ask why, in some cases, the extractives sector has not become the driving sector for the rest of the economy suggests that much greater focus and attention should be given to the role of the sector over time in a country's development strategy. In the light of this, there are four interlinked aspects of extractives-led growth that require greater attention. These are:

*Capacity to diversify*: To use its extractives sector to jump-start growth while allowing enough time for sustainable diversification, a country needs a sufficiently large resource base. Understanding what kind of economic contribution can be expected from reserves over time, as well as whether there is the capacity to channel the revenues into spending and investments that promote self-sustaining growth in other sectors of the economy, will be key to deciding how sensible an option extractives-led growth will be. Such understanding is far from simple to acquire, and factors such as global market risks and trends, including those from increasing carbon constraints and technology shifts, must be taken into account.

*Economic linkages between the new or expanded extractives sector and the rest of the economy*: To achieve these requires in each case a much deeper understanding of what services, jobs, and business opportunities can realistically be generated from the extractives operations and for how long. Energy should be part of this consideration—the policy for allocating and/or pricing fuels for domestic use is a vital dynamic in industrial development, sustainable diversification, and fiscal balance.

*The pace of development*: As suggested above, many low-income producers lack the capacity to take advantage of rapid depletion-led growth and this will in the majority of cases result in entrenched vested interests, imprudent spending patterns, and increased economic inequality. It is time to consider carefully paced development plans (in line with growing national capacity to benefit from the economic opportunities they provide and assuming 'worst case' market scenarios). Key to success will be the institutional arrangements for managing national decisions to hold back from developing certain reserves or to develop them at an appropriate pace against the inevitable pressures from popular, political, and investor (shareholder) interests to ramp up production as quickly as possible.

*Alternative measures of performance*: There is little disagreement that GDP growth is a poor indicator of development, let alone sustainable development. Alternative measures of performance, including the valuation of other natural assets such as clean air, fresh water, and forests, need to be incorporated into the study of the impact of extractives on economies. On a practical level, the right indicators—including the trends in non-hydrocarbon balances—can incentivize countries to balance their portfolio of economic interests.

## References


Africa Progress Panel (2013). 'Equity in Extractives: Stewarding Africa's Natural Resources for All'. Available at: http://www.africaprogresspanel.org/wp-content/uploads/2013/08/2013_APR_Equity_in_Extractives_25062013_ENG_HR.pdf.

Alstine, J. V. (2014). 'Transparency in Resource Governance: The Pitfalls and Potential of "New Oil" in Sub-Saharan Africa', *Global Environmental Politics*, 14(1): 20–39.

Auty, R. (1993). *Sustaining Development in Mineral Economies: The Resource Curse Thesis*. London: Routledge.

Auty, R. (1994). 'Industrial Policy Reform in Six Newly Industrializing Countries: The Resource Curse Thesis', *World Development*, 22(1): 11–26.

Auty, R. (2010). 'Elites, Rent-cycling and Development: Adjustment to Land Scarcity in Mauritius, Kenya and Côte d'Ivoire', *Development Policy Review*, 28(4): 411–33.

Bova, E., P. Medas, and T. Poghosyan (2016). 'Macroeconomic Stability in Resource-rich Countries: The Role of Fiscal Policy'. Washington, DC: IMF. Available at: http://www.imf.org/external/pubs/ft/wp/2016/wp1636.pdf.

Bradley, S., Lahn, G., and Pye, S. (2018). 'Carbon Risks and Resilience: How Energy Transition is Changing the Prospects for Developing Countries with Fossil Fuels'. EER Research Paper. London: Chatham House. Available at: https://www.chathamhouse.org/publication/carbon-risk-resilience-how-energy-transition-changing-prospects-countries-fossil.

Collier, P. and A. Venables (eds) (2011). *Plundered Nations? Successes and Failures in Natural Resource Extraction*. London: Palgrave Macmillan.

Darby, S. (2010). 'Natural Resource Governance: New Frontiers in Transparency and Accountability'. Available at: http://www.transparency-initiative.org/archive/wp-content/uploads/2011/05/natural_resources_final1.pdf.

David-Barrett, E. and K. Okamura (2016). 'Norm Diffusion and Reputation: The Rise of the Extractive Industries Transparency Initiative', *Governance*, 29(2): 227–46.

Dixon, A. D. and A. H. B. Monk (2011). 'What Role for Sovereign Wealth Funds in Africa's Development?' Centre for Global Development. Available at: http://www.cgdev.org/doc/Initiatives/Oil2Cash/Final_Formatted_Monk_and_Dixon_SWF_10.11.pdf.

Dobbs, R., J. Oppenheim, A. Kendall, F. Thompson, M. Bratt, and F. van der Marel (2013). 'Reverse the Curse: Maximizing the Potential of Resource-driven Economies', McKinsey Global Institute. Available at: http://www.mckinsey.com/industries/metals-and-mining/our-insights/reverse-the-curse-maximizing-the-potential-of-resource-driven-economies.

The Extractives Industries Review (2003). 'Striking a Better Balance: The World Bank Group and Extractives Industries', volume I. Available at: http://www.ifc.org/wps/wcm/connect/294e55004ba934bca5adbd54825436ab/01.0+Volume+I+-+The+World+Bank+and+Extractive+Industries,+EI+Review+Report,+ENG.pdf?MOD=AJPERES.

Hilson, G. and R. Maconachie (2008). '"Good Governance" and the Extractives Industries in Sub-Saharan Africa', *Mineral Processing and Extractive Metallurgy Review*, 30(1): 52–100.

Hirschman, A. O. (1977). 'A Generalized Linkage Approach to Development, with Special Reference to Staples', *Economic Development and Cultural Change*, 25: 67–98.

Ibn Khaldun (1967 [1377]). *The Muqaddimah. Translated from Arabic by Franz Rosenthal.* Princeton: Princeton University Press.

IFC (International Finance Corporation) (2003). 'The Chad–Cameroon Pipeline'. Available at: http://www.ifc.org/wps/wcm/connect/region__ext_content/regions/sub-saharan+africa/investments/chadcameroon.

IMF (2007). 'The Role of Fiscal Institutions in Managing the Oil Revenue Boom'. Washington, DC: IMF. Available at: https://www.imf.org/external/np/pp/2007/eng/030507.pdf.

Karl, T. L. (1997). *The Paradox of Plenty: Oil Booms and Petro-States*. Berkeley, CA: University of California Press.

Keblusek, M. E. (2010). 'Is EITI Really Helping Improve Global Good Governance? Examining the Resource Curse, Corruption, and Nigeria's EITI Implementation Experience'. Available at: nidprodev.org/EITI%20-%20Nigeria%20Analysis.pdf.

Lahn, G. and S. Bradley (2016). 'Left Stranded? Extractives-led Growth in a Carbon-constrained World', EER Research Paper. London: Chatham House.

Marcel, V. (ed.) (2013). 'Guidelines for Good Governance in Emerging Oil and Gas Producers'. London: Chatham House. Available at: https://www.chathamhouse.org/publications/papers/view/194059.

Mejía Acosta, A. (2013). 'The Impact and Effectiveness of Accountability and Transparency Initiatives: The Governance of Natural Resources', *Development Policy Review*, 31(1): 89–105.

Myers, K. (2005). 'Petroleum, Poverty and Security'. Africa Programme. London: Chatham House. Available at: https://www.chathamhouse.org/sites/files/chathamhouse/public/Research/Africa/bppetroleum.pdf.

Prebisch, R. (1964). 'Toward a New Trade Policy for Development', in *Proceedings of the United Nations Conference on Trade and Development, New York*, vols I–VIII. New York: UN.

Rosenstein-Rodan, P. (1943). 'Problems of Industrialisation of Eastern and South-Eastern Europe', *Economic Journal*, 53: 202–11.

Shaxson, N. (2007). *Poisoned Wells: The Dirty Politics of African Oil*. Basingstoke: Palgrave Macmillan.

Singer, H. W. (1950). 'The Distribution of Trade between Investing and Borrowing Countries', *American Economic Review*, 40(2): 473–85.

Sovacool, B. K. and N. Andrews (2015). 'Does Transparency Matter? Evaluating the Governance Impacts of the Extractive Industries Transparency Initiative (EITI) in Azerbaijan and Liberia', *Resources Policy*, 45: 183–92.

Sovacool, B. K., G. Walter, T. Van de Graaf, and N. Andrews (2016). 'Energy Governance, Transnational Rules, and the Resource Curse: Exploring the Effectiveness of the Extractive Industries Transparency Initiative (EITI)'. *World Development*. Available at: doi:10.1016/j.worlddev.2016.01.021.

Stevens, P. (2015). 'The Resource Curse Revisited—Appendix A Literature Review'. London: Chatham House. Available at: https://www.chathamhouse.org/sites/files/chathamhouse/field/field_document/20150804ResourceCurseRevisitedStevensLahnKooroshyAppendix.pdf.

Stevens, P. and J. Mitchell (2008). 'Resource Depletion, Dependence and Development: Can Theory Help?' EER Report. London: Chatham House.

Stevens, P., G. Lahn, and J. Korooshy (2015). 'The Resource Curse Revisited', EER Report. London: Chatham House.

Stiglitz, J., A. Sen, and J. P. Fittousi (2009). 'Report by the Commission on the Measurement of Economic Performance and Social Progress'. National Institute of Statistics and Economic Studies, France (INSEE). Available at: http://www.insee.fr/fr/publications-et-services/default.asp?page=dossiers_web/stiglitz/documents-commission.htm.

Talbert, J. (2010). 'Measuring What Matters: GDP, Ecosystems and the Environment'. World Resources Institute blog, 14 April. Available at: http://www.wri.org/blog/2010/04/measuring-what-matters-gdp-ecosystems-and-environment.

World Bank (2008). 'World Bank Statement on Chad–Cameroon Pipeline'. Available at: http://web.worldbank.org/archive/website01210/WEB/0__CO-20.HTM.

# 6

# Political Economy and Governance

*Evelyn Dietsche*

## 6.1 Introduction

More than ten years ago, the World Bank's *Extractive Industries Review* (EIR) marked the beginning of a consensus that 'governance' matters for the developmental outcomes producer countries can achieve from exploiting extractive resources. The independent EIR commission recommended that the World Bank Group and other development finance institutions should refrain from supporting extractive resources projects unless producer countries had 'good governance' in place.

At the time, it was already well understood that producer countries often faced macroeconomic challenges and that, to manage these, they could draw on technical support from the international financial organizations. But there also was growing awareness that the extractive resources sector presented serious political-economic challenges: in the presence of windfall resource rents, self-interested politicians and bureaucrats might be tempted to use these rents for private gain, thus undermining the positive economic and social impacts that the sector could otherwise deliver.

This reasoning gave rise to a second consensus, namely that improving sector governance would mitigate the political-economic challenges. This consensus has experienced a remarkable elevation over the past decade and has nurtured the emergence of several international initiatives and dedicated NGOs diligently working towards improving sector governance.

At the same time, 'governance' as a concept has been subject to much criticism, because it has meant different things to different constituencies (Fukuyama 2013). Hyden et al. (2004) found that comparative political scientists used the term to identify how different rules affect results. International relations scholars looked at how processes deliver rules. Scholars of public administration focused on steering processes towards desired results. And, finally, international development agencies hoped that governance could serve as a tool to deliver positive developmental impacts. For the international development community, the term has provided a reference point to discuss all things political. For practitioners, it has served to reflect upon the political economy of government policies and decision-making and to pinpoint why technical reforms have often failed to deliver expected developmental outcomes. At a more philosophical level, the term has contributed to call attention to the institutional underpinnings of effective economic and political management.

Against the backdrop of this variety of different meanings, it is not too surprising that 'governance' has remained a contested concept. Some argue that things started to go wrong when proponents sought to operationalize 'good governance' to shape policy reform programmes and apply respective 'conditionalities' (Ahrens 2002; Opper 2008; Gisselquist 2012). For example, Mkandawire (2012) criticized that reform programmes introduced under the banner of good governance have supported 'institutional monocropping and monotasking'. The assumption has been that blueprint and standardized interventions can be applied to every country context and deliver similar results. Andrews (2013), who reviewed the mixed outcomes resulting from reforms targeted at improving the governance of public finances, concluded that the concept has been poorly conceptualized in relation to the complexity of insights gained from institutional analyses of the process of institutional change. And, Grindle (2010) has argued that the popularity of the idea of good governance has outpaced its capacity to deliver.

Contrasting these criticisms with the global agenda on the good governance of the extractive resources sector raises the question of whether that agenda has taken proper account of the concerns raised. This chapter argues that the proposition that 'good governance' somehow matters for positive outcomes has been derived from institutional analyses suggesting that 'institutions matter'. But, unfortunately, it is less clear how institutions actually get changed. The chapter suggests reframing the political economy of extractive resources from the negative question: 'How can poor outcomes be prevented?' to the positive question: 'How can positive institutional change for better outcomes be brought about?' This latter question builds on Lahn and Stevens (2017), who ask why extractive resources have not helped countries to grow the rest of their economies.

First, the chapter reviews why it has been found that institutions matter for the political economy of extractive resources and how this finding has informed the resources governance agenda. Second, it explores what institutions are and introduces a framework that highlights where substantial gaps in knowledge remain. It applies this framework to, and assesses the gaps in relation to, the extractive resources sector. Finally, it discusses two critical issues in the political economy of extractive resources.

## 6.2 The Political Economy of Extractive Resources and Institutions

The debate on extractive resources and the developmental impacts of their exploitation rests on the consensus that 'governance matters'. This consensus has developed on the back of the need to explain why many producer countries have experienced negative outcomes, but also why some countries have experienced better outcomes.

### 6.2.1 *Explaining Negative Outcomes*

Research on the political economy of extractive resources dates back to the mid-1990s.[1] Originally, this research drew its inspiration from the desire to explain why many countries reliant on producing such resources had experienced negative economic, political, and social outcomes. A seminal paper by Sachs and Warner (1995) was critical in drawing attention to the question of why resource-abundant countries had not achieved better economic outcomes. These outcomes contradicted the mainstream policy advice of the 1980s that countries could improve their economic situations if they encouraged investment in the resources sector. Initially, confidence had been placed in the ability of technocrats in central banks and finance ministries to manage the macroeconomic risks harboured by economic dependence on the resources sectors—if necessary with help from international financial institutions. However, the observation that many countries had not achieved positive outcomes prompted the search for explanations as to why this was the case.

First, researchers reconfirmed that resource-rich countries had indeed suffered from worse economic performance: the proxies for measuring 'resource dependence' that they had deployed in cross-country statistical analyses could 'explain' variance in economic performance between resource-rich and other countries.

Then, to make sense of this finding, researchers drew on microeconomic analyses to explain the politics of extractive resources. Assuming that politicians and bureaucrats were rational individuals maximizing their self-interests, they conjectured that there could be good reasons why the governments of

[1] There are several literature reviews on this research and these are cited in the longer working paper version of this chapter; see Dietsche (2017).

resource-rich countries were unwilling, or unable, to adopt policies that would counter the macroeconomic risks and maximize the development impacts associated with the resources sectors. They concluded that unproductive rent-seeking behaviour by politicians and bureaucrats caused the poor outcomes. Formal models and regression analysis were used to test more specific hypotheses: for example, why politicians in resource-rich countries tend to over-extract (Robinson et al. 2006), why these countries have ended up with worse political regimes (Jenson and Wantchekon 2004), and whether the presence of a resources sector could be associated with negative social outcomes, such as conflicts and civil wars (Humphreys 2005).

The challenge was that researchers were left with a mixed picture. First, critics pointed out that the quantitative studies on which these findings were based faced some methodological problems, including that results were very sensitive to the proxies used for measuring resource sector dependence (Brunnschweiler and Bulte 2008; Haber and Menaldo 2011). Second, fundamental concerns had been raised about the suitability of the proxies used (Andrews 2013; Arndt and Oman 2006; Van de Walle 2005). And third, some countries had performed better than others, which suggested that the inevitability of negative outcomes could at least be questioned.

### 6.2.2 *Explaining Variance in Outcomes*

The mixed picture suggested that the causal mechanisms responsible for the negative outcomes were not straightforward and that scholars had struggled to penetrate the complexity of causes and effects. In addition, apparent improvements in national-level indicators were sometimes not matched by commensurate improvements in outcomes at the sub-national level, or vice versa apparent improvements at the sub-national level were not reflected in national-level indicators.

#### 6.2.2.1 STRUCTURAL EXPLANATIONS

The proposition that poor outcomes were the result of unproductive rent-seeking behaviour by politicians and bureaucrats who had been left unfettered was soon countered by a second school of thought, taking a closer look at the political-economic power dynamics between different socio-political constituencies and how these affect outcomes. Its proponents suggested that structural characteristics and the political-economic power relationships that these might evoke were conditioning the decisions of politicians, bureaucrats, and other agents. Thus, rather than relying on a priori assumptions about individuals, proponents of structural explanations saw decision-makers embedded in socio-economic and socio-political relationships that, for better or for worse, influenced their behaviour.

The disadvantage for the proponents of microeconomic explanations for bad outcomes over this structural perspective was that the former could hardly appeal to politicians and bureaucrats to change their assumed rational behaviour in order to prevent negative aggregate outcomes. Meanwhile, the structural perspective allowed its proponents to delve deeper into the broader social science research on the role of institutions in economic development. This research had been brought into the development discourse from the mid-1990s, drawing on various strands of institutional analysis that had emerged from around the 1980s to explain differences in the economic, political, and social trajectories of modern market economies.[2] These analyses indicated that positive institutional change does not result from technocratic reforms, but is contingent on dynamic political-economic processes that unfold around the structural characteristics associated with resources.

#### 6.2.2.2 STRUCTURAL CHARACTERISTICS

First among the structural factors identified as likely to affect the political economy of resources are the physical characteristics of different types of resources and the socio-political structures they give rise to. To show this, several authors have drawn on the distinction between *point source* and *diffuse* natural resources, arguing that point source resources are more likely to nurture socio-political structures that support exclusive coalitions of elites holding onto power, which they use to limit non-elites' access to economic opportunities (Auty and Gelb 2001; Isham et al. 2005; Woolcock et al. 2001):

- Characteristic of *point source* resources are capital intensity and concentrated ownership, typically associated with large- and medium-scale oil, gas, and mining projects. Where land rights support large-scale plantations and agro-industrial enterprises producing bulk commodities, such as palm oil, soya beans, sugar cane, corn, or industrial wood, these can also be considered point source resources.
- *Diffuse* natural resources require more moderate capital inputs; economic opportunities are more widely disbursed, capital requirements are less intensive, and ownership is less concentrated. Diffuse resources typically include medium-sized and small farming, but could also include small and artisanal mining.

A related structural source of variance are colonial settlement structures and associated institutional legacies. Acemoglu et al. (2001) observed that European colonial powers developed different colonial-administrative institutions in the

[2] For this literature, see the longer version of this chapter (Dietsche 2017).

territories they occupied, depending on whether the prevalence/absence of tropical diseases discouraged/encouraged European settlers from staying. Where settlers stayed, they typically developed institutions that encouraged investment. For example, they adopted the types of land and mineral rights that they were already familiar with in their mother countries. In turn, where settlers did not stay in large numbers, they developed institutional arrangements designed to extract natural resources and transfer them to the mother country.

Additional sources of institutional variance that researchers have pointed to include social cohesion versus fractionalization and the types of states that producer countries exhibit.[3] For example, socially more cohesive societies have been found to generate greater confidence and trust between governments and citizens, resulting in institutional arrangements that grant broader access to economic and social opportunities (Easterly et al. 2006). Another example is the distinction Auty and Gelb (2001) have drawn between stylized 'predatory states' and 'developmental states', characterizing these by their differences in the relationships between ruling elites, the state apparatus, and society at large. They then tied these characteristics to conditioning structural factors, such as the types of resources exploited, country size, settlement structure, and population density.

Structural explanations for institutional variance have offered a rich source of historical-economic and socio-economic analyses from which insights into the political economy of institutions can be drawn. Yet, they also face two challenges of their own. First, they do not offer immediate solutions to guide policy interventions, because the factors they highlight are often hard-wired; and second, they cannot claim universal applicability because they are typically drawn from single-case and comparative studies.

### 6.2.3 *Guiding Policy Interventions*

The more general turn towards institutions and their role in economic development did not go unnoticed by those who had drawn on rent-seeking arguments to explain poor outcomes. They turned towards 'institutions' to explain variance in outcomes by drawing upon the *rational choice* version of institutional analysis and its basic proposition is that 'good institutions' can interrupt the negative relationship between rational individual actions and poor aggregate outcomes. This led to the conclusion that countries can achieve better outcomes from extractive resources if only they put in place 'good' institutions to constrain self-interested politicians and bureaucrats, who would then no longer be able to capture resource rents for private gain.

[3] See Dietsche (2017) for additional references.

By implication, it has been assumed that adopting 'good' institutions will result in better governance of the sector. To test this hypothesis, proponents looked out for proxies for 'good' institutions to use these as instrumental variables in their multiple regression analyses. Confirming a positive statistical relationship, they developed more or less plausible narratives to connect the quantitative results with a storyline. Collier and Goederis (2007), for example, applied a subjectively chosen threshold to the country rankings of the International Country Risk Guide (ICRG) as a proxy for 'good' governance and on this basis claimed that resource curse does not occur in countries with 'very good institutions'. However, they and the authors of similar other studies have not bothered to explain why the proxies they have instrumented in their analyses actually reflect 'good' institutions.

In conclusion, it has not been straightforward or uncontroversial to create indices for institutional quality. Nevertheless, institutions have come to be seen as a principal variable explaining variance in outcomes across producer countries. This consensus has advocated that improving institutions will improve sector governance and, thereby, outcomes. For those sympathetic to the explanation that unproductive rent-seeking behaviour poses the key challenge, the policy conclusion has been that 'good institutions' are synonymous to 'good governance' and both provide the solution to the 'resource curse'. Thus, improvements in good governance should be encouraged and this would somehow automatically constitute positive institutional change. At the practical level, the consequence has been that countries have been given advice on the institutions they apparently need in order to improve sector governance.[4]

In contrast, those drawing on structural institutional analyses have remained far less ambitious in putting forth policy conclusions and action agendas.[5] Focusing on explaining variance in outcomes, they have looked back and reflected on reasons for the variance in outcomes observed. Perhaps because they are more alert to historically evolved country conditions, respective scholars have taken a more cautious, if not pessimistic, stance towards suggesting that institutional arrangements can be changed by externally imposed advocacy pressures. At the same time, their insights provide a sense-check on how difficult it is in practice to encourage positive institutional change as well as suggesting where levers for encouraging such change might be found.

[4] See Lahn and Stevens (2017) and Gisselquist (2012) for UN organizations, multilateral development banks, and other multilateral organizations embracing the 'good governance' agenda to guide their work on institutional and public policy reforms.

[5] See Dietsche (2014) for a more detailed critical analysis.

## 6.3 What Are Institutions?

As the political economy of extractive resources has narrowed in on 'institutions' to explain variance in outcomes across resource-rich countries and, thus, has come to see institutions as critical to improving the 'good governance' of the resources sector, it warrants taking a step back and asking how social scientists have defined institutions. And, what they have learned about how institutions affect development over time.

### 6.3.1 *Taking Stock of Institutional Analyses*

From the late 1980s onwards, institutional analyses experienced a revival because researchers felt a need to explain observed differences in the economic, political, and social trajectories of industrialized countries. Compared with the rest of the world, these countries had experienced similarly impressive gains in living standards, despite obvious differences in their domestic institutional arrangements that had evolved over decades and centuries. These analyses drew attention to the institutional foundations of market economies, which mainstream neoclassical economics had chosen to ignore by simply assuming that the institutions in place were the most efficient to be had. Thus, there was no need to look at institutions in any great detail, but this task was left to scholars of politics, public administration, economic history, and sociology. Consequently, these disciplines have played a key role in rediscovering institutions, building on the foundations that had already been laid in the first half of the twentieth century.

The challenge for the scholarship on institutions has been that it has comprised a wide spectrum of theories and approaches. When, in 2000, Oliver E. Williamson reviewed the progress economists had made on this subject, he observed that their quest for a better understanding of the role of institutions in economic development had resulted 'in a boiling cauldron of ideas' with diverse and contradictory implications for policy advice (Williamson 2000: 610). This led Williamson to develop a framework that maps out the various strands of institutional economics across four levels of social science analysis.[6] These four levels are further explained below and then applied to the extractive resources sector.

*Level 1: Social embeddedness*: This highest level of institutional analysis has been coined after the work of economic sociologists and anthropologists who have studied social and economic transformations. They underline that

[6] A graphic reproduction of Williamson's framework is included in the earlier WIDER working paper version (Dietsche 2017).

economic systems are ultimately embedded in social networks and relationships and they focus on the most durable social institutions: for example, customs, traditions, norms, and religion. These institutions are not only very powerful but also show a great deal of inertia. Not least, they provide people with the belief systems that shape their understanding of the world around them, and they generate trust and reciprocity among social groups beyond kinship.

*Level 2: Institutional environment*: This second level provides the 'formal rules of the game' for economic transactions, and institutions often relate to the territorial remit of sovereign nation-states. These institutions also include international trade and regional economic unions above the level of nation-states, as well as sub-national states, provinces, or regions where, for example, countries are organized as federations. Thus, this level covers countries' political, bureaucratic, economic, and social systems and associated property rights and legal and regulatory regimes. Level 2 institutions are also not expected to change frequently. But they may change over decades. For example, these institutions can mark distinct political eras.

*Level 3: Governance*: Institutions at this level capture and formalize the commitments of parties involved in economic transactions. Williamson described this level as the 'play of the game', covering the contracts and agreements that underpin market transactions and deal with the risk that parties may cheat or renege on their commitments. Level 3 institutions are typically expected to last for a few months or years.

*Level 4: Efficient allocation of resources*: This level captures conventional market analyses whereby the interaction of demand and supply sets prices, Adam Smith's invisible hand is supposed to do its magic, and economic analyses can focus on the efficient allocation of scarce resources. Institutional arrangements are simply assumed to work efficiently in the background. They are not assumed to change within the time frames in which market participants take their decisions.

The two key points about Williamson's framework which he developed over fifteen years ago are, first, that scholars could well be referring to very different things when they point to 'institutions' as a key determinant of economic outcomes. Meanwhile, those focused on testing theories applying quantitative techniques would be biased towards finding binary proxies of 'institutional quality' that can be instrumented in statistical analyses. And second, while social science scholars have investigated the impact of institutions on economic, political, and social outcomes across all of the three levels from Level 1 to Level 3, mainstream neoclassical economists focus primarily on Level 4.

## 6.3.2 *Gaps in Knowledge*

Next, Williamson used his framework to identify several knowledge gaps in the study of institutions.

### 6.3.2.1 HOW INSTITUTIONS CONNECT ACROSS THE LEVELS

Williamson found that a key gap lies in understanding how the four levels are connected. He maintained that each level imposes constraints on the level immediately below it. Thus, how issues are conceived at Level 1 conditions how they shape the rules put in place at Level 2. These, in turn, shape the 'play of the game' at Level 3, which then affects the allocation of resources at Level 4.

For example, if at the highest level there is a consensus that private property rights are necessary for value creation, this will be reflected in Level 2 institutions where a country's constitution will uphold a regime of private property rights and appropriate legislation and regulations are in place to enforce these. This then provides the basis for the private sector to conclude contracts at Level 3, which in turn will affect the quantities of goods and services produced at Level 4 and the prices at which these are traded.

While Williamson considered the influence from the top to the bottom levels strongest, he was also of the opinion that there sould be feedback from the bottom to the top levels. For example, looking forward it might be assessed at Level 4 that a potential investment project would be financially viable, only if a change were made to institutions at Level 3 or above, such as a change in the fiscal regime resulting in a reduction of costs at Level 4. Or, looking back, it might become clear that the financial assessments of investment projects already undertaken did not take into account actual costs that are now materializing at Level 4 as a result of assumptions made about higher-level institutions and that these assumptions are now proving to be wrong.

A good example to explain this theory concerns the negative externalities associated with greenhouse gas (GHG) emissions. To an increasing number and range of constituencies these emissions now pose risks and increase costs at Level 4: for example, insurance companies, those exposed to extreme weather conditions, and those whose health is being negatively affected, to name but a few. However, to internalize these negative externalities in the price of GHG-emitting energy resources requires institutional changes at all three higher levels. Notably, international climate change policy has been trying to achieve such changes by normatively defining the control of emissions as a 'public good' and developing relevant international agreements and subsequent national legislation, regulations, and contractual relations to transform energy systems and reduce GHG emission levels.

#### 6.3.2.2 INSTITUTIONAL CHANGE AND HOW INSTITUTIONS EVOLVE

Another gap pertains to how institutions evolve across and within the three levels. Williamson conjectured that pressure from above was the most powerful and durable driver of institutional change. Others have concurred that institutional change that originates from the highest levels carries the greatest potential for transforming societies and their economies (Shirley 2008). Economic historians have also pointed to such institutional change to explain where and why major economic transformations have occurred (Tylecote 2015). To them it is clear that transformational institutional change is driven from the upper two levels of institutions.

#### 6.3.2.3 HOW TO BRING ABOUT POSITIVE INSTITUTIONAL CHANGE

A third gap lies in driving positive institutional change. The effectiveness of purposefully conducted institutional change is likely to suffer when such a change results in institutional arrangements that are inconsistent and/or incompatible with institutions at the same level or a level above. For example, it may be very difficult to successfully implement reformed mining or petroleum laws if their contents are at odds with strongly held beliefs about rights to resources and/or with other legislation reflecting vested interests in the status quo. Thus, effective enforcement of changes to specific Level 2 and Level 3 institutions often requires complementary institutional changes to be negotiated in relation to other existing Level 1 and/or Level 2 institutions.

From a historical perspective, economic transformations have typically coincided with fundamental changes to rights granted and enforced at Level 2, with these being brought about by structural changes that have altered the dynamics between socio-political constituencies and challenged the social order that backed the prevailing property rights regime at Level 1. In the past, such structural changes have included technological breakthroughs and major demographic shifts. Notably, climate change and related disruptive innovations may prove to be the next game changer.

#### 6.3.2.4 BUREAUCRACY AND STATE CAPACITY

A fourth gap is the limited attention paid to bureaucracy and state capacity. Yet state capacity plays a critical role in connecting the often informal institutions of Level 1 with the formal political-administrative systems established at Level 2. It is with these systems that nation-states actually exercise their monopoly over power and provide vital public goods and services, such as security and the rule of law. In turn, the 'good governance' agenda has paid scant attention to the question of how the institutions emerge that allow nation-states to accumulate and use power actually, and under what conditions such power is used to support broad-based economic and social development (Andrews 2013).

Williamson concluded that, in general, economists had been rather ignorant about how institutions work and how they emerge. The lowest common denominator is simply that 'institutions matter'. Others have criticized more strongly that 'good governance' has provided a fig leaf for a seemingly 'technical' approach to escaping the dead-end into which mainstream economists believing in market liberalism had been pushed, because of their ignorance vis-à-vis the role of institutions in market transactions (Grindle 2010).

### 6.3.3 *Application to the Extractive Resources Sector*

In applying Williamson's framework to the extractive resources sector, it makes sense to start with Level 4, which captures the market analyses and the economic assessments of projects that resource producers undertake when they evaluate commercial opportunities and risks and decide when and how to invest. Those who are immersed in the world of Level 4 simply assume that the institutions that shape the markets they evaluate are stable. Project models are continuously updated to prepare for and evaluate commercial decisions. Any perceived risks that relate to potential changes in the institutions that underpin these decisions are factored into project economics as additional costs and/or as a mark-up on the discount rate. Thus, the higher such risks are perceived, the more financially rewarding a project has to be.

Extractive resources projects involve large-scale investments in immobile assets that are tied up for long periods. Thus, resource developers strive to lock in the assumptions that shape project economics, using Level 3 instruments such as contracts, agreements, and guarantees. Resource developers typically involve their substantial legal teams and procurement services to negotiate and deliver on such instruments. Notably, the use of Level 3 instruments is based on bilateral relationships. These include, first, the relationships between resource developers and the governments of producer countries, including—where they exist—the latter's state-owned oil, gas, or mining companies. And second, they also include the relationship between resource developers and third parties, namely the companies from which the resource developers purchase goods and services and those to whom they sell their output, and the local constituencies that are impacted by resource exploration and production activities. While a substantial body of sector-specific literature has concentrated on these bilateral contractual relationships, this has come at the expense of neglecting the multifaceted relationships between a national government, its public authorities, and its citizens, and their impact on the enforcement and enforceability of the bilateral contracts and agreements that resource developers rely upon.

These multifaceted relationships play out at Level 2, which covers property rights, political and administrative systems, and sector legislation. Thus, Level 2 institutions set out who formally owns extractive resources and therefore who can issue licences to resource developers to explore and exploit such resources. The sector-specific institutional environment typically includes:

- Constitutional provisions setting out who owns resources in the first instance and how licences to explore for and exploit resources can be granted to resource developers.[7]
- Political and administrative systems determining which government and public authorities (i) develop policies, propose and approve laws, collect taxes and other public revenues, and (ii) administer the enforcement of laws, regulations, and public policies. These systems can be particularly complex in federal and decentralized countries that comprise several decentralized states, regions, or provinces (e.g. Argentina, Australia, Brazil, Canada, Peru, Indonesia).
- Sector-specific and sector-relevant legal, regulatory, and investment-promotion regimes setting out how primary resource rights can be allocated and what obligations come with these, such as domestic market obligations and local procurement and employment targets. These regimes provide the basis for resource developers negotiating the *Level 3* contracts and agreements that underpin their investments, sales, and export decisions.
- Various special fiscal and other institutions that host countries have been advised to adopt as part of good sector governance reforms, such as legislation to establish stabilization and savings funds or sovereign wealth funds, as well as legislation on transparency.

Beyond the sector-specific Level 2 institutions, there is also the wider institutional environment making for the range and the quality of the public goods and services that countries more generally provide to their citizens and the private sector. This environment not only affects the cost structure of the production inputs that resource developers and their supply chains need (e.g. human capital and infrastructure), it also affects the ability of the domestic private sector to leverage the resources sector to induce economic activities in other sectors. This means that Level 2 institutions are most critical for leveraging the potential development opportunities associated with extractive projects, as well as mitigating their negative impacts.

[7] See Scott (2008) for a comprehensive historical review of the legal traditions and historical trajectories that have played a key role in defining primary resource ownership.

At Level 1, sector-specific institutions include long-standing norms and principles, such as the historic principle of *'first come, first served'* applied in the mining sector and the petroleum sector's *'law of capture'*, which in many countries have become firmly and formally embedded in sector legal frameworks. Thus, their universal applicability is often taken as a given.

In addition, Level 1 institutions also include customary rights to resources, such as the rights that indigenous peoples (IP) and other traditional communities may claim. Often, such rights pertain to constituencies that have historically been disadvantaged and for this reason have not been formally embedded in Level 2 institutions. Thus, there is a high risk that they conflict with formally granted exploration and production rights. Sometimes, such as in the case of the Australian Native Title law, efforts have been made to rectify this situation by formalizing these Level 1 institutions with a legal provision that is introduced at Level 2, which in turn then obliges resource developers to take account of IP rights at Level 3 (i.e. a bilateral agreement with such communities) and Level 4 (i.e. absorbing the costs associated with negotiating and concluding such an agreement).

Another example of sector-specific Level 1 institutions is international norms and standards that seek to clarify who and how the externalities associated with resource exploration and production should be internalized across the various institutional levels. To name but a few, these institutions include Convention 169 (1989) of the International Labour Organization, the concept of Free, Prior and Informed Consent (FPIC) set out by the 2007 UN Declaration on the Rights of Indigenous Peoples, the IFC Performance Standards on Environmental and Social Sustainability and the UN Guiding Principles on Business and Human Rights. They also include international norms on resource sector governance, such as Chapter 10 of the EU Accounting Directive of 2013 and sections 1502 and 1504 in the US Dodd–Frank Wall Street Reform and Consumer Protection Act.

### 6.3.4 *Gaps in Knowledge Applied to the Extractive Resources Sector*

Following the same order as sub-section 6.3.2, this sub-section applies Williamson's identified gaps in knowledge to the extractive resources sector.

#### 6.3.4.1 HOW DO INSTITUTIONS CONNECT ACROSS THE LEVELS?

Institutional reforms targeting the extractive resources sector have often focused on specific Level 2 and Level 3 institutions, such as revising the sector's legal and regulatory frameworks and renegotiating contracts and agreements. More recently, however, there has been a shift towards greater recognition of the political settlements traced to levels 1 and 2 and which underpin the societies and economies of resource-producing developing countries (Melia 2015). The aim of this shift has been to better understand variations in the political economy of extractive institutions across (mostly) sub-Saharan African resource-producing countries.[8] Behind these variations lie different institutional legacies that affect the connections between the levels, especially at the intersection of Level 2 with Level 3.

For example, in countries where the legal tradition is based on Common Law, decisions made by judges pertaining to disputes at Level 3 are incorporated into Level 2. In the early phases of petroleum development this meant that the decisions of US judges acting under Common Law set off a path dependency that enticed private landowners to engage in competitive drilling. This, in turn, led to the rapid expansion of fossil fuels as a major source of energy in the United States, with far-reaching consequences well beyond the United States (Daintith 2010).[9]

#### 6.3.4.2 INSTITUTIONAL CHANGE: HOW DO INSTITUTIONS EVOLVE?

Williamson conjectured that pressure from above is the most powerful and durable driver of institutional change across the levels. In the extractive resources sector the trend has been towards establishing new international norms and standards at Level 1 with the intention that their influence percolates downward from there. In some cases, these norms and standards have evoked new Level 2 legislation and regulations, in both producer countries and home countries. An example is Chapter 10 of the EU Accounting Directive which requires EU member states to put in place regulations requiring resource developers to publish details of resources revenues paid to host governments.

In other cases, new Level 1 institutions have surpassed Level 2 and sought to directly target the corporate practices of international resource developers at Level 3, such as in the cases of the IFC Performance Standards on Environmental and Social Sustainability and the UN Guiding Principles on Business and Human Rights.

A question left unanswered is whether, or to what extent, such types of Level 1 institutions have been able to catalyse positive changes to a wider range of Level 2 institutions in the producer countries themselves. It may be suggested that the transformation of social and economic structures in producer countries cannot be brought about solely by improving corporate social performance at Level 3, nor by paying narrow attention to, for example,

[8] Melia (2015) has drawn on three strands of structural-institutional analysis to focus on how social orders established at Level 1 affect the formal rules of the game at Level 2 across various sub-Saharan African countries.

[9] Scott (2008) provides a historical account of the evolution of resource property rights with a focus on mining and other natural resources (e.g. fishing, forestry, grazing, hunting).

the subject of revenue transparency at Level 1. Lahn and Stevens (2017) allude to this in relation to diversifying economies away from extractive resource dependency.

#### 6.3.4.3 BRINGING ABOUT POSITIVE INSTITUTIONAL CHANGE: WHERE ARE THE LEVERS?

Bringing about positive institutional change is about (a) connecting Level 1 with Level 2 institutions and (b) Level 2 institutions complementing and reinforcing each other. The argument is that institutional change cannot be limited to sector-specific institutions if the sector is to contribute positively to wider economic and social development. The latter requires more far-reaching changes to the Level 2 institutions, namely those that underpin the delivery of public goods and services more generally. For other economic sectors to thrive alongside the extractive resources sector requires not only a broadly conducive business environment but also tacit knowledge relevant to the production of non-extractive goods and services. Such knowledge is built by gaining experience over time. Thus, potential levers for positive institutional change are bound to be context-specific and dependent on the ability of relevant stakeholders to forge alliances that can support more comprehensive institutional change across Level 2 institutions. For example, North et al. (2006) developed a conceptual framework spelling out the 'doorstep conditions' and subsequent 'transition steps' that have allowed some countries to transform their social order in such a way as to generate institutions that underpin substantial productivity increases.

#### 6.3.4.4 WHAT ROLE DOES STATE CAPACITY PLAY?

State capacity connects the (often informal) institutions of Level 1 with the (typically more formal) institutions of Level 2 and Level 3. The structural transformation of resource-dependent economies requires coordinated planning and interventions across the wider public sector. North et al. (2006) allude to the institutionalized provision of public goods and services and its importance for broad-based increases in productivity, including the funding mechanisms required to pay for such provision (e.g. taxation). The depth of the challenge of building state capacity is highlighted by recent research into governance in areas of 'limited statehood' and so-called ungoverned spaces (Mingst 2013). Their prevalence stands in stark contrast to the assumption of state sovereignty that underpins 'good governance'. By looking at alternative authority exercised in contested spaces and what economic activities are possible in these, Mingst confirms the importance of state capacity and the institutionalization of public goods and services provision for achieving broad-based developmental outcomes on the back of extractive resources exploitation. This suggest that the 'good governance' agenda needs to reflect upon what it contributes to state capacity building.

## 6.4 Discussion

This sub-section discusses (i) whether institutions are considered an independent or a dependent variable and (ii) what this means for the transformation of social order.

The question whether institutions are considered an independent or a dependent variable is important, because knowing that institutions matter is not the same as understanding how they evolve and how to actively bring about positive institutional change. Even if a particular institutional feature has served as a variable to successfully explain variance in outcomes across resource-dependent economies, it does not necessarily follow that it is a variable suitable for policy interventions. It also does not say much about how a specific institution works in relation to others across and also within each of the three top levels of Williamson's analytical framework.

When institutions are considered an input variable, it is assumed that they can constrain the behaviour of powerful political elites. However, this leaves open the fundamental question as to how these very same elites can be convinced to develop or adopt such institutions in the first place. Elites' push for such institutions is vital for the institutionalized provision of public goods and services. In other words, it takes considerably more than international advocacy and political pressure to convince those exercising authority on the ground that it could be in their interest to use their monopoly over power and public resources to provide institutions that benefit not just themselves, but also others. An additional condition is that they must first hold such a monopoly and be able to raise sufficient public resources.

The consequence of this line of argument is that, first, it takes a conscious shift in mindset from a view that institutions are an *input* for achieving a desired outcome to one where institutions are seen as the *outcome* of socio-political processes. The analytical usefulness of the concept of 'institutions' lies in their serving as a bridge to overcome the fundamental problem of agency-focused theories holding individuals responsible for economic, political, and social outcomes, while structural explanations subject the actions of individuals to more hard-wired conditions. The concept of institutions is useful only in order to identify how and when collective agency makes it to the point of pushing the door open for structural transformation. The issue of transforming social order in such a way that public authorities provide institutions that transcend personal relationships and broaden access to economic opportunities is about the connection and interaction between Level 1 and Level 2 institutions.

What this means for the transformation of social order is, that in order to establish and enforce Level 2 institutions in line with those that exist at Level 1, public authorities must first hold a monopoly over power, and second use this power to institutionalize the provision of public goods and services. Citizens in modern nation-states believe that elected governments and their public authorities hold this power legitimately and are trusted to exercise it fairly and transparently. Using this power, they provide and enforce institutions as a 'public good', thus serving everybody residing with the territorial boundaries of the nation-state. One of the most important public goods is the non-discriminatory enforcement of the 'rule of law' by the public authorities overseen by an elected government. However, it is worth remembering that holding the monopoly over power comes with the ability to threaten and use force to coerce individuals to behave in certain ways.

Enforcing Level 2 institutions also requires sufficient fiscal resources to fund enforcement mechanisms. The cost of enforcement depends on whether Level 2 institutions are underpinned by respective institutions at Level 1. Compliance is most efficiently achieved if individuals internalize norms, principles, and ideas that guide their behaviour and decisions, as opposed to coercing compliance. From this derives the role of public authorities in addressing the issues underlying conflicts and tensions around the exploitation of extractive resources. Positively, this is achieved through political processes that establish a consensus on balancing the costs and benefits associated with such exploitation. Negatively, it is achieved through coercing and oppressing those who must bear the costs imposed on them by the benefits that others are enjoying.

The political economy framework presented by North et al. (2006) describes the complexity of the historical processes that have allowed more advanced countries to transition to social orders that have supported broad-based economic and social development. The countries that achieved such transitions early on were the United Kingdom, France, and the United States, followed by other European, and a few other, countries. These countries have evolved to grant their citizens rights and provide public goods and services that have opened up access to economic opportunities.[10]

Melia (2015) has used part of this framework to describe how Equatorial Guinea, South Sudan, and Chad have become stuck with social orders that *limit access* to economic opportunities to elites. He has also drawn upon two additional political-economic frameworks to describe the relationships and

[10] North et al. (2006) acknowledge that the Netherlands and Switzerland could also be included in the group of early transition countries, but their work does not include them.

alignments between key social constituencies to reflect upon the particular social orders that he sees underpin, respectively, the cases of Mozambique, Angola, and Nigeria and the cases of Peru, Ghana, and Kenya. Contrary to the assumption implicit in the 'good governance' agenda, none of these three frameworks assumes that these various countries are all striving towards some imagined uniform social order where sovereign states 'fully' govern their territories. Rather, each framework highlights that the exercise of statehood is, in some form or another, compromised and/or power is shared between contending parties.

However, the challenge with Melia's static approach is that he can only explain the current status quo that prevails in these countries, but cannot identify what collective agency would be required to move each towards improving their currently unfavourable social orders.[11] In contrast, North et al. (2006) have identified 'doorstep conditions' that precede and lay the foundation for the transition from a social order that *limits access* to one that would *provide open access* to economic opportunities and institutionalize the provision of public goods and services. This chapter does not offer the space to delve deeper into the historical analysis on which these scholars' propositions are built. However, in brief these are:

(i) initial *rule of law for elites*, which refers to the formalization of procedures for arbitration and mediation backed by the establishment of property rights and a judicial system.
(ii) *perpetually lived organizations in the public and private sphere*, which refers to organizations that outlive the individual members who created them and for these organizations to become 'legal persons' bearing rights and duties. Together, the first two conditions create the space for more impersonal relationships among the elites, while at the same time constraining absolute power.
(iii) *consolidated control of the military*, which speaks to the legitimization of the use of violence, as opposed to its arbitrary use.

North et al. (2006) have observed that when these foundations are in place, elites are more likely to find a common interest in transforming their privileges into impersonal rights shared among them and which can then be converted into rights that are granted to citizens more broadly. The conclusion is that transforming social order takes time.

[11] Melia (2015) does discuss 'remedies', referring to macroeconomic assistance, transparency and accountability initiatives, resources for infrastructure deals, direct cash transfers, and focusing on thematic fields for bilateral development engagements on strategic planning, employment creation, local economic development, and cooperation with extractives sector companies.

This also means that, first, the fast development of one extractives project is unlikely to achieve a sustainable outcome. And second, achieving a balance in the benefits and costs derived from the sector is more likely when there are consecutive resource developments that provide a longer time horizon for transforming a country's social order. At the same time, there is still the risk that an unfavourable social order remains entrenched, and elites do not manage to achieve the doorstep conditions that could lead to subsequent structural transformation. The historical literature suggests that structural transformation has much to do with how political and economic elites align between themselves and how their alignment affects non-elites. As institutions of enforcement at Level 2 are conditioned by the social order established at Level 1, positive institutional change requires the transformation of existing social orders.

The bad news is that there is no generic template that shows how to get to a social order that *opens access* to economic opportunities. While North et al. (2006) have provided a conceptual framework that points to the landmarks that pathfinders should look out for on the road to achieving the transformation of social order, Bates (2010) has criticized this framework for rather underplaying the role of those who collectively prepare and walk those paths. Arguably, resource developers and third parties routinely claim that their contributions are supporting positive change.

The longer version of this article provides a graphic reference point for asking whether and how these parties contribute to connecting Levels 1 and 2, building state capacity at Level 2 and achieving consistency and complementarity within and across Levels 1 to 3 (Dietsche 2017). It sets out three modes through which resource developers and third parties make their respective contributions:

- The first mode depicts voluntary delivery and reporting against international standards and social investment in community development, responding to Level 1 institutions that set international norms in respect of companies' social performance and their contributions to the economic and social development of impacted communities. These contributions do not intend to make an explicit difference to Level 2 institutions. Especially in areas of limited statehood and so-called ungoverned spaces, this mode presents resource developers with the only practical course of action to deliver positive outcomes at the community level.
- The second mode depicts contributions that are aimed at particular sector-specific issues expected to improve country-level outcomes. Along the lines of viewing 'good institutions' and 'good governance' as an independent variable, this mode places emphasis on specific Level 2 institutions, such as particular pieces of legislation and/or regulations. Examples include legislation on transparency or on 'local content'.

- The third mode views the extractive resources sector as integral to a country's economy and society and gives concern to the sector's potential contribution towards structural transformation and the diversification of the host country economy. It is premised on positive contributions being possible, if the country's social order evolves towards an institutional environment at Level 2 that broadens access to economic opportunities and institutionalizes better provision of public goods and services. This mode gets close to the doorstep conditions highlighted by North et al. (2006).

Mode 3 holds the most promise for contributing to more comprehensive institutional changes that lead to better outcomes at the country level. Still, a lot remains to be learned about how resource developers and third parties can support collective action and positive realignments among political and economic elites in support of structural transformation.

## 6.5 Summary

This chapter has argued that the challenge of explaining the dismal performance of many resource-rich countries has driven the debate on the political economy of extractive resources. 'Bad' resource sector governance provided one such explanation that led to the argument that institutional constraints are needed to prevent self-interested individuals from using resource rents for personal gain. Those looking towards institutional analyses to provide guidance as to how to improve the governance of the sector often focused on particular types of sector-specific institution. Much hope has been placed on transparency as a means to achieve greater accountability of outcomes. Drawing on the general criticism of the 'governance' concept, this chapter has reviewed the literature on the political economy of extractive resources and institutions. It has argued that too little emphasis has been put on understanding how and why those holding political power might support institutional change that supports positive outcomes.

## References

Acemoglu, D., S. Johnson, and J. A. Robinson (2001). 'The Colonial Origins of Comparative Development: An Empirical Investigation', *American Economic Review*, 91(5): 1369–401.

Ahrens, J. (2002). *Governance and Economic Development: A Comparative Institutional Approach*. Cheltenham: Edward Elgar.

Andrews, M. (2013). *The Limits of Institutional Reform in Development: Changing Rules for Realistic Solutions*. Cambridge: Cambridge University Press.

Arndt, C. and C. Oman (2006). *Uses and Abuses of Governance Indicators*. Paris: Development Centre Studies, OECD.

Auty, R. M. and A. H. Gelb (2001). 'The Political Economy of Resource-abundant States', in R. M. Auty (ed.), *Resource Abundance and Economic Development*, 126–44. UNU-WIDER Studies in Development Economics. Oxford: Oxford University Press.

Bates, R. H. (2010). 'A Review of Douglass C. North, John Joseph Wallis, and Barry R. Weingast's "Violence and Social Orders: A Conceptual Framework for Interpreting Recorded Human History"', *Journal of Economic Literature*, 48(3): 752–6.

Brunnschweiler, C. N. and E. H. Bulte (2008). 'The Resource Curse Revisited and Revised: A Tale of Paradoxes and Red Herrings', *Journal of Environmental Economics and Management*, 55(3): 248–64.

Collier, P. and B. Goederis (2007). 'Commodity Prices, Growth, and the Natural Resource Curse: Reconciling a Conundrum', Department of Economics, University of Oxford (preliminary draft).

Daintith, T. (2010). *Finders Keepers? How the Law of Capture Shaped the World Oil Industry*. Abingdon: Resources for the Future Press/Earthscan.

Dietsche, E. (2014). '"Good Governance" of the Extractive Resources Sectors: A Critical Analysis', PhD Thesis, University of Dundee.

Dietsche, E. (2017). 'Political Economy and Governance', WIDER Working Paper 2017/24. Helsinki: UNU-WIDER.

Easterly, W., J. Ritzen, and M. Woolcock (2006). 'Social Cohesion, Institutions, and Growth', *Economics & Politics*, 18: 103–20.

Fukuyama, F. (2013). 'What Is Governance?' *Governance: An International Journal of Policy, Administration, and Institutions*, 23(3): 347–68.

Gisselquist, R. (2012). 'Good Governance as a Concept, and Why This Matters for Development Policy', WIDER Working Paper 2012/30, Helsinki: UNU-WIDER.

Grindle, M. (2010). 'Good Governance: The Inflation of an Idea', Faculty Research Working Paper Series, 10–023, Harvard Kennedy School.

Haber, S. and V. Menaldo (2011). 'Do Natural Resources Fuel Authoritarianism? A Reappraisal of the Resource Curse', *American Political Science Review*, 105(1): 1–26.

Humphreys, M. (2005). 'Natural Resources, Conflict and Conflict Resolution: Uncovering the Mechanisms', *Journal of Conflict Resolution*, 49(4): 508–37.

Hyden, G., J. Gourt, and K. Mease (2004). *Making Sense of Governance: Empirical Evidence from 16 Developing Countries*. Boulder, CO: Lynne Rienner Publishers.

Isham, J., M. Woolcock, L. Pritchett, and G. Busby (2005). 'The Varieties of Resource Experiences: How Natural Resource Export Structures Affect the Political Economy of Economic Growth', *World Bank Economic Review*, 19(2): 141–74.

Jenson, N. and L. Wantchekon (2004). 'Resource Wealth and Political Regimes in Africa', *Comparative Political Studies*, 37(7): 816–41.

Lahn, G. and P. Stevens (2017). 'The Curse of the One-size-fits-all Fix: Re-evaluating What We Know about Extractives and Economic Development', WIDER Working Paper 2017/21. Helsinki: UNU-WIDER.

Melia, E. (2015). 'The Political Economy of Extractive Resources', GIZ Working Paper. October.

Mingst, K. A. (2013). 'Alternative Governance in Spaces and Territories. Book Review Essay', *International Studies Review*, 13: 285–8.

Mkandawire, T. (2012). 'Institutional Monocropping and Monotasking in Africa', in A. Noman, K. Botchwey, H. Stein, and J. E. Stiglitz (eds), *Good Growth and Governance in Africa: Rethinking Development Strategies*, 80–113. Oxford: Oxford University Press.

North, D. C., J. J. Wallis, and B. R. Weingast (2006). *Violence and Social Orders: A Conceptual Framework for Interpreting Recorded Human History*. Cambridge: Cambridge University Press.

Opper, S. (2008). 'New Institutional Economics and its Application on Transition and Developing Economies', in E. Brousseau and J.-M. Glachant (eds), *New Institutional Economics: A Guidebook*, 389–406. Cambridge: Cambridge University Press.

Robinson, J. A., R. Torvik, and T. Verdier (2006). 'Political Foundations of the Resource Curse', *Journal of Development Economics*, 79: 447–68.

Sachs, J. D., and A. M. Warner (1995). 'Natural Resource Abundance and Economic Growth', Working Paper 5398, Cambridge, MA: NBER.

Scott, A. (2008). *The Evolution of Resource Property Rights*. Oxford: Oxford University Press.

Shirley, M. M. (2008). *Institutions and Development*. Cheltenham: Edward Elgar.

Tylecote, A. (2015). 'Institutions Matter: But Which Institutions? And How and Why Do They Change?' *Journal of Institutional Economics*. DOI: 10.1017/S1744137415000478.

Van der Walle, S. (2005). 'Measuring Bureaucratic Quality in Governance Indicators', Paper for the 8th Public Management Research Conference, Los Angeles, 29 September–1 October.

Williamson, O. E. (2000). 'The New Institutional Economics: Taking Stock, Looking Ahead', *Journal of Economic Literature*, 38(3): 595–613.

Woolcock, M., L. Pritchett, and J. Isham (2001). 'The Social Foundations of Poor Economic Growth in Resource-rich Countries', in R. Auty (ed.), *Resource Abundance and Economic Development*, 76–92. UNU-WIDER Studies in Development Economics. Oxford University Press.

# 7

# New Industrial Policy and the Extractive Industries

*Evelyn Dietsche*

## 7.1 Introduction

Industrial policy is back. Many development economists are now arguing for a new approach to industrial policy to achieve economic and social development in low- and middle-income countries.[1] This revival has been prompted by several developments. First, the pro-poor development agenda of the 2000s has been criticized for failing to develop a sustainable private sector (Whitfield 2012). For example, this is reflected in the Sustainable Development Goals (UN 2015a), placing greater emphasis on economic transformation and sustainable development compared to the more poverty-focused agenda of the preceding Millennium Development Goals.

Second, international climate policy has provided an impetus for a new approach to industrial policy. The climate agreement reached at the Paris COP21 in November 2015 has mandated all signatories to change course and accelerate their move toward more sustainable and greener growth (UN 2015b). This calls not only for cleaner energy, but also for improved resource and material efficiency (OECD 2015). And it exposes the fossil-fuel sector to the risk of 'unburnable carbon' and 'stranded assets' (Lahn and Bradley 2016; Mitchell et al. 2015). 'Green' industrial policy is seen as central to driving the structural transformation towards a more sustainable and greener economic system (Aiginger 2015; Hallegatte et al. 2013; Lütkenhorst et al. 2014; Rodrik 2014). Meanwhile, minerals and metals are seen as critical for a low-carbon future (World Bank 2017).

[1] Dietsche (2017b) provides a fuller version of this chapter, including a broader range of references on new industrial policy.

Third, on both sides of the Atlantic rising disappointment with the social outcomes of economic liberalism and globalization are influencing political agendas towards restrictions on trade and the movement of people (Bailey et al. 2015; Rodrik 2016a, 2016b, 2017; Smart 2017).

Yet, opinions on industrial policy are deeply divided. Associating the term with the interventionist import-substitution policies that were widely applied in the post-World War II and post-independence era, re-introducing industrial policy into economic policy and development debates is seen as encouraging unproductive rent-seeking by politically favoured companies thriving under protectionism.

Proponents of new approaches to industrial policy maintain that investing in productive knowledge and technological capabilities is key to achieving sustainable economic and social development. As explained more fully in Dietsche (2017b), they define industrial policy in broader terms than either the old import-substitution policies, or the inward-looking policy ideas of the new national populism. This means that new industrial policy constitutes a 'broad church' of ideas and people. This chapter asks what the implications of this debate are for the extractive industries and their potential to contribute to inclusive (and sustainable) development.

## 7.2 The Case for Industrial Policy

The theoretical case for industrial policy arises from the problem of transaction costs and how these should be dealt with. Dietsche (2017b) discusses this in more detail by distinguishing between (a) the typical cases of 'market failure' that neoclassical economics identifies on the basis of assuming that transaction costs are zero and (b) the real-life situation where transaction costs are always positive and, therefore, are a problem that is inherent to all markets in which unrelated individuals exchange goods and service.

The general case for industrial policy builds on the argument that there are situations in which markets rely on the public sector to achieve broadly desirable economic and social outcomes. These situations include, but are not limited to, those cases where clear 'market failures' can be identified. In essence, this means that industrial policy sits at the heart of the relationship between markets and states, and it also shapes economies and the social outcomes they produce.

Dietsche (2017b) discusses in some detail the ideal-type situation where transaction costs are assumed to be zero (i.e. case (a)). Even under this assumption, there are 'special' cases of goods and services that warrant interventions because they are marred by the problems of *non-rivalry* and *non-excludability*.

Congruently, interventions are aimed at designing policies and establishing institutions that address *non-rivalry* and *non-excludability*. Dietsche (2017b) also discusses real-life examples to demonstrate that today's production and use of extractive resources are heavily and fundamentally entangled with the characteristics of *non-rivalry* and *non-excludability*. This means that the markets associated with extractive resources are contingent on how institutions have shaped them, not just over the past couple of decades, but at least since the era of the industrial revolution.[2] Equally important is how these markets will work in the future, as not only governments but also social organizations challenge and seek to re-shape existing institutions.

This chapter focuses only on case (b), where the case for industrial policy centres around the question of how transaction costs can be reduced to encourage economic exchanges over distance and across time. With economic historians and institutional economists having spelt out the critical role that 'institutions' play in reducing transaction costs and, thus, in establishing markets (Dietsche 2017a), the probing question is: *Who makes and shapes the institutions that reduce transaction costs to support sophisticated market transactions over distance and across time?* The answer points not only towards collaboration and collective action, but also to the critical role that is played by collective entities with monopoly power over the use of force. In the modern world, this power is held by nation-states, where public authorities provide the institutions that allow the private sector to thrive and, in the best-case scenario, produce broadly acceptable economic and social outcomes.[3]

It is on this discussion that the real-life situation of case (b) builds. The challenge is that the analysis of this situation produces not only more nuanced conclusions but also no consensus on what industrial policy should really entail. A critical initial proposition, well established at the philosophical level, is the case for collective authorities to provide institutions whenever economic agents interact with each other beyond personal relationships. In today's world, nation-states and their public authorities have become the key collective entities that shape and maintain the institutions underpinning economic systems and their economic, social, and political outcomes. Thus, the proponents of industrial policy take the position that public authorities not only *can*, but effectively *do* play a fundamental role in how economies are structured and how they perform. Yet, proponents differ over *how* that role should be played. The analysis in Dietsche (2017b) has identified several themes that run across this ongoing debate.

[2] Relevant here are resource property rights and associated sector legal regimes. On this subject, see also Daintith (2010), Dietsche (2017a), and Scott (2008).

[3] See Dietsche (2017a) for an introduction to some of the relevant literature and a discussion of the role that public authorities play in providing formal institutions.

### 7.2.1 *The Risks of 'State Failure'*

Neoliberal economists have strong reservations against any type of 'policy interventions', because they see 'state failure' as a bigger risk that any type of 'market failure'. This concern stems from their interpretation of some of the negative experiences with interventionist policies prevalent in Latin America as well as in newly independent states in the 1960s and 1970s, which were referred to then as 'industrial policy'.[4]

While some countries found success—notably South Korea and Taiwan (China)—many did not, and by 1980 much of Latin America and many newly independent countries in Africa and Asia faced serious crises. Neoliberal economists concluded that the risks of state failure arising from import-substituting industrialization outweighed any gains from containing market failures. Industrial policy was then seen through the ideological lens of socialism's failure and market liberalism's triumph, reinforced by the decline of the Soviet Union.

Industrial policy came to be seen as a means for creating rents for the benefit of political and economic elites, without delivering many, if any, of the expected economic and social returns. These ideas and experiences were reflected in some new economic theory focusing on rational individuals exploiting their power over public policies to seek unproductive 'rents', thereby maximizing their self-interest.[5] Against the background of these theories, governments were advised to avoid unproductive 'rent-seeking' by keeping their hands away from industrial policy, and instead to open up to foreign investment and trade. To date, 'state failure' and 'rent-seeking' remain two contentious subjects that proponents of the new industrial policy must still confront.

### 7.2.2 *Industrial Policy as a Discovery and Learning Process*

Discovering and learning are core themes in new industrial policy. The argument is that industrial policy should create a network of linkages between public- and private-sector representatives and institutions where processes guide information flows and joint learning. In order to shape the structure of an economy, bureaucrats need to be closer to business. At the same time, the public sector should not simply end up serving particular companies or exclusive business elites. Instead, it must retain its independence from

[4] Specifically associated with such interventionist policies is the concept of 'import substitution industrialization' (ISI), which advocates that trade policy should aim to support domestic manufacturing, in order to reduce imports.

[5] See Congleton et al. (2008) for a review of research on rent-seeking.

individual private interests and focus on the provision of public goods and services that benefit the private sector more broadly.

The question is what role should public authorities play in leading a discovery and learning process that identifies where transaction costs are high and pose a hindrance to productive entrepreneurial activities? There are many views on the specifics. One argument is that the public sector carries a responsibility for coordinating and aligning stakeholders' interests towards reducing transaction costs—and enhancing positive externalities so that the private sector can develop, thrive, and diversify. For example, Rodrik (2007) argues that successfully identifying the true cost structure of an economy takes public policy makers one step closer to removing impediments to 'doing business' and enhancing positive externalities. He suggests a corporatist approach where the public sector brings together and aligns multiple interests across different levels of policy, thereby improving the environment for private-sector activities. Industrial policy is then not the sole responsibility of a separate ministry or department, but a responsibility shared across the public sector. In addition, this view also speaks to encouraging (or nudging) private-sector interests to organize and identify common constraints and impediments that (only) the public sector can address.

There is significant congruence between this view of industrial policy and the proposals by international mining companies to partner with governments, local enterprises, and third parties to enhance mining's positive contributions to development, and minimize any negative impacts on host countries and communities (ICMM 2011; McPhail 2017). Similarly, proponents of the notion of 'shared value'—a concept that has emerged within the debate on extractives-led development—argue that governments, companies, and third parties need to work together to remove impediments that hinder local companies from participating in project supply chains (e.g. bureaucratic, financial sector-related) and to encourage positive externalities by providing training and support to local entrepreneurs and workers (e.g. skills, management, HSE compliance). In addition, industrial policy as a discovery and learning process is also relevant for the challenge of climate change, because it suggests identifying and addressing the impediments and disincentives that discourage the private sector from developing and investing in cleaner and alternative energy sources.

The case for encouraging collaboration for joined-up discovery and learning has been supported by three further observations. First, many Western industrialized countries look back on a long history of experimenting with decentralized approaches to industrial policy, where the development of particular sectors and industries has been supported by fostering innovation and competitiveness through networking across private- and public-sector entities (Keller and Block 2015; Warwick and Nolan 2015).

Second, there are the Asian latecomers to the industrial revolution that successfully supported the structural transformation of their agrarian economies and societies during the second half of the twentieth century. It is now generally accepted that East Asia caught up with the West by actively supporting and networking the export-oriented sectors, industries, and firms—contrary to mainstream policy advice to keep the public sector's hands off the private sector (Amsden 1989; Khan and Jomo 2000; Lin 2012; Lin and Chang 2009; Strom 2017; Wade 1990).

Third, there is the observation that, as countries get richer, so their economies typically diversify (Imbs and Wacziarg 2003). They get richer not by producing more of the same using what resources they already have, but by learning how to produce more of an increasingly diverse range of goods and services. Moreover, more diversified economies have not seen a withdrawal of the public sector: 'learning to produce a more diverse range of goods and services' has often been supported by a higher level of public goods and service provision, supplying inputs that have allowed a broader range of economic activities to prosper. Some see this achievement of structural transformation and diversification as the essence of the development process (Mkandawire 2012; Rodrik 2007).

However, for industrial policy as a discovery and learning process, there is no presumption that the public authorities can always get things right—at least not first time around. It is expected that mistakes will be made and that, as part of learning, these can and will be corrected. What matters is that the incidence of overall successful cases of private–public collaboration eventually exceeds the unsuccessful cases. Indeed, if no mistakes are made it can be taken as a sign that industrial policy is not daring enough and that the country might have missed potential gains. Even in the presence of rent-seeking special interests, leaving market failures unaddressed does not by itself change that order for the better. Stiglitz (2015) argues that not pursuing *any* industrial policy may also serve special interests, namely those interests that cherish the institutional *status quo* and benefit from the absence of state support that broadens access to economic opportunities.

### 7.2.3 *Setting and Pursuing Socio-economic Objectives*

A third theme involves the setting and pursuing of socio-economic objectives. This complements the second theme, because it raises awareness that, first, the process of discovery and learning must serve a purpose and, second, that economic theory on its own cannot decide what the desirable outcomes actually are. Objectives must be guided and set by political processes at various levels. Furthermore, this theme also rejects the binary framing of *whether or not* countries should pursue industrial policies. It replaces it with the much more exploratory question of *how* individual countries can go about *designing* industrial policies in support of the objectives that their respective political leadership has set, as well as *how* their respective public sectors can *implement* such policies efficiently and effectively.

For example, in the context of climate change, a green industrial policy for structural transformation and diversification is often seen as paramount to achieving the global objective of limiting greenhouse gases (GHGs). Altenburg (2011) and Lütkenhorst et al. (2014) see green industrial policy as part and parcel of tackling the daunting challenge of transforming economies and the entire global economic system towards greater energy and resource efficiency and 'beyond GDP' objectives. They also see the past as a problematic benchmark by which to judge contemporary strategies.

### 7.2.4 *Improving Productivity*

A fourth theme is focused on improving productivity. It rests on the argument that, due to positive transaction costs, the private sector is unable to address structural challenges that hold back growth. An example is large-scale investment with high fixed costs, such as the provision of utilities or transport infrastructure that can serve a wide range of economic activities. For these to be built, it may be necessary to incentivize the simultaneous expansion of upstream and downstream activities. For example, in the context of the extractive industries, public investments in infrastructure may not only raise the productivity of a site where extractive resources are produced, but also that of other industries in adjacent areas, such as agricultural and other land-based natural-resources sectors.

However, while in the presence of positive transaction costs public authorities have a key role to play in supporting the private sector to achieve such improvements, this role does not equate to the proposition that the economy should be led or actively run by the state. Nor does it support the proposition that states should own and run industries that are subject to natural monopolies. Rather, the theme highlights the problem that the private sector requires the institutionalization of solutions that address structural market failures and enable positive externalities. Industrial policy becomes the medium to identify the nature of structural market failures and to develop and trial solutions: an example is strengthening education and training systems to deliver highly skilled tradespersons, technicians, and professionals.

Several authors have discussed the role of public authorities in improving productivity with reference to developmentally oriented states, as opposed to 'predatory states' captured by narrow elites (Strom 2017; Wade 2010, 2014). In an earlier UNU-WIDER publication, Auty and Gelb (2001) distinguished between two stylized political-economic models to explain why resource-rich countries had not kept pace with other countries that had started off with similar productive capabilities. These models suggested that economic transformation and diversification on the back of exploiting natural resources hinges on what happens *around* these sectors, as opposed to what happens more narrowly *in* these sectors. A highlighted characteristic of developmentally oriented states is political elites that actively pursue structural transformation and manage to avoid possible capture by economic elites. They also exercise self-restraint in using authoritative power for gains that accrue only to themselves.

However, while proponents of industrial policy agree that improving productivity is a key objective, there is little agreement on how to do it. In this context, an intensely debated question has been whether countries should *comply* with, or *defy*, their comparative advantages based on their existing factor endowments.[6]

### 7.2.5 *Building Comparative Institutional Advantages*

A fifth theme centres on static versus dynamic comparative advantages. A country's *current* comparative advantages are not only, nor even necessarily, down to its endowments of physical resources, but reflect policies and broader institutional arrangements relative to those of other countries. They underpin factor productivity and how it is maintained and improved.

Essentially, this theme emerged from studies tracing the specific characteristics of advanced industrialized countries to their particular institutions and how these have evolved over time as the outcomes of political-economic processes.[7] Historically minded social scientists have focused on the institutional characteristics shared by countries that have managed to transform and diversify their economies. Respective research has included comparative analyses on the economic structures of OECD countries and the 'varieties of capitalism' that these countries display, even though they are all similarly well off.[8] It has also included comparative studies on the successes of East Asian countries.[9]

[6] See the debate between two prominent East Asian economists, who hold opposing views on this (Lin and Chang 2009). This debate is summarized in Box 1 in Dietsche (2017b).

[7] See Thelen (2004) for a comparative analysis of the vocational and skills training systems in Germany, the United Kingdom, the United States, and Japan, and how these relate to the different structures of these countries' economies.

[8] See, amongst others, Eichengreen (2007), Hall and Soskice (2001), Ostrom (2005), and Williamson (1985).

[9] See, amongst others, Chang (2007), Haggard (2004), Khan (2000), and Lin (2012). Some authors, such as Lange (2005), have traced the capabilities associated with such interventions to an institutional legacy of more direct colonial rule, as opposed to the indirect colonial rule exercised elsewhere.

As this research views comparative advantages as the outcome of the comparative *institutional* advantages that a country has built over time, the focus is on the question of how countries have got to where they are, rather than accepting—as a fait accompli—their existing comparative advantages. Importantly, it is not assumed that advantageous institutional arrangements are a given. Hence, this fifth theme links back to the second and third themes regarding discovering and learning, driven by the pursuit of clear socio-economic objectives. Together, these themes indicate that the task is not merely one of quickly designing and announcing industrial strategies and policies, but one of more slowly and purposefully re-shaping existing institutions and building new ones.

### 7.2.6 *The Political Economy of Transaction Costs and Institutional Change*

The last theme is arguably the most controversial: at stake is the issue that—given the positive transaction costs that exist in the real world—there is no guarantee that institutional change per se will reduce these costs. Instead, institutions reflect the political-economic interests of those who were able to shape them in the past and those who are able to re-shape them in the future. Political power-holders may use industrial policy for their own interests and those of their respective political constituencies. Hence, the risk is that of 'rent-seeking' and 'state failure' discussed earlier as the first theme.

However, there are signs that the overtly pessimistic view on the omnipresence of 'state failure' and 'rent-seeking' is being replaced with a more nuanced understanding, which recognizes that the *social order,* underpinning how political power-holders behave and what they see as their legitimate role in society, is critical to outcomes. Thus, this view concludes that what matters is *how* policy makers and state institutions deploy industrial policy: positive institutional change can be a possible outcome, but it is not guaranteed. As a consequence, recommendations on industrial policy should be judged not from an ideological perspective, but on the basis of whether they address systemic challenges that hinder an economy from delivering positive socio-economic outcomes.

Quite clearly, the narrow position that governments should disengage completely from the pursuit of industrial policy is in decline. In part, this is because few still maintain the view that markets are perfect. In addition, industrial policy now has such a wide range of interpretations that economists of different persuasions can subscribe to it. When it comes to the empirics, however, we should expect to see variance in outcomes: the pursuit of industrial policy will fail in some countries, while in others it may well deliver positive results.

## 7.3 Industrial Policy and the Extractives-led Development Agenda

Since the mid-twentieth century, there have been three perspectives on the nexus between extractive industries and industrial policy. These pertain to three periods that have been more fully discussed in Dietsche (2017b).

### 7.3.1 *The Unfulfilled Promise of State-led Industrialization*

From the 1950s to the 1970s, the commonly held view was that state-led industrialization was needed to achieve catch-up with the more advanced economies. Foreign companies had dominated the resources sectors in the colonial era. Generous concessions granted to foreign companies had brought low-cost Middle Eastern oil to the global market and had pushed down prices. But newly independent governments in Asia and Africa sought to take control and transfer the assets of the Western-owned oil majors to newly created state-owned oil and mining companies. The ensuing battle between the governments and the companies prompted key producer countries to actively intervene in the pricing mechanism and encouraged further nationalization. The Organization of Petroleum Exporting Countries (OPEC) was established to tilt market power in favour of the producers. Many governments took control over the sector with a view to using the resulting revenues to finance ambitious industrial development projects as well as the expansion of public provision more broadly.[10]

At the time, some economists raised concerns that it would not be straightforward to industrialize and diversify economies on the back of exploiting and exporting unprocessed commodities.[11] Their fear was that, although countries would be exploiting their current comparative advantages, specialization in primary commodity exports would ultimately lead to declining terms of trade. Economic historians referenced the eighteenth and early nineteenth centuries to point to a mixed picture of potential outcomes.[12]

Analytical tools were also developed to examine the potential connections between the resources-based industries and other economic sectors.[13] For example, drawing on his *linkage theory*, Hirschman (1977) doubted the potential of export-oriented resources sectors to serve broader economic development and diversification.

[10] The history of Chile in these respects is discussed in some detail in Solimano and Guajardo (2017).

[11] See Prebisch (1950) for the Prebisch–Singer hypothesis.

[12] See also Auty (2001).

[13] See Roe and Round (2017 and Chapter 23, this volume) for a more in-depth examination of this subject.

### 7.3.2 *Sector Liberalization in the Ascendant*

By the 1980s many low- and middle-income countries faced fiscal and debt crises, compounded by the commodity price slumps of the time. Some needed to accept IMF and World Bank stabilization and structural adjustment programmes. While state-led industrialization on the back of the resources sectors was not always a failure, it had a high failure rate when political and economic elites abused it for their own advantages. This led to the rise of the microeconomic notion of 'rent-seeking' in the 1980s and 1990s.

This was also the time of the 'Washington Consensus', stressing that state-led industrialization had resulted in state failure. However, its proponents argued that the extractive industries could help countries return to growth by generating much-needed foreign exchange. The implicit industrial policy advice at the time was that countries should attract more foreign investment into the resources sectors, including privatizing and/or selling off their state-owned companies. During this time, almost all national mining and some oil and gas companies in low- and middle-income countries were privatized and sold off, thus removing their troubled finances from the public accounts. In several low- and middle-income countries, significant new private investment and some increased public revenues followed as a result of this approach: for example, in Ghana after 1986 and in Tanzania after 1996.

### 7.3.3 *The Promise of the Extractives-led Development Agenda*

Since the late 1990s, low- and lower-middle-income countries generally have received substantial foreign investment, funding large-scale oil, gas, and mining projects on an unprecedented scale. The construction and operation of such projects has boosted growth rates and per capita income. But they have often not delivered on the expectation that positive trickle-down effects would generate broader benefits, including more diversified economies and more economic opportunities at national, subnational, and community levels. This disappointment has prompted serious questions as to why this expectation has not been met.

Since the early 2000s, this questioning has taken place at two levels. At the community level, the environmental movement has highlighted the discontent of local communities, where negative impacts are felt most immediately and intensively. At this level, extractive companies have been pushed to improve their local impact management and their relationships with local communities.

Meanwhile, at the macro-level many resource-rich countries have done worse than those less well endowed with natural resources (the 'resource curse'). To explain this, researchers initially focused on the macroeconomic and fiscal challenges facing resource-rich countries and identified respective policy responses to counter these. But attention also shifted towards the political economy of resource rents as an explanation of why neoliberal policy-prescriptions had not resulted in broad-based socio-economic development. A consequence of this shift has been the emphasis placed on the 'good governance' of the sector and respectively targeted propositions on 'appropriate' sector governance. International initiatives, such as the EITI and the NRGI, have invested in gathering and disseminating information, and have encouraged greater transparency and better governance in the sector, especially in relation to resource revenues and the legal, regulatory, and contractual arrangements underpinning these.

However, despite these developments it has unfortunately remained rather unclear whether and/or to what extent these efforts have supported countries in diversifying away from the extractives sectors. As underlined in the introduction to this chapter, the very suggestion that it is possible to transform and diversify an economy on the back of extractive projects draws attention to what happens *around* the sector.

## 7.4 What Next on New Industrial Policy and Extractives-led Development?

Unfortunately, because the debate on new industrial policy is so multifaceted, it is also rather inconclusive and does not (yet) provide governments of countries with extractive resources with any sort of road map that could guide them on the strategies most likely to produce long-term sustainable benefits. Worse even, some proponents taking part in the various debates on 'new industrial policy' as outlined above probably regard the extractive industries as an almost irrelevant sideshow to these debates: they would see the focus areas for actions as being the manufacturing sector and its associated technologies.[14] The review so far presented in this chapter—and in more detail in Dietsche (2017b)—has sought to demonstrate that this would be far too narrow a view: several strands in the newer literature suggest important potential roles for the extractive industries in contributing to economic diversification. The remainder of this section sets out four observations on new industrial policy and extractives-led development.

[14] This observation does not mean that all proponents of industrial policy are set on reviving and encouraging manufacturing independently of the extractive industries. For example, Stiglitz (2015) recognizes that an economy based on natural resources can use those resources as a basis for diversification. He points to South Africa's experience of moving from producing earth-moving equipment for the mining sector to producing automobiles.

### 7.4.1 *A Positive Role for the Extractive Industries*

With broader global acceptance of the case for industrial policy and a more nuanced understanding of what such policy might comprise, there are no obvious reasons why considered governments would not choose to devote some part of their resource revenues to support other non-extractive sectors with appropriate policy measures. There is now an increased international understanding of the importance of, first, observing and, second, supporting what happens *around* the sector, as opposed to focusing more narrowly on what happens *within* the sector. This understanding has been reinforced by several years of insights distilled from observing the successes and failures of the extractives-led development agenda to which the UNU-WIDER project on 'Extractives for Development' also seeks to contribute (UNU-WIDER n.d.). This development makes it more likely that any host government that seeks to apply a new industrial policy approach in the interest of long-term sustainable economic development based on diversification will be able to draw on extensive international ideas and increasing support for these, including from development agencies that back respective initiatives such as, for example, the economic growth corridors of the World Bank.

Not least, this suggestion corresponds with the arguments of those proponents who associate new industrial policy with structural transformation, characterized by increased product and service sophistication and the transfer of resources towards more productive economic activities (Felipe 2015). Moreover, it is quite clear that in the period of high investment during the last super-cycle some lower- and middle-income countries have seen the transfer of some resources to higher-productivity activities associated with their increasing dependence on minerals and/or oil and gas. However, the challenge to which there has not yet been a generally good response is how to sustain such initial gains by spreading productivity gains to other sectors.

Contrary to the position that some host governments like to take, such spreading does not happen automatically. In fact, new industrial policy suggests it requires smart policies that support, for example, enterprise development, local skills training, and the promotion of new investments in non-extractive sectors that have the potential to become viable on their own. The more nuanced thinking that is now underpinning the debate provides several ideas on taking forward such policy interventions and developing supportive institutions. Mentioned earlier was the example of Rodrik (2007). Also noted above was that there is significant congruence between this view on new industrial policy and the proposals put forth by international mining companies to partner with governments, local enterprises, and third parties to enhance mining's positive contributions to development, and minimize any negative impacts on host countries and communities (ICMM 2011).

Another line of thought is suggested by Greenwald and Stiglitz (2014, 2017), who have cautioned that the East Asian model of export-led manufacturing growth may not be that relevant for the late developing economies of, for example, sub-Saharan Africa. However, Stiglitz also notes that, in addition to the key role the East African model assigns to manufacturing, several other of its features remain very relevant for new industrial policy. First, as an export-led model, the East-Asian model does not face significant demand constraints from small domestic markets. Second, it addresses the FOREX problem of countries that might otherwise have been limited by that constraint. Third, it provides a convenient basis for learning-by-doing and the absorption of new technologies from abroad, which are critical elements for discovering and learning (see 7.2.2 above). Fourth, the East Asian model provides relatively easily taxable revenues and so helps to boost public spending capacity. Finally, being centred on a discrete number of enterprise units, it provides for a relatively natural system of accountability for those revenues, as compared to the problem of taxing a huge and dispersed set of agriculture-based businesses. A quick glance at this list of key factors suggests their relevance for the extractives-led development model.

On the other hand, there is the question: If not manufacturing, what can serve as the basis for the industrial policy endeavours of the later developers? Stiglitz's answer is that any new industrial policy needs to adopt a multi-sector approach and, with respect to this, in most countries the agricultural sector also needs to play a bigger role. This would require that more policy attention is paid to formalizing land rights, to promoting more advanced technologies, to encouraging non-labour-saving innovations, and to supporting respective skills development and learning across the relevant constituencies. The service sector can also be a key growth sector, but will typically need a lot of support for skills development and the opening up and development of selective tradeable services. Both the agricultural and the service sectors, if better supported by industrial policy, can provide important complementary activities to those of extractive industries. In essence, a multi-sectoral approach needs to be mutually reinforcing and encourage economic activities *around* a buoyant extractives sector in which private operators are willing to invest. Contrary to the view that was commonly held in development circles until recently, there is no reason for the extractives sector to be sidelined and treated as an embarrassing irrelevance alongside calls to industrialize and diversify economies. Instead, host governments should think of the extractives sector as an integral part of a broad-based, multi-sector approach to industrial policy.

### 7.4.2 *Much Theory, Less Practice*

Much of the debate on new industrial policy appears to happen in the intellectual space of economists hypothesizing what countries *ought* to be doing. This comes at the expense of paying less attention to what countries have actually been trying to do and what in the process they have, or have not, achieved.

An example of this is the intellectual debate centred on the arguments about whether countries should be *complying with* or *defying* their comparative advantages,[15] which merits the question of what these arguments hold for the extractives-led development agenda. First, the argument of *complying with comparative advantages* would broadly point to the following recommendation: low-income countries endowed with extractive resources should continue to focus on this advantage and encourage investment in this typically export-focused sector. For example, in the case of sub-Saharan African countries, this would entail pursuing policies that attract those export-oriented, low-skill, labour-intensive industries that China and other emerging market economies are moving on from. The assumption would be that over time both types of foreign investments will somehow contribute to countries advancing their productive knowledge and technological capabilities.

Second, the argument for *defying existing comparative advantages* would suggest that countries should strive to acquire concrete new production experiences via learning-by-doing and cross-sector collaboration. In the context of the extractive sector, this argument would point towards acquiring productive knowledge and technological capabilities that are *transferable* across sectors, so that ancillary sectors with growth potential could gain a basis from which to launch. This is the argument for focusing on selective work packages associated with the development phase of extractive projects to support enterprise and skills development that can be capitalized beyond the sector (such as in relation to civil construction, infrastructure development, or general business services work).

Notably, both arguments are preoccupied with advising on the 'right' policy interventions, but this comes at the cost of paying insufficient attention to how positive institutional change would be brought about as result of such interventions. In addition, both arguments assume the continuation of open trade and cross-border investment.

### 7.4.3 *Local Content as Industrial Policy*

A topic that the extractives-led development agenda has well embraced is 'local content', seen as a means to build linkages between foreign investment in extractives projects and the local economy.[16] But let's face it: local content strategies and policies are industrial strategies and policies by a different name.

[15] Elaborated in more detail in Dietsche (2017b) and Box 1 therein.

[16] Alternative terminology may refer to local procurement and local employment, or to local participation.

And, as there is no universally agreed definition, local content practices can vary widely across countries: for example, some governments focus narrowly on the procurement of goods and services from companies that are owned by nationals, while others consider not only local procurement but also local hiring, skills development, community-based enterprise development activities, or shared infrastructure development (IPIECA 2016). In recent years, many countries hosting extractive industries have put much effort into developing local content strategies, policies, laws, and regulations with the objective of retaining more value in the country and in local communities. At the same time, there is still relatively little research that gathers and investigates empirical evidence on what local content policies have actually delivered.

### 7.4.4 *Secondary Impacts of International Environmental and Social Policies*

Concerns over climate change and associated international, national, and local environmental policies are driving the green industrial policy agenda, which has culminated in the universal climate agreement reached at the Paris COP21 in November 2015. This development has prompted observers to consider the impacts on the extractives-led development agenda. This thinking has shaped new terms, such as 'stranded assets' and 'unburnable carbon', where the low-income producers of extractive resources are often assumed to be the *takers* of the consequences of green industrial policies. At the same time, the extractives-led development agenda is assuming that economic growth will continue to be energy- and material-intensive.

The impacts of green industrial policy vary across the energy and the minerals sector: the former is likely to see a greater shift towards renewables, while at the same time some transformational green economy technologies are reliant on metals and minerals (World Bank 2017). Furthermore, some of the critical green technology minerals are produced as by-products of other, more conventional minerals as well as through various forms of mining: these include not only industrial mining but also artisanal and small-scale mining, which generates a lot of employment but comes with a range of additional environmental and social challenges. In particular, stricter regulations on sourcing minerals from conflict-prone and high-risk areas that are complementing green industrial policies may mean that the mining sector will see the expansion of new forms of so-called *urban* and/or *flexible* mining: that is, the recovery and recycling of secondary materials. Finally, as and when the structural transformation associated with green industrial policy progresses and a more circular economy evolves, there is the influence of the financial sector choosing more carefully what types of primary and secondary extractive resources activities it will invest in, including funding the implementation of a more circular economy.

## 7.5 Conclusions

There are at least three reasons why industrial policy is back on the agenda: the revision of the pro-poor development agenda of the 2000s; international climate policy pushing for a more sustainable and greener economic system; and disappointment with the social outcomes associated with economic liberalization and globalization. Yet, opinions on new industrial policy remain divided, ranging from evoking fears again over the risk of unproductive rent-seeking to populist calls for industrial policy as the panacea for all that might have gone wrong with neoliberal capitalism. Those in the middle of this wide spectrum maintain that investing in productive knowledge and technological capabilities is critical for economic and social development.

This chapter has sought to disentangle the debate on new industrial policy and how it relates to the resources-led development agenda, where the latter builds on the premise that the extractive industries can have a positive developmental impact provided that host governments pursue 'appropriate policies'. The very suggestion that it is possible to transform and diversify an economy on the back of these industries draws attention to what is happening *around* the sector.

The chapter highlighted the challenge that its proponents hold different views on the role that new industrial policy should play. Picking up several themes that run across the ongoing debate, it drew the sobering conclusion that there simply is no consensus on new industrial policy. At best, it is at the philosophical level where a case can be made for the positive role of public authorities in providing institutions that reduce transaction costs. This left a rather incomplete picture of what new industrial policy could bring to the extractives-led development agenda. Looking forward, the chapter drew attention to four observations that re-emphasize some important points of connection between the new mainstream literature on industrial policy and the basic attributes of the extractive industries.

The first observation was that the evolving debate encourages new thinking on a positive role for extractive industries as part of a multi-sector approach to industrial policy. Second, a large part of the new industrial policy debate is focused on what countries ought to be doing and what the right policy interventions are that they should be pursuing. This has come at the expense of paying more attention to the subject of positive institutional change, underpinning the development of comparative advantages for the future.

Third, increasingly popular local content strategies and policies are industrial strategies and policies by another name, but often with a particularly narrow focus and without tracking whether these strategies and policies have actually resulted in the positive institutional change that can sustain gains in productive knowledge and technologies over time.

Finally, the green industrial policy agenda has prompted observers to consider the impact of this agenda on the extractives-led development agenda. With the green industrial policy agenda focusing on the structural transformation of the largest and most industrialized countries and emerging market economies, low-income producers of extractive resources are cast in the role of passive takers of the consequences of these policies. Notably, the impacts vary across the energy and the minerals sector: the former is likely to see a greater shift towards renewables, while the latter, in addition to traditional mining, may see the strengthening of new forms of urban and flexible mining and a shift towards the circular economy concept.

## References

Aiginger, K. (2015). 'Industrial Policy for a Sustainable Growth Path', in D. Bailey, K. Cowling, and P. Tomlinson (eds), *New Perspectives on Industrial Policy for a Modern Britain*, 365–94. Oxford: Oxford University Press.

Altenburg, T. (2011). 'Industrial Policy in Developing Countries', Discussion Paper. Bonn: Deutsches Institute für Entwicklungspolitik.

Amsden, A. (1989). *Asia's Next Giant: South Korea and Late Industrialization*. Oxford: Oxford University Press.

Auty, R. M. (ed.) (2001). *Resource Abundance and Economic Development: UNU-WIDER Studies in Development Economics*. Oxford: Oxford University Press.

Auty, R. M. and A. H. Gelb (2001). 'The Political Economy of Resource-Abundant States', in R. M. Auty (ed.), *Resource Abundance and Economic Development*, 126–44. UNU-WIDER Studies in Development Economics. Oxford: Oxford University Press.

Bailey, D., K. Cowling, and P. Tomlinson (eds) (2015). *New Perspectives on Industrial Policy for a Modern Britain*. Oxford: Oxford University Press.

Chang, H.-J. (2007). *Institutional Change and Economic Development*. London: Anthem Press and UNU Press.

Congleton, R. D., A. L. Hillman, and K. Konrad (2008). 'Forty Years of Research on Rent Seeking: An Overview', in R. D. Congleton, A. L. Hillman, and K. Konrad (eds), *Forty Years of Research on Rent Seeking 1*, 1–9. Berlin: Springer.

Daintith, T. (2010). *Finders Keepers? How the Law of Capture Shaped the World Oil Industry*. London: RFF Press.

Dietsche, E. (2017a). 'Political Economy and Governance', UNU-WIDER Working Paper 2017/24. Helsinki: UNU-WIDER.

Dietsche, E. (2017b). 'New Industrial Policy and the Extractive Industries', UNU-WIDER Working Paper 2017/161. Helsinki: UNU-WIDER.

Eichengreen, B. (2007). *The European Economy since 1945: Coordinated Capitalism and Beyond*. Princeton, NJ: Princeton University Press.

Felipe, J. (ed.) (2015). *Development and Modern Industrial Policy in Practice: Issues and Country Experiences*. Cheltenham: Edward Elgar and Asian Development Bank.

Greenwald, B. and J. Stiglitz (2014). *Creating a Learning Society: A New Approach to Growth, Development, and Social Progress*. New York, NY: Columbia University Press.

Greenwald, B. and J. E. Stiglitz (2017). 'The End of the Manufacturing Export-led Growth Model and its Implications for Development Strategies', presentation at the IEA World Congress, 20 June, Mexico City. Available at: https://www8.gsb.columbia.edu/faculty/jstiglitz/sites/jstiglitz/files/The%20End%20of%20the%20Manufacturing%20Export-Led%20Growth%20Model%20and%20its%20Implication.pdf.

Haggard, S. (2004). 'Institutions and Growth in East Asia', *Studies in Comparative International Development*, 38(4): 53–81.

Hall, P. and D. Soskice (2001). *Varieties of Capitalism: The Institutional Foundations of Comparative Advantage*. Oxford: Oxford University Press.

Hallegatte, S., M. Fay, and A. Vogt-Schilb (2013). 'Green Industrial Policy: When and How', World Bank Policy Research Working Paper No. 6677. Washington, DC: World Bank.

Hirschman, A. (1977). 'A Generalized Linkage Approach to Development, with Special Reference to Staples', in M. Nash (ed.), *Essays in Economic Development in Honor of Bert F. Hoselitz*, 67–98. Chicago, IL: Chicago University Press.

ICMM (2011). *Mining: Partnerships for Development Toolkit*. London: ICMM.

Imbs, J. and R. Wacziarg (2003). 'Stages of Economic Diversification', *American Economic Review*, 93: 63–86.

IPIECA (2016). *Local Content: A Guidance Document for the Oil and Gas Industry*. London: IPIECA. Available at: http://www.ipieca.org/media/1384/local_content_2016.pdf.

Keller, M. R. and F. Block (2015). 'Do As I Say, or As I Do? US Innovation and Industrial Policy Since the 1980s', in J. Felipe (ed.), *Development and Modern Industrial Policy in Practice: Issues and Country Experiences*, 219–46. Cheltenham: Edward Elgar and Asian Development Bank.

Khan, M. (2000). 'Rent-seeking as a Process', in M. Khan and K. S. Jomo (eds), *Rents, Rent-seeking and Economic Development: Theory and Evidence in Asia*, 70–144. Cambridge: Cambridge University Press.

Khan, M. and K. Jomo (eds) (2000). *Rents, Rent-Seeking and Economic Development: Theory and Evidence*. London: Cambridge University Press.

Lahn, G. and S. Bradley (2016). 'Left-Stranded? Extractives-led Growth in a Carbon-constrained World', Chatham House Research Paper. London: Chatham House/The Royal Institute of International Affairs.

Lange, M. (2005). 'British Colonial State Legacies and Development Trajectories: A Statistical Analysis of Direct and Indirect Rule', in M. Lange and D. Rueschmeyer (eds), *States and Development: Historical Antecedents of Stagnation and Advance*, 117–40. New York: Palgrave Macmillan.

Lin, J. (2012). 'From Flying Geese to Leading Dragons: New Opportunities and Strategies for Structural Transformation in Developing Countries', *Global Policy*, 3(4): 397–409.

Lin, J. and H.-J. Chang (2009). 'DPR Debate—Should Industrial Policy in Developing Countries Conform to Comparative Advantage or Defy It? A Debate between Justin Lin and Ha-Joon Chang', *Development Policy Review*, 27(5): 483–502.

Lütkenhorst, W., T. Altenburg, A. Pegels, and G. Vidican (2014). 'Green Industrial Policy: Managing Transformation under Uncertainty', DIE Discussion Paper. Bonn: Deutsches Institute für Entwicklungspolitik.

McPhail, K. (2017). 'Enhancing Sustainable Development from Oil, Gas, and Mining: From an "All of Government" Approach to Partnerships for Development', WIDER Working Paper 2017/120. Helsinki: UNU-WIDER.

Mitchell, J., V. Marcel, and B. Mitchell (2015). 'Oil and Gas Mismatches: Finance, Investment and Climate Policy', Chatham House Report. London: Chatham House/ The Royal Institute of International Affairs.

Mkandawire, T. (2012). 'Institutional Monocropping and Monotasking in Africa', in A. Noman, K. Botchwey, H. Stein, and J. E. Stiglitz (eds), *Good Growth and Governance in Africa: Rethinking Development Strategies*, 80–113. Oxford: Oxford University Press.

OECD (2015). *Policy Guidance on Resource Efficiency*. Paris: OECD.

Ostrom, E. (2005). *Understanding Institutional Diversity*. Princeton, NJ: Princeton University Press.

Prebisch, R. (1950). *The Economic Development of Latin America and its Principal Problems*. Lake Success, NY: United Nations.

Rodrik, D. (2007). 'Industrial Policy for the Twenty-first Century', in *One Economics, Many Recipes: Globalization, Institutions, and Economic Growth*, 99–152. Princeton, NJ: Princeton University Press.

Rodrik, D. (2014). 'Green Industrial Policy', *Oxford Review of Economic Policy*, 30(3): 469–91.

Rodrik, D. (2016a). 'The Politics of Anger'. Project Syndicate. The World's Opinion Page, 9 March. Available at: https://www.project-syndicate.org/commentary/the-politics-of-anger-by-dani-rodrik-2016-03.

Rodrik, D. (2016b). 'No Time for Trade Fundamentalism'. Project Syndicate. The World's Opinion Page, 14 October. Available at: https://www.project-syndicate.org/commentary/protectionism-for-global-openness-by-dani-rodrik-2016-10?barrier=accessreg.

Rodrik, D. (2017). 'Too Late to Compensate Free Trade's Losers'. Project Syndicate. The World's Opinion Page, 11 April. Available at: https://www.project-syndicate.org/commentary/free-trade-losers-compensation-too-late-by-dani-rodrik-2017-04?barrier=accessreg.

Roe, A. and J. Round (2017). 'Framework: The Channels for Indirect Impacts', WIDER Working Paper 2017/79. Helsinki: UNU-WIDER.

Scott, A. (2008). *The Evolution of Resource Property Rights*. Oxford: Oxford University Press.

Smart, C. (2017). 'Industry by Design?' Project Syndicate. The World's Opinion Page, 17 March. Available at: https://www.project-syndicate.org/onpoint/industry-by-design-by-christopher-smart-2017-03.

Solimano, A. and D. C. Guajardo (2017). 'The Copper Sector, Fiscal Rules, and Stabilization Funds in Chile: Scope and Limits', WIDER Working Paper 2017/53. Helsinki: UNU-WIDER.

Stiglitz, J. (2015). 'Industrial Policy, Learning, and Development', WIDER Working Paper 2015/149. Helsinki: UNU-WIDER.

Strom, S. (2017). 'The Political Economy of Industrialisation', *Development and Change*. doi: 10.1111/dech.12281.

Thelen, K. (2004). *How Institutions Evolve: The Political Economy of Skills in Germany, Britain, the United States and Japan*. Cambridge: Cambridge University Press.

UN (2015a). 'Transforming our World: The 2030 Agenda for Sustainable Development. a/Res/70/1', resolution adopted by the General Assembly on 25 September 2015. New York: United Nations.

UN (2015b). *Paris Agreement*. Paris: United Nations Framework Convention on Climate Change.

UNU-WIDER (n.d.). 'Extractives for Development (E4D)'. Available at: https://www.wider.unu.edu/project/extractives-development-e4d.

Wade, R. (1990). *Governing the Market: Economic Theory and the Role of Government in the East Asian Industrialization*. Princeton, NJ: Princeton University Press.

Wade, R. (2010). 'After the Crisis: Industrial Policy and the Developmental State in Low-income Countries', *Global Policy*, 1(2): 150–61.

Wade, R. (2014). *Development Strategies in a Globalized World: Policymaking in an Evolved Framework of Global Governance*. Geneva: UNCTAD.

Warwick, K. and A. Nolan (2015). *Evaluation of Industrial Policy: Methodological Issues and Policy Lessons*. Paris: OECD, Directorate for Science, Technology, Innovation, Committee on Industry, Innovation and Entrepreneurship. Available at: http://dx.doi.org/10.1787/5jz181jh0j5k-en.

Whitfield, L. (2012). 'How Countries Become Rich and Reduce Poverty: A Review of Heterodox Explanations of Economic Development', *Development Policy Review*, 30(3): 239–60.

Williamson, O. (1985). *The Economic Institutions of Capitalism*. New York: The Free Press.

World Bank (2017). *The Growing Role of Minerals and Metals for the Low-carbon Future*. Washington, DC: International Bank for Reconstruction and Development (World Bank).

Part IV

# Policy Challenges in the Macro-management of Extractives

# 8

# The Macroeconomic Management of Natural Resources

*Mark Henstridge and Alan Roe*

## 8.1 Introduction

Managing natural resource wealth requires accommodating to often very large increases in investment, production, exports, and government revenues within the economy of the host country, and setting appropriate macroeconomic policies—especially fiscal, monetary, and exchange rate policies—both to prevent resource wealth from destabilizing the economy and to ensure that its potential for economic development is maximized.

In Part IV of this book we examine this complex challenge from a number of perspectives, both theoretical and practical: the latter based on two contrasting country examples (Ghana and Chile). Before doing this, it is useful to set out a simple framework of ideas to guide the reader through some of the complexity around macro decision-making and policy management. That is the purpose of this short chapter.

At the outset, we make four fundamental points. These follow from the specific and somewhat unusual characteristics of the macroeconomic flows that result from the development of the extractives sector. These four points are: (i) foreign direct investment, production, export, and revenues are often large; (ii) for each project there is a strong degree of uniformity in the sequence of activity from discovery through development to production; (iii) the non-renewable resource is finite, and so are the revenues; (iv) commodity prices are often volatile, which means that extractives' export receipts and public revenues are volatile as well.

First, the scale of extractives follows from the capital intensity of the sector. Oil and mining companies are often large businesses. For low-income countries (LICs) and middle-income countries (MICs) the capital intensity of extractives means that investment and production are frequently non-marginal. This is a point made in Chapter 23 by Roe and Round, in their review of indirect impacts. Indeed, extractives revenues can be so large relative to the size of the host economy that their macroeconomic effects dominate much of what else is happening in the economy.

A second introductory point is the explicit recognition that the conversion of any extractive resource, whether minerals, or oil or gas, into something useful for society (e.g. improved healthcare or a more diversified industrial base) entails a standard sequence of asset transformations. First, the natural resource needs to be discovered, and then it must be 'produced' (i.e. dug up or pumped out and brought above the ground or the ocean) and then sold. This process converts the natural resource into a monetary asset, part of which will (quickly or slowly) be transferred from the extractive companies as a stream of revenue for the host government. That revenue stream takes the form of taxes, royalties, and any claims (through an equity stake or an explicit production-sharing arrangement) that the government has over some part of the production generated by extractives companies. The subsequent transformation into an asset of value to society works through the public finances. This includes public investments in infrastructure as well as education and health services that contribute to the accumulation of human capital, and through any additional private investment which can be stimulated as public projects with high social returns are chosen. In sum, a sequence of macroeconomic flows associated with asset transformation—as illustrated in Figure 8.1—can be anticipated (their scale and character, of course, depending upon whether good or bad policy choices are made).

Third, the economic lifespans of natural assets are finite, ranging from a few decades to several centuries, hence the importance of thinking about their transformation into other assets. Our characterization of the extractives value

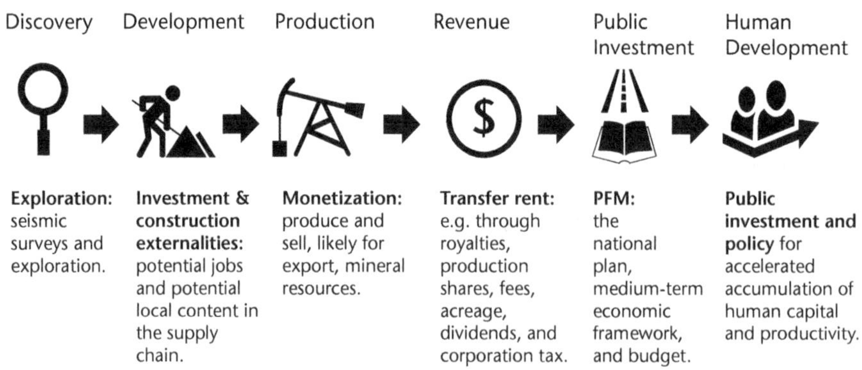


**Figure 8.1.** A chain of asset transformations

*Source*: authors' illustration.

chain as a sequence of asset transformations helps link the finite nature of a natural asset with the set of revenue and spending decisions that constitute the fiscal policy responses to the significant new macroeconomic flows that follow from investment in extractives.

The fourth fundamental observation is that resource revenues are likely to be highly volatile (judged by the historical record of commodity prices over the last hundred years or more). In countries where extractives are large relative to GDP, that volatility translates into significant volatility in export receipts and public revenues. Further, when the fiscal terms for the transfer of rent from extractives companies to a government are progressive—meaning that the government secures a higher share of the rent when prices are higher—then the fiscal terms augment revenue volatility. In other words, contract terms which transfer a greater share of rent at higher prices also transfer a greater challenge of managing volatility to the public finances.

Together these four dimensions of the extractives sector are central to the approach to macroeconomic and fiscal management that any resource-rich country needs to establish. We now discuss the issues in more detail.

## 8.2 The Nature of the Basic Problem

To function well the process of asset transformation needs to be underpinned by a clear set of legal and administrative structures and specific rules within which any policies can be analysed, decided, and then administered. Ideally, these institutions should provide clear legal guidance about the differentiated rights and responsibilities of governments, companies, and citizens. In the absence of such an institutional foundation, backed by a consensus sufficient to sustain it, the potential benefits of extractive activities will be lost. Moreover, the underlying challenges can overwhelm the technical capability of government, leading to adverse development outcomes.

Establishing an appropriate and manageable legal, institutional, and administrative structure is therefore a crucial component of sound natural resource management for all countries. However, the circumstances under which sound institutions are most needed are often also those which are hardest to establish. This is especially so in the LICs. The institutional issues are further illustrated in the case studies on Ghana and Chile in Chapters 10 and 11.

Some dimensions of the institutional structure beyond those involving macroeconomic management are discussed more fully in other parts of the book. One issue discussed is the fact that greater complexity arises because the issues that need to be embraced by this broad structure for policy choices involve many separate parts of government administration: ministries of oil, energy, and mining; industry and environmental regulators, revenue authorities, central banks, and various sector ministries such as labour, health, and education.

Adding further to this complexity is the fact that these different areas of policy responsibility come into play at different times along the asset-transformation chain. Although there is some uniformity in the sequence of discovery, investment, and production, and so predictability in institutional and policy challenges along the chain, the unusually broad range of technical and administrative tasks means that effective management of extractives typically involves a wider range of specialist knowledge and more systematic coordination than do other areas of government policy. Further detail and discussion of these matters can be found in Henstridge and Rweyemamu (2017), AfDB (2015), OPM (2013), and in Chapter 17 by Kathryn McPhail in this volume.

## 8.3 Macroeconomic Management and Dutch Disease

The most familiar way of representing the macroeconomic challenges faced by a country with newly discovered oil or mineral wealth is by reference to the concept of 'Dutch Disease'. This so-called disease arises from a set of circumstances that are commonly associated with the discovery and exploitation of a large natural resource, and the associated asset transformation as described earlier. Specifically, it occurs when a country experiences a large inflow of foreign currency (perhaps in billions of dollars) associated with the new investment needed to produce the extractive resources, and later, after a lag of four to five years or longer, it also experiences the large production revenues from the sale of the extracted resources. These significant foreign inflows can lead to an appreciation of the real exchange rate (RER) which has the effect of making domestic goods (often referred to as 'non-traded' goods) relatively more expensive and international goods (or 'traded' goods such as traditional exports) relatively cheaper. As a consequence, the non-extractive sectors of the economy—such as those in export agriculture or in embryonic industrial production—are likely to experience declining production and employment at the same time as the production of extractives expands.

Why is this problematic for the longer term if not for the immediate future? The answer is in part because natural resources have finite lives and in the longer term any economy needs to look to its non-extractives sector to support a growth path that is genuinely sustainable (see also Chapter 5 by Glada Lahn and Paul Stevens). In addition, any expansion of an enclave (and capital-intensive) sector alone would lead to growth in aggregate GDP, but at the same time a worsening of the income distribution (unless some significant part of the resource rent is redistributed via the fiscal system, either in the form of increased social protection or in public services). Without fiscal redistribution, there is a danger that most people will feel excluded from the fruits of extractives-led economic growth. Public expectations can become seriously inflated merely by the news of resource discoveries, as our case example of the recent experience of Ghana demonstrates. Then any failure to meet these expectations makes the politics of macroeconomic management even more difficult. Managing these problems, and avoiding a sharp RER appreciation, is essential to safeguarding macroeconomic health.

The likely magnitude of the Dutch Disease threat depends on both the size of any new extractive resources relative to the size and absorptive capacity of the host economy, and also the scale of any existing financials flows in and out of the country. But it is fairly clear that Dutch Disease is likely to pose a greater threat to countries with small economies and limited absorptive capacity—especially if the new resource wealth is large relative to the pre-existing macroeconomy.

## 8.4 Fiscal Policy as the Main Instrument

Governments can utilize both fiscal and monetary policy to mitigate the risk of Dutch Disease. However, there is general recognition that in both LICs and many MICs fiscal policy has to bear the brunt of responsibility with monetary policy normally taking the stage as a supporting player.[1] So we proceed for the moment with the proposition that the main bulwark available to governments to avoid or mitigate the threat of Dutch Disease is its own fiscal policy and public finance management (PFM) more generally.[2]

Any host government can in principle choose both the speed at which it scales up its own spending and also the composition of that spending in the light of new extractive revenues. This is a set of choices that need to be considered not only in relation to those revenues that accrue to the government itself, but also in the light of the extra demand pressures caused by the broader economy-wide spending increases associated with the investment and the later production of extractives. Ideally, *total* spending (government plus private) should be managed to grow in line with the absorptive capacity of an economy: that is, the production capacity of the economy to respond,

[1] This is mainly because most LICs and many MICs have only small financial sectors (limited monetary depth and only small or non-existent capital markets), which makes the use of the traditional instruments of monetary policy both difficult and potentially destabilizing (e.g. even small open-market transactions in government debt can cause large movements of interest rates—assuming that these rates are not administratively fixed).

[2] A systematic discussion of how the dangers of Dutch Disease can be addressed by both fiscal and monetary policy can be found in Magud and Sosa (2010).

without excessive shortages, to any extra demand. If total spending outstrips absorptive capacity, this is likely to increase the price of domestic goods (including skilled labour and real estate, both of which are inelastic in supply) and thereby provoke a RER appreciation. But since no government has *direct* control of the *private* component of total spending,[3] it is clear that its major effective instrument is the control of its *own* spending. More specifically, it is commonly argued that increases in total government spending in response to increased extractives revenues need to be moderate and consistent from one year to the next—large and irregular annual increases will make Dutch Disease more likely and more damaging. But government can also mobilize a second useful PFM instrument. This is the pattern or composition of government spending. For example, increased spending on imported goods as well as any government spending that can increase the elasticity of supply of domestic non-tradable goods will reduce the demand pressures on domestic supply and so reduce the risks associated with limited absorptive capacity. One example is increased spending on training, to increase the supply of citizens with the requisite skills. Public spending that reduces the costs of supplying tradable goods can also offset some or all of the negative impact of an RER appreciation on the incentive to produce tradables (e.g. increased spending on transport infrastructure that serves to reduce the costs of exporting may potentially offset any disincentive to export production arising from an RER appreciation).

## 8.5 The Hierarchy of Fiscal Policy Choices

Beyond this basic set of propositions about the role of fiscal policy in managing total spending in the economy, we can conceive of a stylized sequence of policy choices that the fiscal authorities need to address in managing their new-found extractives wealth. This sequence of choices follows the sequence of asset transformations set out above in Section 8.1. These choices have been set out in more detail in AfDB (2015) and are summarized here rather more concisely.

The fiscal policy choices start from any material exploration activity. The investment in exploration will have an early impact on public revenues—from employing people and buying supplies locally—regardless of whether the exploration venture discovers a commercially viable mineral. At the point of declaration of commerciality, expectations of wealth grow, especially in the geographical vicinity of any discovery. Further development will be needed before a 'final investment decision', which would be a material commitment

[3] Some *indirect* control can of course be exerted by the use of interest rate and other components of monetary policy.

in the case of oil and gas (but more like one initial step with more smaller bundles of 'sustaining' capital investment decisions to follow for mining). At the final investment decision point, a reasonable certainty of the likely timing of investment and production can be established.[4] Given expected volumes of production and export, the prospects for future mineral revenue can then be evaluated—based on assumptions about future prices.

The modelling of revenue projections is then a combination of (i) reasonably clear engineering timelines for development and the construction of production and export facilities, (ii) price assumptions (which are subject to uncertainty), and (iii) some assessment of geological and engineering risk, mixed with non-technical risks. These projections condition at least four fiscal decisions:

- Decision 1: Should a government borrow in advance of the arrival of resource revenue streams in the knowledge that those government revenues may be delayed for several years because of (i) a long lag between the construction work involved and production and (ii) a further lag as capital or investment allowances work out and production companies begin to pay significant corporate tax?[5]
- Decision 2: Should a government quickly spend the bulk of its newly acquired revenues or save some part of these, and if so, how large a part?[6]
- Decision 3: Having decided on that part of its own revenues to be spent reasonably quickly, should a government spend on consumption (e.g. increasing the salaries of civil servants or improving operations and maintenance (O & M) activities), or should it instead invest in new infrastructure and other investment goods (e.g. roads, bridges, schools, hospitals)?
- Decision 4: For the saved portion of its revenues, what should government do with those savings? Choices include: saving by depositing funds in general foreign currency reserves managed by the nation's central bank, or establishing some form of sovereign wealth fund (SWF) with or without a stabilization function. In the case of an SWF, there is the further

[4] In practice, commercial operators will normally work with a range of different scenarios, each associated with specific assumptions about uncertain parameters, especially prices.

[5] This is a bigger problem in mining than it is in oil and gas because of the prevalence of production sharing agreements (PSAs) in the latter case.

[6] There is sometimes scope for mixing up the concepts of savings and investment. Here we distinguish between spending, whether on consumption or on investment, and financial savings. This is different from the more formally correct distinction between spending on consumption and savings—where savings can also mean public capital expenditure on physical assets as well as accumulating financial assets. Our reasoning is that an increment to public expenditure financed by foreign revenues is constrained by the absorptive capacity of the economy in a way that parking financial assets overseas is not: see Chapter 9 by Rick van der Ploeg and Tony Venables.

choice of whether the SWF should be mandated to invest funds only outside the country or partly in domestic assets, thereby risking additional local spending pressures as such funds are used for domestic investment.

We now review a few of the considerations that apply to each of these four steps in the sequence of decisions.

## 8.6 Early-stage Borrowing

An early-stage decision for many governments in LICs and MICs is that of whether to borrow internationally to increase their capacity to spend extractive revenues in advance of their arrival.

There are some sound economic reasons why a sensible government might consider borrowing in advance of any receipt of revenue flows. In particular, if the borrowed funds are invested in improving infrastructure and other potentially growth-enhancing activities, this might increase the future tax base, thereby partly self-financing the additional expenditures. The country's growth and development benefits will then arrive earlier, without compromising its ability to make future debt repayments once the revenues from natural resources do arrive. At the same time, all debt-financing incurs a cost for governments in the form of the interest repayments, with the monetary payments involved being in principle usable for other public purposes. Furthermore, for even a well-conceived public good investment (such as a vital new road), it is well documented that the government will only receive a small part of the increased economic production in the form of increased tax revenues with which to service its extra debt (see Buffie et al. 2012).

The *technical* calculus involved here is relatively straightforward: the justification for any additional debt service burden hinges on the costs of the new borrowing, the existing burden of debt service, and above all on whether the borrowed funds are used for productive purposes—their use to finance increased consumption would obviously be problematic. There are also some obvious risks associated with accumulating debt in the expectation of future revenues. Those flows can never be fully certain given the inherent volatility of commodity markets, and the uncertainties about the timing of extractives production. If revenues are smaller than anticipated, then debt levels could become unsustainable, thereby undermining the government's future ability to spend and invest. Venezuela in 2017/18 is an extreme example of this syndrome: one of the world's most richly endowed oil economies is close to debt default. Ghana, discussed by Bawumia and Halland in Chapter 11, is a less extreme example of debt over-accumulation.

Indeed, for many such countries there is a strong temptation to borrow that can override narrow technical factors. The reason is that this decision involves the confluence of three dangerous realities. First, public expectations about a prospective windfall are often hyped up by the media (and often politicians), thereby putting pressure on government to show *early* results. Second, due to the nature of extractives discovery and production, the most significant government revenues (e.g. from corporate tax) may typically take a long time to materialize—sometimes over a decade between discovery, construction, operations, and major revenue streams accruing to governments. Many governments will quite reasonably look for some stopgap financing to limit the time gaps between the announcement of new resource discoveries and any benefits associated with increased spending. The third reality is that international bankers—in search of new investment opportunities—will be only too willing to market new sovereign borrowing opportunities to newly extractive-rich economies—even including some that have had quite recent experience of severe debt-servicing problems (e.g. Ghana before 1996, when the country was granted debt relief under the HIPC and Extended HIPC processes).

One main lesson from Ghana's recent experience (Bawumia and Halland, Chapter 11) is that debt-financed spending in response to short-term political pressures is unlikely to meet longer-term productive goals. It is more likely to result in spending that fails to match the high costs of debt financing, thereby creating a burden for future generations. Ghana illustrates the challenges of establishing a sufficiently comprehensive and credible fiscal framework in circumstances when it is most needed; Mozambique offers a similar and more recent example of this general problem.

## 8.7 To Spend or to Save?

When resource revenues eventually arrive, the government is faced with a second key decision. What part of these revenues should be spent now (for either extra government consumption or public investment) and what part should instead be kept as financial savings for the future?

The classical theoretical answer to this question (referred to as the Hartwick rule) is that since the new revenues represent the monetization of an asset that is being depleted over time as part of a sequence of asset transformations, revenue should be invested fully in commensurate amounts of new public capital—primarily infrastructure and human capital—in order to avoid a net reduction in national wealth (Hamilton and Hartwick 2005). But there is an alternative theoretical viewpoint, namely that natural resource wealth can be regarded as a net addition to national wealth, in which case a portion of the revenues could be used to increase government consumption—in line with the 'permanent income hypothesis' (PIH).

In reality the decision to spend or save and the balance between these will be determined by each particular country's political and economic situation. The theoretical considerations that govern the short-term versus long-term decision (to save or to consume) are reviewed more completely by van der Ploeg and Venables in Chapter 9.

A central and important practical proposition is that many developing economies suffer from a chronic scarcity of the capital needed for economic development. Given this, spending on additional public investment makes sense since the marginal returns from such investment are likely to be high. The most important caveat is of especial relevance to many oil economies such as Nigeria and Venezuela. This is that where the capacity to absorb such spending is limited (e.g. there is perhaps a limited technical ability to design large investment projects or maybe technical limitations on the country's ability to procure, implement, and monitor projects) then there is a need for greater caution. This argues for some planned delays in committing to such investment but also provides a strong case for first 'investing in investing' as recommended in Collier (2010).

Notwithstanding the points made above, at least some of the extractive revenues should arguably be saved in almost all country cases, and for two main reasons. First, the prices of most extractive commodities are highly volatile and so the full spending of all incoming revenues as and when they arrive would likely cause serious negative fiscal consequences. These include: (i) high levels of volatility in government spending characterized by revenue shortfalls in years of low prices and inflated spending in years of high prices; and (ii) pro-cyclical patterns of spending, which would exacerbate the unavoidable booms and busts of commodity price cycles. Second, even when there are demonstrably high returns to extra domestic investment, there are limits in most countries as to how fast government spending can be increased in the short term. As noted earlier, limitations on absorptive capacity together with sudden large increases in total domestic spending (buoyed by high levels of fiscal spend) are likely to inflate prices and appreciate the RER (to the detriment of traded goods production and economic growth). So despite public and political pressures, it would invariably be important to allocate some part of the increased government revenues to increased financial savings, whether to park the revenues in the face of an absorption constraint, or as a more explicit smoothing device rather than spending all of them.

One increasingly common way of avoiding a tendency to volatility in the public finances is to enable inter-temporal expenditure smoothing by adopting clearly defined fiscal rules. Such rules typically set out to stabilize the evolution of revenues over time by requiring savings in periods of higher commodity prices and possible dissaving at other times. They are typically set as numerical limits on budgetary aggregates and can be defined in various different ways: examples include spending rules, budgetary rules, revenue rules, and debt rules—see, for example, IMF (2009); Kopits and Symansky (1998).

The cases of Chile and Ghana (in this volume, Chapters 10 and 11, respectively) illustrate many of the practical issues that arise as countries seek to apply these rules in different and often difficult political-economic contexts.

## 8.8 To Invest or to Consume?

The third decision in our stylized hierarchy of fiscal planning choices relates to the part of total government spending (consumption plus investment) that should be allocated to consumption rather than to investment. Some spending must be used to address immediate short-term needs such as ongoing operations and maintenance (O&M) of existing infrastructure (and also public-sector wages). Some increment to consumption owing to an increase in national wealth is consistent with the PIH. So the available revenues can rarely be committed fully to longer-term investments. But on what basis might the decision on an appropriate allocation be reached? There is no definitive technical answer but merely a few relevant observations.

It is rather appealing—publicly and politically—given capital scarcity to state that investment spending should be favoured as being more visible but also more productive than consumption spending. Hence if the project preparation and management constraints can be addressed, then investment should arguably be the number one priority. However, against this there is much evidence that many LICs also suffer from a serious deficit of adequate O&M of their *existing* capital stock. Indeed, recent economic modelling has shown that the returns to O&M for existing capital investments can be significantly higher than those for new investments (Adam and Bevan 2014).

Further, as noted earlier, building better public-sector capability to select and implement investment projects may be a key contributor to public capital and growth and so some non-investment spending that helps to build this capability is an important complement to actual investments. Ideally the government can strike a balance between consumption spending and investment based on these sorts of considerations as both are needed in most country cases to achieve development objectives. However, in reality the third decision is often hijacked by various political economy pressures (e.g. to hike public-sector wages) as the example of Ghana, in Chapter 11, so clearly demonstrates.

## 8.9 Where Should Any Financial Savings Be Placed?

If and when the government has decided to save some part of its extractive revenues—as in the 'parking fund' outlined by van der Ploeg and Venables, Chapter 9, or as a smoothing instrument, a final choice relates to the decision about where to place ('invest') those savings. Two main options present themselves, namely: (i) save the foreign currency received from the extractive resources as an increase in reserves at the central bank, or (ii) through the creation of a specialized fund—a sovereign wealth fund (SWF).

Saving through an SWF requires initial and ongoing investment. Additionally, both the set-up costs and the ongoing administrative costs (e.g. of fund management) are likely to be large and these costs are only justified if the revenues from natural resources themselves are both large and likely to be sustained over many years. So, for example, the administrative costs of the long-standing Norway SWF, though large in absolute amount, are very small relative to the huge capital fund that has now been accumulated (almost US$900 billion by 2016). By contrast a newly established fund in LICs such as Tanzania and Mozambique would be unlikely to build capital at more than about US$200 million per annum even on favourable assumptions about commodity prices, political will, and investment returns, and so the fixed costs of establishing and then operating that fund could be disproportionately high. More generally if the anticipated revenue streams are small, or saving's main aim is to smooth expenditure over the medium term rather than to accumulate for the long run, then the costs of an SWF may not be justified and saving via the central bank may be preferable. This is likely to be the case in many LICs.

## 8.10 Chapters 9, 10, and 11

The remaining chapters in Part IV develop some of the key points made above and elaborate on them by reference to both theory and some practical case examples.

Chapter 9 by van der Ploeg and Venables considers the *short- versus the longer-term* aspects of the macroeconomic problem. Specifically, they address two of the decisions identified above. The first is how much of any revenue windfall should be used for boosting current household consumption or current government spending, and how much should be used for accumulating assets of some sort: that is, consumption versus saving? The second question is: should any saving be invested in foreign assets (an SWF for short) or domestic assets, and if so, why?

Chapter 9 addresses these problems by first setting out a model framework that captures the trade-offs between the alternative uses of resource revenues. This model is used to confirm that intergenerational efficiency in the saving–consumption choice depends on the return to investment, $r$, and the rate at which society trades off present consumption for future consumption, as measured by the consumption rate of interest.[7] They next use the model to remind us that in an economy that has access to perfect international capital markets (and suffers no capital scarcity), the inter-temporal smoothing of consumption would conform to the pattern suggested by the PIH. This in turn would suggest that all savings and the resulting asset accumulation after a natural resource windfall should be placed in foreign assets, such as an SWF. However, they argue that this theoretical result is largely irrelevant for LICs.

However, that argument is also qualified theoretically by the further point that as capital is accumulated and debt is reduced, the (domestic) rate of return falls, so that a developing economy could in principle converge to look more like an economy with no capital scarcity. Thus, starting from a low base, capital is accumulated, income rises, and the rate of return falls, ultimately reaching a level similar to that in high-income countries where it can support similar levels of income and consumption. This in turn leads to some interesting conclusions about the optimal time path of consumption. Specifically, the consumption increment (relative to that seen in the pre-windfall baseline) is largest immediately after the windfall, and then declines. The intuition behind this is that the current generation is by definition poorer than future generations (assuming that there is positive growth), so that the consumption increment is skewed towards this poorer generation. At the same time, room is made for investments to stimulate growth and development. In brief, once the model framework is modified to include the key developing-country features of capital scarcity and high borrowing costs, the policy messages are significantly different from those of the PIH.

Finally, none of the above takes account of the further developing-country reality of limited absorptive capacity and other bottlenecks that might constrain new investment: e.g. a paucity of high-return projects ready to be implemented even when finance is available. In the public sector, in particular, there is unlikely to be a pipeline of good investment projects. These problems have to be addressed before effective investments can be undertaken. The authors argue that in these cases, it makes no sense to spend natural revenues on inefficient projects. Instead, there is a strong case for

[7] Which in turn is defined as $\rho + \eta g$, where $g$ is the (trend) rate of growth of consumption and the parameter $\eta > 0$ is the inverse of the elasticity of inter-temporal substitution (capturing the rate at which the marginal value of consumption diminishes as individuals become richer).

establishing a 'parking' fund for natural resource revenues until such time as they can be used *efficiently* in the domestic economy.

A central recommendation from Chapter 9 is that resource-rich countries need to be clear about the reasons why they may commit part of any revenue windfalls to invest in foreign assets via some sort of SWF. A commonly mentioned reason is to accumulate financial assets that can be used to generate permanent income for future generations: this implies an investment portfolio comprised mainly of assets with longer-term maturities. However, the relevance of such reasoning in the majority of LIC cases is questionable since it relies on the proposition that returns on additional domestic investment are already close to those available on alternative (foreign) assets. The second possible reason for investing some part of revenue windfalls in an SWF invokes the point that the efficient path of investment in the domestic economy will depend on that economy's absorptive capacity: ideally this should be slowly built up through a programme of 'investing in investing'. In the meantime, it is appropriate to argue that some part of savings might be 'parked' offshore until they can be invested efficiently in the domestic economy: this implies an investment portfolio comprised mainly of assets with relatively short-term maturities. A third possible reason is to insulate the economy from short- to medium-run fluctuations in commodity prices. Given the likelihood of such fluctuations, it makes sense to accumulate precautionary buffers in a *stabilization fund*, especially since it is difficult to hedge away all this risk using financial derivatives.

Chapter 11 by Bawumia and Halland shows how the neat theoretical logic of Chapter 9 tends to unwind when real-world political economy intrudes as it has done with damaging consequences in the case of Ghana over the past decade.

They note that Ghana's discovery of oil in 2007 raised public expectations substantially: for many—the public and politicians—as the prospective revenues appeared to be the long-awaited solution to Ghana's developmental challenges. Regrettably the outcomes that subsequently materialized were sharply at odds with those expectations and at first glance seem to provide support for some versions of the resource curse hypothesis. Between 1983 and 2008 Ghana had been one of the stars of the 'Africa rising' story, the toast of the international development community. Ghana's GDP more than quadrupled in nominal terms between 2001 and 2007 and the country established a track record of macroeconomic stability and fiscal discipline. The chapter provides a detailed picture of how this high-quality track record was gradually established prior to 2007. It also notes that when oil was discovered, the Ghanaian authorities (the governments from both the main political parties) were very assiduous in seeking to learn lessons from the experiences of Nigeria and other oil-rich countries, and to put in place solid institutional arrangements to avoid a resource curse, including apparently strong devices to encourage fiscal discipline.

However, the authors then closely document the set of often relatively small deviations from good practice that cumulatively undermined the initial good intentions—the new institutional arrangements notwithstanding. Their account of this provides an excellent real-world example of the practical messiness of the concept of 'institutions' as elaborated by Evelyn Dietsche in Chapter 6, and how this can so easily rebound to the detriment of macroeconomic management. Specifically, by the end of 2008 Ghana's inflation had risen by 8 percentage points; the exchange rate had depreciated by 20 per cent (relative to 2006); the fiscal deficit had risen to 6.5 per cent of GDP; and international reserves had fallen to only 1.8 months of import cover (versus 3 months in 2006).

Notwithstanding this set of disappointing outcomes, the new post-2008 government continued efforts to build sound institutions to address the known problems of the resource curse: types (ii) and (iii) institutions in Dietsche's terminology. In particular, it developed (with a great deal of international guidance and support) the Petroleum Revenue Management Act (PRMA) and the Petroleum Commission Act, both in 2011. A degree of macroeconomic improvement was also achieved partly as a result of a new IMF programme that ran from 2009 through 2012. The authors document in detail the truly dramatic deterioration in the country's macroeconomic situation that resulted after that and through 2015: a deteriorating current account balance, declining GDP growth, a sharp decline in capital investment, a big increase in external debt and the associated interest payments, a large currency depreciation, and a big increase in the government's borrowing from the central bank.

A critical point from this chapter is what it reveals about the political economy of extractives versus the economic theory of effective macroeconomic management as set out by van der Ploeg and Venables in Chapter 9. In Chapter 11 Bawumia and Halland argue that there were particular factors at work in Ghana's political and electoral system that help to explain the disjuncture as between the theoretical and the practical. For example, Ghana is an effective democracy but one in which only a small margin (in terms of percentages and sometimes actual numbers of votes) has separated the two major parties in recent elections. Consequently, with a four-year term, the pressure on any government to deliver is acute and the loss of even a small number of votes can reverse electoral outcomes. Given this reality, and consistent with the political business cycle literature, incumbent governments in Ghana have generally expanded fiscal policy in election years. If you then introduce a future oil bonanza into the political calculus, political-economy reasoning would suggest that the first government to enjoy resource rents will do all it can to remain in power—even at the risk of macroeconomic instability. A fundamental insight is that, in economic terms, the implicit discount rate used by politicians may exceed the rate of interest by the probability of being removed from office.

Chapter 10 by Solimano and Guajardo adds to this canvas of political-economy issues by documenting the case of Chile and its large mineral exports (mainly copper). Today Chile is usually characterized as a country that has been unusually successful in managing the macroeconomic consequences of its large extractives wealth. But as the authors explain, the historical story has been politically turbulent, the record on macro-management has often been poor, and even today's situation is compromised by a variety of question marks over the practices employed and the manner in which political factors still intrude on decision-making.

Historically Chile has long been dependent on a high level of mineral commodity exports: nitrates from the 1890s and copper from the 1930s until today. However, the magnitude of mineral dependence has varied over time partly as a result of dramatically differing government approaches to industrial policy (ranging from import-substituting dirigisme to aggressively liberal). Historically, the institutional framework of Chile's copper sector has been affected by the dominant role played by mainly US-owned foreign corporations. A variety of disputes over the conduct of that ownership built progressively into a national mood that led first, under the government of Eduardo Frei Montalva, to the Chileanization law no. 16.425 of January 1966 that created copper public–private joint companies. Then, more dramatically, in July 1971 the country's Constitution was modified by the new government, led by Salvador Allende, and the copper mines were nationalized. However, this intense statist approach was then moderated under the military government of Augusto Pinochet that seized power in 1973. That new government moved quickly to establish a far more liberal, outward-looking, and friendly approach to foreign companies, especially in the mining sector.

As the authors point out, this also left in place a somewhat contradictory approach to the mining sector. On the one hand the new mining laws and codes of the early 1980s were explicitly pro-private sector. On the other hand the new Constitution of 1980 reflected much of the pro-state flavour of the 1971 nationalizations. In particular, in 1976 CODELCO was established as a major and distinct state enterprise for managing the still-nationalized copper companies. CODELCO's role then and still today involves it in very significant fiscal transfers to government (largely to finance military purchases by the Chilean armed forces *without* the standard budget oversights).

Over time the Chilean copper sector, CODELCO aside, has been progressively de-nationalized with an ever-larger share of mining assets being held by private companies that also enjoy significant benefits as incumbents. The decree law established in 1974, and still ruling today, allows unlimited repatriation of profits by foreign investors and the option of an invariant tax structure for investments. This has made Chile one of the world's most attractive mining destinations for foreign investors and investment has picked up significantly as a consequence, especially after the final ending of dictatorial rule in 1990.

The chapter then devotes considerable attention to the variety of fiscal stabilization arrangements introduced from the later 1980s to mitigate the impacts of the volatile copper price. It notes that in addition to establishing a set of formal fiscal rules, these and later the revised arrangements from 2006 and later 2011 (the Fiscal Responsibility Law) also established two formal SWFs designed to manage budget surpluses generated at the time of copper price booms. A third relatively hidden fund is constituted by the surplus accumulated from CODELCO by the (secret) copper law which benefits the armed forces. During the commodity price super-cycle, these three funds together accumulated some $30 billion of assets—equivalent to around 10 per cent of GDP. In sharp contrast to what happened in Ghana there is a substantial literature—reviewed by Solimano and Guajardo—that demonstrates how these arrangements have rendered the Chilean economy far more resilient than before to copper price shocks. The formal institutional arrangements have some things in common with those adopted *de jure* in Ghana. But the practical execution of these arrangements has enabled Chile—but not Ghana—to achieve a significant countercyclical fiscal capacity.

But this is not to say that the political-economy dangers are wholly absent in the Chilean case and the authors assess in detail a variety of issues of concern. In particular they suggest that an amount equivalent to 10 per cent of GDP (plus a similar amount in central bank reserves) is likely to be an excessively conservative amount to hold to insure against the risks of copper price volatility. This is particularly the case given Chile's low levels of public spending on education, health, pensions, and other social sectors as a share of GDP relative to those of comparable Latin American economies. Second, they note the high level of discretionary authority still left with the fiscal authorities notwithstanding the rules that are in place. Those rules clearly condition the accumulation of funds but do not offer similar guidance as regards the drawing down of funds. So that decision remains one that can be influenced heavily by political manoeuvrings. Politics is never far out of the picture.

These last points notwithstanding, the Chilean experiences of the past decade demonstrate that a broadly effective macroeconomic management of large extractive revenues and their inherent volatility is possible even though that effectiveness is always under threat from the additional vagaries of political-economy processes.

## References


Adam, C. and D. Bevan (2014). 'Public Investment, Public Finance, and Growth: The Impact of Distortionary Taxation, Recurrent Costs, and Incomplete Appropriability', IMF Working Paper, WP/14/73.

AfDB (2015). 'Delivering on the Promise—Leveraging Natural Resources for Human Development', African Development Bank and Bill and Melinda Gates Foundation. Available at: http://www.NaturalResourcesForHumanDev.org.

Buffie, Edward E., A. Berg, C. Pattillo, R. Portillo, and L.-F. Zanna (2012). 'Public Investment, Growth, and Debt Sustainability: Putting Together the Pieces', IMF Working Paper 12/44.

Collier, P. (2010). *The Plundered Planet*. New York: Oxford University Press.

Hamilton, K. and J. Hartwick (2005). 'Investing Exhaustible Resource Rents and the Path of Consumption', *Canadian Journal of Economics*, 38(2): 615–21.

Henstridge, M. and D. Rweyemamu (2017). 'Managing Hydrocarbon Resources', in C. Adam, P. Collier, and B. Ndulu (eds), *Tanzania: The Path to Prosperity*, 49–85. Oxford: Oxford University Press.

IMF (2009). 'Fiscal Rules—Anchoring Expectations for Sustainable Public Finances', Fiscal Affairs Department, 16 December. Washington, DC: IMF.

Kopits, G. and S. Symansky (1998). 'Fiscal Rules', IMF Occasional Paper No.162. Washington, DC: IMF.

Magud, N. and S. Sosa (2010). 'When and Why Worry about Real Exchange Rate Appreciation? The Missing Link between Dutch Disease and Growth', IMF Working Paper, WP/10/271, December. Washington, DC: IMF.

OPM (Oxford Policy Management) (2013). 'Impact of LNG in Tanzania', report for BG Group presented at a workshop for government, August, Dar es Salaam.

# 9

# Extractive Revenues and Government Spending

## Short- versus Long-term Considerations

*Frederick van der Ploeg and Anthony J. Venables*

## 9.1 Introduction

Resource-rich developing countries often display poor growth performance because they fail to save and invest a sufficiently high fraction of their resource revenues. This problem is exacerbated if they borrow when world prices of their resources are high, fail to put the extra funds to good use, and then get into serious problems with repayment of principal and debt service when world prices collapse. Some have put the revenues from their windfall into sovereign wealth funds (SWFs), but often their objectives are unclear and their management lacks economic discipline, with many of these funds being run down when there is a change of government or times get hard. Furthermore, these funds are often raided for political purposes. Also, countries have often invested in international investments with low returns rather than in domestic investments with potentially higher returns, and have funded SWFs while also issuing expensive sovereign bonds.[1]

[1] An earlier version of this chapter was presented at UNU-WIDER's workshop 'Extractive Industries and Development', led by Tony Addison and Alan Roe, on 11–12 April 2016 in Helsinki. The chapter relies heavily on earlier work: Collier et al. (2010), van der Ploeg and Venables (2011, 2012, 2013), van der Ploeg (2011), van den Bremer and van der Ploeg (2013), Wills (2015a, 2015b), van den Bremer et al. (2016), Venables (2016), and especially Venables and Wills (2016). Thanks to Alan Roe for useful comments. Support from the BP-funded Oxford Centre for the Analysis of Resource-Rich Economies is gratefully acknowledged.

We offer an analytical treatment of how to best manage revenues from non-renewable natural resources, such as oil, natural gas, and minerals, focusing on two key questions. The first is how much of the windfall should be used for boosting current household consumption or current government spending, and how much should be used for accumulating assets of some sort. This concerns the choice between consumption and saving. The second question is whether saving should be invested in foreign or domestic assets. For the time being we refer to foreign assets as SWFs. Domestic asset accumulation means investment in the domestic economy, which might be capital spending by either the public or the private sector, and includes expenditures on education and health that build human capital.

We argue that, in capital-poor developing economies, savings rates out of resource revenues should be high, and the priority should be to invest them in the domestic economy. The case for high savings rates is motivated by both the temporary nature of revenues from exhaustible resources and the lack of sufficient finance to undertake capital projects. There is a clear opportunity cost to placing funds offshore in SWFs, as the case for an SWF turns on three main reasons:

- The first is to transfer part of the windfall to future generations. However, if there is capital scarcity, this is best done by investing in the domestic economy rather than by accumulating assets abroad.
- Second, 'parking' windfall revenue in SWFs is needed if the efficient time profile of domestic investment does not line up with the time path of the resource windfall. Natural resource revenues should be held in SWFs until they can be used most productively in the domestic economy, recognizing that it takes time to build up a high enough quality domestic infrastructure, legal system, educational system, and health system.
- Third, it is necessary to put some of the windfall in a so-called stabilization fund to build up a precautionary buffer to deal with the volatility of commodity prices. This is especially important when hedging is too expensive, as most future markets do not exist or are too thin, or hedging is deemed to be inadvisable from a political point of view.

Section 9.2 sets out our framework and captures the trade-offs between alternative uses of resource revenues. A key feature of our inter-temporal framework, when applied to a developing country, is that the economy starts out with a low capital stock and thus a low level of income per capita. A resource windfall enables the economy to speed up the process of economic development, building up the domestic capital stock by investing at least part of the windfall. Returns to this need to be compared with alternative uses such as consumption or improving the foreign-asset position (cutting sovereign debt or building SWFs).

Sections 9.3 and 9.4 analyse the long-run saving decision, making the case for investing resource revenues in the domestic economy, rather than in SWFs. Section 9.5 investigates the possibility that many developing economies are not—at least in the short run—ready to undertake high-return domestic investments even when they are in dire need of them. Further, such economies have experienced booms and busts as domestic spending has moved in line with natural resource revenues, often rising and falling too abruptly. The domestic investment path must be efficient, so it (and spending more broadly) must be decoupled from current revenue flows. To achieve this, resource revenues may need to be parked in offshore funds, but this should be temporary, until productive domestic investments can be found and funds can be fruitfully and efficiently invested without destabilizing the domestic economy. Section 9.6 discusses issues surrounding the volatility of resource revenues. This volatility impacts government revenues and the macro economy, and there is a case for building a stabilization fund to cushion these impacts. Section 9.7 concludes.

## 9.2 Inter-temporal Efficiency and the Present Value Budget Constraint

Our inter-temporal model has two key ingredients. One is the economy's inter-temporal budget constraint, and the other the efficiency conditions that shape choices within this budget constraint. Starting with the *inter-temporal budget constraint*, we define the wealth of the economy at date $t$, $W_t$, as the value of all assets held by the economy:

$$W_t \equiv F_t + PVY_t + PVN_t \tag{1}$$

Wealth consists of three components. First, net foreign assets, $F_t$ (negative if foreign debt); the rate of return on such assets is $r$ which, for the moment, we take to be constant. Second, the present value of income generated by domestic primary factors minus net investment, that is:

$$PVY_t = \int_t^\infty \left[Y(K_\tau) - \dot{K}_\tau\right] e^{-r(\tau-t)} d\tau \tag{2}$$

The economy's capital stock is $K$ and the production function is $Y(K)$ (depending also on labour, which we hold constant and suppress in the notation). The integral is thus the present value of the economy's output net of investment $\dot{K}$, at all future dates, discounted at rate $r$ to date $t$. The return on domestic capital is the marginal product, $Y'(K)$, decreasing in $K$. Third, the value of subsoil assets: that is, the expected present value of the natural resource revenues, net of extraction costs, which is also commonly referred to as resource wealth. Expected net natural resource revenue at date $\tau$ is denoted $N_\tau$ so that net natural resource wealth at date $t$ is:

$$PVN_t = \int_t^\infty N_\tau e^{-r(\tau-t)} d\tau \tag{3}$$

Discovery of natural resource reserves leads to a path of revenues, $N_\tau$, which we take to be exogenous and determined by the size and nature of the resource deposit and by future prices. Thus, we assume that the extraction path of a given field is not given by Hotelling's rule but determined by geological considerations. For oil, the relevant concern is Darcy's law that describes the flow of a fluid through a porous medium. The discovery—or any change in expected future natural revenues, for example, due to a change in the resource price—shifts the expected present value of resource revenues, $PVN_t$, and hence shifts the resource wealth of the economy.

The inter-temporal budget for the economy constrains the time path of future consumption, $C_\tau, \tau \geq t$ by the economy's total wealth. If the economy has access to perfect capital markets, this too can be expressed as a present value constraint, so:

$$W_t \equiv PVC_t = \int_t^\infty C_\tau e^{-r(\tau-t)} d\tau \tag{4}$$

Hence, at each date, the present value of consumption, $PVC_t$, should not exceed the total wealth of the economy, consisting of foreign assets plus the discounted value of production income plus natural resource wealth. This implies that the 'no Ponzi games' condition is satisfied.

*Intergenerational and inter-temporal efficiency conditions* characterize the efficient allocation of wealth between assets and the efficient rate of asset accumulation: that is, the division of income between consumption and saving. The condition for efficient allocation of wealth is to hold assets yielding the highest return and, where multiple assets are held, choose quantities of each such that their marginal rates of return are equalized. This means dividing domestic and foreign assets, $K+F$, such that $Y'(K)=r$.

Intergenerational efficiency in the saving/consumption choice depends on the return to investment, $r$, and the rate at which society trades off present consumption for future consumption, as measured by the consumption rate of interest. The consumption rate of interest according to the Keynes–Ramsey rule consists of the pure rate of time preference, $\rho$, reflecting inherent impatience and the desire to consume now rather than in the future, and a term that captures changes in society's income through time. To be precise, the Keynes–Ramsey rule implies that the consumption rate of interest is $\rho + \eta g$, where $g$ is the (trend) rate of growth of consumption and $\eta > 0$ is the inverse of the elasticity of inter-temporal substitution (capturing the rate at which the marginal value of consumption diminishes as individuals become richer). The second term, $\eta g$, captures that future generations may be richer than the present generation, so a marginal unit of consumption is less valuable to them than to people currently alive. For a developing country, this can be thought of as saying that, if income is growing ($g > 0$), poverty reduction needs are greater now than in the future, so the future is discounted more heavily, especially if intergenerational inequality aversion is large. Typically, $\rho = 2$ per cent and $\eta = 2$, so if consumption is growing at 3 per cent per annum, the consumption rate of interest equals $\rho + \eta g = 8$ per cent per annum. Developed economies are no longer catching up and typically have lower trend rates of growth and thus lower rates of interest.

The efficiency condition for inter-temporal savings/consumption decisions, recognizing that the consumption rate of interest must equal the rate of return on investment (see Appendix), is thus:

$$r = \rho + \eta g \text{ or } g = (r - \rho)/\eta \tag{5}$$

Thus, given the budget constraint, if the rate of return on assets, $r$, is high, consumption should be relatively low today and high in the future; that is, consumption should grow fast, $g = (r - \rho)/\eta$. The reason is that a lot is saved today to make the most of the high rate of return. If the budget constraint is relaxed (e.g. if a windfall of resource revenue becomes available) and values of $r$, $\rho$, and $\eta$ remain unchanged, overall consumption increases but the subsequent rate of change of consumption is unaffected.

Finally, resource windfalls can last for many decades or even centuries. It is therefore not appropriate for society to base intergenerational welfare comparisons by using data on inter-temporal savings and consumption choices made by individuals. This is why the Keynes–Ramsey rule, as shown in Equation (5), captures the societal ethical trade-offs between generations today and in the future. Following this rule, one should probably use a very small or even zero rate of pure time preference, $\rho$, as it seems unethical to discount the welfare of future, yet unborn generations. A similar stance has been taken by the Stern (2007) review in the context of climate policy. This argument is especially strong for investment projects with a long horizon (Gollier 2013). The coefficient of relative intergenerational inequality aversion should come from deep ethical considerations on how much society is prepared to forsake consumption today to boost consumption of future generations.

## 9.3 Benchmark Management of Windfalls: The Permanent Income Hypothesis

This analysis leads naturally to the celebrated permanent income hypothesis (PIH), which has often been used as the basis for advice to resource-rich economies. It requires the bold assumption that the economy can borrow or lend internationally at a constant (and exogenous) world rate of interest, which we denote $r^*$, and thus that the domestic capital stock adjusts such that $Y'(K) = r = r^*$. This is tantamount to assuming that the economy can borrow from or lend to world capital markets at a relatively low rate and has no shortage of domestic capital because it has invested up to the point where there are no investment opportunities that yield more than the return on foreign assets. Often, a second assumption is made to ensure that the time path of consumption is fully smoothed; namely, that the rate of pure time preference equals the rate of interest, so that $\rho = r^*$. The condition for intertemporal efficiency, $r^* = \rho + \eta g$, given this assumption, is indeed satisfied if aggregate consumption is constant through time, $g = 0$.

Sometimes an alternative, perhaps more realistic, second assumption is made to ensure that the resource dividend per capita rather than aggregate consumption is held constant over time (e.g. van den Bremer and van der Ploeg 2013). This requires that the rate of pure time preference is set to a lower value—that is, $\rho = r^* - \eta n$—where $n$ is the rate of population growth, because this ensures from Equation (5) that $g = (r^* - \rho)/\eta = n$ and thus that *growth* in consumption per capita is smoothed. For ease of exposition, we abstract from population growth in the remainder and suppose that $\rho = r^*$. Many governments in resource-rich developing economies are focused on winning the next election and reaping as much benefit as possible from a resource windfall. Such governments may thus employ a much higher pure rate of time preference, especially when their grip on office is not so strong.

Without resource revenues, an economy satisfying these assumptions will have constant consumption and wealth through time, so:

$$C = r^*W = Y(K) + r^*F \tag{6}$$

The first of these equations comes from integrating the budget constraint, Equation (4), given also that the time path of consumption is fully smoothed. The second is the flow budget constraint stating that consumption equals production income given a constant capital stock.

What is the effect of revenues from a current or future natural resource windfall? A discovery of natural resources, which takes place at date $t = 0$, has present value $PVN_0$. However, it is often the case that revenues may not be earned for a further five or ten years until all exploitation investments have taken place and the pumping or mining starts (e.g. Arezki et al. 2016).

With the assumption that $r^*$ is exogenous, the domestic capital stock, $K$, and hence the level of domestic non-resource output are both fixed by the equation $Y'(K) = r^*$. It follows that a resource windfall has no effect on the optimal choice of domestic capital stock or on investment in the domestic economy. Hence, none of the resource windfall should be invested in the domestic economy because with access to perfect capital markets the level of domestic investment should already be at its optimal level.

Consumption jumps up on impact of the news of the discovery of new resource wealth and well ahead of the revenues pouring in as the news of future resource extraction is fully discounted in efficient markets. Furthermore, with the assumption that $r^* = \rho$ consumption is constant at that new level along the entire optimal consumption path thereafter. Because at the date of discovery wealth jumps by the value of the resource discovery, $\Delta W = PVN_0$, where $\Delta$ denotes the discrete change at the date of discovery, the discrete jump in consumption at the time of the discovery is therefore:

$$\Delta C = r^* \Delta W = r^* PVN_0 \tag{7}$$

Hence, consumption increases by an amount equal to the annuity value of the resource discovery. This increase in consumption occurs at the date of discovery and is maintained in perpetuity thereafter.

These are two very strong results. They stem from two bold assumptions: (i) the domestic capital stock is such that the marginal product of capital equals the world interest; and (ii) $\rho = r^*$, so the time path of consumption is flat. Neither of these assumptions is appropriate for developing economies, and in the next section we look at the consequence of relaxing both of them. Despite this, the PIH gives some important messages for the fraction of natural resource revenues that should be saved and the path of asset accumulation. We turn to these now.

The time profiles of the flows of income and consumption and of the stocks of assets are shown in Figure 9.1, with time on the horizontal axis. For simplicity, we assume that the economy has no foreign assets or debt at the date of discovery, $F_0 = 0$. As noted above, wealth jumps up by $PVN_0$ at the date of discovery, and is constant thereafter. This means that as the resource follows its depletion path (assumed to be exogenous) and $PVN$ declines, so foreign assets, $F$, increase to exactly offset this, $F = \Delta W - PVN$, as illustrated. This is in line with the Hartwick rule which states that any depletion of subsoil wealth must be mirrored by an equal increase in above-ground wealth.

Hence, the increment in consumption is constant, $\Delta C = r^* PVN_0$, that resource income $N$ declines (exogenously; dashed line in Figure 9.1), and that interest earned on other assets, $r^*F$, increases. The sum of these is the post-discovery increment to income, $N + rF$, which, by the flow budget constraint, must equal saving, $S$, plus the increment to consumption, $N + r^*F = \Delta C + S$, as

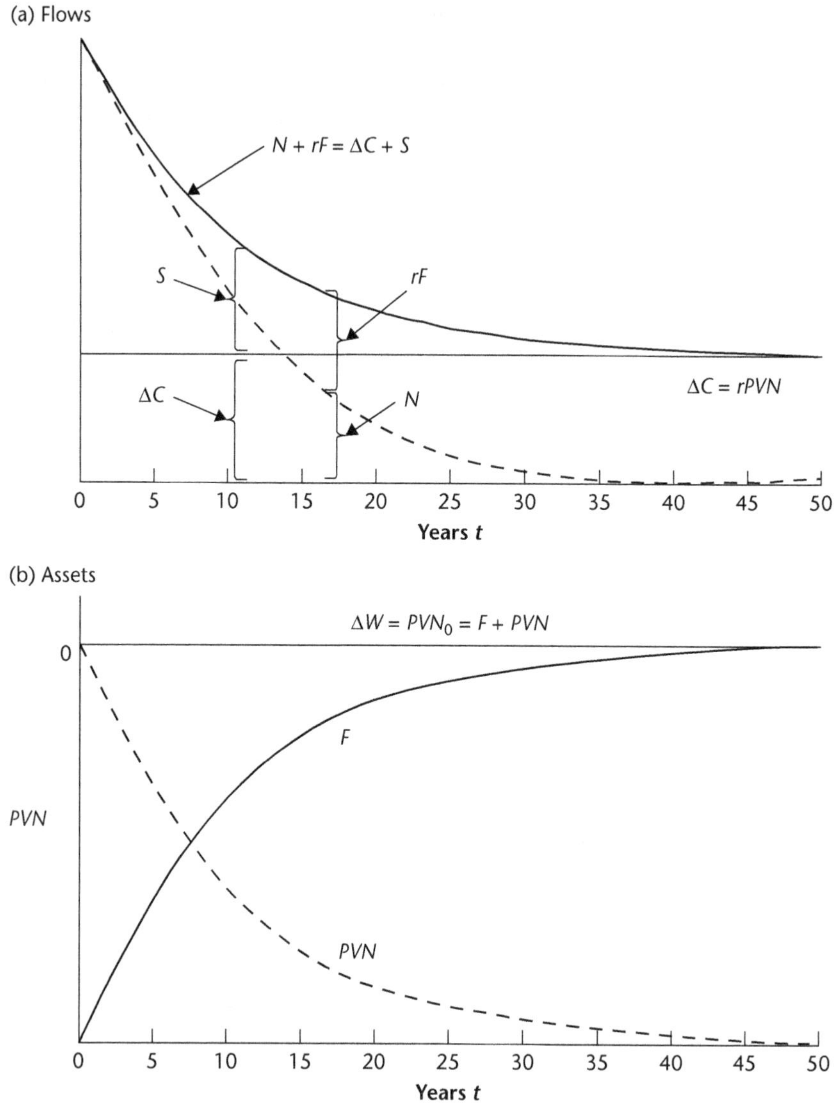


**Figure 9.1.** Managing resource windfalls according to the permanent income hypothesis (PIH): (a) flows; (b) assets

*Source*: authors' illustration based on own calculations.

illustrated by the solid line in Figure 9.1. These savings all go into foreign assets, $S = \dot{F}$, driving the increase in $F$ illustrated in Figure 9.1b.

It is convenient to write these relationships one further way. As $N + r^*F = \Delta C + S$ and $\Delta C = r^*PVN_0 = r^*PVN_0 = r^*\Delta W = r^*F + PVN$, savings must satisfy:

$$S = N - r^*PVN \text{ or } S/N = 1 - r^*PVN/N \tag{8}$$

Equation (8) gives the savings rate out of the flow of resource revenue, $S/N$, a key policy variable. This depends on $PVN/N$, the ratio of the present value of resource revenue remaining to the current flow of resource revenue. For an extremely long-lived stock of natural resources, the present value of natural resource revenue remaining, $PVN$, is large relative to current resource revenue, so the savings rate is low. The shorter lived a resource discovery (the faster the decline in revenues), the smaller is this ratio and the higher the rate of saving. Of course, the PIH implies that saving out of total resource and non-resource wealth is constant. Some further insights follow from some examples.

The simplest one is where the value of resource revenues declines exponentially at the constant rate $\kappa$ (as illustrated in Figure 9.1). In this case, the share of resource revenue saved is:

$$S/N = k/(\mathrm{r}^* + \kappa) \tag{9}$$

because the revenue flow is $N_t = N_0 e^{-\kappa t}$ and the present value of natural resource remaining is $PVN_t = N_0 e^{-\kappa t}/(r^* + \kappa)$. This demonstrates clearly that the savings rate should be higher, the faster the rate of decline of resource revenues, $\kappa$. Thus, for example, if revenues are expected to decline at twice the rate of interest, two-thirds of resource revenue should be saved. The savings rate is constant during the life of the resource, although this is not generally the case, as illustrated in the next example.

Another example is where revenues from depletion are constant until the point of exhaustion is reached at date $T$, so $N = \overline{N}$ for $t < T$ and then drops to zero. As reserves are depleted, $PVN_t$ falls, so the savings rate rises. The present value is $PVN = \left(1 - e^{r^*(T-t)}\right)\overline{N}/r^*$ so the savings rate increases during the lifetime of the resource and goes to unity at the day of exhaustion:

$$S/N = 1 - \left(1 - e^{r*(T-t)}\right)e^{r*(T-t)} \tag{10}$$

The rationale underlying this time profile for the savings rate is that, as time progresses, the stock of the resource left becomes smaller relative to the flow, and thus the windfall becomes more temporary, necessitating more saving.

Figure 9.2 illustrates this for two resource discoveries of equal size. For both, the flow revenue from extraction remains constant over the life of the resource, but the duration and hence the level of the flow differs. Figure 9.2a shows the two revenue flows, with revenue, $N$, on the vertical axis and time on the horizontal axis. Both profiles, shown by the solid and dashed lines, have the same initial present value, $PVN_0 = 1$ (with $r^* = 0.04$), but that depicted by the solid line depletes the resource slowly so that it lasts for forty years, whereas that depicted by the dashed line depletes it more rapidly so that it only lasts for thirteen years. The optimal savings rates for these depletion profiles are illustrated in Figure 9.2b, and follow Equation (10).

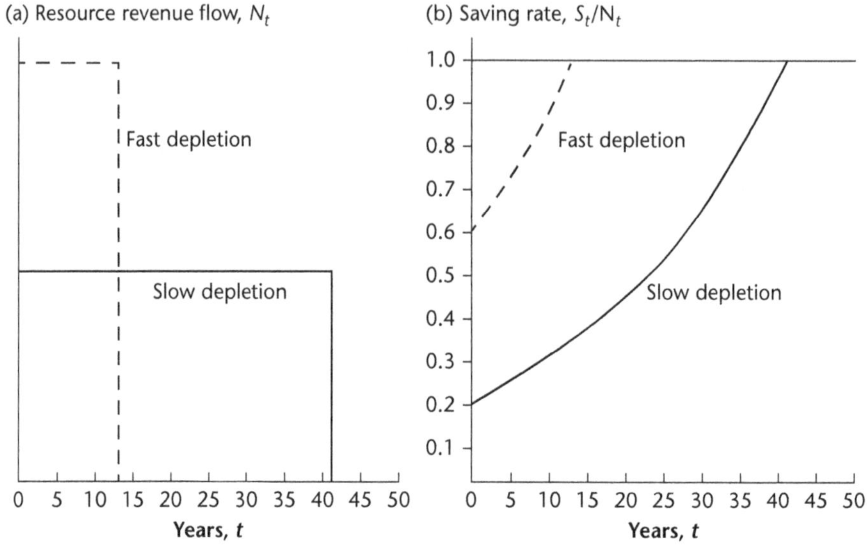


**Figure 9.2.** (a) Resource revenue flow; and (b) optimal savings rates for depletion
*Source*: authors' illustration based on own calculations.

In both cases, the savings rate increases over the course of depletion, as the present value, $PVN_t$, gradually reduces and the flow $N$ and discount rate $r^*$ are constant. Because $PVN_t$ goes to zero as exhaustion approaches, the savings rate eventually rises to 100 per cent. The impact of the rate of depletion is dramatic. In the case of slow depletion, the optimal savings rate rises from 20 per cent in the first year to 100 per cent in the final year. Hence, a policy rule that set a *constant* savings rate would be seriously sub-optimal regardless of the level at which it was set. It is important to have a constant savings rate out of the sum of natural resource and all other forms of wealth, not out of the windfall. The savings path for rapid depletion is strikingly higher than for slow depletion, starting at 60 per cent rather than 20 per cent.

In summary, for a given net present value of a resource discovery, the shorter its duration the higher should be the savings rate, and more surprisingly, for a range of extraction paths the optimal savings rate should rise over time.

## 9.4 The Case for Investing Resource Windfalls in the Domestic Economy

Most developing countries have only asymmetric access to world capital markets: they can save more readily than they can borrow. Typically, as borrowers they are either entirely cut off or they can only borrow at a substantial premium over the world risk-free interest rate to compensate for perceived high risks of default. To capture this reality we retain the model of Section 9.2, but now assume that the interest rate faced by the domestic economy is a function of the stock of foreign assets (or debt) that it holds. We denote this $R(F)$, which implies that the economy has to pay a premium on its foreign debt if it is indebted.[2] If the country is not indebted, then it has perfect access to international capital markets and thus the domestic interest rate relevant for investment decisions, $r$, still equals the world interest rate, $r^*$. If the country is indebted, then it only has imperfect access to international capital markets and thus the domestic interest rate exceeds the world rate, and is increasing in the level of debt. We capture this with the following specification:

$$\text{If } F \geq 0, R(F) = r^* \text{ and if } \quad F < 0, R(F) > r^*, R'(F) < 0 \tag{11}$$

While $R(F)$ is the rate of interest, the marginal cost of borrowing will exceed this if taking on more debt raises the rates the country has to pay on its existing debt.[3] Thus, the marginal social cost of borrowing is $r(F) = R(F) + FR'(F)$, which we assume is also decreasing in $F$ (i.e. increasing in debt). Capital scarcity of this type fundamentally changes the answers to both our questions in Section 9.3—the division of resource revenues between consumption and savings, and the assets that savings should be used to acquire. The policy prescriptions will thus be very different from those emanating from the PIH.

### 9.4.1 *Using the Resource Windfall for Domestic Investment and Foreign Assets*

Total assets in the economy, denoted by $A$, consist of domestic capital $K$ and foreign assets (or debt) $F$: that is, $A \equiv K + F$. If the economy is asset scarce in the sense that the marginal return on domestic capital satisfies, $Y'(A) > r^*$, all assets should be held as domestic investment and none in an SWF. However, if this economy can borrow—that is, choose $F < 0$—it will do this until the return on domestic capital is driven down to the marginal cost of borrowing:

[2] Akitoby and Stratmann (2008) and van der Ploeg and Venables (2011) document empirical evidence for a positive relationship between foreign debt and the domestic interest rate. They find that the effect of resource exports on interest rate spreads is positive, but not significant, suggesting that resources worsen creditworthiness. They also conduct a similar analysis where the cost of borrowing depends on resource revenues, with the main difference being that the cost of borrowing drops, and consumption jumps immediately at the date of discovery of the new natural resource reserves.

[3] It is possible that a resource discovery may reduce the economy's cost of borrowing, as resource revenues are perceived to be a form of collateral for loans. This shifts $R(F)$ downwards.

$$Y'(K) = Y'(A - F) = r(F) \tag{12}$$

If some of the windfall revenue is saved, then the economy's assets increase. To see how much of an increase, $dA$, is allocated to foreign debt reduction, $dF$, and how much to domestic investment, $dK$, we totally differentiate Equation (12) to obtain:

$$dF = \left\{\frac{Y''(A-F)}{Y''(A-F)+r'(F)}\right\}dA, \quad dK = \left(\frac{r'(F)}{Y''(A-F)+r'(F)}\right)dA \tag{13}$$

If the marginal cost of borrowing increases with indebtedness, then $r'(F) < 0$ and the denominators in Equation (13) are always negative. Hence, the bigger the capital scarcity as indicated by the magnitude of $|r'(F)|$, the bigger the fraction of the increment in total assets that is allocated to domestic capital, $dK/dA \in (0,1]$. In the limiting case where the country is completely shut out of capital markets, $r'(F)$ will be equal to minus infinity and the whole of any increase in assets goes to domestic capital formation. At the other extreme, the PIH maintains the assumption that the marginal return or cost of lending and borrowing is given on world capital markets, so $r'(F) = 0$ and hence the domestic capital stock is unaffected by the change in assets, $dK = 0$, and all of the increment in total wealth is allocated to net foreign assets, as we saw in Section 9.3.

Hence, countries that are capital scarce should use resource windfalls to increase the domestic capital stock (including human capital), and pay down foreign debt if investment is sub-optimally low. The case for increasing domestic investment is further reinforced if a resource boom brings its own direct financing needs. For example, infrastructure projects may need to be brought forward and new downstream projects may also be needed, placing a demand on public funds. There may also be general equilibrium effects. The boom is likely to change the structure of the economy, with some sectors booming and others contracting. The 'Dutch Disease' analysis suggests that this will involve a reallocation of economic activity into non-tradable goods and away from non-resource tradables (e.g. Corden and Neary 1982; van der Ploeg and Venables 2013). If the non-resource traded sector is relatively capital-intensive, this will mean that the capital–labour ratio in the economy as a whole will decrease.

### 9.4.2 *Capital Scarcity and the Optimal Choice between Saving and Consumption*

We have thus seen that it is not appropriate for a developing economy to set up an SWF and lend to the rest of the world, but it may be more appropriate to use its newly found wealth to pay down existing debt as well as to invest in the domestic economy. Now we consider total savings out of windfall revenues and their implications for the growth of domestic income and consumption. The inter-temporal efficiency condition and the ethics therein set out in Section 9.2 are fairly standard, but it should be realized that it is a very simple model of development. The inter-temporal efficiency condition (4) now becomes:

$$g = \frac{1}{\eta}\left[r(F) - \rho\right] > \frac{1}{\eta}(r^* - \rho) \text{ with } Y'(K) = r(F) \tag{14}$$

Because the marginal cost of borrowing satisfies $r(F) > r^*$, the cost of capital and thus the consumer rate of interest will be high so that the time path of consumption is rising through time. The rate of return is now higher than the world interest rate. However, as capital is accumulated and debt is reduced, the rate of return falls and the economy converges to one without capital scarcity. Thus, starting from a low base, capital is accumulated, income rises, and the rate of return falls, ultimately reaching a level similar to that in high-income countries and supporting similar levels of income and consumption. This is illustrated in Figure 9.3a showing consumption and Figure 9.3b showing assets; the paths $\tilde{C}$ and $\tilde{A}$ are baseline paths in the absence of natural resource revenues.

A resource bonanza shifts the inter-temporal budget constraint of the economy and allows both current consumption and saving to increase. This speeds up the rate of asset accumulation. More assets at any date mean a lower marginal cost of borrowing, $r(F)$, and hence a slower rate of increase in consumption. The situation is as illustrated by the solid lines $A$ and $C$ in Figure 9.3. Consumption jumps up at the date of discovery but then increases less rapidly. Assets are accumulated more rapidly, more than offsetting the decline in the value of natural resources remaining in the crust of the earth. We also observe the following. First, the consumption increment (the difference between the new path of $C$ and the baseline $\tilde{C}$) is largest immediately after the windfall, and then declines. This is in sharp contrast to the results from the PIH where the increment was constant through time. The intuition is that the current generation is poorer than future generations, so the consumption increment is skewed towards this poorer generation. At the same time, room is made for investments to stimulate growth and development.

If the windfall is very large there may be a 'permanent' element that continues in perpetuity, but for small windfalls the consumption paths illustrated in Figure 9.3 converge (see van der Ploeg and Venables 2011). At the same time as it brings forward consumption, the natural resource windfall also means that accumulated assets are greater at all dates. This brings down the rate of return in the economy as time proceeds. Effectively, the optimal use of the natural resource windfall has the effect of bringing forward development, as can be seen by comparing the consumption and asset accumulation paths

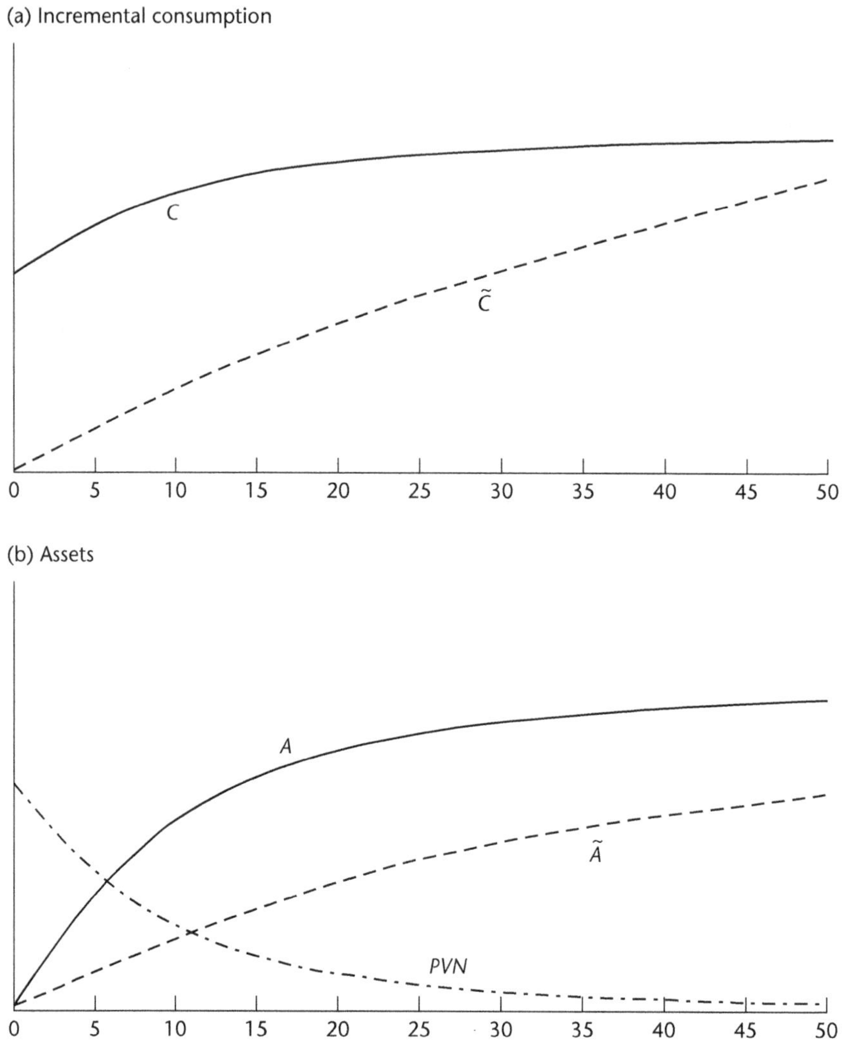


**Figure 9.3.** Optimal revenue management in a developing economy
*Source*: authors' illustration based on own calculations.

with and without the windfall in Figure 9.3. The share of resource revenues that should be saved depends, as before, on the precise path of natural resource revenues. However, the argument of this section tends to reinforce the message that savings should increase through time; the additional effect is the front-loading of the consumption increment.

In summary, allowing for capital scarcity and high borrowing costs, the policy messages are significantly different from those of the PIH. The consumption increment is front-loaded whereas resource revenues should be used principally for domestic investment, which brings forward the development path of the economy.

## 9.5 Absorption Constraints and the Case for a Parking Fund

The prioritization of domestic investment rests on there being domestic projects that yield high returns (social as well as private returns), and on projects being delivered in an efficient sequence and manner. Yet, it is possible that—in the short run—because of absorption constraints and other bottlenecks there may be few high-return projects ready to be implemented. This is a common problem in newly resource-rich developing economies. There will inevitably be a mismatch between the time profile of (efficient) investment and that of resource revenues. Rather than spending revenues on inefficient projects, there is then a case for having a fund that plays the role of 'parking' resource revenues until they can be efficiently used in the domestic economy. The case for parking funds until the economy is able to absorb extra spending efficiently involves both micro- and macroeconomic arguments.

A country seeking to scale up investment will encounter numerous bottlenecks. In the public sector, there is unlikely to be a pipeline of good investment projects. There may be a lack of capacity to design and develop projects; project selection and cost–benefit processes may be weak; and so too may be the ability to procure, implement, and monitor projects. These problems have to be addressed before effective investments can be undertaken, which implies a strong case for 'investing in investing' (Collier 2010). Sequencing of projects also matters because the return to one project depends on other complementary projects. For example, the return to private investment may be low, particularly if public investments in infrastructure and other aspects of productive capacity are lagging.[4] More broadly, additional domestic spending will increase demand for both traded and non-traded goods. As demand increases, the economy will move up its supply curve of non-traded goods, creating both a quantity and a price response. Because the supply curve is less elastic in the short run than in the long run, upwards price pressure is acute if spending increases too rapidly. This is likely to be particularly true for 'home-grown' capital. Although equipment can usually be imported, structures and human capital require domestic capacity (e.g. in the construction and training/teaching professions), all of which take time to develop (van der Ploeg and Venables 2013).

[4] In terms of Equation (14), $Y''(K)$ is particularly large (negative) in the short run. As the marginal product of capital falls more with domestic investment, more should be devoted to foreign assets.

There need not be market failures associated with these bottlenecks and price responses. It may be merely a feature of the supply side of the economy and the political system that projects take time and have to be done in sequence and that it takes time before the necessary political support is obtained. Nevertheless, this timing dictates the rate at which investment can be increased. If extra investment demand is more likely to be met by a price increase than by a quantity response, it is efficient to slow down the rate of investment until supply can catch up. This suggests that investment should be ramped up quite slowly. Hence, resource revenues should be parked in foreign assets until they can be spent in a cost-effective manner on the efficient investment path.

## 9.6 Volatility of Commodity Prices and the Case for a Stabilization Fund

The volatility and unpredictability of commodity prices creates a third argument for using an SWF, namely to function as a stabilization (or liquidity) fund. The optimal size of such a fund depends on how the government and the economy respond, ex post, to shocks that occur. Are shocks sufficiently costly that self-insurance through a stabilization fund is necessary, or can they be managed by other means?

An extreme way of doing self-insurance is the 'bird-in-hand' rule, according to which all revenues are placed in an SWF, and spending is a fixed share of the SWF. Essentially, resource revenues are discounted at an infinite rate, so the spending rule is just a fixed share of above-ground assets and ignores below-ground wealth entirely as done in Norway (e.g. Bjerkholt and Nicolescu 2004; Barnett and Ossowski 2003). Consumption of resource windfalls is thus heavily back weighted, and therefore unable to deliver domestic investment which we have argued is efficient for many resource-rich developing economies. The bird-in-the-hand rule is therefore not appropriate for poor countries because it exaggerates the risk that revenues might terminate earlier than expected, while completely ignoring the much larger risks to low-income societies from deferring growth.

A more general strategy to smooth the impact of commodity price volatility is to save when the price of the resource is high, and dissave or borrow when it is low. The foreign asset holdings will fluctuate with the commodity price but the size of the stabilization fund will fluctuate around zero as the expected value of the fund should be zero. This view needs to be modified for two reasons. First, owing to asymmetry of borrowing and lending costs, borrowing during a sharp downturn in commodity prices is likely to be expensive, if not impossible. Borrowing rates will be high and, following the fall in prices, many a developing resource exporter will likely be shut out of most borrowing options. Second, commodity and in particular oil price shocks are typically long lasting and it is difficult to reject the hypothesis that the oil price follows a random walk. If shocks are permanent or very long-lived, it is less necessary to save for intergenerational reasons but it will be important to build a more sizable stabilization fund (van den Bremer and van der Ploeg 2013). Precautionary savings—that only take into account the utility costs of failing to smooth consumption optimally—stem not from risk aversion but from prudence.

The necessary size of a stabilization fund depends on the costs incurred if, in the event of a commodity price fall, the country is unable or unwilling to borrow. These costs come through both micro- and macroeconomic channels. Precautionary saving only takes into account the volatility costs of failing to smooth consumption optimally, but problems overall are exacerbated by the fact that the fall in revenue principally impacts on government budgets and spending. The consequent and potentially high costs of negative shocks make the case for creating a stabilization fund that gives government the space to undertake countercyclical fiscal policy (e.g. moderate the pace of spending cuts) necessitated by a revenue downturn. Such a fund is even more important in countries with asymmetric borrowing and lending costs. This is because borrowing during a crash in commodity prices is tough.

The management of a stabilization fund, once it is established, opens up a further set of issues. For example, expected rates of return should be adjusted for risk, including political risks such as the possibility that a liquid asset (such as foreign-exchange holdings) might be more easily looted by a future government than illiquid assets (such as domestic infrastructure). Further, a diversified portfolio needs to take into account the correlations between returns on the various assets chosen to hedge against commodity price volatility and the risk inherent in the below-the-ground natural resource wealth (van den Bremer et al. 2016). Rates of return should include the full social costs and benefits of investments and take into account effects of any investment on the domestic macroeconomy, especially if there is limited absorptive capacity.

Finally, we note that an alternative way of insurance is through hedging strategies as used, for example, by Mexico. However, such strategies typically only offer insurance for a relatively short period. We do not discuss them further as future markets are too thin and costly to make this a reliable strategy in most situations. One issue is that the funds may be so large that they could be seen to manipulate commodity prices, and for many countries such hedging strategies are too risky from a political point of view. For most countries, a stabilization fund is a more reliable and pragmatic strategy.

## 9.7 Conclusions

Savings rates should be high and domestic investment should be the priority for developing countries when managing resource windfalls. Management of foreign assets—held perhaps in an SWF—can support this strategy, but it is important that there is clarity on the roles that foreign assets can play.

The first possible role for foreign assets is to put resource windfalls into an *intergenerational savings fund* with the purpose of replacing below-ground resources with above-ground financial assets. The financial assets are used to generate permanent income for future generations. The desirability of such funds depends principally on the long-run investment opportunities (or lack of them) in the domestic economy, and on political economy. Such a fund should focus on assets with long investment horizons, matching the permanent nature of the fund. Such a fund makes sense for countries with a large domestic infrastructure stock already in place and good access to international capital markets. In such a case, the rate of return on further domestic investment is unlikely to exceed the return that can be made on investments through an SWF. Such a fund also makes sense if it is unlikely to be raided by future governments.

These conditions clearly do not prevail in most developing economies. Capital scarcity means that investment in the domestic economy is a priority. This investment can help place the economy on an accelerated growth path, bringing forward economic growth and consumption benefits. If directed towards the non-traded sector, this investment can also limit inflation from resource-financed demand. In developing countries the demographics are such that capital is needed to support jobs for a growing work force, not flows of rent for pensioners. From a political perspective it is also better to sink investments in physical structures and assets rather than leave them in easily drawn-down financial assets. The case for long-run intergenerational funds held in foreign assets is therefore not very strong for developing countries.

To overcome the timing mismatch between receiving resource revenues and investing them efficiently in developing countries, a *parking fund*—the second possible motive for considering investment in foreign assets—is appropriate. The efficient path of investment will depend on the absorptive capacity of the economy, which ideally should be slowly built up through a programme of 'investing in investing'. This requires that savings be parked offshore until they can be efficiently invested in the domestic economy. This fund should invest in assets with a relatively shorter investment horizon than the intergenerational fund.

The third possible reason is to insulate the economy from short-to-medium-run fluctuations in commodity price revenues. Given the likelihood of such fluctuations, it makes sense to accumulate precautionary buffers in a *stabilization fund*, especially as not all the risk of exposure to volatile resource prices can or will be borne by resource extraction firms or can be hedged away with financial derivatives. Therefore, natural resource exporting countries are well advised to self-insure by using stabilization funds. These funds provide an additional source of interest income when commodity prices crash, and a stock of assets that can finance countercyclical fiscal policy when monetary policy is constrained. Stabilization funds, particularly in developing countries, should hold assets that can be liquidated at short notice if the prices of their commodities fall. Their spending rules must permit fast liquidation, contingent on the resource price, and should also include a mechanism for recapitalizing the fund when the price rises again. The establishment and implementation of such rules is no easy task and even in broadly successful cases such as that of Chile there is ongoing controversy about the specific rules for the accumulation and draw-down of funds (see Solimano and Calderón Guajardo 2017).

These three types of funds held in foreign assets have different purposes, and so need separate legal structures and investment mandates. A developed country's intergenerational fund should invest in long-horizon assets and spend a fixed share of above- and below-ground wealth. A developing country's parking fund should invest in shorter duration assets and have a rising limit for the amount spent each year, with this constraint used for domestic investment. Stabilization funds should focus on liquid assets and have spending and recapitalization rules tied to the commodity price. All funds should design their portfolios with subsoil assets in mind—hedging resource price volatility at the start of the windfall and rebalancing towards a diversified portfolio as the resource is exhausted. It may well make sense to manage these funds from within a single government body, but they should remain independent in line with their original aims, with an investment mandate and spending rule that is appropriate for their purpose.

## Appendix: Optimization Problem in the Presence of Capital Scarcity

The social planner's problem is to choose consumption, $C$, and investment, $I$, in the face of exponentially declining oil revenues, $N$, to maximize utility:

$$J(F, K, t) = \max_{C,I} \left[ \int_t^{\infty} U(C_t) \exp\Big(-\rho(\tau - t)\Big) d\tau \right] \tag{A1}$$

subject to the constraints describing foreign asset accumulation:

$$\dot{F}_t = r(F_t)F_t + N_t + Y(K_t) - C_t - I_t \tag{A2}$$

accumulation of physical capital:

$$\dot{K}_t = I_t - \delta K_t \tag{A3}$$

and the time path of natural resource revenues:

$$\dot{N}_t = N_0 \exp(-\kappa t) \tag{A4}$$

(cf. van der Ploeg and Venables 2011; Wills 2015a). We incorporate capital scarcity by assuming that interest rates may be elastic with respect to foreign assets $F$, $r(F_t) > r, r'(F_t) < 0$ for $F_t < 0$ and $r(F_t) = r$ otherwise. Output is produced using domestic capital and a fixed stock of labour, $Y_t(K_t)$ and $\overline{L} = 1$, and it is sold at the constant world price $P^* = 1$. The Hamiltonian for this dynamic optimization problem is:

$$H \equiv U(C) + \lambda_1[r(F)F + N + Y(K) - C - I] + \lambda_2[I - \delta K] + \lambda_3[N_0 e^{-\kappa t}] \tag{A5}$$

This has the optimality conditions:

$$\begin{aligned} &H_C = U'(C) - \lambda_1 = 0, \\ &H_1 = -\lambda_1 + \lambda_2 = 0, \\ &\rho\lambda_1 - \dot{\lambda}_1 = H_F = \lambda_1[r'(F)F + r(F)], \\ &\rho\lambda_2 - \dot{\lambda}_2 = H_K = \lambda_1 Y'(K) - \lambda_2\delta, \text{and} \\ &\rho\lambda_3 - \dot{\lambda}_3 = H_N = \lambda_1, \end{aligned} \tag{A6}$$

and the transversality condition $\lim_{t\to\infty}[\exp(-rt)\lambda_{1,t}F_t] = 0$. The Euler equation or Keynes–Ramsey rule can be derived from the first-order conditions as follows:

$$g = \frac{1}{\eta}[r(F) + r'(F)F - \rho] \tag{A7}$$

where $g$ denotes the rate of growth of consumption. Therefore, consumption will be delayed as long as the country faces a premium on its debt ($F < 0$), so the rate of consumption will be rising with time. The decision of whether to use the income that is not consumed for domestic investment or foreign savings follows from equating the marginal utility of consumption to that of an extra unit of foreign assets or domestic capital:

$$U'(C_t) = \lambda_1 = \lambda_2 \Rightarrow r(F) + r'(F)F = Y'(K) - \delta \tag{A8}$$

Since with capital scarcity the left-hand side is a negative function of $F$, we establish that the optimal stock of capital rises as the stock of foreign debt is curbed.

## References

Akitoby, B. and T. Stratmann (2008). 'Fiscal Policy and Financial Markets', *Economic Journal*, 118(533): 1971–85.

Arezki, R., V. Ramey, and L. Sheng (2016). 'News Shocks in Open Economies: Evidence from Giant Oil Discoveries', *Quarterly Journal of Economics*, 132(1): 103–55.

Barnett, S. and R. Ossowski (2003). 'Operational Aspects of Fiscal Policy in Oil-Producing Countries', in J. Davis, R. Ossowski, and A. Fedelino (eds), *Fiscal Policy Formulation and Implementation in Oil-Producing Countries*, 45–81. Washington, DC: IMF.

Bjerkholt, O. and I. Nicolescu (2004). 'Fiscal Rule Suggestions for Economies with Non-Renewable Resources: Norway and Venezuela', in G. Kopits (ed.), *Rules-Based Fiscal*

*Policy in Emerging Markets: Background, Analysis and Prospects*, 164–79. London: Palgrave Macmillan.

Collier, P. (2010). *The Plundered Planet*. New York: Oxford University Press.

Collier, P., M. Spence, F. van der Ploeg, and A. J. Venables (2010). 'Managing Resource Revenues in Developing Economies', *IMF Staff Papers*, 57(1): 84–118.

Corden, M. and P. Neary (1982). 'Booming Sector and De-industrialization in a Small Open Economy', *Economic Journal*, 92(368): 825–48.

Gollier, C. (2013). *Pricing the Planet: The Economics of Discounting in an Uncertain World*. Princeton, NJ: Princeton University Press.

Solimano, A. and D. Calderón Guajardo (2017). 'The Copper Sector, Fiscal Rules, and Stabilization Funds in Chile: Scope and Limits', WIDER Working Paper 2017/53. Helsinki: UNU-WIDER.

Stern, N. (2007). *The Economics of Climate Policy*. Cambridge: Cambridge University Press.

van den Bremer, T. S. and F. van der Ploeg (2013). 'Managing and Harnessing Volatile Oil Windfalls', *IMF Economic Review*, 61(1): 131–67.

van den, Bremer, T. S., F. van der Ploeg, and S. Wills (2016). 'The Elephant in the Ground: Managing Oil and Sovereign Wealth', *European Economic Review*, 82: 113–31.

van der Ploeg, F. (2011). 'Natural Resources: Curse or Blessing?' *Journal of Economic Literature*, 49: 366–420.

van der Ploeg, F. and A. J. Venables (2011). 'Harnessing Windfall Revenues: Optimal Policies for Resource-rich Developing Economies', *Economic Journal*, 121: 1–31.

van der Ploeg, F. and A. J. Venables (2012). 'Natural Resource Wealth: The Challenge of Managing a Windfall', *Annual Review of Economics*, 4: 315–37.

van der Ploeg, F. and A. J. Venables (2013). 'Absorbing a Windfall of Foreign Exchange: Dutch Disease Dynamics', *Journal of Development Economics*, 103: 229–43.

Venables, A. J. (2016). 'Using Natural Resources for Development: Why Has It Proven So Difficult?' *Journal of Economic Perspectives*, 30(1): 161–84.

Venables, A. J. and S. Wills (2016). 'Resource Funds: Stabilizing, Parking and Intergenerational Transfer', *Journal of African Economies*, 25(Suppl. 2): ii20–ii40.

Wills, S. (2015a). 'Seven Principles for Managing Resource Wealth', OxCarre Working Paper 154. Oxford: Department of Economics, University of Oxford.

# 10

# The Copper Sector, Fiscal Rules, and Stabilization Funds in Chile

## Scope and Limits

*Andres Solimano and Diego Calderón Guajardo*

### 10.1 Introduction

The copper sector constitutes the main export product and the main source of fiscal revenues, foreign exchange, and national income in the Chilean economy. Its price is affected by both changes in productive demand (chiefly in the construction sector and industry in copper-importing countries) and speculation in metals markets. Copper price volatility can have major macroeconomic consequences for the internal economy. Since the 1980s and more aggressively in the 2000s, Chile has developed macroeconomic mechanisms to deal with this volatility.[1]

After the nitrate cycle faded away in the 1930s, copper started to play a pivotal role but mines were owned and exploited by foreign corporations (mostly originating in the United States) under often-generous tax policies and repatriation of dividends arrangements. In this context, 'Chilenization' (mixed public–private partnership) policies were developed in the 1960s to be followed by nationalization in 1971 (approved by the Chilean Congress with the unanimous support of all political parties) that gave the full responsibility of property and management of copper mines to the Chilean state. Those policies were initially maintained by the military regime after the coup of 1973 but then gradually reversed as the country embarked on a neoliberal

[1] The authors thank Alan Roe for the useful comments and suggestions he made on an earlier version of this chapter. Comments from Maximiliano Acevedo and Ciro Giraldez, Chilean Ministry of Finance, are also appreciated.

economic model. As early as 1974 a decree law (DL 600) to attract foreign investment was enacted that provided tax, repatriation, and property protection for foreign corporations. This was followed in the 1980s by a mining code for foreign corporations that paved the way for a steady privatization of the large-scale copper production sector. In the 2000s, the policy regime regarding mining established low royalties and moderate taxes for foreign direct investment along with generous leasing and concession agreements to private investors. However, the prevailing constitution, approved in 1980 and still ruling, maintained several provisos of the nationalization policies of 1971.

At macroeconomic level, the Chilean government created the copper stabilization fund (CSF) in 1987. The copper fund accumulated resources when the current price was higher than a reference price and drew down funds accumulated in previous periods when the current price was below that reference price. This was followed by a fiscal rule (2001) and later by a law of fiscal responsibility in 2006 which created the economic and social stabilization fund (ESSF) that replaced the existing CSF and created the pension reserve fund (PRF).[2] The ESSF is oriented to dampen the effects of changes in international copper prices and copper demand on domestic economic activity (output, investment, and employment), the balance of payments, and the fiscal budget. As the revenues from the copper sector are a main source of fiscal revenues, this would help ensure a more stable stream of revenues for the state and, in principle, avoid cycles of expansion and contraction in public expenditure.

The *fiscal rule* was to operate under the principle that 'permanent' spending should follow 'permanent' revenues (determining what is permanent in an uncertain and volatile world is not easy) and was expected to contain fiscal spending and isolate it from political and social pressures.

The rest of this chapter is organized as follows. Section 10.2 provides a brief historical background of the evolution of the copper sector from the 1930s to the present and focuses on the relationship between GDP and investment cycles and copper price cycles. Section 10.3 documents the evolving institutional regime governing the copper sector from the 1950s until the present. Section 10.4 shows the interplay between fiscal shocks, fiscal rules, and stabilization funds and provides additional detail on their origin, rules of operation, and expected benefits. Section 10.5 evaluates the existing macro framework in Chile to cope with external cycles, highlighting also some open questions regarding the optimal level of resource accumulation of these funds in an economy of high inequality and pending social demands and the prevailing discretion governing the use of funds in bad times.

[2] In addition to the ESSF and PRF there is an opaque fund managed by the defence ministry and Chile's armed forces to finance the acquisition of military equipment (see Section 10.3).

## 10.2 Brief Historical Background of a Mineral-exporting Economy and Cycles Related to the Copper Sector

Historically Chile has been reliant on natural resources such as nitrate, copper, coal, gold, wheat, and other commodities for its development process. Two long cycles of commodity dependence can be distinguished: the *nitrate commodity cycle* of boom and decline that started around 1880 until its eventual decay in the early 1930s, and a *copper cycle* that started in the 1930s and continues until today. In the period 1900–20, nitrate exports accounted for 65–80 per cent of total exports, which, in turn, represented nearly 40 per cent of GDP (Cariola and Sunkel 1982; Meller 2006). The government imposed an export duty on nitrate that financed nearly 50 per cent of total public expenditure. Eventually, the boom of Chilean natural nitrate came to an end when Germany managed to develop synthetic nitrate at a lower cost; as a consequence, the Chilean production of nitrate declined by nearly 75 per cent between 1928 and 1934 (Díaz et al. 2016), prompting an economic and social crisis in the country. Since the 1930s, copper started to replace nitrate as the main export commodity, mainly directed to the American market.

Owing to the adverse effects of the Great Depression in core economies that hit Chile very hard in the first half of the 1930s, the country switched from a commodity export-oriented growth pattern into an import-substitution growth and industrialization strategy. In the period 1950–70, the importance of exports in GDP declined sharply and fluctuated between 7 and 9 per cent. Copper income at that time represented 55–65 per cent of exports and 15–30 per cent of total fiscal revenues (see Table 10.1).[3]

**Table 10.1.** Annual average copper exports from Chile, 1960–2014

| | 1960–9 | 1970–9 | 1980–9 | 1990–9 | 2000–9 | 2010–14 | 1960–2014 |
|---|---|---|---|---|---|---|---|
| US$ millions (2014) | 3,582 | 5,017 | 4,978 | 7,695 | 23,316 | 42,754 | 11,994 |
| % of total exports | 68.9 | 63.9 | 45.8 | 37.8 | 45.7 | 53.9 | 52.5 |
| % of GDP | 8.1 | 9.4 | 9.4 | 8.3 | 14.9 | 16.4 | 10.6 |

*Source*: authors' elaboration based on DIPRES (2016).

[3] In the last fifty-five years, reflecting Chile's effort at diversifying its export base, the share of copper in total exports declined from nearly 70 per cent in the 1960s to 46 per cent in 2000–9, followed by a small increase again by 2010–14 when the share moved slightly above the 50 per cent mark (see Table 10.1). However, as a proportion of gross domestic product (GDP), copper exports doubled from 8.1 per cent in 1960–9 to 16.4 per cent in 2010–14.

### 10.2.1 *The Copper Sector, Policy Shocks, and Macroeconomic Cycles*

In macroeconomic terms, we observe a broad positive correlation between growth and investment cycles on the one hand and the evolution of real copper prices on the other, perhaps acting with some lags (see Figure 10.1). The dominant pattern is that years of slow GDP expansion, including episodes of negative growth (recessions and depressions) have often followed a reduction in real copper prices. These cycles have also been accompanied by adjustments in other variables such as investment, employment, and the real exchange rate.[4] Because of the importance of the copper sector and the country's trade and financial openness, Chile has developed a set of fiscal rules and SWFs to deal, with varying degrees of success, with external shocks, especially volatility of copper prices.[5]

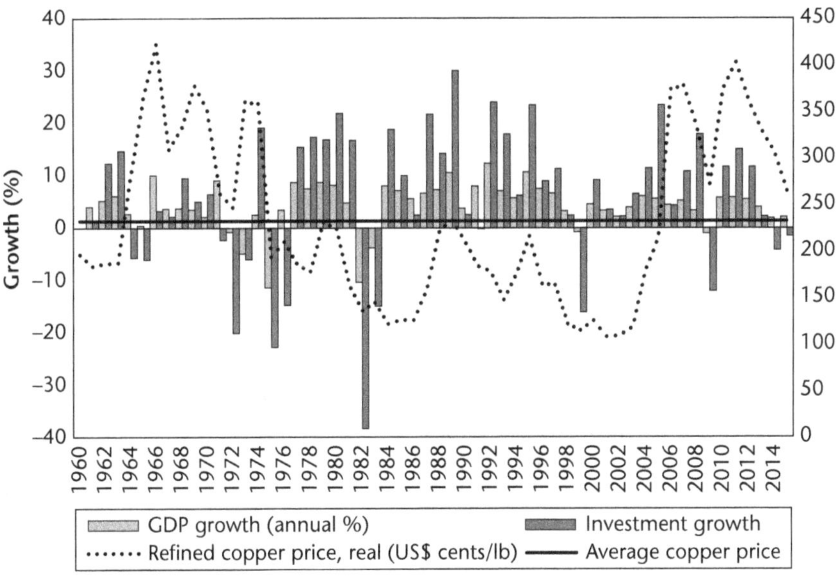


**Figure 10.1.** GDP growth, investment, and real copper price, 1960–2015

*Notes*: Deflator: US producer price index (all commodities).

*Source*: authors' elaboration based on World Development Indicators, Central Bank of Chile data, and Cochilco (2016).

[4] Variations in the price of copper are closely related to external economic activity (chiefly in the construction sector and industry), and also to financial speculation in commodities in what looks like a mean-reversing process.

[5] Several positive evaluations (although somewhat incomplete in their scope) of these rules have indicated that these policies are a successful example of natural resources boom management from a macroeconomic perspective (Céspedes et al. 2014; Frankel 2011). At the same time, these studies make some recommendations regarding forecasting methodology. However, these evaluations tend to overlook the opportunity cost involved in the accumulation of resources in stabilization funds and neglect the lack of rules governing the use of funds. Moreover, in terms of its output

A well-established regularity is that investment is highly pro-cyclical, moving closely with but adjusting more than GDP, particularly in recession years (see Figure 10.1). The two main recessions/depressions in the period 1970–2015 took place in 1975 (associated with anti-inflationary shock policies and a decline in the price of copper) and in 1982–3 (led by a domestic financial crisis following unregulated banking deregulation and a currency crisis) (see Solimano 2012). In these two short but deep 'depressions', GDP declined sharply, on average, between 12 and 16 per cent, and investment contracted further in the range of 20–35 per cent. Then, there were two other years of negative growth—1999 (a follow-up of the effects of the Asian and Russian crises) and 2009 (following the global financial crises)—in which GDP contracted between 1.2 and 1.5 per cent each year and investment declined between 10 and 15 per cent.[6] As could be expected, in these four episodes of recession and depression the cut in investment was several times greater than the decline in GDP. In turn, the durations of the recoveries/expansions after these recessions/depressions were of variable magnitude, the longest being the period 1984–2008 after the depression of 1982–3 and before the recession of 2009.

## 10.3 The Evolving Institutional Framework of the Mining Sector: The Cycle of Foreign Ownership, Nationalization, and de facto Privatization

Historically the institutional framework ruling the copper sector in Chile has been affected by the dominant role played by foreign corporations (mainly US) in the mining sector. Arrangements affecting prices, taxes, royalties, and other factors did not always benefit the Chilean nation. In fact, following World War II (1939–45) and at the start of the Korean War (1950–3), the US government decided, unilaterally, to fix the copper price (at 24.5 cents/lb), which was a level less than half the market price (54.5 cents/lb). As the United States was the main buyer of Chilean copper, this led to losses of export earnings and was a major concern for the Chilean authorities (Millán 2006). A commission was sent to Washington DC to claim retribution for this

composition, the Chilean economy is still heavily concentrated in mining and services with a reduced share of the manufacturing sector in GDP, suggesting tendencies to de-industrialization (Solimano and Schaper 2015).

[6] Historically, the mining industry has been the sector that contributes the most to total investment given its capital intensity and the lumpy nature of investment. In the period 2008–14, mining investments represented nearly a quarter of total gross fixed capital formation. Mining accounted for roughly 30 per cent of total foreign direct investment from 1974 and 2015, equivalent to US$37,700 million.

decision and a Washington Agreement was signed in 1951. This arrangement allowed the Chilean State, through the Central Bank, to dispose of 20 per cent of copper production by buying it from American companies in Chile at the fixed price and then selling it at its market price to other countries. This unfavourable agreement did not appease Chilean demands and in 1955 the New Deal law was established in order to promote copper production and to regulate the tax structure of the mining industry. The shortcomings of this agreement in the following years paved the way for the Chilenización del Cobre ('Chilenization of copper') adopted under the Christian Democrat President Eduardo Frei Montalva (1964–70). The Chilenization law no. 16.425 of January 1966 created copper public–private joint companies in which the Chilean state would buy 51 per cent of the property of the copper mines over a period of fifteen years. This was the first serious, albeit cautious, attempt to recover Chile's natural resources base for national development.

In 1970, the political winds had changed in Chile and a socialist president, Salvador Allende, was elected by popular vote (despite active Nixon–Kissinger covert destabilization efforts, now documented by official United States Congress records). Once in government, Allende sent to Congress a law to nationalize copper mines. In July 1971, the country's Constitution was modified with the unanimous support of representatives of all political parties, from conservatives to communists, and copper mines were nationalized.[7]

On 11 September 1973, the Allende government was ousted by a bloody military coup led by General Augusto Pinochet. Once in power, the military junta started talks with copper companies in the United States (Anaconda and Kennecott) to rapidly compensate American corporations affected by nationalization approved by the previous parliament (that had been closed by the army after the military coup). Friendly policy towards foreign investment in the copper sector and other activities started to develop, although the nationalization of copper mines was not officially reversed by the military government (Solimano 2012). As mentioned before, in 1976 CODELCO was established with the mission of managing the copper companies that were nationalized in the Allende period. Contrary to what happened to other public companies, CODELCO was not included in the several massive privatization programmes undertaken during Pinochet's dictatorship. That same year, a highly controversial *secret* law dating from 1958—the Copper Restricted Law—directed to fund the acquisition of military equipment and weapons out of copper-sector profits was modified. As noted before, in 1976 the new

[7] A technical commission of the Chilean government made an assessment that accumulated earnings of foreign mining companies since their creation were 'excessive' and the Chilean government refused American demands for further compensation. This action reinforced Richard Nixon's determination to destabilize the Allende government.

copper law (maintaining its secret status) stated that '10 per cent of *gross revenues* of CODELCO' were to be directed to finance military purchases by the Chilean armed forces *without* the standard overseeing (formal accountability) by parliament and the general comptroller of the country. This law is not subject to direct congressional oversight in its operation.[8]

Two key elements of the mining legislation in Chile are the Organic Law of Mining Concessions (1982) and the Mining Code (1983). Although the 1980 constitution states that the Chilean state has 'absolute, exclusive, inalienable, and imprescriptible domain over all mines and resources underneath Chilean soil', the combination of the Organic Law and the Mining Code gives, in practice, priority to private property rights and mining concessions (leasing arrangements granted by the state). Leasing rights can be extended indefinitely by the holders of leasing rights (concessioners) by paying very small fees (less than US$10 per hectare per year according to specificities of the surface) with a renovation preference for the incumbent even over the claims of the Chilean state. Indeed, if the Chilean state were to nationalize private mining companies, it would have to compensate private companies with a monetary sum equal to the present discounted value of the estimated future cash flows of the project. In reality, one of the effects of this legislation has been a progressively denationalized/privatized property of mines concentrating increasingly in big private companies that enjoy the incumbent status.

It is important to note that, owing to its corporate governance and tax regime, CODELCO—as with all the public companies—is subject to an extra 40 per cent tax over profits which represents an extra extraction of its net revenues besides the resources transferred to the armed forces through the secret copper law. Currently, with relatively depressed copper prices, this framework has turned CODELCO into a company with a potentially vulnerable financial situation that is affecting its ability to undertake its investment planning for future years. Despite its huge profits, CODELCO's debt grew by 247 per cent between 2004 and 2015 as it transferred to the state treasury an accumulated amount of US$56,000 million in the same period (the highest financial transfer in CODELCO's history).

The history of the mining-sector legislation since the military regime included, as indicated, DL 600, established in 1974 and still in place today, which allows unlimited repatriation of profits by foreign investors and the option of an invariant tax structure for investments made under this regulation. Tax invariability and loopholes have made Chile one of the most attractive mining

[8] Currently, several previous commanders-in-chief of the army and high-ranking generals are under investigation, accused of illicit and corrupt management of funds arising from the secret copper law.

destinations for foreign investors. In response, foreign direct investment in mining started to pick up a few years after the return to democracy (rather than during the period of authoritarian rule itself) and the post-Pinochet governments have maintained, on the whole, the mining legislation enacted by the General.[9]

In 2003, the Chilean Congress appointed a special commission to study the possibility of introducing new private-sector mining taxation arrangements in response to a series of allegations of persistent practices of tax avoidance and after a decade of massive foreign investment in the mining sector.[10] This congressional special commission identified various legal mechanisms used by private corporations to declare negative profits, such as transfer pricing, fraudulent forward manipulations, accelerated depreciation, and

**Table 10.2.** Parametric changes in relevant variables and effects on effective and structural fiscal incomes by sector, 2014

| Variable | Unit of change | Effective income | | Structural income | |
|---|---|---|---|---|---|
| | | US$ million | % of effective income | US$ million | % of structural income |
| CODELCO | | | | | |
| Copper price (CODELCO) | 1 cent (US$) | -36.1 | -1.5 | -0.1 | 0.0 |
| Copper production | 1% change | -17.0 | -0.7 | -21.7 | -0.7 |
| Nominal exchange rate | 1 peso (CLP$) | -4.2 | -0.2 | -5.7 | -0.2 |
| Production cost per unit | 1 cent (US$) | -36.1 | -1.5 | -36.1 | -1.1 |
| Reference copper price | 1 cent (US$) | 0.0 | 0.0 | -36.0 | -1.1 |
| GMP-10 | | | | | |
| Copper price (BML) | 1 cent (US$) | -11.4 | -0.5 | 0.3 | 0.0 |
| Copper production | 1% change | -34.3 | -1.4 | -33.5 | -1.5 |
| Nominal exchange rate | 1 peso (CLP$) | -4.3 | -0.2 | -4.5 | -0.2 |
| Production cost per unit | 1 cent (US$) | -5.1 | -0.2 | -5.1 | -0.2 |
| Reference copper price | 1 cent (US$) | 0.0 | 0.0 | -11.8 | -0.5 |

*Source*: authors' elaboration based on DIPRES (2016).

[9] Accounting schemes (non-penalized) have been used for purposes of tax avoidance. Very favourable conditions for foreign investors were embedded in DL 600: 'It is good to be true, and the companies know it,' said Radomiro Tomic, a copper expert, former Chilean ambassador to the United States, and a prominent political leader who happened to be the Christian Democratic party's presidential candidate in the 1970 elections, running against Salvador Allende and Jorge Alessandri, to *The Washington Post* in 1982, reflecting on the business climate of those years (Diehl 1982).

[10] According to former Senator Jorge Lavandero Illanes (former head of the Mining Commission in Congress), only two out of forty-seven private copper companies paid taxes to the government between 1995 and 2002 (Lavandero Illanes 2004).

debt interest payments. After extended political discussion and lobbying by the mining companies, in 2005, the specific mining tax (SMT)—which taxed profits but not the value of sales or physical production, as in the traditional concept of royalty—was passed into a law.[11] Unfortunately, this SMT contributed only about 1 per cent of total fiscal revenues and 5 per cent of total mining taxes between 2006 and 2015. The fact that the royalty is levied on profits rather than on the value of production, along with the practices to diminish profits, may account for this meagre revenue.

## 10.4 A Macroeconomic Perspective: Orthodoxy, Fiscal Rule, and Stabilization Funds

Chile's economic performance in recent decades is often presented as a case of sound macroeconomic management by international financial institutions such as the World Bank and the IMF and the international financial community. The policies in place provide an almost textbook example of macroeconomic orthodoxy, combining inflation targeting, a fiscal rule, a free-floating exchange rate, and an open capital account and SWFs. The Fiscal Responsibility Law of 2006 is supported by the fiscal rule and two formal SWFs (to manage the budget surplus generated at the time of copper price booms): the ESSF which had accumulated, by mid-2016, nearly US$16 billion and the PRF, running a surplus of US$8.3 billion.[12] A third, relatively hidden fund is constituted by the surplus accumulated by the (secret) copper law aiming to benefit the armed forces, which has accumulated nearly US$5 billion. Thus, over US$30 billion are held in the three stabilization funds, not a minor amount for an economy with US$300 billion GDP.

The literature evaluating these arrangements suggests that they have succeeded in reducing economic volatility stemming from terms-of-trade shocks and other internal and external shocks affecting the macro economy. For example, Franken et al. (2006) show that the reduction in GDP volatility between 1991 and 2003 was associated with a reduction in the volatility of the monetary and fiscal stance. Larraín and Parro (2008) find that roughly 60 per cent of GDP volatility reduction is related to the fiscal rule and the prevailing flexible exchange-rate regime. Furthermore, De Gregorio and Labbé (2011) show that the Chilean economy has become increasingly resilient to copper price shocks since 1985, and especially since the 2000s

[11] Specifically, the specific mining tax is not a royalty because it taxes profits rather than revenues. Therefore, there is a double incentive to underestimate revenues and to overestimate costs.

[12] The ESSF is an amended and extended legal formalization of the Copper Revenue Compensation Fund established in 1987. The PRF is a special fund dedicated exclusively to guaranteeing the sustainability of state-funded pension payments.

owing to its macroeconomic framework. Céspedes et al. (2014) show that the policy mix adopted by Chile allowed a strong *countercyclical* fiscal response to the global financial crisis of 2008–9 (the countercyclical fiscal stance was not pursued after the decline in copper prices in 2013).

An early attempt to manage the impact of the volatility in copper prices was the Copper Revenue Compensation Fund (CRCF) created in 1987. This fund aimed at preventing the occurrence of fiscally induced economic cycles at the time of upward movements in international copper prices (e.g. governments have a propensity to pro-cyclically increase public spending at times of enhanced fiscal revenues). The CRCF was set to accumulate resources (savings) when the copper price was higher than a long-term reference price (set by the fiscal authorities); in turn, accumulated resources could be spent when the copper price was lower than the reference price. According to the Chilean budget office (DIPRES 2016), the CRCF was successful in managing price booms by reducing external debt and unsustainable government spending. The resources of the CRCF were used after the Asian crisis (1998–2003) and before that in 1993 and 1994.

Despite a prudent fiscal policy during the 1990s, a more sophisticated framework was later adopted to tie the hands of fiscal authorities. In 2001, the administration of President Ricardo Lagos created a fiscal rule that implied the calculation of a 'structural balance' (SB). This was intended to operate as a self-imposed policy of fiscal restraint. The idea behind the structural fiscal rule is simple but not without problems of implementation: fiscal spending should follow 'long-run' (*permanent*) values of the price of copper, GDP growth, and other key parameters affecting the budget. However, determining what is permanent can be tricky, particularly in the case of copper prices (a variable with a strong random component), and nor is potential GDP growth a simple concept to define and forecast.[13] It is apparent that a main role of the fiscal rule is of a political–economic nature: it provides a (scientific) disciplinary device, supported by the wisdom of economic experts and aided by the principles of inter-temporal macroeconomic theory, for governments to contain social and political demands for higher public spending, particularly in a highly inegalitarian society such as Chile.

The Fiscal Responsibility Act was approved under the first administration of President Michelle Bachelet in 2006. The SB rule improved its methodology of income measurement and in 2011, under the administration of President Sebastián Piñera, a second generation methodology was applied, giving rise to the cyclically adjusted (fiscal) balance (DIPRES 2011). Again, an advisory fiscal council was convened in 2013 to monitor the operation of the new rule.

[13] Expert committees were appointed and asked to determine the permanent values of key parameters to be used by fiscal authorities.

The structural fiscal balance rule allows government spending expansions only when there is an increase in structural incomes driven by long-term copper prices and potential GDP growth (see Box 10.1 and Table 10.2). If the target for the structural budget is set at 1 per cent of GDP, this means that expenditure for the next year will be 1 per cent of GDP lower than structural income. To obtain structural estimates of potential GDP and long-run copper prices, members of expert teams provide their estimates to the Ministry of Finance. When the effective balance is larger than the SB, the surplus is saved in the ESSF. When the effective balance is smaller than the SB, the deficit is covered with funds from the ESSF. However, the exact value set for the SB in the fiscal rule is still discretional and largely depends on the appraisal of relevant macroeconomic conditions by the Ministry of Finance. The first surplus target for the SB was set by the authorities at 1 per cent of GDP during the period 2001–7. Three arguments to choose this number were given at that time: (i) the public sector was a net debtor (11 per cent of GDP); (ii) there were potential increases in future fiscal liabilities related to pension guarantees, infrastructure investment needs, and potential borrowing constraints; and (iii) it was decided that the Central Bank needed to be recapitalized.[14] Because of progress in funds accumulation during the previous period, the surplus target was diminished to 0.5 per cent of GDP for 2008 and again to 0 per cent in January

**BOX 10.1** FISCAL SENSITIVITY TO CHANGES IN COPPER PRICES AND COPPER PRODUCTION

Chile's fiscal position is strongly affected by changes in both copper price and copper production. Small changes in relevant variables have a significant impact on effective fiscal incomes. However, the use of a medium-term reference price prevents the effects of cyclical shock in structural fiscal incomes. The budget office (DIPRES) simulations show that a change of 1 cent (US$) in the copper price reduces fiscal revenues by US$37.5 million. The reference price is useful to deal with uncertain global conditions, but structural fiscal incomes also depend on internal copper production. A reduction of 1 per cent in copper production of CODELCO has an impact of US$17 million on structural fiscal incomes. The same change in private mining production can reduce fiscal incomes by US$34 million. Production is affected not only by international demand through price changes but also by local conditions such as rising energy costs, water availability, the decreasing ore grade of minerals, and increasing social and environmental issues as a consequence of natural resources activities (see Table 10.2).

[14] Because of the crisis of 1982–3, one of the most severe in Chile, several banks went bankrupt and the government had to bail out banks with a cumulative fiscal cost equivalent to approximately 35–40 per cent of GDP.

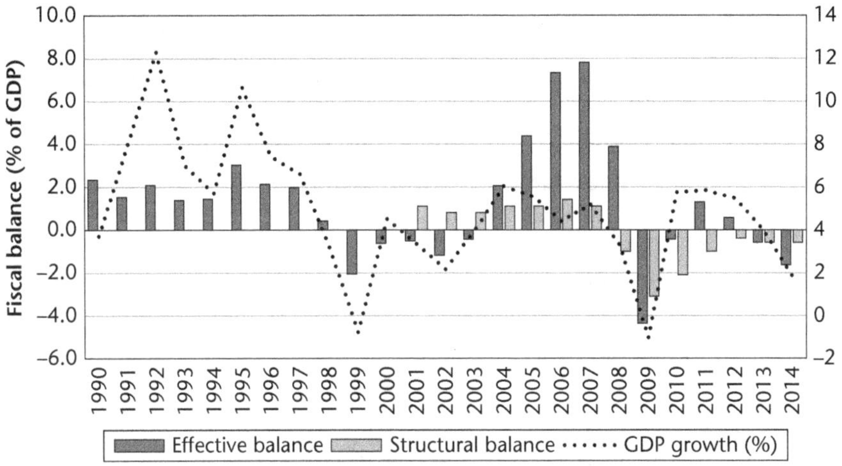


**Figure 10.2.** Fiscal balance, 1990–2014
*Source*: authors' elaboration based on DIPRES (2015a).

2009 to provide fiscal space for a fiscal expansion to counteract the effects of the global financial crisis. This value has been maintained since then.[15]

In Figure 10.2, we can observe the evolution of the SB rule (based on the fiscal balance evaluated at the estimated long-run values of copper prices and GDP) over the business cycle. We can observe, as expected, that the evolution of the SB is smoother than that of the current (effective) balance. The changes (deterioration) in the current (effective) balance were significant during the Asian financial crisis of 1997 (at that time there was only the CRCF) and also during the global crisis of 2008. In 2009, an aggressive countercyclical fiscal expansion was adopted by the fiscal authorities.[16] Between 1990 and 2008, there had been, on average, a surplus in the effective fiscal balance in Chile of 1.9 per cent of GDP. However, by contrast in 2009 the budget ran an effective deficit of 4.4 percentage points of GDP as a consequence of the cyclical decline in revenues and the effect of a large fiscal expansion.

It can be argued that the fiscal rule prevented an unsustainable increase in public expenditure in the period of the copper price boom through 2007, and before the global turbulence of 2008–9. In fact, between 2003 and 2007, when the copper price increased by 220 per cent in real terms, the structural fiscal balance averaged 1.1 per cent of GDP whereas the effective surplus was 4.2 per cent of GDP in the same period. This means that around 3 percentage

[15] The structural balance achieved –3.1 per cent of GDP (deficit) in 2009 and the effective balance was –4.4 per cent of GDP (deficit).

[16] It is worth noting that the IMF at that time was recommending countries with fiscal space to adopt expansionary fiscal policies.

points of GDP were saved by the Chilean state in the SWFs during 2003 and 2007.[17]

As mentioned previously, after the global financial crisis, a combination of countercyclical fiscal policy and expansive monetary policy was adopted in Chile. According to Céspedes et al. (2014), close to US$4 billion from the ESSF (2.8 per cent of GDP) was assigned to the economic recovery in January 2009 (later during that same year more money was spent) with several measures driven by the Central Bank, such as stopping the accumulation of reserves, easing collateral requirement of repo operations, and temporarily loosening rules regarding bank reserves. As we can see in Figure 10.2, the SB has remained consistently negative since then.

In February 2010, a major earthquake followed by a tsunami struck the centre-south of Chile, causing hundreds of deaths and major physical destruction of housing and infrastructure. Reconstruction became the first policy priority of the nation, and public expenditure grew by 5.5 per cent in real terms in 2010. The Piñera administration nonetheless committed to reduce the structural deficit, and a –0.6 per cent of GDP was achieved by the end of the Piñera government in 2014. Since then, the authorities have expressed their intention to return to structural surpluses but predictions are that this may not be achieved before 2020 because of the expected lower price of copper and slower GDP growth.

### 10.4.1 *Asset Management of SWFs*

Because of the copper price boom of 2002–7, the ESSF and PRF both increased their savings in that period, although the government later drew almost US$9 billion from the ESSF throughout 2009 to finance the fiscal expansion of that year. The resources of the ESSF can be used at any time to complement current fiscal incomes in periods of fiscal deficits. They can also be used for the amortization of public debt and to capitalize the PRF. Figure 10.3 presents the accumulation pattern of ESSF and PRF and also the average annual rate of return of each fund. In fact, we can observe that despite its good profitability before the global financial crisis, negative results have been realized in more recent years, mainly in the ESSF. Overall, the total return since 2007 has been in line with a fixed income portfolio with a nominal rate of return of 2.28 per cent for the PRF and 3.14 per cent for the ESSF.[18]

[17] Between 2003 and 2007, average GDP growth was 5 per cent (average GDP growth declined to 3.4 per cent in the period 2008–15).

[18] Rate of return is calculated based on the time-weighted rate of return methodology devised by the Ministry of Finance.

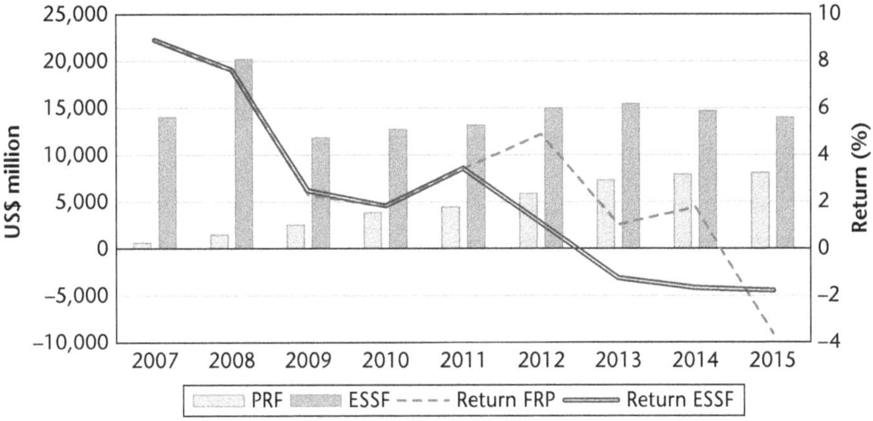


**Figure 10.3.** Sovereign wealth funds
*Source*: authors' elaboration based on DIPRES (2015b).

According to the Fiscal Responsibility Law of 2006, the Ministry of Finance is responsible for both funds and is assisted by a financial committee of experts to define the investment policy and structure of the funds.[19] In turn, the Central Bank, in its role of fiscal agent, is responsible for materializing the investment guidelines defined by the Ministry of Finance and choosing the actual asset allocation and the strategic composition of the funds, and managing the cost related to its administration.

The investment policy of the funds seeks to maximize their market value subject to a certain level of risk exposure. Portfolio composition is tailored to maintain high liquidity, low credit risk, and low volatility investments in order to accommodate future fiscal deficits and avoid losses for the funds. Indeed, the current investment policy of the ESSF and the PRF relies mainly on bonds and fixed-income instruments. Bonds accounted for almost 70 per cent of total assets in both funds in 2010 and 2011. However, following the recommendation of the financial committee, since 2012 in the case of the PRF and 2013 for ESSF, equities have been part of both portfolios in order to increase the profitability of funds, accounting for no more than 18 per cent in the case of the ESSF and no more than 8 per cent for the PRF. All instruments are in reserve currencies such as the Swiss franc, the yen, and the euro but the funds are denominated in US dollars (for a more detailed version of this section see Solimano and Calderón Guajardo 2017).

[19] Decisions and recommendations made by the financial committee are not mandatory for the Ministry of Finance. The president of the committee is chosen by the fiscal authority. The committee also reports to Congress about asset management of SWFs.

### 10.4.2 *Macroeconomic Effects: Growth Volatility and Country Risk Reductions*

As noted earlier, several authors have documented the fact that the macroeconomic framework of Chile has been successful in reducing output volatility since 1990. Indeed, as we can see in Figure 10.4, based on Céspedes et al. (2014), there has been a significant reduction in GDP growth volatility since 1984, with a relatively stable period (lower output volatility) even after the global financial crisis of 2008–9.[20]

The combination of fiscal rules, two SWFs, flexible exchange rates, and an independent Central Bank is said to have contributed to lower output volatility in Chile—certainly relative to most other comparable countries. This may be true, although it is not simple to ascertain the contribution of each component of the policy package to lower output volatility. In addition, it is fair to say that the actual fiscal and monetary policy decisions following the judgement of specific authorities—rather than the pure mechanical operation of supposedly impersonal policy rules—are likely to have contributed also to lower box output volatility.

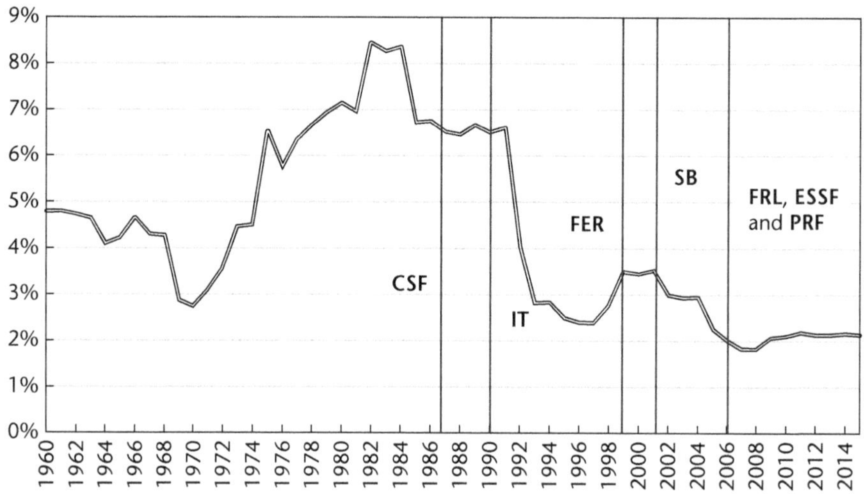


**Figure 10.4.** GDP growth volatility (ten-year rolling window) and macroeconomic policies

*Notes*: CSF, copper stabilization fund; IT, inflation target; FER, flexible exchange rate; SB, structural balance; FRL, fiscal responsibility law; ESSF, economic and social stabilization fund; PRF, pension reserve fund.

*Source*: authors' elaboration based on Díaz et al. (2016) and Central Bank of Chile data.

[20] To measure volatility, Céspedes et al. (2014) compute the ten-year rolling window average of the standard deviation of GDP growth.

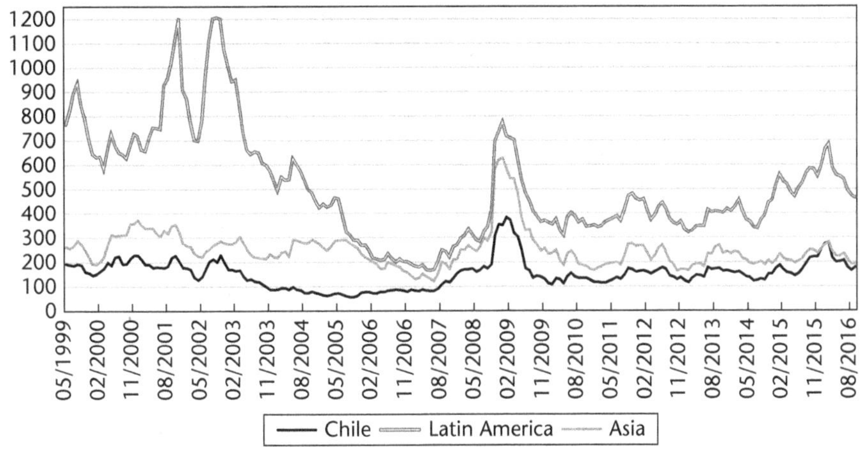


**Figure 10.5.** Country risk spreads (1999–2016, basis points)
*Source*: authors' elaboration based on Central Bank of Chile data.

Another benefit claimed in the literature on the Chilean fiscal rules and SWFs is their contribution to a reduction in the country risk (Larraín and Parro 2008; Marcel 2010; Schmidt-Hebbel 2012). The argument posits that as the government develops its credibility through an independent Central Bank and a solid fiscal position—along with an important implicit source of collateral, namely the ESSF—the risk perception of the economic agents decreases. One manifestation of this, following Marcel (2010), is depicted in Figure 10.5 where we can observe that Chile's sovereign risk spreads have diminished in absolute terms and compared with Latin America and Asia. Although we cannot assume a causal relationship between the adoption of this set of macro rules and the reduction of sovereign risk spreads, we can observe a slight reduction in the sub-period previous to the global financial crisis, and a level considerably lower than in Latin America in the whole period. Moreover, we can see that after the global crisis the level of Chile's country risk spreads returned more rapidly to pre-crisis levels, whereas Latin America continues with its pre-crisis levels.

## 10.5 Conclusions

Historically, macroeconomic cycles in Chile have been led by fluctuations in the prices of its main mineral and agricultural commodities. To counteract this trend, between the 1940s and 1970s the country tried to boost import-substitution industrialization with a productive role attached to the state. However, these industrialization efforts were reversed in the second half of the 1970s as Chile embarked on a neoliberal development strategy (Solimano 2012). As the economy opened to foreign trade more aggressively in the 1970s and 1980s, the relative importance of copper exports as a proportion of total exports declined from nearly 80 per cent in the early 1970s to around 50 per cent nowadays, reflecting the effects of export diversification and a sharp increase in total export volumes. However, in spite of diversification, copper has remained a critical source of fiscal revenues and foreign exchange. Since the 1980s, the authorities have developed various mechanisms to dampen the effects of shocks in the copper price on the domestic business cycle.

A CSF was created in 1987, and then a fiscal rule in 2001. The fiscal rule was expected to constrain possible spending propensities of the fiscal system and delink the domestic economy from the sometimes-disruptive effects of terms-of-trade shocks. In 2006, an ESSF and a PRF were established as part of a broader fiscal responsibility law. There is a relative consensus that these mechanisms have contributed to providing fiscal predictability and have also contained the impact of external shocks on fiscal spending and the business cycle. In addition, it is claimed that this macro framework has contributed to reducing output volatility, thereby reducing borrowing costs and lowering country risk premiums. The IMF and the international financial community often portray Chile as a successful example of the benefits of having fiscal rules, stabilization funds, inflation-targeting, an independent central bank, and flexible exchange rates.

The analysis in this chapter recognizes progress in these areas but also highlights loose ends in the operation of these funds and draws attention to some trade-offs not always recognized in the laudatory evaluations of the Chilean case. In particular, we stress that rules leave substantial space for discretion and their implementation requires non-trivial judgements. Specific features need to be addressed by policy makers. First, the fiscal rule requires predicting, with accuracy, long-run (or permanent) values for key parameters such as copper prices, and potential GDP growth. This is not a simple task and has required a proliferation of expert committees and frequent revisions affecting the operation of these rules. Second, the value for the structural fiscal budget has been changed several times by the authorities since 2001, thereby reducing its anchoring role on expectations and policy predictability. Third, the stabilization funds (particularly ESSF) are asymmetric in their operation rules. On the one hand, there are clear rules of accumulation; on the other hand, no rules exist for *using* the resources of these funds, say to counteract an economic downturn or recession (tied, for instance, to the level of the output gap or the unemployment rate). As of now, the decision to withdraw resources from the SWFs depends on the discretionary judgement of the fiscal authority. In addition, it is not possible to track where the resources are finally spent (e.g. on consumption or investment), and therefore evaluate the efficacy of countercyclical fiscal policies and some of their long-run effects. Fourth, it is unclear what would be the optimal level of resource accumulation in various stabilization funds. Currently, the ESSF, PRF, and defence fund have assets of nearly 10 per cent of GDP (this does not include the international reserves held by the Central Bank, which would add, roughly, another 10 per cent of GDP). This raises the question of a possible tendency towards over-insurance in fiscal management. Fifth, Chile is a country whose levels of public spending in education, health, pensions, and other social sectors as a share of GDP are consistently below those of the OECD and other Latin American economies of middle and large size. This suggests an opportunity cost of over-investing in stabilization funds in terms of over-restricted social spending in a country with high indices of income and wealth inequality (Gini coefficients for income above 50 per cent and for wealth over 70 per cent; Solimano 2016).

Finally, we note that Chile maintains very generous legislation towards foreign direct investment and domestic companies in the mining sector. This includes low royalties along with the perpetual renewal (at almost no cost) of licences granting mining exploitation rights to incumbent private corporations at the expense of forgone revenues for the Chilean state. This has led to a highly privatized and denationalized copper sector. CODELCO, the state-owned mining company, lacking an important degree of financial autonomy from the treasury, is forced by law to provide 10 per cent of its gross revenues to the armed forces for the acquisition of military equipment and weapons. This arrangement affects the ability of CODELCO to finance its capital investment projects with its own internal resources and also reduces the availability of fiscal revenues (transferred from CODELCO to the treasury) that could otherwise be directed to social spending and public infrastructure.

## References

Cariola, C. and O. Sunkel (1982). *Un Siglo de Historia Económica de Chile, 1830–1930. Dos Ensayos y una Bibliografía* [*A Century of Chilean Economic History, 1830–1930: Two Essays and a Bibliography*]. Madrid: Ediciones Cultura Hispánica.

Céspedes, L., E. Parrado, and A. Velasco (2014). 'Fiscal Rules and the Management of Natural Resource Revenues: The Case of Chile', *Annual Review of Resources Economics*, 6: 105–32.

Cochilco (2016). 'Yearbook: Copper and Other Mineral Statistics, 1996–2015'. Available at: https://www.cochilco.cl/Lists/Anuario/Attachments/15/cochilco_anuario_2015.pdf.

De Gregorio, J. and F. Labbé (2011). 'Copper, the Real Exchange Rate and Macroeconomic Fluctuations in Chile', in R. Arezki, T. Gylfason, and A. Sy (eds), *Beyond the Curse: Policies to Harness the Power of Natural Resources*, 203–33. Washington, DC: IMF.

Díaz, J., R. Lüders, and G. Wagner (2016). *La República en Cifras. Chile 1810–2010* [*The Republic in Numbers: Chile, 1810–2010*]. Santiago: Ediciones Universidad Católica de Chile.

Diehl, J. (1982). 'Exxon, Others Find Projects Awry in Chile', *The Washington Post*, 3 October. Available at: https://www.washingtonpost.com/archive/politics/1982/10/03/exxon-others-find-projects-awry-in-chile/590c3c43-4470-4c0a-8751-09641436093c/.

DIPRES (2011). 'Una política fiscal de balance estructural de segunda generación para Chile' ['A Fiscal Policy of Structural Balance for Second Generation Chile']. Public Finance Studies Series 18. Chile: Ministerio de Hacienda. Available at: http://www.dipres.gob.cl/572/articles-81713_doc_pdf.pdf.

DIPRES (2015a). *Informe de Finanzas Públicas: Proyecto de Ley de Presupuestos del Sector Público para el año 2016* [*Report of Public Finances: Project of the Law of Public Sector Budgets, 2016*]. Chile: Ministerio de Hacienda.

DIPRES (2015b). *Informe Anual de los Fondos Soberanos* [*Annual Report of the Sovereign Wealth Funds*]. Chile: Ministerio de Hacienda.

DIPRES (2016). 'Evolución, administración e impacto fiscal de los ingresos del cobre en Chile' ['Evolution, Administration and Fiscal Impact of Copper Revenues in Chile']. Public Finance Studies Series 23. Chile: Ministerio de Hacienda.

Frankel, J. (2011). 'A Solution to Fiscal Procyclicality: The Structural Budget Institutions Pioneered by Chile', NBER Working Paper 16945. Cambridge, MA: NBER. Available at: http://www.nber.org/papers/w16945.pdf.

Franken H., G. Le Fort, and E. Parrado (2006). 'Business-cycle Responses and the Resilience of the Chilean Economy', in R. Caballero, C. Calderón, and L. F. Céspedes (eds), *External Vulnerability and Preventive Policies*, 71–108. Santiago: Central Bank of Chile.

Larraín, F. and Parro, F. (2008). 'Chile menos volátil' [A Less Volatile Chile]. *El Trimestre Económico*, 75(299): 563–96.

Lavandero Illanes, J. (2004). *Una política para el Cobre chileno* [*A Policy for Chilean Copper*]. Santiago: Impresos Prólogo. Available at: https://recuperaciondelcobre.files.wordpress.com/2011/04/una-polc3adtica-para-el-cobre-chileno.pdf.

Marcel, M. (2010). 'The Structural Balance Rule in Chile: Ten Years, Ten Lessons', Discussion Paper IDB-DP-289, June. Washington, DC: Inter-American Development Bank. Available at: http://services.iadb.org/wmsfiles/products/Publications/37889337.pdf.

Meller, P. (2006). *Un Siglo de Economía Política Chilena (1890–1990)* [*A Century of Chilean Political Economy, 1890–1990*]. Santiago: Uqbar Editores.

Millán, A. (2006) *La minería metálica en Chile en el siglo XX* [*Metal Mining in Chile in the Twentieth Century*]. Santiago: Editorial Universitaria.

Schmidt-Hebbel, K. (2012). 'Fiscal Institutions in Resource-rich Economies: Lessons from Chile and Norway', Documento de Trabajo 416. Santiago: Instituto de Economía, Pontificia Universidad Católica de Chile. Available at: http://epge.fgv.br/conferencias/commodity-prices/files/KlausSchmidt-Hebbel.pdf.

Solimano, A. (2012). *Chile: The Neoliberal Trap. The Post Pinochet Era*. Cambridge: Cambridge University Press.

Solimano, A. (2016). *Global Capitalism in Disarray: Inequality, Debt, and Austerity*. Oxford: Oxford University Press.

Solimano, A. and M. Schaper (2015). 'The Paradoxes of Chilean Economic Development: Growth, Inequality, Deindustrialization and Sustainability Risks', in A. Hansen and U. Wethal (eds), *Emerging Economies and Challenges to Sustainability*, 162–76. London: Routledge.

Solimano, A. and D. Calderón Guajardo (2017). 'The Copper Sector, Fiscal Rules, and Stabilization Funds in Chile: Scope and Limits', WIDER Working Paper 2017/53. Helsinki: UNU-WIDER.

# 11

# Oil Discovery and Macroeconomic Management

## The Recent Ghanaian Experience

*Mahamudu Bawumia and Håvard Halland*

## 11.1 Introduction

Much of the extensive literature on the 'resource curse' phenomenon suggests a negative correlation between national resource endowments and economic growth (Sachs and Warner 1995, 2001), and literature subsequent to Sachs and Warner has concentrated on identifying various mechanisms, including real exchange rate appreciation, through which natural resource wealth impacts on growth.

Ghana's situation since 2007 provides a unique opportunity to consider the resource curse hypothesis: it also provides limited support for some versions of this hypothesis, but ultimately tells a simpler story of insufficient fiscal and monetary discipline. Crucially—and contrary to, for example, Botswana—Ghana was not able to manage expectations sufficiently after oil was discovered: consistent with van der Ploeg's (2011) 'anticipation of better times' hypothesis. It is also consistent with Atkinson and Hamilton (2003), who conclude that the combination of natural resource rents and high government consumption provides an explanation for the curse, and with Collier and Hoeffler (2009), who argue that the combination of rents from the extractive industries and open democratic systems is associated with slower growth unless there are enough checks and balances. Ghana's large increase in sovereign debt after the discovery of oil reflects the concerns of Mansoorian (1991), who suggests that an abundance of natural resources may encourage countries to assume unsustainable levels of debt.

Tightly fought elections in 2008 and 2012, in the context of expected or newly available oil revenues, generated a situation where electoral promises trumped the need to manage expectations, generating spending pressures that ultimately could not be contained. Institutional weakness, often proxied by aggregated indexes of institutional quality, took on very precise forms in Ghana. These specific forms are discussed in detail later.

Ghana's case also corresponds well with arguments from the political economy of macroeconomics, surveyed by Persson and Tabellini (2000). A fundamental insight from this literature is that the implicit discount rate used by politicians may exceed the rate of interest by the probability of the politicians being removed from office. Hence if a political faction expects to be expelled from office in the near future, it will extract oil or minerals much faster than is socially optimal and will also borrow against future oil revenues (van der Ploeg 2011).

Bleaney and Halland (2016) do not find evidence that natural resource wealth in general promotes fiscal indiscipline. However, their econometric model, based on the most disaggregated fiscal data currently available, does not explain the performance of some outliers such as Ghana and Mongolia. The Ghana case example therefore extends what can be learned about the resource curse based on the prevailing econometric analysis.

Section 11.2 of this chapter elaborates some of the key developments in the period leading up to the discovery of oil, in 2006. Section 11.3 provides the context of the policy choices faced by the government in the subsequent period, after oil had been discovered and the legal framework was put in place to manage oil revenues. Sections 11.4 and 11.5 examine the deterioration of public finances and the large accumulation of debt following the discovery of oil. Section 11.6 analyses the decline in capital expenditure following Ghana's oil discovery. Section 11.7 examines Ghana's declining economic growth following the oil discovery, while Section 11.8 examines the possible increase in corruption post-oil discovery. Section 11.9 places the developments in Ghana in the context of the political system and in particular the specific context of narrowly won elections. Section 11.10 asks the question 'What could Ghana have done differently?' Section 11.11 concludes.

## 11.2 Ghana before Oil

From the year 2000 until around 2012, Ghana was one of the stars of the 'Africa rising' story, the toast of the international development community, and a benchmark for other African countries in the areas of democracy and development. In particular: (i) it had enjoyed stellar GDP growth between

2001 and 2008; (ii) it had received substantial HIPC debt relief to the tune of $4.2 billion after reaching the HIPC completion point in 2004; (iii) it had benefitted from the experiences of countries such as Nigeria, and had put together a strong framework to avoid the oil curse: especially the Petroleum Revenue Management Act (PRMA) 2011 (Act 815, Republic of Ghana 2011) which sets the key parameters for the accounting and collecting of petroleum revenues; and (iv) it had a good record of democratic governance, with a free press, rule of law, and reasonably strong institutions.

One would therefore have expected Ghana to be on the list of sub-Saharan African countries able to avoid the natural resource curse. However, even before oil production began in 2011, and increasingly thereafter, Ghana found itself embroiled in all the problems of the curse. In August 2014 it requested a bailout from the IMF.

In a recent earlier paper (Bawumia and Halland 2017) we summarized in some detail the disastrous record of macro management in Ghana in the pre-liberalization period from 1960 through 1983, and the recovery through 1999. This recovery was followed by a period of excessive fiscal expansion, high inflation, and currency depreciation in the run-up to the presidential and parliamentary elections in 2000. The immediate focus of the new government of President John Agyekum Kufuor, elected in peaceful democratic elections that year, was to restore macroeconomic stability. This would consolidate many of the reforms initiated in 1993 by the Economic Recovery Program (ERP). The term of the Kufuor government also coincided, from 2001 to 2006, with adherence to an IMF programme that was required to obtain HIPC debt relief. In a sense, Ghana during this period was operating under a quasi-fiscal rule: a major shift in macroeconomic policy, from one of monetary accommodation, to one of fiscal stringency and monetary discipline.

A major pillar of this macroeconomic stabilization process was old-fashioned fiscal consolidation, involving a medium-term fiscal framework; the stabilization of domestic public debt; robust revenue mobilization; and prudent spending. For example, the 2003 fiscal programme set a target of zero net domestic financing of the public-sector borrowing requirement for 2003.

The government's fiscal policy strategy from 2001 also focused on debt reduction. Under the Multilateral Debt Relief Initiative (MDRI), Ghana's debt relief was estimated at US$4.2 billion in nominal terms. After the enhanced HIPC initiative was implemented in 2004 and Ghana qualified for the MDRI, Ghana's external debt decreased significantly, from 156.3 per cent of GDP in 2000 to only 17.2 per cent of GDP by 2006 (Table 11.1). Thanks to the HIPC relief, Ghana's external debt declined from US$6.02 billion in 2000 to US$2.17 billion by 2006. Furthermore, the proportion of exports used to service Ghana's debts declined from 28.1 per cent in 2000 to only 4.5 per cent by 2006 (Table 11.1).

**Table 11.1.** Ghana, selected economic indicators, 2000–8

| | 2000 | 2001 | 2002 | 2003 | 2004 | 2005 | 2006 | 2007 | 2008 |
|---|---|---|---|---|---|---|---|---|---|
| Annual percentage change unless otherwise stated | | | | | | | | | |
| Real GDP | 3.7 | 4.2 | 4.5 | 5.2 | 5.7 | 5.8 | 6.4 | 6.3 | 7.3 |
| Inflation | 40.5 | 21.3 | 15.2 | 23.6 | 16.4 | 13.9 | 10.9 | 12.7 | 18 |
| Broad money | 46.5 | 41.4 | 50.0 | 35.8 | 25.9 | 14.3 | 39.1 | 36.3 | 37.0 |
| Reserve money | 52.6 | 31.3 | 42.6 | 28.2 | 18.5 | 11.2 | 32.3 | 30.6 | 34.1 |
| Ex-rate depreciation | 49.8 | 3.7 | 13.2 | 17.3 | 2.2 | 0.9 | 1.1 | 4.8 | 20.1 |
| 91-day T-bill | 38.0 | 27.0 | 26.6 | 19.6 | 17.1 | 11.8 | 9.6 | 10.6 | 24.7 |
| BOG prime rate | 27.0 | 27.0 | 24.5 | 21.5 | 18.5 | 15.5 | 12.5 | 13.5 | 17.0 |
| Gross reserves (US$m) | 233.4 | 364.8 | 640.4 | 1,425.6 | 1,732.4 | 1,894.9 | 2,269.8 | 2,836.7 | 2,036.0 |
| Months of imports | 0.84 | 1.2 | 2.2 | 3.2 | 3.3 | 3.5 | 3.0 | 2.7 | 1.8 |
| Overall balance (US$m) | –116 | 8.6 | 39.8 | 558.3 | –10.5 | 84.3 | 415.1 | 413.1 | –940 |
| Crude oil (US$ per barrel) | 25.93 | 25.5 | 29.9 | 29.5 | 41.7 | 56.8 | 58.1 | 94.1 | 98.5 |
| NPLs % | 11.86 | 19.6 | 22.7 | 18.3 | 16.1 | 13.0 | 7.9 | 6.9 | 7.6 |
| External debt (US$m) | 6,021 | 6,025 | 6,131 | 7,548 | 6,447 | 6,347 | 2,172 | 3,590 | 3,871 |
| Debt services/exports | 28.1 | 16.4 | 10.1 | 4.9 | 7.2 | 7.7 | 4.5 | 4.6 | 4.3 |
| As percentage of GDP | | | | | | | | | |
| Budget deficit | 8.6 | 7.7 | 4.9 | 3.2 | 3.2 | 2.0 | 4.8 | 4.9 | 6.27 |
| External debt | 156.3 | 115.9 | 105.9 | 101.0 | 73.1 | 59.6 | 17.2 | 24.9 | |
| Debt services | 14.1 | 5.9 | 3.5 | 1.7 | 2.2 | 2.0 | 1.3 | 1.3 | 0.4 |
| External debt services (US$m) | 544.8 | 306.6 | 204 | 126 | 194.9 | 215.2 | 166.7 | 192.5 | 52.2 |
| Current account balance | 10.16 | 10.7 | 4.3 | 1.3 | 9.7 | 12.6 | 13.1 | 16.1 | 24.2 |

*Note*: BOG = Bank of Ghana; NPLs =non-performing loans; T-bill = treasury bill.

*Source*: data from Bank of Ghana.

Monetary policy at this time was underpinned by the adoption of an inflation-targeting framework, with the central bank taking advantage of the statutory independence that parliament had enshrined in the Bank of Ghana Act (BOG 2002).

This new fiscal and monetary policy framework caused a decline in inflation and inflationary expectations and strengthened Ghana's external payments position. Headline inflation declined from 40.5 per cent in 2000 to 12.7 per cent by 2007 (Table 11.1), and between 2001 and 2007 the overall balance of payments was in surplus (except for a deficit of US$10.5 million recorded in 2004. Gross international reserves increased from US$233 million (less than one month of import cover) in 2000 to US$2.84 billion (approximately three months of import cover) by 2007.

Additionally, after an initial sharp depreciation of the cedi against the US dollar in 2000, relative exchange rate stability was restored in 2001, when the depreciation was just 3.7 per cent. This was followed by 13.2 per cent and 17.3 per cent nominal depreciation in 2002 and 2003 respectively, 2.2 per cent in 2004, 0.9 per cent in 2005, 1.1 per cent in 2006, and 4.8 per cent in 2007 (anchored by declining inflation expectations). Between 2004 and 2007 the cedi depreciated by an average of 2.25 per cent annually against the dollar (Table 11.1). This contrasted with the cedi's historical instability and its 50 per cent depreciation in 2000.

## 11.3 Discovery of Oil

In June and August 2007, the United Kingdom-based firm Tullow Oil, and its US partners Kosmos Energy and Anadarko Petroleum, announced two significant oil discoveries off Ghana's coast. Some initial reports conveyed the impression that Ghana's reserves were comparable in size to those of Nigeria and Angola (Modern Ghana 2007b). With oil production projected to begin in 2010, the Ghanaian government was optimistic. In particular, there was a sense among government officials and the population that Ghana would be freed from the clutches of donors and international financial institutions, to pursue a more independent and growth-based development agenda. With an election on the horizon for 2008, the incumbent New Patriotic Party (NPP) government was eager to maximize the political benefit of the oil discovery, and in the process arguably did not manage expectations sufficiently.

However, impacted soon afterwards by increased domestic government spending and the global oil and food crisis, Ghana's economic environment took a turn for the worse as early as 2007–8. The government then faced the choice of fiscal contraction, full cost recovery for utilities, postponement of some already committed expenditures, or continued fiscal expansion.

Government factions argued that as the world was heading towards a recession, Ghana needed to adopt counter-cyclical policies, and therefore aggregate demand needed to be increased in the interim. It was argued that public finances could then be balanced later when the crisis was over, and when the oil was expected to start flowing in 2010.

For a government facing an imminent election in 2008, this was a persuasive argument. There was a sense that some reduction in the price trends and the forthcoming oil revenues would soon abate Ghana's economic difficulties (Bawumia 2010). The natural resource discovery gave a false sense of greater fiscal space than was the case.

The expansionary fiscal policy of 2007–8 had predictable consequences, and the economy quickly suffered a setback: inflation increased from 10.9 per cent at the end of 2006 to 18.1 per cent at the end of 2008, and the exchange rate depreciated by 20.1 per cent in 2008 compared with only 1.1 per cent in 2006. The budget deficit increased from 4.8 per cent of GDP in 2006 to 6.5 per cent of GDP by 2008. Gross international reserves declined from US$2.27 billion (three months of import cover) in 2006 to US$2.04 billion (1.8 months of import cover) in 2008 (Table 11.1). Thus the beginnings of the natural resource curse were evident in Ghana long before the first drop of oil was produced.

Nevertheless, the government of Ghana expressed determination to make sure that Ghana's oil resources would be managed well so as to avoid the dreaded curse (Amoako-Tuffour and Ghanney 2013). The government convened stakeholders to learn what pitfalls to avoid and which best practices to follow from the diverse experiences of countries such as Nigeria and Norway. To underpin good governance in the oil sector, and based on these discussions, public consultations, and international best practices, Ghana passed both the PRMA and the Petroleum Commission Act in 2011.

## 11.4 Deterioration of Public Finances

Notwithstanding these efforts, Ghana's public finances began to deteriorate quickly following the oil discovery in 2007. As Figure 11.1 indicates, public finances were generally sound from 2001 to 2007. For example, in 2005 there was a deficit of only 2 per cent of GDP. But this subsequently increased to 6.5 per cent of GDP in 2008, as the government increased expenditure in that election year in anticipation of oil revenues. Following the onset of oil production in 2010, the fiscal deficit narrowed, but only temporarily, to 4 per cent of GDP in 2011.

The 2008 elections brought a change of government, from the NPP to the National Democratic Congress (NDC). Notwithstanding this change, increased pre-election spending also occurred ahead of the 2012 elections.

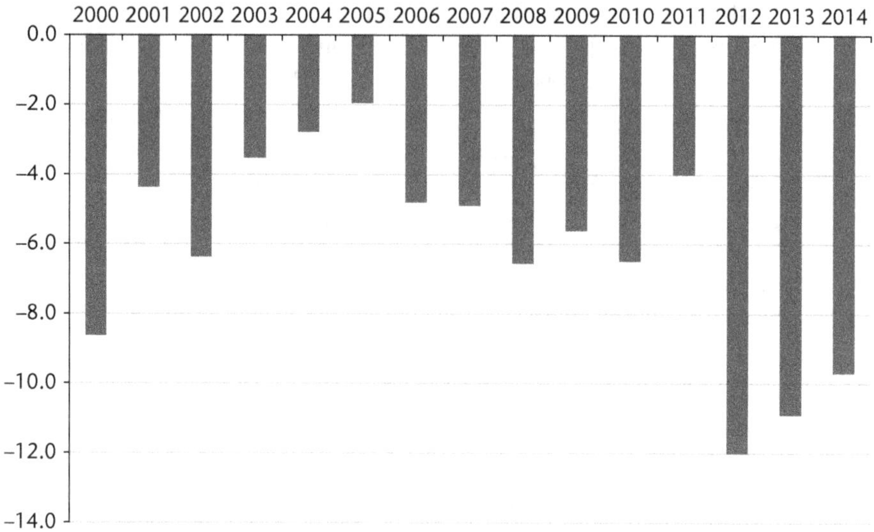


**Figure 11.1.** Fiscal deficit as percentage of GDP, 2000–14
*Source*: authors' illustration based on data from Ministry of Finance and Economic Planning, Ghana.

Ghana signed on to a further IMF programme in 2009 for the period 2009–12.[1] The programme restored macroeconomic stability, with inflation, the fiscal deficit, the exchange rate, and real GDP growth generally moving in the direction of stability. However, the IMF programme ended in mid-2012, and notwithstanding the progress that had been made, and with an election approaching, the authorities decided not to renew it.

As the 2012 presidential and parliamentary elections drew closer, the incumbent NDC government dramatically increased expenditures relative to 2011–14 revenues. While government tax revenue averaged approximately 18.9 per cent of GDP between 2011 and 2014, government expenditures increased from 20.1 per cent of GDP in 2011 to no less than 34.5 per cent in 2012—before declining to 28.2 per cent at the end of 2014 (IMF 2014). In the 2012 election year, Ghana's budget deficit reached GH₵8.7 billion, and amounted to some 11.6 per cent of GDP. This was the highest-recorded budget deficit in Ghana's history.

To make matters worse, the bulk of the increase in government expenditure between 2011 and 2014 (94 per cent) was on recurrent expenditure (mainly wages and interest payments). At the end of 2008 the government wage

[1] This was a Poverty Reduction and Growth Facility arrangement that built on Ghana's second Poverty Reduction Strategy. The main goal of the programme was to eliminate Ghana's large fiscal imbalances by 2011 and put in place strengthened institutions for public financial management (IMF 2009).

bill represented some 46 per cent of total tax revenue but it had increased to 72.3 per cent of tax revenue by December 2012 (Government of Ghana 2013). This fiscal stance resulted in double-digit fiscal deficits of 11.6 per cent and 10.9 per cent of GDP in 2012 and 2013 respectively, and a provisional 9.5 per cent of GDP in 2014 (Figure 11.1). This was the first time in independent Ghana's history that the country had double-digit fiscal deficits two years in a row. It resulted in the government being cash-strapped and unable to meet some of its obligations on statutory payments (for example, for health, education, and local government) as well as some non-statutory payments.

## 11.5 Rising Levels of Public Debt

The debt relief that Ghana obtained under the HIPC initiative had reduced the country's debt burden significantly. However, with the onset of oil production, the public debt stock rose dramatically to GH₵76.1 billion (67.1 per cent of GDP) by the end of 2014, versus 30 per cent in 2008 (Figure 11.2).

At the same time government borrowing became much more expensive, as illustrated by the successive increases in the risk-free treasury bill rate; see Bawumia and Halland (2017: Figure 3). Such large-scale borrowing—made possible by the large expected oil revenues—and the ensuing high returns offered to investors by investing in risk-free treasury bills, had the effect of

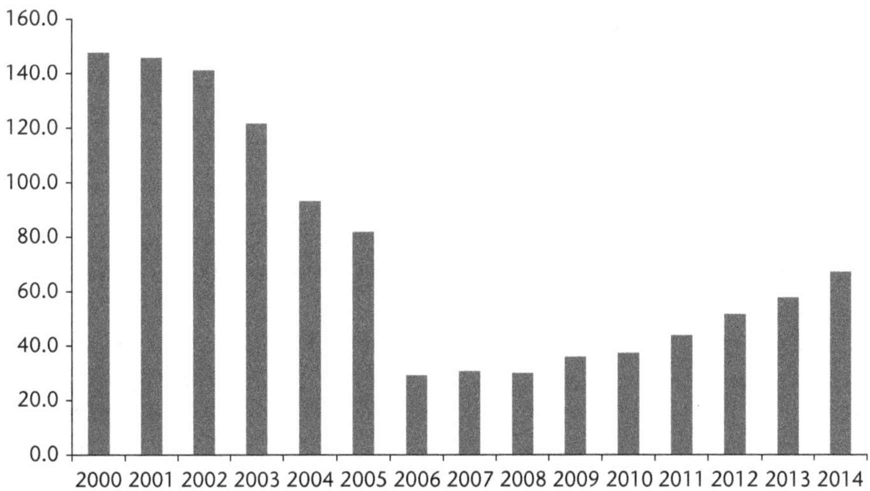


**Figure 11.2.** Total debt as percentage of GDP, 2000–14

*Source*: authors' illustration based on data from Ministry of Finance and Economic Planning, Ghana.

crowding out the private sector, which was increasingly unable to borrow. The interest burden of this increased and high public debt stock was also considerable, growing from 2.8 per cent of GDP in 2008 to 7.1 per cent in 2015; see Bawumia and Halland (2017: Figure 4).

This high interest burden, combined with a rising wage bill, left Ghana's government with very little money for other critical areas. To put this in perspective, the government of Ghana's 2015 budget allocations (excluding internally generated funds and donor contributions) to the eight most important spending ministries including education, health, and transport ministries amounted to about GH₵952 million, whereas interest payments on Ghana's public debt stock in the same year were GH₵9.5 billion—that is, ten times the combined allocations to these critical ministries.

This situation was reminiscent of Ghana's situation before the HIPC debt relief, when the debt burden had reduced the critical fiscal space that could have enhanced capital and social expenditure. At 67 per cent of GDP, by 2015 Ghana's debt stock had crossed the critical 60 per cent level: a particular concern to developing countries with limited access to capital flows. In fact, Ghana's debt by then was right back on the unsustainable track that had led to its HIPC relief.

How could Ghana have so misjudged its capacity to borrow so soon after obtaining HIPC relief just a few years earlier? Apart from the oil discovery heightening expectations about ability to pay, it should also be noted that Ghana's GDP was rebased in 2010, thereby statistically increasing Ghana's GDP by 60 per cent from 2007. This made the debt-to-GDP ratios look satisfactory on the surface. What policy makers may have overlooked was the fact that the GDP was rebased without an attendant increase in foreign exchange liquidity. In this situation, taking comfort from an apparently low debt-to-GDP ratio was potentially very misleading.

## 11.6 Declining Capital Expenditure

According to Hartwick's rule (Hamilton and Hartwick 2005), resource-rich countries, in order to maintain wealth and build strong foundations for economic growth, should offset the depletion of their natural resources by commensurate levels of investment in produced capital—primarily infrastructure and human capital. But in the case of Ghana, notwithstanding the new oil revenues and the massive increase in the debt stock, capital expenditure as a percentage of GDP has actually been in decline since 2007. From an average of 12 per cent of GDP between 2004 and 2008, capital expenditure declined to 4.8 per cent by 2014; see Bawumia and Halland (2017: Figure 5). This meant that before the oil discovery Ghana was spending a much higher proportion of

**Table 11.2.** Annual budget allocations to four priority areas, 2012

| Priority areas | GH₵ millions |
|---|---|
| Expenditure and amortization of loans for oil and gas infrastructure | 100.00 |
| Roads and other infrastructure | 232.41 |
| Agricultural modernization | 72.47 |
| Capacity-building | 111.95 |
| Total | 516.83 |

*Source*: Government of Ghana Budget 2013.

its income on capital investment than it has been since the oil discovery. This decline in infrastructure investment runs counter to what was anticipated and what good practice recommends.

In accordance with Section 21(5) of the 2011 PRMA, four priority expenditures were approved by parliament in the annual budget funding amount (ABFA): (i) expenditure and amortization of loans for oil and gas infrastructure, (ii) roads and other infrastructure, (iii) agricultural modernization, and (iv) capacity-building (including oil and gas). Table 11.2 shows the allocations to these areas for the 2012 budget year.

Even though on paper the oil revenue allocation is skewed towards infrastructure, the overall decline in capital spending means that the total allocation to infrastructure is likely to have declined. In the earlier paper (Bawumia and Halland 2017) we document several other related aspects of the macroeconomic deterioration that followed the discovery of oil. These include: (i) a loose monetary policy associated with excessive Bank of Ghana financing of the government (2017: Figure 6); a sharp increase in the rate of inflation through the 2008 elections and beyond (2017: Figure 7); a severe deterioration in the country's external payments position and its international reserve holdings (2017: Figures 8 and 9); and a progressive depreciation of the cedi currency against the US dollar (2017: Figure 10).

## 11.7 Declining Real GDP Growth

While the discovery of oil initially boosted economic growth, real GDP growth in Ghana has declined significantly since 2011. Data from the Ghana Statistical Service (Government of Ghana 2011) show that real GDP growth increased from 8.4 per cent (without oil) in 2008 to 15 per cent in 2011 (when it ranked among the highest in the world). Since 2011, however, real GDP growth (including oil) has slowed down—to 7.9 per cent in 2012, and then to around 4.2 per cent in 2014 (Figure 11.3). The 2015 budget projected further declining growth—to 3.5 per cent, at a revised oil price

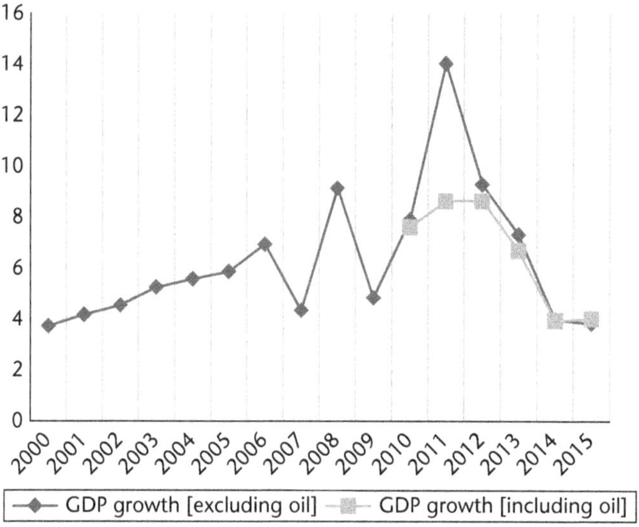


**Figure 11.3.** Real GDP growth, 2000–15, %

*Source*: authors' illustration based on data from Ghana Statistical Service.

of US$52.8 per barrel.[2] Non-oil GDP growth has similarly declined from 9.4 per cent in 2011. The actual growth rate in 2015, at just under 4 per cent, was only marginally higher than in the year 2000, and less than half the rate achieved in 2008 without oil.

The decline in real GDP growth can be attributed to a number of factors including: (i) the increase in current expenditure at the expense of capital expenditure; (ii) the greater fragility of both the private sector (partly due to a reliance on public contracts that could not be honoured given the need for adjustment post-elections) and the financial sector (given the accumulation of arrears and related non-performing loans); (iii) an increasing debt burden that has reduced critical fiscal space; and (iv) reduced macroeconomic policy credibility affecting investment decisions.

## 11.8 Corruption, Post-discovery

We now turn to consider some aspects of Ghana's political economy that arguably contributed to the macroeconomic malaise. An important strand of the relevant literature attributes the natural resource curse to the political and social factors at play in a given country (Rosser 2006; Schrank 2004; Snyder and Bhavnani 2005; van der Ploeg 2011). Such factors include, in particular,

[2] The initial budget projection was based on a price of US$99.3 per barrel.

corruption and rent-seeking (Acemoglu et al. 2004; Isham et al. 2002; Leite and Weidmann 1999; Sala-i-Martin and Subramanian 2003).

The basic argument is that natural resources create 'rents', and political elites compete for their control. Natural resource wealth provides incentives for administrations to stay in power at any cost—by bribing voters, supporting patronage-based and unproductive investments, encouraging a shift from productive entrepreneurial activity to unproductive rent-seeking, weakening institutions, and resisting efforts to further accountability, transparency, and modernization. Corruption imposes a fiscal burden on the economy through the waste of resources that could otherwise have been more productively spent or saved.

In the case of Ghana, there are some indications that corruption may have been on the rise since the discovery of oil. The Afrobarometer Survey conducted in 2014 by the Ghana Centre for Democratic Development (CDD 2014) found that public perceptions of corruption had significantly increased, and there was a similar perception of an increase in corruption in the run-up to the 2008 elections under the NPP government (CDD 2008). According to the 2014 survey, three quarters (75 per cent) of respondents said corruption had increased over the previous year.

## 11.9 Ghana's Political System in the Context of Oil

Ghana practises what has been described as a hybrid system of governance. The president is chosen directly by the people (unlike in the parliamentary system), and is head of government and head of state with full executive powers. But unlike the system operating in the United States, for example, the Fourth Republican Constitution, under which Ghana operates, makes it mandatory for the president to select a majority of ministers from parliament.

This means that power is significantly skewed towards the executive branch of the government. The executive branch also plays a large part in judicial duties. Not only is the executive responsible for framing and presenting all bills sent to parliament, but the president wields the power to appoint justices to the High Court and Appeals Court (acting on the advice of the Judicial Council) as well as Supreme Court, and to appoint the chief justice (in consultation with the Council of State and with prior approval of parliament). In practice, therefore, Ghana's political system is a presidential one (Banful 2011).

Ghana is a multiparty democracy, but of the twenty-three political parties registered in 2014, very few are active across the nation. The strongest, most vibrant parties are the NDC, the NPP, the Convention People's Party, the People's National Convention, and the Progressive People's Party. However, the relative dominance of the NDC and NPP in various general elections since 1992 has turned Ghana's democracy into a virtual two-party system. All general elections since that year have been closely contested between the two parties. Only a small margin (in terms of percentages and sometimes actual numbers) separates the two. The NDC won the presidential elections and a majority of parliamentary seats in the 1992, 1996, 2008, and 2012 general elections, while the NPP won the presidential elections and a majority of parliamentary seats in the 2000 and 2004 general elections.

In the 2000 presidential elections, after both candidates failed to secure the required votes in the first round, the elections went to a run-off. NPP candidate John Agyekum Kufuor secured 56.9 per cent of the votes. In the 2004 presidential elections, he secured 52.5 per cent of the votes to defeat John Atta Mills of the NDC, who received 44.6 per cent of the votes. The 2008 election saw Mills defeat Nana Akufo-Addo of the NPP by the narrowest of margins—41,000 votes (of about 12 million registered voters and nine million actual votes)—in another run-off poll. This was the closest election in Ghana's history, and the first to follow the discovery of oil. It speaks to the functioning of Ghana's democracy that the incumbent government was defeated by the opposition, even if by the narrowest of margins. The 2012 election was also fiercely fought: the NDC's John Dramani Mahama received 50.7 per cent of the votes, while NPP candidate Nana Akufo-Addo received 47.7 per cent.

Such close margins indicate that for a four-year political term, the pressure on any government to deliver is high. A loss of the support of even a small number of voters can turn the next election. When oil was discovered in 2007, the incumbent NPP government raised public expectations to a high pitch (Modern Ghana 2007a). President John Agyekum Kufuor declared, 'It is a great time to be Ghanaian' (Abissath 2008). As discussed above, the main economic indicators deteriorated sharply ahead of both the 2008 and 2012 elections: the first time while the NPP was in government, the second time while the NDC was the incumbent party. Consistent with the political business cycle literature, incumbent governments in Ghana have generally expanded fiscal policy in election years. In a joint review of public expenditure, Republic of Ghana et al. (2011) found that in election years (1992, 1996, 2000, 2004, and 2008) the fiscal deficit (on a cash basis) as a percentage of GDP was 1.5 per cent higher than the year before. The 2012 fiscal outcome following the oil discovery therefore cannot be wholly attributed to the discovery, even though the discovery may have amplified the political business cycle.

Much of the literature on the political economy of resource booms predicts that the first government to enjoy resource rents will do all it can to remain in power. In the case of Ghana, this drive may be further facilitated by a skew in political power towards the executive branch. Ghana's experience may therefore support some researchers' hypothesis that presidential systems are more prone to this sort of tendency than parliamentary systems (Andersen and Aslaksen 2008; Persson and Tabellini 2003).

## 11.10 What Could Ghana Have Done Differently?

The objective of this chapter is not to provide policy recommendations for Ghana, but rather to make available an empirical account of Ghana's experience to guide other recent or upcoming oil and mineral producers. In that context, it is useful to briefly discuss what Ghana might have done differently.

### 11.10.1 *Anchoring Fiscal Discipline*

While the discovery of oil provided Ghana with the fiscal space to increase its borrowing, there was a clear and present danger amid a lack of actual value. In the case of Ghana, plans for how the loans were to be spent—and the transparency of their terms—were arguably insufficient. In sum, institutional mechanisms were insufficient to check unsustainable fiscal expansion. Could fiscal rules of the type employed in Chile and discussed by Solimano and Guajardo (2017) help in such a context?

### 11.10.2 *Fiscal Rules and Budget Institutions*

Fiscal rules commit governments, usually but not always by legislation, to numerical targets—most often budgetary aggregates pertaining to debts, deficits, expenditures, and revenues. The aim is to tie the hands of the fiscal authorities in a bid to achieve greater fiscal discipline. Research on the effectiveness of fiscal rules does not indicate that fiscal discipline follows in a straight line. Ossowski and Halland (2016) note that fiscal rules seem to work best in countries with:

- a prior commitment to fiscal discipline
- strong institutions
- political commitment and consensus
- policy credibility
- strong public financial management capacity
- fiscal transparency
- robust monitoring.

It is therefore possible that a country lacking fiscal rules but possessing a commitment to fiscal discipline and good institutions will outperform a country with well-designed fiscal rules but with less commitment to fiscal discipline, less political consensus, and weaker institutions.

Chile provides a good example of the success of fiscal rules amid dependence on resource revenues (Frankel 2011). There, the structural budget balance is targeted to allow some cyclical flexibility. The medium-term equilibrium price of copper and the output gap are decided on by two non-partisan expert panels. The Chilean mechanisms are discussed in greater detail in Solimano and Guajardo (2017).

What is clear from the literature is that political economy considerations are critical to the success—or otherwise—of fiscal rules in resource-rich countries. For many low-income countries, infrastructure needs are high, as is the demand for public-sector wages. Governments facing elections may attempt to respond to these two demands by embarking on infrastructure projects without conducting value-for-money audits, and by increasing public-sector wages (Gelb 1986; Medas and Zakharova 2009). Similarly, it is very difficult for politicians in low-income countries that have just discovered a resource to make an argument to increase savings and pursue some sort of inter-temporal optimization, as argued in a recent paper by van der Ploeg and Venables (2017). This is especially the case when elections are hard fought and close in margin.

The application of fiscal rules in Ghana is new territory: the history of the country's fiscal management is one of fiscal indiscipline interspersed with short periods of fiscal discipline. In Ghana even a fiscal rule enshrined in law might lose out to the promise of electoral victory. Judging from successive administrations' very poor observance of the 2002 Bank of Ghana Act and the 2011 PRMA, fiscal rules by themselves are not sufficient to guarantee fiscal discipline.[3] What is needed is a commitment to, and a consensus on, fiscal discipline across the political divide, policy credibility, and a building up of strong public financial management capacity. With these requirements in place, a fiscal responsibility law might work.

### 11.10.3 *Anchoring Monetary Discipline*

In Ghana, fiscal dominance following oil discovery and production was also accommodated by the central bank. Central bank financing of the fiscal deficit increased fourfold between 2007 and 2008 (in the run-up to the 2008 elections), and subsequently increased sixfold between 2008 and 2014. In the context of an inflation-targeting framework, such an increase in liquidity was bound to compromise the central bank's ability to maintain price and exchange rate stability.

[3] A detailed account of how Ghana's PRMA has been breached one way or another can be found in NRGI (2015).

### 11.10.4 *Bank of Ghana Act and Central Bank Independence*

The case for central bank independence is based largely on the argument that central banks need to be protected from political interference to deliver on the goal of price stability (Alesina and Summers 1993; Bade and Parkin 1982; Parkin 1987). Yet some research (including in developing countries) indicates mixed economic performance even where central banks are independent (Berger et al. 2001; Eijffinger and de Haan 1996; Klomp and de Haan 2007).

In the case of Ghana, the 2002 Bank of Ghana Act was a landmark piece of legislation that established the independence of the Bank of Ghana. The Act states that the Bank shall support the general economic policy of government and promote economic growth and effective operation of banking and credit systems in the country, *'independent of instructions from the Government or any other authority'* (BOG 2002, emphasis added). This provision makes the Bank of Ghana, on paper at least, one of the most independent central banks in the world (Ayensu 2007).

This historic Act gives operational independence to the Bank of Ghana and specifies, inter alia, that: 'Government borrowing from the central bank in any year will be limited to 10 per cent of its current year's revenue.'

To achieve its primary objective of maintaining price stability, the Bank of Ghana in 2002 formally adopted an inflation-targeting monetary policy framework. In this context, the central bank should be able to choose instruments independent of political pressure and have minimal responsibility for financing government deficits. Fiscal reforms to maintain a broad revenue base, and thus reduce the need for seigniorage revenue,[4] then become crucial (Tuladhar 2005).

So why is it that central bank financing has grown so dramatically since the oil discovery, and why has the 2002 Bank of Ghana Act not been able to restrain this increase in central bank financing? The answer most likely lies in the inability of the central bank to withstand pressure from the government. There are no sanctions for breaches of the Act. Also, under the Act, the Bank of Ghana does not report to parliament. Furthermore, the ceiling on lending to the government is based on the government's estimate of revenue collection in the current year, which provides an incentive for an upward bias in revenue estimates. Instead, such legislation might have set a ceiling on the bank's lending to the government that is based on the government's *actual* revenue collection in the previous year.

The experience of Ghana shows that while monetary policy matters, fiscal policy matters even more. Ghana's monetary policy framework could not

[4] Seigniorage is the difference between the face value of money (i.e. coins or notes) and its production cost.

withstand sustained pressure from the fiscal authorities in 2000, 2008, and 2012—all election years. It is clear that governments must commit to fiscal discipline beyond the electoral cycle. In the context of Ghana's most recent (April 2015) IMF programme, the central bank will be required to reduce its financing to the government to 5 per cent in 2015 and zero in the 2016 election year (IMF 2015). The conditions necessary for fiscal rules to be effective are probably the same as those necessary for central bank independence: policy credibility, fiscal discipline, political commitment, and political consensus, among others.

### 11.10.5 *Transparency*

To enhance transparency in the management of natural resources, Ghana joined the EITI in 2003 and became compliant with EITI in 2011. Ghana also set up the Public Interest Accountability Committee (PIAC) in 2011 under Section 51 of the 2011 PRMA, to conduct independent assessment and monitor the utilization of oil revenues to ensure accountability by government.

How much have PIAC and EITI contributed to enhancing transparency in the management of Ghana's oil resources? The Institute of Economic Affairs Ghana, a leading public policy think tank, has developed the Petroleum Transparency and Accountability Index Project to monitor transparency and accountability in the oil sector. Since 2011, their index has focused on four key areas: revenue transparency, expenditure transparency, contract transparency, and the Ghana Petroleum Funds. The 2015 report found that steady progress had been made with regard to the transparency of revenue expenditure and the Ghana Petroleum Funds. Contract transparency showed the least improvement, lacking public disclosure of contracts and the associated process (IEA 2015). An earlier paper (Bawumia and Halland 2017) discusses in more detail how the example of the Atuabo gas project—the largest project yet undertaken in the sector—illustrates the remaining weaknesses of the approval and transparency processes in Ghana.

Furthermore, contracts for prospecting or exploration rights in Ghana's oil and gas sector are issued through an administrative process, rather than through competitive tendering. Ghana also does not have any mandatory contract disclosure provisions, and therefore details of oil contracts that do not come before parliament can only be disclosed by a ministerial directive.

Notwithstanding Ghana's compliance with EITI, civil society organizations have called for increased disclosure of information on beneficial ownership in mineral and oil contracts (which is voluntary under EITI). Furthermore, there have been claims that the oversight responsibility of PIAC may have been curtailed, with the record over the last two years indicating that PIAC has had difficulties in carrying out its mandated activities, partly because of under-funding. According to Adam (2014), in 2014 PIAC received only 14 per cent of its proposed budget.

### 11.10.6 *Strengthening the Law on the Management of Oil Revenues*

Ghana's PRMA includes the following provisions: a limit of 70 per cent utilization of expected revenues in any financial year; payment of excess revenues into the Petroleum Holding Fund; a transfer of at least 70 per cent of the excess revenue into the Petroleum Stabilization Fund; and the balance to be paid into the Heritage Fund (an endowment fund to support development for future generations when petroleum reserves have been depleted). However, and significantly, the oil revenues due to the budget can be used as collateral for loans for the first ten years of oil production.

But in the face of dwindling foreign exchange reserves, the Petroleum Stabilization Fund was drawn down well before oil prices fell and before there was any actual shortfall in petroleum revenues. As the funds transferred to the Stabilization Fund increased in 2012 and 2013, the government set a cap on the amount to be transferred into the Stabilization Fund. The cap was set at US$250 million at a time in 2013 when the balance on the account stood at US$426 million. This provided fiscal space to spend the excess amount of US$176 million. Civil society organizations have argued that the withdrawal of funds from the Stabilization Fund was illegal, since this was not made public and was first brought to the attention of the public by a whistle-blower (ACEP 2014).

The lesson from Ghana's experience thus far is that while stabilization and savings funds can help smooth spending, they can certainly not prevent slippages. In the case of Ghana's PRMA, this is compounded by the fact that the collateralization of future oil revenues was allowed for the first ten years, thereby undermining the initial objective to limit excessive spending permitted by oil discoveries.

### 11.10.7 *Forecasting Government Oil Revenues*

The literature contains persuasive evidence that official forecasts of revenues during resource booms tend to be overly optimistic (Forni and Momigliano 2004; Frankel 2011; Jonung and Larch 2006). This is arguably the case in Ghana.

According to Adam (2014), Ghana's 2011 PRMA as designed gives an incentive to governments to project higher revenues than are likely realizable, because 70 per cent of the projected benchmark oil revenue is allocated to the budget. In 2011, for example, the government projected that it would receive GH₵1.2 billion from oil. The total amount actually earned, however, came down to GH₵667 million: a shortfall of GH₵583 million. The benchmark revenue for 2012 was over GH₵1 billion,[5] but the amount actually received came down to GH₵562.4 million. In 2011, forecasted oil revenues from corporate taxes of GH₵600 million (GH₵384.1 million for 2012) were arguably unlikely to materialize (Adam 2014). This is because the Jubilee partners[6] are entitled to capital cost recovery under the 1987 Petroleum Income Tax Law (PNDC Law 188). The over-projection was arguably motivated by the provisions of the 2011 PRMA, by which the proportion of the oil revenues that accrues to the budget (ABFA) is based on projected benchmark oil revenues.

In the 2015 government budget presented to parliament in November 2014, oil revenues were estimated at GH₵4.2 billion (3.1 per cent of GDP) based on an oil price of US$99.3 per barrel, although global oil prices at the time were actually close to US$50 per barrel. It was argued that the use of the US$99.3 price was dictated by the PRMA legislation (specifying a seven-year moving average price) even though this over-optimistic forecast implied a revenue shortfall of GH₵2.7 billion (2 per cent of GDP). Prudence would have dictated that, despite the law, no budgeted expenditures be made in relation to unrealizable revenues.

The government later, at the prompting of the IMF (in the context of negotiations for an IMF bailout), presented a revised oil revenue estimate of GH₵1.5 billion (1.1 per cent of GDP) using the price of US$52.8 per barrel, even though the original budget was approved by parliament in January 2015.[7]

To enhance safeguards against fiscal indiscipline, adequate legislation and methodologies for revenue forecasting are imperative. Frankel (2011) has recommended, following Chile's example, that resource-rich countries consider establishing independent non-partisan expert panels to forecast resource revenues for the budget and the extent to which prices (of oil, in the case of Ghana) deviate from their long-run averages.

## 11.11 Conclusions

The discovery of oil in Ghana in 2007 raised public expectations quickly and substantially. For many, oil appeared to be the long-awaited solution to

[5] Based on a projected oil price of US$90 per barrel.

[6] Tullow, Kosmos, Anadarko, and Sabre, which are currently producing Ghana's oil from the Jubilee Field (offshore).

[7] Statement to parliament on 'Implications of the fall in crude oil prices on the budget' submitted by the Minister of Finance and Economic Planning, 12 March 2015.

Ghana's developmental challenges. This hope was reflected in the 2008 and 2012 election campaigns, which saw political parties promise much in the areas of education, infrastructure, and health, among others. Fiscal discipline was given only low priority, as parties focused on winning the elections.

As a consequence, the public finances deteriorated, debt returned to unsustainable levels, current account deficits ballooned, foreign exchange reserves dwindled, the exchange rate depreciated rapidly, interest rates rose, inflation rose, and real GDP growth declined. These are the hallmarks of a resource curse. Ghana's experience suggests that symptoms of the resource curse that have dimmed its prospects since the discovery of oil could have been avoided in the presence of:

- a broad-based political commitment to fiscal and monetary discipline
- strong institutions, including the ability to uphold relevant fiscal and monetary legislation
- strong public financial management capacity
- transparency in the management of oil resources
- alignment of expenditure with realizable revenues in the context of independent oil revenue forecasts.

The case of Ghana may not fit smoothly into the classic resource curse narrative because several of the usual transmission channels—such as an appreciation of the real exchange rate and the volatility of commodity prices, or armed conflict—are missing. Nevertheless, as discussed in the introduction to this chapter, the Ghanaian story in part reflects the insights provided by the literature on the political economy of development, and more specifically the political economy of the resource curse. The immediate cause of the lack of fiscal and monetary discipline in Ghana can be found in policy decisions.

Importantly, Ghana's story allows us to move beyond vague notions of 'institutional quality', measured by aggregate indexes, when considering the resource curse and how to address it. The Ghanaian experience confirms that there are *specific* institutional checks and balances that must be upheld if a country is to avoid the resource curse. From a fiscal and monetary perspective, these include early management of expectations, a broad-based political commitment to fiscal discipline as opposed to a reliance on fiscal rules, full and real (as opposed to nominal) independence of the central bank, as well as the establishment of means to isolate from political pressures any sovereign wealth fund and the government entity responsible for oil revenue projections. Ultimately, these factors are likely to be critical in determining whether the discovery of natural resources will adversely affect economic growth or not. In the case of Ghana, they were key factors in the regrettable shift from boom to gloom just four years after the discovery of oil.

## References


Abissath, M. K. (2008). 'Preparing for Oil Boom in Ghana: Any Lessons from Others?' *Abissath Features*. Available at: abissathfeatures-mawu.blogspot.co.uk/2008/06/preparing-for-oil-boom-in-ghana-any.html.

Acemoglu, D., J. A. Robinson, and T. Verdier (2004). 'Kleptocracy and Divide-and-Rule: A Theory of Personal Rule', *Journal of the European Economic Association*, 2: 162–92.

ACEP (2014). 'Press Release on the Illegal Withdrawal of Funds from the Petroleum Stabilization Fund, June 2014'. Accra: Africa Centre for Energy Policy.

Adam, M. A. (2014). 'Three Years of Petroleum Revenue Management in Ghana', Public Interest Report 2. Accra: Africa Centre for Energy Policy.

Alesina, A. and L. Summers (1993). 'Central Bank Independence and Macroeconomic Performance: Some Comparative Evidence', *Journal of Money Credit and Banking*, 25: 151–62.

Amoako-Tuffour, J. and M. Ghanney (2013). 'Ghana's Petroleum Revenue Management Law: A Social Contract for Good Economic Governance and Possible Challenges', in K. Appiah-Adu (ed.), *Governance of the Petroleum Sector in an Emerging Developing Economy*, 35–64. Farnham: Gower Publishing.

Andersen, J. J. and S. Aslaksen (2008). 'Constitutions and the Resource Curse', *Journal of Development Economics*, 87: 227–46.

Atkinson, G. and K. Hamilton (2003). 'Savings, Growth and the Resource Curse Hypothesis'. *World Development*, 31(11): 1793–807.

Ayensu, E. (ed.) (2007). 'Commemoration of Fiftieth Anniversary'. Accra: Bank of Ghana.

Bade, R. and M. Parkin (1982). 'Central Bank Laws and Monetary Policy', unpublished paper. London, ON: University of Western Ontario, Department of Economics.

Banful, A. B. (2011). 'Do Formula-Based Intergovernmental Transfer Mechanisms Eliminate Politically Motivated Targeting? Evidence from Ghana', *Journal of Development Economics*, 96: 380–90.

Bawumia, M. (2010). *Monetary Policy and Financial Sector Reform in Africa: The Experience of Ghana*. Bradenten, FL: BookLocker.

Bawumia, M. and H. Halland (2017). 'Oil Discovery and Macroeconomic Management: The Recent Ghanaian Experience', World Bank Policy Research Working Paper No. 8209. Washington, DC: World Bank.

Berger, H., J. de Haan, and S. Eijffinger (2001). 'Central Bank Independence: An Update of Theory and Evidence', *Journal of Economic Surveys*, 15(1): 3–40.

Bleaney, M. and H. Halland (2016). 'Do Resource-rich Countries Suffer from a Lack of Fiscal Discipline?' Policy Research Working Paper No. 7552. Washington, DC: World Bank, Governance Global Practice Group.

BOG (2002). 'Bank of Ghana Act (Act 612)'. Accra: Bank of Ghana.

CDD (2008). 'How Ghanaians Rate the Performance of the NPP Administration', Afrobarometer Briefing Paper 49. Accra: Ghana Center for Democratic Development.

CDD (2014). 'Afrobarometer Survey: Dispatch 6'. Accra: Ghana Center for Democratic Development.

Collier, P. and A. Hoeffler (2009). 'Testing the Neocon Agenda: Democracy in Resource-rich Societies', *European Economic Review*, 53(3): 293–308.

Eijffinger, S. and J. de Haan (1996). 'The Political Economy of Central Bank Independence', Special Papers in International Economics 19. Princeton, NJ: University of Princeton, International Finance Section.

Forni, L. and S. Momigliano (2004). 'Cyclical Sensitivity of Fiscal Policies Based on Real Time Data', Economic Working Paper No. 540. Rome: Bank of Italy, Economic Research Department.

Frankel, J. A. (2011). 'A Solution to Fiscal Procyclicality: The Structural Budget Institutions Pioneered by Chile', NBER Working Paper No. 1694. Cambridge, MA: NBER.

Gelb, A. (1986). 'Adjustment to Windfall Gains: A Theoretical Framework', in J. P. Neary and S. van Wijnbergen (eds), *Natural Resources and the Macroeconomy*. Cambridge, MA: MIT Press.

Government of Ghana (2011). *Auditor General Report 2011*. Accra: Government of Ghana.

Government of Ghana (2013). *Budget Statement 2013*. Accra: Government of Ghana.

Hamilton, K. and J. Hartwick (2005). 'Investing Exhaustible Resource Rents and the Path of Consumption', *Canadian Journal of Economics*, 38(2): 615–21.

IEA (2015). 'Petroleum Transparency and Accountability (P-TRAC) Index Project'. Accra: Institute of Economic Affairs.

IMF (2009). 'Ghana: 2009 Article IV Consultation and Request for a Three-Year Arrangement under the Poverty Reduction and Growth Facility—Staff Report; Staff Supplement; Staff Statement; Public Information Notice and Press Release on the Executive Board Discussion; and Statement by the Executive Director for Ghana', IMF Country Report 09/256. Washington, DC: IMF. Available at: www.imf.org/external/pubs/ft/scr/2009/cr09256.pdf.

IMF (2014). 'Article IV Consultation: Staff Report', IMF Country Report 14/129. Washington, DC: IMF.

IMF (2015). 'Ghana Request for a Three-Year Arrangement under the Extended Credit Facility'. Washington, DC: IMF.

Isham, J., M. Woolock, L. Pritchett, and G. Busby (2002). 'The Varieties of Rentier Experience: How Natural-resource Export Structures Affect the Political Economy of Economic Growth', Economics Discussion Paper 03–08. Middlebury, VT: Middlebury College.

Jonung, L. and M. Larch (2006). 'Improving Fiscal Policy in the EU: The Case for Independent Forecasts', *Economic Policy*, 21(47): 491–534.

Klomp, J. G., and J. de Haan (2007). 'Inflation and Central Bank Independence: A Meta Regression Analysis'. Mimeo. Groningen: University of Groningen.

Leite, C. and J. Weidmann (1999). 'Does Mother Nature Corrupt? Natural Resources, Corruption, and Economic Growth', IMF Working Paper WP/99/85. Washington, DC: IMF.

Mansoorian, A. (1991). 'Resource Discoveries and "Excessive" External Borrowing', *Economic Journal*, 101(409): 1497–509.

Medas, P. and D. Zakharova (2009). 'Primer on Fiscal Analysis in Oil-Producing Countries', IMF Working Paper 56. Washington, DC: IMF.

Modern Ghana (2007a). 'Oil: Huge Field Discovered at Cape Three Points', *Modern Ghana*, 19 June. Available at: www.modernghana.com/news/137981/1/oil-huge-field-discovered-at-cape-three-points.html.

Modern Ghana (2007b). 'Oil! Oil! Oil!' *Modern Ghana*, 27 June. Available at: www.modernghana.com/news/138430/1/oil-oil-oil.html.

NRGI (2015). 'In Ghana, Fiscal Responsibility Remains Elusive Even as Oil Flows'. Accra: NRGI.

Ossowski, R. and H. Halland (2016). *Fiscal Management in Resource-rich Countries: Essentials for Economists, Public Finance Professionals, and Policy Makers*. Washington, DC: World Bank.

Parkin, M. (1987). 'Domestic Monetary Institutions and Deficits', in J. M. Buchanan et al. (eds), *Deficits*, 67–85. Oxford: Basil Blackwell.

Persson, T. and G. Tabellini (2000). *Political Economics: Explaining Economic Policy*. Cambridge, MA: MIT Press.

Persson, T. and G. Tabellini (2003). *The Economic Effects of Constitutions: What Do the Data Say?* Cambridge, MA: MIT Press.

Republic of Ghana (2011). 'Petroleum Revenue Management Act 2011 (Act 815)'. Accra: Ghana.

Republic of Ghana, World Bank, DANIDA, and KfW (2011). 'Joint Review of Public Expenditure and Financial Management'. Accra: Ghana.

Rosser, A. (2006). 'The Political Economy of the Resource Curse: A Literature Survey', Working Paper 368. Brighton: Institute of Development Studies.

Sachs, J. D. and A. M. Warner (1995). 'Natural Resource Abundance and Economic Growth', NBER Working Paper 5398. Cambridge, MA: NBER.

Sachs, J. D. and A. M. Warner (2001). 'The Curse of Natural Resources'. *European Economic Review*, 45(4–6): 827–38.

Sala-i-Martin, X. and A. Subramanian (2003). 'Addressing the Natural-resource Curse: An Illustration from Nigeria', IMF Working Paper WP/03/139. Washington, DC: IMF.

Schrank, A. (2004). 'Reconsidering the "Resource Curse": Sociological Analysis versus Ecological Determinism'. New Haven, CT: Yale University, Department of Sociology.

Snyder, R. and R. Bhavnani (2005). 'Diamonds, Blood, and Taxes', *Journal of Conflict Resolution*, 49(4): 563–97.

Solimano, A. and D. Calderón Guajardo (2017). 'The Copper Sector, Fiscal Rules, and Stabilization Funds in Chile', WIDER Working Paper 2017/53. Helsinki: UNU-WIDER.

Tuladhar, A. (2005). 'Governance Structures and Decision-making Roles in Inflation Targeting Central Banks', IMF Working Paper WP/05/183. Washington, DC: IMF.

van der Ploeg, F. (2011). 'Natural Resources: Curse or Blessing?', *Journal of Economic Literature*, 49(2): 366–420.

van der Ploeg, F. and A. J. Venables (2017). 'Extractive Revenues and Government Spending: Short- versus Long-term Considerations', WIDER Working Paper 2017/45: Helsinki: UNU-WIDER.

# Part V

# National Institutions of Extractives Management

# 12

# The Regulation of Extractives

## An Overview

*Tony Addison and Alan Roe*

## 12.1 The Basic Frameworks

Every resource-abundant country faces the need to regulate the sector in the public interest, given its significant social, economic, and environmental impact. This is a daunting challenge, requiring not only the design and enactment of the necessary laws and regulations but also the assignment of responsibilities for implementation across ministries and often numerous government agencies. The regulatory system must be: comprehensive (while avoiding unnecessary complexity, which is a disincentive to investment); transparent (to avoid manipulation and regulatory 'capture'); and implemented to a high standard. These are demanding tasks for any country, not least a low-income country (LIC) with a resource windfall, yet they are essential for the extractives sectors to yield economic and social improvement.

Three frameworks are available to help understand the nature and complexity of the regulatory challenge: (i) a life-cycle approach, (ii) an institutional approach, and (iii) an organizational-responsibility approach.

The first is based upon the *life cycle* of mining as well as oil and gas projects. That cycle begins with the discovery of the asset after geological exploration. This is followed by its extraction, entailing often large investments by a company or companies which has the effect of monetizing the previously inert asset, then the taxation of the project which transfers some of the monetary benefits to the public authorities during its operational phase, and then a closure process. At the beginning of the life cycle, a licensing process is required for the exploration stage and for the project's eventual investment and operating stages. Contracts must be drawn up with the companies involved, some of which may be based upon general law, while others will be case-specific agreements requiring negotiation. There are also likely to be a wide range of regulatory issues associated with, for example, the release of land for the project, and for the re-settlement and compensation of resettled populations.

Over the life of the project, which may last decades, not only do the taxation arrangements need to be clearly defined and implemented (and possibly adjusted to changing circumstances) but ongoing arrangements are also needed to ensure that the project complies with national environmental and social standards as well as employment regulations. Any social obligations that the commercial companies take on towards local communities need to be implemented as agreed and monitored. This may involve legislation and the enforcement of mandatory obligations. Any public projects to catalyse the economic development impacts of the extractives activities must be legislated for and implemented. Finally, project closures can be demanding, especially as environmental and social standards are now higher than historically: these require additional regulatory processes that need to be actioned effectively. At all stages of the project's life cycle, there may be breaches of standards, regulations, and local laws which require the application of appropriate sanctions.

A second useful framework is that based upon Oliver Williamson's 'economics of institutions' (Williamson 1985, 2000), which is discussed in detail in Chapter 6 by Evelyn Dietsche on institutional dimensions in this volume. This framework reminds us that any regulatory system must be aware of, and sensitive to, four layers of institutions. The first layer consists of *social, cultural, and religious norms* which provide the core of the country's social networks and relationships, and which exert a high level of influence over economic decisions. These norms differ across countries. Therefore, transplanting a regulatory system that works well in one country context into another does not automatically result in success. The second layer consists of the *formal institutional structure* that has been created with or without the presence of an extractive industry. This includes the institutional arrangements that govern relationships between the central government and regional and local governments, as well as relations between countries: for example, membership of a customs union. Again, a country's political, bureaucratic, and economic arrangements will be unique in some respects, and so not amenable to a simple transplant of another country's regulatory system for extractives.

The third layer involves those institutions that Williamson refers to as the '*play of the game*'. These include the agreements and contracts negotiated to deal with certain specific and relatively narrowly defined issues pertaining to an extractives project. These agreements may or may not be consonant with the general laws of the country that would be included in Williamson's second layer. Finally, any country's *market system* will have its own characteristics, including the completeness or otherwise of its markets, and its definition of the appropriate role for government policy. The design of regulatory systems for extractive industries should consider how national markets systems operate given the other three institutional layers.

The third useful framework relates to the range of *different organizations* that input to the overall regulation of an extractives sector; government ministries, parastatal organizations, and other public and quasi-public agencies. Although it is often assumed that regulatory responsibility resides narrowly with a country's ministry of mines or petroleum, the reality is invariably far more complex. This is evident in the wide range of issues that we have already listed in discussing the first and second frameworks. A stylized and somewhat over-simplified example of this complexity is presented in work by Oxford Policy Management (OPM 2013) which is reproduced as Figure 12.1. It maps the main stages of the life cycle (shown in the top row of the diagram) to the main legal and organizational arrangements that most countries have in place for industry regulation. The lower part of this figure labelled 'administration' identifies just a few of the many public agencies that are typically involved in addressing those aspects that are described in the other two sections of the figure. Later in this chapter we provide a specific country example that fills out the detail of this more fully.

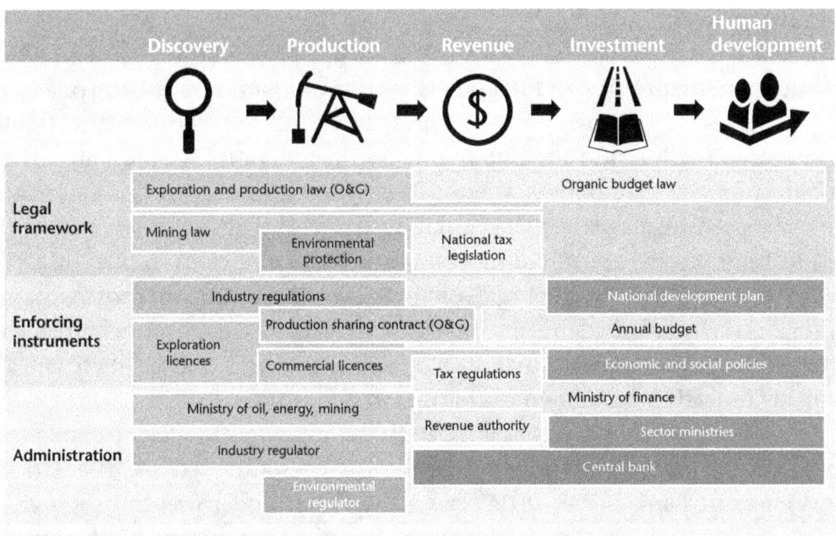


**Figure 12.1.** The regulatory framework: a typical structure

*Note*: O&G refers to oil and gas.

*Source*: OPM (2013: Figure 3); reproduced here by permission.

## 12.2 The Regulatory Challenge in Low-income Countries

Multiple layers of time sequences, institutions, and organizations make it difficult to effectively regulate the extractive industries in any country. LICs with incomplete and fragile institutions, and even some middle-income countries (MICs) with limited capacities, typically encounter significantly worse difficulties. When a new mineral or energy resource is discovered, the country may have had very limited previous exposure to the technical challenges of regulation in the extractives sector, a weakness often compounded by a politics that favours short-term gains for the governing elite and their commercial associates rather than building regulatory systems that protect the public interest into the longer term. There is a large literature on the regulatory deficiencies—and consequent disasters, especially environmental—experienced in the developing world, especially (e.g. see Chapter 16 by Bell).

There are generic reasons that help to explain weak regulation and the even weaker enforcement of established regulations. Many LICs lack capacity in technical areas such as tax administration and environmental protection to effectively enforce regulation. Often the data—for example, on production levels and the quality of particular minerals—are extremely poor, thereby undermining those elements of regulation that rely on hard information. Some legal systems are highly vulnerable to corruption and other abuses of power: indeed, the motive to obtain high government office is sometimes driven by the expectation of large personal gains from the extractive industries themselves.

In other cases, active state ownership in the production of an extractive resource—perhaps through full ownership of a production company or a large equity stake—can easily result in conflicts of interest between the public regulatory function and the public ownership function. Additionally, when public employees are poorly paid, fines for breaching regulations can sometimes be used to supplement personal incomes. The ultimate sanctions that would truly punish an offending company, including closure, may often be avoided because they would impose a high level of hardship on those working in, or otherwise dependent on, the extractive activity. Lighter penalties will often be regarded as a normal cost of doing business and as such will be unlikely to lead to fundamental changes in behaviour.

These generic issues are associated with the many more specific challenges that host governments face in regulating their resource sectors. The African Development Bank (AfDB, 2015) has drawn attention to several such challenges, among which three are perhaps the most significant. First, can the recipient countries find enough of the specialist personnel and the related technical resources needed to design, implement, and manage a comprehensive regulatory system? This is a particularly difficult challenge not least because the industries of mining and oil and gas are so highly specialized with many technical dimensions: the skills available to the regulatory bodies should ideally match those available to the mining and oil and gas companies themselves. Even though highly skilled specialists can be bought in—for example, to help negotiate contracts—they may not be easy to find and certainly they are expensive. Yet, governments still need some minimum basic skills to identify, recruit, and work effectively with any external specialists who are hired.

Second, can governments manage the effective coordination of regulatory activity across the wide range of ministries and other state agencies that ideally should be involved in different aspects of extractives regulation? This is a profound and under-estimated problem not least because few governments acquire such a high level of cross-agency coordinating capability in relation to most other aspects of their work. Tanzania is an example of the complexity involved: see Box 12.1. The Tanzanian experience demonstrates that coordination across such a wide range of actors is difficult and that failures can arise at many different points. Effective coordination requires that different agencies must recognize the specialist roles of other agencies

**Box 12.1** WHO IS INVOLVED IN EXTRACTIVES REGULATION IN TANZANIA?

The complete answer to this question involves a plethora of ministries and agencies that ideally need to work together to deliver the 'all of government' approach as advocated by Kathryn McPhail in Chapter 17. The Vice President's Office is nominally the lead coordinator on overall policy but often needs to defer to the Office of the President. The other ministries and agencies involved include the Ministry of Energy and Minerals (MEM) as the lead policy and administrative institution; the Tanzania Petroleum Development Corporation or TPDC (now the NOC) as the designated national partner in all petroleum-related ventures; the Petroleum Upstream Regulatory Authority that has recently taken over the regulatory functions of TPDC; the Energy and Water Utility Regulatory Authority (EWURA) whose role is confined to midstream and downstream regulation; the Ministry of Finance (MOF), as the lead policy institution in setting royalty and profit-sharing terms at the project level; the Attorney General's office that is also involved in negotiation and devising contracts; the Tanzania Revenue Authority that collects income taxes from gas companies. MEM collects the large share of non-tax revenues from petroleum activities via TPDC, including royalties, licence fees, application fees, and annual rent, and the government profit share from oil and gas, with MOF collecting revenues from equity holdings, and local authorities collecting a local service levy from mining companies. If this was not already a daunting enough coordination challenge, the National Environment Management Council in coordination with local authorities manages environmental issues; the Ministry of Labour and Employment leads on the formulation of labour, labour market, social security, and employment policies; while the Ministry of Lands, Housing and Human Settlements Development has to approve land allocations for extractives use.

*Source*: Roe (2017: Box 1); reproduced here by permission.

on some sub-topics and defer to them in those cases in the national interest rather than compete for supremacy. But it is often difficult for a powerful ministry or minister to defer to another agency, although there will invariably be a need for some sacrifice of power if the system is to work well. Tanzania and other cases also illustrate how important it is to have some over-arching national vision of the long-term potential of the extractives sector that can guide the policy and regulatory interventions. In the absence of such a vision effective coordination is likely to be compromised (Roe 2017).

Third, in developing that vision and the policies needed to support it, how do governments resist the inevitable and loud public pressures to see quick material gains from any newly discovered extractive resource: pressures that are encouraged and amplified by the often-exaggerated media coverage which accompanies discoveries? The problem here is that the clamour for quick gains stands in stark contrast to the realities of the lengthy delays and inherent uncertainties that are associated with natural resource discoveries. Production never starts immediately and even when it does start there is always a further lengthy lag before government revenues begin to rise commensurately. This fact is well documented in many case studies but is invariably poorly communicated to the public at large. Consequently, governments are often pressured into poorly devised policies that are seriously sub-optimal and effective only in pandering to populist sentiments. But set against public anxiety for quick results is the reality that high-quality strategic decision-making—about, for example, supporting investments—needs long-term and carefully built institutional arrangements. The recognition that the *long term* necessarily spans the lives of several governments—assuming they are democratically elected—requires not only skills in the civil service but also some consensus across different political parties if a regularly changing and unpredictable regulatory structure is to be avoided.

## 12.3 Outline of Part V

The five chapters that follow in this part of the book address merely a sample of the critical issues in the regulation of the sector; they do, however, serve to illustrate the points made here about the generic problems and challenges of this task. So, they include a perspective on the problem contributed by a senior practitioner—Toni Aubynn—who has overseen mining regulation in a large African economy, followed by detailed reviews of the regulatory challenges in relation to three key areas of regulation. These are the *taxation* of mining operations, the operation of *state-owned oil companies*, and the wide range of issues associated with *environmental regulation*. This is not an all-encompassing list but it does cover a set of the most important regulatory areas. Other areas of regulation are addressed in earlier and later chapters of the book (for example, *macroeconomic* regulation) and some aspects of regulation as applied to local content, *downstream activities*, and *community development* are to be found in the relevant chapters on those topics.

## References


AfDB (2015). *Delivering on the Promise: Leveraging Natural Resources to Accelerate Human Development in Africa*. Abidjan: African Development Bank Group and Bill and Melinda Gates Foundation.

OPM (2013). *DFID PEAKS Topic Guide: Extractive Industries, Development, and the Role of Donors*. Oxford: Oxford Policy Management.

Roe, A. R. (2017). 'Tanzania—from Mining to Oil and Gas: Structural Change or Just Big Numbers?' WIDER Working Paper 2017/175. Helsinki: UNU-WIDER.

Williamson, O. E. (1985). *The Economic Institutions of Capitalism*. New York: The Free Press.

Williamson, O. E. (2000). 'The New Institutional Economics: Taking Stock, Looking Ahead', *Journal of Economic Literature*, 38(3): 595–613.

# 13

# Regulatory Structures and Challenges to Developmental Extractives

## Some Practical Observations from Ghana

*Toni Aubynn*

## 13.1 Introduction and Background

It is now well established that resource-rich low-income countries, even though they produce less minerals than their high-income counterparts, depend more on the extraction of natural resources for their economic development and transformation (McKinsey Global Institute 2013). This places natural resources extraction at the heart of the socio-economic development and transformation of resource-endowed LICs. Central to the transformative and developmental power of natural resource extraction is an effective regulatory framework that guides the promotion of investments in the sector and the establishment of effective procedures responsible for the extraction and management of the utilization of such resources. The role of an effective regulatory regime in promoting economic growth and development has excited considerable interest among researchers and practitioners in recent years (Jalilian et al. 2007; World Bank 2004).

One key question is how do regulatory structures contribute to the harnessing of the natural resource extraction for socio-economic development and transformation? Another is how can regulation be improved to ensure that it plays its expected role in the attraction of investment and the management of the utilization of natural resources in LICs?

Ghana is currently a lower-middle-income economy with per capita income of about US$1,100 and a population of 24 million (Government of Ghana 2013). The country is endowed with significant amounts of ferrous and non-ferrous minerals including, gold, bauxite, manganese, diamonds, and iron ore as well as hydrocarbons. It has been mining gold for over a century and currently ranks second only to South Africa in gold production in Africa. By virtue of its former name, Gold Coast, Ghana has historically been touted as being adorned with gold and other natural resources. The country's mining sector has also undergone various reforms including a regulatory transformation that is believed to have resulted in the improvement in the mining sector of the country.

Drawing largely on the case of Ghana, this chapter seeks to share the experience of regulation in the extractives sector and offers some perspectives on the purposes, content, and challenges of the practical regulation of an extractives sector in a resource-endowed lower-middle-income economy. The chapter looks at both the design and content of a regulatory system in the mining sector of Ghana and throws light on the practical challenges (technical and political) of implementation. In the light of the increasing allure of resource nationalism the chapter also takes a brief navigation into the manner in which relationships are established and maintained by the regulatory bodies with both large multinational companies and artisanal and small-scale (ASM) mining operations. An earlier related paper (Aubynn 2017) also examined the manner in which Ghana's regulatory standards have been influenced by various international initiatives including those of the following agencies: the IFC, EITI, and the ICMM, to highlight their impact on domestic regulations. Conclusions are drawn to underscore the importance of effective and collaborative regulations in maximizing the transformative potential of resource extractions in less developed resource-endowed countries.

There is an abundance of literature on natural resource extraction and its impact on resource-rich countries, and a growing body of work that focuses on natural resources governance (Auty 1993, 1998, 2000; Collier and Venables 2011; Roberts 2013) which informs the background to this chapter. The chapter also relies quite heavily on the author's own experience in the field as well as documentary materials at the Ghana Minerals Commission, the Ghana Chamber of Mines, the Ghana Statistical Services, the Bank of Ghana, and the ICMM.[1]

The rest of this chapter is organized as follows. Section 13.2 provides a brief but general examination of the role of regulatory structures in mineral resource development. Section 13.3 focuses on Ghana's case with a brief

[1] The author was until August 2017 the Chief Executive Officer of the Minerals Commission of Ghana, with nearly twenty years of previous experience in the extractives sector including oil and gas. Before he joined the Commission, he was the Chief Executive Officer of the Ghana Chamber of Mines. He has written a number of academic papers on mining-related subjects.

look at the trajectory of the country's regulatory structures since the colonial era. A recent related paper (Aubynn 2017) spells out more fully the contrasts between the policies of the early post-independence years; the reformed policies of the post-1983 period; and the gradual evolution of policies subsequently. The post-1980s regulatory reforms reveal an interesting dynamic in the contribution of mining to the economy and so this period is examined more fully in Section 13.4 in terms of both the regulatory reforms themselves and their impact on the various benefit streams within the economy. A key emerging outcome of recent regulations in resource economies has been the more active promotion of local content and value addition frameworks. Section 13.5 provides a brief examination of Ghana's local content policy framework and its implications for the integration of the country's mining sector into the economy. Section 13.6 concentrates on the regulatory challenges from the practical perspective of the regulator and provides some suggestions on how the extraction of Ghana's mineral resource can play a true catalytic role in Ghana's development. Finally, Section 13.5 draws some conclusions on the role of regulatory structures in enhancing the contribution of mineral resource extraction to the socio-economic development of the country.

## 13.2 The Role of Regulatory Structures in Mineral Resource Development

This section deliberately avoids the temptation of engaging with the intricacies of the definition of 'policy' and 'regulation'. Suffice to say that regulation policies are the 'rules of the game' governing decision-making by socio-economic actors—such as individual firms as producers or consumers. They are a purposeful course of action designed and implemented mainly by government agencies with the objective of shaping future outcomes in ways that are more desirable than would otherwise be expected. Furthermore, regulations have the force of law, prescribed by a superior or competent authority, relating to the actions of those under that authority's control. Regulation policies can take many forms and the specific forms of regulation policy adopted in developing countries have changed over time (Jalilian et al. 2007; Minogue 2006). Roberts (2013) asserts that regulatory variability seen in sub-Saharan Africa is based on the push and pull of domestic elite-driven governance orientations, including among others, post-conflict and constitutional transitions, transnational influences and impositions. Ghana like many African countries since independence has experienced regulatory variability ranging from command and control to relatively liberal structures, which also reflect the governance orientation and, to a greater extent, external influences and impositions (including influences and impositions, and conditions arising out of the World Bank and IMF). The outcome of a regulatory system can be assessed against the yardsticks of effectiveness and efficiency. To assess the effectiveness of regulation one needs to assess the extent to which it achieves the goals set by the government for the regulatory authority.

According to Parker (1999), a well-functioning regulatory system is one that balances accountability, transparency, and consistency. Similarly, effective regulatory agencies must not only operate within their legal powers but also be accountable for the consequences of their actions. *Transparency* relates to regulatory decisions being reached in a way that is revealed to all interested parties. Regulatory *consistency* ensures predictability and the 'level playing field' which also underpins legitimacy. Inconsistent regulatory decisions undermine public confidence in a regulatory system and could also create uncertainty for investors. One area that undermines regulatory consistency is political interference, which often alters the regulatory rules of the game for short-term political advantage. This makes a strong case for some kind of 'independent' regulator.

Jalilian and his colleagues (2007) argue that between the 1960s and the 1980s, market failure was used to legitimize direct government involvement in productive activities in many developing countries as they promoted industrialization through import substitution, invested directly in industry and agriculture, and extended public ownership of enterprises. However, the role of state regulation has since been redefined and narrowed to reflect the requirement for an undistorted policy environment in which efficient markets could thrive. This change of course has resulted from the apparent success of market liberalization programmes in some developed countries, and the evidence of the failure of state-led economic planning in developing ones (World Bank 1995).

## 13.3 Trajectory of Mining Regulations in Ghana

Since the date of Ghana's independence in 1957, several variants of mining and allied regulation policies have been implemented with the aim of ensuring the most effective exploitation of the country's mineral resources and the maximization of their benefits. The recent regulatory framework in the mining sector of Ghana was part of the deregulation framework adopted under the Structural Adjustment Programmes (SAP) of the early 1980s, which in a way reflects the role of external factors in grooming the regulatory framework in the country. The key driver for the chosen mode of regulation of the mining sector under the SAPs was the disappointing record of investment and the imperative to drive that higher. The central objective of regulation in the mining sector since the early 1980s has been to promote sustainable development and poverty reduction through the extraction of the country's mineral resources. Efficiency and effectiveness have also been key drivers of mineral sector regulation to ensure the greater attraction of foreign direct investment into the sector, increased mineral production and revenues, and from that enhanced foreign exchange earnings and greater employment.

However, prior to the 1980s and tracing back to colonial days,[2] other regulatory paradigms were adopted to try to ensure the sustainable and beneficial exploitation of the country's mineral resources. Two of those systems—the pre- and immediate post-independence policies, and those in place during the period of mining's decline from the 1960s—are examined in detail in Aubynn (2017). The analysis there underscores the point that the last three decades have seen significant progress in minerals production and benefits due largely to the implementation of relatively effective mining regulation policies.

## 13.4 The Regulatory Structures of the 1980s and Beyond

The steeply declining fortunes of mineral production during the 1970s and early 1980s have been blamed less on the absence of ore than on the overall impact of the macroeconomic malaise of the country, as well as production constraints of the sector (Hutchful 1996; Barning 1990). The unfavourable macroeconomic environment coincided with increased concerns over political unrest and the possibility of expropriation or sudden changes in taxation policies, as well as the prevailing laws on repatriation of profits. These factors combined precluded any new large-scale direct foreign investment (Warhurst and Bridge 1997: 2):

> The political and economic exigencies of the time, including the continuing decline in most economic indicators, made a compelling case for the government of the Provisional National Defence Council (PNDC) to begin negotiations with the International Monetary Fund (IMF) and the World Bank for economic assistance and policy guidance, which culminated in the adoption the Structural Adjustment Programme (SAP), locally christened the 'Economic Recovery Programme' (ERP) in 1983.

The policies and regulations introduced from that date typified the free market and private property ownership advocacy of neoliberalism. This stood in stark

[2] It is not known exactly when mining began in the country. Tsikata (1997) suggests that the existence of large mineral resources including gold, diamonds, bauxite, and manganese was among the key reasons for Britain's imperial adventures in Ghana during the nineteenth century, even though the country's influential position in mining dwindled steadily over that period (Acquah 1995; Dumett 1999).

discord with the previous state control of mining, which saw the state as dirigiste or neo-patrimonialist. As Ghana is a resource-rich developing country, it was no surprise that special attention in the Ghanaian reforms was paid to the mining sector to enable it to play a leading role in the revival of the ailing economy through the conversion of mineral assets into other national wealth. More specifically, Ghana developed a set of new mining codes that have come to best illustrate the so-called 'first generation of mining codes' in Africa since the 1980s (Campbell 2004; *African Agenda* 2003). According to Campbell (2004), Ghana's mining codes of the 1980s (Minerals and Mining Law 1986, PNDCL 153) reflect a new generation of liberalization approaches, including the privatization of state enterprise and very extensive deregulation. Numerous mining operations previously owned by the state were privatized and new private investment in the sector was actively encouraged. Section 13.4.1 takes a closer look at the key legislative and institutional reforms that were introduced as part of this major change.

### 13.4.1 *Mining Legislation*

The key legislative framework for mining in Ghana since the mid-1980s is laid down in the Minerals and Mining Law, 1986, PNDCL 153 (Law 153) as amended by the Minerals and Mining Amendment Act 1993, Act 475 (Act 475), and since 2007, amended and replaced by the Minerals and Mining Law 2006 Act 703. Within the current legal framework, all minerals in Ghana are vested in the president on behalf of and in trust for the people of Ghana. Thus, regardless of who owns the land upon or under which minerals are situated, the state is the owner of all minerals occurring in their natural state within Ghana's land and sea territory, including its Exclusive Economic Zone (EEZ).[3] According to the law governing mining in Ghana, the exercise of any mineral right requires a licence to be granted by the government minister responsible for mining, who in turn acts as an agent of the state for the exercise of powers relating to minerals.

The World Bank (2002: 152) notes quite correctly that building effective regulatory structures in developing countries is not simply an issue of the technical design of the regulatory instruments but also involves the quality of the supporting regulatory institutions and their capacity. During this period new institutions were established, such as the Minerals Commission and the Environmental Protection Agency (EPA), set up in 1986 and 1994 respectively to

[3] EEZ is a sea zone prescribed by the United Nations Convention on the Law of the Sea, over which a state has special rights regarding the exploration and use of marine resources, including energy production from water and wind. It extends no more than 200 nautical miles from the territorial sea baseline.

provide regulatory effectiveness. The Minerals Commission in particular has the authority under the constitution to regulate, manage, and advise the government on the utilization of mineral resources and coordinate policies in relation to minerals. It also seeks to ensure a one-stop shop for investors in the minerals sector in order to reduce the problems of complex bureaucracy and administrative inertia.

### 13.4.2 *Investment*

In order to attract venture capital into the mining industry in Ghana, the government also undertook a review of its tax and incentive regimes in respect of mining and introduced what may be described as relatively generous investment incentives for mining companies, including:

- Reduction of corporate tax from 50–55 per cent prior to SAP, to 45 per cent in 1986; later scaled down further to 35 per cent in 1994. In 2002, the corporate tax rate was further reduced to 25 per cent but was reversed to its 1994 level of 35 per cent in 2012 (Government of Ghana Budget 2012).
- Exemption from customs duties in respect of plant, machinery, equipment, and accessories imported specifically for mining.
- Sweeping changes, in 2012, in the fiscal imposts for mining. For example, capital allowances were increased from 20 per cent in the first year of production and 15 per cent per annum subsequently, to 75 per cent and 50 per cent respectively in 1986; further revised in 2002 to 80 per cent in the first year and 50 per cent subsequently. Capital allowances have since 2012 been changed to a flat rate of 20 per cent for a fixed period of five years.
- A negotiable foreign exchange retention regime of a minimum of 25 per cent.

The government, through SAP, also provided finances for the purchase of spare parts and materials, and for the rehabilitation of infrastructure such as roads and railways. This new climate of support led to significantly increased investment in the mining sector, making it a major springboard for the economic recovery of the 1980s up to the mid 1990s. What then has been the impact of these policy and regulatory changes on minerals production and the benefit streams for the economy since the 1980s? Section 13.5 briefly examines the various contributions of the mining or minerals sector to the economy of Ghana. An earlier paper (Aubynn 2017) looks at these same contributions in greater detail.

## 13.5 Contribution and Benefits Stream

The mining industry globally has been generally criticized for not contributing 'enough' to its host countries. In Africa, and Ghana in particular, the criticism has been heightened in recent decades when commodity prices, particularly for gold, have experienced an unprecedented bull rally (United Nations Economic Commission for Africa 2011; Africa Union Commission and United Nations Economic Commission for Africa 2009). The impression created, especially by anti-mining NGOs and certain sections of the media, is that mining is not only environmentally destructive but also does not make any meaningful contributions to the socio-economy of the country. This impression of extreme environmental degradation and limited benefits associated with the mining industry has often overshadowed the contribution it makes to the economy. In Section 13.5.1 I briefly review the sector's contributions to investment and associated capital inflows to the economy, and on domestic revenue, foreign exchange earnings, and employment.

### 13.5.1 *Capital Injection and Production of Minerals*

During the decade prior to 1985, the mining industry had virtually stagnated. There were no significant new investments into the country in general and the mining sector in particular. However, largely as a result of the policy measures described above, coupled with favourable world market prices for some minerals, a substantial re-capitalization for the mining industry, notably gold, has occurred. Specifically, between 1983 and 2016 the Ghanaian mining sector experienced cumulatively over US$16 billion in new foreign direct investment (FDI) for exploration, the establishment of new mines, and the expansion and rehabilitation of already existing ones (Aubynn 2017). This investment boom has led to the doubling of production of all minerals since 1983. Gold, by far the most important mineral, has increased more than tenfold. Production from the ASM sector has increased from less than 1 per cent in 1990 to over 30 per cent of total gold produced by 2013.

Furthermore, these investments have also provided enormous spin-off benefits including the provision of supporting infrastructure (roads, houses, schools, hospitals, and the extension of electricity), on-the-job training opportunities for Ghanaians and the transfer of technology.

#### 13.5.1.1 DOMESTIC REVENUE GENERATION

One of the major areas of benefits from mining has been and remains the generation of domestic revenue to the state. Details of the substantial increases in the various fiscal contributions are shown in Table 13.1.

**Table 13.1.** Contributions of the mining sector to domestic revenue, 1990–2012

| Year | Corporate tax | Mineral royalties | PAYE | Reconstruction levy | Total IRS (GRA) | % mining to total |
|---|---|---|---|---|---|---|
| 1990 | 2,825,941,158 | 1,893,436,000 | | | 52,818,068,300 | 8.94 |
| 1991 | 821,844,979 | 3,021,277,000 | | | 61,485,625,496 | 6.25 |
| 1992 | 455,051,883 | 4,545,804,000 | | | 74,931,531,366 | 6.67 |
| 1993 | 4,393,447,293 | 7,485,121,000 | 2,649,306,000 | | 113,236,997,000 | 12.83 |
| 1994 | 7,214,082,000 | 12,783,689,000 | 4,810,802,000 | | 166,595,941,000 | 14.89 |
| 1995 | 20,392,973,000 | 20,911,926,000 | 7,951,763,000 | | 275,513,201,000 | 17.88 |
| 1996 | 9,160,528,000 | 35,527,027,000 | 16,834,543,000 | | 424,491,908,000 | 14.49 |
| 1997 | 9,868,796,000 | 34,594,950,000 | 25,022,023,000 | | 605,782,577,000 | 11.47 |
| 1998 | 14,450,773,000 | 49,841,242,000 | 31,016,506,000 | | 785,436,693,000 | 12.13 |
| 1999 | 31,117,108,000 | 48,620,419,161 | 27,839,260,000 | | 901,663,758,000 | 11.93 |
| 2000 | 15,789,167,000 | 118,736,935,173 | 59,243,800,000 | | 1,409,445,273,000 | 13.75 |
| 2001 | 24,812,893,000 | 127,358,386,430 | 76,111,678,000 | 4,251,467,579 | 1,950,162,751,000 | 11.92 |
| 2002 | 23,501,158,000 | 153,452,471,032 | 101,457,668,000 | 26,474,633,878 | 2,757,747,781,032 | 11.06 |
| 2003 | 68,137,702,000 | 194,387,579,429 | 141,049,450,000 | 16,785,882,702 | 3,824,078,389,429 | 10.99 |
| 2004 | 100,331,114,000 | 215,743,706,000 | 134,357,711,000 | 36,346,622,100 | 5,333,114,704,000 | 9.13 |
| 2005 | 269,889,639,000 | 235,951,903,000 | 194,058,939,000 | 22,957,004,700 | 6,446,385,048,000 | 11.21 |
| 2006 | 404,361,775,000 | 316,254,789,000 | 216,525,776,000 | 11,085,262,400 | 7,333,916,866,000 | 10.20 |
| 2007* | 47,415,690 | 40,882,042 | 34,587,597 | – | 901,242,340 | 14.42 |
| 2008* | 73,554,697 | 59,004,892 | 47,139,242 | – | 1,222,272,177 | 15.32 |
| 2009* | 124,600,880 | 90,415,902 | 103,061,985 | – | 1,731,633,034 | 18.21 |
| 2010* | 241,578,780 | 144,697,000 | 132,469,710 | – | 2,441,331,842 | 21.29 |
| 2011* | 649,902,536 | 222,024,706 | 161,822,107 | – | 3,746,024,194 | 27.61 |
| 2012* | 893,773,828 | 359,392,853 | 207,495,934 | – | 7,461,202,977 | 20.72 |

*Note*: * Contributions for 2007 to 2011 are in GH¢; all others are in cedis.

*Source*: Minerals Commission, Statistical Overview of Ghana's Mining Industry (1990–20); Internal Revenue Service/GRA (2004–11).

The collection of these taxes and royalties over the years has positioned the mining sector as the leading contributor to the domestic public revenue of Ghana. Between 1995 and 2003, for example, the total annual average contribution by the sector to Ghana's domestic revenue was the equivalent of approximately US$40 million, representing an average of approximately 10 per cent of total government domestic receipts. However, the drastic decline in gold and other commodity prices since around 2011, coupled with the expansion in domestic revenue sources, especially resulting from the novel inclusion of revenues from oil production,[4] has meant a dip in the sector's contribution to overall domestic revenue generation to a low of 16 per cent in 2016 (Aubynn 2017). That notwithstanding, the mining sector has maintained its leadership as the single most important revenue generator for Ghana.

#### 13.5.1.2 FOREIGN EXCHANGE

In terms of foreign exchange or receipts from merchandised export for Ghana, mineral exports have maintained a consistent leadership as the country's number one earner since 2000. The share of minerals in the total foreign exchange earnings of Ghana has increased from 14 per cent in 1990 to an average of 41 per cent since 1993, outpacing cocoa as the country's most important foreign exchange earner (ISSER 1996–2010; Bank of Ghana 2010). As shown in Figure 13.1, in 2012, mining earned over US$4 billion or 42 per cent of gross merchandise exports value and has sustained a similarly high percentage contribution in spite of the inclusion of oil in Ghana's basket of exports. Gold has contributed between 86 per cent and 95 per cent of the earnings from the country's total mineral exports since 1986.

Critics of the industry have questioned the extent to which the high foreign exchange earned from mineral exports is retained in Ghana (Akabzaa 2001; Akabzaa and Darimani 2001). Available records at the Minerals Commission, the Ghana Chamber of Mines, and the Bank of Ghana, however, indicate that on the average, mining companies have returned more of their export proceeds than is statutorily required (GCoM 2012, Aryee 2014). This is contrary to the commonly held view that mining companies retain their earnings largely in offshore accounts.[5]

Table 13.2 shows that since 2009, mining companies have kept an average of about 27 per cent of their exports in overseas accounts and returned approximately 72 per cent to Ghana. Similar ratios pertained in earlier years. For instance, between 1998 and 2000, a quarter of export earnings (US$177 million)

[4] Ghana began the production and export of oil in commercial quantities during the last quarter of 2010.

[5] The former Governor of the Bank of Ghana, Paul Acquah, confirmed this in a speech he delivered at the 79th Annual General Meeting of the Ghana Chamber of Mines; see Ghana News Agency report on 9 June 2006; ghanadistricts.com.

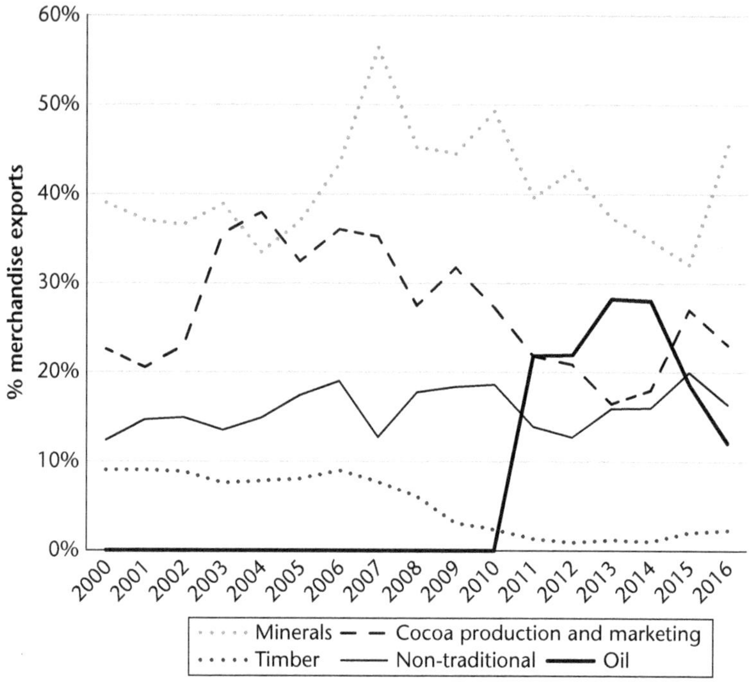


**Figure 13.1.** Trends in total exports by key sectors
*Source*: author's illustration based on Minerals Commission data.

**Table 13.2.** Total mineral revenue and mineral revenue returned, 2009–12

| Year | Total mineral revenue (US$) | Mineral revenue returned (US$) | % of mineral revenue returned |
|---|---|---|---|
| 2009 | 2,384,836,583 | 1,812,255,608 | 76 |
| 2010 | 3,290,792,703 | 2,222,901,896 | 68 |
| 2011 | 4,245,370,284 | 3,173,491,961 | 75 |
| 2012 | 4,525,657,336 | 3,268,084,143 | 73 |

*Source*: data from Ghana Chamber of Mines (2013).

was repatriated to Ghana. But by the end of the commodity price boom in 2012, the sector was returning approximately US$3.2billion, representing 73 per cent of total mining exports receipts.

#### 13.5.1.3 EMPLOYMENT GENERATION

In terms of employment numbers, the massive retrenchment occasioned by the implementation of SAP from 1983 onwards had an initial debilitating effect on employment in the mining sector. Before the 1980s, the sector was estimated to have directly employed about 27,000 Ghanaians. By 1987 the number is believed to have shrunk to 15,000 largely due to the massive retrenchment exercise at the long-established Obuasi Mines. However, total direct employment in the mineral sector surged back to about 23,000 by 1995. By 2004, this number had reduced to approximately 18,000 Ghanaians, attributable to the closure of the Teberebie Mines and redundancies at the Damang Mines.[6] According to the 2012 Annual Report of the Ghana Chamber of Mines, the sector at that time directly employed approximately 23,000. The Minerals Commission put it significantly higher at 27,000 to include other mining contractors. But of great significance is the fact that analysts including the Chamber of Mines have long estimated that indirectly the mining industry offers employment of about five times this number through local purchases and the award of various kinds of contracts (Ghana Chamber of Mines 2001). These estimates, which relate only to large-scale mining, are substantially increased when we also consider the large army of youth employed in the informal, less-regulated artisanal mining sector of Ghana, which has been variously estimated at 500,000–1,000,000 (e.g. McQuilken and Hilson 2016).

Critics of the industry argue that open-pit mining lacks the capacity for significant employment generation, due largely to its capital intensity and short lifespan.[7] This is true to some extent and examples to make the point are shown in Aubynn (2017). But, flipped to the other side, one can argue that the breath of the newer technology of open-pit mining in the early 1990s has been rather timely, allowing for new operations in areas such as DamangWassa, Akyempim, Iduaprim, Tebeberbe (in the Western Region), Amansie, Chiraano, Kenyaasi, Nkawkaw, Akyim, and so on, which were not previously known to produce gold, to be brought on stream and actively benefit from the production of gold. This development provided a conduit to absorb the mass of underground miners who would otherwise have been retrenched following the evidently poor performance of the country's underground mines.

From the foregoing, it is clear that the prioritization of the mining sector under SAP for major support has paid good dividends in terms of employment, foreign exchange earnings for the country, and capital injections. It remains to be seen how these dividends have translated into social improvement and sustainable development. That question is discussed in Section 13.5.1.4.

[6] This estimate includes employees in the exploration area, contractors as well as suppliers to the large-scale mining companies. It does not include artisanal and small-scale miners.

[7] Open-pit gold mining has become common in Ghana only in recent years (i.e. since the 1990s) relative to underground mining, as the fortunes of the latter, which was then dominant in the country, began to dwindle.

#### 13.5.1.4 GDP AND INTEGRATION INTO THE ECONOMY

The data suggest that the contribution of mining to GDP has seen steady improvements since the implementation of the SAP in the 1980s. In 1991 the industry contributed only a little over 1 per cent to Ghana's GDP. Since 1992, mining has contributed an average of 5 per cent to Ghana's GDP, and like the other indicators, this has steadily improved since the 2000s, reaching a high of 14 per cent in 2012 (ISSER 2005–12).

However, in an earlier paper, I observed a rather poor linkage of the sector with the rest of the national economy (Aubynn 2015). The relatively low, but increasing contribution to GDP, coupled with low linkage of mining with the economy, can partly be explained by the absence of an industrial base for the manufacture of inputs required by the rather technology-intensive industry, leading to most being imported. The steady increases in the mining GDP contribution may, however, reflect the gradual increase in local participation in mining support services such as drilling, exploration and explosives, cement, lime, and high density pipe (HDPE) manufacturing and supplies in the sector. The topic of local content is critical in this respect and Aubynn (2017) looks in greater detail at the regulatory challenges to which this matter gives rise.

### 13.5.2 *Local Content: The Evolution of Regulatory Approaches*

In Ghana, both the Minerals and Mining Act Law, 2006 (Act 703) and its predecessor, Act 153 of 1986, sought to promote a localization policy and facilitate improved production linkages. Indeed Clause 50(3) of Act 703 calls for the eventual 'localization' of mining staff. In particular, the passage of the Minerals and Mining (General) Regulative Instrument (LI), 2012-LI 2173, marked the beginning of the implementation of local content policies in the mining sector.

The LI 2173 provides a clear interpretation of Act 703 and focuses on three areas, namely: (i) the employment and promotion of a local workforce, (ii) the procurement of locally produced goods and services, and (iii) additional licensing and reporting requirements. Of most interest for our present purposes are the first two, as discussed in greater detail in Aubynn (2017).

#### 13.5.2.1 THE STATE OF IMPLEMENTATION

Since the regulation came into force in 2013, the Minerals Commission has worked with the Ghana Chamber of Mines for its practical implementation. Most of the major mining companies submitted both their localization and procurement plans to the Commission for approval by the end of 2014 and 2015. These plans have been reviewed by a committee. Between 2012 and 2013, the Commission undertook a process of broad consultations based on which an initial list of eighteen items earmarked for local procurement was pruned down to eight. While this may seem a significant reduction, in reality, the eight items constitute over 60 per cent of the value of the total local procurement budget of the companies concerned. In 2014, following the successful implementation of the first edition of the local procurement list, the latter was increased by an additional eleven items to a total of nineteen. So far an average of 80 per cent compliance has been recorded by the Commission with a total of US$183 millions worth of goods and services procured at home in Ghana in 2015. These were purchases that were otherwise likely to have been made from abroad.

With respect to the localization of employment, all the major mining companies submitted their plans, which were reviewed by a committee established at the Commission and that includes representation from the Immigration Services. This means that approval for residence and work permits for expatriate staff of holders of mineral rights will depend on satisfactory compliance with the localization regulation.

## 13.6 Key Regulatory Challenges

The previous sections have underscored the very impressive change to the key indicators of mining and the macro and socio-economy of Ghana in the past three decades. The assumption has been that the impressive performance of the mining sector has been due in part to the regulatory changes introduced since the reforms starting in 1983. However, as a student and practitioner of mining policies and sector regulation in Ghana for close to two decades, one cannot help but observe a number of regulatory challenges and bottlenecks which need to be addressed if Ghana is to achieve the full benefits of its minerals extraction. The following four are prominent amongst the challenges and are discussed in some detail in the ensuing sections:

- multiple regulations and inter-institutional conflicts
- capacity of regulators
- political will and interference
- the balancing act of dealing with large-scale multinational and local small-scale mining sectors.

### 13.6.1 *Multiple Regulations and Inter-institutional Conflicts*

The mining sector is, arguably, not only the most regulated sector in Ghana, but also the sector whose regulation engages by far the largest number of regulatory institutions. The 1992 Constitution and the Minerals and Mining

Law of 1986, as amended in 2006, mandate the Minerals Commission to regulate and manage the utilization of Ghana's mineral resources, including the coordination of policies in relation to them.[8] However, currently at least six other institutions, namely the EPA, Forestry Commission, Water Resources Commission, Ghana Revenue Authority, Ministry of Finance, and the Bank of Ghana, exercise additional strong and direct regulatory authority over the sector. The resulting complexity of regulations and interpretation as well as conflicts among regulators have in many cases led to undue delays and bureaucracy in permit processing. A classic example of the institutional conflict is the confusion between the Minerals Commission (the Inspectorate Division) and the EPA as to who has the ultimate authority to regulate tailings dam construction. Both institutions claim the authority to regulate the construction of any tailings dam, to the effect that while the EPA insists on the plastic lining of all tailings, the Minerals Commission argues that the imperative of lining of tailings dams with plastic sheeting should be dependent on the nature of the soil. It argues further that, where proven to be technically feasible, clay lining would achieve the same outcome. This has created serious concerns among mining companies, which continue to seek clarity in the area. Similarly, in October 2015, the Bank of Ghana announced certain regulatory changes in respect of the export of gold in the small-scale mining sector. The areas of new regulations were deemed to be outside the scope of the Bank of Ghana. This led to the association of Licensed Gold Exporters (LGE) taking the matter to court. In the event, the government found it necessary to temporarily suspend the granting of new export licences and renewal of expired ones. Even though the matter has been successfully resolved out of court, one can surmise that the loss of revenue and foreign exchange during the period of the suspension has been considerable.

These multiple regulations and their attendant inter-institutional conflicts not only defeat the purpose for which the Minerals Commission was established, i.e. to be a one-stop shop for investment in the mining sector, but also create needless layers of bureaucracy and institutional tensions. The recent merger of the hitherto stand-alone Mines Department with the Minerals Commission as the latter's Inspectorate Division, and the subsequent relocation in the same office premises, has clearly improved regulatory efficiency and effectiveness. Thus, the original idea of a one-stop-shop Commission modelled on the Australian Department of Minerals and Mining is not only achievable but also relevant today if regulatory efficiency and the effectiveness of the Minerals Commission are to be enhanced.

[8] Minerals Commissions Act 1993, Act 450 section 2(1).

### 13.6.2 *Capacity of Regulators*

Institutions and regulations are only as good as the capacity of their human resources. After over a century of mining in Ghana and with relatively good training institutions such as a dedicated university (University of Mines and Technology) for the training of mining professionals, Ghana can be said to have relatively good-quality human resources in the area. As indicated earlier, Ghana is endowed with substantial mineral resources and its predominant mineral, gold, appears to spread almost across the nation, leading to a number of major mines and numerous small-scale (legal and illegal) mining activities across the country. As of September 2016, the Minerals Commission had total staff strength of 220, less than 40 per cent of which is technical. Most of the technical staff have solid academic backgrounds (mostly second degree holders) with significant industry experience and are generally highly competent. However, more than half of the Commission's technical staff are in their fifties, posing potential challenge for succession.[9] Similarly, the sheer volume of monitoring, inspections, and extension work required makes it impossible for fewer than a hundred staff to fully and effectively accomplish all the tasks that are required. In particular, the absence of adequate staff at district level has been blamed as one of the reasons for the increasing spate of illegal mining operations in the country. The inadequacy of staff has also meant insufficient inspection and monitoring of the operating mines. For example, the current situation in which the country's flagship Obuasi mine has had to be placed under care and maintenance for a revised programme of development has also been partly blamed on a failure of the Inspectorate Division to effectively monitor the development of the mine to pick up early warning signs.[10] The lack of sufficient staff with the relevant skill sets is not dissimilar to the generally unattractive remunerations and other conditions of employment that most public sectors face—particularly in African countries. The Commission is currently undertaking the refurbishment of some of its regional and district offices to make these more attractive and also to enable them to accommodate the anticipated increase in the number of staff.

Similarly, equipment and other resources that technicians need for their work to be effective raise further significant capacity challenges. Apart from the Machine Laboratory, which has been described as being in good state, the Environmental and Assay laboratories currently require major investment in order to meet the challenges of a modern mining industry.

[9] The statutory retirement age in Ghana is sixty.
[10] The Obuasi mine is estimated to have over 7 million ounces of gold reserves.

### 13.6.3 *Political Will and Interference*

One of the key features of the post-SAP regulatory structures in Ghana has been the reduction in the previous undue bureaucracy and political interference in the minerals title-application processes. For example, the application for minerals title is now based on a first-come-first-served principle and should begin at the Minerals Commission. The law endows the minister responsible for mining with the ultimate power to grant or terminate any mining title. Yet, the same law ensures that the power is not capriciously exercised. The regulations enjoin the minister to seek the advice of the Minerals Commission and the minister's decisions are largely based 'on the advice' of the Minerals Commission.[11] This principle has largely held in Ghana. However, there have been a few, even if outlying, instances where ministers have instructed that specific concessions be processed by the Commission for some companies, contrary to the advice offered. Similarly, it is common knowledge, backed by anecdotes, that some individual politicians (not necessarily representing the position of government or parties) have used their influence to intervene in the seizure of equipment used by illegal miners in their operations. Even though this practice may be the exception rather than the rule, it needs to be checked to avoid gradual escalation and also to maintain the integrity of regulations and their implementation. To overcome the temptation of political interference and to deepen transparency as well as reduce other human intervention in the minerals titling (granting of concession) processes, the Commission is establishing a digitized mineral cadaster system that allows applications for mineral titles to be made via the Internet.[12] This, combined with an open tender system, will hopefully reduce to the barest minimum the incidence of interference and human intervention in the grant of minerals titles and improve the efficiency and transparency of the system.

Another area of concern is the creeping 'regulator uncertainty' that occurs in the aftermath of political change. It is fast becoming a norm that new political parties, which win general elections, begin office with a significant dose of suspicion of the regulator and often question the regulator's loyalty to the new government.[13] In the event, some regulatory agencies become the punch-bag for needless public ridicule and opprobrium. During this period of 'sparring' for trust between political appointees and technocrats, instances of

[11] The law enjoins the minister to seek the advice of the Minerals Commission, and a minister who disagrees with the Commission must provide reasons for the disagreement.

[12] This process is expected to 'go live' by April 2017.

[13] This situation may have arisen and perhaps have been made worse by instances in which some public servants have shown open bias and declared support for certain political parties.

direct intervention in the day-to-day activities of regulators become common. One example, widely publicized in the local and social media, is the recent action (in May 2017) of the minister of Lands and Natural Resources announcing the interdiction of nine district mining officers of the Minerals Commission and their assistants without recourse to the policies of the Commissions. Similarly, normal processes of recruitment by the Commission have had to be halted at the instance of the sector minister for an indefinite period, compounding the capacity challenges intimated earlier in this chapter. These burgeoning attempts at interfering and truncating the activities of regulatory agencies during political transitions have occurred under the governments of both of the dominant political parties in Ghana, the NDC and the NPP. This has tended to directly affect not only the image and credibility of public institutions but also the psyche of employees who work in them and who feel unjustifiably victimized as a result of political change.

### 13.6.4 *The Balancing Act of Dealing with Large-scale Multinational and Local Small-scale Mining Sectors*

Ghana has two broad categories of mining operations—the large-scale, operated largely by multinational mining giants such as Newmont, Gold Fields, and Anglogold Ashantis; and a large mass of small-scale miners (SSM), including many unlicensed operators locally known as the *galamseys*. Ghana is among the few countries where small-scale mining is legalized and regulated (Small-Scale Mining Law 1989, PNDC Law 218). Yet, it is also an area where an estimated 70 per cent of the over one million operators work outside the confines of the law.

As noted in earlier sections of this chapter, the large-scale mining (LSM) companies produce the bulk of Ghana's gold (an annual average of 3 million ounces in 2011–15). Gold production by the SSM cannot be dismissed as unimportant, however, as the contribution from them has grown from negligible levels in the 1990s to over a million ounces (since 2013). Laws and regulations exist for both categories of miners and these are managed mainly by the Minerals Commission and the EPA. Regulation of LSM operations has been largely formal and relatively simple.

However, the same cannot be said of the SSM. The mass of SSMs, dominated by the galamseys, have been associated with the devastation of the physical environment of some areas, including the massive pollution of a number of river bodies in Ghana such as the Ankobra, Tano, and Abirem. The galamseys operate substantially in utter defiance of the law. Interestingly, their negative operations are known to be carried out in the full glare of law enforcement institutions and in disregard of any real sense of sound environmental management. There have been instances, often reported in the media, where unregistered mining operators have invaded the concessions of some LSM companies. The most recent (February 2016) case in point is the invasion of AngloGold Ashanti property at Obuasi by a large mass of illegal miners (estimated at five thousand). Clearly, while the LSMs are held to strict regulatory standards, the same cannot currently be said of the SSMs.

The question that has often been raised is why SSMs are not, in fact, held to the same regulatory standards as the LSMs. Several reasons have been assigned to explain their defiance of the law in spite of the existence of clear laws and regulations to guide their operations. These reasons include the long and complex application process, lack of finance in an activity which has increasingly become capital intensive, lack of knowledge about the licence acquisition process as well as the lack of capacity fully to enforce the law due to the sheer numbers of people (estimated at 1.5 million) involved in SSM in the country vis-à-vis the available resources for policing.

While these stated reasons are no doubt very important and potent, the major challenge, in my view, is the lack of political will by successive governments to resolutely address the challenges posed by illegal SSM. The culprits here are the two biggest political parties, the NDC and NPP, which are the only political parties to have run the affairs of government under the Fourth Republic constitution since 1992. As noted in two previous papers (Aubynn 2006, 2009), successive governments have considered the political economy of enforcing the SSM regulations potentially too costly to their electoral fortunes by virtue of the number of SSMs representing a potential source of significant electoral votes and the lack of alternative employment opportunities.[14] They have had to engage in a difficult balancing act of attracting and protecting FDI in the country while at the same time ensuring the availability of opportunities for local citizens, including the opportunity to engage in mining. In the event, politicians of successive governments have often competed in their overt or tacit support for the operators of illegal mining.[15] This lack of political will to enforce the laws and regulations on SSMs has not only strengthened the hands of these unregistered operators but also emboldened them to invade the concessions of some large-scale mining companies. Some analysts have attributed the recent (February 2016) invasion of AngloGold Ashanti property at Obuasi by illegal miners as an

[14] This becomes more pronounced during national election years as major political parties compete in their tacit or overt declaration of support for galamseys and their assurance of assistance to them once they win power.

[15] In 2012, the arrest of a group of illegal mining operators by the National Security in the Obuasi area was described by a member of parliament as 'state terrorism'. Some people have also interpreted the failure of the government to act decisively in the face of the illegal occupations of the Obuasi mine as a lack of political will to act in the face of impending elections. It must, however, be said that there have been significant efforts by the current NPP government to fight the menace of galamseys in a more comprehensive manner.

eloquent expression of this 'emboldenment' occasioned by the lack of political will to enforce the laws. It is imperative that a bipartisan approach is found to the enforcement of the country's minerals and mining laws if a race to the bottom, with devastating environmental consequences, is not to take place. There should be a collective understanding that illegal mining is a 'no-go' while encouraging and supporting regularized SSMs.

## 13.7 Conclusions

The role of natural resources extraction in the socio-economic development and transformation of resource-endowed LICs is no longer a myth. Evidence abounds, in the literature, about the potential of resource-endowed countries to harness their resources to propel growth and development. What remains to be fully understood, however, is the role that a regulatory framework can play in promoting and accelerating such transformation and development. Drawing on the case of Ghana, this chapter has discussed in some detail the powerful influences that regulatory structures can have in changing the dynamics of resources extraction to benefit a country's growth and development. The study did not attempt to establish a direct correlation between specific regulations and their impact on mining and economic growth. However, significant improvements in all the macroeconomic indicators on mining, including FDI inflows, production, domestic revenue, and foreign exchange, since the implementation of Ghana's post-1980s liberalized mining regulatory framework, provides ample evidence of the potential of a well-managed liberalized regulatory regime to catalyze minerals-driven economic growth in the country. The recent (2013) implementation of new localization and content policies is a conscious effort to more fully integrate mining into the local economy.

The ability of the state to provide effective regulatory institutions with high levels of technical and other capacities is key to how well regulatory regimes can drive the transformation and economic development catalyzed by mineral extraction. The establishment of institutions such as the Ghana Minerals Commission and the EPA, with relative independence, during the 1980s and 1990s respectively, and the relative *efficiency* of the regulatory policies and instruments that they are asked to operate, are likely to have contributed in no small measure to the relative success of Ghana's case. Collaboration among regulatory agencies as well as regulated organizations is also imperative for the efficiency and effectiveness of regulation. The practice of regulation by 'policing' and insistence must give way to regulation by collaboration and assistance. Similarly, the implementation and adaptation of international standards have proven to be effective tools for addressing some of the capacity challenges of regulatory institutions, including the transfer of internationally acceptable practices to developing countries in general and Ghana in particular.

There are, however, some critical challenges that need to be addressed if the benefits of regulatory reforms are to be harnessed fully. For example, creeping political interference in the affairs of these regulatory institutions, inadequate human resource capacity, multiple regulations, and inter-institutional conflicts threaten to reduce or significantly erode the gains made by the new regulatory structures in Ghana.

## References

Acquah, P. (1995). 'Natural Resources Management and Sustainable Development: The Case of the Gold Sector of Ghana'. United Nations Conference on Trade and Development (UNCTAD). Available at: http://unctad.org/en/Docs/pocomd41.en.pdf.

Africa Union Commission and United Nations Economic Commission for Africa (2009). *Africa Mining Vision*. Addis Ababa: AUC-UNECA.

Akabzaa, T. (2001). *Boom and Dislocation: The Environmental and Social Impacts of Mining in the Wassa West District of Ghana*. Accra: Third World Network Africa.

Akabzaa, T. and Darimani, A. (2001). 'Impact of the Mining Sector Investment in Ghana: A Study of the Tarkwa Mining Region', *Structural Adjustment Participatory Review International* (SAPRI). Available at: https://pdfs.semanticscholar.org/db7d/5a2fe09a7c14bf82dddee1bd5cb6ea32d003.pdf.

Aryee, B. N. (2014). 'Strengthening the Impact of the Mining Sector on Development: Supporting More Ghanaian Mining Companies to Industrialization and Structural Transformation Agenda', A presentation by Mineral Commision at the National Policy Dialogue on Economy-wide Consequences of Ghana's Dependence on Minerals. Accra: Third World Network Africa, March.

Aubynn, A. (2006). '"Live and Let's Live"—The Relationship between Artisanal/Small-scale and Large-scale Miners in Ghana: The Abosso Goldfields Experience', in G. Hilson (ed.) *Small-Scale Mining, Rural Subsistence, and Poverty in West Africa*. Rugby: Practical Action Publishing.

Aubynn, A. (2009). 'Sustainable Solution or a Marriage of Inconvenience? The Coexistence of Large-scale Mining and Artisanal and Small-scale Mining on the Abosso Goldfields Concession in Western Ghana', *Resources Policy*, 34(1–2): 64–70.

Aubynn, T. (2015). 'Mineral Resource Policy Dynamics and the Contribution of Mining to Ghana's Development', in K. Appiah-Adu and M. Bawumia (eds), *Key Determinants of National Development: Historical Perspective and Implication for Developing Economies*, 187–211. Surrey: Gower Publishing.

Aubynn, T. (2017). 'Regulatory Structures and Challenges to Developmental Extractives: Some Practical Observations from Ghana', UNU WIDER Working Paper 179/2017. Helsinki: UNU-WIDER.

Auty, R. M. (1993). *Sustaining Development in Mineral Economies: The Resource Curse Thesis*. New York: Routledge.

Auty, R. M. (1998). *Resource Abundance and Economic Development*. Helsinki: UNU-WIDER.

Auty, R. M. (2000). 'How Natural Resources Affect Economic Development', *Development Policy Review*, 18: 347–64.

Bank of Ghana (2010) *Annual Report 2010*. Accra: Bank of Ghana.

Barning, K. (1990). 'A Review of Mineral Exploitation Activities in Ghana 1984–1990', in K. Barning (ed.), *Symposium of Gold Exploitation in the Tropical Rainforest Belt of Southern Ghana*, 6–22. Accra: Minerals Commission.

Campbell, B. (ed.) (2004). 'Regulating Mining in Africa: For Whose Benefit?' Discussion Paper 26. Uppsala: Nordiska Afrikainstitutet.

Collier, P. and T. Venables (eds) (2011). *Plundered Nations? Success and Failures in Natural Resource Extraction*. Basingstoke: Palgrave Macmillan.

Dumett, R. (1999). *El Dorado in West Africa: The Gold Mining Frontier, African Labour and Colonial Capitalism in the Gold Coast, 1875–1900*. Athens, OH: Ohio University Press.

Ghana Chamber of Mines (2001). *Annual Report of the Ghana Chamber of Mines*. Accra.

Ghana Chamber of Mines (2012). *Annual Report of the Ghana Chamber of Mines*. Accra.

Ghana Chamber of Mines (2013). *Annual Report of the Ghana Chamber of Mines*. Accra.

Government of Ghana (2013). 'Public Expenditure and Financial Accountability (PEFA) Performance Review', Final Report, ECORYS, June. Accra.

Hutchful, E. (1996). 'Structural Adjustment in Ghana 1983–1994', in P. Engberg-Pedersen, P. Gibbon, P. Raikes, and L. Udsholt (eds), *Limits of Adjustment in Africa: The Effects of Liberalization, 1986–94*. Copenhagen and Oxford: Centre for Development Research.

Institute of Statistical and Social Research (ISSER) (1996–2012). *The State of Ghana's Economy*. Accra: University of Ghana.

Jalilian, H., C. Kirkpatrick, and D. Parker (2007). 'The Impact of Regulation on Economic Growth in Developing Countries: A Cross-country Analysis', *World Development*, 35(1): 87–103.

McKinsey Global Institute (2013). 'Reverse the Curse: Maximizing the Potential of Resource- driven Economies'. London: McKinsey & Company.

McQuilken, J. and G. Hilson (2016) *Artisanal and Small-scale Gold Mining in Ghana: Evidence to Inform an 'Action Dialogue'*. London: IIED.

Minogue, M. (2006). 'Apples and Oranges: Comparing International Experiences in Regulatory Reforms', in M. Minogue and L. Carino (eds), *Regulatory Governance in Developing Countries*. Aldershot: Edward Elgar.

Parker, D. (1999). 'Regulation of Privatised Utilities in the UK: Performance and Governance', *International Journal of Public Management*, 12(3): 213–35.

Roberts, C. W. J. (2013). 'Extractive Sector Regulations in Africa: Old Practices and New Models for Change', in H. Besada (ed.), *Governing Natural Resources for Africa's Development*, 64–91. Ottawa: North-South Institute Forum.

Tsikata, F. S. (1997). The Vicissitudes of Mineral Policy in Ghana. *Resource Policy*, 23(1–2): 9–14.

United Nations Economic Commission for Africa (2011). *Minerals and Africa's Development: The International Study Group Report on Africa's Mineral Regimes*. Addis Ababa: Economic Commission for Africa.

Warhurst, A. and G. Bridge (1997). 'Economic Liberalisation, Innovation and Technology Transfer: Opportunities for Cleaner Production in the Minerals Industry', *Natural Resources Forum*, 21(1): 1–12.

World Bank (1995). *Bureaucrats in Business: The Economics and Politics of Government Ownership*. Oxford: Oxford University Press for the World Bank.

World Bank (2002). *World Development Report, 2002: Building Institutions for Markets*. Washington, DC: World Bank.

World Bank (2004). *Doing Business in 2004: Understanding Regulation*. Washington, DC: World Bank.

# 14

# The Taxation of Extractive Industries

## Mining

*James M. Otto*

## 14.1 Mineral Taxation: A General Overview

### 14.1.1 *Introduction*

Governments have probably been taxing mines ever since governments began taxing economic enterprises. What have governments learned over time? Basically you need to achieve balance—if you tax too heavily in the short term, over the long run the tax base is reduced because investors will shy away from new investment, and if you tax too lightly, society may feel slighted that it is not receiving its fair share of benefits. Though the above principle is simple, its implementation is not, and tax policy makers grapple with how to create a balanced tax system on an ongoing basis. Mineral prices fluctuate widely, more so than many other commodities, which complicates achieving a balanced approach. While there are tax tools that assist in accommodating wide price fluctuations, many governments simply adjust rates and bases to react to commodity price cycles. This chapter examines some of the issues that are of current interest to tax policy makers.

### 14.1.2 *Mineral Rents*

The academic literature is replete with the treatment of the concept of economic rent—a surplus of income that can theoretically be taken away from an investor without altering its economic behaviour. Economists have long toyed with the concept that the taxation of a mine can be adjusted to capture its economic rent. In practice, few nations attempt to capture all economic rent, but all taxes, and especially taxes such as income tax, excess profits taxes, and additional profit taxes do appropriate at least some economic rent for the benefit of the state. It is beyond the scope of this short discourse on mine taxation to delve into the extensive literature on the application of economic theories concerning mineral rents to the design of taxation policy in the mineral sector.[1] While tax policy makers may grasp the theoretical concept of economic rent, most are more concerned about achieving a real-world balance that satisfies investors and society.

### 14.1.3 *Evolution of Mineral Taxation Schemes*

Mineral-sector taxation systems tend to evolve, reflecting the current state of the business cycle. When mineral commodity prices are high, mineral-led economies tend to adjust their tax systems to capture more of the perceived excess profits, and when prices are low, systems are adjusted to allow mines to remain economically viable. The most common way to adjust the mineral-sector taxation system is through the manipulation of royalty schemes, because such schemes do not affect other types of taxpayer. Recently, as a result of the commodity super-cycle commencing around 2002, there was a resurgence of interest in self-adjusting taxation approaches designed to appropriate 'excess profits' based on either rate-of-return measures (Australia, Liberia, Malawi, Solomon Islands, and Zimbabwe) or annual profit-to-cost ratio principles (Chile and Peru).[2] Tax schemes imposed during times of rising prices are vulnerable to repeal when prices fall or investors shy away. For example, Australia imposed a resource rent-type tax when the super-cycle became apparent but repealed it shortly thereafter when prices dropped. Likewise, new taxes driven by the super-cycle in both Zambia and Mongolia were short-lived, and other nations have reconsidered their royalty rates.

Governments have several direct ways of participating in the sharing of income generated by mines. The simplest way is through taxation. Another route is for the government to establish a state mining enterprise that itself operates mines, or to enter into some sort of joint-venture interest arrangement with a private investor that entitles it to a share of profit or dividend distributions. However, the trend over the past several decades has been to emphasize taxation and few nations today directly mine minerals or take a majority equity position in mines. Worldwide, the trend has been to privatize state-owned mines (for example, in Malaysia, Peru, Poland, and Russia) and today their governments look to taxes, not dividends, for fiscal revenues.

[1] For an introduction to mineral rent theory and taxation see Tilton and Guzman (2016), Boadway and Keen (2013), Land (2010), Otto et al. (2006), Cordes (1995), and Land (1995).

[2] For a history of resource rent types of tax applied to mining see Land (2010).

Other nations that formerly owned mines have partially privatized them by reducing the state's equity to a minority share, such as ZCCM Ltd. in Zambia.

Over the past two or three decades there has been a trend to rely on laws rather than negotiated agreements to set out taxation schemes for large projects. For example, Indonesia, which developed the 'Contract of Work' that has been widely emulated worldwide in mining agreements since the early 1970s, has abolished such agreements and now miners are subject to tax under the general tax laws. Negotiated mining agreements played an important role in many mineral-led economies that lacked an adequate system of laws to accommodate large-scale projects, but as legal systems have matured, the need for them has waned in many nations. In this author's opinion, the time for such agreements has passed, and nations that still use them should concentrate their efforts on strengthening the underlying system of laws, rather than spending the considerable time and effort it takes to negotiate them.

When considering fiscal reform, whether through changes to taxes or equity options, it is important to examine the system as a whole, not just a component part. For example, when comparing royalty rates in different nations, one should not be too concerned if the rate is higher or lower than in another nation because other tax types may offset the difference in royalty (for example, a lower or higher income tax rate). Today, many nations make use of project fiscal modelling to aid with tax reform policy analysis. Mine fiscal models allow a holistic assessment, taking into account the impact of all taxes on a typical or model mine, and provide the ability to carry out a sensitivity analysis of the impact of various scenarios on measures such as internal rate of return and total effective tax rate. For example, Peru was interested in changing its approach to mine taxation and it commissioned fiscal modelling to assess its competitive position with other countries,[3] conducting fiscal scenario modelling exercises before moving forward with two rounds of mineral-sector-related fiscal legislation. Such analysis has become more routine and accessible with the availability of tools such as the Fiscal Analysis of Resource Industries Methodology and supporting software developed by the IMF (IMF 2016). Many governments that lacked confidence in their capacity to conduct fiscal system analysis have sought outside assistance from organizations such as the IMF, the Commonwealth Secretariat or specialist consultants, such as this author. In this author's experience, nations that analyse the impact of a variety of tax scenarios using modelling have more stable fiscal systems and are better able to attract investment on a continuing basis.

While various tax rates and bases have changed over the past several decades, the basic mineral-sector taxation methods have remained more or less

[3] See Otto et al. (2000) and Otto (2002).

the same, with the exception of the introduction of general value-added taxes (VAT) in many nations and a new form of additional profits mining tax in Chile and Peru. In this author's opinion, the greatest strides recently have been in tax administration, where the capacity of governments has strengthened, particularly in mineral-led developing economies. Many governments have established specially trained large-taxpayer compliance units. For nations that still have weak tax administration capabilities, simpler taxes remain preferable to more complex types. For example, the author, when meeting with officials in the tax authority of an African nation, learned that none of them knew how a new excess profits tax (crafted by external consultants and passed into law) was calculated or were familiar with the concept of internal rate of return.

Unfortunately, the progress made by tax authorities to stem fiscal leakages resulting from transfer pricing practices remains slow in both developed and developing economies. While input and output transfer pricing mechanisms are well known, the ability of governments to address these practices has remained weak. In this author's opinion, most nations today have developed their mineral-sector tax systems to achieve a 'theoretical' fair balance between national and investor interests, but transfer pricing linkages remain a major challenge that distorts actual revenue collection.

### 14.1.4 *Revenue Distribution*

Fiscal revenues generated by mining enterprises can be substantial, and where they form a large part of government's revenue, their distribution can be controversial. Subnational governments obtain their revenues from two principal sources: through the annual allocation of the national budget and from earnings that they obtain from the collection of various taxes that they are empowered to collect. Often, the respective tax powers of central and subnational governments are set out in the nation's constitution. In many nations, there is tension between the various levels of government regarding revenue distribution and taxing powers.

Subnational governments in areas that have mines often profess that since the minerals come from lands within their jurisdictional boundaries, they should receive preferential distribution. Subnational governments in areas without mines take a more egalitarian view—mineral revenues should be distributed equally or in such a manner that will do the greatest number of beneficiaries the greatest good. Otto (2001) has observed that in practice, there is often a bias in mineral-sector fiscal systems towards those areas in which minerals are mined either through statutorily defined distribution mechanisms or through the devolution of taxing powers to subnational government. In many developing nations there is a trend for subnational governments to acquire a greater portion of the tax or royalty take, either through statutory allocation of certain tax receipts or through direct taxation, as they gain administrative capacity. For example, over the past several decades, subnational governments have seen a greater portion of tax taken in nations as diverse as Brazil, Peru, Madagascar, Papua New Guinea, and Indonesia.[4] In Brazil, the royalty is split as follows: 23 per cent to the state in which the mine is located; 65 per cent to municipalities; 2 per cent to the national fund of scientific and technological development; and 10 per cent to the mining ministry, of which 2 per cent is earmarked for environmental protection in the mining regions.

The degree to which taxing authority is devolved among various levels of government is usually set out in the national constitution and in the various organic laws of the nation. Most mines are subject to a variety of taxes and fees levied by different levels of government. Typically, the largest tax cost is at the national level with lesser amounts being paid to subnational governments. Table 14.1 indicates the author's assessment of whether a certain type of tax is amenable to being assessed at the national, provincial, or local level.

In a previous study by this author (Otto 2001), an analysis of the mining fiscal systems in over twenty nations indicated that mining sector taxation in developing economies was highly centralized at the national level. Today, in

**Table 14.1.** Fiscal methods and their amenability to fiscal decentralization

| Tax type | National govt. | Provincial govt. | Local govt. |
|---|---|---|---|
| Income or profits-based tax | Y | P | N |
| Import duty | Y | N | N |
| Export duty | Y | N | N |
| Royalty (profit-based type) | Y | P | N |
| Royalty (ad valorem type) | Y | Y | P |
| Royalty tax (unit type) | Y | Y | Y |
| Royalty tax collected nationally and % distributed | Y | Y | Y |
| Licensing fees | Y | Y | Y |
| Surface rental or land use fees | Y | Y | Y |
| Withholding taxes on loan interest, dividends, services | Y | N | N |
| VAT on goods and services | Y | P | N |
| Sales and excise tax | Y | P | P |
| Stamp duty | Y | Y | Y |
| Property tax (on book or assessed value) | Y | Y | Y |
| Payroll-based taxes | Y | P | N |
| Surtaxes | Y | Y | Y |
| User fees | Y | Y | Y |

*Note*: Y = Yes, well suited; P = possibly suited; N = not a good fit
*Source*: Otto (2001).

[4] A variety of distribution mechanisms used in Australia, Canada, Chile, Ghana, Indonesia, Peru, and Tanzania are described in ICMM (2009).

the author's experience, that remains the same. Nations with a federal system of government, such as Australia, Brazil, Canada, and the United States, were more likely to devolve significant taxation power to provincial governments, most often through a jurisdictional income tax, royalty, and sales tax. The study revealed that few governments devolve significant taxing power to local government, except occasionally a tax on the value of commercial property. The World Bank Group collects and makes available detailed data on fiscal decentralization[5] but unfortunately mining is not broken out as a separate category for analysis.

While there are often good reasons to consider fiscal decentralization, it is this author's experience that achieving such decentralization is in practice quite difficult. Taxation powers are often inflexible either because of constitutional constraints or because central authorities are reluctant to devolve taxation power to subnational governments. The more common scenario is for national government to maintain taxing authority but to mandate a certain portion of fiscal revenues for distribution to subnational governments. For example, in Peru there was substantial pressure for the central government to share more of the mining fiscal take at the local level and the operations of some mines were affected by acts of civil disobedience. In response, over the course of a decade changes were made to the fiscal laws and now a portion of both royalties- and mines-derived income tax is collected centrally but distributed to subnational governments.[6]

### 14.1.5 *Special Aspects of the Mining Industry and the Tax Policy Response*

Tax policy makers grapple with the issue of uniqueness. Should a uniform taxation system be applied to all economic sectors or should a more complicated system be developed that takes into account the unique attributes of each sector? Each sector in an economy has different cost, revenue, and related profit attributes. Additionally, government objectives for various sectors may differ—for example, maximizing employment, providing security of supply, generating revenue, and so forth. The willingness of investors to invest in a sector may be impacted by how that sector is taxed, and if the taxation system is different for each sector of the economy, will preferential taxation cause detrimental distortions in the economy? Additionally, if the tax system is non-uniform, it will be inherently more complex and relatively more difficult to administer.

[5] Database access is available at: http://www1.worldbank.org/publicsector/decentralization/fiscalindicators.htm.

[6] For a description of the distribution system see Neyra (2011).

Otto (2001) has observed that most governments afford the mining sector special tax treatment, taking into account those attributes that distinguish it from other sectors. This section briefly examines several mineral-sector tax discrimination issues and related policy approaches. The treatment below is not exhaustive, and it is not uncommon to witness other discriminatory taxation practices.

*Discrimination by mineral type*: Because operational economics may differ from mineral type to mineral type, many nations statutorily define groups of minerals that are subject to different royalty rates. For example, diamond mines have the potential to generate profit levels that may not be obtainable by a gravel mine, and the royalty rate for diamonds will be set higher than that for gravel mines.

*Discrimination by level of investment*: Mines come in many sizes and some are on a commercial scale while others are artisanal. Many nations exempt artisanal miners, whether licensed or not, from paying a royalty, in recognition of the fact that tax enforcement is unrealistic. SSMs may enjoy a reduced royalty or royalty exemption and may also be subject to a low-income tax rate if the system is graduated. Very large mines and mines owned by the political elite may be offered the opportunity to negotiate unique tax terms in a special agreement with the state. Large, expensive-to-build, long-lived mines may be offered the ability to stabilize all or some types of tax for a defined time period. To entice investment in a large project or to aid in project financing, a few nations, such as Indonesia and the Philippines,[7] provide a 'holiday' from one or more types of tax for qualifying projects.

*Discrimination by nationality*: Bilateral investment and double taxation treaties offer special tax treatment for investors from the signatory nations; this may be unavailable to investors from non-treaty countries.

*Discrimination relating to costs*: Exploration expenses. Exploration expenses will be incurred before taxable income is available. Governments provide special provision for how pre-production (pre-income) exploration expenses are handled for future income tax purposes. Many nations allow exploration expenditure to be carried forward to the first year of mineral production, when the accumulated expenditure is either expensed or amortization is commenced.

Mine development and equipment. The development of a mine is dependent on specialized equipment and the developer will initially need to import large quantities of equipment from specialized foreign suppliers. Many nations exempt mining equipment from import duty and VAT during at least the initial development period. Other nations provide refunds or apply zero-rating schemes that have the same impact as an exemption. Mines are

[7] Details are reported in PWC (2012).

capital-intensive and large expenditures are required before operations can commence. Many governments recognize the capital intensity of the industry and provide various means to accelerate recovery of capital costs (i.e. depreciation) once production commences.

*Discrimination relating to export sales*: Many mine products are often destined for highly competitive international markets. Many governments impose no (or very low) export duties on minerals and provide a means whereby VAT on export sales is either not applied or applied in a way that allows for a refund or credit.

*Discrimination relating to commodity price cycles*: Mines produce raw materials used to make other things and are vulnerable to substantial price changes on a periodic, business-cycle-related basis. Some countries allow royalties to be waived or deferred from time to time for projects experiencing short-term financial stress and most countries provide for the carrying forward of losses.

### 14.1.6 *Transparency*

The disclosure of tax payments by commercial enterprises is a delicate fiscal issue in most nations. In a competitive world, information on a rival's payment of its various taxes can provide a competitive advantage. Many governments accept this principle and hold such payment information confidential. On the other hand, the public may be concerned that payments to government, perhaps to the executive branch or to a tax authority controlled by the political elite, can result in corrupt or questionable practices. One of the challenges faced by tax policy makers is how to balance the need for commercially necessary confidentiality with the public's need to hold the government accountable for how it receives and uses fiscal remittances.

Mineral-sector taxation transparency is an issue that has seen rapid change in recent years. This is largely attributable to two factors: the EITI and the shift away from secretly negotiated agreements (arranged with individual companies) towards standardized licencing and/or the public disclosure of negotiated agreements. In today's interconnected world, even if a nation and company desire to keep a negotiated agreement confidential, it may be leaked, and once available on the Internet, it is difficult to then control its unlimited dissemination. Specialized 'agreement websites' make it possible for the public to access a wide range of mining agreements.[8]

The EITI is a relatively new initiative that can trace its political origins to 2002 when the United Kingdom's then prime minister Tony Blair released a speech that called on companies and governments to publicly disclose their

[8] For example, visit http://www.resourcecontracts.org/, http://www.atns.net.au/, and http://repository.openoil.net/wiki/Main_Page to access a wide variety of mining and petroleum agreements.

fiscal payments and receipts. The following year, a Statement of Principles to increase the transparency of payments and revenues in the extractives sector was prepared and then agreed by over forty institutions. In 2004, under the auspices of the IMF and the World Bank, a World Bank-administered Multi-Donor Trust Fund (MDTF) was created to lend technical assistance to governments that wished to promote extractives sector fiscal transparency policies. Over time, the 2003 principles laid the groundwork for an international standard of reporting: governments report on the various taxes they receive from the mineral industry and companies report the amounts that they pay to government; the amounts paid and received are then subject to reconciliation, and the results are normally made public. A nation's EITI process is periodically assessed to determine whether it is in compliance with the standards or not. By 2016, over fifty nations, including many mineral-led economies, were attempting to implement EITI, of which about 60 per cent were largely compliant with the standards.[9] Being a fairly new initiative, it remains to be seen whether the EITI will prove effective in stemming corrupt practices over the longer term, but the disclosure of fiscal payments and receipts required by the EITI promotes accountability and is certain to make fiscally related corrupt practices more difficult to implement.[10]

### 14.1.7 *Revenue Stabilization Funds*

Mineral commodity prices are prone to large fluctuations attributable to many factors but such changes are most often related to global business cycles affecting the mineral supply and demand balance. Nations whose annual budgeting is dependent on mining revenues are vulnerable to rapid increases and decreases in commodity prices. It can be a challenge to match annual expenditure to annual fiscal receipts, particularly where a large part of the budget is relatively inelastic, such as in the payment of government workers, the provision of essential health, education, and security services, debt servicing, and so forth.

One approach used to deal with revenue instability is to implement a revenue stabilization scheme whereby 'excess' revenues received during times of high prices are saved and then expended when revenues drop. Such schemes have been implemented by a number of nations and they can

[9] A history of EITI, a listing of nations that are implementing it and a copy of the Standards are available at https://eiti.org/eiti/history.

[10] For an introduction to transparency issues see 'Emerging Economies: Transparency and Accountability in the Extractive Industries', Working Paper Series published by the NRGI available at http://www.resourcegovernance.org/analysis-tools/publications/emerging-economies-transparency-and-accountability-extractive-industries.

effectively help to smooth the revenues available for the budgeting process.[11] However, such schemes are vulnerable to political risk—a system adopted by one generation of lawmakers may be quickly repealed by the next generation of lawmakers who may want to dissolve the fund to spend the saved funds immediately (perhaps to fulfil election promises or to service debt). For example, Papua New Guinea implemented a mineral revenue stabilization fund scheme commencing in 1974 that worked well for several decades before its implementing act was repealed in 1999 by a new generation of lawmakers.[12]

### 14.1.8 *Sovereign Wealth Funds*

A SWF is a state-owned fund or entity that usually receives its capital infusion partially from fiscal revenues and partially from its earnings on investments. Such funds are becoming increasingly common and over seventy funds now exist or are in the planning stage.[13] For natural resource-led economies, such funds have at least four main purposes—to provide revenue smoothing (such as when used in part as a revenue stabilization fund), to provide for future sustainable income, to diversify income, and to limit available government expenditure based on the absorptive capacity of the economy. Key issues for policy makers when setting up such a fund include: who controls or manages the fund, the type of investments that can be made by the fund, and how and when earnings on those investments will be disbursed and to whom.[14] Many SWFs are set up so that annual disbursements to the government or its designated recipients are capped in some way in order to preserve the fund's capital base. Like revenue stabilization funds, SWFs are vulnerable to political action by lawmakers and governments who may over time seek to redirect a fund's earnings and distributions. While some SWFs operate with a goal of public transparency, others are highly secretive and may be subject to potential abuse and corrupt practices. In an attempt to promote transparency, good governance, and accountability standards, several dozen SWFs have jointly developed a voluntary code of principles, the 'Generally Accepted Principles and Practices', also known as the

[11] For an example, Chile maintains an Economic and Social Stabilization Fund whose primary purpose is to partially cover cyclical reductions in fiscal revenues. A description of the fund is available at http://www.hacienda.cl/english/sovereign-wealth-funds/economic-and-social-stabilization-fund.html.

[12] For a discussion of the fund and an accounting of its inflows and outflows see Otto and Dorian (1989) and Lum et al. (1995).

[13] A listing of SWFs and a brief description of their main attributes are available at http://www.swfinstitute.org/sovereign-wealth-fund-rankings/.

[14] An analysis of the issues involved in setting up a SWF within the context of a non-renewable resource-led developing economy (Papua New Guinea) is available at http://www.treasury.gov.pg/html/misc/Sovereign%20Wealth%20Fund%20Discussion%20Paper.pdf.

'Santiago Principles'. However, an analysis by Behrendt (2010) concludes that adherence to the principles is generally falling short.

Bauer (2014) and Ang (2010) have documented that many SWFs are successful in meeting their objectives but others are not. For example, the Nauru Phosphate Royalties Trust was a SWF set up by the Republic of Nauru, funded through revenues derived from phosphate mining.[15] At its peak the trust was worth more than one billion Australian dollars, which were to be used for the future benefit of Nauru's population of around 10,000 people. Abusive spending and mismanagement of the fund by officials eventually led to its bankruptcy and it was dissolved in 2014.[16]

Ang (2010) has proposed that for SWFs to work well they must meet four critical benchmarks: legitimacy, intent, performance, and endurance. Bauer (2014) recommends six steps that promote good SWF governance:

- Set clear fund objective(s) (e.g. saving for future generations, stabilizing the budget, and earmarking natural resource revenue for development priorities).
- Establish fiscal rules—for deposit and withdrawal—that align with the objective(s).
- Establish investment rules (e.g. a maximum of 20 per cent can be invested in equities) that align with the objective(s).
- Clarify a division of responsibilities between the ultimate authority over the fund, the fund manager, the day-to-day operational manager, and the different offices within the operational manager, and set and enforce ethical and conflict-of-interest standards.
- Require regular and extensive disclosures of key information (e.g. a list of specific investments, and names of fund managers) and audits.
- Establish strong independent oversight bodies to monitor fund behaviour and enforce the rules.

Unfortunately, while some SWFs meet the Ang benchmarks and the Bauer recommended steps criteria, many do not.

### 14.1.9 *Double Taxation Treaties*

Many nations have entered into 'double taxation treaties' in order to promote investment. Such bilateral conventions between two countries seek to eliminate the double taxation of income arising in one country and being paid to

[15] Nauru Phosphate Royalties Trust Act 1968.
[16] RONWAN Finalisation Act 2014 (No. 19 of 2014).

residents of another country. Otto and Cordes (2002: 6–17) have observed that the content of such treaties varies widely but that they usually contain a number of investment-related incentives useful to the mining industry such as: reduced dividend withholding tax; reduced interest withholding tax; crediting of income taxes paid by a home country taxable entity in the host country; and the recognition of deductions allowed in the host country as a valid deduction for host country crediting. Of particular importance to foreign-owned mines are treaty provisions that reduce withholding tax rates below the normal statutory rate. Withholding taxes are often one of the largest fiscal costs incurred by a mine and treaty rates are often set substantially below the statutory rate. Many double taxation treaties have been deposited with the United Nations Secretary General and are available at its online United Nations Treaty Series collection website.[17]

## 14.2 Direct Taxes and Indirect Taxes

### 14.2.1 *Direct Taxes*

For the purposes of this chapter, 'direct taxes' means all taxes, fees, and other imposts that are paid by a mine directly to government. The most common types of direct tax applied to mining include:

- income tax
- turnover tax
- withholding tax on dividends
- withholding tax on loan interest payments
- withholding tax on salaries
- withholding tax on payments for foreign-sources services
- royalties
- value-added tax or sales tax
- customs duties
- taxes based on land area (such as rent)
- taxes based on asset value (such as property tax)
- provincial and local taxes
- stamp duties/transaction fees
- transport-related taxes (user fees)
- employee or payroll related taxes, fees, contributions, pension schemes
- progressive, excess profits, and additional profits taxes.

[17] To access the United Nations Treaty, see https://treaties.un.org.

Of these direct tax types,[18] perhaps the types that are of most interest to the current generation of tax policy makers are excess/additional-profit-type taxes. Most nations do not impose taxes specifically designed to capture 'excess profits'. However, when the business cycle pushes mineral commodity prices up, governments may act to impose a special tax to capture some of the perceived increased rent and then may repeal the tax when prices descend.[19] For example, during the price boom that commenced around 2004 Australia, Mongolia, and Zambia imposed new 'excess profits' taxes but repealed them at the end of the price-cycle boom. Other nations, such as Chile, Peru, and Liberia, imposed new excess profits taxes but have maintained them even as prices have fallen. A well-designed excess profits tax should be self-adjusting to take into account changing profitability. However, the reality is that excess profits taxes are highly vulnerable to political processes linked to the business cycle.

In practice, there are three principal types of tax designed to capture excess profits: progressive tax; rate-of-return-triggered additional profit tax; and profitability-ratio-triggered additional profit tax:

- With a progressive tax, when the magnitude of annual profits (net income) or prices goes up, the tax rate goes up. The rate increase may be tied to a progressive corporate income tax (such as in the United States and Venezuela) or to a royalty where the royalty rate is linked to the commodity price (such as in Mongolia).
- A rate-of-return-triggered additional profit tax is favoured by many economists because of its linkage to the concept of economic rents. When a statutorily set rate of return is exceeded, an additional tax is imposed. The idea is not to capture all economic rent, but to instead impose an additional tax when a rate-of-return calculation suggests that there are substantial economic rents. Typically the rate-of-return trigger rate is in the range of 20–25 per cent. The rate-of-return-based calculation is based on the statutorily defined cash flow of the mine to date. It is possible for the tax to be triggered in some years, but if the rate of return falls below the trigger rate, the additional profits tax is not paid. Historically, rate-of-return-triggered additional profit taxes were considered by many governments as too administratively difficult to calculate, but new accounting and calculation approaches developed by the IMF (such as IMF-drafted provisions incorporated into the income tax acts of Liberia, Malawi, and other client countries) have addressed this problem.

[18] For a description of each direct tax type and its trend see Otto (2017).

[19] For additional information see Otto (1992a); Otto (1992b); Land (1995); Cordes (1995); Otto and Cordes (2002); Otto et al. (2006); Land (2010).

- Unlike rate-of-return-based taxes, another form of additional profits tax is based not on the cash flow of the mine over time, but instead on the ratio of profits to costs in the current tax year. If the ratio exceeds a limit, the additional profits tax is triggered and it is applied to either gross sales revenues (a form of royalty) or a statutorily defined operating income. In Chile and Peru, which impose this latter type of tax, the effective rate of the additional tax is progressive—as the profit ratio increases, the effective additional profits tax rate increases.

In this author's opinion, rate-of-return and additional profits taxes can be a good option for some governments, provided that they have adequately trained tax officials. Any sort of tax whose calculation takes into account costs is more difficult to implement, monitor, and enforce than one that is solely based on revenues (such as an ad valorem type royalty). The introduction of these types of complex tax is best suited to nations with a strong, well-funded, and educated tax authority.

### 14.2.2 *Indirect Taxes*

In addition to taxes that result in a direct monetary payment to government, governments can impose requirements on miners that result in higher costs. Several examples of indirect taxation include:[20]

- free equity or carried interest state participation
- mandatory local sourcing (goods, services)
- downstream processing
- local community development requirements
- social infrastructure development requirements
- foreign exchange requirements.

In most parts of the world, mining by wholly state-owned enterprises is no longer practised. However, some nations, predominantly in the West Africa region and in former 'Soviet sphere' nations, still require a minority state participation interest in large mines. The means by which such an interest is created varies.[21] Daniel (1995: 175) describes state equity as 'the acquisition by the government (or a state-owned enterprise) on behalf of the state of a participating interest or a claim on after-tax profits in a joint venture with

[20] For a detailed description of various types of cost that can be considered indirect taxation see Otto (2017).

[21] For a more complete treatment see McPherson (2010), Otto and Cordes (2002), and Radetzki (1985).

privately owned companies'. Today, most equity interests provided to governments fall into one of the following categories:

- *Free equity*: This term means that the state receives an ownership interest at no cost to itself. The ownership interest does not oblige the state to participate in the paying of any costs associated with the project. It does, however, entitle the state to a share of distributed dividends or profits. Free equity will have a large impact on an investor's profits, and a requirement for free equity is considered a major detriment by most potential investors. The rationale usually offered for such a requirement is that the state's contribution to the project is the mineral endowment being exploited. Free equity interest requirements have waned over the past several decades, except in the West Africa region. This is probably attributable to the realization by governments that free equity requirements had a substantial negative impact on potential fiscal revenue because little investment was occurring. Even in the West Africa region where many nations have a 10 per cent free equity requirement in their mining code, proposed mining projects can sometimes be exempted from the equity requirement (such as in Ghana and Guinea).
- *Working interest equity*: This type of equity is sometimes referred to as 'participating equity'. The equity interest is purchased by the state according to an agreed or statutorily defined pricing scheme. After purchasing its interest, the state must contribute to the expenditure needs of the project in proportion to its shareholding. Most investors do not view this requirement as a major investment impediment as long as the purchasing procedures are clearly defined. Some investors welcome working equity participation because it can reduce certain types of risk. The mining laws of Botswana, Indonesia, and Papua New Guinea require that the government has an option to acquire a working equity interest in new mines, and these nations have been successful in attracting foreign investment into their mining sector.
- *Carried interest equity*: This is a specialized form of working equity and has all the attributes of working equity described above, except for the way in which the ownership interest is paid for. Instead of a cash-based transaction, the state's share is paid for from the future dividends or profit distributions that would have been distributed to that ownership interest. The amount due to the investor from the state accrues interest. The result is that for many mines, the state will not receive any dividend distribution for many years, if ever. In practice, the amount due to the investor may increase steadily over the years. In effect, but not in form, this type of equity can be thought of as similar to a loan extended by the investor to pay the state's equity purchase cost and ongoing participation costs.

Carried interest requirements are typically not required by statutory mandate, but arise instead in a negotiated agreement where a government desires an equity stake but is not able to currently pay for it. For example, the government of Mongolia entered into an agreement with foreign investors to develop a large copper mine, and a related shareholder's agreement provided the carried interest equity provisions.[22]

- *Free carried interest equity*: This type of equity is a special form of carried interest equity but with the important difference that interest does not accrue on the amount payable to the investor by the state. In effect, this is a form of interest-free loan.

If the private-party investor is required to provide free equity or free carried interest equity, many investors would view this as a form of indirect taxation.

## 14.3 Tax Incentives and Income Tax Adjustments

Governments that seek to attract mineral-sector investment sometimes provide special provisions within their tax legislation or in negotiated agreements. Otto (1992a, 2000) has identified ten particular tax incentives that some nations offer to mineral-sector investors. Based on the author's experience, these tax incentives and income tax adjustments are still in use today[23] and include:

- tax holiday from one or more tax types
- loss carry forward
- exploration cost expensing
- social cost expensing
- capitalized cost deductions (such as amortization/depreciation)
- loan interest deduction
- depletion allowance
- tax credits
- tax stabilization.

It should be noted that the generosity of nations offering tax incentives to potential investors tends to vary with the business cycle. With the advent of the so-called super-cycle that commenced around 2002 and that resulted in mineral commodity price increases and an abundance of exploration and mine development funding, some nations pulled back on incentives and many

[22] The agreement between the state, Rio Tinto, and other parties is available at http://www.resourcecontracts.org/contract/ocds-591adf-MN2079301876RC-1658/view.

[23] For a description of current trends in tax incentives see Otto (2017).

raised taxes. With the subsequent fallback in prices, it can be expected that the tax incentives will be one tool used by governments to compete for scarce capital. It is interesting to note that a major mining tax study conducted by ICMM (2009) concluded that many companies consider tax incentives as less important than tax disincentives, such as high rates.

In many developing nations, a key issue is that of tax stabilization. Large mines can require enormous initial investment. Because the resource on which they depend is location specific, that capital is immobile, and the captive investment is subject to the risk that once the mine is built, the taxation system may be amended to the investor's disadvantage. Investors and their lenders may thus require an assurance that the fiscal system will remain stable, at least for a certain period for time, before they will invest. In its tax study, ICCM (2009) reported that the two most important tax issues for mining companies are stability and predictability. Governments with a history of political or fiscal system instability may find it necessary or useful to provide the stabilization of one or more taxes in order to meet investor needs. There are two principal ways in which governments approach fiscal stabilization: by terms negotiated in an agreement with the state[24] and by provisions set out in a law. The former approach allows flexibility on a project-by-project basis while the legislated approach provides for standardization. Fiscal policy makers, whether negotiating an agreement or drafting a law, have several key issues to address: Which taxes are to be stabilized? Is just the tax rate stabilized or is the tax basis also stabilized? For what time period is stabilization allowed? Can the state impose new tax types on a project once it is built?

Governments that provide tax stabilization have increased administrative challenges. Each mine that operates under a stabilization scheme becomes an exception to the existing fiscal system when that system evolves. Over time, many separate 'tax systems' may result as mines are granted tax stabilization at different points in time. The administration of many mine tax schemes, each operating under a different stabilized regime, is a challenge for governments. As an example, in discussions with a tax authority in a South American nation that allows tax stabilization, the author was told that over fifty different taxing schemes had resulted for mines in that nation. In this author's experience, standardized limited-term tax stabilization, such as the non-negotiated optional schemes set forth in the tax laws of Chile and Peru, can be a valuable tool to attract investors, but should be restricted to very large projects in order to avoid the administrative challenge.

[24] For a description of tax stabilization provided in mining and petroleum agreements, see Daniel and Sunley (2010).

## 14.4 Transfer Pricing and Other Tax Minimization Schemes

Transfer pricing is a major problem for governments, and a particularly vexing challenge for mineral-led economies. Kar and Spanjers (2015) estimate that developing and emerging economies lost US$7.8 trillion from illicit financial flows from 2004 through to 2013, of which about 83 per cent was attributed to the fraudulent invoicing of trade.

There are many ways in which taxpayers can reduce their tax burden, either legally or otherwise. Most tax minimization schemes are linked to either increasing reported costs (input manipulation) or reducing reported revenues (output manipulation). Often, the goal is to transfer a portion of taxable income from the host nation in which the mine is located to a low or zero income tax rate jurisdiction.

An example of input manipulation is when a foreign investor sets up a company in a no- or low-tax nation (i.e. a tax haven) and that company then provides management, marketing, or other services and/or loans to the company in the host nation at a cost higher than what might have been charged had the two firms not been affiliated. The effect is to increase tax deductions in the host nation, thus lowering taxable income. For services, it is difficult if not impossible for a tax authority to determine whether input costs are artificially inflated. For loans, the task is somewhat simpler but is still difficult to identify. For this reason, some nations have developed loan interest deduction limits based on thin capitalization rules[25] and a cap based on a percentage of earnings before income tax (for example, in Germany and Norway). If the cap is breached, a portion of loan interest may be non-deductible or treated as a dividend. In this author's experience, tax minimization schemes based on input cost manipulation are commonplace.

The output manipulation of revenues has long been a problem for tax authorities. A simple practice is for the producing company to sell its output to an affiliated company located in a no- or low-tax jurisdiction, reducing its gross revenues in the host nation and thus reducing the amount of income tax and royalty. The affiliated company in the low-tax nation then on-sells the output at the full market price. Some nations still lack basic provisions in their mining and tax laws to control output price manipulation. In mining laws recently drafted with the help of this author, sections to assist tax authorities have been added, such as: a requirement that all sales must be on an arm's-length basis; all sales to an affiliate must be reported; all sales are subject to royalty revision if the authority determines that the sale was not on an arm's-length basis; penalties apply in case of a violation; and repeated violations can

[25] For an in-depth analysis of thin capitalization rules, see Blouin et al. (2014).

result in the mining concession being cancelled. Even with adequate legal provisions, detecting transfer pricing can be a major challenge because determining whether a fair price has been paid is a challenge. For many commodities a reliable reference price is not available. For example, many metals are sold in the form of a multi-metal concentrate (such as one containing lead, zinc, and silver) that contains varying levels of deleterious substances (such as sulphur and arsenic). There are market reference prices for refined lead, zinc, and silver but not for concentrates. Tax authorities often lack access to information (for example, sales contracts from across the globe) that would allow them to determine a fair price for such concentrates. An additional challenge is that each sale for delivery may be different because the mineral being mined may be non-homogeneous—the coal from one part of a mine may have a different heating value than the coal from another part of the mine and thus a different monetary value.

Most developing-nation tax authorities are ill equipped to deal effectively with transfer pricing schemes and even well-trained tax authorities in developed nations struggle.[26] International 'standards' such as the OECD Transfer Pricing Guidelines for Multinational Enterprises,[27] in this author's opinion, have been ineffectual as applied to the mineral sector.

Worldwide, governments continue to work on improving legislation to control transfer pricing and other tax minimization practices. For example, it is fairly simple to amend a mining law to require that any mineral sales or any service type contract with an affiliate be reported to government along with a copy of any associated invoice or contract. Arm's-length pricing requirements are suitable for both a mining law and the tax act and thin capitalization rules are appropriate for the tax act.

## 14.5 Conclusions

The taxation of the mining industry varies considerably from nation to nation and evolves over time, reflecting the response of politicians and tax authorities to a variety of factors including commodity price cycles, levels of investor interest, and changing national objectives. Today, governments have access to tools and expertise that can assist them in fiscal reform efforts, including software that can allow the holistic testing of tax scenarios on project economics. While the types of taxes and investment incentives have remained

[26] The World Bank Group has published a mineral sector transfer pricing reference guide for use by tax authorities. See Guj et al. (2017).

[27] OECD (2010).

fairly static for the past several decades, annual profit-ratio-based additional profits taxes newly introduced by Chile and Peru may be of interest to some nations—particularly nations with a strong, well-funded, and educated tax authority. Historically, some nations sought mineral-sector-derived revenues from both taxation and direct participation by the state in mining, either through a state mining enterprise or through some sort of joint venture or share participation arrangement. That is less common now, except in the West Africa region, as many nations have concluded that they can achieve the objectives of control and risk-free revenue more effectively through legislation. There has also been a move away from negotiating mining agreement fiscal terms towards a reliance on standardized requirements set out in the general tax laws. In this author's opinion, the time for such agreements has passed, and nations that still use them should concentrate their future efforts on strengthening the underlying system of laws, rather than spending the considerable time and effort it takes to negotiate agreements that are specific to individual companies or projects. However, for very large projects, standardized limited-term tax stabilization, achieved either by a non-negotiable tax stabilization agreement or by statute, can be a useful tool to attract investment and requisite financing. Progress by tax authorities to stem fiscal leakages that result from transfer pricing practices remains slow in both developed and developing nations. While input and output transfer pricing mechanisms are well known, the ability of governments to address these practices has remained weak. In this author's opinion, most nations today have developed their mineral-sector tax systems to achieve a 'theoretical' fair balance between national and investor interests, but transfer pricing linkages remain a major challenge that distorts actual revenue collection.

The emphasis of this chapter is on how mines are taxed, not on how tax revenues should be distributed and used. However, it is noted that one of the influencing factors that drives fiscal system policy evolution is the issue of whether subnational governments should have additional taxing power or be given special revenue dispensation. It is the author's experience that devolving additional tax power to subnational governments is often not possible because of constitutional constraints or because central authorities wish to maintain control. The more common scenario is for national government to maintain taxing authority but to statutorily mandate a certain portion of fiscal revenues for distribution to subnational governments.

In summary, this chapter has provided an introduction to the various taxation approaches applied by governments to the mining sector and it includes a description of the principal tax types and investment tax incentives. It has briefly described the main policy issues pertaining to mineral-sector taxation. Its primary message is that governments, when devising mineral-sector fiscal systems, should carefully assess their fiscal options in a holistic, balanced approach that anticipates commodity price cycles, and that mining companies should anticipate fiscal system changes that reflect the evolution of the political economy in which they operate.

## References


Ang, A. (2010). 'The Four Benchmarks of Sovereign Wealth Funds'. New York: Colombia Business School. Available at: http://ssrn.com/abstract=1680485.

Bauer, A. (ed.) (2014). 'Managing the Public Trust: How to Make Natural Resource Funds Work for Citizens'. New York: Revenue Watch Institute and Vale Columbia Center on Sustainable International. Available at: http://www.resourcegovernance.org/sites/default/files/NRF_RWI_Complete_Report_EN.pdf.

Behrendt, S. (2010). 'Sovereign Wealth Funds and the Santiago Principles: Where Do They Stand?' Beirut: Carnegie Endowment for International Peace. Available at: http://www.carnegieendowment.org/files/santiago_principles.pdf.

Blouin, J., H. Huizinga, L. Laeven, and G. Nicodème (2014). 'Thin Capitalization Rules and Multinational Firm Capital Structure'. Washington, DC: IMF. Available at: https://www.imf.org/external/pubs/ft/wp/2014/wp1412.pdf.

Boadway, R. and M. Keen (2013). 'Rent Taxes and Royalties in Designing Fiscal Regimes for Non-Renewable Resources'. Oslo: University of Oslo Department of Economics. Available at: http://www.sv.uio.no/econ/english/research/news-and-events/events/guest-lectures-seminars/ofs-seminar/documents/Boadway-Keen.pdf.

Cordes, J. (1995). 'An Introduction to the Taxation of Mineral Rents', in J. Otto (ed.), *Taxation of Mineral Enterprises*, 25–46. London: Graham and Trotman/Martinus Nijhoff.

Daniel, P. (1995). 'Evaluating State Participation in Mineral Projects: Equity, Infrastructure and Taxation', in J. Otto (ed.), *Taxation of Mineral Enterprises*, 165–86. London: Graham and Trotman/Martinus Nijhoff.

Daniel, P. and E. Sunley (2010). 'Contractual Assurances of Fiscal Stability', in P. Daniel, M. Keen, and C. McPherson (eds), *The Taxation of Petroleum and Minerals: Principles, Problems and Practice*, 404–24. New York: IMF/Routledge. Available at: http://s1.downloadmienphi.net/file/downloadfile3/161/1394930.pdf.

Guj, P., S. Martin, B. Maybee, F. Cawood, B. Bocoum, N. Gosai, and S. Huibregtse (2017). 'Transfer Pricing in Mining with a Focus on Africa: A Reference Guide for Practitioners'. Washington, DC: World Bank Group. Available at: http://documents.worldbank.org/curated/en/801771485941579048/Transfer-pricing-in-mining-with-a-focus-on-Africa-a-reference-guide-for-practitioners.

ICMM (2009). 'Minerals Taxation Regimes: A Review of Issues and Challenges in their Design and Application'. London: ICMM. Available at: https://www.icmm.com/en-gb/publications/society-and-the-economy/minerals-taxation-regimes-a-review-of-issues-and-challenges-in-their-design-and-application.

IMF (2016). 'Fiscal Analysis of Resource Industries (FARI) Methodology'. Washington, DC: IMF. Available at: https://www.imf.org/external/pubs/ft/tnm/2016/tnm1601.pdf.

Kar, D. and J. Spanjers (2015). 'Illicit Financial Flows from Developing Countries: 2004–2013'. Washington, DC: Global Financial Integrity. Available at: http://www.gfintegrity.org/wp-content/uploads/2015/12/IFF-Update_2015-Final-1.pdf.

Land, B. (1995). 'The Rate of Return Approach to Progressive Profit Sharing in Mining', in J. Otto (ed.), *Taxation of Mineral Enterprises*, 91–109. London: Graham and Trotman/Martinus Nijhoff.

Land, B. (2010). 'Resource Rent Taxes: A Re-appraisal', in P. Daniel, M. Keen, and C. McPherson (eds), *The Taxation of Petroleum and Minerals: Principles, Problems and Practice*. New York: IMF/Routledge. Available at: http://s1.downloadmienphi.net/file/downloadfile3/161/1394930.pdf.

Lum, J., J. Dorian, and A. Clark (1995). 'Managing Cyclical Mineral Revenues: Papua New Guinea's Stabilization Fund', in J. Otto (ed.), *Taxation of Mineral Enterprises*, 189–209. London: Graham and Trotman/Martinus Nijhoff.

McPherson, C. (2010). 'State Participation in the Natural Resource Sectors: Evolution, Issues and Outlook', in P. Daniel, M. Keen, and C. McPherson (eds), *The Taxation of Petroleum and Minerals: Principles, Problems, and Practice*, 263–88. New York: IMF/Routledge. Available at: http://s1.downloadmienphi.net/file/downloadfile3/161/1394930.pdf.

Neyra, C. (2011). 'Mining Taxation: Principles, Tools and their Application in Peru'. Lima: GIZ. Available at: http://www.gobernabilidad.org.pe/buen_gobierno/galleries/115510511_022-Mining%20TaxationprinciplesTools%20AndTheir%20AplicationIn%20Peru.pdf.

OECD (2010). *Transfer Pricing Guidelines for Multinational Enterprises and Tax Administrations*. Paris: OECD. Available at: http://www.ilsole24ore.com/pdf2010/SoleOnLine5/_Oggetti_Correlati/Documenti/Norme%20e%20Tributi/2011/02/istruzioni-uso-societa-perdite-fiscali/ocse-linee-guida-2010-prezzi-trasferimento.pdf.

Otto, J. (1992a). 'Mineral Sector Taxation Methods: A Global Review', in *Minerals Industry Taxation Policies for Asia and the Pacific*, 11–21. New York: United Nations.

Otto, J. (1992b). 'Taxation Methods That Have Been Used in Selected Mineral-Producing Nations', in *Minerals Industry Taxation Policies for Asia and the Pacific*, 33–55. New York: United Nations.

Otto, J. (2000). 'Tax Incentives to Promote Mining Sector Development'. Tokyo: Metal Mining Agency of Japan.

Otto, J. (2001). 'Fiscal Decentralization and Mine Taxation'. Washington, DC: World Bank. Available at: https://commdev.org/userfiles/files/1403_file_Fiscal_Decentralization.pdf.

Otto, J. (2002). 'Position of the Peruvian Taxation System as Compared to Mining Taxation Systems in Other Nations'. Lima: Ministry of Economy and Finance. Available at: https://www.mef.gob.pe/contenidos/pol_econ/documentos/perufinalreport_280504Otto.pdf.

Otto, J. (2017). 'The Taxation of Extractive Industries: Mining', WIDER Working Paper WP2017/75. Helsinki: UNU-WIDER. Available at: https://www.wider.unu.edu/publication/taxation-extractive-industries.

Otto, J. and J. Cordes (2002). *The Regulation of Mineral Enterprises: A Global Perspective on Economics, Law and Policy*. Westminster, CO: Rocky Mountain Mineral Law Foundation.

Otto, J. and J. Dorian (1989). 'Metals Trade and Industry Protection: The Metals-Exporting Developing Countries', in J. Cordes and T. Torries (eds), *Surplus Capacity in the International Metals Industry*, 67–82. Englewood, CO: Society of Mining Engineers.

Otto, J., C. Andrews, F. Cawood, M. Doggett, P. Guj, F. Stermole, J. Stermple, and J. Tilton (2006). 'Mining Royalties'. Washington, DC: World Bank. Available at: http://siteresources.worldbank.org/INTOGMC/Resources/336099-1156955107170/miningroyaltiespublication.pdf.

Otto, J., M. L. Batarseh, and J. Cordes (2000). *Global Mining Taxation Comparative Study Second Edition*. Golden, CO: Colorado School of Mines.

PWC (2012). 'Corporate Income Taxes, Mining Royalties and Other Mining Taxes: A Summary of Rates and Rules in Selected Countries'. London: Price Waterhouse Coopers Ltd. Available at: https://www.pwc.com/gx/en/energy-utilities-mining/publications/pdf/pwc-gx-miining-taxes-and-royalties.pdf.

Radetzki, M. (1985). *State Mineral Enterprises: An Investigation into their Impact on International Markets*. Washington, DC: Resources for the Future Press.

Tilton, J. and J. I. Guzman (2016). *Mineral Economics and Policy*. New York: Resources for the Future Press.

# 15

# Doubling Down

## National Oil Companies as Instruments of Risk and Reward

*Patrick R. P. Heller*

### 15.1 Introduction

State-owned oil companies have proven themselves capable of reaching tremendous heights—as generators of public revenues and stewards of core industries—and devastating lows—as squanderers of national resources and agents of corruption. Sometimes, both sides of this dichotomy can be illustrated by the same company. In recent years, no company has provided a more dramatic example than Brazil's Petrobras.

Coming of age during Brazil's period of military rule, Petrobras became an increasingly skilled and effective company throughout the 1960s and 1970s, developing particular expertise in deep-water exploration and production. The Brazilian oil sector was opened to competition in the 1990s, and Petrobras continued to thrive, generating secure petroleum resources and revenues for the state, helping stimulate a thriving Brazilian private sector in oil services, and expanding its operations overseas (de Oliveira 2012). By the 2010s, industry 'experts' were hailing Petrobras as a model of commercial efficiency (Heller et al. 2014).

The company's positive image came crashing down in 2014, with the public revelation that Petrobras had sat at the centre of a multi-billion-dollar bribery and price-fixing scheme. Petrobras directors had conspired with a cabal of Brazil's leading construction companies to rig massive procurement processes and award contracts to participants in the conspiracy at inflated prices. Corrupt officials were rewarded for their efforts with bribes upward of US$2 billion. The impact of this 'Operation Car Wash' scandal on Petrobras was dire. The company had to write down US$17 billion in direct and indirect losses from the scandal, and saw its share price fall by 80 per cent from its 2014 high point to the end of 2015. The scandal devastated Brazil's economy and rocked its political system to its core, resulting in the indictments of hundreds of public officials and industrial titans, undermining confidence in the economy, and contributing heavily to the political destabilization that culminated in the impeachment of President Dilma Rousseff (IMF 2016; Leahy 2016).[1]

At its best, Petrobras illustrated the reasons that so many countries have invested in national oil companies (NOCs) as centrepieces of their strategies for developing their oil and gas sectors. The world's most successful NOCs, including Norway's Statoil, Saudi Arabia's Saudi Aramco, and Malaysia's Petronas, have maintained vigorous exploration programmes, delivered strong returns on public resources, decreased long-term reliance on costly private partners, and/or helped promote the rise of a technocratic class of private businesses and professionals.

But Petrobras' fall from grace underscores a set of risks common to NOCs, which in many cases have resulted in a net negative contribution to oil-sector management. At the core, these enterprises' positions at the intersection of public policy, commercial ambition, massive economic rents, and networks of elites leave them particularly vulnerable to being used as vehicles for patronage. In many cases, weak incentives or management structures also render them ineffective developers of petroleum reserves, which can mean they waste significant portions of the public resources they are entrusted with overseeing.

NOCs can increase their governments' share of long-term rewards from the oil and gas sector. In virtually all cases they expose their governments to risk, both financial (via the reinvestment of petroleum revenues back into the unpredictable industry itself) and political (via the kinds of patronage mechanisms to which Petrobras so spectacularly fell prey and via principal–agent problems that sometimes lead the company's goals to diverge from the government's). Rather than examining the question of whether or not a state should create an NOC, this chapter takes as a given that almost all non-OECD oil producers already have one. The study focuses instead on constructing a

[1] The formal rationale for Rousseff's impeachment was improper manipulation of the budget, not the 'Car Wash' scandal. But the scandal was one of the major flash points that led to the crisis in confidence in the government.

risk–reward lens oriented around an attempt to inform the kinds of practical questions that governments face in managing their NOCs; namely:

- *what* commercial investment strategies are most likely to maximize return on investment
- *how* public oversight and corporate governance measures can enhance performance incentives and reduce the most serious risks
- *how much* public revenue should be entrusted to the NOC to manage in order to balance reinvestment in the sector against other public expenditure needs.

An environment of low global oil prices poses a special set of challenges to which NOCs must respond. Well-managed NOCs have been able to develop projects efficiently even when prices are high and money flows freely. But many companies become bloated during boom times, failing to sufficiently take advantage of opportunities and deliver sufficient benefits to their countries. Many of these companies face crises when volatile oil prices fall, as has happened at the time of the writing of this chapter. Lean times force NOCs to cut costs and seek efficiency gains. As such, a moment of crisis can offer an opportunity for some NOCs to implement reforms with long-term benefits, but only if they focus on systemic corporate governance and accountability commitments that are robust even to the pressures of plenty that will return if prices rise again.

After this introduction, Section 15.2 provides a brief review of the key literature on state-owned enterprises (SOEs), emphasizing a typology of different types of NOCs and some of their fundamental characteristics. Section 15.3 constructs the central analytical lens of the piece, classifying the most important risks and modes of assessing potential rewards. Section 15.4 offers a practical response, outlining some of the strategies that have been demonstrated to maximize the chances that rewards will be realized and to mitigate risks. Finally, Section 15.5 provides concluding thoughts, with a particular emphasis on the specific challenges surrounding NOC governance during times of low petroleum prices.

## 15.2 SOEs: Context and Core Characteristics

There is a rich scholarship on the role that SOEs play in the management of modern economies, and about some of the problems associated with entrusting key sectors to SOEs. These enterprises are often given a privileged role in critical sectors of the economy based on the multiple ambitions of enabling the state to meld the management of strategic industries to policy priorities (particularly in highly regulated sectors); generating positive externalities that benefit citizens; building up a targeted class of skilled managers and technocrats; and generating financial returns to the treasury (Megginson and Netter 2001; Victor et al. 2012).

Much of the empirical work around SOEs generally has focused on their commercial performance vis-à-vis that of private companies, with the preponderance of studies showing that private companies tend to be more efficient and to generate stronger financial returns than SOEs (Boardman and Vining 1989; Megginson and Netter 2001; Shirley and Walsh 2000). Several inherent factors contribute to these performance challenges, most of which, following Shirley and Walsh, can be categorized as 'incentive effects' (stemming from the weak incentives for SOE managers) and/or 'information effects' (stemming from a weak relationship between data on performance and decisions/imperatives for SOE actions). Specific problems include the following:

- *Mixed mandates*: The goal of private-sector firms is clear: profit maximization. SOEs are often called upon to pursue profits while simultaneously promoting other public goods, which can impede clear strategy, hinder rigorous performance monitoring, and create incentives to distort markets (Sappington and Sidak 2003).
- *Politicized and bureaucratized decision-making*: Politicians and bureaucrats—often with limited commercial skills and experience—can play a disproportionate role in SOE management by virtue of shareholding rights and, in many cases, their appointment and remuneration powers.
- *Reduced competition*: Some SOEs enjoy total monopolies in their sectors; others are subject to some competition but benefit from systemic advantages—via access to markets, inputs, or financing. This means that these enterprises can continue to operate (and often to dominate) even in the presence of substantial inefficiencies.
- *Difficulty in accessing replenishment capital*: The fiscal relationships between SOEs and the state are often complex and unpredictable, with SOEs sometimes being used as 'cash cows' by the treasury. This can make it difficult for these enterprises to have the financial predictability to plan effectively and address problems when they arise.

SOE efficiency problems can result in the wasting of public resources invested in them, and poor provision of the public goods they are expected to provide. Beyond these efficiency concerns, many SOEs have been involved in corrupt activities and have further weakened economic governance through self-dealing or abuse of the power they occupy at the juncture of public oversight and market participation (Kane and Christiansen 2015).

**Table 15.1.** NOC average commercial performance vs. private oil companies, 2002–4

| | Revenue per employee (US$) | Revenue per reserves (US$) |
|---|---|---|
| NOCs | 1,000 | 5 |
| Major IOCs | 2,865 | 15 |
| Other IOCs | 1,629 | 11 |

*Source*: based on Eller et al. (2007).

Nonetheless, SOEs have played an important role in the management of national petroleum sectors across the world since a wave of nationalizations swept the Middle East and Latin America in the 1970s. Today, more than 90 per cent of the world's top per capita oil and gas producers have an NOC, with the major exceptions being OECD countries such as the United States, Canada, and the United Kingdom (Myers 2015). At a global level, NOCs have been estimated to control 90 per cent of global oil reserves and 75 per cent of production (Tordo 2011).

Like SOEs in other sectors, the performance of NOCs has been chequered. Some NOCs have generated strong financial returns and other public goods, including innovation, stimulation of a domestic class of oil-sector technocrats, and effective promotion and oversight of the sector. But on average, research indicates that NOC commercial performance lags significantly behind that of private-sector international oil companies (IOCs) on most industry-standard measures (Eller et al. 2007; Victor 2007; Wolf 2009; see Table 15.1). Studies that have examined NOC performance in the light of their broader range of objectives have presented mixed conclusions, with strong successes such as Statoil (Norway) and Saudi Aramco balanced against failures such as Venezuela's Petróleos de Venezuela (PDVSA) and the Nigerian National Petroleum Corporation (NNPC) (Heller et al. 2014; Tordo 2011; Victor et al. 2012).

SOEs in the oil sector are subject in many cases to exacerbated versions of the general challenges facing SOEs in other sectors, particularly vis-à-vis corporate governance. Oil is a classic 'rentier' sector, whereby the state has the power to manage and distribute huge flows of public wealth without facing public pressure to stimulate innovation or maintain the consent of the governed (Karl 1997; Ross 2012). Because of the massive revenues that can be generated by petroleum, and the complexity of the processes necessary to get it out of the ground, NOCs often find themselves managing larger flows of public revenues than SOEs in other sectors. Their objectives are shaped by the state's distributional power, the company's own ambitions for growth, and a swirling pool of private interests trying to gain access to this lucrative and strategic sector. As such, many NOCs have experienced weak incentives for efficient performance, and strong temptations for patronage. NOCs have been key players in large-scale corruption scandals in countries ranging from Nigeria to Brazil to Russia.

With some exceptions, NOCs have not been subjected to the kinds of strong oversight and reporting requirements that promote strong corporate governance in the private sector. The Resource Governance Index, which measures accountability mechanisms in oil- and mineral-producing countries worldwide, found that only twenty-six of the seventy-four SOEs in its sample demonstrated satisfactory governance (NRGI 2017; see Figure 15.1).[2]

## 15.3 Analysing Risk and Reward

### 15.3.1 *Key Choices and Lenses of Analysis*

The significant ambitions ascribed to NOCs, and their frequent failure to meet these ambitions, point to the need for governments to engage in rigorous analysis as they set up, structure, and manage their NOCs. This need is particularly pertinent in some of the new oil and gas countries that are only now beginning to exploit large reserves (examples include Tanzania and Kenya). The question in most countries is not *whether* to have an NOC—the vast majority of current and prospective oil producers, with limited exceptions, already have one (or several), and few countries are likely to eliminate them. Rather, the question is *what kind* of NOC can best achieve the country's goals, and *how* to create the right incentives for performance.

Three questions are particularly critical. First, the government should decide what kind of *commercial mandate* the NOC will be empowered to pursue. NOC commercial mandates vary dramatically, as is appropriate given the wide divergence in capacities and oil-sector prospects across countries (Heller and Marcel 2011). The aspirations of Saudi Aramco are thus necessarily of a different order of magnitude than those of a company like the National Oil Corporation of Kenya; Figure 15.2 elaborates.

Second, the government should decide what, if any, kinds of *non-commercial* roles to empower NOCs to carry out. While some analysts suggest that an enterprise should refrain altogether from non-commercial activities, the reality is that almost all NOCs engage in some degree of non-commercial work, and in some cases this mixing of responsibilities has been compatible with

[2] The 2017 Resource Governance Index evaluated the governance of the extractive sectors in eighty-one countries. It assessed seventy-four SOEs across the oil and mining sectors. Each country's score is assessed via a 149-question survey divided into three components: value realization, revenue management, and enabling environment. For an explanation of the overall methodology of the index, see http://resourcegovernanceindex.org/about/methodology. The ranking for SOEs displayed in this graphic is based on a subset of questions addressing each SOE's transparency, oversight, and clarity of responsibilities. See http://resourcegovernanceindex.org/data/both/issue?category=1&region=global&subcategory=4.

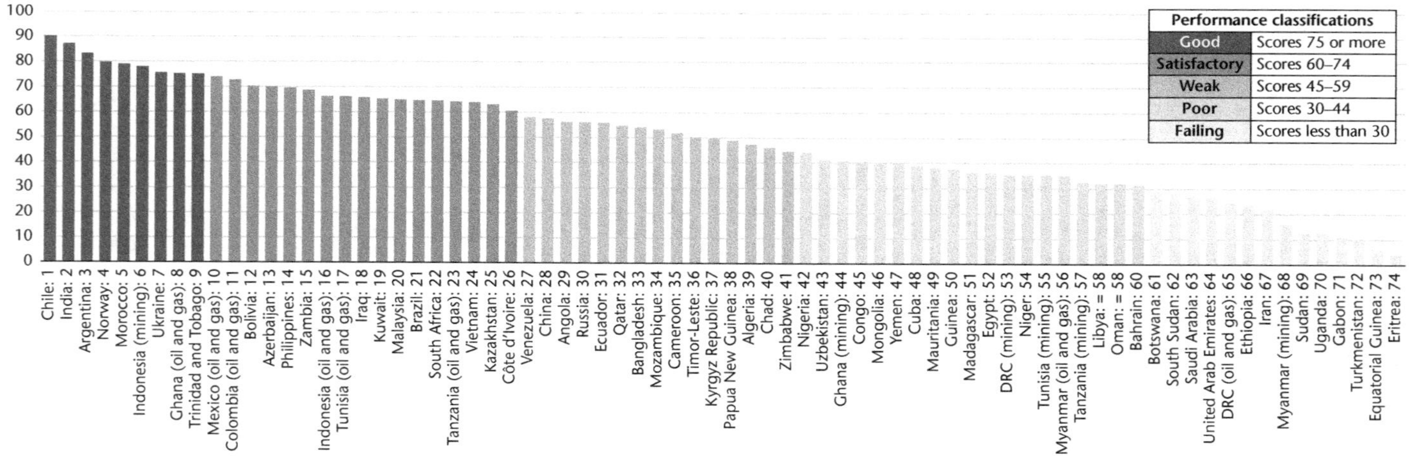


**Figure 15.1.** State-owned enterprise governance: Resource Governance Index score

*Source*: NRGI (2017); reproduced here by permission.

Most complex

| Commercial role | Example |
|---|---|
| Operator of large-scale or complex upstream fields | Statoil (Norway) |
| Operator of smaller or simpler upstream fields | Sonangol (Angola) |
| Major equity holder in joint venture group, but non-operating | Nigerian National Petroleum Corporation (Nigeria) |
| Manager of refineries or downstream distribution networks | Petrotrin (Trinidad & Tobago) |
| Investor or manager of subsidiaries or service providers | Sonatrach (Algeria) |
| Sale of government share of oil/gas | Société Nationale des Hydrocarbures (Cameroon) |
| Minority equity investor | Ghana National Petroleum Corporation (Ghana) |
| Data sales and fees | National Oil Company of Liberia (Liberia) |

Least complex

**Figure 15.2.** Spectrum of NOC commercial roles

*Note*: many NOCs play several of these roles simultaneously. The inclusion of a company as an example of a particular role on this list does not imply that this is the only role the company plays. Angola's Sonangol, for example, is simultaneously an operator of smaller/simpler upstream fields, a major non-operating shareholder in some fields, the manager of refineries and distribution networks, the owner of subsidiaries, the seller of government oil and gas, and a minority equity investor.

*Source*: author's illustration.

efforts to build skills and knowledge over time (Thurber et al. 2011). Table 15.2 illustrates some of the most common tasks. The assignment of some non-commercial responsibilities can enable a new producer to begin to increase capacity by concentrating resources in one public institution, and can thereby support the delivery of important public goods to citizens (Marcel 2016a). But such assignments can damage NOCs' commercial efficiency, by saddling them with costly burdens that divert resources away from the core commercial goals. They can impede effective performance monitoring by muddying fundamental goals and benchmarks. And they can create conflicts of interest by ostensibly requiring NOCs to enforce regulations against themselves or to choose between themselves and other prospective contractors.

Third, the government should decide how to allow the NOC to finance its mandate. These companies can be stewards of huge portions of public revenues—in countries ranging from Azerbaijan to Angola more than half of all government revenue passes through the hands of the NOC. This creates a dilemma. In order to execute their commercial and non-commercial programmes, some NOCs have to spend extensively, on operating costs and investments. The leadership of these companies often lays claim to hold on to large shares of the revenues that they capture. But if the NOC is allowed to control too great a share of public revenues, it risks becoming a sort of parallel

**Table 15.2.** NOC non-commercial roles

| Role | Purported justification | Potential drawbacks | Example |
|---|---|---|---|
| *Regulatory* | | | |
| Leaders of efforts to draft laws/regulations | NOC can be most knowledgeable/skilled government body | Conflict of interest, over-extension of company, may distort orientation of sector rules | National Oil Company of Liberia (Liberia) |
| Licensing of rights to private companies | NOC is knowledgeable/skilled, may be well positioned to choose its partners | Conflict of interest | Petronas (Malaysia) |
| Approving and monitoring company work programmes, de facto or de jure enforcement of legal rules | NOC has closest access to projects | Conflict of interest, over-extension | Sonangol (Angola) |
| *Quasi-fiscal* | | | |
| Providing large-scale public employment | NOC may have resources to cushion labour-market stresses, deliver income to population | Cost burden, damage to company efficiency, potential for corruption | China National Petroleum Corporation (China) |
| Building public infrastructure | NOC may be most competent, financially strong government entity | Cost burden, potential for corruption, distracts from core mission | Saudi Aramco (Saudi Arabia) |
| Providing social services | NOC may be most competent, financially strong government entity | Can weaken other government institutions, cost burden, potential for corruption, distracts from core mission | PDVSA (Venezuela) |
| Managing subsidized fuel programme | Nexus between extraction and consumer fuel needs | Cost burden, potential for corruption | NNPC (Nigeria) |

*Note*: Many NOCs play several of these roles simultaneously. The inclusion of a company as an example of a particular role on this list does not imply that this is the only role that company plays.

*Source*: author's illustration.

treasury that reinvests the lion's share of public gains from the sector back into oil and gas, or spends on public goods outside of the ordinary institutional procedures of government. This can subvert effective and accountable fiscal management.

In making these decisions, this chapter posits that many governments would be well served to conduct more concentrated risk–reward analyses of available options, and to share the results of these analyses with the public. Such an approach can lead to a better matching of strategy to the state of the country's oil sector, quality of its institutions, level of dependence of oil revenues, and ability to withstand failure.

### 15.3.2 *Categorization of Risks*

Government officials sometimes focus on the upside potential of major investments in an NOC—and the national pride that the company can amplify—without systematically assessing how likely the upside is to be realized in practice or how costly failure would be. Following are two basic categories of risks that countries face when investing in NOCs.

#### 15.3.2.1 COMMERCIAL/FINANCIAL RISK

Making an investment in an NOC can mean that the government ends up with fewer economic benefits than it would have had absent such an investment. For purposes of this discussion, it is useful to think in terms of a hypothetical upstream oil project, and presume two stylized options. Under the first option (referred to as the 'British Option') the state does not involve an NOC in the project at all. Instead it awards a licence to a private company and extracts financial benefits purely by taxing that company.

Under the second option (the 'Saudi Option'), the NOC has exclusive dominion over the project, exploring and operating the field itself (and hiring any contractors it may need to get the job done). Here, the idea is that the NOC will invest upfront as necessary to execute the project effectively, and that the state will ultimately derive financial benefits via taxes, dividends, or other transfers from the NOC to the treasury, and also from any other non-commercial benefits that may accrue (public employment, ancillary infrastructure, etc.).

Over the long run, if the government chooses the Saudi Option, it may hope for stronger overall returns. By cutting out the middleman (private company), a larger share of rents from the project can stay in the country. But in order to generate this potential, the government needs to pay, today. It will itself have to spend on building up a set of core competencies and staff—hundreds or thousands of skilled technical staff members—in order to become a major operational player (Marcel 2016b). And it will have to invest in project-specific capital and operating costs.[3] In executing a project in this manner, there are several elements of financial risk:

- *Geology*: This is the most significant risk. Exploration is expensive, and usually fails. Research by a global energy research firm showed that only 8 per cent of wells drilled in 'frontier' countries—those without previous discoveries or production—yielded a commercially viable play from 2011

[3] Note that if we deviate from the stylized all-or-nothing hypothetical, an NOC can defer some of these project costs by having its interests in a project 'carried' through the exploration or development stage by private partners. This reduces risk meaningfully, but the government is still responsible for repaying its share of costs out of any revenue stream ultimately generated by the project.

to 2015. Even in mature oil producers, the success rate was only 36 per cent (Myers 2016). In general, the more promising are the oil plays that exist within a country, the better an NOC may be able to spread its geological risk. When a state opts for an NOC to be responsible for exploration—as opposed to assigning that responsibility to a private company—the burden of any exploration failure falls on the public purse.

- *Market fluctuations*: Extractive industries are notoriously volatile, and periods of low prices can devastate the returns on investments.
- *Performance divergence*: For the reasons discussed above, an NOC may be less efficient than private-sector oil companies, which may mean that the government choosing the Saudi Option will be less likely to see the project carried through successfully to production, and more likely to experience management inefficiencies that wipe out many of the benefits deriving from the state's exclusive control.
- *Opportunity cost*: Revenues reinvested in the oil sector via NOC projects are not available for immediate public investment, including in growth-promoting sectors such as infrastructure or education. Sometimes a cycle of reinvestment lasts for years or decades, with a strong share of NOC revenues being returned into the company in a continual bet on future returns. Table 15.3 shows that many oil producers have tied up large sums of national assets in their NOCs (Manley et al. 2016). For comparative purposes, the table shows what NOC asset holdings represented as a percentage of total government spending in that year.[4] A period of low prices or of repeated project failures can mean that years of sector revenues are wasted without an appreciable impact on national development.

**Table 15.3.** Public assets held by 100% NOCs, 2014

| Country (company) | Total assets (US$ million) | State assets as percentage of annual government expenditure |
|---|---|---|
| Angola (Sonagol) | 54,496 | 103 |
| Azerbaijan (SOCAR) | 30,684 | 135 |
| Indonesia (Pertamina) | 49,507 | 30 |
| Malaysia (Petronas) | 164,531 | 181 |
| Mexico (PEMEX) | 160,119 | 44 |
| Qatar (Qatar Petroleum) | 110,031 | 164 |
| Venezuela (PDVSA) | 226,760 | 104 |

*Source*: company annual reports; author's calculations.

[4] This is an admittedly crude proxy for the opportunity cost of the allocation of assets to SOEs, as the accumulation of assets takes place over the course of years or decades. But it is listed here to show the relative size of NOC assets in terms of the total public sector. Upcoming research by the author examines the relationship between NOC assets as revealed in financial statements and various measures of public assets and government activity.

The British and Saudi options represent the two extreme poles of the spectrum of commercial roles assigned to NOCs, and the choices that a government makes about what kind of role the company can play. The key is that a dollar invested in the NOC is a dollar of public funds put at risk. The more complex a role that the government assigns the NOC (i.e. the further towards the top the company sits among the range of options shown in Figure 15.2), the greater is the risk that the government is undertaking with public revenues.

#### 15.3.2.2 GOVERNANCE RISKS

In addition to the financial risks associated with putting public revenues into play via a national company with commercial aspirations, poor management of (or by) NOCs can disrupt the governance of the oil and gas sector or even the economy as a whole. These risks are elevated when the NOC is assigned significant regulatory or quasi-fiscal responsibilities. These can take several forms.

First, a country can discourage private investment if would-be participants in an oil market perceive that the playing field is uneven and that NOCs are accorded special privileges. In order to generate the confidence necessary to invest the huge sums associated with exploration and development, oil companies want assurances that commitments will be honoured. Oil executives routinely cite the fear that assets will be 'nationalized' (i.e. handed over to the NOC) or that project decisions will be forced upon them midstream by NOC partners as one of the biggest risks their companies face (Ernst & Young 2013). At an extreme, uncertainty about excessive grants of NOC access to the best geological resources or the power of NOCs to force project decisions can cause oil companies to avoid a market altogether. Venezuela has been one of the strongest recent examples of this risk—strong fears of dramatic retrenchments of private oil company contractual rights in favour of PDVSA have made the country too risky for some oil companies. At a minimum, oil companies price these concerns into their assessments of the value of a project and the fiscal terms they will accept.

Second, the enforcement of generally applicable rules can be weakened significantly if the NOC is not subject to strict oversight. In order for its oil sector to be managed effectively and in the interests of the public, a country needs strong legal rules covering issues ranging from fiscal oversight to environmental protection to health and safety. Many countries have enforced these rules weakly or inconsistently vis-à-vis the NOC, either because the NOC enjoys *formal* regulatory power (and is thus in the conflicted position of ostensibly overseeing itself) or because it exercises *informal* political power that protects it from the strictures of the rules. This increases the likelihood that oil projects will fail to deliver promised economic results or, worse, that dangerous incidents may occur without redress. It also damages the prospects for the development of strong public institutions. In countries such as Nigeria, the NOC is widely perceived to play outside of the rules, exercising power through a tapestry of informal levers. This has dramatically hindered the development of strong systems of financial and environmental control, resulting in continually disappointing financial returns from the NOC to the state and a persistent state of conflict and environmental damage.[5]

Third, for the reasons cited above, NOCs have too frequently been at the centre of large corruption scandals, many of which have spread to the broader economy or polity. The Petrobras pay-for-play scandal is the most dramatic recent case, but other examples abound. In Congo-Brazzaville, the government has been accused of selling public oil through an intermediary company controlled by the head of the NOC Société Nationale des Pétroles du Congo (SNPC). A Norwegian company was sanctioned both in Norway and in the United States for paying US$5 million into the accounts of the director of the National Iranian Oil Company in order to secure access to a major Iranian gas field.

### 15.3.3 *Assessment of Potential Rewards*

As is noted above, some NOCs have contributed extensively to their countries' efforts to develop an efficient and profitable oil sector that delivers benefits to their countries, including, but not limited to, extensive revenue flows. But in order to make a country's investment in an NOC a net-positive proposition, the government needs to map the precise benefits that it hopes to derive from its NOC, and the conditions necessary for those benefits to come to pass. This step is sometimes overlooked by governments, as is the need to assess trade-offs between different goals.

Among the benefits that NOCs can deliver to their states and the conditions for their achievement are the following:

- *Larger revenues*: For countries at a relatively early stage of oil-sector development, relying heavily on an NOC as a commercial player rarely generates greater revenues to the state than would relying on private companies and taxing them effectively. Minority equity stakes do not generally produce financial returns beyond those that could otherwise be captured via other fiscal mechanisms (NRGI 2014). And especially during early years, assigning responsibility to an NOC to manage oil fields itself is

[5] For an understanding of the frequency and scope of oil spills in Nigeria, see the *Nigerian Oil Spill Monitor* at https://oilspillmonitor.ng.

likely to result in efficiency losses as the company develops its competence from an initially low base. But over time, if an NOC succeeds in developing itself as a skilled and efficient project manager, the state can successfully extract a greater share of revenues deriving from production, without having to rely heavily on private contractors. Various NOCs in the Middle East, Latin America, and Asia have demonstrated the long-term potential of these enterprises as effective revenue generators. In order for this goal to be realized, a country needs a sufficient supply of oil or gas to provide for a long enough period of production for the NOC to build economies of scale and develop its capabilities (Heller and Marcel 2011).

- *Greater control over the pace of oil-sector development*: Governments often want to manage the development of an oil sector strategically, timing the pace of exploration and extraction according to national macroeconomic goals, a desire to maximize long-term production levels, or the aim of building up national capacities/systems before extracting large quantities of petroleum. Managing this pace of development can be difficult if a government is working overwhelmingly with private contractors, which are driven by their own market-oriented time frames. Planning projects principally through an NOC can enable a government to exercise greater control (see Lahn and Stevens 2017; Stevens et al. 2015).
- *Building a nexus of domestic expertise in the sector*: Countries such as Malaysia and Angola have used the NOC as the place where human and financial resources are concentrated in the early stages of oil-sector development. NOCs are often able to attract the country's best talent, benefitting from more flexible compensation systems than government ministries. Some countries have chosen to 'mass forces' in the NOC as a way to build a body most capable of developing and enforcing a strategy for the sector. This sort of concentration approach carries a greater risk of conflict of interest, but those risks have been deemed acceptable by governments in many new producer countries (Thurber et al. 2011).
- *Promoting local content and positive economic spillovers*: Many governments see their NOCs as the most natural champions of efforts to use the oil and gas sector as an incubator for small and medium enterprises that service the oil sector and that ultimately can develop into dynamic companies in their own right. Such responsibilities can be imposed upon private/foreign oil companies, but embedding them in the core mandate of an NOC is often seen as a clearer route to place local content at the core of the business. Further, the successful development of a local content strategy can be a key to transforming exhaustible petroleum resources into something more sustainable.

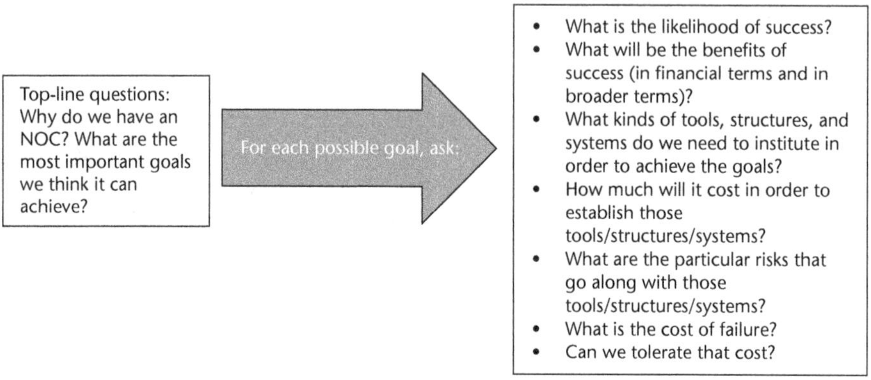


**Figure 15.3.** Key questions for NOC goal-setting
*Source*: author's illustration.

### 15.3.4 *Putting the Pieces Together*

The various considerations that have been set out above work together to determine the type of NOC structure and systems that will be most suitable in any particular country context. When a government makes decisions about how to structure its NOC, or indeed when the NOC's own leadership sets out its strategy, it needs ideally to examine the foregoing elements of risk and potential reward side by side. Doing so can provide an opportunity to be more systematic and realistic, and to tailor plans more effectively. Asking a basic sequence of questions such as those set out in Figure 15.3, and conducting the in-depth analysis necessary to answer those questions, can also aid effective policymaking and execution. Figure 15.3 illustrates the basic sequence.

In practice, some NOCs—ranging from commercially sophisticated giants such as Statoil and Qatar Petroleum to more limited-mandate companies such as tiny Staatsolie of Suriname—have engaged in this kind of strategic and systematic assessment. However, such concerted strategic decision-making remains the exception rather than the rule. Other NOCs have a mixed decision-making record, executing thorough strategic planning on several aspects of their business while occasionally bowing to the idiosyncratic demands of the political system. And still others are more purely reactive to the ebbs and flows of national politics and have failed to develop coherent strategies.

## 15.4 Mitigating Risks and Making Rewards More Likely to Materialize

The precise contours of the risk–reward analysis discussed above will vary based on the characteristics of the oil endowment of each country, the quality of institutions, and the state of economic development. But there are certain strategies for how to manage an NOC that have been demonstrated to reduce risks and increase the chances of success across a range of countries.

Perhaps most importantly, it is important to define the roles and responsibilities of the NOC as clearly as possible. Major problems arise where the contours of what the NOC is allowed and required to do are unclear, or where the lines of authority between the NOC and other government agencies are blurred. This is particularly problematic in countries where the NOC engages in non-commercial activities. Blurred lines of responsibility and scope can damage the NOC's commercial bottom-line, because they saddle the company with non-core responsibilities that lack clear limits. They also impede good governance, by obscuring channels of accountability and requiring stakeholders to navigate complex systems. Across a range of different types of NOCs, with different sorts of mandates, the countries that have reduced ambiguity and inefficient overlaps in responsibility have tended to see better performance. Colombia's Ecopetrol, for example, has over time been increasingly absolved of what were once a large set of non-commercial activities, and allowed to focus energy and resources on its economic bottom-line. Even some NOCs with mixed mandates—such as Malaysia's Petronas—have been able to develop commercial units with clear mandates and benchmarks that have kept focused on business success.

Given the tensions around revenue flows, with NOCs advocating for larger shares of revenue flows and ministries of finance pushing for these companies to pay more to the treasury, instituting a clear system governing revenue transfers is particularly important. When payment flows between the company and the state are subjected to the whims of the government or of company executives, the results can seriously impede planning, both within the company and for the national budget. In a worst-case scenario, the company becomes a state within a state that plays an outsized role in public spending. A clear and easy-to-understand system not only enhances public planning, but can also make it easier for the public to track the decisions that the NOC is making on the use of public resources. The easiest systems to comprehend are those that subject NOCs to the same rules that apply to private taxpayers, such as in Norway and Brazil. But these systems may not deliver adequate shares of oil revenue to the state in some countries, so where special rules apply to the NOC, they should be transparent and easy to track.

The most successful countries have developed governance mechanisms to promote strong incentives for effective NOC performance. These mechanisms can broadly be grouped into three categories:

- *Corporate governance* tools within the NOC to promote accountable and effective management, including strong standards for board membership, merit-based hiring and promotion, internal ethical standards, anti-corruption training, and whistleblower protections for employees who report malfeasance. Additionally, the conduct of rigorous and independent audits by highly skilled, independent auditors is perhaps the most important mechanism to promote sound financial management. Examples of companies that have implemented such tools are Malaysia's Petronas and several Persian Gulf NOCs.

- *Intragovernmental checks and balances*, such that the executive and the legislature can hold company leadership accountable for their performance. Among these measures are the establishment of clear goals and performance targets, and the assessment of company leadership against those targets; consistent reporting by the NOC to its 'shareholder' ministry within government; and the requirement that the company provide detailed reports to the legislature and appear before the legislature when summoned. Norway and Colombia are often considered the countries that have developed the strongest such checks and balances.
- *Public reporting systems* that can give citizens a clear picture of how NOCs are managing national resources. NOCs should publish clear annual reports with detailed financial information—including income statements, balance sheets, and cashflow statements. The public also benefits from detailed information about the NOC's plans, budgets, operational activities, and internal management policies. Several prominent NOCs—including many in Latin America and Asia-Pacific—provide very extensive public reports. Middle Eastern and African NOCs—including giants such as Saudi Aramco and Nigeria's NNPC—have had weak public reporting to date.

## 15.5 Conclusions

Much of the recent scholarship on NOCs has been developed with tacit or explicit framing within the context of the high-price era, with a concomitant focus on helping countries maximize the opportunities presented by their oil and gas wealth via NOCs. At the time of writing, the world was ensconced in what appears likely to be a sustained period of lower oil and gas prices. The price fall has hit NOCs particularly hard, and across the world these companies have been forced to shelve projects, cut costs, or take on new debt in order to keep going.

This low period in the decades-long cycle of oil-price volatility has laid bare some of the risks inherent in government choices to invest heavily in NOCs as a core of their sector development strategy. By doubling down on the country's dependence on the oil sector—that is, reinvesting revenues in the sector, via the NOC—governments give themselves the opportunity to extract larger shares of the benefits during boom times. But they leave themselves doubly exposed when times are more difficult, with the negative impact being felt in the national budget and with struggling entities needing to be propped up as national 'flagships'.

But in the midst of this period of crisis for many countries may lie the opportunities for reform that can help reduce the risk of similar swings in the future. NOCs from across the world—from major producers such as Azerbaijan and Malaysia to countries that are not yet producing oil, such as Liberia—have announced a range of measures to try to strengthen their balance sheets, including reducing personnel budgets, selling their interests in risky exploration plays, and eliminating investments in non-core or money-losing ventures. The elimination or reduction in fuel subsidies in a range of countries—including such seemingly unlikely places as Saudi Arabia and Venezuela—could also prove a boon to the finances of NOCs that have carried a heavy burden.

These responses to market challenges could have a long-term impact, helping NOCs to become leaner, better managers of public resources, if they are sustained. But that is a big 'if', and past efforts to tighten NOC operations during tough times have tended to be pushed aside by patronage and bloat when prices rise again, reducing the benefits that accrue to the companies' public shareholders. The key for the NOCs that have been able to remain focused and efficient in times of plenty—including Colombia's Ecopetrol, Norway's Statoil, and Malaysia's Petronas—has been a strong commitment to corporate governance, transparency, and oversight. If they wish to take advantage of this period of fiscal crisis to implement durable reforms, governments and NOCs in other countries can aim for systemic reforms that commit the companies to stronger technocratic management and internal decision-making, tougher performance incentives and protections for whistleblowers, and better public communication. They would also be well served to use this moment to reflect carefully on the risk–reward calculus, and adapt strategies to hit an optimal balance for the long term.

## References

Boardman, A. and A. Vining (1989). 'Ownership and Performance in Competitive Environments: A Comparison of the Performance of Private, Mixed and State-Owned Enterprises', *Journal of Law & Economics*, 32(1): 1–33.

de Oliveira, A. (2012). 'Petrobras: Strategy and Performance', in D. G. Victor, D. Hults, and M. C. Thurber (eds), *Oil and Governance: State Enterprises and the World Energy Supply*, 515–56. Cambridge: Cambridge University Press.

Eller, S., P. Hartley, and K. Medlock (2007). *Empirical Evidence on the Operational Efficiency of National Oil Companies*. Houston, TX: Rice University James A. Baker III Institute for Public Policy.

Ernst & Young (2013). *Business Pulse: Exploring Dual Perspectives on the Top 10 Risks and Opportunities in 2013 and Beyond—Oil and Gas Report*. London: Ernst & Young.

Heller, P. and V. Marcel (2011). *Institutional Design in Low-Capacity Oil Hotspots*. New York: Revenue Watch Institute.

Heller, P., P. Mahdavi, and J. Schreuder (2014). *Reforming National Oil Companies: Nine Recommendations*. New York: NRGI.

IMF (2016). 'Transcript of the Press Conference on the Release of the World Economic Outlook Update', 19 January. New York: IMF.

Kane, K. and H. Christiansen (2015). *State-owned Enterprises: Good Governance as a Facilitator for Development*. Paris: OECD.

Karl, T. (1997). *The Paradox of Plenty: Oil Booms and Petro-States*. Berkeley, CA: University of California Press.

Lahn, G. and P. Stevens (2017). 'The Curse of the One-size-fits-all Fix: Re-evaluating What We Know about Extractives and Economic Development', WIDER Working Paper WP2017/21. Helsinki: UNU-WIDER.

Leahy, J. (2016). 'What Is the Petrobras Scandal That Is Rocking Brazil?' *Financial Times*, 31 March.

Manley, D., J. Cust, and G. Cecchinato (2016). 'Stranded Nations? The Climate Policy Implications for Fossil Fuel-Rich Developing Countries', Oxcarre Policy Paper 34. Oxford: Oxford Centre for the Analysis of Resource Rich Economies. Available at: www.oxcarre.ox.ac.uk/images/stories/papers/PolicyPapers/oxcarrepp201634.pdf.

Marcel, V. (2016a). *Guidelines for Good Governance in Emerging Oil and Gas Producers 2016*. London: Chatham House.

Marcel, V. (2016b). *The Cost of an Emerging National Oil Company*. London: Chatham House.

Megginson, W. and J. Netter (2001). 'From State to Market: A Survey of Empirical Studies on Privatization', *Journal of Finance*, 49(2): 403–52.

Myers, K. (2015). 'Getting the Most from State-Owned Enterprises', Presentation at NRGI–Blavatnik School of Governance Executive Course on Oil, Gas, and Mining Governance, Oxford.

Myers, K. (2016). 'Introduction to the Oil and Gas Sector', presentation at New Petroleum Producers Discussion Group, Georgetown, Guyana, on file with author.

NRGI (2017). *2013 Resource Governance Index*. New York: NRGI. Available at: http://resourcegovernanceindex.org/.

NRGI (2014). *Natural Resource Charter*, 2nd edition. New York: NRGI.

Ross, M. (2012). *The Oil Curse*. Princeton, NJ: Princeton University Press.

Sappington, D. and J. Sidak (2003). 'Incentives for Anti-competitive Behaviour by Public Enterprises', *Review of Industrial Organization*, 22(3): 183–206.

Shirley, M. and P. Walsh (2000). *Public vs. Private Ownership: The Current State of the Debate*. Washington, DC: World Bank.

Stevens, P., G. Lahn, and J. Kooroshy (2015). *The Resource Curse Revisited*. London: Chatham House.

Thurber, M., D. Hults, and P. Heller (2011). 'Exporting the Norwegian Model: The Effect of Administrative Design on Oil Sector Performance', *Energy Policy*, 39: 5366–78.

Tordo, S. (2011). *National Oil Companies and Value Creation*. Washington, DC: World Bank.

Victor, D., D. Hults, and M. Thurber (eds) (2012). *Oil and Governance: State Enterprises and the World Energy Supply*. Cambridge: Cambridge University Press.

Victor, N. (2007). *On Measuring the Performance of National Oil Companies*. Programme on Energy and Sustainable Development Working Paper 64. Stanford, CA: Stanford University.

Wolf, C. (2009). 'Does Ownership Matter? The Performance and Efficiency of State Oil vs. Private Oil (1987–2006)', *Energy Policy*, 37: 2642–52.

# 16

# Protecting the Environment during and after Resource Extraction

*Ruth Greenspan Bell*

## 16.1 Introduction: Resource Extraction Has High Potential for Immediate and Long-term Environmental Damage

Even responsible extraction of natural resources imposes environmental costs. Digging minerals, metals, and hydrocarbons from the ground inevitably disrupts the landscape. Beyond the actual mine site, extraction brings the construction of roads to bring in equipment and bring products to market, power transmission corridors in order to power the operations, facilities to house mine workers and often their families and all the attendant services that must be provided, and waste disposal both from the mining and from other human activities, among a stream of impacts. The activity of building roads and then the use of those roads and equipment introduces fuels and exhaust fumes.[1,2]

The mining and extraction operations themselves often use hazardous chemicals including cyanide, sulfuric acid, and solvents to separate ore from rocks. In addition, beyond chemicals which are considered appropriate and for which technologies exist to capture and clean or separate, many mining operations continue to use illegal chemicals. One example is mercury. Despite strong evidence that mercury does not break down in the environment and continues to cycle between water, land, and the atmosphere, travelling large distances from the original application or breach, and building up in humans

[1] Greg Jehle, a JD/MA candidate at the Georgetown University Law Center and Center for Contemporary Arab Studies, an expert in international project finance, environmental law, and law and development, provided invaluable research support.

[2] Detailed supporting references and footnotes have been removed to fit space demands; these can be found at the UNU-WIDER website: https://www.wider.unu.edu/publication/protecting-environment-during-and-after-resource-extraction.

and animals and in the food chain, neither regulation nor even bans are sufficient to keep it from being used in various parts of the developing world and in so-called artisanal mining.

Water is often an essential component of mining. The contaminated water might be discharged or stored in large tailings ponds that can reach the size of small lakes. As explained in more detail in Bell (2017), the impacts of water contamination are broad and often unpredictable; weak communities have little power against powerful mining interests.

Today, the search for valuable resources has taken mining and extraction into increasingly remote areas of the undeveloped world, such as the Amazon rainforest (see Kimerling 2000), reflecting the demands of the increasing number of people on the globe with the resources and the interest in purchasing upscale goods and in engaging in large-scale building (for more detail, see Bell 2017).

The areas marked for extraction can include fragile ecosystems previously difficult to access. For a number of reasons, the true nature of the trade-offs between development and environmental protections may not be properly understood.

The general point is that mining, including hydrocarbon operations, can and likely will contaminate land, water, and air. They can cause health problems for workers and people living near mines. They may, but do not necessarily, create off-setting local wealth. Further, when mining operations are opened up in natural areas, there are consequences for the local biodiversity—impacts on organisms and species in the local ecology, the way water is captured and stored (or not) by trees and vegetation, and on other natural processes that support both local and distant human activity. People and their livelihoods are displaced, and traditions and cultures disrupted. In some instances, long-term interests are sacrificed for short-term returns (Cronin 2009).

These factors illustrate the need for effective regulation to ensure that the benefits obtained by a country by exploiting its natural resources are not undercut and eroded, if not entirely destroyed, by an exploitation that leaves behind irreparable damage. Regulation is here defined as 'a process of setting and then assuring the implementation of requirements that protect, to the extent necessary to meet local needs and expectations, the landscape and the people working in the mines and living in the general vicinity or dependent on natural processes of the exploited area'. Regulation provides a broader perspective so that immediate needs are balanced by longer-term considerations, namely preservation to the extent possible of natural and other values. It can not only address the conditions under which a country's natural resources are extracted and exploited but also set rules for what happens when these resources are depleted and the companies exploiting them leave.

## 16.2 Countries with Weak Institutions and Poor Success Rates in Addressing their Other Environmental Challenges Face Significant Challenges in Regulating Extraction

Reconciling extraction and protection of the natural world is an ongoing challenge in the highly developed world, even in countries with well-established regulatory systems and long-standing adherence to the rule of law. Every day brings evidence of the complexity of holding to account companies that pollute (see McQuaid 2009). Some impacts are obvious and visual; we have all seen stark photos of, for example, table top removal to extract coal. Others are more difficult to track and comprehend. When pollution enters an ecosystem, tracing its source and attributing responsibility is a challenge, even in countries with vast resources, empowered citizens with the tools and legal recourse to play a watchdog role, and well-functioning systems of law (see Rich 2016).

The challenge is exponentially greater in countries of the developing world with fragile institutions where experience in compliance and enforcement is weak, where environmental issues are not considered a priority, and where there are correspondingly poor success rates in addressing other environmental and societal challenges. There is a vast literature on the regulatory deficiencies in the developing world and many of the disasters that have ensued (see, for example, Puvimanasinghe (2009)). The challenge of regulatory deficiency may be further compounded by the way environmental matters are managed in concession and extraction agreements.

Developing countries that host extraction often have the necessary *formal* structures, such as environmental laws and agencies. But for a variety of reasons—ranging from seemingly simple issues of capacity to corruption and malicious intent—these laws and agencies are frequently either ignored, inconsistently invoked, or in the worst cases, used as a weapon for purposes totally unrelated to rectifying environmental harm. Russia offers examples where environmental laws were invoked against businesses and their leaders to discipline or punish insufficient loyalty, retrieve financial gains, or tie up assets, rather than to stop pollution.

There are multiple explanations for weak or inconsistent enforcement. Countries may simply lack adequate systems, reflecting insufficient resources and/or inadequate experience or training, even if they have the will to act. In cases where an industry is owned in part or entirely by the state, the inherent conflict of one arm of the state prosecuting another discourages enforcement against polluting facilities (see Bell 1994: footnotes 33 and 34).

Poland as it emerged from the Soviet bloc provides an example of a country with inadequate enforcement tools. It could impose fines and criminal penalties and threaten facility shut-downs. But these are a 'nuclear option' and when actually used, shut-downs were often merely symbolic and imposed for only a day or two rather than correcting the circumstances that were causing the pollution. Fines were either weak or ignored. More perniciously, where the enterprise was state owned, fines were a line item in the industry's annual state-provided budget; a fine for a state enterprise would simply be offset by an increased subsidy through the budget or by borrowing from state banks (Bell 1992). Poland, in short, was missing tools that could nudge, push, and incentivize facilities to come into compliance with laws and regulations.

The depth of commitment to enforcement is another barrier. Requirements that appear sensible on the books often languish without implementation when only officials can bring enforcement cases. Where citizen enforcement provisions, or meaningful opportunities for citizens to come forward with evidence, are lacking, there is generally no one to step into officials' shoes to ensure that environmental requirements are met. Numerous countries including the United States have variations on citizen involvement in environmental enforcement. Elsewhere, ecological associations, legislative bodies at different levels of government, trade unions, workers and/or local self-governments can initiate enforcement actions (see May 2003). Provisions like these are particularly useful when violations are taking place in remote locations, where government officials may be few and far between and affected citizens may be in the best position to observe and document harmful environmental impacts.

Finally, there is the issue of corruption. The resource curse is often expressed as the concern that exploitation of natural resources in countries with weak governance and fragile institutions often benefits a small number of powerful people who siphon the proceeds for their own needs (see Lawson-Remer and Greenstein 2012). Some mechanisms for this diversion of proceeds are described in Bell (2017). Consequently, environmental requirements and their enforcement are high on the list of restraints that are jettisoned in many countries in favour of quick profits. In power struggles over such issues, environment ministries are typically among the weakest bodies in government; and environmental regulators and the people who depend on them hardly have a chance.

The bottom line is that enforcement is essential to making laws and requirements work, but it should not be assumed that the existence of enforcement agencies or paper commitments are a sufficient guarantee that pollution control and clean-up requirements will be honoured.

## 16.3 Commonly Used Approaches to Work around Domestic Legal, Institutional, and Experiential Deficiencies

Experts debate how to manage these systemic deficiencies in environmental enforcement that parallel similar obstacles to financial and other oversight of extraction and its consequences. This section begins by considering the regimes under which mineral and other rights are granted and administered, and specifically the distinction between 'licensing' vs. 'contractual' approaches for building in protective requirements.

'Licensing' is shorthand for a process whereby agreements to extract rely on generally applicable laws to define rights and obligations. The contract between the extraction company and the government is controlled by established law to the extent possible. The virtue of this, as stated by the Columbia Center on Sustainable Investment (CCSI), is that the process of considering and writing laws, hopefully including public participation in the legislative process in democratic states, 'provides a venue for incorporating the public's concern for the sector and decreases the likelihood of political volatility"' (internal citation omitted, CCSI 2015: 7). The laws apply equally, the system is easier to administer and opportunities for corruption are more limited.

Licensing is preferable from a policy point of view because requirements are relatively consistent across the board and transparent rather than being ad hoc and specifically negotiated. In principle, one need only read the law (rather than be privy to specific negotiations) to understand how the particular requirements on any one extraction operation were derived. Further, requirements reflect policy judgements that take into account wider considerations than the needs of a specific project and are hopefully formulated through democratic processes.

Nevertheless, some countries with well-established laws have difficulty in their implementation, and acutely so with respect to environmental requirements. Accordingly, although the relationship between the government and the extraction company might seem to be securely rooted in established law/policy, in fact the parties' obligations might be formalities rather than tenets that guide practice (Bell 1994). The literature commonly distinguishes between 'laws on the books' and 'laws in practice'. Weak institutions, corruption, and lack of political will, as outlined earlier, can undermine any kind of agreement on rights and obligations, including those that rely on existing law.

Another weakness is the contract negotiation process itself. Many developing countries cannot match the legal and commercial sophistication of the extractives companies, or believe they lack the bargaining power to push for certain requirements. More damagingly, as in Poland before a hard-fought memorandum of understanding introduced environmental expertise into privatization teams, the country's environmental experts are frequently excluded from the negotiating table and sometimes are not even consulted. If they were involved, engagement in negotiations would improve environmental results as well as help build their expertise over time and make them better negotiators and regulators.

One means of substituting for weak institutions and deficient or even non-existent legal structures is to negotiate contracts that specifically spell out the obligations of the company and the government in the context of a specific project (Affolder 2013). In this model, 'mineral licenses and the accompanying rights and obligations are negotiated for specific projects with each individual company' (CCSI 2015: 7). These types of arrangements are clearly more flexible, but that flexibility is also their weakness. They represent decisions made within the context of the specific proposed project. For that reason, it is harder to integrate more general or long-run environmental (or other) aspirations, incorporating the wider implications of any specific development or injury to the environment and a much longer time frame extending even to post-extraction remediation. After all, environmental laws are, among other things, societal decisions balancing development against preservation.

Decisions made through such contracts, involving matters that inevitably impact the long-term health and resilience of a country are made case by case, presumably in a less coherent fashion than would happen when policy emerges from generally applicable laws. Additionally, when requirements are handcrafted to a specific site and/or a specific entity, the burdens of monitoring and implementation imposed on a fragile bureaucracy are even greater than is the case with across-the-board laws. Accordingly, overstretched and under-resourced environmental authorities must police the very specific terms agreed to, which can differ project by project.

Accepting nonetheless that contracts will continue to be an important part of setting rights and responsibilities in many countries, several organizations have refined the contracting approach by developing guides that identify issues and standardize solutions. These include the Environmental Law Alliance Worldwide's (ELAW) *Natural Resource Contracts: A Practical Guide* (ELAW 2013) and the International Institute for Environment and Development's (IIED) *Investment Contracts and Sustainable Development: How to Make Contracts for Fairer and More Sustainable Natural Resource Investments* (Cotula 2010). Each offers practical tips for drafting natural resource contracts more favourable to environmental protection and suggest ways to structure investment contracts that maximize the investment's contribution to sustainable development. Further details on some of these are presented in Bell (2017).

Analysts assessing the contractual approach have found it difficult to draw clear-cut conclusions on the adequacy of this approach for protecting environmental values. On the one hand, Kimerling (2001) argued that the privatization of environmental regulation of oil extraction in the Ecuadorian Amazon through the use of contractual environmental provisions and an environmental management plan (EMP), in lieu of robust state regulation, represented an abdication of state responsibility and a threat to democratic accountability. She expressed strong concerns about the lack of transparency in the Ecuadorian government's dealings with the United States-based oil company, Occidental Petroleum, and called for a 'credible, independent and transparent audit... to evaluate the company's environmental standards and performance' (Kimerling 2001: 394). Her concerns about lack of consultation with affected communities, especially indigenous peoples, predate the use of the IFC Performance Standards, Equator Principles, and other private-sector initiatives that have evolved in recent years and are discussed in the extraction literature. Affolder (2013) is ultimately agnostic as to the usefulness of the contractual approach, arguing that it is a technique of global environmental governance that represents a particular policy choice (one that tends to favour the growth of extractive industries). She notes the 'contractual webs' created through the development of large extractive industry projects (e.g. loan agreements, insurance agreements) as an example of how regulation through private agreements can either strengthen (as in community benefit agreements or the IFC Performance Standards) or undermine environmental protection at the host country level (Affolder 2013).

In part because of some of the hazards of one-off negotiations, negotiations often now rely on model agreements and other industry codes of conduct including the Equator Principles and the IFC Performance Standards. While these have an aspect of being industry or lender-led initiatives, their development responded to a growing public concern about the impacts of extraction. Any number of activist and advocacy groups have publicized the need for greater attention to stewardship concerns domestically within the United States and internationally. Prominent examples include the Rainforest Action Network in the 1980s (focused on clear-cut logging, from 1996 through 2003); anti-extraction advocacy US-based Project Underground which published Drillbits and Tailings to alert readers to problematic extraction operations; and other such advocates. Eventually NGOs focused on the role of the financial services sector as a point of control over the activity of extraction companies. These initiatives, and incidents like the Bhopal accident discussed later, helped accelerate a shift in the way 'financial decision-makers perceive[d] that environmental concerns could affect their company's profit margin' (Jordan and Chamberlain 2001).

This NGO engagement with the financial services sector in its role as lender to large-scale extractive projects with potentially harmful environmental impacts led to the creation of several private-sector (voluntary) codes of conduct to be followed in all projects funded by participating financial institutions. These principles supplement the purely contractual approach by indicating to governments where negotiation is appropriate (see Langer 2013).

The Equator Principles (Mongoven 2006) help participating banks and other financial institutions (currently over seventy) to determine, assess, and manage environmental and social risk in projects. The objective is that they should not provide loans to 'projects where the client will not, or is unable to, comply with' the principles (Mongoven 2006). Although, in theory, the principles should allow project lenders to serve as a lever to improve environmental performance through the threat of reduced access to funds, in practice the results have been somewhat mixed (see Lance 2013; Sarro 2012). However, the Equator Principles have effectively given environmental and other NGOs a voice in major international lending decisions: a major change.

Similarly, the IFC Performance Standards, a primary inspiration for the Equator Principles, are intended to provide guidance to clients of the IFC 'on how to identify risks and impacts, and are designed to help avoid, mitigate, and manage risks and impacts as a way of doing business in a sustainable way, including stakeholder engagement and disclosure obligations of the client in relation to project-level activities' (IFC 2012).

These are powerful models, in part because of the status of the IFC, a member of the World Bank Group, as the self-described 'largest global development institution focused exclusively on the private sector in developing countries', willing to 'engage in difficult environments' and providing 'leadership in crowding-in private finance' (Freestone 2013). Adhering to these principles is a requirement of obtaining IFC support and is intended to set standards that the IFC client must meet 'throughout the life of an [IFC] investment' (IFC 2012). They have widely been copied as a model for other financial institutions (Langer 2013).

Another slice at this problem is found in the corporate social responsibility (CSR) approach. This is a self-regulatory approach predicated in the belief that companies can increase long-term profits, reduce business and legal risk, and improve shareholder trust by cultivating positive public relations and a reputation for high ethical—including environmental—standards. This belief came out of hard experience. Episodes like the Bhopal incident heightened corporate sensitivity about the reputational impacts of industrial accidents and loss of life, even in faraway places. In 1984 in Bhopal, a remote part of India, a major industrial accident with lasting consequences for local health and safety led to close scrutiny of Union Carbide's practices in the developing world (Verma 2014). Bhopal was heavily covered by the international press and, whether or not justified, came to represent corporate irresponsibility with continued argument to this day about specific responsibility (see Rogge 2001).

Some companies take implementation beyond compliance and engage in 'actions that appear to further some social good, beyond the interests of the firm and that which is required by law' (Wikipedia n.d.). For example, Nike has committed to monitoring factories in its global supply chain that produce its products under contract and that have historically subjected workers to substandard working conditions (Connor 2010).

In a similar vein, the ICMM, a voluntary organization of mining and metals companies, established another business-led self-regulatory initiative in 2003. The twenty-three member companies joined with thirty-four national and regional mining associations and global commodity associations to establish ten industry-focused principles of sustainable development (O'Callaghan 2011). Member companies also seek to comply and reinforce guidelines established by other organizations including the OECD and the World Bank.

Finally, environmental requirements are often incorporated into the contractual process through the conduct of an environmental (generally joined with social) impact assessment. Environmental impact assessments first came into use when the US Congress passed the National Environmental Policy Act (NEPA) in 1969. NEPA requires that all proposed major actions of the US federal government be accompanied by an examination of its environmental impacts including direct, indirect, and cumulative impacts of the alternatives being considered, detailed examination of any adverse environmental effects that cannot be avoided, and the means to mitigate adverse environmental impacts. Although the United States has no requirement that a project be rejected if the EIA shows adverse environmental impacts (Hershowitz 2008), merely requiring the analysis has often averted the worst projects and facilitated improvements of others.

EIA laws and policies were adopted by countries across the globe in response to the global environmental movements of the 1960s and 1970s, particularly following the 1972 United Nations Conference on the Human Environment in Stockholm, and further accelerated through (non-binding) international declarations such as the 1992 Rio Declaration on Environment and Development. The latter calls upon states to require EIAs for activities likely to have a significant adverse impact on the environment (Fromherz 2013). Many of these national laws have come to serve a somewhat different purpose than the US model, in some cases having a licensing or even a regulatory component (see Crippa 2008; Gray 2000).

In their application, EIA analyses ideally are intended for the 'project design phase'. The analysis can identify problems, consider how they might be mitigated, and incorporate this planning into the project's contractual structure (Kohn 2002). The difference from the US NEPA model is that these EIAs often are used to set terms going forward, particularly to satisfy contractual obligations to project lenders. This is quite a different role than an assessment of alternatives for a federal decision-maker (who might be functioning in a licensing or approval posture, for example) or to inform public review of potential projects (Lawrence and Thomas 2004).

In the German Federal Ministry for Economic Cooperation and Development's review, EIAs are used in both licensing and contract regimes. In the former, this is mostly as a result of domestic legislation; consequently, the assessment is 'evaluated as part of the approval process for granting the license'. There is also a mixed use of EIAs with some requirements included in mining agreements to supplement gaps in the law, in which case they can 'provide an avenue for remedy in the case of a violation not covered under the ESIA [Environmental and Social Impact Assessment] provisions of the law' (internal citation omitted, CCSI 2015: 36–7).

The use of the EIA tool in contracts may be problematic in part because it comes to substitute for missing legal elements: the EIA sometimes becoming another issue to be negotiated in the absence of uniform requirements. Timing of the conduct of the review can also be problematic, especially if 'only conducted after an agreement is concluded, often making them perfunctory exercises' (CCSI 2015: 36).

### 16.3.1 *How a Combination of Approaches Was Put to Work in a Liberian Mining Contract*

Special contracts with mining companies were used in Liberia to work around the legal and institutional inadequacies in Liberia following the huge disruptions of its brutal civil war (see Bell (2017: 9) for full details). These try to establish workable standards and strategies for implementation, and in Liberia's circumstances, it is not clear what more could be done. However, given Liberia's chronic regulatory capacity problems, it is also easy to be sceptical about those contracts' potential for putting serious environmental restraints on the mining companies.

### 16.3.2 *How Effective Are Voluntary Standards for Improving Environmental Performance?*

Debate rages about the effectiveness of approaches built on lending practices and reputational considerations. Certainly, failure to commit to the Equator Principles and the IFC Performance Standards can make certain kinds of important financing off limits. The acid test is whether the requirements imposed through these various means have been translated into meaningful change on the ground in the form of genuine environmental protections. Meyerstein (2013) concludes that the Equator Principles have been somewhat successful in the project finance sphere. He cannot answer satisfactorily whether this has been translated into meaningful change on the ground as his research focuses on institutional practices rather than ultimate environmental outcomes (Meyerstein 2013).

CSR and other reputational approaches depend on mutual accountability ('peer pressure') and pressures from the close-watching public. Research reviewing these has reached mixed conclusions. Hilson (2012) is dubious, concluding that CSR's effectiveness in any location requires a 'foundation of robust regulations and enforcement...for it to complement', rather than replace or undermine domestic law (Hilson 2012: 136). Oshienbo (2009) says poor implementation has made the World Bank's environmental and social policies inadequate in ensuring sustainable development in the extractive industries.

Sethi and Emelianova's (2006) fairly damning conclusion is: 'in the absence of standardized measures of performance, uniformity in reporting, and transparency, in full disclosure, the notion of voluntary codes of conduct is rendered almost worthless since it lacks any assurance of credibility and accuracy' (Sethi and Emelianova 2006: 235).

They struggle to answer whether CSR's objectives are lofty to the point of being unrealistic; is CSR window dressing or even actively undermining the role of government? While Sethi and Emelianova (2006) are not focused principally on environmental requirements, the parallel concern would be 'greenwashing', a form of spin where seemingly environmentally friendly commitments are deceptively used to promote the perception over a more grim reality.

By contrast, Spense's view as an economist (2011) is that it is in corporations' and shareholders' best interests to act in an environmentally and socially responsible manner; especially in the absence of strong host country environmental protection laws. Yakovleva and Vazquez-Brust (2012) interviewed stakeholders in the Argentine government, civil society, international financial institutions, and mining industry who posit that the most critical element of CSR in the mining context is 'to do what is safe for the environment' (Yakovleva and Vazquez-Brust 2012: 191), again without consideration of how CSR policies are translated into meaningful change on the ground. Livermore (2014), agreeing with Dashwood (2012), concluded that 'the ICMM's authority will continue to influence the CSR decisions of mining companies and support them to be directed at developmental outcomes' (Dashwood 2012). Livermore (2014), relying heavily on the World Bank's 2014 report, agrees that 'standards only matter to the extent they are complied with' but is convinced that 'the pressure for compliance with the ICMM's standards is...amplified by factors in members' larger political and financial environment' (Livermore 2014: 38). In other words, many of these analysts believe that companies acting in their own self-interest will do the right thing.

My own experience, including as an American environmental regulator and with considerable on-the-ground experience in transitional economies and the developing world, makes me cautious about voluntary standards. Reputation is important but companies are rarely monoliths. Rather, they are complex widely dispersed organizations, with numerous, sometimes conflicting perspectives. Leadership can care a great deal about public perceptions but still have reasonable differences about many of the important details concerning pollution and its generation and dissemination: for example, differences about the impacts of pollution and what is dangerous; after all, similar disagreements can take place among scientists concerning what is a safe exposure level. Agreements made at one level of the company (e.g. at HQ) might not be translated fully into action, especially when the actual activities must be carried out in remote places. There can be substantial differences between asset managers and sustainability teams, or management and on-the-ground personnel or, even more importantly, poor communication and/or oversight. Divisions of responsibilities can result in shaky on-the-ground commitments because lines of communication are blurred. Any number of institutional and human factors can confound legitimately good intentions and break the chain of responsibility that is assumed by the supporters of reputational approaches. The passage of time and possible conveyance of project ownership or assignment of a company's rights may obscure obligations that were incurred at the time of the initial transactions. Sometimes even good faith negotiators need the bludgeon of enforcement to compel action.

Finally, piecemeal approaches may not address systemic issues or address the wider significance of development (see Affolder 2013). In contrast, laws, in the best case, attempt to think through a wider vision and write rules consistent with that vision (see Caruso 2009). A country may strive for development and at the same time want to preserve important domestic values: for example, balancing agrarian vs. industrial objectives. A sophisticated regulator might understand that mining in a faraway mountain range has consequences for the quality of water consumed in cities. There is a reason for independent bodies that set and then enforce standards and obvious reasons why specific projects should fit into this overall vision, rather than the other way around (Sethi and Emelianova 2006: 230).

Among other things, Sethi and Emelianova point out that any code must cover 'issues that are of concern to the community' not just industry (Sethi and Emelianova 2006: 230), including a governance structure that assures external input 'from independent experts who have the respect and confidence of all parties involved', not just reliance on industry promises (Sethi and Emelianova 2006: 230). Despite industry resistance to external monitoring and independent checks on their compliance performance, theirs is not and should not be the only relevant voice; external monitoring and compliance verification is also necessary (Sethi and Emelianova 2006: 230).

Company CSR policies understandably are aimed at specific projects: it is not their job to manage bigger picture issues such as longer-term impacts on a nation's entire biodiversity, and even less how pushing on one part of closed-system earth might have knock-on effects elsewhere, including outside the country boundaries. The argument for laws and general principles over CSR is based on the need for looking through a wider and more inclusive lens and for adding a public dimension to the inquiry.

Approaching environmental protection through the EIA approach, either contractually or through licensing schemes (with the dimensions of the EIA being possibly set by law), has the virtue of injecting a greater degree of specificity than can emerge from CSRs or voluntary codes. A comprehensive EIA considers many factors, including all environmental impacts and a range of alternatives to the proposed action and their own impacts (including air, water, endangered species, historic and cultural sites, social and economic impacts, and the costs for each alternative including mitigation). It could also consider a financial plan for the proposed action, identifying the sources of secured funding, a mitigation plan, and where necessary additional documentation involving other relevant environmental requirements (in the United States this can include state and local permits and reviews) (see Kohn 2002).

Even so, the EIA tool has received mixed reviews. For example, despite Peruvian requirements that all projects submit an EIA supervised by the Ministry of Energy and Mines, *The Economist* (2016) concludes that their implementation undercuts environmental protection. *The Economist* quotes an NGO that works with communities affected by mining saying that 'people don't believe in the rigour of EIAs'. Local rigour in enforcing requirements is undermined because half of a mine's corporate income tax is devolved to regional and local governments in the area under Peru's *canon minera*. So 'showering' some mining districts with more money than they can spend fosters pro-mining corruption.

Thomsen et al. (2001) recommend 'ongoing and long-term evaluation or monitoring of environmental and social impacts at the ecosystem level... to ensure that the area's ecological and social integrity was upheld' for hydrocarbon exploration in sensitive and 'poorly explored' ecosystems (Thomsen et al.: 90). They fear that EIAs, as they are traditionally carried out, are insufficient to provide adequate biological and ecological information. More information must be available for on-going project management:

> managers of oil and gas exploration projects must factor in the need for ongoing data gathering into project time frames and allow enough flexibility in the project execution phase to take full advantage of such data. (Thomsen et al. 2001: 109–10)

### 16.3.3 *The Limits of CSR*

A wild card in the CSR discussion is the growing significance of investors from countries with poorly established environmental requirements who do not face the same kinds of domestic, international cultural, or economic pressures in their overseas business activities that contributed to the development of voluntary approaches. China is actively exploiting mineral resources in Africa, Asia, and Latin America at the same time its domestic commitments to and implementation of pollution controls have been relatively weak and questionable (see, for example, Bell 2003), raising questions about China's overseas environmental behaviour.

How significant is Chinese investment? Gonzalez (2012) says that China is vying with the World Bank and the Inter-American Development Bank to become a major lender in Latin America. Chen et al. (2015) calculate that China's FDI in Africa was only about 3 per cent of total continental FDI. UNCTAD's *World Investment Report 2015* (UNCTAD 2015) puts that figure at 4.4 per cent of the total in 2013–14, behind the EU countries (led by France and the United Kingdom), the United States, and even South Africa. Nevertheless, it is where they are investing that is important:

> China's investment is more visible in the poor rule-of-law countries because China *has* invested in those locations whereas Western investment generally stayed away from them... Angola, Burundi, the Central African Republic, the Democratic Republic of the Congo, Eritrea, Guinea, and Zimbabwe. (Chen et al. 2015)

Struggling to assess China's influence, Tan-Mullins (2014) assesses the feasibility of 'socializing' China towards adopting current CSR global norms, particularly given the large number of state-owned enterprises participating in the extractive sector (and the Chinese government's corresponding unwillingness to embrace initiatives based on transparency and stakeholder consultation) (Tan-Mullins 2014).

Jansson et al. (2009) are more optimistic, concluding that Chinese extractive companies are receptive to the principle of transparency and general CSR standards; positing that 'there are, in fact, few differences in the operating procedures between Chinese corporations per se and other international actors engaged in Africa'. Their report focuses on transparency in general (rather than environmental impacts) but contains useful information and case studies about Chinese companies' environmental practices in Africa (including the use of EIAs) (Jansson et al. 2009).

### 16.3.4 *Falling Commodity Prices: Some Consequences for Environmental Regulation*

Environmental controls, investor interest in accommodating local requirements, and country-level will to identify, agree, and implement them have been further complicated by the so-called commodity price slump that began in approximately 2014 (Pakiam and Katakey 2015). Until around 2012, the unusual profits associated with the so-called super-cycle of commodity prices were a key driver of decisions about extraction. Many countries, contrary to the advice from several sources to use those revenues to help diversify their economies, instead became increasingly dependent on extractives. With declining revenues, what are the new challenges to implementation of environmental requirements? Are countries arguably now even less likely to impose restrictions and/or to enforce them? Will they try to increase production or otherwise change the rules to favour extraction companies and so try to compensate for the impacts of the commodity price falls?

In Peru, for example, the government's response to lower growth and tax revenues has been toward policies to relax the conditions of mining and oil contracts, in order to increase production (Ballivián 2013). The increased export volume would in their view 'compensate' for the losses brought by the end of the super-cycle. Campodonico (2015) finds this policy relaxation particularly noticeable in the areas of previous consultation and EIAs.

In Bolivia, investment is now permitted in previously off-limits Protected Natural Areas including allowing access to land owned by communities and indigenous people to 'encourage' investment (Campodonico 2015).

Clearly, where these revenues are central to the functioning of a poor country, the temptation will be to make extraction easier and less onerous. *The Economist* (2016) noted 'ironically' that 'the end of the boom may increase both government and public support for mining'.

## 16.4 A Bigger Tool Box: Alternatives to Improve Environmental Performance in Extraction

The general principles identified in the literature point in the right direction. Good practice standards should provide 'guidance on how to identify risks and impacts... and be designed to avoid, mitigate, and manage risks and impacts as a way of doing business in a sustainable way, including stakeholder engagement and disclosure obligations of the client in relation to project-level activities' (IFC 2012).

Tools like contracts, lender and industry-created principles, CSR initiatives, and EIAs each try in their own way to achieve this goal in difficult venues but the experience of uneven results has led to even more ideas.

Oshienbo (2009) offers a mixed bag of suggestions to improve implementation of the World Bank Group's own environmental standards. These include: a) incorporating project borrowers' past environmental performance into funding decisions; b) imposing financial sanctions (i.e. fines); c) involving independent environmental experts at all stages of project development, including EIA review; d) training and other environmental capacity-building at the borrower/host government level; and e) community and NGO consultation. Some of this suite of recommendations are, of course, already part of the compliance toolkit.

Freeman (1997) argues for environmental insurance in transnational lending as a means to improve enforcement of international environmental norms. Unlike EIAs, insurance can both assess and enforce compliance with international environmental standards because insurance companies require as a contractual condition that the insured adheres to specified prudent behaviours and procedures. As with domestic insurance practices, implementation would create a private legal regime that enforces public (environmental) values.

Many commenters elaborate on the community consent approach. Based in part on anthropological research conducted among affected communities in Peru, Laplante and Spears (2008) recommend engaging in consent processes with communities and groups directly affected by extractive projects with a view to obtaining their free prior and informed consent (FPIC). They argue that FPIC-compliant companies will have a competitive advantage (avoiding later community protests) with communities obtaining enforceable agreements granting them control over their natural resources and environment (Laplante and Spears 2008). Recent reporting from Peru suggests, with some exceptions, that forms of community consent have helped prevent conflicts, and that the lower commodity prices and resulting delays or postponements of some projects 'potentially offer more time for consultations' (*The Economist* 2016). On the other hand, *The Economist* reports Colombia examples in which prior consultation became a 'means to extort money from companies'.

Baker (2012) proposes an expansive interpretation for the IFC's Performance Standard 7: project developers would enter into 'Environmental and Social Risk Agreements' with affected communities incorporating explicit contractual allocation of environmental and social risks among all stakeholders. This would help to shift environmental risks from indigenous communities to project developers.

Seck would make home states responsible for exercising extraterritorial jurisdiction over extraction activities of their nationals abroad. Jurisdiction would be based on 'territorial points of control', including 'stock exchanges, financial institutions, and enabling corporate laws', as well as services provided by export credit agencies and trade commissioners (Seck 2008: 188–9), modelled on Canada's and Australia's extraterritorial regulation of their extractive industries. The nexus of corporate law would impose such regulations: for example, requiring all companies listed on home state stock exchanges to comply with minimum environmental standards (i.e. consultation with affected communities).

Seck's idea can be scrutinized through a legal or a political lens. For the legal case, Bernhard concludes that Seck's arguments 'alternatively raise concerns about the implications of imposing one country's legal regime on another or appealing to the tenants of universal jurisdiction, which are typically reserved for the most heinous crimes' (Bernhard 2014: 212). Instead, Bernhard suggests bringing such cases to an International Corporate Criminal Court (Bernhard 2014). Looking at this suggestion through a political lens is even less encouraging, as incorporating such ideas would require massive amounts of political will in the face of often powerful domestic interests. Some similarly frustrated transparency advocates argue for the utility of Section 1504 of the US Dodd–Frank Act, which requires extractive companies to disclose the amounts they pay to foreign governments for licences and other approvals. However, the future of this rule, which was made an early target of the new Trump administration, is unclear at the time of this writing (see Rubenfeld 2015).

'Naming and shaming' is another potential tool, relying on independent monitors collecting information about environmental and social performance or extracting the requisite information from CSR commitments, the Equator Principles, the IFC Performance Standards, and the like. Equator Principles information provides 'ammunition [to NGOs and others who might lodge protests or organize shaming opportunities], as well as additional avenues to contact and interact with the project borrowers' (Lance 2013: 198). This strategy adds a significant public component, assuring that the information is made available where it counts—directed to investment vehicles, shareholders, the press, the NGO community, or concerned governments.

Sunshine on activities in far-distant places that would otherwise be hard to track and evaluate and that might otherwise escape public notice can be a powerful disinfectant. Williams would use this to expose multinational corporations to reputational harm in the event that they fall short of their social or environmental standards (Williams 2004). She argues that civil society demands for CSR are changing the social and business paradigms under which some multinational corporations operate, and in turn are also changing norms of appropriate industry action in environmental matters, creating global standards of action.

The three significant constraints on naming and shaming are information overload, the limited number of response tools for those who are outraged, and the complexity of creating effective campaigns around information.

Naming and shaming assumes that reports will be read and will generate actionable embarrassment for the malefactors, and particularly that there is a motivated audience ready to act—often a huge assumption in this time of information overload. Poor environmental practices are likely to be found in many countries and in many types of extraction involving any number of extraction companies and/or corrupt or distracted domestic officials. This by definition blurs the focus.

The hope in such campaigns is that investors and consumers will be motivated to act, for example, by lobbying for new or improved laws and regulations. This has worked in the regulation of forestry products with the enactment of the Lacey Act, the European Union Timber Regulation, and the Australian Illegal Logging Prohibition (Nogueron and Cheung 2014) which have in turn made it easier for consumers to organize product boycotts and move investments towards more socially responsible companies.

But successful campaigns for new laws and boycotts generally require considerable organization and coordination, and they work better with a human face or story to give life to the facts. The strategy of stigmatizing has proven more effective in boycotting diamonds than have restrictions on purchases of materials (such as oil or copper) that are processed and then combined with other materials to make a variety of products (Haufler 2015).

Another approach is to tweak existing elements of the contracting tool chest to make them more robust and trustworthy: for example, improving audits of compliance with contract and EIA obligations and assuring that they are truly independent, a tough task. Wara's (2008) research on the Kyoto Protocol demonstrated the unreliability of those so-called independent audits; name brand auditing firms were found to have provided cut-and-paste identical language in numerous supposedly independent reviews. Genuinely independent auditing is probably obtainable but not without considerable effort and cost. A legitimate audit is an important enforcement tool and, when shared with the public, can provide important feedback to a variety of stakeholders.

Another approach is to enhance the skills of domestic environmental experts and give them a seat at the table in contract negotiations. Often, the domestic environmental authorities are presented with a fait accompli or asked to contribute too late in the process.

A final idea is to connect the extraction agreement and its climate-related terms to each country's Intended Nationally Determined Contributions (INDCs). Countries that are party to the United Nations Framework Convention on Climate Change (UNFCCC) articulated the steps each would take within their own countries to address climate change. These INDCs represent specific emissions reduction plans, taking into account domestic circumstances and capabilities. The UNFCCC is working on a plan to review these INDCs and keep them current. This would tie environmental commitments to a more general global review process that gets a fair amount of global attention. While it might seem odd to tie extraction agreements to the UNFCCC INDC process, INDCs can encompass a variety of country greenhouse gas control commitments.

## 16.5 Conclusions

The challenges reviewed above are not new. Considerable commentary over several decades concludes with the perfectly reasonable yet frustratingly nonspecific exhortation that countries need clear and specific laws and 'the development of an environmental protection ethos'. Smith and Kormos (2001), for example, conclude that:

> legislation and contracts need to establish fiscal regimes that encourage, and do not discourage, environmentally sound mining and drilling practices and... bidding procedures that include environmental protection criteria.... there must be environmental protection provisions that specify best practices for specific projects, effective environmental protection monitoring, and investment in training of environmental protection personnel. (Smith and Kormos 2001: 241)

The Natural Resource Charter similarly urges companies to 'take steps that go beyond minimum legal requirements to respect the highest environmental... standards' (NRGI 2010). The Charter's drafters recognize 'there is no guarantee that rules will be followed or capable institutions will work for the country's benefit'. Their suggested antidote is 'strong accountability'.

The Chatham House Report on conflict and coexistence in the extractive industries (Stevens et al. 2013) recommends to 'go slow', with an emphasis on building capacity to regulate companies and manage these complex responsibilities. Indeed, Chatham House's preferred option is 'delaying development for Afghanistan and Somalia, for example, given the combination of political instability, conflict, and environmental stress they are currently facing' (Stevens et al. 2013).

It is not that these recommendations (such as the first above, which is from a book published in 2001 but could equally have been written today) are wrong. Rather, such authors rarely explain how the results they recommend are to be achieved in practice.

The challenges in making environmental requirements actionable, much like other governance challenges surrounding resource extraction in fragile countries, have no easy solutions. Lindblom's (1959) pragmatic insight proposed more than sixty years ago might be most realistic—that no single tool can resolve deficiencies in local institutions and a lack of political will. A process of accretion, Lindblom's 'successive limited comparisons', might be more realistic, if it builds expertise and deepens experience over time (see Lindblom 1959: 81). Even in ideal situations, as Lindblom says:

> Policymaking is a process of successive approximation to some desired objectives in which what is desired itself continues to change under reconsideration... Making policy is at best a very rough process... if [the policy maker] proceeds through a *succession* of incremental changes, he avoids serious lasting mistakes in several ways. (Lindblom 1959: 86)

As a practical matter, gaining incremental knowledge and learning by doing is in fact what is happening in the real world. A variety of tools have been developed and tested, and the best contracts draw on more than one of these.

In the best cases, domestic skills are (or should be) developed as contracts are negotiated. What happens after negotiation is also important—ideally a variety of genuinely independent eyes (including independent auditors, international and domestic NGOs, and domestic institutions) will monitor developments and assess whether obligations are honoured. Newer agreements must learn from what has previously been decided in order for this incremental learning process to take place.

Finally, all of this assumes that countries do not treat the extraction contracting process as a mere formality, but rather as an opportunity to fulfil a real desire to develop the kind of local expertise and interest that will contribute sustained attention to the environmental consequences of extraction.

## References


Affolder, N. (2013). 'Beyond Law as Tools: Foreign Investment Projects and the Contractualization of Environmental Protection', in P. Dupuy and J. E. Vinuales (eds), *Harnessing Foreign Investment to Promote Environmental Protection: Incentives and Safeguards*, 364–66. Cambridge: Cambridge University Press.

Baker, S. (2012). 'Why the IFC's Free, Prior, and Informed Consent Policy Does Not Matter (Yet) to Indigenous Communities Affected by Development Projects', *Wisconsin Journal of International Law*, 30(3): 668–705.

Ballivián, E. Q. (2013). 'Political Risk Management and Sustainability in Latin American Natural Resource Projects', Rocky Mountain Natural Resource Law Institute Paper 3A. Westminster, CO: Rocky Mountain Natural Resource Law Institute.

Bell, R. G. (1992). 'Environmental Law Drafting in Central and Eastern Europe', *Environmental Law Reporter*, 22(9): 10597–606.

Bell, R. G. (1994). 'EPA's International Assistance Efforts: Developing Effective Environmental Institutions and Partners', *Environmental Law Reporter*, 24: 10593–99.

Bell, R. G. (2003). *Choosing Environmental Policy Instruments in the Real World.* Paris: OECD Global Forum on Sustainable Development: Emissions Trading.

Bell, R. G. (2017). 'Protecting the Environment during and after Resource Extraction', WIDER Working Paper, 2017/164. Helsinki: UNU-WIDER.

Bernhard, A. (2014). 'Response: Sara L. Seck, Home State Responsibility and Local Communities: The Case of Global Mining', *Yale Human Rights and Development Law Journal*, 11: 206–15.

Campodonico, H. (2015). 'Towards Full Recognition of Tenure Rights in Latin America: Challenges and Opportunities in the Context of Colombia and Peru as New Emerging Economies', paper presented at the 2015 Bern Conference: From Rhetoric to Action: Scaling Up Community and Indigenous Peoples' Land and Resource Rights. 1 October.

Caruso, D. (2009). 'Private Law and State-making in the Age of Globalization', *NYU Journal of International Law and Politics*, 39(1): 2–74.

CCSI (2015). *Natural Resource Contracts as a Tool for Managing the Mining Sector*. Hanover: German Federal Ministry for Economic Cooperation and Development.

Chen, W., D. Dollar, and H. Tang (2015). 'China's Direct Investment in Africa: Reality v. Myth', Brookings Institution, 3 September. Available at: http://www.brookings.edu/blogs/africa-in-focus/posts/2015/09/03-china-africa-investment-trade-myth-chen-dollar-tang.

Connor, M. (2010). 'Nike: Corporate Responsibility at a "Tipping Point"', *Business Ethics*, 24 January. Available at: http://business-ethics.com/2010/01/24/2154-nike-corporate-responsibility-at-a-tipping-point/.

Cotula, L. (2010). *Investment Contracts and Sustainable Development: How to Make Contracts for Fairer and More Sustainable Natural Resource Investments*. London: IIED.

Crippa, L. (2008). 'Cross-cutting Issues in the Application of the Guatamalan "NEPA": Environmental Impact Assessment and the Rights of Indigenous Peoples', *American University Law Review*, 24: 104–40.

Cronin, R. (2009). 'Natural Resources and the Development-Environment Dilemma', in R. Cronin and A. Pandya (eds), *Exploiting Natural Resources: Growth Instability and Conflict in the Middle East and Asia*, 63–81. Washington, DC: Stimson.

Dashwood, H. (2012). *The Rise of Global Corporate Social Responsibility: Mining and the Spread of Global Norms*. Cambridge: Cambridge University Press.

*Economist, The* (2016). 'Mining in Latin America: From Conflict to Cooperation', 6 February. Available at: http://www.economist.com/news/americas/21690100-big-miners-have-better-record-their-critics-claim-it-up-governments-balance.

ELAW(2013). *Natural Resource Contracts: A Practical Guide*, November. Available at: https://www.elaw.org/system/files/Natural_Resource_Contracts_Guide.pdf.

Freeman, P. K. (1997). 'Environmental Insurance as a Policy Enforcement Tool in Developing Countries', *University of Pennsylvania Journal of International Economic Law*, 18: 477–8.

Freestone, D., (2013). *The World Bank and Sustainable Development: Legal Essays*. Available at: http://booksandjournals.brillonline.com/content/books/9789004202337.

Fromherz, N. (2013). 'From Consultation to Consent: Community Approval as a Prerequisite to Environmentally Significant Projects', *West Virginia Law Review*, 116(1): 110–96.

Gonzalez, C. (2012). 'China's Engagement with Latin America', in E. Blanco and J. Razzaque (eds), *Natural Resources and the Green Economy: Refining the Challenges for People, States and Corporations*, 37–79. Leiden: Martinus Nijhoff Publishers.

Gray, K. (2000). 'International Environmental Impact Assessment: Potential for a Multilateral Environmental Agreement', *Colorado Journal of International Environmental Law and Policy*, 11(1): 83–128.

Haufler, V. (2015). 'Shaming the Shameless? Campaigning against Corporations', in H. R. Friman (ed.), *The Politics of Leverage in International Relations: Name, Shame, and Sanction*, 185–200. London: Palgrave Macmillan.

Hershowitz, A. (2008). 'A Solid Foundation: Belize's Chalillo Dam and Environmental Decisionmaking', *Ecology Law Quarterly* 35(1): 73–105.

Hilson, G. (2012). 'Corporate Social Responsibility in the Extractive Industries: Experiences from Developing Countries', *Resources Policy*, 37: 131–7.

ICMM (2012). *Community Development Toolkit.* London: ICMM.

IFC (2012). 'Performance Standards on Environmental and Social Sustainability'. Available at: http://www.ifc.org/wps/wcm/connect/115482804a0255db96fbffd1a5d13d27/PS_English_2012_Full-Document.pdf?MOD=AJPERES.

Jansson, J., C. Burke, and W. Jiang (2009). 'Chinese Companies in the Extractive Industries of Gabon and the DRC: Perceptions of Transparency', EITI. Available at: https://eiti.org/document/chinese-companies-in-extractive-industries-of-gabon-drc-perceptions-of-transparency.

Jordan, L. and C. Chamberlain (2001). 'Partner or Pariah, Public Perceptions and Responses to the Extractive Industries', in I. Bowles and G. Prickett, (eds), *Footprints in the Jungle: Natural Resource Industries, Infrastructure, and Biodiversity Conservation,* 37–52. Oxford: Oxford University Press.

Kimerling, J. (2000). 'Oil Development in Ecuador and Peru: Law, Politics, and the Environment', in A. Hall (ed.), *Amazonia at the Crossroads: The Challenge of Sustainable Development.* London: University of London Press.

Kimerling, J. (2001). 'International Standards in Ecuador's Oil Fields: The Privatization of Environmental Law', *Columbia Journal of Environmental Law,* 26: 289–398.

Kohn, D. (2002). 'Setting a Standard: Environmental Impact Assessment Policies of Multilateral Development Banks and Export Credit Agencies', *Environmental Lawyer,* 9: 281–306.

Lance, J. (2013). 'Equator Principles III: A Hard Look at Soft Law', *North Carolina Banking Institute,* 17(1): 175–99.

Langer, M. J. (2013). 'Key Instruments of Private International Finance', in P.-M. Dupuy and J. E. Viñuales (eds), *Harnessing Foreign Investment to Promote Environmental Protection: Incentives and Safeguards,* 131–75. Cambridge: Cambridge University Press.

Laplante, L. and S. A. Spears (2008). 'Out of the Conflict Zone: The Case for Community Consent Processes in the Extractive Sector', *Yale Human Rights and Development Law Journal,* 11: 69–116.

Lawrence, R. F. and W. L. Thomas (2004). 'The Equator Principles and Project Finance: Sustainability in Practice', *Natural Resources and Environment,* 19(2): 20–6.

Lawson-Remer, T. and J. Greenstein (2012). 'Beating the Resource Curse in Africa: A Global Effort', Council on Foreign Relations, August. Available at: https://www.cfr.org/blog/battling-africas-resource-curse.

Lindblom, C. (1959). 'The Science of 'Muddling Through', *Public Administration Review,* 19(2): 79–88.

Livermore, K. (2014). 'Mining Companies as Partners in Development: The Role of the International Council on Mining and Metals in Institutionalising the Development Activities of Mining Companies', dissertation, London School of Economics. Available at: https://www.ucl.ac.uk/anthropology/mongolian-economy/articles/K_Livermore_Dissertation.

May, J. (2003). 'Now More Than Ever: Trends in Environmental Citizen Suits at 30', *Widener Law Review,* 10(1): 39–41.

McQuaid, J. (2009). 'Mining the Mountains', *Smithsonian Magazine,* January (updated 2010). Available at: http://www.smithsonianmag.com/ecocenter-energy/mining-the-mountains-130454620/?no-ist.

Meyerstein, A. (2013). 'Transnational Private Financial Regulation and Sustainable Development: An Empirical Assessment of the Implementation of the Equator Principles', *NYU Journal of International Law and Policy*, 45: 487–98.

Mongoven, B. (2006). 'The Equator Principles: The Next Stage for Activists', *Stratfor*, 14 July. Available at: https://www.stratfor.com/article/equator-principles-next-stage-activists.

Nogueron, R. and L. Cheung (2014). 'Sourcing Legally Produced Wood: A Guide for Businesses', World Resources Institute. Available at: http://www.wri.org/publication/sourcing-legally-produced-wood-guide-businesses.

NRGI (2010). 'Natural Resource Charter', NRGI. Available at: https://resourcegovernance.org/approach/natural-resource-charter.

O'Callaghan, K. (2011). 'Corporate Social Responsibility: A Framework for Understanding the Legal Structure', *Rocky Mountain Mineral Institute*, 57: 17A.

Oshienbo, E. (2009). 'World Bank and Sustainable Development of Natural Resources in Developing Countries', *Journal of Energy and Natural Resource Law*, 27(2): 193–227.

Pakiam, R. and R. Katakey (2015). 'Commodities Slump to Sixteen-year Low on Mining, Oil Stocks', *Bloomberg*, 23 August.

Puvimanasinghe, S. (2009). 'Towards a Jurisprudence of Sustainable Development in South Asia: Litigation in the Public Interest', *Sustainable Development Law and Policy*, 10(1): 41–9.

Rich, N. (2016). 'The Lawyer Who Became DuPont's Worst Nightmare', *The New York Times Magazine*, 6 January. Available at: https://www.nytimes.com/2016/01/10/magazine/the-lawyer-who-became-duponts-worst-nightmare.html.

Rogge, M. (2001). 'Towards Transnational Corporate Accountability in the Global Economy: Challenging the Doctrine of *Forum Non Conveniens* in *In Re*: Union Carbide, Alfaro, Sequihua, and Aguinda', *Texas International Law Journal*, 36(2): 299–318.

Rubenfeld, S. (2015). 'SEC Pushes Extractive Disclosure Rule Decision to 2016', *Wall Street Journal*, 30 March. Available at: https://blogs.wsj.com/riskandcompliance/2015/03/30/sec-pushes-extractive-disclosure-rule-decision-to-2016/.

Sarro, D. (2012). 'Do Lenders Make Effective Regulators? An Assessment of the Equator Principles on Project Finance', *German Law Journal*, 13(12): 1522–55.

Seck, S. (2008). 'Home State Responsibilities and Local Communities: The Case of Global Mining', *Yale Human Rights and Development Law Journal*, 11(1): 177–206.

Sethi, S. P. and O. Emelianova (2006). 'A Failed Strategy of Using Voluntary Codes of Conduct by the Global Mining Industry', *Corporate Governance*, 6: 226–38.

Smith, D. and C. F. Kormos (2001). 'Conservation and Concession Contracts: Environmental Issues in Mineral Extraction Agreements', in I. Bowles and G. Prickett (eds), *Footprints in the Jungle: Natural Resource Industries, Infrastructure, and Biodiversity Conservation*. Oxford: Oxford University Press.

Spense, D. (2011). 'Corporate Social Responsibility in the Oil and Gas Industry: The Importance of Reputational Risk', *Chicago-Kent Law Review*, 86(1): 59–85.

Stevens, P., J. Kooroshy, G. Lahn, and B. Lee (2013). *Conflict and Coexistence in the Extractive Industries*. London: Chatham House.

Tan-Mullins, M. (2014). 'Successes and Failures of Corporate Social Responsibility Mechanisms in Chinese Extractive Industries', *Journal of Current Chinese Affairs*, 43(4): 19–39.

Thomsen, J., C. Mitchell, R. Piland, and J. Donnaway (2001). 'Monitoring Impacts of Hydrocarbon Exploration in Sensitive Terrestrial Ecosystems: Perspectives from Block 78, Peru', in I. Bowles and G. Prickett (eds), *Footprints in the Jungle: Natural Resource Industries, Infrastructure, and Biodiversity Conservation*, 90–112. Oxford: Oxford University Press.

UNCTAD (2015). *World Investment Report 2015: Reforming International Investment Governance*. UNCTAD. Available at: http://unctad.org/en/PublicationsLibrary/wir2015_en.pdf.

Verma, S. (2014). 'Photos: Living in the Shadow of the Bhopal Chemical Disaster', *Mother Jones*, 2 June. Available at: http://www.motherjones.com/environment/2014/06/photos-bhopal-india-union-carbide-sanjay-verma-pesticides-explosion/.

Wara, M. (2008). 'Measuring the Clean Development Mechanism's Performance and Potential', *UCLA Law Review*, 55: 1759–803.

Williams, C. A. (2004). 'Civil Society Initiatives and "Soft Law" in the Oil and Gas Industry', *NYU Journal of International Law and Policy*, 36(2/3): 457–502.

Yakovleva, N. and D. Vazquez-Brust (2012). 'Stakeholder Perspectives on CSR of Mining MNCs in Argentina', *Journal of Business Ethics*, 106(2): 191–211.

# 17

# Enhancing Sustainable Development from Oil, Gas, and Mining

## From an 'All of Government' Approach to Partnerships for Development

*Kathryn McPhail*

## 17.1 Overview

The sustainable development objectives of low- and middle-income countries have been set out, since 2000, in the Millennium Development Goals and, since 2015, in the Sustainable Development Goals. Action to deliver these goals is led by national governments. And the private sector has a key role to play. Why? Over a similar period (1996–2014), more low- and middle-income countries became dependent on the natural resources sector. By 2014, seventy-two such countries obtained 30 per cent (or more) of export earnings from oil, gas, and mining—in some cases up to 90 per cent. This export dependence is on an upward trend, despite the recent commodities price collapse, with sustained increases in most countries since 1996.[1] Looking ahead, investments in oil and gas and minerals may need to increase at more than double historical rates to meet new demand (e.g. Paris Agreement) and replace existing supply.[2] New investment could potentially reach $17 billion by 2030.

Work undertaken since 2000 has shown that dependency on oil, gas, and/or mining does not automatically translate into broader-based economic and social benefits. It requires companies to invest in lower-income countries to

[1] https://www.wider.unu.edu/publication/it-or-not-poor-countries-are-increasingly-dependent-mining-and-oil-gas.

[2] McKinsey Global Institute (2013).

design and operate their projects to make a genuine contribution to the sustainable development of their host societies. And such investments need to be supported by appropriate host country policies to catalyse broader economic development.

This chapter draws on detailed country case studies to identify actions companies and governments can take so that resource-driven countries[3] can realize their full potential. It begins by focusing on the need for an agreed set of data and analysis showing the current and potential future contributions of the natural resources sector at the national and local levels. Many line agencies of government have roles and responsibilities in addition to the more traditional actors: ministries of finance, economy and planning, energy, and mining. All need to engage, and work together, to facilitate an 'all of government' approach, which is critical to success.[4]

That the evidence base to guide effective policymaking (and advocacy) is not always widely known, shared, or indeed even agreed upon, across different line ministries and other players, is the first challenge to overcome. This is especially true of the forward-looking projected data that guide many policies. This sobering fact has been demonstrated in a number of analytical country case studies and assessments on which the chapter will draw.

The chapter next addresses a further regrettable truth that, in some countries, there is a lack of trust among stakeholders: for example, between companies and civil society organizations (CSOs), between companies and government, between government and CSOs, and between federal, state, and municipal government. Finding ways to bring all these interests together in various ways is an important dimension of the overall governance challenge associated with the natural resources industries. The chapter uses the example of multi-stakeholder workshops (based on prior in-depth evidence around the industry) to show how a robust evidence base can form the basis for prioritizing common objectives and agreeing actions by each of the stakeholder groups. Results can enhance the natural resources sector's contributions, and at the same time help to build trust.[5]

[3] For the definition of 'resource-driven', see McKinsey Global Institute (2013).

[4] 'All of government' includes all of national government (the presidency and the various line ministries) as well as all of subnational government in terms of the agencies and political-administrative arrangements at the regional and local levels (which are particular for any given country). Understanding this arrangement, including where/by whom policy is set and where/by whom it is implemented, is key. In the absence of the full range of agencies and players working together and using an agreed set of facts, sub-optimal policy outcomes can be the result.

[5] My grateful thanks are given to various colleagues who worked with me, or commented on aspects of this chapter and on other papers on which this chapter depends. They include Kate Carmichael, Alan Roe, Tom Butler, and Catherine Macdonald. The chapter also benefitted from comments and suggestions at an 'Extractive Industries and Development' project meeting hosted by UNU-WIDER in Helsinki, 11–12 April 2016. My thanks go also to Joseph Laredo for his most helpful editorial suggestions.

Lastly, the chapter probes related ways to mitigate the negative impacts of resource development and to enhance its potential positive contributions, particularly at the local level and where the governance context is often weak. Although 'good governance' is clearly critical in enhancing the benefits of extractive industries, 'governance' as such is a difficult and complex topic to pin down. Other studies in this project (for example, Dietsche 2017) set out this very complex process. Furthermore, the common features of good governance are impacted by the presence of large-scale natural resource investments. Hence, leadership and coordination across government must be both horizontal and vertical: necessarily extending to the subnational level. Partnerships for Development can help to fill the so-called 'governance gaps'.

In conclusion, the chapter briefly recaps the three steps outlined: (i) establishing an evidence base to facilitate cross-government coordination; (ii) building trust through multi-stakeholder dialogue; (iii) creating Partnerships for Development—and considers the unfinished agenda.

## 17.2 Evidence Base to Facilitate 'All of Government' Coordination

### 17.2.1 *Context/Issue*

Why is it important that the evidence base to guide effective policymaking is widely known, shared, and agreed? The potential economic and social contributions of the mining, oil, and gas industries are manifold but poorly understood and not automatic—particularly at the local level.[6] In many countries, policy makers and others do not have access to comprehensive datasets on the resources sector's current contribution, nor are aggregated forward-looking data (projections) available.

The impacts of the oil, gas, and mining industries on the more visible, national-level indicators (such as on foreign direct investment, exports, GDP, and government revenue) are, in a growing number of low- and middle-income countries,[7] extremely large and significant in macroeconomic terms—see Figure 17.1 for typical numbers in mining. In parallel, however, there are few direct benefits at the local and community levels given the capital-intensive nature of such investments and their inherent export orientation (Figure 17.1 also stylistically illustrates this). Yet it is here that the negative disruptions

[6] This is a key finding of a multi-year research project initiated by the ICMM in collaboration with UNCTAD and the World Bank Group, which resulted in the Mining: Partnerships for Development (MPD) Toolkit (ICMM 2011a)—see www.icmm.com/mpd.

[7] See Roe and Dodd (2017).

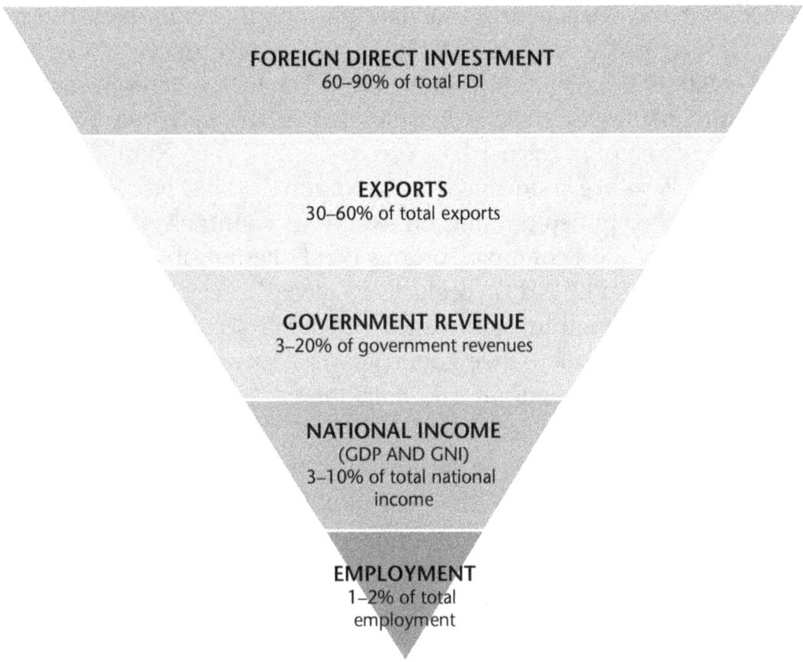


**Figure 17.1.** Typical macroeconomic contributions of mining to low- and middle-income economies

*Source*: based on ICMM (2009), reproduced here by permission.

from resource development are most likely to be felt. It is this disconnect between costs and benefits as between national and local levels that is absolutely central to the controversies surrounding the role of extractive industries: such investments have an impact on local communities and their perceptions that is often compounded by national decisions about how to use the 'national' resource rents.

In 2004, ICMM began the Resource Endowment initiative (REi) in collaboration with UNCTAD and the World Bank Group with the specific objective of identifying good practices in how mining can contribute to sustainable development. The starting point was to first identify and then understand 'success cases' and explanations of 'success' amongst some thirty-three low- and middle-income countries that in 2004 were dependent on mineral resources. This was done by constructing an analytical framework, which was then used to identify the multifarious factors that seemed to help or hinder mining's contribution to economic and broader social development in any particular country. This was followed by a second phase of its work, which applied the analytical framework to a series of country case studies, in order to collect evidence that was then used to validate the premises on which the more recent 'Mining: Partnerships for Development' (MPD) process is based. The ICMM Toolkit itself has been fully implemented by ICMM in seven countries,[8] and by third parties in another five. As part of the initiative, ICMM also undertook the Mapping Partnerships exercise, which covered nineteen countries. Both the Resource Endowment initiative and the MPD process as a whole, therefore, provided in-depth results on twenty-six countries.

In 2011, the World Economic Forum's (WEF) Responsible Mineral Development Initiative (WEF 2011) conducted a global survey that involved 145 representatives from mining companies, the public sector, NGOs, academia, and civil society in thirty-three countries. The survey concluded that the priority is to 'conduct rigorous and collaborative socio-economic studies'. Two-thirds of all stakeholders considered rigorous socio-economic studies 'very' or 'extremely' helpful to advancing responsible development. This action is valued across all countries, reflecting the benefits of a solid, reliable fact base, regardless of national development or industry maturity.

The REi demonstrated that the prospects for positive impacts from mining depend on a number of factors, which necessarily involve many different government agencies. These are set out below:

1. sound national macro-economic management (which is typically the responsibility of the ministry of economy, central bank)
2. revenue transparency (tax authorities)
3. reasonable standards of national governance (president/prime minister's office, attorney general)
4. responsible behaviour by natural resource companies (companies, ministries of resources/energy, regulatory authorities, environment, and community development)
5. the capacity of government to design, implement, and monitor policy provisions (ministries of planning, industry, labour, education, health, and community development)
6. a sound mineral fiscal regime (ministry of finance)
7. implementation of key international initiatives, such as the EITI[9] (all of government, all of industry, and all of civil society), and the UN Voluntary Principles on Security and Human Rights (e.g. armed forces, police, and public and private security).

[8] All of ICMM's MPD country case studies received the endorsement of host countries before any results were published.

[9] See Section 17.2.3.

These can be further enhanced by:

8. the quality of governance at the subnational level (regional and local institutions)
9. the quality of collaboration between government, companies, development partners, and civil society organizations to enhance impacts.

Given the large number of government agencies (and others) involved, this necessarily raises the question of how the roles and responsibilities are defined and how leadership and coordination are effected to ensure that resource investments create development. As the World Bank has noted:

> The EI sector, more than many others, depends for its efficient functioning on a complex ecosystem of governmental institutions and functions... The multifaceted character of the sector is reflected by the involvement of a large number of ministries and public entities whose coordination may be highly complex. Efficient extractive-based economic development requires the effective cooperation of these public entities while drawing on the specialized capacity of each. Yet, cooperation often suffers as individual entities seek to maintain control of their share of the extractive portfolio—and revenues. (Halland et al. 2015: 3)

Cross-government coordination is also key to realizing greater benefits from oil, gas, and mining industries in some of the functional areas that are priority concerns of large private-sector operations (e.g. recruitment and procurement), as well as government priorities—even where these policies are not specifically about resource development (e.g. education policy, industrial policy, improving the enabling environment for private-sector investment). On one side, companies may seek to address issues around procurement and recruitment without engagement with corresponding government agencies—just as governments may pursue education and skills development policies that do not take into account the needs of the industry.

### 17.2.2 *Objectives*

- To develop a shared understanding of the evidence base of the potential economic and social contribution to support informed decision-making.
- To facilitate the cross-government coordination required for successful natural resource development.

### 17.2.3 *Action in Support of the Objectives*

#### 17.2.3.1 SHARED UNDERSTANDING OF THE EVIDENCE BASE

From the outset, the REi was consultative and inclusive on the basis that developing the relevant knowledge of mining's full social and economic costs and benefits, and how countries can escape the so-called resource curse, must be built on a comprehensive understanding of all stakeholder positions. Developing a Toolkit that can be applied in a standardized way across countries, with input from all affected stakeholders, allows the identification of good practice with broader applicability.

The WEF's Responsible Mineral Development Initiative (RMDI) started in 2010, with the objective of identifying the key challenges facing responsible mineral development. In 2011, the Initiative's continuing work led to the identification of six building blocks that together provide a constructive framework for the mining sector and a neutral multi-stakeholder platform for the discussion and development of ideas capable of unlocking the potential socio-economic benefits of mining. During 2012, the RMDI focused on 'Mineral Value Management' (MVM), a perception tool it developed for enhancing the understanding of drivers of value in mining so as to trigger frank and open discussion on the issues that affect, unite, and divide stakeholders in the mining sector. Similarly, the Commonwealth Scientific and Industrial Research Organisation (CSIRO) has surveyed public perceptions of mining on a national scale, building on regional-scale work undertaken 2008–11.[10]

The Extractive Industries Transparency Initiative (EITI), which began in 2003, is an international standard to promote open and accountable management of natural resources by publishing information on tax payments, licences, contracts, production, and other key elements of resource extraction. It is designed to improve accountability and public trust in the revenues paid and received for a country's resources. It asks companies to publish what they pay for oil, gas, quarrying, and mining, and governments to disclose what they receive from oil, gas, quarrying, and mining. These figures are audited by an independent administrator and published along with contextual information in the EITI report. In each implementing country, the EITI process must be supported by a coalition of government, companies, and civil society working together as a multi-stakeholder group and provides a shared understanding across different interests.

#### 17.2.3.2 CROSS-GOVERNMENT COORDINATION

In China, government has taken a strategic approach to helping the nation's resource-oriented companies to follow standardized sustainable business practices abroad. The Chinese Chamber of Commerce of Metals, Minerals and Chemicals Importers and Exporters (CCCMC) issued Guidelines for Social Responsibility in Outbound Mining Investments in October 2014. This initiative was supported by the National Development and Reform Commission

[10] See Moffat et al. (2014) for an example for Chile.

(NDRC), working with eight government ministries (CCCMC 2014). One year later, in order to promote implementation of the Guidelines, the CCCMC, again supported by the NDRC, published a draft Chinese Outbound Mining Investment Industry Social Responsibility Action Plan (CCCMC 2015).

In Lao PDR, the National Economic Research Institute (part of the Ministry of Planning and Investment) led the MPD Toolkit analysis. All key government agencies attended the workshop, which provided simultaneous translation into Lao, Mandarin, and English, thereby facilitating the engagement of participants. This succeeded in bringing together a larger and broader cross-section of key government stakeholders than had been achieved previously.

### 17.2.4 *Early Results*

This section presents case examples of efforts led by companies such as BG Group (Tanzania) and by governments (Lao PDR).

#### 17.2.4.1 RESULTS

In Tanzania (ICMM 2009), the national and local debate on mining has become increasingly intense and contentious since large-scale mining began. ICMM took the view that a starting point for dialogue would be to analyse the evidence base regarding mining in Tanzania—for example, its contribution to FDI, government revenues, and exports as a whole. This led to the development of an innovative life-cycle analysis[11] as part of the MPD Toolkit. This brought companies and government together by providing, on a pooled basis, realistic projections by the companies of their future investment activities, human resources requirements, taxes, and other payments, thereby allowing the government to get an understanding of the full project cycle. A shared understanding of the potential economic and social contribution enabled the government to plan its own role in developing the country's human resource needs, training, income, and expenditures, and contributed to developing a consensus across 'all of government'.[12]

[11] See ICMM (2009). This analysis can help 'all of government' understand the choice between a policy regime that favours, for example, high rates of growth vs. an approach consisting of raising revenues in the short term at the risk of lowering sectoral growth in the future. By focusing on production, government revenues, and employment over a 15–20-year timeframe, it provides policy makers with the big picture.

[12] Previously, the debate had been informed only by a very short period of historical data—beginning with the resumption of large-scale mining investments in the late 1990s. In a May 2009 cross-government workshop supported by the Tanzanian Chamber of Minerals and Energy, the government was able to understand better the interrelationships between decisions made in different ministries and consider introducing a broadened and integrated sector strategy. As one example, the Tanzania Ministry of Power's inaction regarding mining companies' proposal to extend the power grid meant that operating costs remained higher and royalties lower than

This same MPD framework successfully transitioned to the offshore oil and gas industry in 2013. BG—a FTSE 100 exploration and production company specializing in gas—funded a forward-looking life-cycle analysis based on a realistically assessed 10 Tcf of deep-water offshore gas reserves, using the same methodology. The objective was to determine how the development of liquefied natural gas (LNG) could impact the country's economy, in terms of FDI, exports, government revenues, contribution to GDP, and employment. This provided a comprehensive picture of the potential direct, indirect, and induced risks and benefits associated with LNG development over an extended period of time. This enabled the government to understand better what the future could look like and the implications for planning and immediate action.

Findings were presented to senior government officials at a workshop convened by the Uongozi Institute of African Leadership for Sustainable Development and chaired by the chief secretary to the president. One important outcome was the establishment of a 'government team of negotiators for natural gas and oil'. These 'twenty-five carefully picked experts from different backgrounds [are] to ensure that all the relevant issues for building a sustainable oil and gas economy are addressed'.[13] The team were supported by a six-month capacity-building programme organized by the Institute of African Leadership for Sustainable Development,[14] supported by the Columbia Center for Sustainable Investment and the International Senior Lawyers Project.

Making an agreed-upon dataset and analysis available also facilitates informed policymaking. In Lao PDR, the government prioritized the hydropower sector. The application of ICMM's MPD Toolkit (ICMM 2011a) demonstrated that mining was in fact the largest industrial sector—contributing 80 per cent of FDI, 45 per cent of exports, 12 per cent of government revenues, and 10 per cent of national income. Similar evidence has been referenced in the UN Economic Commission for Africa–African Union's 'African Mining Vision', in Natural Resource Charter datasets, and in national government and donor agency communications.[15]

might otherwise have been the case (the mining companies' operating costs were five times the global average for gold mining because they had to import fuel).

[13] The institutions represented include the Ministries of Mining, Energy and Minerals, Finance, Labour, and Industries and Trade, regional administrations, and local governments, The Attorney General's Chambers, Tanzania Electricity Power Company, Tanzania Petroleum Development Corporation, the Planning Commission, the State-owned Mining Company, Prime Minister's Office, Bank of Tanzania, Tanzania Revenue Authority, and the National Environmental Management Corporation.

[14] See http://www.thecitizen.co.tz/News/Business/State-launches-team-of-negotiators-for-gas-deals/1840414-2707728-11rsv6f/index.html.

[15] For example, a Lao PDR government official presenting key findings at the China Mining Congress 2011, Tianjin; and World Bank Sector Policy Reviews (see Schuler and Lokanc 2015).

A 2013 independent evaluation[16] found that the majority of stakeholders interviewed had 'very positive views' of the MPD Toolkit (ICMM 2011a). 'Furthermore, there was strong support across all stakeholder groups for ICMM to continue to apply the MPD Toolkit. The Toolkit is widely seen as a unique tool that generates high quality data about mining's contribution to development and also provides concrete steps for how partnerships that increase the opportunity for development can be established.'

#### 17.2.4.2 CHALLENGES

A common challenge is consistency and availability of data. For example, in Zambia, there are widely divergent perspectives and disputes within and between different arms of government and industry about key datasets. The MPD process in Zambia identified that in 2012, revenue from the mining industry was 5.9 per cent of GDP and 32 per cent of tax revenue, representing a significant increase since 2008. However, much smaller numbers from earlier years were still being used in official communications as recently as 2013. This is now being addressed to some degree by the Zambia-EITI process, which has strengthened transparency on revenues from mining through its credible multi-stakeholder process.

The importance of a national 'champion' to follow up and ensure implementation was a finding of the 2013 evaluation (footnote 11) of the MPD Toolkit: 'ICMM's future MPD work should revisit existing applications, ensure recommendations are properly progressed following a Toolkit application, and step up the delivery of Toolkit applications in new countries.' The evaluation noted that independently commissioned studies have used the MPD Toolkit methodology in a further seven countries and, in order to scale up, regional partners should be considered to deliver additional applications and support in-country implementation.

## 17.3 Building Trust through Multi-stakeholder Dialogue

### 17.3.1 *Context/Issue*

This section will address a further regrettable truth that, in some countries, there is a lack of trust among stakeholders: for example, between companies and CSOs/NGOs (e.g. in Ghana), between companies and government (e.g. in Tanzania and Zambia), between government and CSOs/NGOs (e.g. in Lao PDR and Mongolia) and between federal, state, and municipal government (e.g. in Ghana, Brazil, and Peru).

[16] Independent Evaluation of ICMM Mining: Partnerships for Development Toolkit (CSRM 2013).

The WEF noted: 'In 2012, significant resource-related disagreements flared up in almost every significant mining region, from Mongolia to Chile and from South Africa to Indonesia. Some, notably in South Africa, escalated sufficiently to migrate from the financial to the news sections of the national and global media.' This was the genesis of the Responsible Mineral Development Initiative because 'events of the last few years have heightened the need for stakeholders involved in mineral development to find common ground to understand each other's needs, perceptions, and priorities' (WEF 2013: 6).

The same WEF report highlighted the importance of national dialogue platforms. It found that these have the potential to 'offer consistent, inclusive dialogue and collaboration among stakeholders on a countrywide sector basis. The aim is to enable responsible development, find synergies and align stakeholders, and devise action plans for longer-term working partnerships. It should mean that every project begins with structures for engagement throughout its life cycle already in place.' Sixty-nine per cent of all stakeholders surveyed considered that a national dialogue platform would be 'very' or 'extremely' helpful. Responses were strongest where it was felt that a platform could significantly enhance dialogue, particularly if there were limitations in existing national infrastructure.

### 17.3.2 *Objectives*

- Create a neutral space to voice different perspectives on the impacts of mining, oil, and gas.
- Identify challenges and opportunities (e.g. conflict management, management of revenues, economic linkages, local development) as input for policy development.
- Develop a shared understanding of, and accountability for, responsible investment.

### 17.3.3 *Action in Support of the Objectives*

#### 17.3.3.1 CREATE A NEUTRAL SPACE TO VOICE DIFFERENT PERSPECTIVES

Dialogue processes are important as a way to convene local stakeholder opinions in the formulation of specific policy recommendations. Such views can best be identified and assessed through an inclusive on-the-ground process. The examples below are drawn from a series of multi-stakeholder workshops held in several countries as part of ICMM's MPD process.

Multi-stakeholder dialogue processes at the subnational level are equally important. In Peru, beginning in 2012, copper and zinc prices fell dramatically and local employment at mine sites dropped, as did the government's local and regional budgets. Consequently, at the subnational level, a number of mining companies realized that they needed to work with local government in taking the lead on promoting development. Despite significant expenditures in the past, the local populations remain poor, with low Human Development Index rankings. Antamina (a mining company in Peru) developed 'The Multi-Stakeholder Development' approach to focus on building dialogue with local government and other stakeholders. This also reduced the potential for special interests to derail development processes. Within this framework, Antamina, the local governments, and stakeholders are addressing:

1. how to obtain a sufficient share of mining taxes and royalties for local governments from the central government
2. how to build capacity for planning and implementing development activities
3. how to respect the rule of law and put in place transparent processes and accountability to end corruption by local governments.

#### 17.3.3.2 IDENTIFY CHALLENGES AND OPPORTUNITIES

The disconnect mentioned earlier—between costs and benefits as between national and local levels—which is absolutely central to the controversies surrounding the role of extractive industries, is evident in a number of countries.

In Ghana, while new mining investment (from the mid-1980s) had helped to turn around the national economy and reduce poverty at national and local levels, local communities did not perceive that they were receiving sufficient economic benefits. Companies also needed to improve their management and delivery of broader economic contribution. A 2008 workshop was explicitly structured to tackle some of the main challenges highlighted by the research (ICMM 2008b). It was informed by a mapping exercise of existing initiatives in Ghana against six priority areas. These had been identified as benefitting from collaboration or partnerships between governments, companies, and others that seemed to have the greatest potential for enhancing the social and economic returns from mining. That mapping exercise found that there were a lot of partnerships and initiatives around poverty reduction, but far fewer for regional development planning and revenue management—although both of these activities critically impact the effectiveness of what is done in the other areas. As a result, disputes between communities and companies were more numerous and more confrontational than they needed to be (ICMM 2008b).

The success of the 2008 Ghana workshop was due in part to broad and comprehensive consultations by the research team over a four-month period, with a wide range of stakeholders in Accra (and the localities of Tarkwa), in

Ahafo, and to a limited extent also in Obuasi (where earlier project fieldwork had been concentrated). These consultations were designed to obtain a wide range of views about the strengths and weaknesses of present arrangements for mining in Ghana. The number of invitees was intentionally limited and focused in order to ensure the best possible debate on the day. It was the agreed wish of the meeting that workshop minutes be circulated also to organizations and individuals who were likely to play a significant policy-influencing role. The summary notes were posted on the websites of the Chamber of Mines and ICMM.

In Zambia, the issues and opportunity areas identified during ICMM's November 2013 workshop included: (i) the need for reliable data to support a better informed debate; (ii) scope for an improved investment climate in Zambia to support greater local content and job creation; and (iii) scope for more effective social investment by mining companies, through improved coordination between companies and with the priorities of local government.

#### 17.3.3.3 DEVELOP A SHARED UNDERSTANDING OF, AND ACCOUNTABILITY FOR, RESPONSIBLE INVESTMENT

Negative impacts on local communities are often concentrated in the immediate vicinity of a mining or oil and gas operation. In many countries, resources available to communities to seek redress are more limited at the subnational than at the national level, in part due to weak government capacity and limited fiscal resources. This can lead to growing discontent at the local and community levels, which in turn can percolate up to the national level and undermine support for the industry.

There are many practical tools for understanding and managing community relations around large-scale investments, based on recognized international good practice. IFC's CommDev[17] is focused on enhancing benefits to communities by providing documents, tools, case studies, training materials, presentations, and resources produced by the IFC as well as by others, to guide how benefits from infrastructure can be shared with local communities. The World Bank's *EI SourceBook*, which identifies key levers for enhancing success in the natural resource industries, also includes links to many other guides and case studies for action at the subnational level. ICMM's Community Development Toolkit gives practical guidance on how to responsibly address issues of community engagement, develop baseline studies, set up community-level agreements, foster their implementation, and establish dispute resolution mechanisms (ICMM 2012a).

[17] https://commdev.org/.

### 17.3.4 *Early Results*

Efforts to build trust via multi-stakeholder dialogue processes have been led by NGOs (Myanmar), ICMM (Ghana, Lao PDR, and Brazil) and industry (Peru).

#### 17.3.4.1 RESULTS

In Myanmar, a consultation draft of a pioneering oil and gas sector-wide impact assessment (SWIA) formed the basis of multi-stakeholder workshop consultations. The dual objective was to:

1. discuss the key draft findings and ensure their relevance and completeness (i.e. have key issues been missed or misunderstood?)
2. allow multi-stakeholder participation in developing recommendations for actions by the government and companies and other stakeholders (local and foreign). The purpose was to improve both the outcomes of oil and gas projects for the benefit of Myanmar society, and the framework for responsible investment.[18]

Following publication[19] in 2014, the Myanmar Centre for Responsible Business (MCRB) held a 'Multi-Stakeholder Workshop on Community Engagement in the Extractive Industries' in Yangon in January 2015 to discuss international best practice in strategic community investment and engagement, including how to handle grievances. The workshop was attended by over 100 representatives from government departments, oil, gas, and mining companies, and CSOs from across Myanmar, as well as international NGOs and donor organizations. These workshops were a complement to the multi-stakeholder dialogue on revenue transparency taking place as a result of Myanmar's candidacy for the EITI. On the objective of building trust, the director of the MCRB said: 'our assessment found that engagement, information, and genuine two-way communication by business with stakeholders, particularly local communities and national NGOs, has historically been almost completely absent in Myanmar, leading to mistrust, misunderstanding, and occasionally conflict.'

#### 17.3.4.2 CHALLENGES

In Brazil, an MPD workshop was held in Brasilia in 2012 and attended by 120 people, representing a broad cross-section of government, industry, and

[18] See www.myanmar-responsiblebusiness.org/news/oil-and-gas-swia-draft-for-consultation.html.

[19] Institute for Human Rights and Business, Myanmar Centre for Responsible Business, The Danish Institute for Human Rights: Sector Wide Impact Assessment: Myanmar Oil & Gas Sector Wide Impact Assessment, 2014, http://www.myanmar-responsiblebusiness.org/swia/oil-and-gas.html.

civil society (ICMM 2012b). Discussion focused on the ability of public institutions to scale up and replicate company-initiated partnerships. Three challenges were identified:

1. Brazil has a complex system of public revenue management at all three tiers of government, namely federal, state, and municipal. There is a significant overlap of responsibilities between these three tiers. A graphical description that indicates the complexity of these arrangements is shown as Figure 2 in McPhail (2017). One consequence of 1988 fiscal decentralization was huge revenue variations across municipalities.
2. Municipal government has limited capacity to allocate and spend funds effectively. Often the challenge is not lack of funds.
3. National government policies often do not seek opportunities to link oil, gas, and mining operations into the region as a whole. This will require leadership thinking through a bigger regional economic integration picture, which in turn requires clearly defined public- and private-sector roles. For example, the public-sector education system's chronic undersupply of scientists and engineers with skills that are useful to industry constrains participation in the opportunities that mining creates. Taxation incentives to encourage companies to invest are only a 'sticking plaster' on a system of education in need of overall reform.

Dialogue platforms inviting participatory problem-solving around resources development are increasingly common. A recent example is the CSO-led Groupo de Dialogo operating at national level in some Latin American countries.[20] At the global level, the OECD's globally convened Policy Dialogue on Natural Resource-based Development[21] is a multi-interest group that has identified a useful 'Operational Framework on Public–Private Collaboration for Shared Resource-based Value Creation'.[22] This provides a strong framework for progress to the next step: country-specific evidence to be collected and debated in-country as a basis for building trust and consensus around action needed.

## 17.4 Partnerships for Development

### 17.4.1 *Context/Issue*

This section explores how to mitigate the negative impacts of resource development and to enhance its positive contributions.

[20] See http://www.dialogolatinoamericano.org.
[21] See http://www.oecd.org/dev/natural-resources.htm.
[22] See http://www.oecd.org/development/public-consultation-pd-nr.htm.

The REi found that there are certain common features of governance structures and institutions across better performing mining countries. Where these institutions are in place, economic development—and, in some cases, the social development of these countries—has been enhanced by new sustainable private-sector activities that complement the impacts of mining itself. Some of these features are shown in Figure 3 of McPhail (2017), which depicts, in general terms, the link between natural resource extraction and economic and social development. This simple formulation shows that the minerals or the oil and gas industries cannot be treated in isolation from the rest of a country's institutional and governance structures. Their presence has an impact on these structures and calls for leadership and coordination across the different ministries and agencies with companies and others. This illustrates the multifaceted nature of 'good governance'. Where this is missing, Partnerships for Development can help fill the so-called governance gaps.

Why partnerships? Where the features shown in Figure 3 (McPhail 2017) are weak or missing, in-depth research has shown that resource governance can be strengthened through partnership. For example, there is a role for development partners in supporting developing host countries that embark on prescribed reforms. The private sector has a critical role to play in generating incomes and employment. CSOs can work directly with local communities and can demand accountability. Faith-based groups can communicate costs and benefits of natural resource development to citizens and help manage expectations.

Given these different roles and responsibilities of resource governance between the many actors involved, mainstreaming partnerships will require ongoing engagement and dialogue. Partnerships cannot always provide magic bullets, or be deployed in all situations. They can drive action and enhance accountability. When combined with clear goals, such as the MDGs/SDGs—as represented in top-down policy recommendations—they can enhance development outcomes.

Partnerships can also help to uncover the highly local contextual factors driving the economic and social impacts of resource projects. For example, ICMM research findings confirm the negative impacts around operations, and cover a large number of issues. This suggests the need for better local knowledge and evidence about the impacts of such projects, which will depend on close engagement with local actors (partnerships) (ICMM 2008a, 2008b).

Often much of the failure to capture equitable benefits from resource extraction among developing countries is due not to the absence of top-level policies, but rather to the absence of capacity to effectively implement these policies and hold policy makers to account. This can apply at both national and subnational levels. For example, where companies provide social investment (e.g. health, education, and infrastructure), capturing the full benefits from such investments may be undermined if the regional development planning processes are weak. Experience has shown that partnerships—including those involving industry as well as donors and civil society—complement central government policy in enhancing capacity to deliver benefits.

ICMM's MPD work suggests that there are six recurring themes where partnerships can be used to enhance development outcomes: poverty reduction, revenue management, regional development, local content, social investment, and disputes resolution.

### 17.4.2 *Objectives*

- Enhance economic and social outcomes by filling national and local governance capacity gaps.
- Strengthen accountability.
- Uncover local contextual issues that, if unchecked, can create social tensions and political pressures.

### 17.4.3 *Action in Support of the Objectives*

#### 17.4.3.1 ENHANCE ECONOMIC AND SOCIAL OUTCOMES BY FILLING GOVERNANCE CAPACITY GAPS

With regard to the subnational level, this partnership challenge has been taken up at the local and regional levels by a number of resources companies, with promising early results.

In Colombia (ICMM 2015), the government in La Guajira was responding in only a limited way to the needs of communities. One of the mining companies, Cerrejón, decided that, rather than respond individually to local demands, it would focus on engaging with the local as well as the central government. Through its Institutional Strengthening Foundation, Cerrejón provided technical assistance to the planning committees and local officials to improve planning, management, implementation, and evaluation. This has also helped municipalities to overcome their fiscal deficits. There is now improved accountability due to a focus on improving reports to the central government on the use of financial resources by La Guajira. Progressively, the municipalities have improved their performance, and these mining-affected local administrations are now at the top of national rankings. The La Guajira Department occupies second place in performance.

#### 17.4.3.2 STRENGTHEN ACCOUNTABILITY

Between 2008 and 2012, the Revenue Watch Institute (RWI) supported governments and CSOs in Ghana, Indonesia, Nigeria, and Peru to improve subnational mineral revenue management (Pellegrini and Venugopal 2013). Early results that it documented included:

- improved revenue tracking and facilitated investment in sustainable development. In Peru, the Arequipa and Piura regional governments used RWI's forecasting tool to accurately predict their entitlements and formulate multi-year budgets. Overall regional government spending effectiveness (the percentage of the allocated budget that is actually spent) increased from 89.2 per cent in 2009 to 93.7 per cent in 2011 in Arequipa, and from 79.9 per cent in 2009 to 84.5 per cent in 2011 in Piura.
- the development of lasting mechanisms for participation and accountability. Work in Indonesia led to the creation of multi-stakeholder steering committees that promote oil industry transparency in Blora and Bojonegoro. In both districts, these multi-stakeholder groups were formally acknowledged and the government allocated funds from local budgets to support their continued operation. In Nigeria, the RWI-funded Bayelsa Expenditure and Income Transparency Initiative (BEITI) developed a multi-stakeholder platform to track state revenues, transfers, and expenditures.

#### 17.4.3.3 UNCOVER LOCAL CONTEXTUAL ISSUES

In Zambia, an emerging priority for action was improved mining company investment in the social areas. The dialogue process agreed that this is a primary responsibility of companies, which should also collaborate both across the industry and with key stakeholders, including communities and local government. Both CSOs and mine managers noted that successful outcomes are contingent on government's contribution: for example, local authorities require improved financial and technical resources to develop stronger local development plans. In the absence of these plans and partnership with other stakeholders, company social investment has no clear framework in which to operate and cannot deliver real local benefits.

### 17.4.4 *Early Results*

This section presents case examples of efforts to identify partnerships in several countries and briefly considers where the challenges remain. Efforts highlighted here have been led by the mining industry alongside government and development partners (Brazil, Ghana, and Lao PDR).

#### 17.4.4.1 RESULTS

In Brazil (ICMM 2012c), Vale and other large mining companies (such as Alcoa and Hydro) are working strategically and in partnership to drive positive developmental outcomes in south-east Pará—one of the least developed regions of Brazil. All major mining companies in the state are committed to a broad range of voluntary and mandatory initiatives to help address the challenges of infrastructure, human capacity limitations, and public administration constraints in their areas of operation. Vale's approach to economic development of the regions in which it operates is the most highly formalized and involves all levels of government.[23]

Priorities are set in a matrix of intersecting processes:

- consultation with, and agreement from, the environmental licensing process
- voluntary in-depth socio-economic diagnostic studies conducted to provide regular forward projections of demand (e.g. for schools and other public services) based on expected future mining investment, the induced future demand for public services that may result, and deficits in capacity to meet these demands
- connecting municipalities with federal government agencies through a 'Public–Private Social Partnership' framework (see Figure 17.2).

This is collated in 'Letters of Agreement' that Vale signs with all municipalities where it operates, setting out the roles and responsibilities of different actors in delivering social projects. These are public documents that enhance both the accountability and the commitment of all partners. Vale's social investment framework illustrates the power of combining mandatory and voluntary processes of stakeholder alignment.

In Ghana, the Chamber of Mines and the Minerals Commission organized a May 2009 workshop to follow up on a 2008 multi-stakeholder workshop action plan, which had prioritized compensation payments and small-scale mining (ASM).[24]

Although local content was not prioritized in the 2008 Ghana workshop,[25] nevertheless a partnership was formed between the Ghana Minerals Commission, the IFC, and the Ghana Chamber of Mines to assess local supplier development needs and create linkages between mines and the local economy.

[23] See ICMM (2012b).

[24] Less progress appears to have been made on engaging companies in the implementation of District Development Funds (DDF)—where there was already good coordination between donors and local governments for budget support to the districts both for project financing and for capacity-building; as well as on the transparency agenda. The 2008 workshop noted that this had been developed well in Ghana already but not nearly enough had been done to provide the relevant information down to local levels of society. The initial EITI reports in Ghana highlighted that many payments by companies and distributions by national government were not being made according to the requirements, or not being properly accounted for and used at a local level (GHEITI 2015).

[25] ICMM (2007) finds that of the US$110 million spent annually by the company on local procurement, 47 per cent was sourced in Ghana, while in Peru, of the total annual spend of US$229 million, almost 78 per cent was sourced in Peru.

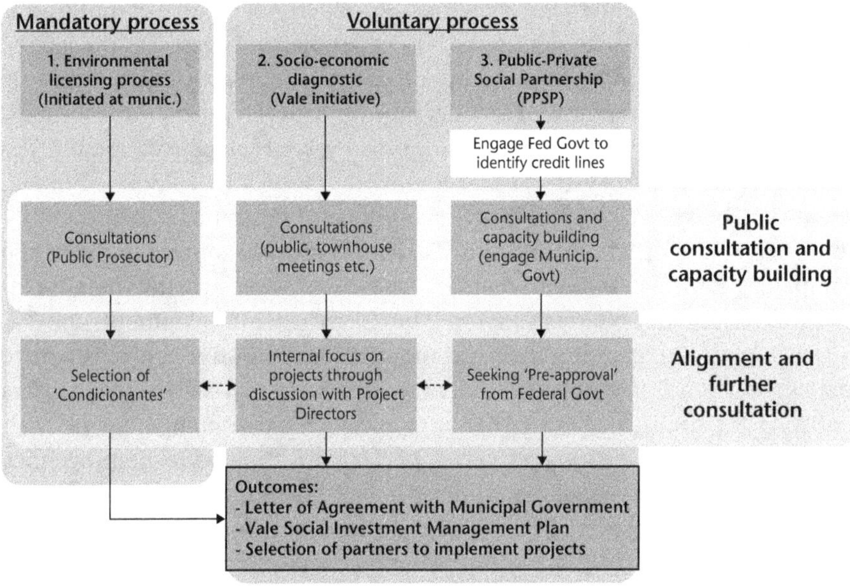


**Figure 17.2.** Vale Foundation social investment framework, Brazil
*Source*: ICMM (2012c: 52); reproduced here by permission.

The objective of the partnership is to build the capacity and productivity of up to thirty-five suppliers over three years. The result is expected to be an increase in the share of items produced and bought in Ghana by identifying twenty-eight new production lines.

Results to date are slow in coming. In 2014, ICMM was invited to return to Ghana and conduct a forward-looking analysis of mining's future contributions. The analysis indicated that, recent efforts to enhance local procurement notwithstanding, there had been little change in the proportion of company procurement spend in Ghana over the decade. The prize is clear. Increasing local procurement by 25 per cent across certain categories would create about US$50 million annually in value-added to the economy. Indirect employment would be even further increased. An April 2015 multi-stakeholder workshop again prioritized local content, with calls for a national action plan.

#### 17.4.4.2 CHALLENGES

No matter how well intentioned (or well funded), no individual stakeholder (including a national government and a company) can 'go it alone' in seeking positive outcomes. For companies, a failure to engage broadly through partnerships clearly increases company risk. In Tanzania[26] mining has made an

[26] Roe (2016).

impressive contribution to the economy since 2000, yet the role of mining has been underrated and often criticized. One possible explanation is the lack of linkages. Did companies miss the opportunity to seize a vision that mining can be used as a focal point for broader economic development? The location of the major mine investments is in a broadly contiguous area around the southern part of Lake Victoria and is shown graphically in McPhail (2017: Figure 5). The very close proximity of millions of dollars of mining activity could have supported a significant cluster of new economic activity if the companies had first worked together—and then engaged with the government to ensure that policies supported broader cluster development.

Developing a shared objective and maintaining collective action is a huge challenge when different stakeholder interests have markedly different time horizons for decision-making. For example: extractive companies in their forward-looking projections can take a 20–100-year time horizon; indigenous peoples take a multi-generational perspective; governments face a 3–5-year electoral cycle; investors look to quarterly results; communities often have an immediate need for jobs. And prices are constantly changing. Measuring and tracking progress and attributing results across different partners (which can be required by investors—whether donors, government agencies, or companies) is difficult both conceptually and methodologically.

Some of the most interesting data on local economic and social impacts in the ICMM Lao PDR study[27] resulted from the innovative use of biannual household surveys by the companies operating the country's two large-scale mines. MMG Sepon undertakes two yearly household surveys of thirty-four villages around the mine site with a total population of 8,500. Quantitative information includes population growth, food sources, household possessions, and income, and qualitative opinions are sought on land use and operations of the mine. The surveys found that the average annual per capita income in the villages had increased considerably since 2001. Specifically, it grew from US$64 in 2001 to US$436 in 2009/10. Interestingly, per capita income increased despite a rapidly growing population. For example, the number of inhabitants in the immediately affected communities doubled from around 1,100 in 2001 to 2,200 in 2009/10. Moreover, income inequalities declined over the same period. Looking across villages, the Gini coefficient in 2001 was 27 (meaning that 27 per cent of the total income would need to be redistributed to attain perfect equality across villages). By 2009/10 it had fallen to 12 per cent. Looking within villages, in every case the Gini coefficient fell from 2001 to 2009/10 (on average from 50 per cent to 34 per cent). This is a significant fall in the Gini coefficient. The remaining inequality within the

[27] ICMM (2011b).

villages is largely a function of family structures—where there are elderly couples and young couples with multiple infant children who have not yet benefitted as much as families with adults of working age.

## 17.5 Conclusions

Work over a decade shows that investments in the natural resources sector do not automatically result in broad-based development, which requires:

- comprehensive understanding of the sector's full economic and social impacts—positive and negative, quantitative and qualitative, national and local—and the existing sector governance framework
- a vehicle for sharing this understanding across government and stakeholders, to (i) recognize the complexity of sector governance and (ii) explore how to diversify their economies from, and citizens' expectations of, dependence on the sector
- a platform for developing collaborative partnerships between government entities, companies, development partners, and CSOs to fill gaps in the governance framework.

This is an unfinished agenda. At the national level, for example, coordination within government, industry, or others, and between these players, is often embryonic. The MPD Toolkit provides some tools to achieve this. What are the best leadership and coordination models? There are recent examples of ways to improve intra-government coordination, effectiveness, and learning in Peru[28] and in China (CCCMC 2014). Many Chambers of Mines or Energy can lack technical and outreach capacity and need strong engagement and support from company members.

At the national and regional levels, a key priority of many governments is infrastructure development. However, there are few examples of shared-use infrastructure (roads, railways, power plants, etc.) either within countries or across borders. The World Bank considered the potential and challenges of power–mining integration in the sub-Saharan Africa context (Banerjee et al. 2014). The challenges highlighted include aligning incentives (regulatory and commercial), commodity price volatility, and political instability.

[28] For example, in Peru, the Humala government came into office in 2011 and prioritized an approach to improving social cohesion that centres on achieving greater coordination between government entities and horizontal accountability across the public sector. A new Ministry of Development and Social Inclusion was conceptualized to improve the inter-ministerial coordination and the efficiency of existing social and other public policies. In addition, a presidential decree created a National Office and a High Commissioner for Dialogue and Sustainability, situated in the Prime Minister's Office (see ICMM 2013).

At the global level, the three steps set out in this chapter could support implementation of the Sustainable Development Goals and the Paris Agreement on Climate Change. In December 2015, 195 nations signed an agreement to combat climate change and implement 'nationally determined contributions' towards a low-carbon future. This is to be achieved by keeping the global temperature rise this century to below 2°C, and driving efforts to limit the temperature increase even further to 1.5°C above pre-industrial levels. The Paris Agreement came into force in November 2016 in one of the fastest ratification processes in climate negotiation history. It provides a framework for country-driven action by 'non-state actors', including cities, states, and regions, companies, and investors.

## References

Banerjee, S. G., Z. Romon, G. McMahon, P. Toledano, and P. Robinson (2014). 'The Power of the Mine: A Transformative Opportunity for sub-Saharan Africa'. Washington, DC: World Bank. Available at: https://openknowledge.worldbank.org/handle/10986/21402.

CCCMC (2014). 'Guidelines for Social Responsibility in Outbound Mining Investments'. Beijing: CCCMC. Available at: http://www.eic.cat/promocio/normativa/new_xina/Draft_Guidance_PublicConsultation_ENG_20140310.pdf.

CCCMC (2015). 'Chinese Outbound Mining Investment Industry, Social Responsibility Action Plan (2016–2018)' (Draft). Beijing: CCCMC.

CSRM (2013). *Independent Evaluation of ICMM Mining: Partnerships for Development Toolkit*. Brisbane: Centre for Social Responsibility in Mining, University of Queensland.

Dietsche, E. (2017). 'Political Economy and Governance', WIDER Working Paper 2017/24. Helsinki: UNU-WIDER.

GHEITI (2015). *Annual Activity Report for Ghana Extractive Industries Transparency Initiative*. Accra: Ministry of Finance, Government of Ghana.

Halland, H., M. Lokanc, and A. Nair (2015). 'The Extractive Industries Sector: Essentials for Economists, Public Finance Professionals, and Policy Makers', World Bank Studies. Washington, DC: World Bank. Available at: https://openknowledge.worldbank.org/bitstream/handle/10986/22541/The0extractive0s00and0policy0makers.pdf.

ICMM (2007). 'Peru Country Case Study—The Challenge of Mineral Wealth: Using Resource Endowments to Foster Sustainable Development'. London: ICMM.

ICMM (2008a). 'Ghana: Sowing the Seeds for Action. Action Learning through Partnerships. The Challenge of Mineral Wealth: Using Resource Endowments to Foster Sustainable Development', Spotlight series 11. London: ICMM.

ICMM (2008b). 'Peru: From Boom to Broad-based Development? The Challenge of Mineral Wealth: Using Resource Endowments to Foster Sustainable Development', Spotlight series 12. London: ICMM.

ICMM (2009). 'Mining in Tanzania: What Future Can We Expect?' London: ICMM.

ICMM (2011a). 'Mining: Partnerships for Development Toolkit'. London: ICMM.

ICMM (2011b). 'Utilizing Mining and Mineral Resources to Foster Sustainable Development of the Lao PDR'. London: ICMM.

ICMM (2012a). 'Community Development Toolkit'. London: ICMM.

ICMM (2012b). 'The Mining Sector in Brazil: Building Institutions for Sustainable Development', Spotlight series 17. London: ICMM.

ICMM (2012c). 'The Mining Sector in Brazil: Building Institutions for Sustainable Development', Report. London: ICMM.

ICMM (2013). 'Responsible Mining in Peru'. London: ICMM.

ICMM(2015). 'Maximising the Potential Benefits of Mining: Understanding Local Government Capacity Issues', MS, report and tool. London: ICMM.

McKinsey Global Institute (2013). *Reversing the Curse: Maximizing the Potential of Resource-driven Economies*. Available at: https://www.mckinsey.com/~/media/McKinsey/Industries/Metals%20and%20Mining/Our%20Insights/Reverse%20the%20curse%20Maximizing%20the%20potential%20of%20resource%20driven%20economies/MGI_Reverse_the_curse_Full_report.ashx.

McPhail, K. (2017). 'Enhancing Sustainable Development from Oil, Gas, and Mining: From an "All of Government" Approach to Partnerships for Development', WIDER Working Paper 2017/120. Helsinki: UNU-WIDER.

Moffat, K., N. Boughen, A. Zhang, J. Lacey, D. Fleming, and K. Uribe (2014). 'Chilean Attitudes to Mining: Citizen Survey 2014 Results'. Australia: CSIRO. Available at: https://www.csiro.au/en/Research/Mining-manufacturing/CSIRO-Chile/Other-initiatives/Chilean-attitudes-to-mining.

Pellegrini, M. and V. Venugopal (2013). 'From Conflict to Collaboration: Lessons Learned from RWI Work at the Sub-national Level'. New York: Revenue Watch Institute. Available at: www.resourcegovernance.org/sites/default/files/From%20Conflict%20to%20Collaboration%20-%20SHORT20131203.pdf.

Roe, A. R. (2016). 'Tanzania and Extractives—Structural Change: Another Twist?' WIDER Working Paper 2016/79. Helsinki: UNU-WIDER.

Roe, A. R. and S. Dodd (2017). 'Dependency on Extractive Industries in Lower-income Countries: The Statistical Tendencies', WIDER Working Paper 2017/98. Helsinki: UNU-WIDER.

Schuler, P. M. and M. Lokanc (2015). 'Zambia Economic Brief: Making Mining Work for Zambia', Issue 5. Washington, DC: World Bank Group. Available at: http://documents.worldbank.org/curated/en/2015/06/24646516/making-mining-work-zambia-zambia-economic-brief-making-mining-work-zambia.

WEF (2011). 'Responsible Mineral Development Initiative Report'. Geneva: World Economic Forum. Available at: https://www.weforum.org/reports/responsible-mineral-development-initiative.

WEF (2013). 'Responsible Mineral Development Initiative: Mineral Value Management: A Multidimensional View of Value Creation from Mining'. Geneva: World Economic Forum.

World Bank Group (2016). *EI SourceBook*. Washington, DC: World Bank. Available at: www.eisourcebook.org.

Part VI

# International Regulatory Concerns and Structures

# 18

# Towards Contribution Analysis

*R. Anthony Hodge*

## 18.1 Introduction

The mining industry's boom–bust pattern is well recognized. During the last twenty years, the swings have been particularly extreme: the 2002–12 super-cycle was followed by a precipitous fall.[1] By late 2016, recovery seemed evident. 2017 has been marked by much global uncertainty and has seen fluctuations, depending on commodity. Societal attitudes towards mining are nested in these swings. However, they are more fundamentally governed by the evolution of two factors: (1) deeply held values related to human and ecological well-being, and (2) the perceived role of various human activities (such as mining) in contributing or not to that well-being.[2]

Through the past fifty years, the mining industry has found itself subject to increasing levels of criticism, accusations of unfairness, and resulting regulatory and public scrutiny. The industry has been ill prepared for these 'perfect storm' conditions and for many, it has brought discomfort and a defensive reaction; see Hodge (2017) for a more detailed discussion of these trends.

At the turn of the millennium, the industry mounted an unprecedented initiative. This was the Global Mining Initiative, with its flagship project, Mining, Minerals and Sustainable Development. Over two years, multi-interest

[1] See Humphreys (2015: 1) for a clear description of mining's 2002–12 super-cycle.

[2] The ideas presented in this chapter have evolved over the last thirty years, and along the way many people have contributed, including Al Freeze, Ron Rice, David Brooks, Alex Michalos, Britt Banks, and numerous friends and colleagues with whom I have shared professional assignments. Though these others remain unnamed here, I am deeply appreciative of the push and pull of debate along the way. The following reviewed this chapter, offering many helpful comments: Tony Addison, Chris Anderson, Toni Aubynn, Jim Cooney, Robert Court, Lois Craig, Rodrique Djahlin, Liesel Filgueries, Bob Gibson, David Humphreys, Anne Johnson, Mike McPhie, Lee Merkhofer, Alan Roe, Andrew Roman, Ian Thomson, and Mark Wade. Much of the richness is due to their input, while the weaknesses that remain are solely due to the author.

discussions about mining practices were convened across the world. Many thousands participated in an examination of mining practices. An action plan for improvement was developed and the International Council on Mining and Metals (ICMM) established to serve as a change agent in its implementation and further development (MMSD 2002).

As part of this evolution, a remarkable number of initiatives *in addition to many changes in formal law and regulation* in countries across the world have been mounted to improve industry performance. Table 18.1 lists forty-five such initiatives, nine from within the industry, thirty-six from outside. There may be others. The majority of these are multi-interest collaborations involving companies, civil society organizations, and government.

The obvious question that arises is whether all this effort is making any difference. Is mining and metals' contribution to human and ecosystem well-being more aligned to the values and expectation of society as a result? These questions are largely unaddressed: there is no accepted, systematic way to assess and track a mine operation's contribution through the full project life cycle, let alone the whole industry. Given this gap, it is not surprising that industry champions and detractors disagree, sometimes vehemently, on the nature of mining and metals' contribution to host communities and countries. Questioning of the legitimacy of current project approvals processes continues for communities, governments, and project proponents alike. Filling this gap is the challenge taken up by this chapter—how to understand, assess, and track mining's contribution to human and ecosystem well-being over the full project and product life cycles. As will be seen, the task is far from trivial: the ecological, social, cultural, political, and economic setting of mining is global and complex, and requires a synthesis of many perspectives.

A values-driven conceptual foundation is offered that addresses: (1) both *ends* and *means*; (2) *inputs, outputs*, and *results*; and (3) not only the *substance* of decisions and actions (the *what*), but also the *process* used in implementation (the *how*). Importantly, a *four-part generic assessment cycle* is introduced that explicitly links measurement, story, synthesis/judgement, and communication. Central to the approach is a conceptual shift from the current deeply entrenched practice of focusing on the identification and mitigation of negative impacts (or effects) to a 'higher test' based on the achievement (or not) of a net positive contribution to human and ecosystem well-being over the long term.

Implementation will build on more traditional environmental and social impact assessment including cumulative effects assessment. However, contribution analysis will require a significant evolution in the culture of decision-makers, proponents, and assessment participants.

Some important steps towards contribution analysis have already been taken by communities, by governments, by companies, by academia. In sum, a start

**Table 18.1.** Forty-five initiatives aimed at improving the performance of the mining and metals industry over and above the formal legal system

a. Multi-interest performance-enhancing initiatives within the industry

| Date | Initiative |
|---|---|
| 1991–2001 | 1. International Council on Metals and the Environment; 29-company international coalition aimed at strengthening mining's environmental performance. |
| 1992–4 | 2. Multi-interest Whitehorse Mining Initiative; Leadership Accord to govern mining practices |
| 2000–2 | 3. The Global Mining Initiative (GMI) with its central element, the Mining Minerals and Sustainable Development (MMSD) and subsequent creation of the ICMM |
| 2003–ongoing | 4. ICMM Principles and Sustainable Development Policy framework (with further position statements: 2003—Mining and Protected Areas; 2003/5/6/9—Transparency and Mineral Revenues; 2004/10—Mining: Partnerships for Development; 2006/9/11—Climate Change; 2008/13—Mining and Indigenous Peoples; 2010—ICMM public reporting first completed including independent assurance and consistent with GMI reporting requirements) |
| 2003, 2009–ongoing | 5. Prospectors & Developers Association of Canada (PDAC) e3 Plus, a web-based inventory of recommended good practice for environmental and social management of mineral exploration projects |
| 2004–ongoing | 6. Towards Sustainable Mining initiative of the Mining Association of Canada |
| 2006–ongoing | 7. Principles of Enduring Value established by the Mining Council of Australia based on the ICMM Principles, revised in 2015 |
| 2009–ongoing | 8. Aluminium Stewardship Initiative |
| 2011–ongoing | 9. Conflict-Free Sourcing Initiative (CFSI) Smelter Program (CFSP) |

b. Performance-enhancing initiatives outside the industry aimed at influencing industry practices

| Date | Initiative |
|---|---|
| *Broadly focused, multi-interest initiatives from outside the industry* | |
| 1976–2011, ongoing | 1. OECD Guidelines for Multinational Enterprises, Annex to OECD Declaration on International Investment and Multinational Enterprises |
| 1995–ongoing | 2. AccountAbility AA1000, Assurance Standard |
| 1997–ongoing | 3. Social Accountability SA8000 |
| 1998–2000; 2014–15, ongoing | 4. The United Nations Millennium Development Goals (MDGs) adopted in 2000 and then the Sustainable Development Goals (SDGs) adopted in 2015 |
| 1999–ongoing | 5. Dow Jones Sustainability Index (DJSI) |
| 2000–ongoing | 6. United Nations Global Compact (First Leader's Summit convened in 2004) (voluntary initiative based on chief executive officer (CEO) commitments to implement universal sustainability principles (human rights, labour, environment, anti-corruption) and to take steps to support United Nations goals) |
| 2000–ongoing | 7. United Nations Voluntary Principles on Security and Human Rights (designed explicitly to guide extractive industry companies in maintaining the safety and security of their operations within an operating framework that encourages respect for human rights) |
| 2000–ongoing | 8. Global Reporting Initiative (mining supplement issued in 2011) |
| 2001–ongoing | 9. FTSE4 Good Index |

*(continued)*

**Table 18.1.** Continued

b. Performance-enhancing initiatives outside the industry aimed at influencing industry practices

| Date | Initiative |
|---|---|
| 2002–4 | 10. Extractive Industries Review by the World Bank Group concluding that extractive industries can contribute to poverty reduction and that World Bank Group involvement can positively influence industry standards |
| 2003–ongoing | 11. Equator Principles established in 2003, revised in 2006 (a risk management framework, adopted by financial institutions, for determining, assessing, and managing environmental and social risk in projects) |
| 2003–ongoing | 12. Extractive Industries Transparency Initiative (EITI) principles agreed on in 2003; Tripartite (governments, companies, and civil society organizations) International Advisory Board established and six key criteria established marking the beginning of implementation in 2005; Secretariat established in 2007. Current 'Standard' launched in 2013 |
| 2004–ongoing | 13. Electronic Industry Citizenship Coalition (EICC) |
| 2006–12, ongoing | 14. International Finance Corporation Principles and Guidance on Social and Environmental Sustainability established in 2006, revised in 2012 |
| 2006–ongoing | 15. Initiative for Responsible Mining Assurance (IRMA) |
| 2007–ongoing | 16. Devonshire Initiative |
| 2009–ongoing | 17. Equitable Origin, *EO100™ Standard* |
| 2010–ongoing | 18. Responsible Mineral Development Initiative, RMDI (WEF) |
| 2010–ongoing | 19. Free, Prior and Informed Consent (FPIC) Dialogue (The Forests Dialogue) |
| 2010–ongoing | 20. Natural Resource Charter launched (a set of principles for governments and societies on how to best harness the opportunities created by extractive resources for development); now part of the NRGI |
| 2013–ongoing | 21. OECD Due Diligence Guidance for Responsible Supply Chains of Minerals from Conflict-Affected and High-Risk Areas |
| 2015–ongoing | 22. Responsible Mining Index, under development by the Responsible Mining Institute |
| *Issue or commodity focused multi-interest initiatives from outside the industry* | |
| 2001–ongoing | 23. Global Business Coalition on HIV–AIDS, tuberculosis, and malaria—GBC Health |
| 2003–ongoing | 24. Kimberley Process Certification Scheme (KPCS) |
| 2004–ongoing | 25. Business and Biodiversity Offsets Programme (BBOP) |
| 2004–ongoing | 26. Alliance for Responsible Mining (ARM) |
| 2005–ongoing | 27. Partnering Against Corruption Initiative (PACI) of the World Economic Forum |
| 2005–ongoing | 28. Responsible Jewellery Council |
| 2005–ongoing | 29. International Cyanide Management Code (ICMC) for gold mining |
| 2005–ongoing | 30. Diamond Development Initiative (DDI) |
| 2011–ongoing | 31. Public–Private Alliance for Responsible Minerals Trade |
| 2011–ongoing | 32. Solutions for Hope |
| 2012–14 | 33. Conflict-Free Tin Initiative |
| 2013–ongoing | 34. Better Gold Initiative |
| 2016–ongoing | 35. New mining code from the World Initiative of Mining Lawyers (WIOML) |
| 2016–ongoing | 36. World Bank Social and Environmental Framework |

*Source*: this compilation builds on those compiled in Resolve Solutions Network and the World Economic Forum (2016). There may be others.

has been made. But significant distance remains on the path ahead. This chapter attempts to go beyond that start.

## 18.2 Conceptual Foundation

As with most aspects of human activity, in the arena of mining–society relationships, it is not only *what* is done that matters, but also *how* it is done. For assessing contribution, both substance and process are essential considerations. In the paragraphs below, a conceptual foundation for each is summarized.[3]

### 18.2.1 *Substance*

The origins and definitions for 'sustainability', 'development', and 'sustainable development' that are used in this chapter are provided in Box 18.1.

In summary, the three definitions described in Box 18.1 translate to the use of a value set that is best stated as 'parallel care and respect for the ecosystem and people within' (Hodge et al. 1995: xiii). To judge mining's success at achieving a positive contribution requires testing mining for its compatibility with that value set.

The higher test of achieving a 'positive contribution to sustainability' was first articulated by Gibson (2000, 2002). MMSD North America (2002: 7) points out:

> The above 'positive contribution to sustainability' criterion is different from, though built upon, the 'mitigation of adverse effects' criterion that is the focus of traditional environmental and social impact assessments. The implications of the shift are twofold. On one hand, the positive orientation opens the door to a much fuller treatment of the benefits that result from mining activities than has traditionally been the case with impact assessment approaches. On the other, the same positive orientation sets the assessment bar higher.

These assertions do not negate the fact that mining causes impacts, or that human and/or ecosystem well-being might be degraded, and permanent ecosystem or social change might occur. However, when the full life cycle of projects/operations and products is considered, a net positive contribution to human and ecosystem well-being should be realized. If not, the mining/mineral activity will not be contributing positively to sustainability.

[3] More detail can be found in Gibson (2000, 2002), Gibson et al. (2006), Hodge (1995, 2011, 2017), Hodge et al. (1995), and MMSD North America (2002).

**Box 18.1** APPLIED SUSTAINABILITY: ORIGINS AND THREE KEY DEFINITIONS

During the 1980s, the United Nations Secretary General convened three linked international enquiries. Willy Brandt, former chancellor of West Germany, was asked to lead a review of North–South inequities (Brandt Commission on North–South Issues 1980); Olaf Palme, former president of Sweden, was asked to lead a review of international security (Palme Commission on Security and Disarmament 1982); and Gro Harlem Brundtland, prime minister of Norway, was mandated to integrate the results and collaboratively develop a strategy that would reduce North–South inequities while increasing international security (WCED 1987). Brundtland called her strategy sustainable development. It is this concept that provides the anchor for this chapter.

### Three key definitions

The following three definitions serve as the starting point of this discussion.

**Definition 1. Sustainability**: the persistence of certain necessary and desired characteristics of both people and the enveloping ecosystem (of which people are a part) over a very long time—indefinitely (modified from Robinson et al. 1990).

**Definition 2. Development**: to expand or realize the potentials of; bring gradually to a fuller, greater, or better state (modified from Daly 1989).

**Definition 3. Sustainable development**: an ongoing process in which humans take actions leading to development (as described above) that meets the needs of the present without compromising the ability of future generations to meet their own needs. This is the classic Brundtland Commission definition which has stood the test of time (WCED 1987). It elegantly captures the idea that humankind is on a continuous quest for certain necessary and desired characteristics of both humans and the enveloping ecosystem over a very long time—indefinitely. The needed test of achieving success is whether or not human actions (such as mining) contribute positively to human and ecosystem well-being over the long term—thus meeting current without compromising future needs.

A full discussion of the underlying ideas and implications of these definitions is provided in Hodge (2011, 2017) and MMSD North America (2002).

From an engineering design perspective, the achievement of a positive contribution to human and ecosystem well-being over the long term serves as a kind of two-dimensional design criterion. From the perspective of 'results-based management', achievement (or not) of human and ecosystem well-being is the result or 'end' that is to be sought and designed for, and whose achievement (or not) is to be tested over time, and publicly reported on.

MMSD North America (2002) points out that these ideas veer sharply away from seeking a 'trade-off' between human and ecosystem well-being. There are obviously many small trade-offs in any practical application: between interests, between components of the ecosystem, across time, and across space. And Gibson et al. (2006: 130–41) propose a carefully considered set of six rules for dealing with such trade-offs. However, in an overarching sense, the idea of sustainability calls for both human and ecosystem well-being to be maintained or improved over the long term. One without the other subverts the concept.

Contributing positively to that 'two-dimensional' improvement provides the conceptual starting point for assessing the substantive aspects of mining's success. Figure 18.1 draws these ideas together to provide the fundamental conceptual framework for assessing contribution.

In Figure 18.1, the two upper bubbles show the sought *results* or *ends* and the lower bubble captures the *means*. The central bubble summarizes essential considerations: (1) both process and substance; (2) both short- and long-term time horizons; and (3) the effectiveness of systems that are in place to facilitate continuous learning.

The means we have at our disposal include: (1) actions within the formal economy; (2) actions outside the formal economy (a broad range of unpaid work (e.g. housework) and volunteer activities); and (3) governance systems which span all the written and unwritten rules—sometimes entrenched in law, sometimes not. In assessing these means for a mine, these 'means' activities must be assessed for the benefits they offer to people and ecosystems and the stresses (costs and risks) that are imposed.

Figure 18.1 is a broad systems analogy of the results-based management approach to tracking and assessing inputs, outputs/outcomes, and results/ends.

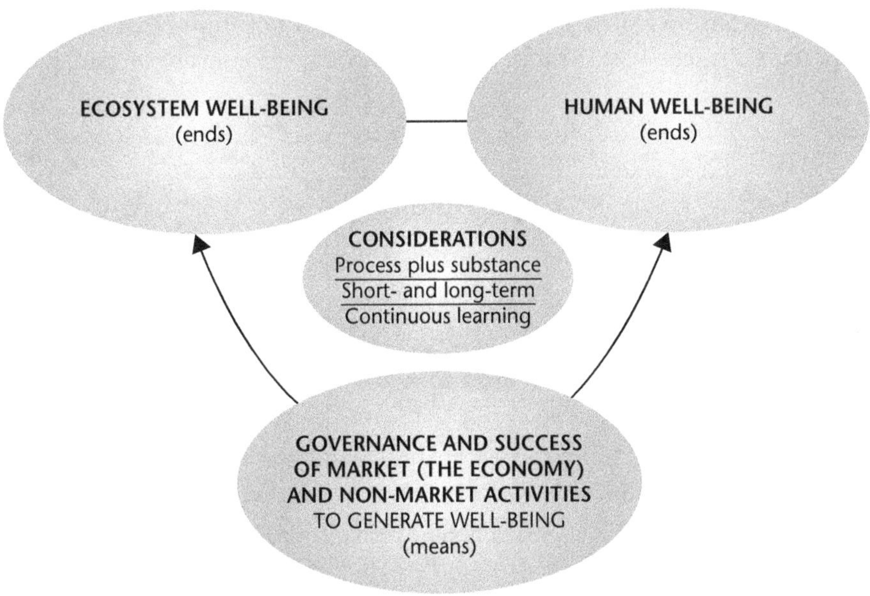


**Figure 18.1.** Conceptual framework for assessing contribution
*Source*: Hodge (2014a).

For any organization (company, community, civil society organization, government) or society, each of these is important for different reasons:

1. *inputs* to establish commitment for action and set the needed level of resources (budget)
2. *outputs/outcomes* to track the success or not of meeting commitments and holding those responsible to account
3. *results/ends* to test whether the overall goal and the specific objectives are being met—and if not, to signal a need for adjustments.

Combined with an effective feedback mechanism, these processes together facilitate continuous learning and performance improvement so efficiencies can be achieved over time.

In contribution analysis, the concept of *sustainability* is seen to be much more than environmental protection in another guise. It is a positive concept that addresses both: (1) achieving well-being for people and ecosystems; and (2) reducing stress or mitigating impacts.[4] Or, as articulated by Gibson et al. (2006: 165), 'the aim is to deliver multiple, lasting, mutually re-enforcing gains rather than just mitigation of environmental damage'. Dietsche (this volume) similarly calls for a transition to focus on the positive, not just the negative.

Figure 18.2 translates the concepts of Figure 18.1 into the form of a high-level hierarchical system. It maps elements from the most general (above) to more detailed (below). At the lowest level, each element can be further disaggregated. Note that such a hierarchical form does not imply that the most important element is at the top. Rather, following the concept of 'panarchy'

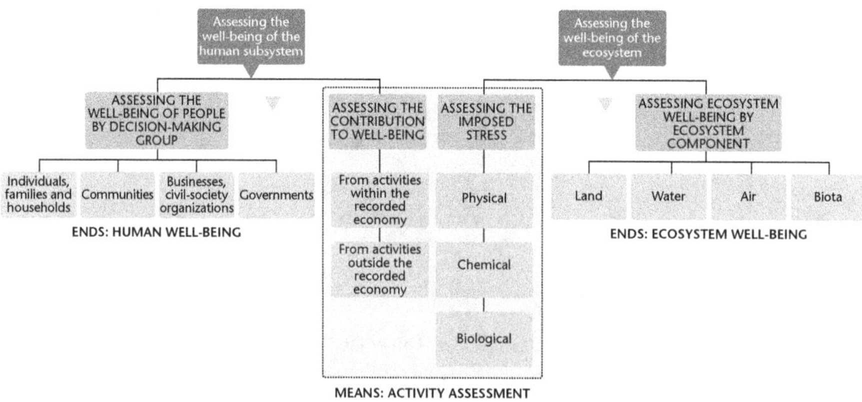


**Figure 18.2.** Assessing contribution—hierarchy of data and information
*Source*: author's illustration.

[4] Hodge (1995); Hodge and Taggart (1992); Hodge et al. (1995); NRTEE (1993); Prescott-Allen (2001).

proposed by ecologists Gunderson and Holling (2009), the most important element at any given time may turn up anywhere in the 'hierarchy'. For project-level application, the above general conceptual approach translates into four questions:

#### 18.2.1.1 MEANS ASSESSMENT

1. What is the project's contribution to human well-being over the full project life cycle, taking into account the full suite and distribution of benefits, costs, and risks?
2. What is the project's contribution to ecosystem well-being over the full project life cycle, taking into account the full suite and distribution of benefits, costs, and risks?

#### 18.2.1.2 ENDS ASSESSMENT

3. What is the resulting state and distribution of human well-being (recognizing that there are other contributors)?
4. What is the resulting state and distribution of ecosystem well-being (recognizing that there are other contributors)?

Figure 18.2 diaggregates these four questions in a high-level hierarchy of data and information that must be considered. In fact, this hierarchy is generic and would apply to all scales and foci of analysis, though the details of each would vary depending on the scale, boundaries of enquiry, and application.

Human well-being is categorized by 'decision-makers'.[5] This approach is particularly useful because: (1) if the well-being of these four decision-making groups is high or is threatened together or individually, it closely reflects the overall state of human well-being (regardless of scale from local through regional to national); (2) each of these groups or interests (to varying amounts depending on a given society) plays a critical role in bringing change and improvement; and (3) data and information are often gathered and stored by these categories.

Ecosystem well-being is disaggregated in four parts: (1) land, (2) water, (3) air, and (4) biota (living organisms). This categorization is useful in that it reflects the regulation of human activity and the organization of many relevant departments of government, business, and civil society organizations. However, ecologists have rightly pointed out that such a compartmentalization of the ecosystem does not effectively capture the state of structure,

[5] This categorization was first suggested by Alex Michalos (personal communication 1992), the founding editor (in 1972) of the *Journal of Social Indicators Research*. It was adopted by the Sustainable Development Reporting Task Force of the Canadian Prime Minister's National Round Table on the Environment and the Economy (NRTEE 1993).

function, and diversity. Nor does it address systemic flows (e.g. of energy, nutrients, and materials) within the ecosystem or the holistic nature of the ecosystem itself. It is thus deficient from an ecological perspective. Another option is to use this categorization (to facilitate a link to governance and decision-making) but include an additional form of analysis and synthesis that speaks to ecologists' concerns.

The third, means or activity assessment, captures the benefits, costs, and risks produced by the project. Here it is articulated as both 'support for' well-being and 'stress on' well-being. Human activities always generate some balance of 'support' and 'stress' (the positive and the negative) that add up to some degree of contribution to human and ecosystem well-being. Not to be forgotten, however, is that the natural ecosystem not only provides the nourishment needed by human society but also generates significant stress from extreme natural events such as drought, extreme rainfall, earthquakes, tsunamis, volcanic eruptions, and explosive diseases (Berger 2007).

Assessing benefits includes not only the typical measured quantities (jobs and taxes) but also the citizen/societal perspective on the very fabric of their family and community lives. Many of these aspects are outside the market economy but contribute massively to well-being. Examples include housework, volunteer/not-for-profit activities, cultural activities, and so on that may be touched (positively or negatively) by the project. A useful description of the nature and richness of volunteer/not-for-profit activities is provided by the United Nations/Johns Hopkins University classification of not-for-profits (United Nations 2003).

An approach that greatly facilitates navigation of this maze can be borrowed from the auditing profession who use concrete and specific questions for which answers can be sought in practical terms. This approach is championed in the 2003 multi-interest work *Seven Questions to Sustainability: How to Assess the Contribution of Mining and Metals Activities* (MMSD North America 2002). Figure 18.3 provides the high-level seven-question framework that emerged in the MMSD North America work.

For each question, an ideal answer is offered and then data and information sought that facilitate assessing how close to the ideal is being achieved. A full detailing of the approach described above is provided in MMSD North America (2002), including a catalogue of various sub-elements, indicators, and metrics. A synthesis of the answers to all seven questions can then serve as the foundation for a multi-interest assessment of contribution based on overlapping values.

To be effective, the process of assessing contribution needs to be ongoing from project inception through the full life cycle, not simply to be undertaken once to obtain a legal or social licence to operate. A concrete assessment and reporting schedule is required that matches the project time horizon. This

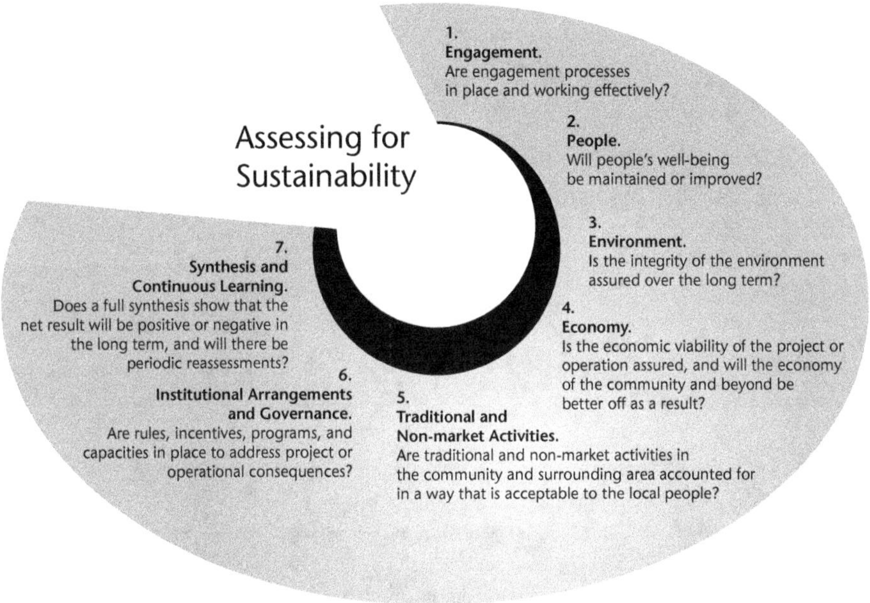


**Figure 18.3.** The MMSD North America framework: 'seven questions to sustainability—how to assess the contribution of mining and minerals activities'
*Source*: MMSD North America (2002); reproduced here by permission.

ongoing process of assessing and reporting facilitates the continuous learning and adjustment that is so essential—see Buckley, McCulloch, and Travis (Chapter 27, this volume).

Figure 18.4 shows the mine project life cycle by phase and level of activity. The different phases are marked by significant variations in activity levels and different implications for human and ecosystem well-being, all of which needs careful and rigorous consideration. It is only since the millennium change that this full project life cycle has been integrated into the thinking of the industry and regulators.

### 18.2.2 *Process*

Assessing process effectiveness is an essential part of contribution analysis. Below, the idea of 'overlapping consensus', introduced by Rawls (1971, 1987) some forty years ago, is offered as a conceptual starting point for doing so. In the context of a mining operation–community relationship, a state of overlapping consensus would be achieved if community, company, and government each agreed on the same way forward based on their own perspective and moral position (Wenar 2013). In other words, it would be a consensus based on overlapping values.

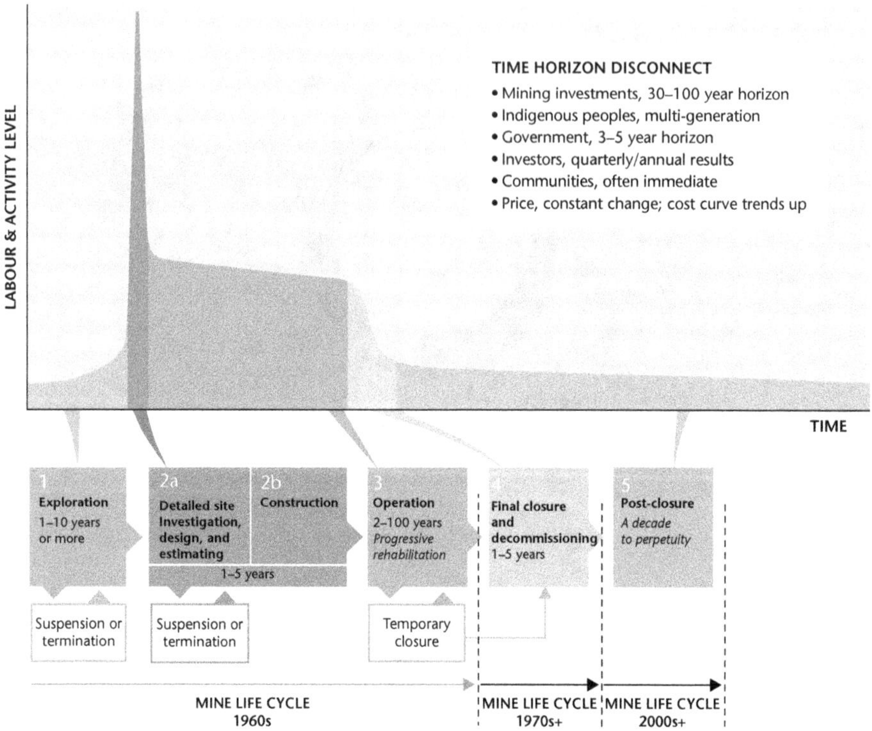


**Figure 18.4.** The mine project life cycle with indicative activity levels
*Source*: Hodge (2011), ICMM (2012); reproduced here by permission.

Because it is based on each interest's own beliefs and values, it is each interest's best option. Such a consensus stands in stark contrast to a compromise that arises through an explicit trade-off or 'deal' in the face of others' power. The weakness of such deals is that a power shift can trigger a loss of social stability if the position of one of the parties to the 'deal' changes. Such a shift can occur when a change in ownership of a mining property or operation leads to reneging on company–community agreements made between the community and the former owner.

In summary, a values-based, overlapping consensus offers greater and longer-lasting stability and security, exactly what applied sustainability concepts call for. It is true that an ideal state of overlapping consensus will not always be possible to achieve or, once established, to endure. However, it is the best route to social stability that a free society can hope to attain.

In practice, success can be assessed by comparing the reality on the ground with this Rawlsian ideal. A scale can be created with the 'ideal' providing an end-point and current success scored accordingly. The ideal may not be achieved, but continuous improvement towards that ideal can be sought as learning takes place. Clearly the scaling process emerges as a key to implementation.

Undertaking such an assessment collaboratively with all interests (community, company, government thus bringing different values to the table) provides a special opportunity to develop a sense of respect, trust, integrity, fairness, justice, authenticity, and overall security for community, company, and government that the right thing will be done in the right way. In short, the assessment process itself can be used to re-enforce social stability and strengthen a sense of security for all interests.

Relationships and values evolve over time. Thus, a collaboratively-built, values-based assessment is not a task that can be theoretically addressed in a licensing hearing and then set aside when approval is received. Continuous effort by all parties through the full project life cycle is essential.

Examples of successful ongoing engagement can be found in ICMM's 'Mining Partnerships for Development' initiative (ICMM 2016). Another is the Ahafo Social Responsibility forum in Ghana, now in its tenth year of seeking overlapping consensus in mining-related issues (Anderson, personal communication, 2016). See also discussion in McPhail (Chapter 17, this volume). Not surprisingly, broad consensus-seeking engagement of key interests is one of the success factors that has emerged in achieving a 'social licence to operate'.[6] Effective engagement is now a central component of most mining law and regulation; it is a key part of every progressive company's policies. At the same time and despite powerful examples of success, many examples of ineffective engagement processes across the mining industry continue to occur. An evolving compilation of examples of good and bad practice would greatly facilitate learning.

## 18.3 Generic Assessment Cycle

In recent years there has been much emphasis devoted to generating clear *measured* indicators of change. This interest has been felt within many aspects of human endeavour, including mining. It has been reinforced by the common saying 'you manage what you measure, so best measure what you want to manage'. In fact, the real challenge is not simply to measure but to understand and track change or success with rigour and integrity using all the means available—including but not limited to measurement. And while there is

[6] The 'social licence to operate' is a concept first proposed by James Cooney, vice president of Placer Dome Inc. in 1997. See Joyce and Thomson (2000) for an early treatment of the proposition and Owen and Kemp (2013) for a useful cautionary critique.

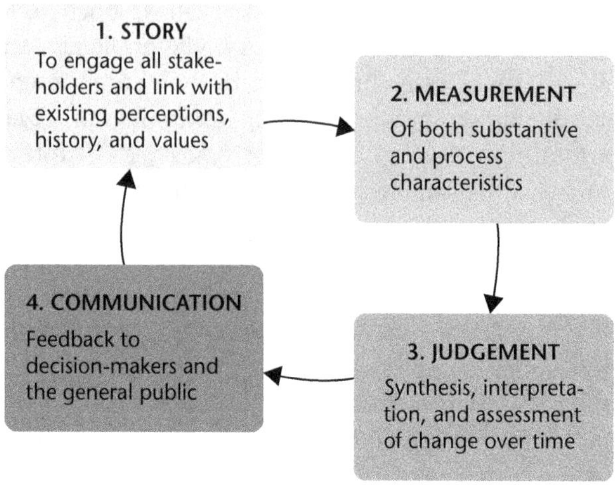


**Figure 18.5.** Generic assessment cycle
*Source*: Hodge (2006); reproduced here by permission.

power in measurement, measured indicators serve as only one element of a more complex generic assessment cycle (Hodge 2006). Figure 18.5 shows that cycle and its four essential elements: (1) story; (2) measurement and indicators; (3) judgement; and (4) communication.

### 18.3.1 *Story*

Many aspects of contribution have not been or cannot be measured. But they are known about by those who carry the institutional and community memory. Real insight emerges through the 'story' that these people can offer. And importantly, seeking the stories of various interests offers a powerful and respectful means of engagement.

Over the past several decades, there has been a major resurgence of interest in story as a key element of policy development, decision-making, and organizational behaviour—for example, see Denning (2001) and Fischer and Forester (1993). However, formal quality control techniques of 'story' remain less developed than for measurement.

Gathering stories requires strong listening and hearing skills, often intercultural in scope. These capacities have not typically been strengths of the mining industry. In general, the technically oriented people who comprise most of mining industry personnel have a high comfort level with numbers and counting, and a low comfort level with storytelling, particularly when the issues to be described range into social, cultural, political, economic, and environmental concerns that are outside their immediate training—but critical to understanding the values of community and government, and therefore critical to any operation.

In many communities, the multi-generational livelihoods brought by mining are important not only because of jobs, wages, and cash contributions to local charities and organizations, but more importantly because of the stability, respect, and confidence they bring to families and the community during and after the life of the mine. There are many examples of where this strengthening of community fabric and the psychological security based on hope in the future combined with the absence of fear far outweighs the significance of what may be only short-term mining income. While this side of mining's contribution defies capture in easily measured indicators, over the long run it may be more important than all the 'quantities' that are typically measured in the annual reporting cycle demanded by managers and investors. To get a sense of this part of mining's contribution requires building relationships with people so they will share their true feelings based on a mutual sense of respect and integrity.

At the same time, storytelling includes the dark side—the negative experiences. And mining, like every facet of human activity, has its share of negatives. These might include loss of life from accidents or poor design, labour strife and violence, environmental disaster, unintended contribution to a substance-abuse culture or the spread of sexually transmitted disease, the wrench of ill-prepared-for closure, or overall community discontent about lost land uses, displacement of people from their lands, unassigned responsibilities, lack of accountability, and unfairness in the distribution of mining-related benefits, costs, and risks. By honestly sharing the bad along with the good stories, a sense of honesty and integrity can be earned that contributes to finding solutions to these challenges while building the long-term trust that is so essential. A significant barrier to such openness and transparency is the attitude of legal advisers that any admission or apology that might infer liability—even by association—is to be avoided.

### 18.3.2 *Measurement*

In the broad sweep of history, trying to capture the nature of change over time has been a topic of interest since at least the ancient Greeks. However, the last half of the nineteenth century and through the twentieth century saw an increasing focus on counting. Lewis Mumford (1934) voiced a concern about this when he wrote:

> The new attitude toward time and space infected the workshop and the counting house, the army and the city. The tempo became faster, the magnitudes became greater; conceptually, modern culture launched itself into space and gave itself

> over to movement. What Max Weber called the 'romanticism of numbers' grew naturally out of this interest. In time-keeping, in trading, in fighting, men counted numbers, and finally, as the habit grew, only numbers counted.

Even Albert Einstein added a note of caution with a sign hung on his Princeton University office door that read: 'Not everything that can be counted counts and not everything that counts can be counted.'

The specific role of 'measured' indicators is to provide a firm 'backbone' for assessing contribution. They serve to: (1) confirm and give confidence to the observations that are often first recognized through 'story', and (2) provide counter-intuitive insight that can trigger whole new perspectives. However, the overall power of counting can only be realized by seeing it as part of the larger generic cycle that is needed to recognize, track, record, and understand change and, in the context of this chapter, the contribution of mining and metals activities.

### 18.3.3 *Synthesis and Judgement*

In tracking contribution, whether the scale be that of the project, organization, community, industry, or political jurisdiction, the greatest challenge is to bring together the power of both qualitative and quantitative analysis—story and measurement—in a synthesis that can be respected across all interests.

Even when 'measured' indicators have been compiled and nested in the story of place or issue, the task remains of weighing the elements and judging the significance of the gathered evidence. This means establishing the relative significance of both measured indicators and articulated story and combining the result in a synthesis that allows an overall assessment of contribution and progress. This task is far from trivial. It is the classic problem of combining apples, oranges, and socks.

Gibson et al. (2006: 165–79) offer a set of criteria drawn mainly from the environmental assessment literature (which includes social and economic as well as ecological effects) for assessing the significance of a project's effects. However, they point out that little of the available literature on this topic has been developed with sustainability criteria in mind.

There is much relevant experience in courts of law across the world where significance must be assigned to different pieces of evidence and those pieces weighed to draw a judgement.[7] Critically, 'reasons for decisions' are carefully developed and made public to serve as a continuously growing body of knowledge from which to draw. In this way, continuous learning concepts can be

[7] A similar 'weight-of-evidence' approach was adopted long ago by the government of Canada for addressing a broad range of social and environmental issues; see Canada (1993).

applied to the assessment processes. An important result is the enhancement of people's sense of integrity in and trust for such assessment processes.

### 18.3.4 *Communication*

All is for naught without effective communication of assessment results. Communication is a two-way process. So the starting point of this is *listening*, not telling—listening to the stories, listening to what is important to others and, through that, developing a sensitivity to others' culture and values. If that is done, the chances of being able to then *tell* in a way that will be heard is greatly enhanced.

### 18.3.5 *Explicitly Bringing Values into 'Technical' Assessment and Decision-making*

The generic assessment cycle described above is dependent on careful and rigorous application of each component. An ongoing challenge, however, is the integration of values—often those held by indigenous people, community members, and government representatives—into the technical assessment and decision-making process. One powerful technique available to do this is 'multi-attribute utility analysis' which comes from the field of 'Decision Analysis'. It has been taught in engineering and operations research programs for many years and applied in major industries to best characterize risks when making high-stakes decisions. However, it has only rarely been used to address sustainability-driven decisions and the kind of contribution analysis championed in this chapter. An example application involving an assessment of closure options of a major mine project is detailed in IRP (2007) and Hodge and Merkhofer (2008), and summarized in Hodge (2017: Box 3). A second example is detailed in Ben-Eli et al. (2004) which describes how alternatives were assessed for managing high-level nuclear waste over the long term in Canada.

The technique involves careful definition of objectives, specification of alternatives, identification of the hierarchy of factors, uncertainties, and relationships that influence the performance of each alternative under each objective, rigorous assessment of performance using carefully crafted performance scales, weighting the importance of each objective, synthesizing the result, and lastly, completing a sensitivity analysis. The whole process is completed collaboratively and may include periodic 'negotiations'. It is a powerful approach to assessment and decision-making that effectively combines the facts, judgements, and probabilities provided by the technical experts with the value judgements provided by the participating interests.

### 18.3.6 *Need for an Interdisciplinary Team*

Each aspect of the assessment cycle requires a special skillset. Further, those expert in story may not be expert in measurement, and neither may be proficient in synthesis, weighing evidence, and making judgements about significance. Further, none of these groups are necessarily effective at communicating. The inevitable conclusion is that a multidisciplinary team is essential for success.

## 18.4 The Tough Challenge of Attribution

A mining/metals operation is one of many contributors to the state of social, cultural, economic, and ecological conditions at a given location. Other industries (large and small), government facilities, and households all play a role. In fact, a significant issue that commonly arises is when boundaries of responsibility between community, company, and government become blurred. When there is such confusion and multiple interests share responsibility for social and ecological conditions, a central question is then how to apportion contribution—positive and negative—among interests.

For measurable indicators such as the following, clear-cut answers to the attribution issue can be established, provided accurate statistical systems are in place and maintained:

- standard project key performance indicators—production cost numbers relative to the expected mine life, financial numbers tracked by company management
- supplies and services drawn from the local community and region
- employment numbers (primary, secondary, tertiary)
- household members associated with employees and their participation in community activities
- payment of taxes, royalties, licences, and fees
- contributions to local infrastructure, charities, and a range of civil society organizations (sport, cultural, recreational, advocacy, education)
- discharges of contaminants to surface or groundwater
- loss of biodiversity in a region.

Unfortunately, the above numbers do not get at the feeling of a community about the presence of a mining project, their sense of fairness about the distribution of benefits, costs, and risks, and, most importantly, the feelings of a community about itself—its own sense of current and future security and well-being. This part of contribution can only be accessed by talking to people and drawing out their stories in a way that is acceptable to local cultural norms.

On this front, all interests—company, community, and governments—are at an early stage of learning about how best to proceed. There is no formula to draw on for apportioning contribution on this front when it is essential to do so—only the use of common sense and courtesy in processes of open and collaborative assessment that are marked on all sides by respect and integrity.

## 18.5 Steps towards Contribution Analysis

This section describes three concrete examples of steps that have been taken towards contribution analysis.

### 18.5.1 *The 1997–2002 Assessment and Approval of the Labrador-based Voisey's Bay Nickel Project Based on Assessment of Projected Contribution to Sustainability*

Assessment and approval of the massive Voisey's Bay nickel project occurred through a landmark environmental assessment process. For the first time, 'contribution to sustainability' was used as the basic test of acceptability. Details of this process are provided in Voisey's Bay Mine and Mill Environmental Assessment Panel, 1997. The panel determined that progress toward sustainable development would require:

1. the preservation of ecosystem integrity, including the capability of natural systems to maintain their structure and function and to support biological diversity
2. respect for the right of future generations to the sustainable use of renewable resources
3. the attainment of durable and equitable social and economic benefits.

Adoption of the sustainability-based decision criteria based on achieving a net positive contribution changed how the main issues were addressed, how the project was designed, and what was approved. Further, it shifted the focus from the mitigation of negative effects during the life of the mine to attaining net gains over the long term. Critically, it entrenched a collaborative and ongoing process for monitoring and reporting on project performance against the sustainability-based criteria. Now in 2018, the project is well into operation. A strong foundation has been established for long-term success, but the critical test will come when it enters the closure and post-closure phases.

### 18.5.2 *A First Nation's Assessment of Mining's Contribution: Past, Present, and Future*

In the spring of 2003, the Tahltan First Nation, whose territory covers some 100,000 km$^2$ in north-western British Columbia set out to:

1. assess the relationship—past (say, fifty years ago), present, and sought future (including specific actions for various interests)—between themselves, their land, and the mining industry, taking into consideration activities on their traditional territory from exploration through operation, closure, and post-closure
2. build a strategy to guide that relationship in the future.

To guide this assessment, they used the 'seven questions to sustainability' assessment framework (MMSD North America 2002) described in Figure 18.3. The full process and results are described in Tahltan Nation and the International Institute for Sustainable Development (2004).

The facilitated, but community-driven, process involved collaboratively assessing past, current, and sought future experience for each of (1) exploration, (2) operation, and (3) closure/post-closure activities in (or projected to be within) their traditional territory. Available data and experiential anecdotes of elders and others in the community where compiled as a starting point for the assessment.

The Tahltan Nation made it very clear that they were prepared to support mining activities if: (1) mining interests and government respected their values and concerns; and (2) a fair distribution of benefits, costs and risks, responsibilities, and accountabilities could be achieved. The collaborative assessment process combined with the seven-question framework provided a means to fairly and openly treat not only direct and indirect employment and local procurement of services, but also more sensitive and difficult issues that included:

1. innovative collaborative approaches to the overall management/co-management of mining-related activities
2. how to move towards a fairer distribution (considering all participating interests) of all benefits, costs, and risks
3. more effective approaches for addressing environmental and health/social/cultural implications of mining/mineral activity that in their view continue to receive inadequate attention.

### 18.5.3 *Strengthening Mining's Contribution to Poverty Alleviation*

In November 2015, the United Nations General Assembly adopted seventeen Sustainable Development Goals. Highest priority among these is poverty alleviation. This example weaves together threads of two thought processes that focused on strengthening mining's contribution to poverty alleviation.

In September 2013, the Vatican's Pontifical Council on Peace and Justice convened a day of reflection involving some twenty-five individuals drawn from the mining world and fifteen from the Church and related civil society. Stemming from this initiative, the Integrity of Creation Working Group, of the Justice, Peace and Integrity of Creation Commission of the International Union of Superiors General, convened a second meeting a few days later. It involved some twenty individuals representing fifteen religious orders of the Catholic Church along with the president of the ICMM. In both cases, the focus of discussion was on how the contribution of mining could be strengthened for host communities and nations (Hodge 2014).

The Integrity of Creation Working Group met again in May 2014, this time focusing entirely on the priority issue of mining's role in poverty alleviation. They did so recognizing: (1) the potential significant and positive results of a contribution from mining on this front; and (2) a sense that to gain trust, the mining industry needs to send a concrete signal that it is not simply continuing 'business-as-usual' on the poverty alleviation front, despite many fine words being expressed to the contrary.

Simultaneously with these events, Jim Cooney, a former vice president of Placer Dome Inc. (a large Canadian gold-mining company that subsequently merged with Barrick Inc.) was also reflecting deeply on mining's role in poverty alleviation. He recognized that poverty has both an absolute and a relative dimension. And while mining contributes demonstrably to economic growth in host countries and local areas of operations, it also tends to widen the income gap. This is because mine-generated wealth is more likely to be directed into the hands of those with the greatest ability to capture and capitalize on that wealth. For their part, not only are the poor less able to take advantage of it, but the greater numbers of poor compared to rich means that the share of wealth that is captured is diluted. In short, the rich get richer faster than the poor get richer. This widening income gap can lead to a sense of growing poverty among the poor (Cooney 2014).

Cooney outlined four simple steps for mining companies that would greatly strengthen mining's contribution to poverty alleviation:

1. Develop a comprehensive and in-depth understanding of poverty around their operations.
2. Work collaboratively with others to narrow the income gap in local communities and seek a more equitable and sustainable distribution of the benefits from mining.
3. Regularly assess the changes in the income gaps in local communities, and if widening is evident, adopt corrective interventions.

4. Embrace a Poverty Action Plan for the marginalized, working with local civil society institutions that can assert and advance the best interests of the marginalized.

The Integrity of Creation Working Group adopted these steps as the core of an action plan. In addition, they called for: (1) ensuring that all subsidiaries and service providers were included in the envelope of any given operation, and (2) industry participation in the four-year, nation-by-nation United Nations' Universal Human Rights Periodic Review (Integrity of Creation Working Group 2014).

Strengthening mining's contribution to poverty alleviation rather than serving to exacerbate it is central to the idea of sustainability. The action plan itself is nothing more than what any company would do to address other priorities.

## 18.6 Progress Made, Progress Needed

The examples in Section 18.5 provide strong evidence for the evolution underway towards the use of an integrated and ongoing assessment of success at strengthening mining and metals' contribution to human and ecosystem well-being. The elements needed for assessing contribution are available, the conceptual foundation is clear, and effective practice of various steps along the way has been demonstrated. In short, much progress has been made. However, the full step has not yet been taken. The task urgently needs addressing.

Most companies, communities, and governments are not yet driving decision-making with an assessment of the full contribution of a project through its full life cycle, even though the practical benefits of doing so are clear. Instead, they continue to focus their attention on the identification and mitigation of 'impacts'—the negatives of human activity—generally at a fixed point in time such as at licensing or when an accident occurs. Bizarrely, it is like making a cost–benefit analysis thinking only about part of the costs. This limited perspective is contributing to the ongoing tensions that exist between communities, companies, and government.

The chapter makes the following seven assertions:

1. Contribution assessment must be driven by the interlinked ideas of sustainability and sustainable development. At the core is a value set best described as parallel care and respect for people and the enveloping ecosystem, particularly over the long term.
2. Following ideas of results-based management, assessing contribution should similarly be based on understanding and tracking: (1) human

well-being (result), (2) ecosystem well-being (result), and (3) the support generated, and the stress imposed, by the activity on human and ecosystem well-being (means).

3. To be consistent with the ideas of sustainability and sustainable development, a project should achieve a net positive contribution to human and ecosystem well-being over the long term.
4. At an operational level, achieving a 'net positive contribution to human and ecosystem well-being over the long term' should be adopted as the fundamental two-dimensional criteria for designing, constructing, operating, and closing mining operations. Thus, it should be similarly used as the test for project approvals and ongoing success. The shift will require a change of culture on all sides.
5. For effective assessment of contribution, a generic assessment cycle should be used involving compilation of insight from both story and measurement, a rigorous form of synthesis and judgement, and careful communication that includes listening and respectful sharing of results.
6. Contribution has both substantive and process dimensions. Both must be addressed.
7. Contribution analysis for mining must span the full project and product life cycles. Significant unknowns are inevitable; a commitment to continuous learning and adaptive management is essential.

Extractive industries are *bridging activities* that are taking us from now into the future (MMSD North America 2002). They are not 'sustainable' in and of themselves but their contribution to human and ecosystem well-being should be—and needs to be tested for, tracked over time, and with so many unknowns, a learning culture instilled if we are to achieve the kind of insights for low- and middle-income countries that this project is seeking.

## References

Anderson, C. J. (2016). Personal communication. Principal, Yirri Global LLC, Greenwood Village, Colorado.

Ben-Eli, M., J. Neate, J-A. Facella, A. Hodge, T. Isaacs, W. Leiss, M. Margolick, K. Moshanas Cole and F. Roots (2004). 'Assessing the Options: Future Management of Used Nuclear Fuel in Canada', NWMO Assessment Team Report, June. Toronto: Nuclear Waste Management Organization. Available at: https://www.nwmo.ca/~/media/Site/Files/PDFs/2015/11/04/17/33/1092_9-1assessingtheoptions_nwmoass.ashx?la=en.

Berger, A. R. (2007). 'Where Is Sustainability When Landscapes Change Rapidly?' LESTARI Public Lecture. Bangi: Institute for Environment and Development (LESTARI), University of Kebangsaan, Malaysia.

Brandt Commission on North–South Issues (1980). *North–South: A Programme for Survival*. London: Pan Books.

Canada (1993). *Canada's Response to Recommendations in the Sixth Biennial Report of the International Joint Commission*. Ottawa: Minister of Supply and Services Canada. Available at: www.ijc.org/files/tinymce/uploaded/documents/reportsAndPublications/Canada_s%20Response%20To%20The%20Sixth%20Biennial%20Report.pdf.

Cooney, J. P. (2014). 'Mining within the Context of a "Preferential Option for the Poor"', paper presented at the 2014 Annual Meeting of the Canadian Institute of Mining and Metallurgy (CIM), March.

Daly, H. E. (1989). 'Sustainable Development: From Concept and Theory towards Operational Principles', paper presented at the Hoover Institution Conference, Population and Development Review. Also published in H. Daly (1991). *Steady State Economics*, 2nd edition. Washington, DC: Island Press.

Denning, S. (2001). *The Springboard: How Storytelling Ignites Action in Knowledge-era Organizations*. Woborn, MA: Butterworth-Heinemann.

Fischer, F. and J. Forester (1993). *The Argumentative Turn in Policy Analysis and Planning*. Durham, NC: Duke University Press.

Gibson, R. (2000). 'Favouring the Higher Test: Contribution to Sustainability as the Central Criterion for Reviews and Decisions under the Canadian Environmental Assessment Act', *Journal of Environmental Law and Practice*, 10(1): 39–54.

Gibson, R. (2002). *Specification of Sustainability-based Environmental Assessment Decision Criteria for Determining Significance in Environmental Assessment*. Hull, Quebec: Canadian Environmental Assessment Agency.

Gibson, R. with S. Hassan, S. Holtz, J. Tansey, and G. Whitelaw (2006). *Sustainability Assessment: Criteria and Processes*. London: Earthscan.

Gunderson, L. H. and C. S. Holling (2009). *Panarchy: Understanding Transformations in Human and Natural Systems*. Washington, DC: Island Press.

Hodge, R. A. (1995). 'Assessing Progress toward Sustainability: Development of a Systemic Framework and Reporting Structure', PhD dissertation, School of Urban Planning, Faculty of Engineering, McGill University, Montreal.

Hodge, R. A. (2006). 'Tracking Progress toward Sustainability: Linking the Power of Measurement and Story', *Mining Engineering*, 58(9): 63–8.

Hodge, R. A. (2011). 'Mining and Sustainability', in P. Darling (ed.), *SME Mining Engineering Handbook*, 3rd edition, 1665–88. Englewood, CO: Society for Mining Metallurgy and Exploration Inc.

Hodge, R. A. (2014). 'Bridging the Language of Mining, Sustainability, and the Social Doctrine of the Catholic Church', unpublished discussion paper prepared for the Integrity of Creation Working Group and the Pontifical Council for Justice and Peace, The Vatican. London: ICMM.

Hodge, R. A. (2017). 'Towards Contribution Analysis', WIDER Working Paper 2017/19. Helsinki: UNU-WIDER.

Hodge, R. A. and L. Merkhofer (2008). 'Faro Mine Closure, Assessing the Alternatives: An Application of Multi-attribute Utility Analysis', Final Report of the Faro Closure Assessment Team. Whitehorse: Assessment and Abandoned Mines Branch Department of Energy Mines and Resources, Government of Yukon.

Hodge, R. A. and I. Taggart (1992). 'Reporting on Sustainability: Human Well-being within Ecosystem Well-being', unpublished report prepared for the Policy and Research Committee, Ontario Round Table on Environment and Economy, Toronto.

Hodge, T., S. Holtz, C. Smith, and K. Hawke-Baxter (eds) (1995). *Pathways to Sustainability: Assessing Our Progress*. Ottawa: National Roundtable on the Environment and the Economy. Available at: http://www.publications.gc.ca/collections/collection_2016/trnee-nrtee/En134-2-11-1995-eng.pdf.

Humphreys, D. (2015). *The Remaking of the Mining Industry*. Basingstoke: Palgrave Macmillan.

ICMM (2016). 'Mining: Partnerships for Development Toolkit'. Available at: www.icmm.com/mpd.

Integrity of Creation Working Group (2014). 'Statement on the Impacts of Mining: Solidarity for the Common Good and Poverty Reduction', unpublished note. Rome: Integrity of Creation Working Group.

IRP (2007). 'Review of the Remediation Alternatives for the Anvil Range Mine Complex', final report prepared for Deloitte and Touche Inc., Faro Mine Closure Planning Office (FMCPO), Indian and Northern Affairs (INAC), Yukon Government, Selkirk First Nation at Pelly Crossing, Kaska Tribal Council represented by the Ross River Dena Council. Whitehorse: IRP.

Joyce, S. and I. Thomson (2000). 'Earning a Social Licence to Operate: Social Acceptability and Resource Development in Latin America', *Canadian Mining and Metallurgical Bulletin*, 49(1037): 49–53.

MMSD (2002). *Breaking New Ground: Mining, Minerals and Sustainable Development*. London: IIED.

MMSD North America (2002). *Seven Questions to Sustainability: How to Assess the Contribution of Mining and Metals Activities*. Winnipeg: International Institute for Sustainable Development. Available at: www.iisd.org/pdf/2002/mmsd_sevenquestions.pdf.

Mumford, L. (1934). 'The Monastery and the Clock', in D. L. Miller (1986). *The Lewis Mumford Reader*. New York: Pantheon Books.

NRTEE (1993). 'Toward Reporting Progress on Sustainable Development in Canada: Report to the Prime Minister'. Ottawa: National Round Table on the Environment and the Economy. Republished as Part I in Hodge, T., S. Holtz, C. Smith, and K. Hawke Baxter (1995). *Pathways to Sustainability: Assessing Our Progress*. Ottawa: National Round Table on the Environment and the Economy.

Owen, J. R. and D. Kemp (2013). 'Social Licence and Mining: A Critical Perspective', *Resource Policy*, 38(1): 29–35.

Palme Commission on Security and Disarmament (1982). *Common Security: A Programme for Disarmament*. London: Pan Books.

Prescott-Allen, R. (2001). *The Wellbeing of Nations: A Country-by-Country Index of Quality of Life and the Environment*. Washington, DC: Island Press.

Rawls, J. (1971 [1999]). *A Theory of Justice*, revised edition. Cambridge, MA: Harvard University Press.

Rawls, J. (1987). 'The Idea of an Overlapping Consensus', *Oxford Journal of Legal Studies*, 7(1): 1–25.

Resolve Solutions Network and the World Economic Forum (2016). 'Voluntary Responsible Mining Initiatives: A Review'. Geneva: WEF Forum. Available at: http://www3.weforum.org/docs/Voluntary_Responsible_Mining_Initiatives_2016.pdf.

Robinson, J. G., G. Francis, R. Legge, and S. Lerner (1990). 'Defining a Sustainable Society: Values, Principles, and Definitions', *Alternatives*, 17(2): 36–46.

Tahltan Nation and the International Institute for Sustainable Development (2004). 'Out of Respect: The Tahltan, Mining, and the Seven Questions to Sustainability', report of the Tahltan Mining Symposium, 4–6 April 2003. Winnipeg: International Institute for Sustainable Development. Available at: www.iisd.org/pdf/2004/natres_out_of_respect.pdf.

United Nations (2003). *Handbook on Non-profit Institutions in the System of National Accounts*. New York: United Nations. Available at: http://unstats.un.org/unsd/publication/seriesf/seriesf_91e.pdf.

Voisey's Bay Mine and Mill Environmental Assessment Panel (1997). 'Environmental Impact Statement Guidelines for the Review of the Voisey's Bay Mine and Mill Undertaking, 20 June 1997'. Ottawa: Canadian Environmental Assessment Agency. Available at: www.ceaa-acee.gc.ca/default.asp?lang=En&n=0A571A1A-1&offset=2&toc=show.

WCED (1987). *Our Common Future: Report of the World Commission on Environment and Development*. London: Oxford University Press.

Wenar, L. (2013). 'John Rawls', in E. N. Zalta (ed.), *The Stanford Encyclopedia of Philosophy*, winter 2013 edition. Available at: http://plato.stanford.edu/archives/win2013/entries/rawls.

# 19

# The Role of Governance and International Norms in Managing Natural Resources

*James Cust*

## 19.1 Introduction

Countries rich in natural resources face a series of development challenges. As custodians of resource wealth, they must make difficult policy choices to maximize the benefits for citizens. However, many of these choices are shaped or constrained by the institutional or governance context.[1] The high rents often associated with resource extraction place a premium on effective policy but can also act to exacerbate weaknesses in governance. The resource curse describes what can happen when the interaction of governance and resource wealth goes wrong.[2]

Good or bad governance has great repercussions for poverty and inequality within resource-rich countries. However, it can also have spillover effects, affecting governments, companies, investors, and citizens both inside and outside of producing countries. These externalities give rise to a public goods problem—good governance of natural resources may have a larger benefit than the private benefit captured by any national government or single administration. Further, according to various theories of the resource curse, the private benefits accruing from conducting bad governance may be greater still for a small group of elites.

[1] This chapter follows the definition of institutions and governance used by much of the literature on this topic, from North (1990); as institutions defined as the rule of the game in a society, constituting laws, norms, policies, and codes of conduct.

[2] I thank Tony Addison, Alan Roe, Alexis Ballesteros, David Manley, David Mihalyi, and participants of the UNU-WIDER 'Extractives for Development' workshop in Helsinki, May 2016, for helpful comments, and staff at NRGI for their inputs.

In response to these externalities a variety of international norms and initiatives have emerged. They cover a range of aspects of resource governance, but typically have commonality in their aim to codify successful historical experience and best practices around effective policy design and implementation. Further, they may seek to guide decision-makers in their choices, and, in some cases, to provide external standards as accountability mechanisms that countries can bind themselves to.

This chapter provides a review of different types of governance codes and standards for extractive-sector performance and the role of international efforts to support governance reforms. The chapter offers some critical reflections on the conceptual foundations as well as the practical experience of these efforts. Specifically, it examines government-targeted approaches including the EITI, the Kimberly Process, the Natural Resource Charter (NRC), and the Africa Mining Vision (AMV).

While oil and minerals prices have fallen from the sustained highs of the commodity price super-cycle of the 2000s and first half of the 2010s, the stakes remain high for LICs and MICs. Historically almost 90 per cent of mineral and oil investments have taken place in upper-middle and HICs. However, by 2030 an estimated US$11–17 trillion of new investment is needed in mineral, oil, and gas projects in LICs (Dobbs et al. 2013). Research suggests that as investment has moved towards the Global South, the role of institutional quality has increasingly come to the forefront of investors' minds and has shaped the pattern of investments around the world (Cust and Harding 2014; Arezki et al. 2016). As well as shaping investment choices, governance and the quality of institutions also shapes outcomes, and is generally considered to be the crucial determinant in whether a country succumbs to the resource curse or not (Barma et al. 2012; Mehlum et al. 2006; Robinson et al. 2006).

The chapter provides a critical review of governance initiatives targeting governments as the key agents of change. We find a mixed picture of success, where uptake has been strong, for example, in terms of the number of countries engaging with or signing on to normative frameworks. However, evidence for causal impact remains weak and sometimes limited to anecdotal evidence. We offer some critical reflections on the challenges these initiatives have faced and potential ways forward to build on the lessons and achievements of the past decade.

## 19.2 What Is the Rationale for Improving Governance in Resource-rich Economies?

Natural resources such as oil, gas, and minerals present a unique set of policy dilemmas to government. The potential for large revenues or rents arising from their extraction means that big government, in a fiscal sense, is hard to avoid. In nearly all countries, governments are charged by their citizens to manage resource wealth on their behalf, which invariably entails contracting resource companies and taxing the revenues derived from the extraction process. Governments receiving these sizeable tax receipts must then navigate a series of policy choices on how to manage and distribute these revenues to maximize economic development gains, while minimizing the risks of misappropriation, patronage, and plunder. Getting it wrong can mean countries succumbing to the so-called resource curse—see Ross (2014); van der Ploeg (2011); also Dietsche, Chapter 6, this volume.

Getting it right, and creating outcomes that are robust to politics, has proved challenging. Putting the right rules and government institutions in place is a difficult task, and one that is exacerbated by the challenge of doing so in a highly politicized context. Political cycles can often endanger reform agendas, and populist politics can create pressures for wasteful spending or inequitable distribution strategies; see Collier (2017) and Bawumia and Halland (Chapter 11, this volume). The record in recent years has not been good (Venables 2016).

However, governance does matter: a wide range of empirical studies demonstrate the important role for institutional setting, policy choices, and good governance in shaping the risks and rewards associated with resource wealth (e.g. see Mehlum et al. 2006; Robinson et al. 2006). This can operate via private-sector channels, whereby good governance can reduce the risks or lower the operating costs faced by investors. Similarly, good governance can ensure prudence and efficiency in the managing of resource revenues for the ultimate benefit of citizens, such as via fiscal rules, SWFs or expenditure policies. Countries that have successfully navigated these challenges include Chile, with fiscal rules; Norway, with its savings fund; and Botswana, with its promotion of broad-based human development in spending choices.

How we should define good governance, and what actions matter in this regard, are widely debated.[3] A definition taken from the Worldwide Governance Indicators provides an expansive basis for understanding governance: the traditions and institutions by which authority in a country is exercised. This includes: (a) the process by which governments are selected, monitored, and replaced; (b) the capacity of the government to effectively formulate and implement sound policies; and (c) the respect of citizens and the state for the institutions that govern economic and social interactions among them.[4]

Countries with relatively weak governance—such as a lack of constraints on executive power—run the heightened risk of various resource curse challenges.

[3] This is an issue discussed in greater detail in Dietsche, Chapter 6, this volume.

[4] http://info.worldbank.org/governance/WGI/#faq.

First, elite capture prior to deals, or signing poor deals with extractives companies, can limit the overall scope for revenues received by government. This in turn will limit spending options and otherwise implicitly subsidize extraction by private agents. Upon receipt of revenues governments may be at additional risk of elite capture, myopic fiscal policies, and dilemmas regarding the inter-generational equity of spending and saving decisions. In the worst cases, resource wealth in the context of poor governance can exacerbate corruption and drive conflict.

There is a growing consensus in the academic literature of a *conditional* resource curse (Torvik 2002, Mehlum et al. 2006; Robinson et al. 2006), moving beyond the earlier unconditional view of the resource curse (Sachs and Warner 1997, 2001; Auty 1994, 2001). By this, it is meant that conditional on the overall level of governance in a country, resources can result in either a curse or a blessing. This stands in sharp contrast to the conventional resource curse hypothesis, where the impact of resources is attributed unconditional blame for the economic under-performance of the average country defined as resource-rich.

According to the conditional variant of the resource curse hypothesis, below some critical governance threshold, resources may be more likely to lead to negative outcomes relative to a counterfactual of no-resource discovery. On the other hand, countries enjoying strong prior institutions are more likely to be able to harness resource wealth for long-term prosperity. For example, Collier and Goderis (2008) show that while the first few years of commodity price booms benefit resource-rich African countries, the long-term effects reduce output by around 25 per cent relative to the counterfactual. They estimate that for a resource-rich country to benefit from its newfound wealth, it must possess the governance levels and institutional quality equivalent to Portugal in the late 1980s. The alternative scenario increases the prospect of plunder and mismanagement (Collier 2010). Indeed, recent empirical work has argued that critical thresholds exist in polices and institutions that determine whether resource abundance leads to perverse political choices (Casselli and Tesei 2016) and whether major oil discoveries lead to short-run growth disappointments or not (Cust and Mihalyi 2017).

### 19.2.1 *The Emergence of Governance Initiatives*

The international community has built various rules and guidelines, with notable successes, originating from the challenges faced in OECD countries, such as the Basel standards for banking, and the WTO for trade (Collier 2008). Recent years have seen the wider creation and adoption of codes and standards that are more obviously of relevance to the specific challenges faced in development and by developing countries. Examples include the

International Court of Justice, the international monitoring of elections, as well as environmental codes and more recently best practice guidelines for the extractives sector.

Since the beginning of the most recent commodity price super-cycle, starting in roughly 2002, a multitude of guidelines, codes, best practice manuals, scorecards, and assessment tools have been proposed.[5] Many of these have explicitly targeted the governments of resource-rich countries, seeking to provide them with additional tools to navigate policy choices and otherwise act to improve governance in the resource sector.

A pioneer in the extractives sector was the Kimberly Process Certification Scheme (KPCS). This international certification and monitoring initiative emerged in response to concerns around resources being used to trigger and sustain violent conflict, in the hope of regulating the trade of so-called blood diamonds. Subsequent to KPCS, the most widely adopted has been the EITI, initiated in 2002, which encourages action by both government and companies to release information about the tax revenues received and paid in the course of extraction.

Since KPCS and EITI, others have followed such the NRC and the AMV. In some cases, these movements have spawned legislative steps, such as US resource revenue transparency legislation in section 1504 of the Dodd–Frank Act 2010 (Dodd–Frank 2010), which, while repealed under the Trump administration, has led to similar actions across the European Union, Canada, and Norway.

The codes and standards discussed in this chapter share several common features. They seek to support domestic political institutions and policy makers via normative benchmarks or thresholds for countries to meet. For example, the EITI and KPCS give you member status if you meet their standards, which are commonly applied across the globe. NRC and AMC provide principles that countries can integrate into their own reforms but which they are not required to adopt in part or in full. NRC and AMV are more wide ranging, speaking to the entire chain of decisions countries face when seeking to harness their resource wealth. EITI and KPCS, for example, have been more focused on specific challenges or areas of weakness in the governance of resources. All, however, speak to the role and actions of government in the process of resource exploitation, and as the custodian of resource wealth on behalf of a country's citizens.

Beyond good governance initiatives there are a plethora of efforts relating to corporate standards, such as the ICMM Principles laying out good practice guidelines for mining companies, and IFC Performance Standards for companies

[5] An extensive list of these is included in Hodge (2017).

receiving IFC financing. Others address pressing environmental and social risks. While important, these lie beyond the scope of this chapter.

## 19.3 What Are the Governance Challenges Posed by Resource Wealth?

Resource wealth creates several additional challenges for developing countries and their governments; it makes governing harder, it makes strengthening institutions harder still, and in the meantime it can erode institutions, potentially fuelling a move towards autocracy.

First, the presence of resource wealth may make aspects of governance harder, regardless of the institutions the country inherits prior to discovery. Drivers of this challenge arise from the rents or revenue windfalls associated with resource extraction. Such rents can often be a magnet for elevated public expectations of public service delivery. These expectations may be hard to fulfil, particularly within time frames that may not align closely to the production life cycle of the extractive sectors. Relatedly, resource extraction and rents raise difficult challenges of balancing inter-temporal and intergenerational priorities. The non-renewable nature of such resources makes the inter-temporal trade-offs much sharper than is the case with the proceeds from other public assets. Finally, populist policies can become more tempting and viable when large revenue windfalls are available for distribution, even at the expense of these inter-temporal trade-offs. Combating such populist tendencies is a difficult task for governments of all stripes and competencies.

Second, improving institutions is made more challenging in the face of resource wealth, especially for countries possessing relatively weak institutions at the time of discovery. And yet, strengthening institutions over time remains a core component of driving prosperity. Resource wealth can make governance reforms harder to implement and ultimately less likely to succeed or survive political cycles and changes of government. This is due to the increased vested interests around current ways of managing the sector, compounded by the scale of rents available to those engaged in the sector—either via the private or public sector. Furthermore, any reforms that seek to disrupt these vested interests may face forceful challenges and ultimately be overturned when governments change (or, in the worst cases, are overthrown).

Third, resource wealth can serve to erode the very institutions that need to be strong and strengthened during the period of resource exploitation. Governments seeking to meet public expectations or political commitments for increased spending, including on public goods, may be tempted to pressure central banks to play a more political role—such as occurred in Ghana after its discovery of oil in 2007; see Bawumia and Halland (Chapter 11, this volume).

Such pressure can undermine the independence of such institutions, weakening their strength to pursue their main objectives of monetary prudence and price stability more generally. Similarly, resource windfalls can create pressure to flout fiscal rules and other checks and balances on government spending that might not otherwise have been tested. In some cases, institutional quality can actually deteriorate in the face of resource wealth (Ross 2001).

## 19.4 What Is the Rationale for Codes and Standards?

The resource sector—and the various challenges discussed above—exemplifies why there is a potential role for standards and codes that are overseen by various external agencies such as the EITI. The special characteristics of the resource sector and how it relates to political and economic institutions mean that the potential usefulness of codes and standards is higher than elsewhere in the economy.

First, due to the scale of government revenues in resource-dependent countries, the demand for overseas development assistance (ODA) may be lower than for countries otherwise similar in income or poverty characteristics. Aid tends to be associated with institution-building—at least in its intent with respect to multilateral aid and bilateral aid from OECD Development Assistance Committee (DAC) donors. By contrast, inflows of resource revenues have little institution-building imperative (including better regulation of the sector itself) and therefore may be more open to abuse and less likely to generate such outcomes.[6] Taken together this means conventional channels of support via aid flows are less useful to resource-rich countries, while less conventional global public goods such as codes and standards may now be relatively more valuable.

Second, the governance challenges and risks posed by resource extraction mean that there is likely to be a premium on getting rules and safeguards in place early (and preferably prior to exploration and exploitation) when policy decisions may already have been made that influence whether the resource sector has a positive or negative economic (and social) impact over the longer term. Making a bad rather than a good deal in the very early stages of developing a resource sector can cast a long shadow—with negative consequences persisting far into the future. Here, therefore, guides that address what lessons other counties have learned, or standards that reflect expertise

[6] It is important to note that both resource revenues and ODA as sources of external finance may be seen to delay or deter governments building a tax base among the general population, which in turn may weaken the social contract. Authors such as Besley and Persson (2013) argue for such mechanisms.

on how a country might proceed can be valuable at a time of competing interests, especially if a country has little prior experience of navigating these new interest groups and new technical challenges.

Third, the process of exploiting resource wealth is far from smooth: the number and size of deposits discovered and then exploited is 'lumpy'; many years may elapse between investments, so that current policy makers have little accumulated experience. Such time frames are compounded by the number of economic agents involved (who may have considerably more experience than governments), all implying that governments may have very few 'rounds of the game' in which to 'experiment' and get policy right. This limits the opportunities for learning by doing, and for mid-course corrections based on experience. These characteristics of the sector magnify the *ex ante* value of peer learning as well as drawing from international and historical experience.

Finally, good or bad resource governance has spillovers both internationally and inter-generationally. The social benefits to good national resource governance may exceed the gains to any individual government administration. Meanwhile bad governance can lead to regional instability, expropriation of international investors, or disruption in resource supply chains relied upon by resource-importing countries and consumers worldwide. Such features create a case for provision of global public goods in the form of efforts to bolster strong governance in resource-exporting countries. Here there may be a particularly high payoff to investments in such public goods; if such tools can avert bad policy choices early in the cycle of resource exploitation, the impact on poverty reduction and economic growth may substantially dwarf the initial cost of such efforts.

### 19.4.1 *The Policy Makers' Dilemmas*

Policy makers also face a series of dilemmas. Policy mistakes during a boom are costly, but with slow feedback loops. For example, the policy lessons from the 1970s commodity boom only emerged once that boom was over in the 1980s. Likewise, the full implications and lessons from the recent commodity price super-cycle may not be fully known for several years, limiting the ability of governments to react and respond in a timely manner. This is especially the case in those countries where the resources profile is limited in duration to a decade or two. Strengthening governance is a slow process. The payoff to reforms likely exceeds the length of the political cycles of many countries under the normal conditions of multi-party democracy, leading to delay and underinvestment in the necessary institution-building. Furthermore, the needs of the private sector may run counter to citizen priorities. Decision-makers must grapple with how to balance these trade-offs: for example, the trade-off over the speed of the investment and the jobs and revenue it brings, versus the need to protect the rights of the community as well as the natural environment and future generations.

Given this context, external codes can bring credibility and rigour. They can provide a general reference for best practice. This in turn can lower the costs of understanding and information by helpfully collating experience and viewpoints into a single location, for governments to then draw upon. This can be particularly valuable for governments in countries with weak technical capacity in key ministries; they do not have to start from scratch when a well-developed body of knowledge and practice and associated technical guidance is available to them. Furthermore, by providing a general reference that draws on international expertise and experience, these external codes can provide a credible source for best practice—thereby cutting through politicized domestic arguments and allowing every country to benefit from countries which have followed a similar but earlier path.

In addition to providing a useful point of reference, external codes can also provide a potential anchor for political commitment. Decision-makers can 'bind themselves to the mast' of an external code, or codes, thus restricting their own government's or successive governments' ability to deviate from particular policy choices. In technical jargon this can help overcome the time-inconsistency associated with many policy dimensions of resource management.

External codes can seek to induce changes in government behaviour—such as the EITI which requires countries to make transparent the tax receipts and other information associated with the extractives sector. However, such steps are not costless—they can upset interest groups or even impose direct 'costs' on elected officials who may find channels of earning illicit money harder to access. Such costliness can operate as a positive feature rather than as a limitation; in the language of political economy scholarship, it acts as a signal of 'type', whereby the 'cost' of compliance is greater for politicians engaged in corruption, compared to the honest politicians who are not. Thus, the signal—such as signing up to EITI—can be observed by voters who can then infer that honest politicians are more likely to pursue such steps since it is less 'costly' for them to do so. This may help build political coalitions that support honest politicians, leading to real improvements over time.

## 19.5 What Is at Stake?

The governance challenges of harnessing non-renewable resources begin with the decision to open the country for exploration and prospecting, and then span all the way through to choices around the spending of revenues and to the development path of the country beyond the lifetime of the resource wealth. Furthermore, the challenges relate to the public sector, to the private sector, as well as to the role of international companies and to cross-border challenges. Guides such as the NRC discuss the trade-offs and choices countries face (Cust and Manley 2014). Historical experience points to both successes and failures at each stage of the so-called decision chain, beginning with the decision by a country to explore and extract, through to how the proceeds will be used to support sustainable development.

First, normative guides typically point to the high cost of failure. Policy missteps and mismanagement of extractive wealth has carried a significant price for some developing countries. Between 1997 and 2002, for example, around US$4.2 billion of revenue 'disappeared' in Angola—more than the country's total ODA receipts for that decade (IMF 2012). For the period 2007–10, the IMF found US$32 billion in unaccounted government funds, equivalent to a quarter of Angola's annual GDP. Despite its oil wealth, Angola ranks 148th on the Human Development Index. Similarly, numerous reports of missing money emerge from Nigeria on a regular basis. Between 2010 and 2012 the Africa Progress Panel (2013) reported a missing US$6.8 billion lost because of corruption and mismanagement of fuel subsidies. Meanwhile, the Petroleum Revenue Special Task Force identified losses of US$29 billion (FMPR 2012).

On the other hand, key policy levers can yield big benefits when they are implemented well. The opportunities for gains are no less significant in revenue and public expenditure management, when prudent and sometimes technically complex policies are pursued. Codes and standards can help capture successful policies or reforms, spreading their lessons elsewhere. One example of a policy success was in Mexico. The government was able to save $5 billion in potentially lost revenues during a price dip in 2009 by means of careful hedging against future oil prices (Thomson 2012). This amount was equivalent to 7 per cent of the government's revenue, and just under half of total FDI (US$11.4 billion). Nigeria's reform of public procurement delivered similarly major benefits in the efficiency of public spending. Prior to 1999 the government lost an average of about US$300 million each year from corrupt practices in public procurement (Okonjo-Iweala and Osafo-Kwaako 2007). Since the reform of the procurement system the federal government has saved an estimated US$1.5 billion over the period 2001–7 in the form of better contract prices, while the initial prices quoted by various government contractors have also declined significantly.

The stakes are high but the potential gains can be higher. Between the late 1960s and 2000, Botswana, for example, has recorded an average growth rate of 7 per cent per annum. This has supported average growth in primary school enrolment of 4.8 per cent per year, every year. In contrast, the DRC collected only US$92 million in minerals' taxes and tariffs on estimated mineral exports of US$2 billion in 2008 (Bauer and Quiroz 2013). Cameroon has been estimated to collect roughly 12 cents on the dollar for its oil; Norway 78 cents on the dollar (Bauer and Quiroz 2013). Such gaps may emerge from different hidden costs to companies, arising from political risk or instability. On the other hand, governments may be able to catch more revenue via better-negotiated deals or fiscal terms.

## 19.6 Examples of Codes and Standards

During the recent commodity super-cycle a wide range of external codes, standards, and best practice guides were created and recommended for adoption by resource-rich counties. Some have focused on economic policy-making, such as the AMV or NRC. Others have focused on anti-corruption and transparency, such as EITI, while a final category, such as the ICMM Principles or the IFC Performance Standards, have concentrated on the private sector's role and responsibilities as an effective development partner to government. Our focus in this chapter is on the codes, standards, and best practice guides addressed primarily to governments and to those engaged in policymaking and design.

### 19.6.1 *The Benefits of Transparency*

Transparency's benefits as a potential pathway to increased accountability are widely espoused. Indeed, much effort has been put into changing the transparency behaviours of key actors in the resource sector such as governments and operating companies. There is good evidence that behaviour change has occurred. However, scientific evidence for the impact of changing transparency—such as the role it may play in delivering a reduction in corruption and conflict or increasing accountability—has proved elusive.

International voluntary standards such as the EITI have proved effective in changing practices around the disclosure by governments and companies of payments made for resources. Between 2003 and 2008, over 150 companies and their affiliates reported under the EITI process in Africa, covering over US$130 billion in revenues paid to governments. Such efforts have spurred binding 'publish what you pay' legislation now passed in Norway, Canada, and the European Union and new listing requirements now required on the Hong Kong Stock Exchange. The Dodd–Frank Act (section 1504) which (before its repeal in 2017) required disaggregated reporting of payments would have applied to 90 per cent of the world's largest internationally operating oil and gas companies, as well as eight of the world's ten largest mining companies.

Some of the best evidence we have is related to transparency and the cost of capital. Glennerster and Shin (2008) find that countries that choose to become more transparent experience lower borrowing costs. Henisz et al. (2014) find that close to two-thirds of the market capitalization of twenty-six publicly traded gold mining companies are a function of on-the-ground stakeholder engagement practices and political risk, whereas only 37 per cent is a function of the value of gold they control.

## 19.6.2 *Extractive Industries Transparency Initiative*

EITI was created on the back of growing civil society pressure for transparency in the resource sector. The idea was first floated by Tony Blair, the then UK prime minister, at a conference on sustainable development. A coalition of international and domestic NGOs took shape in 2003, and operated under the name of Publish What You Pay. The formation of EITI followed soon after. EITI's original mission and mandate was to help alleviate corruption, both perceived and real, in the extractive sector, by revealing to the public the amount companies were paying in taxes to government and to reconcile those numbers against what governments reported receiving.

The first countries announcing their intention to join EITI came in 2003. To achieve EITI 'compliance' countries had to successfully complete Reconciliation Reports which compiled company disclosures with government reported receipts and reconciled the two sets of information. This was conducted by an independent administrator.

In 2013 a new EITI Standard was adopted which updated the rules of the initiative, broadening the focus to encompass a wider set of transparency measures. The initiative has seen a broadening of focus to include transparency on other dimensions of resource sectors including more contextual information about sectors, production, contracts, beneficial ownership of licences, and so forth.[7] Between 2003 and 2017, some fifty-two countries announced implementation; and are at various stages towards compliance with the rules of EITI.[8] Of those, forty-six have published revenue numbers in some form, while the number of fully 'compliant' countries stood at twenty-six in 2016.

### 19.6.2.1 HAS EITI HAD A POSITIVE IMPACT?

Measured in terms of uptake EITI has become a widespread standard for resource-rich countries. This has led to behaviour change in many as EITI

[7] https://eiti.org/document/standard.

[8] The withdrawal of the United States from EITI in late 2017 has subsequently reduced this number to fifty-one, see https://eiti.org/explore-data-portal.

reports contain published revenue numbers that would otherwise not be available in the public domain in the same form.

While setbacks such as the 2017 withdrawal of Azerbaijan and the United States demonstrate that backsliding can occur, new countries continue to join. The swelling number of members indicates a strong appetite and momentum for signing up to EITI's goals. The number of countries publishing revenue numbers is also high, reflecting more than mere tokenism in the intention to implement EITI rules. However, evaluating positive attributable impacts to EITI is a tricky task and one that a burgeoning number of evaluations are attempting to take on.

There are several examples of potentially attributable impacts of EITI membership—that is, those that would not have occurred in EITI's absence. For example, Nigeria's 2005 EITI report shows that over US$500 million in oil taxes had not been collected or had gone missing; this is more than seven times the amount the government spent on agriculture that year (EITI 2008). In 2012, a Nigerian EITI report uncovered $800 million in discrepancies between what companies said they paid and what the government received, an amount exceeding the individual budgets of the ministries of health and education. Of the US$800 million, US$560 million proved to be a shortfall in company payments (EITI 2012). According to De Sa (2013) EITI reports between 1999 and 2008 identified a loss of US$2.6 billion in revenues and a failure by oil companies to pay US$9.9 billion in royalties. These findings have enabled Nigeria to recover US$2.4 billion to date. In Tanzania, the EITI analysis showed that companies paid lower taxes for extraction than the withholding tax paid on income payments to employees in 2008. These findings are supported by other anecdotal stories. For example, the interest spread on Nigerian government bonds (above US Treasury bonds) dropped from 13 to 6.9 per cent between 2002 and 2003 and continued to decline steadily through 2006 as Nigeria implemented the EITI and adopted strong transparency measures.

EITI's supporters argue that examples of discovered discrepancies—which are potentially recoverable—may alone justify EITI's global financial expense, as well as the political capital expenditure it entails. However, such isolated cases are hard to falsify or evaluate scientifically—for example, how much of these discrepancies could and would have been found by other less costly means remains unknown. A further challenge posed for those seeking to evaluate the impact and success of EITI is the lack of a clearly articulated theory of change—an expression of how EITI might be expected to make its impact and therefore to identify some measures of success or failure more rigorously.

A useful activity would be to ground future EITI activities and strategy in a framework that facilitates both rigorous testing of impact, but also the potential for falsification of the underlying and implicit theories of impact. The expansion of the EITI Standard may offer opportunity to stagger rollout of certain reporting features, allowing for an experimental testing of impact between countries.

#### 19.6.2.2 HAS EITI IMPROVED THE AVAILABILITY OF INFORMATION ABOUT PUBLIC REVENUE FROM EXTRACTIVE RESOURCES?

One metric by which EITI might reasonably be evaluated is a proximate operational measure of success such as the actual availability of published government revenue data from extractives. A metric such as this may not connect directly to impacts or outcomes, but might be a good gauge of the healthiness of the initiative, especially in a sector where policy impacts and desired outcomes, such as reduced corruption, are likely to be very hard to measure directly and may in any case take a long time to manifest themselves.

The availability of published government revenue data from countries adopting EITI is one of EITI's core requirements, and a chief mechanism by which EITI may, in principle, achieve its impact. As noted earlier, forty-six countries had by 2015 published government revenue numbers in some form during the period under review. A key question is whether this is a big number or a small number relative to what information may already be available via other means. Other questions are what additional value such publishing might bring, if any, and whether it is likely that such additional disclosures may have an impact on behaviour.

The case of Africa is a good example to examine in addressing these questions since this is a continent characterized by relative opacity in the extractives sector and poor data collection systems in some countries (Jerven 2013). Taking 2010 as a sample year, for which there is coverage from three different sources, we see resource revenue figures within sub-Saharan Africa from twenty-one countries in EITI reports, eighteen countries in IMF research, and twelve countries in the International Centre for Tax and Development (ICTD) database, with only limited overlaps.[9] For additional analysis, including comparisons for 2014, please also see Mihalyi and Fleming (2017).

This implies that EITI is reaching countries that otherwise do not have available data. Additionally, it improves our picture of government revenue data for the Africa region as a whole. However, there are several caveats to consider. First, these revenue numbers remain non-comparable across countries due to inconsistencies in the EITI reporting formats. Steps are now being taken to align reporting fields with the IMF Government Finance Statistics

[9] http://www.ictd.ac/datasets/the-ictd-government-revenue-dataset.

(GFS) methodology and may yield greater consistency across countries and time.[10] The second challenge is data quality, which remains a challenge in EITI reporting countries due to the approach of self-reporting and reconciliation; this does not align to IMF data collection standards that would make such data suitable for inclusion in official statistics. Finally, the reporting is too infrequent and unpredictable to lend itself to time-series analysis or comparison for EITI reporting countries, although progress is being made to reduce the time lag of reporting. Sierra Leone and Myanmar, both EITI member countries, have latest data available only from 2014 in 2018.

The IMF and World Bank are yet to include EITI reported numbers in their statistical databases, while few governments currently report EITI revenues as part of their official government statistics. It is hoped that improvements to methodological consistency and alignment, alongside improvements in timeliness and data quality, can elevate EITI data to a status equivalent to that of the data from international institutions. Until this happens it remains a significant unfulfilled potential that EITI is yet to provide resource revenue data considered sufficiently authoritative to gain widespread usage.

#### 19.6.2.3 HOW HAS GOOD GOVERNANCE EVOLVED IN EITI ADOPTING COUNTRIES?

Adoption of EITI does not occur in a vacuum. As such it is worth reflecting on the wider governance environment surrounding EITI. Various hypotheses around EITI adoption have been considered. For example, it might be argued that adoption is associated with a reforming government, in which case we might expect to observe other governance indicators improving concurrently to EITI. This would not necessarily imply causality from EITI, but may be reflective of the more positive governance steps being taken at the same time. In contrast, we might hypothesize that EITI implementation could be accompanied by backsliding on other governance indicators, whereby governments use such initiatives as an international 'fig leaf' of reform, meanwhile neglecting more impactful or costly reforms, or even reversing good governance progress.

Kaufmann (2014) analyses trends in various Worldwide Governance Indicator variables pre- and post- entry as a EITI member country. His observed trends do not show any strong evidence for backsliding, nor do they show significant improvement in two broad-based measures of governance: voice and accountability, and the control of corruption. If anything, this provides more support for the hypothesis that EITI adoption is not part of a broader reform package but may even be used domestically to delay or substitute for

[10] https://eiti.org/document/eiti-summary-data-template.

alternative domestic reforms. This in turn may give pause for reflection on the way EITI may complement or substitute for other reforms and whether one might expect improvement in other indicators to be associated with its adoption as a broader metric of success.

#### 19.6.2.4 REFLECTING ON THE FIRST DECADE OF EITI

Various studies have been conducted with the objective of evaluating the success and impact of EITI. These have ranged from qualitative studies with a focus on operational and managerial performance, through to quantitative studies seeking to identify evidence for cross-country trends or impacts resulting from EITI's adoption.

Most recently, Rustad et al. (2017) have examined the objectives and successes of EITI. They note that 'in many ways the EITI has succeeded in terms of reaching its institutional goals and some of its operational goals, in particular when it comes to producing annual reports. The EITI has through several measures engaged the civil society groups, particularly through the Multi-Stakeholder Group, but seems to have failed to empower the public to hold the governments and companies to account.' However, they caution that evaluations of EITI may not have thus far used the right criteria to measure success.

As noted in a recent GIZ review of EITI (GIZ 2016) a number of authors express concern regarding the lack of results-based monitoring and evaluation systems and/or demand their improvement (e.g. EITI Scanteam 2011, MSI Integrity 2013, and Scanteam 2015). The same GIZ study concludes that it was difficult to establish the views held by stakeholders outside of those directly participating in the initiative. Supporting evidence for the EITI's impact is largely either anecdotal or based on the perceptions of stakeholders belonging to the 'in-group'—those NGOs, donors, private-sector actors and other experts involved via either the EITI board or membership of EITI multi-stakeholder groups, or otherwise engaged in extractives transparency directly.

As noted earlier, over the period 2003–17, fifty-two countries announced implementation and have reached various stages towards compliance with the rules of EITI. Of those, forty-six have published revenue numbers in some form, while the number of fully 'compliant' countries stood at twenty-six in 2015. This constitutes a significant scaling up of the initiative as a truly global effort, spanning fragile states, LICs and MICs, as well as high-income OECD countries. EITI's expansion accompanies, and may in part be responsible for, a spreading norm of transparency in the extractives sector. For example, contracts are increasingly being disclosed. This trend began before its inclusion in the new EITI Standard as a reporting recommendation; however, it has received wider momentum since its inclusion. While it remains only a recommendation rather than a requirement under EITI, many hope that its presence in EITI may further raise the profile of this emerging norm and help it spread to other jurisdictions.

Many evaluations and commentators praise EITI for its role in providing a platform for dialogue among stakeholders in countries where other avenues for such dialogue may have been difficult or even impossible. Rich and Moberg (2017), drawing on their experiences running the EITI international secretariat, highlight the multi-stakeholder platforms created under EITI as an important component and strength of the initiative. Providing a platform for citizen dialogue was not the primary aim of EITI; however, it may constitute a valuable ancillary benefit and could yet emerge as its key contribution to improving governance.

An emphasis on promoting transparency in poor governance environments may be risky according to leading academics researching the role of governance in economic development. Acemoglu and Robinson (2013) note that marginal reforms of the resource sectors such as EITI might even make things worse. One issue they point to is that the feasible path out of poverty for most countries with poor institutions is a messy one involving much compromise and many bargains. In such circumstances, a feasible model of developmental patrimonialism may well involve tolerating some amount of extraction of resource rents and other aspects of patrimonialism by the elite in exchange for allowing institution-building elsewhere in the state: for example, allowing the development of other fiscal bases such as a value-added tax and perhaps the building of a bureaucracy that can be a constraint on worse abuses in the future. In certain circumstances Acemoglu and Robinson argue that 'an obsession with the application of EITI might be counter-productive'.[11]

Looking ahead, experts are now increasingly pointing to the challenge of 'managing expectations' and to the task of building an informed citizenry (Collier 2017) to hold government accountable and to demand good governance in resource-rich countries. Here a norm of transparency, platforms for dialogue, and a vibrant civil society that can help citizens understand choices made in their extractives sector may all prove invaluable tools in the challenges ahead.

### 19.6.3 *The Regulation of Conflict Minerals*

One of the resource curse mechanisms identified in the research literature is the role of conflict. Here resources may act as a trigger but also as a means of sustaining the financing of conflict, thus contributing to both conflict incidence and duration (Berman et al. 2017; Collier and Hoeffler 2004; Ross 2014).

[11] http://whynationsfail.com/blog/2013/9/12/what-could-be-wrong-with-extractive-industries-transparency.html.

Valuable minerals produced by small-scale mines, so-called point source minerals, can be easily appropriated by despotic governments, warlords, and militia, compared to resources such as oil which require more capital-intensive and technically sophisticated extraction, or agricultural products which have relatively few rents associated with them, and where any rents are more geographically disbursed. Furthermore, minerals with high value to weight ratios are also well suited to illicit trade and can be used to buy arms. As such, certain minerals are particularly susceptible to being connected to conflict (Lujala et al. 2005). These include gold, diamonds, silver, coltan, gemstones, and some other precious metals. Most famously, 'blood diamonds' financed armed conflicts in West Africa and led to the establishment of the Kimberley Process, a cooperative venture between governments, the diamond trade, and NGOs, under which the conflict-free origin is certified.

The KPCS was established by the UN in 2003. It is supposed to ensure that revenues from the diamond trade are not financing violence by rebel movements and undermining legitimate governments. KPCS also requires member countries to collect and publish data on mining and international trade in diamonds. In addition to the Kimberley Process, governments and the international community have sought other means to decouple resources and conflict via a variety of policy mechanisms and market instruments. The United States was the first to legislate (the Dodd–Frank Act); but the European Union is following closely behind. In 2010 the United States passed the Dodd–Frank Act Wall Street Reform and Consumer Protection Act which included provision 1502 aimed at stemming the trade in conflict minerals (a broader definition than in the diamond trade covered under Kimberley). Here the rules put the onus on companies rather than via a certification scheme, with the objective to improve supply-chain management to ensure minerals were not being sourced from conflict-affected regions, in particular the Eastern DRC and neighbouring countries.

What has happened? Since the introduction of these schemes, the buyers of valuable minerals have been deterred from sourcing their raw materials from the DRC and neighbouring countries. This has reduced demand in these countries which has impacted the ability to finance conflict, but may have also harmed livelihoods for miners and their families. Further consequences have included a deterrence effect for sourcing from artisanal miners, who may be less able to validate their supply chains compared to larger-scale mines or traders. Debate continues as to whether the improvement in security has been significant, and if it has, whether it exceeds the damage to livelihoods of some of the poorest people in the world (Seay 2012; Parker et al. 2016). The OECD Due Diligence Guidance for Responsible Supply Chains of Minerals from Conflict-Affected and High-Risk Areas may also support efforts to mitigate the trade in illicit minerals and conflict minerals.

### 19.6.4 *Best Practices and Economic Policy Guides for Policy Makers*

Several initiatives have focused on the challenges faced by policy makers with particular reference to the economic governance of resources such as the African Union's AMV (African Union 2009) and the NRC (Cust and Manley 2014). They have focused more on the synthesizing and sharing of knowledge based on historical experiences and expert consensus on how countries might act to make the most of resource wealth. However, like standards-based clubs such as EITI, they also face some of the same shortcomings, including complex and untested theories of change as well as measures of success that are hard to conceptualize or quantify.

#### 19.6.4.1 AFRICA MINING VISION

Created by the members of the African Union (AU) in 2009, the AMV[12] is a policy framework designed to lay down principles for individual countries to harness their mineral wealth for economic development.[13] It is intended to provide a rubric for promoting broad-based benefits from mineral exploitation, and to help shift the sector beyond enclave development to becoming a more integrated sector to the rest of the economy. An action plan followed in 2011 that laid down the need for a specialist agency to oversee continental implementation of the AMV and nine clusters of activity across different policy priorities.

The AMV covers various policy considerations along the mineral value chain, from issues around exploration and project development, through taxation and safeguards, to revenue management, downstream value addition, and linkages to the rest of the economy. The AMV emphasizes resource-driven development, pointing to the potential positive benefits from resource exploitation in contrast to the more pessimistic view of resources' impact on prosperity identified in the academic literature on the resource curse. The AMV does, however, note the centrality of good governance to achieving the goals laid out, including support for transparency and certification schemes, thus internalizing the conclusions of the conditional resource curse thesis; namely that conditional on getting governance right, resource wealth can be a blessing (van der Ploeg 2011; Ross 2014).

Country Mining Vision (CMV) is the mechanism by which the AMV is intended to be translated into practical policy prescriptions at the country level. Intended to bring together dialogue among different stakeholders (government, industry, and civil society), the CMVs are intended to be an inclusive and deliberative policy-setting process. The African Mineral Development

[12] http://www.africaminingvision.org/.

[13] The AMV was developed by the United Nations Economic Commission for Africa on behalf of the African Union Commission and was universally adopted by AU Heads of State.

Center was created to support the implementation of CMVs. Eight years after its inception, questions remain over the slow pace of implementation, and whether it is meeting the purposes for which it was created (Ushie 2017). Despite being endorsed by all fifty-four member states of the AU, by early 2017 only one country, Lesotho, has fully adopted the AMV through the development of a CMV. Oxfam points to the lack of engagement with civil society as one of the shortcomings of the approach.

#### 19.6.4.2 NATURAL RESOURCE CHARTER

The NRC,[14] launched in 2009, is an exercise to synthesize best practice lessons from country experience and academic research. The twelve precepts of the NRC identified the key 'links in the chain' countries needed to get right to successfully transform subsoil resources into economic prosperity, without compromising sustainability.[15] The Charter recognizes the fragility of the process of resource exploitation and that any weak link in the chain could push countries onto a pathway towards the resource curse. It therefore emphasizes a holistic policy perspective towards addressing gaps and key shortcomings to ensure the whole process is managed carefully (Cust 2013).[16] In addition, it emphasizes the important role of an authorizing environment that spans traditional government ministries, noting that the challenges of resource wealth span ministries of energy, minerals, finance, and environmental agencies, tax administration and those responsible for national development strategies.

Spanning issues from exploration, contract negotiation, taxation, revenue management, and investing for sustainable development, the Charter identifies key recommendations as well as policy trade-offs that countries need to consider. It does not provide a one-size-fits-all prescription, but rather a compilation of the policy ingredients that have proved successful in different country contexts. The precepts capture the high-level general principles of successful resource management that can be achieved via different means.

The NRC empathizes the role for both careful policy design and engagement with an informed citizenry. It recognizes the need for governments to get decisions right early and continuously, but that the only longer-term protection

[14] https://resourcegovernance.org/approach/natural-resource-charter.

[15] The Charter was led by a distinguished Oversight Board, chaired by Ernesto Zedillo, the former President of Mexico, and accompanied by Mo Ibrahim, Shengman Zhang, Luisa Diogo, and Abdulatif Al-Hamad. Its content and recommendations were developed and maintained by a Technical Advisory Committee chaired by Michael Spence, Nobel Laureate in Economics, and joined by leading academics and practitioners including Paul Collier, Tony Venables, Karin Lissakers, Joseph Bell, Bob Conrad, and Michael Ross.

[16] In the interests of full disclosure, it should be noted the author of this chapter served as the founding staff member and led the secretariat of the Natural Resource Charter during 2009–13.

against the resource curse comes from an environment of accountability to citizen demands and responsiveness of the state.

The implementation of the NRC at the country level is applied via the NRC Benchmarking Framework. This toolkit provides the means for countries to self-assess against the twelve precepts of the NRC to identify policy gaps and shortfalls. This then allows the formulation of a policy action plan that can help countries lay out the policy priorities. The NRC benchmarking framework has now been applied in a range of resource-rich countries—including Nigeria, Tanzania, Sierra Leone, and Myanmar. In 2013, the NRC was merged with the Revenue Watch Institute to form NRGI—NRGI is now a leading organization in the field of improving the governance of natural resource wealth in developing countries. It is also the main implementing agent for the NRC; however, it also provides an open licence for other NGOs and governments to use the benchmarking toolkit independently.[17] As with other good governance efforts during this period, causal attribution of the benefits to countries using the NRC have proved difficult to articulate and quantify. Further, some countries that pursued and integrated the principles espoused by the NRC in their own reforms—such as Ghana in its prudent approaches to oil revenue management legislation in the period following 2009—have since proved less able to protect the country from negative price shocks and populist spending pressures than many had hoped.

## 19.7 Critical Reflections

The purpose of the international codes, standards, and initiatives discussed here is to foster and support the adoption, implementation, and embodiment of policies and institutions for the good governance of extractive resources.

Good governance is widely agreed to play a fundamental role in determining the economic performance of countries; however, less is known regarding the steps and time it takes for a country to move from a situation of weak governance to one of stronger governance. In the case of resource-rich countries, this challenge is exacerbated by the role such resources play in shaping and distorting economic and political incentives. Indeed, extractive resources can fuel a situation where governing well is made harder by the rents they provide to governments. This in turn can create incentives that serve to undermine or weaken institutions, such as circumventing checks and balances, or fostering a culture of corruption and patronage. In turn, the resources that

[17] https://resourcegovernance.org/analysis-tools/tools/natural-resource-charter-benchmarking-framework.

are hard to govern may also make other political and economic reforms harder and less likely to succeed.

Various proposals have been advanced for normative measures that may help countries strengthen governance of the resource wealth, while avoiding policy missteps that have been made in other countries. These include increasing transparency in the resource sector (EITI), creating supply-chain checks against minerals sourced from conflict (Kimberley Process), and pursuing policy choices and development strategies that have proved robust in other settings (NRC, AMV).

It might be unreasonable to expect good governance initiatives to yield large and observable effects in a relatively short space of time. It might even be unreasonable to expect them to succeed with a high degree of probability. Neither factor invalidates the importance of their pursuit; the payoffs might be slow to emerge but are potentially large when realized even with a low probability of success in any given country. However, the extent of the challenge, or the pace of underlying change, should also not preclude a rigorous approach towards measuring and reporting impact, as well as articulating a testable and falsifiable theory of change.

While tangible success on changing policy and embedding best practice has been hard to identify, this does not imply the focus is incorrect. Given the time profile of extraction—requiring upfront policy choices that determine the path for years to come—the need for urgency is inherent. The policy choices made by governments today can have long-lasting repercussions for current and future generations. Therefore, efforts to support improved policymaking today, even in the context of other institutional weaknesses, may be justified on the grounds of these long tails of effect and the magnitude of money involved.

### 19.7.1 *Downsides of Codes and Standards*

However, such external guides can have downside risks too. Successful policies and institutions as operated in one setting are not always straightforward to replicate in other contexts; indeed, in some cases what has proven successful in one country may prove inappropriate or unsuccessful in another. Countries must carefully navigate the path between what lessons can reasonably be drawn from international experience and what is likely to be feasible and effective in a domestic setting. Doing so in the face of strong political and populist pressures unleashed by resource wealth can be especially difficult.

Learning what works, and what does not, also takes time. Given the uncertainty associated with any given reform or policy action to improve governance, a learning process is likely to play an important role. There is, however, a high opportunity cost to initiatives that require significant investment of money—sometimes donor aid—and political will.

A recurring concern regarding successful governance initiatives is their ability to learn from and adapt and respond to emerging evidence and changing circumstances on the ground. The Kimberley Process has struggled with controversy around its effectiveness in stemming the flow of conflict diamonds, but has also struggled to update definitions of conflict diamonds given the evolving role of diamonds that may be profiting state actors engaged in conflict. The EITI has likewise faced challenges in terms of expanding scope and whether reforms to the standard agreed centrally are feasible or desirable for all member countries to adopt. The link from some improvements in transparency to better development outcomes or increased accountability remains tenuous and hard to identify.

### 19.7.2 *Looking Ahead*

Norms ranging across transparency, competitive licence allocation, getting the right saving–spending balance, and building strong institutions for checks and balances such as fiscal rules and accountability bodies, are spreading internationally. The pace of their adoption and retention can feel slow, but such governance reforms are likely to take time and are at constant risk of backsliding and being undermined in the face of price declines, such as those seen after 2011, and rising populism. There is emerging praise for the role governance initiatives have played in creating spaces for civil society dialogue, while a greater shift towards open data that are accessible and used by domestic actors may further entrench ideas of public scrutiny and debate around how the resource sector is governed.[18]

In addition, the gradual spread of the 'good governance' norms, the process of learning about what works and what does not, is well underway. Several countries have recently experienced resource booms for the first time, and some of these are learning their own lessons of what to avoid in the future. Similarly, international initiatives have the advantage of being able to share and spread lessons and experiences with relatively low transaction costs as such reforms to the global approach can permeate quickly to the country level: one example is the new EITI Standard.[19]

Finally, the end of the recent super-cycle of commodity prices, and the prospect of potentially permanently lower prices for fossil fuels due to global

[18] More recent developments in the space of resource governance have included the creation of the Resource Governance Index (NRGI) which ranks countries according to normative measures of good governance. Additionally organizations such as the World Bank and Inter-Governmental Forum have been developing their own tools to support governance in resource-rich countries.

[19] This centralized reform of EITI rules had almost immediate implications for all fifty-plus participating countries, which would now be expected to work towards compliance with the new, more expansive, reporting requirements rather than the rules they originally signed up to.

decarbonization, creates potential positive opportunities for improved governance. First, low prices reduce the rents that have fuelled rent-seeking, supported vested interests, and spurred populist promises and elevated expectations. Second, fiscal budgets are squeezed so there is increased incentive to tighten governance loopholes and crack down on corruption and other forms of waste. Further, many governments are now looking to take steps to end fuel subsidies and to correct other 'poor' governance choices during high price periods, such as a lack of savings provisions and bad deal-making. The pause in new foreign investment into extractives in some countries buys those countries time to continue the process of institution-building to prepare for future new investments, extraction, and revenues.

The future may, however, look quite different from the past when it comes to the governance of extractive resources. First, commodity prices may not recover to the same highs seen during the super-cycle, or the length of such price cycles may not match the recent one. Second, the opportunity for greenfield developments (in terms of non-resource-rich countries entering resource richness) is shrinking as many new producers entered the market during the recent boom. This may have led to the locking-in of poor policy choices and weak institutions that may be harder to fix for next time, not least since vested interest groups have had the chance to entrench their positions. Finally, the urgent threat of climate change and technological breakthroughs in carbon-zero technologies mean that the demand and hence the price for fossil fuels (coal, but also oil and gas) may never recover, or the period of higher prices may be curtailed in the near future by technology substitution.[20] Similarly, depletion of resources during this boom may mean that some countries have few new deposits left to develop.[21]

As such, developing countries may have to prepare themselves for a post-extractive future, one where the diversification of the economy and a strengthening of governance may be no less vital, and may even prove easier in the absence of such large resource rents. For those countries rich in fossil-fuel resources, the gradual decarbonization of the global economy may have long-term implications for their economies. While those specialized in minerals and metals may see strong demand extending long into the future, the temporary slowdown in demand across some of these commodities may create a prime opportunity for reflection and lesson-learning, and ultimately an opportunity to strengthen governance.

[20] The link between the climate change agenda and the situation facing extractive industries (both mining and oil and gas) is discussed in detail in Addison, Chapter 22, this volume, who also explains the important differences that are likely to apply to minerals as opposed to fossil fuels.

[21] However, many have significant resources left untapped. Among fossil-fuel-rich developing countries, the median time to depletion exceeds forty-five years at current rates of production.

## References

Acemoglu, D. and J. Robinson (2013). 'What Could Be Wrong with Extractive Industries Transparency Initiative?' 12 September. Available at: http://whynationsfail.com/blog/2013/9/12/what-could-be-wrong-with-extractive-industries-transparency.html.

Africa Progress Panel (2013). 'Equity in Extractives: Stewarding Africa's Natural Resources for All'.

African Union (2009). *Africa Mining Vision*. Addis Ababa: African Union Commission and United Nations Economic Commission for Africa. Available at: http://www.africaminingvision.org/amv_resources/AMV/Africa_Mining_Vision_English.pdf.

Arezki, R., R. van der Ploeg, and F. Toscani (2016). 'The Shifting Natural Wealth of Nations'. CERDI Working Paper.

Auty, R. (1994). 'Industrial Policy Reform in Six Large Newly Industrializing Countries: The Resource Curse Thesis', *World Development*, 22(1): 11–26.

Auty, R. (2001). 'The Political Economy of Resource-driven Growth', *European Economic Review*, 45(4–6): 839–46.

Barma, N., K. Kaiser, T. Le, and L. Viñuela, (2012). *Rents to Riches? The Political Economy of Natural Resource-led Development*. Washington, DC: World Bank.

Bauer, A. and J. C. Quiroz (2013). 'Resource Governance', in A. Goldthau (ed.), *Handbook of Global Energy Policy*, 244–64. Chichester: Wiley Blackwell.

Berman, N., M. Couttenier, D. Rohner, and M. Thoenig (2017). 'This Mine Is Mine! How Minerals Fuel Conflicts in Africa', *American Economic Review*, 107(6): 1564–610.

Besley, T. J.and T. Persson (2013). 'Taxation and Development'. CEPR Discussion Paper No. DP9307. Available at SSRN: https://ssrn.com/abstract=2210278.

Caselli, F. and A. Tesei (2016). 'Resource Windfalls, Political Regimes, and Political Stability', *Review of Economics and Statistics*, 98(3): 573–90.

Collier, P. (2008). 'Laws and Codes for the Resource Curse', *Yale Human Rights and Development Law Journal*, 11(1): article 2.

Collier, P. (2010). *The Plundered Planet: How to Reconcile Prosperity with Nature*. London: Penguin.

Collier, P. (2017). 'The Institutional and Psychological Foundations of Natural Resource Policies', *Journal of Development Studies*, 53(2): 217–28.

Collier, P. and B. Goderis (2008). 'Commodity Prices, Growth, and the Natural Resource Curse: Reconciling a Conundrum'. Available at SSRN: https://ssrn.com/abstract=1473716.

Collier, P. and A. Hoeffler (2004). 'Greed and Grievance in Civil War', *Oxford Economic Papers* 56:4.

Cust, J. (2013). 'The Natural Resource Charter in Africa: A Tool for National Strategy and Evaluation'. NSI-CA. Available at: http://www.nsi-ins.ca/wp-content/uploads/2013/05/6.-Jim-Cust.pdf.

Cust, J. and T. Harding (2014). 'Institutions and the Location of Oil Exploration.' OxCarre Working Paper 127.

Cust, J. and D. Manley (2014). 'Natural Resource Charter', 2nd edition. NRGI. Available at: https://resourcegovernance.org/analysis-tools/publications/natural-resource-charter-2nd-ed.

Cust, J. and D. Mihalyi (2017). 'Evidence for a Presource Curse? Oil Discoveries, Elevated Expectations and Growth Disappointments'. World Bank Policy Research Working Paper WPS8140.

De Sa, P. (2013). 'From Transparency to Accountability'. Washington, DC: World Bank. Available at: http://www.worldbank.org/en/news/feature/2013/12/19/from-transparency-to-accountability.

Dodd–Frank Wall Street Reform and Consumer Protection Act (2010). 'Public Law 111–203', *US Statutes at Large* 124 (2010): 1376.

Dobbs, R. et al. (2013). 'Reverse the Curse: Maximizing the Potential of Resource-driven Economies'. McKinsey Global Institute.

EITI (2008). 'EITI Nigeria Financial Audit Report 2005'. Available at: https://eiti.org/document/2005-nigeria-eiti-report.

EITI (2012). *Nigeria EITI: Making Transparency Count, Uncovering Billions*. Oslo: EITI. Available at: http://eiti.org/document/case-study-nigeria.

Federal Ministry of Petroleum Resources (2012). 'Report of the Petroleum Revenue Special Task Force', Federal Ministry of Petroleum Resources, Nigeria. Available at: https://www.premiumtimesng.com/docs_download/Report_of_the_Ribadu_led_Petroleum%20Revenue%20Special%20Task%20Force%202012.pdf.

Glennerster, R. and Y. Shin (2008). 'Does Transparency Pay?' *IMF Staff Papers*, 55(1): 183–209.

Henisz, W. J., S. Dorobantu, and L. Nartey (2014). 'Spinning Gold: The Financial Returns to External Stakeholder Engagement', *Strategic Management* Journal, 35(12): 1727–48.

Hodge, R. A. (2017). 'Towards Contribution Analysis'. WIDER Working Paper 19/2017. Helsinki: UNU-WIDER. Available at: https://www.wider.unu.edu/publication/towards-contribution-analysis.

IMF (2012). 'IMF Country Report: Angola: 2012 Article IV Consultation and Post Program Monitoring', No. 12/215. Available at: http://www.imf.org/external/pubs/ft/scr/2012/cr12215.pdf.

Jerven, M. (2013). *Poor Numbers: How We Are Misled by African Development Statistics and What to Do about It*. New York: Cornell University Press.

Kaufmann, D. (2014). 'Presenting NRC-RWI'. Presentation to the Natural Resource Charter annual conference, June. Oxford.

Lujala, P., N. P. Gleditsch, and E. Gilmore (2005). 'A Diamond Curse? Civil War and a Lootable Resource', *The Journal of Conflict Resolution*, 49(4): 538–62.

Mehlum, H., K. Moene, and R. Torvik. (2006). 'Institutions and the Resource Curse', *The Economic Journal*, 116(508): 1–20.

Mihalyi, D. and A. Fleming (2017). 'Countries Struggling with Governance Manage $1.2 Trillion in Resource Wealth'. Blogpost. Available at: https://resourcegovernance.org/blog/countries-struggling-governance-manage-12-trillion-resource-wealth.

Natural Resource Charter (2010). 'Natural Resource Charter', 1st edition. NRGI. Available at: https://resourcegovernance.org/sites/default/files/NRC_Eng_2011.pdf.

North, D. (1990). *Institutions, Institutional Change and Economic Performance*. Cambridge: Cambridge University Press.

Okonjo-Iweala, N. and P. Osafo-Kwaako (2007). *Nigeria's Economic Reforms: Progress and Challenges*. Washington, DC: Brookings Institution.

Parker, D. P., J. D. Foltz, and D. Elsea (2016). 'Unintended Consequences of Sanctions for Human Rights: Conflict Minerals and Infant Mortality', *Journal of Law and Economics*, 59(4): 731–74.

Rich, E. and J. Moberg (2017). *Beyond Governments: Lessons from the Extractive Industries Transparency Initiative*. New York: Routledge.

Robinson, J. A., R. Torvik, and T. Verdier (2006). 'Political Foundations of the Resource Curse', *Journal of Development Economics*, 79(2): 447–68.

Ross, M. (2014). 'What Have We Learned about the Resource Curse?' *Annual Review of Political Science*, 18: 239–59.

Ross, M. L. (2001). 'Does Oil Hinder Democracy?' *World Politics*, 53(3): 325–61.

Rustad, S. A., P. Le Billon, and P. Lujala (2017). 'Has the Extractive Industries Transparency Initiative Been a Success?' *Resources Policy*, 51: 151–62.

Sachs, J. and A. Warner (1997). *Sources of Slow Growth in African Economies*. Oxford: Oxford University Press.

Sachs, J. and A. Warner (2001). 'The Curse of Natural Resources', *European Economic Review*, 45(4–6): 827–38.

Seay, L. (2012). 'What's Wrong with Dodd–Frank 1502? Conflict Minerals, Civilian Livelihoods, and the Unintended Consequences of Western Advocacy'. Working Paper 284, Center for Global Development, Washington, DC. Available at: https://www.cgdev.org/publication/what%E2%80%99s-wrong-dodd-frank-1502-conflict-minerals-civilian-livelihoods-and-unintended.

Thomson, A. (2012). 'Mexico Reveals $85 a Barrel Oil Hedge', *Financial Times*, 31 January. Available at: https://www.ft.com/content/c07f47e6-4c35-11e1-bd09-00144feabdc0.

Torvik, R. (2002). 'Natural Resources, Rent-seeking and Welfare', *Journal of Development Economics*, 67(2): 455–70.

Ushie, V. (2017). 'From Aspiration to Reality: Unpacking the Africa Mining Vision'. Oxfam Briefing Paper. Oxfam International, March.

van der Ploeg, F. (2011). 'Natural Resources: Curse or Blessing?' *Journal of Economic Literature* 49(2): 366–420.

Venables, Anthony J. (2016). 'Using Natural Resources for Development: Why Has It Proven So Difficult?' *Journal of Economic Perspectives*, 30(1): 161–84.

# 20

# Oil and Gas Companies and the Management of Social and Environmental Impacts and Issues

## The Evolution of the Industry's Approach

*Kathryn Tomlinson*

### 20.1 Introduction

In the 1990s, the environmental and social impacts of extractive industries on communities increasingly became debated in various forums, and extractive companies faced mounting criticism in the media and from NGOs, academics, and wider civil society. In the oil and gas industry, this situation partly came about due to the high-profile negative impacts of certain oil and gas operations, such as the Exxon Valdez oil spill in Alaska (1989), the clash between Royal Dutch Shell (Shell) and Greenpeace over the disposal of the Brent Spar oil rig in the North Sea (1995), the execution of Ken Saro-Wiwa, who was campaigning against Shell in Nigeria (1995), and BP's and Talisman's alleged complicity in security-related human rights abuses in Colombia (1997) and Sudan (1999) respectively. In the mining sector, it was cases such as the tailings dam failure of BHP's Ok Tedi mine in Papua New Guinea (1984), Placer Dome's Marcopper mine in the Philippines (1996) and the subsequent ongoing pollution, and the conflict in Bougainville sparked by Rio Tinto's mining operations (1989) that drew the most attention and criticism. As a result of these cases and others, damning reports were published and targeted actions undertaken, such as the picketing of companies' head offices, service stations, or shareholder meetings.[1]

[1] This chapter represents a social practitioner's perspective on social and environmental performance in the oil and gas industry. In terms of methodology, although I refer to the

Specific campaigns against oil and gas companies started to gain wider momentum during this period. The criticisms levelled at the industry covered a wide range of issues, including allegations that oil wealth fuelled corruption and conflict, and propped up repressive governments;[2] suggestions that the companies were either complicit in or benefitted from human rights abuses committed by host governments and security forces (such as in the Shell Nigeria and BP Colombia cases);[3] complaints about negative environmental impacts from oil and gas projects (such as oil spills and flaring), leading to environmental damage, and health and livelihood impacts on local communities;[4] and rising awareness of the wider environmental question of the industry's role in climate change and how it collectively approached the subject, including allegations that the industry was actively promoting climate change denial. Within the wider 'resource curse' debate, it was pointed out that although oil and gas projects were generating vast sums of wealth for low- to middle-income countries, many communities affected by these projects, such as in Nigeria or Angola, continued to live in poverty, and had in fact become worse off than previously due to the presence of oil and gas projects. Finally, many critics of the industry decried what they saw as a lack of transparency in the industry, and a propensity for active corporate lobbying against social and environmental legislation.[5]

As the campaigns against the industry intensified by the late 1990s, some companies started to respond to the criticisms, and began to address social and environmental issues in their operations. Since then, broadly speaking, there has been an evolution of the management of social and environmental issues in the extractive industries, from a general neglect of these issues coupled with ad hoc philanthropy to a risk management approach modelled on specific international standards. Today, the principles of social and environmental management in the extractive industries can be broadly summarized as 'do no harm' and 'provide benefits' at the local community level, and a number of tools and approaches have been developed and are variously implemented

academic literature on corporate social responsibility in the oil and gas industry, the chapter is largely based on a review of primary materials such as company sustainability reports and NGO reports, as well as my own observations and experiences of working within the extractive industries, including many conversations with colleagues in the oil and gas industry, over the last ten years. I would particularly like to thank Ramanie Kunanayagam for sharing with me her ever-insightful perspectives, which have helped shape my views on social performance in the industry over the years, as well as my arguments for this chapter. I would also like to thank Alan Roe for his feedback on earlier drafts. I am extremely grateful to Evelyn Dietsche for reading earlier drafts of the chapter and giving me very helpful editorial and content feedback, which helped strengthen the chapter. Last but not least, I would like to thank my husband, Ronald Ranta, for reading and commenting on various drafts, as well as holding the fort while I was immersed in the writing.

[2] See e.g. Global Witness (1999).

[3] See e.g. *The Economist* (1997); Human Rights Watch (1999).

[4] *The Economist* (1997); Human Rights Watch (1999).

[5] See e.g. Christian Aid (2004).

across the industry to address these two high-level aims. This chapter focuses on the evolution of the management of social and environmental issues in the oil and gas industry specifically, with some comparisons with the mining industry. It also presents how the industry has responded to external criticisms around social and environmental issues, and how it is now approaching the management of these issues.[6] The chapter examines some of the key international standards that have shaped this area, breaks down what the management of social and environmental performance actually means in practice, and looks at some of the gaps and ongoing debates in the implementation of social and environmental performance in the industry. Ultimately, the chapter argues that the area of social and environmental performance in the oil and gas industry is often misrepresented in the broader literature under the umbrella term 'corporate social responsibility', which in turn has led to general misunderstandings around the drivers of this management area, as well as of the nature of social performance itself.

## 20.2 The Initial Industry Response

Faced with mounting criticism from NGOs and activists, by the late 1990s oil, gas, and mining companies had come under increasing pressure to reform the ways they operated. At the time, most large Western extractive companies had some kind of internal corporate environmental policy and management system in place, but very few had explicit social performance policies or were strategically managing social and community issues (McPhail and Davy 1998: 57). Interestingly, in responding to this pressure the mining industry took a different trajectory from the oil and gas industry. At the time, several chief executive officers (CEOs) of international mining companies decided

[6] It is worth noting that it is difficult to speak about the oil and gas industry as a whole, as it can be roughly divided into international oil companies (IOCs) and national oil companies (NOCs). The IOCs comprise seven 'supermajor' publicly owned multinational companies (ExxonMobil, Chevron, BP, Shell, Total, ENI, and ConocoPhillips), and numerous medium-sized to small independent oil and gas companies (e.g. Hess, Marathon, OMV, BG Group, Anadarko, Tullow, Woodside). Currently the largest oil and gas companies in the world based on revenue are made up predominantly of NOCs (e.g. Saudi Aramco, Gazprom, PetroChina, Petrobras, Sinopec, Kuwait Petroleum Corporation), and these are increasingly competing with IOCs in oil and gas development. However, it is mainly IOCs that have experienced external pressure from NGO campaigns, adverse media reports, and shareholder activism, and as a result it is predominantly IOCs that have pushed and developed the agenda of social and environmental management. NOCs, on the other hand, have often been used by governments as vehicles for local development, and therefore tend to have a view that equates social performance with corporate philanthropy (e.g. PEMEX in Mexico, PDVSA in Venezuela, Petrobras in Brazil, and Saudi Aramco in Saudi Arabia). Furthermore, in my experience, there is a variety of practices and conceptions of what social and environmental management in oil and gas operations should look like. Nevertheless, this chapter will look at some of the broad trends in the industry, whilst also trying to point out the differences in policy and practice.

that their industry needed to address the issues of mining and sustainable development in a more coordinated manner (Littlewood 2000). As a result of this decision, nine of the major mining companies grouped together to form the Global Mining Initiative (GMI), to spur a major study of mining and sustainability entitled 'Mining, Minerals, and Sustainable Development' (MMSD). The aim of the MMSD project was to carry out a comprehensive survey of the sector and identify a path towards more responsible behaviour and improved reputation (MMSD 2002). The GMI eventually led to the creation of the International Council on Mining and Metals (ICMM), which was founded in 2001 to improve sustainable development performance in the mining and metals industry (Franks 2015). The ICMM was set up as a CEO-led industry association with a mandate based on the core recommendations of the MMSD project, and over the years it helped both to standardize the mining industry's approach to social and environmental issues and to promote good practice norms through the development of industry standards and good practice guidance (see Hodge 2017 for a more in-depth discussion of the mining industry and sustainability).

The oil and gas industry, on the other hand, did not collectively respond to these external pressures. No comprehensive sustainable development review of the industry took place, and there was no industry-wide response to sustainable development issues, let alone one led by company CEOs. Though there was already a global oil and gas industry association for environmental and social issues called IPIECA,[7] its influence on the industry was limited.[8] Indeed, during this period, change around social and environmental management amongst IOCs was mainly being led by individual companies, in particular Shell and BP, while the rest of the industry lagged behind on these issues. Utting and Ives (2006) have elaborated on this issue of the 'leaders' versus the 'laggards' with regard to social and environmental management in the oil and gas industry. They argue that this difference in approach, between proactive IOCs such as BP and Shell, and others such as Exxon, Chevron, and Total, can be explained by a complex mixture of differences in the pressures

[7] When IPIECA was originally set up in 1974, the acronym stood for the International Petroleum Industry Environmental Conservation Association. However, in 2002 IPIECA stopped using the full title and now only refers to itself as 'IPIECA, the global oil and gas industry association for environmental and social issues'.

[8] IPIECA was formed in 1974 following the launch of the United Nations Environment Programme, and its mandate was, and continues to be, to help 'the oil and gas industry improve its environmental and social performance by: developing, sharing and promoting good practices and solutions; enhancing and communicating knowledge and understanding; engaging members and others in the industry; working in partnership with key stakeholders' (http://www.ipieca.org/, accessed 28 December 2016). Unlike the ICMM, IPIECA has not developed binding standards or policies for members, and it is not CEO-led. It also works on a significantly smaller budget, with membership fees a fraction of those of the ICMM. As a result, the extent to which IPIECA has driven good practice in social and environmental performance in the industry is questionable.

applied to companies by NGOs, which tended to focus on specific countries and companies at the expense of others—such as Shell in Nigeria—and the pressures exerted by shareholders and governments, as well as by internal leadership (Utting and Ives 2006: 23–6). As with many of the mining companies that spearheaded the GMI, Shell and BP's approach to social and environmental issues in the late 1990s was driven by their experiences of social and political crises in Nigeria and Colombia respectively, and of the resulting civil society pressures, as well as by internal senior leadership.[9]

Throughout the 1980s and 1990s, the oil and gas industry mainly focused on improving health and safety, and to a certain extent environmental performance. There was very little focus on social performance and impacts on communities other than the usual philanthropic community development projects and donations. From the late 1990s, Shell and BP started to engage more with external stakeholders, in particular various civil society groups and NGOs, as well as to set up internal social performance units to drive better social management throughout the business. The mandate of these new social performance departments was to push for better social performance throughout the business to ensure impact management, proper consultation with local communities and more widely impacted stakeholders, and more sustainable community social investment projects linked to the business (see e.g. Fossgard-Moser 2005). With regard to environmental management, the new approach involved tackling and engaging with wider environmental issues beyond local permitting regulations, such as climate change, gas flaring, and biodiversity. Notably early compared with its peers, in 1997 BP became the first oil and gas company to publicly recognize the risks of climate change and push for industry to take a role in finding solutions.

In terms of company standards, Shell took the first step in 1997 to update its business principles to include commitments to manage its social impacts and be a 'good neighbour' to the communities in which it operated. In 1999, after the merger with Amaco, BP updated its business policies to include commitments on ethical conduct, employees, relationships, and health, safety, and environmental performance. Both companies publicly expressed their support for human rights, and also took an active role with the UK and US governments and various human rights NGOs in bringing about one of the

[9] John Brown, CEO of BP between 1995 and 2007, was known to promote the importance of good corporate citizenship, and during his tenure at BP took a very involved leadership role in pushing for better social and environmental performance in the company (Bader 2014). Within Royal Dutch Shell, Mark Moody-Stuart, chairman of the company between 1998 and 2001, reportedly believed that 'sustainable development and financial performance are inextricably linked' (*The Independent* 2000).

key voluntary human rights standards for the industry, namely the Voluntary Principles on Security and Human Rights (VPs).[10]

In effect, Shell and BP during this period started to shift the field away from a 'business as usual' approach—characterized by the attitude that a company's role is to focus on its core commercial business, comply with local laws, pay its taxes, and provide a few ad hoc benefits to local communities—to a 'good corporate citizen' approach. This new approach was characterized by the view that strict legal compliance is often not enough, and that social and environmental impacts can be complex and multifaceted, necessitating more comprehensive management focused on impact management, benefit enhancement, and relationship-building with external partners. Part of this approach involved developing internal corporate codes of conduct and establishing corporate functions dedicated to what became known as 'social performance' issues, as well as publishing annual reports detailing how they were managing social and environmental issues, joining various international initiatives aimed at improving social and environmental standards in industry, and engaging more openly with some of the NGO critiques.

## 20.3 Laggards Catching Up: The Impact of International Standards

Despite the variations in initial company responses to social and environmental management in the oil and gas industry, most IOCs eventually started to take the same direction in their approaches. This incremental change across the industry can be partly explained by companies influencing each other's behaviour, as well as by growing societal expectations around company conduct, and an emerging group of social and environmental professionals sharing and implementing practice across the industry. However, the development of international norms has played a major role in pushing the wider industry at least partly along the same trajectory as BP and Shell took in the late 1990s.

[10] Established in 2000, the VPs were designed as a set of voluntary principles specifically for the extractive industries, with the aim of guiding companies in maintaining the safety and security of their operations within an operating framework that encourages respect for human rights (http://www.voluntaryprinciples.org, accessed 28 December 2016). The VPs initiative is a tripartite multi-stakeholder group, currently made up of ten governments, thirty extractive companies, and thirteen NGOs. As of 2016, all of the five major international oil and gas companies are signatories of the VPs (Shell, BP, Exxon Mobil, Chevron, and Total), as well as a number of large to medium-sized international oil and gas companies (Conoco Phillips, Hess Corporation, Marathon Oil, Occidental Petroleum Corporation, Repsol, Statoil, ENI, Pacific Exploration & Production Corporation, Premier Oil, Seven Energy, Woodside Energy, and Tullow Oil), but no national oil and gas companies.

In the last fifteen years, a number of international standards related to the management of social and environmental issues in the extractives or the business sector more widely have been developed, more often than not through multi-stakeholder processes. For example, the above-mentioned VPs are a standard that has widely shaped the way IOCs manage the security of their operations. Similarly, the UN Guiding Principles on Business and Human Rights, developed over several years and unanimously approved by the UN Human Rights Council in 2011, have arguably led to most major and medium-sized IOCs developing internal human rights policies and focusing more on their community grievance mechanisms. But the standard that has had the most impact on the oil and gas industry and how it approaches social and environmental issues is the International Finance Corporation (IFC) Environmental and Social Performance Standards.

The IFC Performance Standards came about as the result of years of pressure from civil society on the IFC, as well as controversy around IFC projects such as the Pangue Hydroelectric project in Chile (Franks 2015: 4). Building on the World Bank Social and Environmental Safeguard Policies and the recommendations of the independent review report on the Pangue Hydroelectric project, the Performance Standards were developed in the first part of the 2000s, formally published in 2006, and updated in 2012. Applying to IFC private-sector clients operating in emerging markets, there are eight IFC Performance Standards: Assessment and Management of Environmental and Social Risks and Impacts; Labour and Working Conditions; Resource Efficiency and Pollution Prevention; Community Health, Safety, and Security; Land Acquisition and Involuntary Resettlement; Biodiversity Conservation and Sustainable Management of Living Natural Resources; Indigenous Peoples; and Cultural Heritage. These standards cover most aspects of the management of social and environmental issues facing oil and gas projects from early project development through to construction, operation, and closure.

The Performance Standards can be seen as ground-breaking for their time, presenting for the first time a comprehensive set of environmental and social standards, which detailed real steps a private-sector project needs to take to manage its environmental and social risks and impacts. Importantly, these standards went beyond the vague policy statements of guidelines such as the Global Compact. Overall, the IFC Sustainability Policy, as embodied in the Performance Standards, can be summarized as 'do no harm' and 'enhance positive development outcomes on the ground' (IFC 2012: 2). Since their inception, the IFC Performance Standards have become the international best-practice benchmark for social and environmental risk management in the extractive sector, widely referred to by social and environmental practitioners, even for projects that are not seeking IFC funding. Furthermore, the Equator Principles—an environmental and social risk management framework adopted by most major international banks and applied to private-sector projects seeking financing from these banks—are modelled on the IFC Performance Standards, therefore furthering the influence of these standards. As a result, the IFC Performance Standards have become a significant contributing factor to a gradual shift towards wider adoption of social and environmental risk management amongst IOCs, and a majority of IOCs now model their internal environmental and social standards on the IFC Performance Standards to a certain extent.

The growing interest of mainstream investors, from both international banks and private equity investment funds, in environmental, social, and governance (ESG) investment criteria to assess oil and gas firms' activities is an additional factor in shaping the industry's approach to social and environmental issues (BSR 2012; Crosse and Horaks 2014). In particular, investors with a long-term focus on investments such as pension funds and SWFs have been increasingly integrating ESG criteria across their portfolios and applying pressure on IOCs to improve their performance in this regard (BSR 2012). The recent example of Norway's Sovereign Wealth Fund putting pressure on ExxonMobil and Chevron to report in more detail on the risks of climate change, and in 2013 threatening to sell shares in IOCs operating in Equatorial Guinea because of the clear lack of benefit to the local populations, is a case in point (Doyle and Fouche 2016; Fouche 2013).

The overall approach to social and environmental management amongst IOCs today is therefore one of risk management based on the overall aim to 'do no harm' and 'provide benefits' to local communities. What we have seen in the multinational oil and gas industry, over a trajectory of twenty years or so, is a gradual shift from a 'business as usual' approach—characterized by very little engagement with the social arena other than through philanthropy and small social investment projects in neighbouring communities, and a legal compliance approach to environmental management—to an environmental and social risk management approach modelled on various international standards. Most IOCs today are members of the VPs; have human rights policies or statements that claim to be consistent with the UN Guiding Principles; have social and environmental policies and management frameworks modelled to a certain extent on the IFC Performance Standards; and adhere to international environmental and health and safety standards.[11] Section 20.4 looks at what this approach actually means in practice, and at some of the gaps and issues that have emerged in its implementation.

[11] e.g. ISO 14001 and OHSAS 18001.

## 20.4 Environmental and Social Performance in the Industry Today: The Realities and Problems of Practice

Environmental management in practice broadly entails impact identification and management. Most IOCs (in common with mining companies) will prepare environmental impact assessments (EIAs) early on in project developments, and then implement environmental management plans to manage environmental issues throughout operations.[12] Specifically, the major IOCs report that they implement measures to limit impacts on and protect biodiversity in areas where they operate; minimize water consumption and discharges; implement preventative measures to avoid spills, and measures for rapid spill response in case of spills; reduce damaging air emissions such as volatile organic compounds that have negative impacts on human health; and reduce greenhouse gas (GHG) emissions in order to address the risk of climate change through improved energy efficiency in operations, as well as reduced flaring, venting, and fugitive emissions.[13] Some companies, such as ExxonMobil and Statoil, report that they are actively exploring carbon capture storage techniques as well. In terms of quantifiable data, all of the major IOCs report on the volume of operational spills, GHG, and other emissions from gas flaring and refineries, water use, and waste disposal. From a health and safety perspective, the common approach across the industry is to implement policies and management systems, and to report annually on fatalities, injuries, and process safety incidents. In sum, in terms of actual performance outcomes, it is clear from the data that the environmental and health and safety performance of large IOCs has improved over time, with overall reductions in fatalities, oil spills, freshwater consumption, and air and GHG emissions.[14]

However, despite this overall improvement in environmental and health and safety performance in IOCs, there are clearly still ongoing environmental issues and unmitigated impacts amongst IOCs as well as across the industry as whole. Furthermore, the push to find new resources is leading companies to engage in higher-risk types of operation in more remote places, such as deepwater drilling, or drilling in the Arctic. These types of operation usually involve the potential for more significant environmental impacts. The 2010 BP Deepwater Horizon rig explosion and subsequent oil spill in the Gulf of

[12] BP (2015); Chevron (2015); ENI (2015); ExxonMobil (2014); Royal Dutch Shell (2015); Statoil (2015); Total (2014).

[13] BP (2015); Chevron (2015); ENI (2015); ExxonMobil (2014); Royal Dutch Shell (2015); Statoil (2015); Total (2014).

[14] BP (2015); Chevron (2015); ENI (2015); ExxonMobil (2014); Frynas 2010; Royal Dutch Shell (2015); Statoil (2015); Total (2014).

Mexico is a high-profile example demonstrating this problem. This particular environmental and health and safety disaster resulted in eleven fatalities and the largest offshore oil spill in US history. Subsequent investigations demonstrated a history of cost-cutting and risk-taking beneath the rhetoric of improving environmental and health and safety performance (Lyall 2010; Morrison 2014: 5–9). Recurring oil spills, such as in Russia or the Niger delta, and ongoing gas flaring in the Niger delta and other parts of sub-Saharan Africa, are other examples of poor environmental performance in the industry (Greenpeace Russia 2014; Luhn 2015). The fact remains that although the oil and gas industry has improved its environmental performance over the last couple of decades, in large part due to the development and tightening of national environmental regulation as well as increased public scrutiny, it remains an industry with significant negative impacts on the environment, in particular with regard to oil spills and GHG emissions.

In terms of social performance, the general practice amongst IOCs is one of managing social impacts at the local community level through social impact assessments (SIAs) and social impact management plans, with a particular focus on critical issues such as impacts on indigenous peoples, resettlement, or cultural heritage. This overall approach to social performance also includes consulting and engaging with communities in a systematic manner, and providing benefits through community development projects (termed in the industry 'social investment' (SI)) or local content programmes. However, it is difficult to ascertain actual social performance across the industry in these areas, as there is a general lack of social indicators in company and project reporting, and when these do exist they tend to be 'process' rather than 'outcome' focused. For example, in terms of social data, IOCs only report on annual SI expenditure, which gives no indication of how effective these SI projects actually are, or how well they are managing their social impacts. Other than SI expenditure, the rest of social reporting tends to focus on case studies or processes, which do not necessarily give a good overall indication of the strength of social performance in the industry, or of actual outcomes. Although IOCs have clearly come a long way in their approach to managing social risk at the community level, a number of issues and trends are emerging that point towards implementation failures, as well as shortcomings in this social risk management framework of social impact mitigation and management, community consultation, and social investment.

SIAs, for example, are now commonly implemented by IOCs in early project development, as a tool to manage social risk as well as part of larger EIAs to gain local environmental permits. The underlying philosophy of SIAs is to minimize negative impacts and enhance positive impacts, through ongoing social impact management (Vanclay and Esteves 2011: 5). For example, in a typical oil and gas project, an SIA might seek to minimize potential community resettlement by moving planned project infrastructure to different areas, and maximize potential jobs available to local community members by advising the implementation of a local content policy. The social element of impact mitigation and planning was historically omitted from national permitting processes. However, in the last decade more countries have mandated SIA as part of the EIA permitting process (although the level of requirements on the social side still tends to be much lower than on the environmental side), and SIAs are mandated in most projects that seek international financing.

Nonetheless, the implementation of SIAs has been widely criticized by NGOs and communities as ineffective. One of the problems with SIAs is the lack of common methodology or criteria, creating a lack of consistency and often of quality, a view backed up by Bruce Harvey in his foreword to a book on emerging trends in SIA (Harvey 2011: xxx). Other issues include a lack of information disclosure and consultation with impacted communities during the SIA process; a failure to adequately deal with cumulative impacts or positive impact enhancement measures; and regular omission of wider macroeconomic and institutional issues, such as the distribution of taxes and royalties (Harvey 2011; Vanclay and Esteves 2011). Ultimately, as Vanclay and Esteves argue, SIAs too often become box-ticking exercises, with little real analysis or integration of the SIA into cross-functional social management plans (Vanclay and Esteves 2011). Finally, SIAs in the oil and gas industry are usually only developed for new project developments, leaving many older operations—developed before SIAs became the norm—without adequate impact management procedures.

A related emerging issue is a common failure of oil and gas proponents to engage in meaningful consultation or proper information disclosure with impacted communities. On the one hand, consultation with communities is much more common and systematic in extractive projects now than in previous decades, when government arguments of 'national interest' prevailed and the exclusion of local communities from project decision-making processes was the norm (Harvey 2011: xxviii–xxix). Most IOCs have made public commitments to engage in an ongoing manner with impacted communities, through both formal and informal mechanisms; and the importance of consultation to gain a company 'licence to operate' is commonly accepted across the industry. Furthermore, there is a growing trend of governments developing legislation around community consultation, particularly in the cases of extractive projects (e.g. Bolivia's Consultation and Participation law in the case of hydrocarbon projects), with which companies have to comply. The IFC Performance Standards and other international financing standards also stress the importance of community consultation in all project phases.

However, despite these advances, oil and gas companies frequently fail to consult properly with communities, and as a result consultation processes are often seen not to be meaningful (Wilson et al. 2016). It is clearly as a result of these failings that NGOs pushed for the promotion of 'meaningful consultation' throughout the draft Environmental and Social Standards in a recent review and update of the World Bank Safeguard Policies (Wilson et al. 2016: 9). The World Bank definition of 'meaningful consultation' stresses that consultation should take place early and throughout the project cycle, be inclusive and free from coercion, and provide timely disclosure of relevant and understandable information (Wilson et al. 2016). Similarly, it could be argued that the push from civil society organizations for companies to apply processes of 'free, prior, informed consent' (FPIC) to all communities, rather than just to the indigenous communities from whose rights FPIC derives, is another example of the increasing dissatisfaction with current practices of consultation in oil and gas projects.[15]

Finally, ongoing criticisms of company SI projects still prevail. This is not new, and various academics have written about the problems of company SI projects over the years (e.g. Frynas 2003; Frynas 2005; Hilson 2012; Utting and Ives 2006). The problems highlighted have ranged from the outright failure of SI projects, to their inadequacy to address truly sustainable development challenges, to SI projects creating tensions and conflicts within communities (Utting and Ives 2006: 20). Many authors have ultimately questioned whether companies should be engaging in community development projects at all (e.g. Hilson 2012). However, the fact remains that in MICs and LICs, communities are often marginalized from an already weak public administration system. As a result, companies are usually expected to step into the gap and contribute to local community development. Not only is this widely expected by communities themselves, it is now increasingly becoming mandated by national governments in permitting requirements or production-sharing contracts (McNab et al. 2012).

Broadly speaking, over the last two decades, IOCs have gradually changed their approach to community SI, from ad hoc philanthropic donations to an approach that has sought to deliver better development impacts at the local level (Kunanayagam and Dietsche 2014; McNab et al. 2012). This approach involves developing partnerships with delivery-oriented NGOs or development organizations to implement projects (Kunanayagam and Dietsche 2014; Utting and Ives 2006: 20). One of the objectives of this newer approach for companies has been to use SI 'to mitigate social risk, protect their corporate social licence to operate, and address growing societal expectations' at the

[15] See e.g. Oxfam's campaign for FPIC from all communities affected by extractive projects (Katz 2015). See also Tomlinson (2017) for an overview of the extractive industry's approach to FPIC.

local community level (McNab et al. 2012: 2). However, one of the drawbacks of this approach is that the industry has failed to deal with the social risks it faces through wider resource governance issues around poor macroeconomic performance and rent-seeking in many LICs and MICs (Kunanayagam and Dietsche 2014: 23–4). As a result, even the best thought-out SI projects actually often fail to meet growing societal expectations of benefits from oil and gas projects.

Although most IOCs have improved the transparency of their payments to governments and the management of bribery and corruption through membership of the EITI, as well as through compliance with legislation such as the UK 2010 Bribery Act, national regulations developed on the back of Chapter 10 of the EU Accounting Directive, and/or the US Foreign Corrupt Practices Act, increasing transparency is only the first step towards improving government accountability and public policies (Kunanayagam and Dietsche 2014: 24). Kunanayagam and Dietsche, amongst others, suggest that the way to address these problems is for companies to take a more strategic approach to SI based on four elements: an upfront macro and micro socio-economic and political analysis; a strategy to seek areas of investment 'where public policies authorities are currently not sufficiently addressing the drivers of social risks and opportunities'; delivery of SI projects through 'delivery-oriented development partners'; and implementation of pilot SI projects to test the ground for the potential to leverage and scale up these projects through collaboration with government and donor agencies (Kunanayagam and Dietsche 2014: 25). This type of 'strategic SI' is slowly being taken up by some IOCs, has more of a direct link with the business of oil and gas, and usually includes projects that aim to enhance local content through supplier or skills development initiatives. However, most IOCs have yet to fully grasp the overall strategic importance of this type of approach to SI, particularly for gaining not only the local community 'licence to operate' but also, in many LICs and MICs, the government and wider society 'licence to operate'.

IOCs' approach to social and environmental management has evolved over the last twenty years, leading to developing practice and improvements in performance. All IOCs today publicly claim to be managing social and environmental issues, and report on how they are doing this in practice through their annual sustainability reports. However, despite this overall improvement in performance, gaps in the implementation of standards and poor practices still continue in the industry. This section has broken down what social and environmental management means in practice, and has discussed some of the gaps and issues that are emerging. Section 20.5 looks at how social and environmental management in the oil and gas industry is often misunderstood, leading to a lack of understanding of the nature and drivers of social and environmental performance.

## 20.5 Moving beyond Corporate Social Responsibility and Common Misunderstandings

In academia, and among NGOs and policy makers, the extractive industries' approach to dealing with the social and environmental aspects of corporate operations has generally been categorized under the umbrella term 'corporate social responsibility' (CSR).[16] The overall idea of CSR is that companies should behave responsibly towards society and contribute to sustainable economic development, and that they can expect to achieve this by adhering to various voluntary standards (i.e. self-regulation). Over the years, there have been a number of academic articles and NGO reports outlining some of the problems of CSR in the oil and gas industry.[17] By and large, the criticisms have been that although CSR programmes in companies have had some positive impacts, they are often little more than badly thought-out community development programmes with a corporate public relations aim, for an industry that is still having negative impacts on communities and the environment, and that is actively resisting any kind of regulation of its activities (e.g. Christian Aid 2004; Hilson 2012).

One of the problems around much of the writing on CSR in the oil and gas industry is that not only are the drivers of CSR commonly misunderstood and misrepresented, but what the oil and gas industry is actually doing in terms of environmental and social management also does not fit with the concept of CSR. More often than not, CSR is conflated with community social development programmes, therefore omitting the important aspect of management of impacts. A recent World Bank study on the extractive industries sector, in which 'social safeguards' in the extractive industries are equated with sharing the benefits of projects at the local level through community development funds, is a case in point (Halland et al. 2015: 82). As discussed in this chapter, corporate contributions to community development (or social investment) are but one aspect of social performance in the oil and gas industry.

Another common misunderstanding is that the term CSR has come to be equated with 'voluntary' self-regulation, as opposed to formal regulation. One of the main criticisms of CSR is that since it is a voluntary approach, companies cannot be penalized for poor or absent implementation. In fact, some NGOs have even critiqued CSR's entire *raison d'être* as a corporate-led approach 'to promote self-regulation as a substitute for regulation at either national or international level' (Christian Aid 2004: 5). According to some

[16] It should be noted that there is no single agreed definition of CSR and the term is often used to describe a number of different approaches, from corporate philanthropy, to corporate partnerships in community development, to the broad management of social issues. See e.g. Christian Aid (2004); Frynas (2005); Frynas (2010); Hilson (2012).

[17] See e.g. Christian Aid (2004); Gilberthorpe and Banks (2012); Hilson (2012).

authors, this situation is compounded in developing countries where 'the drive to legislate and enforce regulations is lacking', as opposed to the situation in developed countries (Hilson 2012: 136).

However, this is not an accurate overall depiction of the field of social and environmental management in the oil and gas industry. Although some companies in certain instances have no doubt lobbied against certain pieces of environmental or social legislation (such as corporate lobbying against the US Alien Tort Claims Act), the drivers of social and environmental performance in the industry are often not entirely voluntary. For example, there is a growing body of national legislation around community consultation and mandated company social investment in Latin America and Africa,[18] and companies are increasingly having to comply with national legislation around environmental and social issues, as well as with international standards to access financing. Furthermore, increasing community expectations and pressure are also pushing companies to manage these issues from a risk perspective, in order to gain their 'social licence to operate'. Therefore, the simple dichotomy between voluntary self-regulation and mandated regulation is not in fact an adequate way to understand the complexities of the field of social and environmental management in the industry.

One of the main aims of this chapter has been to give an overview of how oil and gas companies are actually approaching and implementing social and environmental performance across their operations. It is only by unpacking and ultimately discarding terms such as CSR, and by truly examining what oil and gas companies are actually doing in social and environmental management, that one can begin to answer the question of whether and how companies can generate positive and sustainable economic and social outcomes for local populations, or perhaps more importantly—considering the state of performance across the industry—why they are failing to do so in many instances.

While implementing strong social and environmental performance can be challenging and complex, the tools for implementation are there, and are well articulated for the industry. For environmental performance, this involves implementing EIAs and management plans, and ultimately minimizing and avoiding the well-known environmental impacts of oil and gas operations around air emissions, oil spills, water discharges, and water consumption. For social performance, these involve implementing SIAs and social management plans, engaging in a meaningful manner with stakeholders, and implementing strategic investment projects that address both macro and micro socio-economic issues in partnership with government, donor, or NGO parties. To successfully implement these tools and achieve good social and environmental

[18] See e.g. Otto (2017).

performance, a company needs strong and engaged leadership, qualified and experienced professionals to implement the work programmes, mandatory standards, well-designed management systems, and a clear assurance process to drive performance across the company.

The question therefore remains: why is performance still so uneven, both across IOCs and within individual companies? Part of the answer may lie in the challenges of driving good performance across a large multinational company and poorly developed internal management processes. While most IOCs have environmental management systems that are usually quite clearly articulated, social management systems and standards can be a little vague, and often are not performance-based. This potentially leaves different arms of a company with too much autonomy to implement social performance according to their own interpretations, with only limited checks possible from any assurance framework. However, despite gaps in management frameworks, poor environmental and social performance often ultimately comes down to senior leadership.

Although social and environmental management in the oil and gas industry is now largely about risk management and legal compliance, the manner in which a company approaches this is fundamentally down to leadership and culture. It is no surprise that one of the main reasons that BP and Shell took a proactive approach to social and environmental issues, at a time when most other IOCs were not doing so, was largely due to internal leadership. It is also widely known that when these engaged leaders stepped down, the emphasis on social and environmental performance changed; see, for example, Bader (2014); Morrison (2014: 7). The presence of strong social and environmental performance in a company is almost always because the person put in charge of these areas has been given a clear mandate by, and has a direct link to, the CEO and/or the board. Unfortunately, industry leadership on these issues today is uneven, and too often IOCs are seen to be taking a legalistic or public relations approach to solving complex social and environmental conflicts. Chevron's approach to the Texaco oil pollution lawsuit in Ecuador is a telling example. It is interesting that in Daniel Franks's account of the formation of the CEO-led GMI and the subsequent MMSD project in the late 1990s, he recounts that initially the CEOs of the different mining companies present were broadly split into two camps: those who thought that the solution to the growing anti-mining movement was to 'sell' a better message (i.e. better public relations), and those who thought that the problems were more fundamental, and that systematic poor social and environmental performance in the industry had to be addressed (Franks 2015: 8). This same tension—between selling a better message and actually addressing poor performance and taking a more sustainably driven strategic approach—remains an ongoing and unresolved issue in the oil and gas industry today.

## 20.6 Conclusions

This chapter has sought to give an overview of social and environmental management in the oil and gas industry. It has outlined the evolution of IOCs' approaches over the last twenty years, and has reviewed what social and environmental management amongst IOCs means in practice, and what some of the emerging issues are today. The chapter has argued that much of the academic writing looking at social and environmental performance in the oil and gas industry uses the outdated concept of CSR. This is a concept that is no longer widely used by social and environmental practitioners in the industry, and one that also leads to misunderstandings of what social and environmental management actually consists of, and of what the drivers behind its application are. The complexity and variety of these drivers—ranging from compliance with international standards to the need to access financing, to local legislation, the risks of the trend of developing local legislation, reputational issues, and gaining a 'social licence to operate'—is the reason why the portrayal of social and environmental management in the industry as a 'voluntary' approach is somewhat inaccurate and misleading. While it is clear that the manner in which a company chooses to approach social and environmental management remains at least partly voluntary and is usually dependent on leadership and culture, no oil and gas company today can operate without addressing these issues to a certain level. This is particularly compounded by the growing national trend to develop social and environmental legislation related to extractives. As a result, oil and gas companies increasingly have to balance international standards, local legislation, and growing community expectations. It is precisely because of this reality that companies need durably strong internal policies and management systems, as well as adequate human and financial resources to manage these challenges.

The chapter also describes how, despite a certain variety of approaches between IOCs in addressing social and environmental issues, the broad trend across the industry is to approach these issues through a risk management approach, broadly modelled on international standards such as the IFC Performance Standards and the VPs. Overall, this risk management approach entails an underlying philosophy of 'do no harm' and 'provide local development benefits'. However, although IOCs have clearly evolved their approaches and performance in relation to social and environmental impacts and issues, actual performance is still uneven across the industry, with many instances of unmitigated social and environmental impacts, and a lack of sustainable development benefits for many local communities affected by oil and gas projects. While some of this poor performance comes from the challenges and complexities of implementing good social and environmental performance and driving standards throughout a large company, the usual root cause of poor performance lies in a lack of senior leadership engagement on these issues, leading to companies that either avoid international best practice in social and environmental performance, take unnecessary risks to increase profits, or lack the internal capacity to implement standards and approaches.

## References

Bader, C. (2014). *The Evolution of a Corporate Idealist: When Girl Meets Oil*. Brookline, MA: Bibliomotion.

BP (2015). 'Sustainability Report 2015'. Available at: https://www.bp.com/content/dam/bp/pdf/sustainability/group-reports/bp-sustainability-report-2015.pdf.

BSR (2012). 'Trends in ESG Integration in Investments: Summary of the Latest Research and Recommendations to Attract Long-term Investors, BSR Report, August 2012'. Available at: https://www.bsr.org/reports/BSR_Trends_in_ESG_Integration.pdf.

Chevron (2015). 'Corporate Responsibility Report 2015'. Available at: https://www.chevron.com/corporate-responsibility/reporting.

Christian Aid (2004). 'Behind the Mask: The Real Face of Corporate Social Responsibility'. Available at: http://www.eldis.org/go/home&id=14595&type=Document#.V17GkscsH8s.

Crosse, G. and M. Horaks (2014). 'Financing Criteria for Extractives: A Changing Dimension'. *Ethical Corporation*, September: 9–13.

Doyle, A. and G. Fouche (2016). 'Norway Wealth Fund Turns up Climate Heat on Exxon and Chevron'. Reuters, 3 May. Available at: https://www.reuters.com/article/us-exxon-mobil-norway-swf-idUSKCN0XU1BF.

*The Economist* (1997). 'BP at War'. 17 July. Available at: https://www.economist.com/node/151558.

ENI (2015). 'Sustainability Report 2015'. Available at: http://www.eni.com/docs/en_IT/enicom/sustainability/eni_for_2015_report_eng_.pdf.

ExxonMobil (2014). 'Corporate Citizenship Report 2014'. Available at: http://read.uberflip.com/i/514689-exxonmobil-2014-ccr-fullreport-digital-final.

Fouche, G. (2013). 'Norway Wealth Fund Mulls Selling Oil Firms That Work in E. Guinea', Reuters, 24 April. Available at: https://www.reuters.com/article/exxonmobil-norwayfund-idUSL6N0DB3F820130424.

Fossgard-Moser, T. (2005). 'Social Performance: Key Lessons from Recent Experiences within Shell', *Corporate Governance*, 5(3): 105–18.

Franks, D. M. (2015). *Mountain Movers: Mining, Sustainability and the Agents of Change*. London: Routledge.

Frynas, J. G. (2003). 'Royal Dutch/Shell', *New Political Economy*, 8(2): 275–85.

Frynas, J. G. (2005). 'The False Developmental Promise of Corporate Social Responsibility: Evidence from Multinational Oil Companies', *International Affairs*, 81(3): 581–98.

Frynas, J. G. (2010). 'Oil Industry's Increasing Focus on CSR: Are Oil Companies Acting Responsibly Yet?' *Petroleum Economist*, February: 18–20.

Gilberthorpe, E. and G. Banks (2012). 'Development on Whose Terms? CSR Discourse and Social Realities in Papua New Guinea's Extractive Industries Sector', *Resources Policy*, 37: 185–93.

Global Witness (1999). 'A Crude Awakening: The Role of the Oil and Banking Industries in Angola's Civil War and the Plunder of State Assets'. Available at: https://www.globalwitness.org/en/archive/crude-awakening/.

Greenpeace Russia (2014). 'A Brief Overview of the Oil Spill Problem in Russia'. Available at: http://www.greenpeace.org/russia/Global/russia/report/Arctic-oil/GPRussia_Oil_spills_briefing_ENG.pdf.

Halland, H., M. Lokanc, and A. Nair, with S. P. Kannan (2015). *The Extractive Industries Sector: Essentials for Economists, Public Finance Professionals, and Policy Makers*. Washington, DC: World Bank Group. Available at: https://openknowledge.worldbank.org/bitstream/handle/10986/22541/The0extractive0s00and0policy0makers.pdf?sequence=1&isAllowed=y.

Harvey, B. (2011). 'Foreword: SIA from a Resource Developer's Perspective', in F. Vanclay and A. M. Esteves (eds), *New Directions in Social Impact Assessment: Conceptual and Methodological Advances*, xxvii–xxxiii. Cheltenham: Edward Elgar.

Hilson, G. (2012). 'Corporate Social Responsibility in the Extractive Industries: Experiences from Developing Countries', *Resources Policy*, 37: 131–7.

Hodge, T. (2017). 'Towards Contribution Analysis', WIDER Working Paper 2017/19. Helsinki: UNU-WIDER.

Human Rights Watch (1999). 'The Price of Oil: Corporate Responsibility and Human Rights Violations in Nigeria's Oil Producing Communities'. Available at: https://www.hrw.org/reports/1999/nigeria/.

IFC (2012). 'International Finance Corporation's Policy on Environmental and Social Sustainability'. Available at: http://www.ifc.org/wps/wcm/connect/7540778049a792dcb87efaa8c6a8312a/SP_English_2012.pdf?MOD=AJPERES.

*The Independent* (2000). 'The Man Who Was Sure of Shell', *The Independent*, 11 April. Available at: www.independent.co.uk/news/people/profiles/the-man-who-was-sure-of-shell-281050.html.

Katz, M. (2015). 'More Consultation, Less Exploitation: The Extractive Industries and Community Consent'. Available at: http://policy-practice.oxfam.org.uk/blog/2015/09/more-consultation-less-exploitation-extractives-and-community-consent.

Kunanayagam, R. and E. Dietsche (2014). 'Managing Social Risks in the Extractive Resources Sector', *Ethical Corporation*, September: 19–25.

Littlewood, G. (2000). 'The Global Mining Initiative'. Address to Mining 2000 Conference, Melbourne, 20 September.

Luhn, A. (2015). 'Russia's Rosneft Charged over Pipeline Leak that Caused Oil to Come out of Taps', *The Guardian*, 30 June. Available at: https://www.theguardian.com/environment/2015/jun/30/russias-rosneft-charged-over-pipeline-leak-that-caused-oil-to-come-out-of-taps.

Lyall, S. (2010). 'In BP's Record, a History of Boldness and Costly Blunders', *The New York Times*, 12 July. Available at: http://www.nytimes.com/2010/07/13/business/energy-environment/13bprisk.html?_r=0.

McNab, K., J. Keenan, D. Brereton, J. Kim, R. Kunanayagam, and T. Blathwayt (2012). *Beyond Voluntarism: The Changing Role of Corporate Social Investment in the Extractives Sector*. Brisbane: Centre for Social Responsibility in Mining, University of Queensland. Available at: https://www.csrm.uq.edu.au/publications/beyond-voluntarism-the-changing-role-of-corporate-social-investment-in-the-extractive-resources-sector.

McPhail, K. and A. Davy (1998). 'Integrating Social Concerns into Private Sector Decision-Making: A Review of Corporate Practices in the Mining, Oil, and Gas Sectors', Discussion Paper 384. Washington, DC: World Bank.

MMSD (2002). 'Breaking New Ground: The Report of the Mining, Minerals and Sustainable Development Project'. Available at: http://pubs.iied.org/9084IIED.html.

Morrison, J. (2014). *The Social License: How to Keep Your Organization Legitimate*. Basingstoke: Palgrave Macmillan.

Otto, J. (2017). 'How Do We Legislate for Improved Community Development?' WIDER Working Paper 2017/102. Helsinki: UNU-WIDER.

Royal Dutch Shell (2015). 'Sustainability Report 2015'. Available at: http://reports.shell.com/sustainability-report/2015/.

Statoil (2015). 'Sustainability Report 2015'. Available at: http://www.statoil.com/no/InvestorCentre/AnnualReport/AnnualReport2015/Documents/DownloadCentreFiles/01_KeyDownloads/2015_Sustainability_report.pdf.

Tomlinson, K. A. (2017). 'Indigenous Rights and Extractive Resource Projects: Negotiations over the Policy and Implementation of FPIC', *International Journal of Human Rights*, doi: 10.1080/13642987.2017.1314648.

Total (2014). 'Sustainable Growth Report 2014'. Available at: http://www.total.com/sites/default/files/atoms/files/rapport_croissance_durable_va.pdf.

Utting, P. and K. Ives (2006). 'The Politics of Corporate Responsibility and the Oil Industry', *Stair* 2(1): 11–34. Available at: http://users.ox.ac.uk/~stair/2_1/utting&ives.pdf.

Vanclay, F. and A. M. Esteves (2011). 'Current Issues and Trends in Social Impact Assessment', in F. Vanclay and A. M. Esteves (eds), *New Directions in Social Impact Assessment: Conceptual and Methodological Advances*, 3–19. Cheltenham: Edward Elgar.

Wilson, E., S. Best, E. Blackmore, and S. Ospanova (2016). 'Meaningful Community Engagement in the Extractive Industries: Stakeholder Perspectives and Research Priorities'. London: IIED. Available at: http://pubs.iied.org/pdfs/16047IIED.pdf.

# 21

# The Role of Gender in the Extractive Industries

*Catherine Macdonald*

## 21.1 Overview

### 21.1.1 *Scope*

The extractive industries are a major source of revenue for many resource-rich economies around the world, and are central to their economic growth and social development. Many studies have revealed that the extractive industries have different impacts upon men and women, in a variety of ways. It is important to try to understand those impacts and to determine whether mitigation policies and programmes are needed. Further, the full social benefit of resources development can only be realized if women and girls are able to participate as fully as males in all aspects of resources activity and consequent economic development and social progress. This requires that the principles of gender equality are embedded within policies and practices applied to resources and associated development.

In recognizing that women's participation and gender equity is a precondition for achieving the best development outcomes, some extractive industries companies have committed to integrating gender equality, inclusion, and women's economic empowerment into aspects of their operations, but others have not. Examples are provided that demonstrate what leading companies are doing to integrate gender concerns into their corporate and social/community policies. This chapter considers how widespread these practices are and whether they are effective or need improvement.

### 21.1.2 *Content*

Research for this chapter encompassed a survey of the wide-ranging literature that discusses matters of gender in the extractives sector over the past five years or so. Several themes have emerged from that survey, and these form the basis for the discussion in the rest of the chapter, with a section focusing on each. These main themes are:

1. the understanding of gender in the extractives sector and how this has changed over time
2. the gendered impacts of the extractive industries and whether women are, indeed, passive victims of the sector rather than active participants
3. the nature of extractives-associated sex work and gender-based violence in various settings
4. women's role in resisting the expansion of extractives projects
5. industry efforts towards achieving gender balance and equity in the sector.

This chapter is not based on primary research, although it is informed by fieldwork and interviews that the author has undertaken for other research over many years. It is primarily a synthesis of the most important evaluations of the situation of gender in the extractives sector as undertaken by an impressive array of scholars and practitioners, many of whom have undertaken primary fieldwork and have also made recommendations for improving women's status. It includes an assessment of industry efforts and a summary of recommendations for future action.

## 21.2 The Many Facets of Gender and the Extractive Industries

There is a wealth of literature about women and gender in the extractive industries published over the last five years, and much that goes further back. The more recent articles analyse a variety of different issues and geographical locations and are often unrelated to each other, although a number of common themes emerge from the entirety of the scholarship. Some articles deal with the negative impacts experienced by women in extractive industries project areas; others deal with women's active engagement in the extractives field and the choices they make to improve their position in life; some focus on informal mining (artisanal and small-scale mining (ASM)), while many are concerned with large-scale industrial extraction projects. Certain analyses demonstrate that the business case for diversity in the workplace is fundamentally harnessed by the industry to perpetuate the status quo—which is of a highly masculinized sector—while presenting a modern and acceptable face to society. Much of the research features statistics and indicators about women's involvement in extractive industries, both formal and informal, and whether the numbers are increasing towards any semblance of equality;

still others describe women's participation in battles against extractive projects in many different countries. Other articles discuss the persistently masculine culture of the extractive industries, which continues to work against gender equity despite efforts to transform it. The disparate nature of the scholarship makes it difficult to distil a clear picture of 'gender and the extractives' as a single topic. However, this variety also lends a great richness to the field and reveals the many facets of women's interactions with and activities in and around the extractive industries.

### 21.2.1 *Women and Gender: What Do We Mean?*

Gender is a complex and contested term and tends to be assumed to be 'natural', in the sense that men are seen as 'masculine' and women as 'feminine', and we all think we know what that means.[1] In discussions of gender and the extractive industries, the term 'gender' is often used synonymously with women. Academics writing about women and the extractive industries prefer to use the term 'gender' to indicate that they are referring to a cultural construct, not a biological descriptor; in practice, most of these articles are referring only to women—the gender of men is not considered. A very few scholars (Laplonge 2014, 2016; Mayes and Pini 2014) have recently started to point out that this conflation of the term 'gender' with women is enabling the industry to remain highly masculinized in spite of considerable efforts to achieve 'gender balance' and 'gender diversity'.[2]

It is important to describe the actions of women in and around the extractives sector, both currently and historically. So, much of the discussion in this chapter focuses on women, although the discussion about recommendations for change will return to the discussion about gender, including masculinity.

### 21.2.2 *Aspects of Gender in the Extractives Sector*

Although many scholars describe the masculinity of the extractives sector, Lahiri-Dutt (2015) posits the development of a 'feminization of mining'. She bases this on the increasing presence of women in the *formal* mining sector, the large numbers of women involved in the growing *informal* mining sector, and an evolving debate on the nature of sex work associated with mining settlements. All of these are themes that recur throughout the literature and are discussed at greater length below. She makes a strong case for

[1] See Macdonald (2017: 2–5) for a fuller discussion of the scholarship on gender constructs in the extractives sector.

[2] In this chapter, gender equality is defined as providing equal chances and opportunities to women and men, and gender equity as the process of being fair to both men and women (which might require compensation for past disadvantage in order to level the playing field).

recognizing women's agency in the mining sector, thereby moving beyond the portrayal of women as victims of the negative impacts of mining: an opinion shared by a number of other scholars, such as Mahy (2011), O'Faircheallaigh (2013), and Bryceson et al. (2013a, 2013b). This is an important contribution to the debate about gender in the mining sector.

Women have been active in the mining sector in various ways and in many countries for a very long time. Geographically, the presence of women in the extractives field has varied from place to place. Lahiri-Dutt and Burke (2011) demonstrate the essential roles played by Asian women miners throughout history, specifically in Japan and India. Although they describe the many challenges and obstacles that have existed and continue to exist for women, they nonetheless paint a compelling picture of Asian women's active role in mining history. Murillo (2013) concludes that women were essential to the development of the Mexican silver industry in the seventeenth and eighteenth centuries. Although she was unable to find evidence that women actually worked in mines, she demonstrates that they were crucial to the supporting economy. It is quite common for observers to comment upon the supportive roles played by women around extractives projects, often because they are not able to be directly involved, but Murillo takes the case further with her conclusion that the historical Mexican silver mining industry would not have been able to exist without women's actions.

O'Faircheallaigh makes a strong case for the recognition of indigenous women's influence in the development of agreements with mining companies in both Canada and Australia, especially at the end of the twentieth century and the beginning of the current one. However, he notes also that many scholars have tended to overlook this, stating that the 'dominant view in the academic and activist literature is that women are bypassed in agreement negotiations, and as a result are often excluded from the benefits of mining while continuing to experience its economic, social, and environmental costs' (O'Faircheallaigh 2013: 1790). Although he concedes that women are sometimes (but not always) excluded from negotiating teams, for a variety of reasons, he contends that they are usually actively involved in processes leading up to the formal negotiations, influencing the shaping of agendas and objectives of negotiations and in the institutional structures that implement the provisions of agreements, thus ensuring that matters of importance to them are not forgotten. In his experience as a legal adviser to various aboriginal groups in Australia and Canada before, during, and after their negotiations with resource developers, he has observed that women exerted significant power over the negotiations and the content of the agreements. He gives examples of how indigenous women have ensured that gender equality provisions were included in agreements, particularly concerning income distribution, and also that sufficient attention and resources were allocated to sustainable investments and the recognition of and respect for cultural traditions. He notes that indigenous men did not oppose these measures, but that it was women's influence that insisted upon them.

Thus, although the extractive industries are very largely masculine, nonetheless women have played and continue to play a wide range of significant roles in the sector. Although there are negative impacts of extractives projects upon women, as noted elsewhere in this chapter, it is important to also acknowledge the strong and active roles taken by women in the mining and oil and gas industries throughout history.

### 21.2.3 *Gendered Impacts*

Many discussions of the impacts of the extractive industries state that these effects are differentiated by gender and are more likely to affect women negatively than they do men. For example, the following statement from the World Bank's 2009 guidance volume on gender and the extractive industries states that:

> Men have most access to the benefits, which consist primarily of employment and income, while women and the families they care for are more vulnerable to the risks created by Extractive Industries, which consist of mostly harmful social and environmental impacts. (Eftimie et al. 2009: 1)

While much of the discussion in this chapter emphasizes the ways in which women are actively engaging with the extractive industries in order to derive benefits from them, it is important to also note that there are detrimental impacts coming from the sector that seem to fall disproportionately upon women.

One of the main ways in which women are negatively affected by extractives projects is in the impacts of involuntary resettlement and also by environmental damage. The loss of land and waters that they rely upon to grow food for their families when they are forced to move by an extractives project will typically have a greater impact upon the women of a community, as they frequently take the greater responsibility for subsistence farming. In a similar way, as women are often the most fully engaged in subsistence farming activities, they are also most affected by any pollution arising from extractives projects, because they are more directly exposed. Men are also affected by displacement and environmental damage but, as noted by the World Bank above, they often have better access to alternative incomes and the ability to move to other locations to seek alternative and often better opportunities.

In the case of an iron-ore mining area of Goa in India, D'Souza et al. (2013) describe the detrimental effects of badly managed physical displacement of subsistence farming families, when men took the little compensation paid by the companies but wasted it on short-term personal consumption, while the women lost the land on which they could grow food for their children. They also lost access to clean water, necessitating much more time-intensive efforts to obtain water for the home. These negative impacts, among others, were shown to take a great toll on women's social and physical well-being, as well as their economic situation. This case study, which focused on the health impacts of mining on women, showed that poorly mitigated impacts of extractives projects can have major deleterious effects upon women and their families.

A number of articles (Mukherjee 2014; Omeire et al. 2014) describe the way in which indigenous women in developing countries (here India and Nigeria) are disproportionately affected by mining and oil and gas projects because they lose access to forest, fields, and fisheries, which they have previously used for food to feed their families. Men may receive compensation and jobs when an extractives project moves into their area, but women generally do not, so there is no replacement for their lost incomes. Articles by Omeire et al. (2014) and Oluduro and Durojaye (2013) assert that in the Niger Delta women are also more affected by pollution and gas flaring because their livelihoods are dependent upon the land and the water, so they are more exposed. They also say that the loss of land-based livelihoods drives more women to sex work in the absence of alternatives. Some of these outcomes are not actually well demonstrated—they are merely stated, but there is clearly some basis for the claims.

Other scholars have researched the varied roles that women have played in the extractives sector in Africa, most often in informal mining, but some have also looked at women in the formal sector. For example, Lauwo (2016) presents an interesting post-structuralist analysis of the essentially masculine mining discourse, even when discussing gender balance, and then relates it to the poor performance of gender equality measures in the Tanzanian mining industry. This theme, of the disconnect between rhetoric and reality, occurs throughout the literature, and is observed from a number of theoretical perspectives. Mayes and Pini (2014) demonstrate that the use of gender equity targets and discussion of the business case for diversity in the Australian extractives industry discourse are actually effective in maintaining the status quo: perhaps an unintended consequence. The business case, central to these efforts, incorporates a normative role for women, namely that of an idealized 'civilizing' woman, but fails to mention equality. Instead, women's difference is used to ensure that the mining space remains male. Indeed, the public efforts at 'gender balance' are used as a rationale for resisting regulation and structural change. Lauwo's effort at radical post-structuralist feminist analysis arrives at the same conclusion—that talk of gender balance in the Tanzanian mining industry bears no significant result (Lauwo 2016).

### 21.2.4 *Gender-based Violence and Sex Work*

Much of this section on sexual relationships around extractives projects is concerned with the ASM sector, but not all. Studies by Lockie (2011) in Australia and Cane et al. (2014) in Mongolia focus on large-scale mining projects. Mahy's (2011) Indonesian research was also concerned with the impacts of a large-scale rather than an artisanal mining project. These examples apart, the cases referred to in this section focus on ASM. That is not surprising, as the informal mining sector is reputed to provide many more opportunities for women than does the formal mining sector, which is heavily masculinized. That is not to say that the ASM sector typically treats women fairly, or that there is any form of gender balance or equality operating there. Just the same, there is much more scope and flexibility for women to participate in the economic benefits associated with informal mining, so it is important to discuss gender relations in the ASM context. Much of the work available for women in mining areas involves sexual aspects, but not all of it. Further, a growing list of scholars is demonstrating that the sex work associated with mining is often a preferential choice for women workers, rather than a case of victimization.

Mahy (2011) and Bashwira et al. (2014) point out that mining-related prostitution can benefit women who choose to earn money in this way and that it is crucial not to assume that all women taking up this profession are 'victims' of mining, especially in the ASM sector. Mahy's research among women sex workers near a large mine in Indonesia revealed that they should be viewed neither as victims nor as heroines, but rather as 'women who are pursuing a livelihood opportunity within their wider socio-economic context' (Mahy 2011: 53). She points out that, for many of the women, their income-earning capacity and lifestyle options are greater in the mining town than they would have been in their home villages, and this includes a variety of work possibilities for single women, among which sex work is often the most lucrative, especially for those with limited education. The range of options may be limited, but engaging in sex work in a mining settlement can be a rational livelihood choice rather than a question of force or victimization.

In a far more conflict-ridden setting, that of the Democratic Republic of Congo (DRC), two separate teams of scholars—Bashwira et al. (2014) and Kelly et al. (2014)—criticize the use of the high incidence of gender-based violence (GBV) as a reason for controlling women's ability to access income in the ASM sector. They particularly oppose the tendency of development programmes, implemented by donor agencies and NGOs, to promote women's exit from the mining sector as the best means for protecting them from violence and exploitation. They argue instead that programmes to improve women's situation within the industry would be more beneficial. They agree with Mahy that women's agency and active engagement in the mining sector need to be recognized and understood better by policy makers. Given the paucity of alternative means of income and the choice of many women to become involved in the ASM sector as an attractive livelihood alternative among the limited options available, they claim that legislation aimed at protecting women by excluding them from mining areas is actually counterproductive. They state that 'instead of serving as an instrument to protect women miners' health, it is currently being abused as a tool to consolidate the male-dominated nature of the ASM industry in eastern DRC' (Bashwira et al. 2014: 112). They exhort policy makers to work at empowering women within the industry, tackling gendered power relations and structural inequalities rather than excluding women from the field.

Some of the most interesting recent scholarship documenting the 'complex interplay of competing sexual desires, emotional needs, social status, daily practicalities, and economic security objectives' (Bryceson et al. 2013a: 50) of women in informal mining settlements has been produced by Bryceson et al. (2013a, 2013b) in Tanzania. They reject the assumption that most women's roles in these communities revolve around prostitution, and have instead developed the term 'wifestyles' to describe the many different relational forms pursued by women and men in these places. Through their fieldwork, they have concluded that women seek to become steady girlfriends or wives of miners, as this status provides some material and emotional security, in exchange for a range of domestic services. Although many of these relationships are not long term, in the light of the fluid nature of mining work and settlements, some certainly are. This in turn highlights the need to avoid blanket statements about women's welfare in artisanal mining settings. Another key point they make is that the relationships in ASM communities are financially and emotionally interdependent, and of benefit to both men and women, as long as they last:

> What is important to stress is that it is not only women depending on men's income. Miners, constrained by erratic income-earning, fall back on girlfriends/wives' income-earning as well. It remains to be seen who provides the bulk of shared income but whatever the case, the income exchange is likely to be vital for continued habitation in the settlement. (Bryceson et al. 2013a: 51)

Reflecting upon Tanzanian women's views of themselves and their life choices ('wifestyles') is an essential part of understanding the role of gender in the ASM sector. Perhaps the strongest point made about women's agency and choices by Bryceson et al. is that:

> Women in Tanzanian mining settlements generally do not perceive or portray themselves as victims of sexual oppression. No longer subject to the control of their elders, they have migrated to the mining settlements, engaged in sexual relationships, and pursued productive and reproductive paths of self-making in or out of relationships with men. (Bryceson et al. 2013b: 102)

This research demonstrates the inadequacy of statements assuming that women engaged in a range of relationships and activities in communities surrounding mining settlements are victims, as prostitutes or in some other way. Some may be, but many women choose to engage in this sphere as their best chance for self-advancement at a certain stage of their lives.

Bashwira et al. (2014) suggest the mainstreaming efforts of the Mongolian government and the Swiss Agency for Development Co-operation (SDC) in the Sustainable Artisanal Mining Project as a good example of how to improve gender equality in ASM communities. This deliberate approach to include women in training to build their capacity is anticipated to yield better results than the approach of finding alternative income sources as was proposed in the DRC case (Purevjav 2011: 209). The SDC project website (www.sam.mn) notes that more than 40 per cent of Mongolian ASM workers who registered in social and health insurance schemes in 2015 were women, although they accounted for only 30 per cent of registered ASM miners, indicating that women were embracing opportunities to improve their living standards while engaging in artisanal mining. Although the issues are different between the formal and informal sectors, Cane et al. (2014) reported in a study of GBV undertaken in communities in the proximity of two large-scale mines in southern Mongolia, that the Mongolian government and a private mining company entered into a partnership to make the mining area community a safer and more family-friendly place (Cane et al. 2014: 35), thus altering the gendered behaviour status quo.

In-depth studies of GBV in relation to extractives projects are not yet common, but are likely to increase in importance as current campaigns about gendered violence in mining-focused countries like Australia and Canada gain currency, along with rising expectations that companies must act to ensure human rights in their spheres of influence, as promoted by the Ruggie Principles (United Nations Office of the High Commissioner for Human Rights 2011). Often, all that is available in the public domain are news articles and announcements of company programmes, such as those made by the Canadian gold miner Barrick in response to authenticated allegations of sexual assault against community women perpetrated by its security forces at two of its mines, one in Papua New Guinea and one in Tanzania (Barrick 2011a, 2011b).

One of the limited examples of detailed research was undertaken by Lockie in Queensland; he found that intimate partner violence was no worse in mining towns than in the general population, but that other factors, such as family finances and alcohol and drug use, were far more likely than mining-related elements to exacerbate family and sexual violence (Lockie 2011). By contrast, Cane et al. found that the social changes provoked by the transient populations associated with mine-related transportation and the temporary population growth during construction led to an increase of domestic violence and sex work. The preponderance of men spending long rostered periods away from their families contributed to anti-social behaviour and increased GBV (Cane et al. 2014). Both studies conclude that much more work needs to be done on the subject of GBV in mining areas, as their own work is just the beginning. As different results have been found in different contexts, it is certainly clear that additional research is still required and also that generalizations based on what we already know may be misleading.

### 21.2.5 *Women's Resistance to Extractives Projects*

One important sphere in which gendered modes of action seem to predominate is that of environmental activism against extractives projects. Much of the literature discusses the gendered representations that commonly occur among women anti-extractives activists. This is often labelled as 'eco-maternalism', or the belief that women, as mothers and nurturers, care more about conserving the planet than do men, and are therefore more prepared to fight for it. In a mining context, the literature reveals the important but previously understudied role of women as anti-mining activists in Andean Peru and Ecuador. Jenkins (2014) highlights the active role played by women in the anti-mining campaigns there, based on their acute concerns about the environmental degradation of the land and waters that they rely upon to grow food for their families. For many women activists, their engagement has entailed great personal sacrifice and risk. They experienced significant opposition from within their own families, especially from senior men, but also intimidation from pro-mining groups, both within their communities and outside, and from mining companies (Jenkins and Rondón 2015).

Andean women continued with their activism in spite of this high level of pressure, as they had become used to 'experiences of violence, harassment, and intimidation which have become part of their daily lives over a number of years' (Jenkins and Rondón 2015: 419). This echoes the experiences reported by the US women shale energy activists described below. The Andean women activists also share the 'eco-maternal' perspective of their North American sisters. As Jenkins records when discussing the Ecuadorian women activists, in 'explaining their activism, the women identify themselves with Pachamama, reflecting broader (and by no means unproblematic) eco-feminist maternalist tropes of women as Earthmothers and guardians of nature' (Jenkins 2014: 451). This use of Pachamama—the indigenous Mother Earth figure of Ecuadorean culture—can be seen as strategic essentialism, appealing to a strong cultural basis as justification for their actions.

Another example of this occurs in the shale energy industry, which has catalysed a high level of resistance among women in the United States. Willow and Keefer (2015: 114) observe that women activists opposing hydraulic fracturing in Ohio are 'fashioning a new (but still highly gendered) relationship to motherhood'. Women activists they interviewed stated that political action had become more important than traditional care-giving tasks in order for women to protect the future of their families. They explained that women were used to being disempowered in their personal lives and therefore having to stand up to authority figures, who no longer impressed them. Having less to lose from the system, they were more likely than men to challenge it. They had come to view activism and care-giving as complementary goals. This self-portrayal of women as 'eco-warriors' enables them to represent their non-traditional, even anti-social, behaviour as good mothering, giving strength to their movement (Willow and Keefer 2015). Although clearly an essentialist position, presuming that all women are better than all men in some respects, this self-empowering depiction of women's abilities to force social and economic change has emerged as a strong gender model in certain extractives contexts.

Similarly, in the Niger Delta, women have played very specific and active roles in the resistance movement fighting the oil and gas extraction industry: roles that prescribe clear limitations for women's actions, although these actions are crucial for the cultural and practical survival of the male soldiers (Oriola 2012). On one hand, post-menopausal women play an essential spiritual role in performing cleansing rituals for male insurgents in the Delta's creeks, and supply them with various herbal preparations designed to protect and sanctify the warriors. On the other hand, non-menopausal women are forbidden from entering the creeks as they are believed to defile the area with their presence. Oriola states that the Delta insurgents do not seem to perpetrate sexual violence against women, unlike the Nigerian security forces, so perhaps the strong belief in women's spiritual potency, for good or evil, has influenced the behaviour of the male insurgents towards women. He also states that female Delta insurgents do not engage in prostitution, although other women 'for instance, professional prostitutes and girlfriends or sex partners of male insurgents' (Oriola 2012: 550) are encouraged to undertake 'soft prostitution': that is, to befriend oil workers and security service officers in order to gather intelligence. Women insurgents are also active in smuggling arms and ammunition and 'benefit from the gender stereotypes and chivalry displayed by security operatives' (Oriola 2012: 549). Although this range of specific gendered roles had given women in the Niger Delta insurgency a range of positive positions, Oriola admits that this freedom disappeared once the insurgents were offered an amnesty by the state. Thereafter, 'women's participation in the insurgency and the rehabilitation exercise seems devalued and relegated to the fringes' (Oriola 2012: 551). Women insurgents were not offered rehabilitation programmes until after their male counterparts and were often viewed with suspicion in their home communities, as they had transgressed traditional gender boundaries and no longer behaved in the ways expected of women. So, although the insurgency had provided some women with opportunities for unprecedented freedom of action, they were later punished for this liberty, and not rewarded as heroes like their brothers were.

Presenting a highly critical view of the effects of oil development upon gender equality, Etkind (2014) describes the Russian oil oligarchy as *petromachismo* and claims that it is gender discriminatory, as well as anti-human rights in many ways. He agrees with Ross's (2008) finding that oil-fed development decreases women's employment opportunities, stating that the 'synergy between the oil and gas trade and security services creates a hypermasculine, cynical, and misogynistic culture: *petromachismo*, as I prefer to call it' (Etkind 2014: 161). He also highlights women's activism:

> Promoting archaic values of aggressive masculinity, the post-Soviet overreliance on natural resources and security services denies the role of women as the critical drivers of human capital. Victims of the regime, they become leaders of the resistance. At the turning points of the protest movement, rebellious femininity confronts the overbearing masculine state, with symbols of female sexuality acting as powerful, liberating political messages. (Etkind 2014: 167)

One of the examples he refers to is that of the radical Russian women's rock group, Pussy Riot, notoriously jailed for performing a protest song in a Russian Orthodox cathedral in 2012. Etkind demonstrated the inherently anti-feminist stance of the Russian state when he noted that the judge, a woman, commented when making her judgment, that: 'Though feminism is not a crime, it is incompatible with Orthodoxy, Catholicism, or Islam . . . Feminists violate the sphere of decency and morality' (Etkind 2014: 167). Of course, the Pussy Riot 'Punk Prayer' does not refer to the oil business, and Etkind does not clearly demonstrate a link between the extractive industries and Russian chauvinism, which may well be cultural as Rorbaek (2016) asserts that chauvinism in oil-rich Muslim countries is. Etkind does, however, emphasize that one of the largest employment sectors in Russia is the security industry (including the armed forces and police), which employs about 10 per cent of the population, almost all male, revealing an inherent 'hypermasculinity' already existing in the Russian state. A similar condition certainly seems to exist in Russia's oil and gas sectors.

By contrast, Rorbaek (2016) has challenged the assertion that a 'resource curse' in the Middle East has had a negative influence upon women's rights by making it easier to confine women to the home—an idea advanced by Ross in 2008. He concludes instead that it is Islamic culture that is detrimental to women's rights and that although the possession of oil wealth may bolster the degree of repression of women, it does not actually cause it. He cites Afghanistan and Pakistan as examples of repressive Islamic regimes which do not possess oil wealth. Thus he concludes that, although the eleven OAPEC[3] nations are definitely oil-rich and repressive, so too are many other Muslim countries without similar wealth. So he argues the explanation must be more than economic, and he analyses historical and cultural reasons for his findings.

Whether or not it is possible to prove that the oil industry, whether in the Middle East or in Russia, leads to increased repression of women, it is clear that the presence of large-scale oil development is certainly not bringing about an increase in gender equity and sexual equality in these host countries. This begs the question of whether oil companies should be making greater efforts to bridge the gap by working towards gender equality themselves when host states are clearly not doing so. Although it may be that the cultural subjugation of women in Islamic countries has resulted in less female activism than is apparent elsewhere, which may benefit extractives companies, their human rights policy commitments should lead them to encourage host governments towards greater gender equity. If the industry provided more positive opportunities for women worldwide, it is possible that their motivation to oppose it might be lessened.

### 21.2.6 *Industry Efforts towards Gender Balance*

In spite of much research producing general agreement about what needs to be done to make the extractive industries more gender-balanced and equal, and major efforts by a number of resources companies to try to implement the recommended changes, there has not been much visible improvement, with female employment rates still not exceeding 20 per cent even in the more advanced economies. See Macdonald (2017) to view available statistics on women's participation in the sector and for further detail on the gender pay gap and gender segregation by occupation. Some of the more radical scholars (Laplonge, Mayes, and Pini) claim that industry efforts to date have done little more than perpetuate the status quo while making cosmetic changes around the margins. By frequently restating its commitment to gender equality policies and programmes, the extractives sector arguably can avoid regulation or intervention even if the actual results of its actions to date are negligible.

In Laplonge's view, gender 'is not about what men and women are; rather it needs to be seen as what men and women do' (Laplonge 2014: 36). This then enables a discussion of behaviour rather than nature, which can lead to options for cultural change. His research has shown that women in mining

[3] i.e. Organization of Arab Petroleum Exporting Countries.

'often consciously make an effort to not act like girls when on site' (Laplonge 2014: 70) and women managers in mining do not like to associate their success with feminism or women's rights for fear of alienating their mostly male colleagues. Williams et al. also found this in the American oil and gas industry when they discovered that women managers were actually subconsciously discriminating against the promotion of women in their departments for fear of appearing gender-biased in favour of women. They observe that in such a situation 'the only way to prove one is neutral and objective is to hire a white man for a position' (Williams et al. 2014: 455). Thus, although many of the women geoscientists interviewed stated a preference for having more women and greater opportunities for women in their industry, most of them were opposed to any form of preferential policies for women in case it undermined the positions that they themselves had achieved.

Laplonge and Mayes and Pini both criticize the belief that women will 'civilize' the workplace, stating that there is no proof either that this works in practice or that women want this responsibility (Mayes and Pini 2014; Laplonge 2014). Laplonge notes that there is often an assumption that if the numbers of women in the workplace reach an arbitrary 'critical mass', often stated to be 30 per cent, then the cultural balance will be tipped. He argues that there has never been any proof that the critical mass assertion is true, therefore women should not bear the sole responsibility for bringing about this cultural change (Laplonge 2014). Mayes and Pini describe the assumption that women will change the workplace and work behaviour by their very presence and innate femininity as an expectation that they will perform 'unpaid civilizing work' (Mayes and Pini 2014: 542). Laplonge declares that the only way to truly effect cultural change in the extractives sector that may eventually lead to sexual equality in employment is to challenge both male and female gender stereotypes and the associated sets of behaviour, an approach that would be supported by Mayes and Pini.

## 21.3 Conclusions

For the enhancement of gender equality in the ASM sector, Kelly et al. propose efforts 'to promote rights and education to ensure safe and fair working conditions for those doing work in and around mining tunnels' (Kelly et al. 2014: 103). In their view, campaigns and programmes should target the education of government and customary leaders in how to assist women miners to organize and represent themselves, rather than trying to persuade them to leave the industry they have chosen. Better medical services, especially for those engaged in transactional sex, would benefit women workers, as would microcredit services. These practical programmes have been recommended to improve women's status in the ASM sector.

On the formal extractives side, although efforts have been put into increasing the representation of women in leadership positions, especially on boards, these efforts have yet to bear much fruit.[4] PWC's studies over the past three years have produced the following conclusion:

> Mining is still perceived by both genders to be a male-dominated industry where women do not possess equality of opportunity to advance. The only way that an organisational culture change of this nature can be effective is if it is led from the top. The boards and executive teams of mining companies need to understand and champion the business imperative to promote and support women within their organisations. They need to drive cultural change within their own companies to create a more profitable and sustainable industry. (WIM (UK) and PWC 2015: 11)

It is doubtful that sufficient change will emerge from these worthwhile but only advisory research efforts, and the researchers themselves estimated that it would take decades to achieve notable change at the current rate (WIM (UK) and PWC 2015: 10–11). One of the leading movements for corporate gender transformation in Australia is called Male Champions for Change, but only one mining company representative, from Rio Tinto, has become engaged with this group, and even in that case not at CEO level. This demonstrates a general lack of commitment to change from the top of the extractives sector. While it is true that change can only occur if driven by management, change at the top is not necessarily reflected by change throughout an organization, and the expectation of the PWC study that the achievement of 30 per cent women on extractives company boards will bring about a transformation is not supported by evidence (Laplonge 2014). Nor are the common statements by successful women that women themselves must bring about change.

Continuing efforts to recruit more women certainly make the industry more appealing to them. Company efforts to make more positions available to women in the full range of extractives occupations, at equal pay, will surely be of benefit to some women. However, these corporate initiatives can only bring about superficial changes to the masculine nature of the extractives sector without there being a root-and-branch, industry-wide cultural transformation focused on men as much as women. Laplonge accuses the industry of refusing to 'investigate the practices of masculinity which dominate in the

[4] A few women have reached positions at the top of major extractives companies, but very few. Cynthia Carroll was CEO of AngloAmerican plc from 2007 to 2013, when she stepped down, leaving Kay Priestly, CEO of Turquoise Hill Resources, as the only woman CEO in the top 100 resources companies. Cheryl Carolus, of Gold Fields Limited, was the sole female chair of a major resources company in 2015. In the larger group of the top 500 resources companies women fared a little better, with twelve being CEOs, an increase from seven in 2014 (PWC 2015: 11).

business of mining and which continue to ensure that most women . . . will find it hard to succeed in this industry' (Laplonge 2014: 70).

Thus, what is needed is change at the top, driving cultural change throughout each extractives organization and primarily focused on the men who make up the majority of the workforce. Some work has been done in this field, and the Western Australian government has sponsored the development of a training course along these lines (Department of Mines and Petroleum 2012). Others may want to follow this example. Fundamentally, unless the men who run the extractive industries decide to make these changes, which will not be easy, there is unlikely to be much improvement in the gender balance of the sector in the foreseeable future.

## References


Barrick Gold Corporation (2011a). 'Statement by Barrick Gold Corporation Concerning the North Mara Mine, Tanzania', Toronto, 30 May.

Barrick Gold Corporation (2011b). 'Statement by Barrick Gold Corporation in Response to Human Rights Watch Report', Toronto, 1 February.

Bashwira, M.-R., J. Cuvelier, D. Hilhorst, and G. van der Haar (2014). 'Not Only a Man's World: Women's Involvement in Artisanal Mining in Eastern DRC', *Resources Policy*, 40: 109–16.

Bryceson, D. F., J. B. Jonsson, and H. Verbrugge (2013a). 'For Richer, for Poorer: Marriage and Casualized Sex in East African Artisanal Mining Settlements', *Development and Change*, 45(1): 79–104.

Bryceson, D. F., J. B. Jonsson, and H. Verbrugge (2013b). 'Prostitution or Partnership? Wifestyles in Tanzanian Artisanal Gold-Mining Settlements', *Journal of Modern African Studies*, 51(1): 33–56.

Cane, I., A. Terbish, and O. Byambasuren (2014). *Mapping Gender-based Violence and Mining Infrastructure in Mongolian Mining Communities*. Brisbane: Sustainable Minerals Institute, University of Queensland.

D'Souza, M. S., S. N. Karkada, G. Somayaji, and R. Venkatesaperumal (2013). 'Women's Well-Being and Reproductive Health in Indian Mining Community—Need for Empowerment', *Reproductive Health*, 10(24): 1–12.

Department of Mines and Petroleum. (2012). *Gender and Safety in Mining: Workshop Planner Series*. Perth: Resources Safety, Department of Mines and Petroleum, Western Australia.

Eftimie, A., K. Heller, and J. Strongman (2009). *Gender Dimensions of the Extractive Industries: Mining for Equity*. Washington, DC: World Bank.

Etkind, A. (2014). 'Post-Soviet Russia: The Land of the Oil Curse, Pussy Riot, and Magical Historicism', *Boundary 2*, 41(1): 153–70.

Jenkins, K. (2014). 'Unearthing Women's Anti-Mining Activism in the Andes: Pachamama and the "Mad Old Women"', *Antipode*, 47(2): 442–60.

Jenkins, K. and G. Rondón (2015). '"Eventually the Mine Will Come": Women Anti-Mining Activists' Everyday Resilience in Opposing Resource Extraction in the Andes', *Gender and Development*, 23(3): 415–31.

Kelly, J. D. T., A. King-Close, and R. Perks (2014). 'Resources and Resourcefulness: Roles, Opportunities and Risks for Women Working at Artisanal Mines in South Kivu, Democratic Republic of the Congo', *Futures*, 62: 95–105.

Lahiri-Dutt, K. (2015). 'The Feminization of Mining', *Geography Compass*, 9(9): 523–41.

Lahiri-Dutt, K. and G. Burke (2011). 'Gender Mainstreaming in Asian Mining: A Development Perspective', in K. Lahiri-Dutt (ed.), *Gendering the Field: Towards Sustainable Livelihoods for Mining Communities*, 213–30. Canberra: ANU Press.

Laplonge, D. (2014). *So You Think You're Tough: Getting Serious about Gender in Mining*. Perth: Factive.

Laplonge, D. (2016). 'A Toolkit for Women: The Mis(sed) Management of Gender in Resource Industries', *Journal of Management Development*, 35(6): 802–13.

Lauwo, S. (2016). 'Challenging Masculinity in CSR Disclosures: Silencing of Women's Voices in Tanzania's Mining Industry', *Journal of Business Ethics*, doi: 10.1007/s10551-016-3047-4.

Lockie, S. (2011). 'Intimate Partner Abuse and Women's Health in Rural and Mining Communities', *Rural Society*, 20(2): 198–215.

Macdonald, C. (2017). 'The Role of Gender in the Extractives Industries', WIDER Working Paper 2017/52. Helsinki: UNU-WIDER. Available at: https://www.wider.unu.edu/publication/role-gender-extractives-industries.

Mahy, P. (2011). 'Sex Work and Livelihoods: Beyond the "Negative Impacts on Women" in Indonesian Mining', in K. Lahiri-Dutt (ed.), *Gendering the Field: Towards Sustainable Livelihoods for Mining Communities*, 49–66. Canberra: ANU Press.

Mayes, R. and B. Pini (2014). 'The Australian Mining Industry and the Ideal Mining Woman: Mobilizing a Public Business Case for Gender Equality', *Journal of Industrial Relations*, 56(4): 527–46.

Mukherjee, S. (2014). 'Mining and Women: The Case of the Maria of Chhattisgarh', *Social Change*, 44(2): 229–47.

Murillo, D.V. (2013). 'Laboring above Ground: Indigenous Women in New Spain's Silver Mining District, Zacatecas, Mexico, 1620–1770', *Hispanic American Historical Review*, 93(1): 3–32.

O'Faircheallaigh, C. (2013). 'Women's Absence, Women's Power: Indigenous Women and Negotiations with Mining Companies in Australia and Canada', *Ethnic and Racial Studies*, 36(11): 1789–807.

Oluduro, O. and E. Durojaye (2013). 'The Implications of Oil Pollution for the Enjoyment of Sexual and Reproductive Rights of Women in Niger Delta Area of Nigeria', *The International Journal of Human Rights*, 17(7–8): 772–95.

Omeire, E. U., A. A. Aveuya, O. C. T. Muoneme, A. Gold, U. Akudo, and C. A. Omeire (2014). 'Between the Devil and the Deep Blue Sea: Niger Delta Women and the Burden of Gas Flaring', *European Scientific Journal*, 10(26): 151–62.

Oriola, T. (2012). 'The Delta Creeks, Women's Engagement and Nigeria's Oil Insurgency', *British Journal of Criminology*, 52: 534–55.

Purevjav, B. (2011). 'Artisanal and Small-Scale Mining: Gender and Sustainable Livelihoods in Mongolia', in K. Lahiri-Dutt (ed.), *Gendering the Field: Towards Sustainable Livelihoods for Mining Communities*, 197–212. Canberra: ANU Press.

Rorbaek, L. L. (2016). 'Islamic Culture, Oil, and Women's Rights Revisited', *Politics and Religion*, 9: 61–83.

Ross, M. L. (2008). 'Oil, Islam, and Women', *American Political Science Review*, 102: 107–23.

United Nations Office of the High Commissioner for Human Rights (2011). *Guiding Principles on Business and Human Rights: Implementing the United Nations 'Protect, Respect and Remedy' Framework*. Geneva: OHCHR. Available at: http://www.business-humanrights.org/specialrepportal/home.

Williams, C., K. Kilanski, and C. Muller (2014). 'Corporate Diversity Programs and Gender Inequality in the Oil and Gas Industry', *Work and Occupations*, 41(4): 440–76.

Willow, A. J. and S. Keefer (2015). 'Gendering Extraction: Expectations and Identities in Women's Motives for Shale Energy Opposition', *Journal of Research in Gender Studies*, 5(2): 93–120.

Women in Mining (UK) and PWC (2013–15). *Mining for Talent*. Annual Report. London: WIM (UK) and PWC.

# 22

# Climate Change and the Extractives Sector

*Tony Addison*

## 22.1 Introduction

Climate change is one of the world's most complex and urgent global problems; indeed, many would argue that it is *the* greatest challenge.[1] Climate change adaptation and mitigation are fundamental to the evolution of our economies and societies over the rest of the twenty-first century and beyond. The extractive industries are in many ways at the heart of the climate change challenge. The extractives sector must support national and international efforts to respond to climate change, by adjusting exploration and production to shifting patterns of demand for energy and minerals—as policies and new technologies encourage progress along low-carbon pathways.[2] None of this is easy, and success is not assured.[3]

This is a large topic, and the task of this chapter is to set out some of its main issues as they relate to the extractives sector. The next section summarizes some of the technologies and policies that shape our energy future. Adoption of renewable energy technologies is accelerating, but without major technological breakthroughs fossil fuels will continue to dominate the immediate energy future. The chapter then discusses three vital issues for the transition: natural gas as an energy bridge; carbon sequestration; and energy efficiency. After that, the chapter considers the policy dilemmas inherent in the 'stranding'

[1] The Intergovernmental Panel on Climate Change (IPCC) provides the most exhaustive review of the state of research on climate change through its three Working Groups on the Physical Science Basis (WG 1); Impact, Vulnerability, and Adaptation (WG 2); and Options for Mitigating Climate Change (WG 3). See https://www.ipcc.ch/report/ar5/.

[2] While the chapter uses the shorthand terms 'low-carbon pathways' and 'net zero-carbon' future, emissions of other greenhouse gases such as methane must be cut as well.

[3] I thank John Drexhage and Nadia S. Ouedraogo for their comments, contributions, and suggestions. Any errors remain my own.

of fossil-fuel resources, and the responses to climate change by extractives companies. The chapter's second half considers the implications for minerals and metals of the transition to low-carbon pathways. The evidence points to a conclusion that achieving a 'net zero-carbon' future will be considerably *more* materially intensive than existing fossil-fuel technologies. While low-carbon pathways offer considerable opportunities for developing countries with the right mineral resources, there are major concerns as well, including the distribution of the benefits within what are often highly fragile nations. The chapter concludes that with greenhouse gas (GHG) emissions reaching their highest ever levels, recent progress in sustainable development and poverty reduction is in imminent danger, posing a great challenge for the extractive industries and their host countries.

## 22.2 Looking to the Energy Future

Peering into the energy future is fraught with difficulties (see Stevens, Chapter 4, this volume). Forecasts by fossil-fuel and mining companies have been repeatedly wrong (especially on prices) despite their strong incentives to get forecasting right. The authoritative forecasts of the International Energy Agency (IEA) are regularly revised to incorporate the impact of new developments in technologies and markets. The governments of resource-abundant countries frequently stick to erroneous assumptions (especially in their revenue projections) long after the markets have turned, and despite the vital importance of the extractives sector to their economies and public finances.

This is an especially difficult time to look ahead. While technology continually shapes the energy future, the world is now in an era of 'renewables acceleration' driven by falling costs.[4] For solar photovoltaic (PV) the IEA's 2017 estimate for capacity growth to 2022 is *one-third* higher than its 2016 estimate (IEA 2017a). Renewables now account for almost two-thirds of the world's additional (net) power capacity (IEA 2017a). By 2022, renewables could be providing 30 per cent of global electricity generation, ahead of natural gas, and equivalent to the total power consumption of China, India, and Germany (IEA 2017a).

But such a fast growth in renewables still leaves fossil fuels (coal, oil, and gas) with a dominant, if reduced, share of global energy production. Innovation in battery storage is the game-changer to watch. The power from renewables such as solar, wind, and tide fluctuates. Cheaply capturing and storing

[4] On uncertainty and carbon pathways, see Hughes et al. (2013) and Usher and Strachan (2012). Rhodes et al. (2014) discuss the drivers of energy innovation, and the respective roles of the private and public sectors.

their energy in large-capacity batteries would eliminate this drawback. The technology of 'flow batteries' holds promise for large-scale storage. But they remain expensive (pricey vanadium is used). Grid energy storage together with the shift to electric vehicles now underway will require lots of cobalt, lithium, and other materials to the benefit of the producing countries (see later discussion). Research will eventually replace the expensive materials with cheaper alternatives, heralding a new era of large-scale (and cheap) grid storage. But we are not there yet.

Public policy also shapes the energy future; regulations, taxes, and subsidies have profound effects on incentives. Fossil-fuels subsidies amount to billions of dollars annually. Estimates of their annual size range from US$200–500 billion and upwards, depending on whether subsidies to production are included, or just those to consumption.[5] When environmental externalities are accounted for, fossil-fuel subsidies amount to trillions of dollars. One IMF study calculates that if all the externalities of fossil fuels are included (such as harm to human health) then fossil-fuel subsidies amounted to US$5.3 trillion in 2015, equivalent to 6.5 per cent of global GDP (Coady et al. 2017). In sum, fossil-fuel subsidies far exceed subsidies to renewables in power generation, which amounted to US$140 billion in 2016 (Shirai and Adam 2017).

Coal is abundant and relatively cheap, and historically the price has not reflected its many negative environmental externalities. Coal accounts for well over half of all electricity generated in China and India (IEA 2017b). Globally, about US$750 billion was invested in new coal power plants in 2000–10 alone. Hundreds of coal-fired facilities could still be operational in the coming decades, and more than 2,400 new coal-fired power stations are already planned for construction by 2030 (van Breevoort et al. 2015). Coal has been called the 'zombie fuel'—dying, but far from dead. Proponents of 'clean coal' even argue that carbon sequestration, discussed later, will breathe new life into the use of this fossil fuel.

That coal will be around for a long time, even in a much-reduced role, is a worrying scenario for human health let alone planetary health. More than half the world's deaths from ambient air pollution occur in China and India, and coal is a big culprit.[6] Irrespective of climate change mitigation, the

[5] On the methodologies for measuring fossil-fuel subsidies see McCulloch (2017). Estimates vary depending on whether consumption and/or production subsidies are counted. Some eleven European countries and the EU itself provided at least €112 billion in annual subsidies over 2014–16 towards the production *and* consumption of fossil fuels (Gençsü et al. 2017). Production subsidies at the US federal and state levels were US$20.5 billion in 2015, with 80 per cent going to oil and gas, and 20 per cent to coal (OCI 2017: 5). Broadening the definitions of subsidies to include the harm that fossil fuels do to the environment and human health (respiratory illnesses etc.) yields much larger estimates (Coady et al. 2017).

[6] Lancet Commission on Pollution and Health (http://www.thelancet.com/commissions/pollution-and-health).

leaderships of China and India want to cut air pollution, and are now willing to trade some growth for cleaner air. Emissions of nitrogen dioxide and sulphur dioxide (the major air pollutants) from gas-fired plants are a fraction of those of coal-fired plants (de Gouw et al. 2014). Stronger environmental regulations are squeezing the use of coal to the benefit of gas. China, which consumes half the world's coal, has cut its own production (by an amount equal to the total production of South Africa, the world's fifth largest coal exporter) and is curbing imports as well (IEA 2017b). China is today responsible for around half the global take up of solar PV, as it increasingly switches to renewables (IEA 2017b).

Carbon pricing shifts incentives in electricity generation away from coal and towards natural gas (which emits much less $CO_2$ than coal in power generation) and, best of all, towards renewables. The UK's 'carbon price support', a tax paid by its coal and gas generators, helped reduce coal's share to a record low of 2 per cent by mid-2017, when the United Kingdom had its first day without coal-fired power generation since the Industrial Revolution (Aurora Energy Research 2017). Some forty national and twenty-five subnational jurisdictions now put a price on carbon, a doubling over the last decade (World Bank/ECOFYS 2017: 2). Official carbon pricing covers some 15 per cent of the world's GHG emissions (World Bank/ECOFYS 2017: 2). This will rise to around 25 per cent when China introduces a national carbon market by the end of 2017, making it the world's largest scheme. In Canada, the federal government plans to introduce a carbon-pricing regime to support existing provincial carbon-pricing initiatives.

Both regulation and market-based instruments like carbon pricing alter business behaviour. In its principles for climate change policy design, the International Council on Mining and Metals (ICMM) states that: 'without a clear price signal, members' ability to plan and make sound decisions is at risk' (ICMM (2011: 4).[7] As more key mining jurisdictions move to carbon-pricing systems (e.g. Chile and South Africa) companies will need to adjust their operations accordingly.

The 2015 UN Paris climate agreement opens a new chapter in the global energy future, one in which the shift from fossil fuels to renewables should accelerate.[8] However, progress is not assured. In 2017 President Trump announced US withdrawal from the Paris Agreement, a decision that is,

[7] The ICMM (2011) principles for climate change policy design were re-affirmed by ICMM members in 2015 prior to the Paris Climate Conference. For the oil and gas industry, IPIECA (2015) also affirms the importance in climate policy of long-term price signals and market certainty. In expectation of increased regulation on emissions (and carbon pricing), more companies are using an internal carbon price in their business planning (CDP 2017: 5).

[8] The Paris Agreement was reached at the 21st session of the Conference of Parties to the UN Framework Convention on Climate Change (UNFCCC) in December 2015.

however, 'open to review' and one that faces strong domestic opposition at state and city government levels (meanwhile other signatories remain committed, notably the EU, China, and India). Although the European Union's emissions trading scheme (ETS) was the world's first and most comprehensive scheme, the price for its carbon credits is now too low to deter Europe's heavy emitters: gas has not replaced coal on the scale initially expected in Europe (Stern 2017). Australia has scrapped its carbon tax, replacing it with a weaker ETS-style scheme. The United Kingdom's carbon price must now rise, otherwise coal-fired power generation will return. Some fossil-fuel companies continue to lobby against increased regulation, carbon pricing, and, indeed, the Paris Agreement itself.

### 22.2.1 *Natural Gas, Carbon Sequestration, and Energy Efficiency*

Even under very optimistic scenarios for emissions reductions, fossil fuels will still be in use for many decades while renewables grow their market share as the costs of the technologies fall. This tension between climate goals and the pace of technology could be eased by three sets of measures: the first is heavily promoted (natural gas), the second is untested at scale (carbon capture and storage (CCS) or 'carbon sequestration') and the third has very wide support (energy efficiency).

Champions of natural gas, including (not surprisingly) the producing companies and the countries with the resource, argue that it is a vital 'energy bridge' from the era of hydrocarbons to the low-carbon future. High efficiency gas-powered plants using combined cycle technology (CCGT) emit less than half as much $CO_2$ as the typical coal-fired plant per unit of energy produced (de Gouw et al. 2014). Gas is now the fastest-growing fossil fuel (IEA 2017c). Liquefied natural gas (LNG) is showing especially strong growth: forty countries now import LNG, up from seventeen a decade ago (BP 2016). New gas producers, like those in East Africa, have high hopes of selling LNG to Asia, especially to China as it moves on from coal.

However, the notion of gas as an energy bridge is contested. The main constituent of natural gas is methane ($CH_4$). This is emitted into the atmosphere via deliberate venting as well as from leaky gas wells and pipelines (with corroded iron pipes and other old infrastructure leaking more). Methane is a more potent GHG than $CO_2$, and contributes about 17 per cent of radiative forcing (the warming impact on climate) (NOAA 2016). Whether gas has a better climate footprint than coal depends on how much methane emissions can be reduced. A methane leakage rate of 1.3 per cent of gas production (the EPA's leakage estimate for the US) is associated with a 46 per cent emissions reduction (when efficient gas plants substitute for coal power) whereas a 5 per cent leakage rate (found in other US estimates) implies only a 24 per cent reduction (Lazarus et al. 2015: 4). After reviewing the US evidence, Brandt et al. (2014) conclude that methane leakage is substantially underestimated. Shale gas is of special concern; Howarth (2015) concludes that when methane emissions are properly measured, the greenhouse footprint of shale gas exceeds that of conventional gas, coal, and oil.

The necessary technology exists to cut methane emissions (Balcombe et al. 2017). US regulation is now tighter, and the US industry has successfully reduced its methane emissions (Schwietzke et al. 2016). Is this feasible elsewhere, especially in Eurasia and the Middle East which account for around half the world's methane emissions? Gould and McGlade (2017) calculate that it is cost-effective to avoid three-quarters of current global methane emissions (in part because captured methane can be sold). If gas is to act as an energy bridge, then a great deal hinges on putting methane abatement measures in place globally, especially for shale gas.

There are additional concerns around natural gas that call into question the optimism of its champions. These include the environmental costs of its extraction (local pollution from shale gas 'fracking' is a worry) and a concern that switching to cheap gas, with all its attendant investment in production and distribution infrastructure, will delay the adoption of near-zero-carbon technologies. Some modelling exercises find only a small reduction in $CO_2$ emissions and indeed, in some scenarios, an increase in climate forcing (McJeon et al. 2014). Several studies argue that gas can only act as a bridge fuel (to displace coal) until around 2030 (a shorter time than many policy makers expect); renewables must then take the lead if the world is to achieve the target of keeping the increase in global temperature within 2°C above its pre-industrial level (the 2DS goal) (Banks and Taraska 2013).[9]

We are left with a conundrum: how to maintain energy supplies while keeping to global emissions goals? Carbon sequestration and improved energy efficiency might both help.[10]

Advocates of carbon sequestration see it as a transitional mitigation technology allowing the continued use of fossil energy (particularly gas) as the share of renewables grows. More controversially, proponents of coal see it as means to keep, and indeed grow, coal's share in energy generation. This is a major concern for critics of CCS: by keeping coal alive, CCS could divert resources from clean energy technologies—CCS will absorb a lot of resources

[9] The 2DS goal was adopted at the 16th session of the Conference of the Parties to the UN Framework Convention on Climate Change (in 2010). To keep the rise in the global temperature within 2°C above preindustrial levels, the concentration of atmospheric GHGs must be stabilized within 450 parts per million (ppm) $CO_2$ equivalent by 2050.

[10] Space limitations preclude a full discussion of carbon sequestration and energy efficiency. The reader is instead referred to Sovacool et al. (2016) and Sovacool (2017).

as it has large initial investment costs—and reduce the incentive to adopt renewables and improve energy efficiency.

Many climate models find that CCS will be essential to achieving international targets (IPCC 2014: 82). Industrial sectors such as iron, steel, and cement have few if any alternatives for achieving deep emission reductions (Florin and Fennell 2010). Their technologies have already seen large gains in energy efficiency (DECC 2015). Moreover, the production of cement, chemicals, and steel, as well as petroleum refining, generate significant additional $CO_2$ emissions as a by-product of the chemical reactions inherent in the industrial process; these cannot be eliminated (ZEP 2013).

How much could CCS achieve in emissions reduction? Much is unknown. On some estimates, CCS could deliver 13 per cent of the cumulative reduction in emissions required by 2050 to meet the 2DS goal (IEA 2015a). But for the present, CCS projects are mostly small scale, and there are strong concerns regarding operational safety and risks in transporting large volumes of liquefied $CO_2$. Fundamentally, there is little experience with long-term $CO_2$ storage and certainly no proof that storage can be managed or safely guaranteed for centuries.

In sum, CCS will have a role in the industrial sectors. In the power sector, CCS may have a role in coal decarbonization in the short term—provided it does not delay the energy transition—and perhaps a vital role in decarbonizing natural gas use in the medium and long term, as gas replaces coal in electricity generation. But at what scale this occurs is uncertain for technical, economic, and legal reasons.

Continuing to improve energy efficiency will therefore be imperative if 2DS is to be met: it must have one of the biggest roles in reducing emissions, especially in construction, industry, and transport (IEA 2015b; UNEP 2011). Some US$14 trillion in energy efficiency investments are needed up to 2035 (IEA 2014). This implies more demand for the minerals and metals that go into the technologies that achieve greater energy efficiency; in lighting and temperature control in buildings and in public transport, for example. This demand will be additional to that for energy storage and electric vehicles. We return to the implications for the mining sector later in this chapter.

### 22.2.2 *Stranded Assets*

Cutting methane emissions from gas infrastructure, together with carbon sequestration and improved energy efficiency, will reduce some of the pressure to leave fossil fuels unused and 'in the ground'. Nevertheless, a radical readjustment in the world's GHG emissions to meet the Paris commitments will render a substantial portion of the known stock of fossil-fuel reserves unusable, and they will become, in effect, 'stranded assets' (Cust et al. 2017;

van der Ploeg 2016). One much-cited study concludes: 'Our results suggest that, globally, a third of oil reserves, half of gas reserves and over 80 per cent of current coal reserves should remain unused from 2010 to 2050 in order to meet the target of 2°C' (McGlade and Elkins 2015: 187).

Fossil fuels are of growing concern to the investment community. Norway's SWF (which invests the country's oil wealth); the heirs to the Rockefeller oil fortune; French bank BNP Paribas; the World Council of Churches; and universities such as Stanford are among the investors and public and private organizations that are selling all or part of their fossil-fuel investments (coal mining, oil and gas extracted from shale and tar sands, and Arctic oil and gas exploration and production feature in the disinvestments). This does not for the present deprive fossil-fuel companies of capital as environmentally insensitive investors will buy the assets. But disinvestment does reinforce the political message that action on climate change is urgent.

Eventually fossil-fuel companies will face a rising cost of capital as the value of their assets declines, the consequence of ever-cheaper renewable energy. As we discussed earlier, firm implementation of the Paris Agreement will make obsolete billions of dollars of existing and planned investments in oil, gas, and coal as these resources are stranded. For the largest publicly traded oil and gas producers, Carbon Tracker Initiative (2017) reckons that one-third of potential capital investment up to 2025 could become obsolescent. This has become a concern for financial regulators as well.[11]

Large adjustments must be made by fossil-fuel companies, but even more so by countries with the oil, gas, and coal resources. Companies can gradually shrink the size of their businesses, and diversify into renewables: 'companies could, if they wanted to, run down their existing reserves in less than 15 years... most countries must wait 45 years on average to liquidate their fossil-fuel wealth (Cust et al. 2017: 47–8). Countries can only convert their resource wealth into revenue once it is developed and produced. This becomes tougher for countries with high production costs as prices weaken and they become less attractive to companies (relative to lower-cost producers). Their dreams of large revenues to finance ambitious economic and social development plans may never be fulfilled.

Consider natural gas for instance. At present, the global market is glutted.[12] As a result, companies are keenest on long-standing low-cost suppliers like Qatar with its large reserves as well as new and large low-cost producers like the United States (whose exports, supplied by the shale gas boom, could

[11] The Bank of England has a task force on climate-related financial disclosures (https://www.fsb-tcfd.org). See also Zenghelis and Stern (2016).

[12] After the Fukushima nuclear disaster, Japan's import price for LNG tripled between the millennium and 2012, but by 2017 it was back around its 2000 level.

match Qatar's). Existing producing countries, and the companies operating in their jurisdictions, are keen to grow their exports while it remains remunerative to do so. New producers such as Tanzania, which hopes to develop LNG for export to Asia, are entering a buyers' market for gas. One consequence is that a more attractive tax package is now necessary to encourage companies to invest than in the years of high prices when countries expected a revenue and foreign-exchange bonanza.[13]

Prices could recover, but international climate action and the shift to renewables are strong headwinds for producers of fossil fuels. Governments would be well advised to reduce their dependence on revenues from oil and gas (and especially coal) by broadening their tax bases and systems of tax collection to raise more non-resource revenues (Addison et al. 2018). They also need to build fiscal buffers by saving more revenue in readiness for unexpected earnings shocks, including future breakthroughs in renewable energy technologies that eat into the market shares of oil, gas, and coal. Some fossil-fuel-rich countries also have reserves of those metals and minerals that are in high demand as the world transitions onto low-carbon pathways (see the discussion in Section 22.3). Developing these would help reduce their existing over-dependence on one or two fossil fuels for revenues (and foreign exchange). Above all, investing resource revenues into high value-added activities in agriculture, manufacturing, and services will diversify the economies of countries with abundant fossil fuels. This in turn requires an active 'industrial policy' (Dietsche, Chapter 7, this volume). None of this will be politically or technically easy but countries need to be well prepared to take the necessary action.

Phasing out generous subsidies to consumers of gasoline and other fuels as well as subsidies to exploration and production will further reduce the exposure to carbon-market risk of fossil-fuel-abundant countries (Cust et al. 2017). This is also politically difficult, as consumers, the transport sector, and businesses are used to cheap fuels, while state-owned oil and gas companies are vigorous champions of more investment in their sector. Low-income countries (LICs) and middle-income countries (MICs) will need financial and technical assistance to adjust (including accelerated investment in renewable energy). So far, aid donors have done little to encourage development partners to cut fossil-fuel subsidies (McCulloch 2017). The United Kingdom's DFID, for instance, gives twice as much support to projects relating to fossil fuels as it does to renewables (Wykes and Scott 2017: 4).

Given the internationally agreed temperature target (2DS etc.), there will come a time when humanity has used up its cumulative GHG budget (IPCC 2015). That could be within a few decades. Caney (2016) poses three vital

[13] Scurfield and Manley (2017) assess Tanzania's options in the current global gas market.

questions: (i) For a given GHG budget, how is it decided which producing countries have the best case to continue production? (ii) Do the countries with unused fossil fuels merit compensation? (iii) If so, who should pay? From the perspective of distributive justice, LICs and MICs have a stronger case for continuing to produce than do high-income countries (HICs), which owe a good portion of their present prosperity to the use of their own fossil-fuel resources.

Politically, it is hard for LICs and MICs to agree to strand their fossil fuels. Their resource revenues are generally much larger than any aid flows, and 'aid fatigue' prevails among donors—so full compensation (including that from new streams of climate finance) may not appear credible to them. Many of today's largest producers of fossil fuels are HICs. The United States is now the world's third largest oil producer, and is closing fast on the leading oil and gas exporters, Saudi Arabia and Russia. President Trump is also trying to revive the US coal industry. Would any US administration agree to curtail US fossil-fuel development for the benefit of poorer countries like Mozambique (which has natural gas and abundant coal)? Whether wealthy or poor, every fossil-fuel-abundant country wants to avoid catastrophic climate damage. Yet political short-termism works against HICs taking on the burden of responsibility and stranding their own fossil-fuel reserves. In sum, distributive justice would favour an international agreement. But in its absence, resolution of the question of which countries get the revenues, and which are left with the stranded fossil-fuel assets, will be left to hard global politics—in which the LICs have the least power.

### 22.2.3 *Implications for Extractives Companies*

In a warming world, extractives companies must adapt their operations to higher temperatures and rising sea levels, and more frequent droughts, floods, heatwaves, and storms (ICMM 2013a). Every stage of the extractives cycle is at risk from physical damage as well as breakdowns in supply chains and vital public services: the exploration and discovery of resources; the construction of the infrastructure to exploit the deposit; the operation of the asset over its life (often spanning decades); and the reclamation of the site at closure and after. Other climate risks include disrupted supplies of power, water, materials, and equipment to the site, and delays in transporting products to refiners, manufacturers, and end-markets. Companies operating in politically fragile nations, where the state has only limited capacity to respond to disaster, are most vulnerable to extreme weather events. Disasters constitute commercial risks for companies and economic risks for host countries (especially to the public finances when revenues are lost from production stoppages).

Mines, oil/gas rigs, and associated infrastructure require engineering excellence and high-level risk management, not least in extreme and isolated environments. The extractive industries now have the tools to assess their risks.[14] Effective risk management requires an ecosystem perspective that is unique to each location. Mining as well as oil operations inevitably disrupt the landscape, and potentially stress the ecosystem in ways that accentuate its vulnerability to climate change (and thereby endanger the livelihoods of local communities). Land degradation can also compromise the structural integrity of mine sites as well as oil and gas wells together with their supporting water and energy and transport infrastructure (ICMM 2013a). For these reasons, some companies have stepped up the protection of the ecosystem around their operations, as well as to meet their sustainability objectives.

Inevitably, there are business and financial consequences: higher operating costs (to raise safety margins); unplanned capital expenditures (to climate-proof older infrastructure); as well as reputational and financial costs arising from litigation, regulatory non-compliance, and adverse publicity (ICMM 2013a, 2013b). These will affect the value of both private and publicly listed companies. Consequently, there is more pressure on extractives companies to incorporate climate risk into their strategic plans, environmental and social impact assessments, safety management systems, and engineering practices. That pressure comes from within the industry itself (and industry associations), host governments (including sector regulators who must protect the public interest), campaigning NGOs (both national and international), and increasingly from investors such as pension funds, as well as regulators of financial markets.

Minimizing mining's footprint is important since forest clearance for mining and associated infrastructure contributes to GHG emissions (and biodiversity loss). Some mining companies are investing to 'green' their operations by use of renewables to power their operations and measures to reduce their carbon footprint. The use of renewable energy, particularly solar PV and wind power, is scaling up in mining projects, notably in Canada and Australia. However, there are many GHG emissions from mining, particularly in transport, that are not under the industry's direct control (ICMM 2013b). This is a major challenge to the industry given the expected scale of mining's expansion over coming decades.

[14] On corporate risk management and climate change in the oil and gas industry see IPIECA (2013) and in mining see (ICMM 2013a). Though not climate related, the Exxon Valdez and Deepwater Horizon oil spills, as well as the Samarco (Beno Rodriques) tailings-dam collapse, illustrate the damage that major accidents can cause.

## 22.3 Impact of a 'Net Zero-carbon' Future for Minerals and Metals

Achieving the necessary reduction in emissions alongside the UN Sustainable Development Goals (SDGs) represents a massive and unprecedented challenge. The five areas where emission reductions will be most critical are in: power; transportation; buildings; industry; and land-use management. Constructing the necessary new technologies and infrastructures requires minerals and metals for non-carbon power providers (hydro and nuclear), renewable power technologies (wind turbines, solar PV, and tidal power), zero-emission buildings, hybrid/electric transportation vehicles, and alternative transportation modes (especially rail). To reach a 'net zero-carbon future' the IEA estimates that annual investments in low-carbon technologies and energy efficiency need to more than double by 2020 to US$790 billion and to increase by about six times to reach US$2.3 trillion by 2035 (IEA 2014).

It is difficult to exaggerate the potential implications for the growth in demand for key minerals and metals if the world does move successfully towards clean energy and onto low-carbon pathways (IRP 2017). The most comprehensive and recent assessment is provided in a study undertaken by the World Bank with support from ICMM (World Bank 2017). This predicts future demand for metals in the transition to a low-carbon future, based on the IEA's Energy Technology Perspective (ETP) scenarios (IEA 2015b). The IEA study examines the implications for renewable technologies of meeting the 2°C (2DS), 4°C (4DS), and 6°C (6DS) goals for global temperature warming (IPCC 2014). Renewables (solar, wind hydropower, biomass etc.) rise in the ETP scenarios from 14 per cent of the present energy mix to 18 per cent in the 6DS scenario, and 44 per cent in the 2DS scenario (IEA 2015b).

The World Bank study concentrates on wind, solar, and energy storage batteries as they are critical to future energy supplies in a carbon-constrained world, while recognizing that there are many other important technologies and transmission systems (including power, transportation, buildings, industry, and land-use management) (World Bank 2017). It identifies what materials will be required, and how much, as these technologies are scaled up under the various climate goal scenarios (especially 2DS). Of course, forecasting metals demand is highly contingent, not least on intra-technology choices (which types of wind technology are adopted, for example) and whether the Paris Agreement holds.

For mineral-abundant counties and mining companies, the World Bank's main conclusion has great significance: 'The report clearly shows that the technologies assumed to populate the clean energy shift—wind, solar, hydrogen, and electricity systems—are in fact significantly MORE material intensive in their composition than current traditional fossil-fuel-based energy supply systems' (World Bank 2017: xii). This is a striking result. Inevitably the demand for metals for use in low-carbon technologies is much greater under the ambitious 2DS scenario than the 4DS scenario; this is especially so for aluminium, cobalt, iron, lead, lithium, manganese, and nickel.

Similar conclusions emerge from other studies. Vidal et al. (2013) examine the material requirements necessary for the world to achieve 100 per cent renewable energy by 2050, a target of the World Wildlife Fund for Nature (WWF 2011). Vidal et al. (2013) conclude that the material requirements of solar and wind installations greatly exceed those of conventional energy infrastructure. Relative to an equivalent installation from traditional (fossil-fuel) energy technologies, renewable technologies require fifteen times more concrete, ninety times more aluminium, and fifty times more iron, copper, and glass. To reach the WWF goal, this would mean 3,200 million tonnes of steel, 310 million tonnes of aluminium, and 40 million tonnes of copper; this amounts to an increase of 5–18 per cent in annual production over the next forty years (Vidal et al. 2013).

Clean energy is very demanding of materials, a message underscored by Kleijn et al. (2010) who explore potential resource constraints in a hydrogen-centred economy with a scenario whereby renewables generate all energy. This study estimates an increase in iron/steel production of six times to service wind turbines and forty times should specially constructed pipelines be required. If electricity is the main carrier/transmitting agent for clean power, it would mean an increase of copper production by seventy times from its current levels.

Wind technologies require copper, iron ore (steel), and neodymium (Kleijn et al. 2010: 2790). For solar PV and concentrated solar power systems (CSPs) it is expected that demand for the 'rare earths' could substantially exceed production capacity, thereby encouraging the development and adoption of synthetic substitutes (Kleijn et al. 2010: 2789). Smart grid systems integrated with solar and wind installations will raise the demand for base metals even further.

In the case of zero-carbon vehicles, strong growth in fuel cells implies dramatic growth in platinum demand and production (possibly exhausting current reserves) and neodymium production levels would need to increase by 200 per cent (Kleijn et al. 2010). Lithium, copper, nickel, and platinum will see considerable demand growth under a net zero-carbon scenario for transport (García-Olivares et al. 2012: 567). CCS at scale will require significant amounts of nickel and molybdenum (Kleijn et al. 2011: 5647).

### 22.3.1 *Supply Concerns*

The metals required for a low-carbon future are experiencing increasing demand and higher prices. The evolution of the future pattern of demand will depend upon the mix of technologies adopted and how these evolve over time as new scientific breakthroughs are made. Finding cheaper substitutes for the most expensive metals in batteries is a pressing concern (Dawkins et al. 2012).

Increasing recycling can reduce some of the pressure on supplies. Creating a closed-loop or 'circular' material economy can save energy (compared with producing new material from ore, biomass, or oil) and contribute to reducing GHG emissions. Increasing rates of steel recycling would reduce the demand for primary steel production and save about 75 per cent of emissions for every ton of scrap recycled (Climate Strategies 2014: 5). There are, however, many technical challenges in recycling (Reck and Graedel 2012). Moreover, the potential for recycling is constrained by the volume of material available (Allwood et al. 2013; Pauliuk et al. 2013). The availability of material from end-of-life products is limited by the time delay between initial production and the discard of products. If the global demand for materials and metals continues growing, then recycling will no longer be able to create a closed-loop or 'circular' material economy. Older developed economies (like the UK) could possibly operate a closed-cycle for steel, whereas emerging economies such as China and India cannot do this until their stocks have grown further (Müller et al. 2011).

Consequently, there is an extremely promising development opportunity for mineral-abundant countries and an excellent commercial opportunity for many miners. Investors in metals have taken note and speculation drove the price of lithium up by 80 per cent in 2016, and cobalt's price rose by 80 per cent over 2017 (making it the year's best performing financial asset) as investment funds accumulated large stockpiles in anticipation of growing demand. Cobalt and lithium were already in demand for smartphone and laptop batteries; and demand is now growing even faster—stimulated by electric vehicle manufacture and associated battery technologies, which are expected to result in a 39 per cent growth in cobalt demand through to 2022.[15] China is central to this dynamic: 80 per cent of its cobalt imports are used to manufacture rechargeable batteries (USGS 2017: 53). Car makers are trying to lock in cobalt supplies, but in 2017 Volkswagen failed to secure five years of supply at a fixed price.[16] As grid-storage technologies become commercially viable they will add greatly to the demand for cobalt, lithium, and other metals.

### 22.3.2 *Development Impacts and Concerns*

Our knowledge of the scale and distribution of the world's mineral and metal resources remains patchy. This is especially so for the rare earths; there are no

[15] 'Five Charts That Matter for Investors', *Financial Times*, 12 October 2017.

[16] 'VW Fails to Secure Long-term Cobalt Supply for Electric Vehicles', *Financial Times* (15 October 2017).

data on production and reserves for Africa and, in the developing country group, only Brazil, China, India, Malaysia, and Thailand have data (World Bank 2017: 50, citing the US Geological Survey). From what we do know, Latin America is especially well placed for the low-carbon energy transition. It has many of the mineral resources central to the transition, notably lithium for batteries; Chile and Argentina (the world's second and third largest producers, respectively) together with Bolivia constitute Latin America's so-called 'lithium triangle'. India has iron and steel and titanium, Indonesia, Malaysia, and the Philippines have bauxite and nickel, while Oceania's small islands also have resources (New Caledonia's has 10 per cent of the world's nickel reserves, for instance). For the rare earths, China has the largest production and reserves with Australia coming a distant second in production (World Bank 2017: 50).

Africa is mineral rich, and its metals will see growing demand from the energy revolution. The Democratic Republic of the Congo (DRC) has between 47 per cent and 60 per cent of the world's cobalt reserves, and supplies more than half the world's total production (USGS 2017: 53).[17] There are at present no other sources of cobalt that match those of the DRC. Despite considerable mining investment over the last decade, Africa's potential remains largely untapped. Africa's shares of world production are often far below its shares of world reserves; Guinea, for example, has 26 per cent of known bauxite reserves but accounts for only 6.5 per cent of global production (World Bank 2017: 26).

In sum, the developing world is potentially well placed to benefit from the material needs of the various low-carbon pathways. However, there are at least four concerns (in addition to uncertainty around the Paris Agreement, as well as the difficulty that countries face in managing resource booms in general).

First, those countries and companies expected to meet increasing demands for base and rarer metals and minerals face an uphill task in reducing their own GHG emissions and the material footprint of their operations (discussed earlier in this chapter). Increasing extraction and production levels, as well as decreasing ore grades, imply that the relative GHG, energy, and water intensity of operations will also increase. Increasing the share of renewables in the power supplied to mines, and in the transport system which serves them, will be vital. Achieving this greatly depends on shifting the host country's power system towards renewable energy.

Second, countries with mineral wealth cannot enjoy the associated earnings and revenues unless they mine. Here there is a marked gap between the success of emerging economies like Brazil, Chile, and China, with their high levels of production relative to reserves, and those of the LICs (and some

[17] The DRC's cobalt reserves are reported as 60 per cent of total world reserves in many sources, but the USGS figure is 47 per cent (Yager 2014).

MICs) where the gap between their reserves and production is often very wide (Guinea's bauxite is an example). This requires large investments, supporting infrastructure (especially reliable power systems and good railways and ports), and a supportive and stable policy framework. This is sadly lacking in many countries.

Third, who will benefit in the producing countries from any resource boom resulting from the low-carbon transition? Some 42 per cent of the reserves of metals and minerals required for a clean energy future are in countries with 'good' or satisfactory resource governance, while 37 per cent are in countries with weak scores according to analysis by Tilley and Manley (2017) which runs the World Bank (2017) data against the NRGI's Resource Governance Index. The countries with weak scores are often fragile states, with high poverty rates. Global poverty is falling, yet many resource-rich countries are exceptions, including mineral-rich countries like the DRC, Zambia, and Zimbabwe. In these countries both the number of poor people, and their share of the total population, are rising (Ferreira et al. 2017). The DRC's resources include copper, cobalt, and nickel—all essential to clean energy technologies—but the country's politics are highly unstable and exclusionary, and over 60 per cent of the population is poor (Nanivazo and Mahrt 2016: 421–45). Zimbabwe is the world's fifth largest producer of lithium, with deposits which may be among the world's largest (USGS 2017: 101). Yet Zimbabwe's mineral wealth has enriched its ruling elite and funded an oppressive security apparatus (Global Witness 2017). Will any lithium boom play out differently? It seems unlikely until Zimbabwe's politics improves.

Fourth, it would be unacceptable if scaling up mining to facilitate the low-carbon transition results in large environmental and social costs around mine sites and for local people. The supply chain for cobalt is especially worrying, as it relies on the DRC which provides over half the world's supply. A handful of industrial miners dominate the DRC's cobalt mining sector, which is attracting more foreign investment.[18] Environmental damage together with human rights abuses in unregulated artisanal mining (including child labour, lack of workplace safety, and forced relocation of villages) are widespread (Amnesty International 2016; Frankel 2016; SOMO 2016). To curb the funding of Congolese militias, US law requires companies registered with the US Securities and Exchange Commission (SEC), and which sell products containing gold, tantalum, tin, or tungsten, to disclose whether these are sourced from the DRC or adjoining countries (subject to independent audit).[19] While cobalt is not implicated in militia-funding, many campaigners (and some companies)

[18] China Molybdenum bought the Tenke copper and cobalt mine in the DRC in 2016.

[19] This is a provision in the Dodd–Frank Wall Street Reform and Consumer Protection Act, passed into law by Congress in July 2010.

argue that cobalt's inclusion in the list would help to curb human rights abuses in the DRC's mining (Frankel 2016).

Batteries for electric vehicles, together with the much larger batteries of grid-storage, require far more cobalt than IT devices. The renewable energy, electric vehicle (and IT industries) all face a serious dilemma, as their technologies greatly depend on the DRC's cobalt. A leading battery maker, LG Chem, stopped buying DRC-sourced minerals, including cobalt, in 2015.[20] Car manufacturer Tesla is trying to replace DRC cobalt with North American supplies for its electric cars. This will be difficult as combined Canadian and US production is less than one-sixth of the DRC's, and the DRC's reserves are more than three times those of Australia, which has the second-largest reserves (USGS 2017: 53). Battery makers are searching for substitutes (nickel is favoured, but iron, manganese, and silicate may be feasible). Research results are promising, but moving from the laboratory to commercial manufacture is some years ahead. Meanwhile, the technologies underlying the renewables (and information) revolutions depend on a mineral supplied by one of the world's most fragile nations.

Cobalt is only one example of the many important minerals that are sourced from fragile states. Embedding these supply chains in a proper framework of governance and transparency (that also avoids environmental damage and human rights abuses) is essential if this resource wealth is to serve inclusive development and help countries end their fragility (Ali et al. 2017).[21] Strengthening state effectiveness and accountability is essential as well (Addison and Brück 2009: 15–30). Weak resource governance deters the investment necessary to develop the sector and provide the revenues (e.g. Zimbabwe's output of lithium is far below other producing countries, despite its having much larger reserves). Fragile states like the DRC risk missing the many development benefits of a future in which low-carbon pathways generate a rising demand for their metals and minerals.

## 22.4 Conclusions

In 2016, atmospheric concentrations of $CO_2$ reached their highest level in 800,000 years, and atmospheric methane also reached new highs (WMO 2017). Since 1990 there has been a 40 per cent increase in radiative forcing by all GHGs (NOAA 2016). To avoid more than a 2°C temperature rise will require GHG concentrations to not exceed more than 430–450 ppmv; this

[20] Apple is also working to reform its cobalt supply chain (Frankel 2016).

[21] The OECD has a set of guidelines on responsible supply-chain management in conflict-affected countries (OECD 2013). Other actions are outlined in RESOLVE (2010).

means that GHG emissions must peak within the next few decades and the world must then achieve zero net-emissions (IPCC 2015).

The shape of the energy future and of the societies that it underpins, as well as the prospects for producers of fossil fuels and manufacturers of low-carbon technologies (and the metals and minerals that go into them), depend on a myriad of decisions by private and public actors, whose decisions interact. Billions of dollars of private investment in extractives (together with investments in renewable energy, electric vehicles, energy-efficient buildings etc.) hinge on the parties to the Paris Agreement fulfilling their commitments and supporting the transition with publicly funded research—thereby encouraging the private sector to shift decisively towards delivering low-carbon pathways for economies and societies. This includes companies in the extractives sector who need to reduce their carbon footprint while, in the case of miners, increasing the production of the metals and minerals essential to the technologies of the low-carbon future.

The cost of renewables is falling at an unexpectedly fast pace yet fossil fuels will dominate power generation for many more years, even as societies improve their energy efficiency. Eventually, much of the world's remaining reserves of fossil fuels must remain unused (unless there is a massive scaling up of carbon sequestration, which is itself problematic). Many of the LICs and MICs still hoping for rapid prosperity based on their oil, gas, and coal reserves face the stranding of these assets in a carbon-constrained future. Although there may be scope for prioritizing their supplies while maintaining a fixed global carbon budget, this requires an international agreement that will be tough to achieve.

At the same time, those LICs and MICs which have reserves of the materials and metals essential to the construction of wind, solar, electricity transmission, and public transportation face a brighter future than that of fossil-fuel producers. If international climate action, including implementation of the Paris Agreement, is robust then minerals and materials that should see sustained demand-growth include: copper, nickel, high-grade steel, aluminium, zinc, molybdenum, platinum, chromium, cobalt, lithium, silver, and a range of rare earths. Renewables have a much greater need for metals and materials than fossil-fuel power generation. This represents a promising market opportunity for producing countries. However, they will face great challenges in simultaneously meeting their Paris goals (to reduce their own GHG emissions) and their SDG goals (to reduce the material footprint of their economic activities). International assistance is vital, especially for the poorer countries.

The story of extractives and climate makes for a complex and intriguing narrative. Success is vital to humanity's continued progress and to reducing the dangers posed by climate change to sustainable development, to poverty reduction—and to humanity's very existence.

## References


Addison, T. and T. Brück (2009). 'Achieving Peace, Participation and Prosperity', in T. Addison and T. Brück (eds), *Making Peace Work: The Challenge of Social and Economic Reconstruction*, 15–30. Basingstoke: Palgrave Macmillan for UNU-WIDER.

Addison, T., M. Niño-Zarazúa, and J. Pirttilä (2018). 'Fiscal Policy, State Building and Economic Development', WIDER Working Paper 2018/5. Helsinki: UNU-WIDER.

Ali, S. H., D. Giurco, N. Arndt, E. Nickless, G. Brown, A. Demetriades, R. Durrheim, M. A. Enriquez, J. Kinnaird, A. Littleboy, L. D. Meinert, R. Oberhänsli, J. Salem, R. Schodde, G. Schneider, O. Vidal, and N. Yakovleva (2017). 'Mineral Supply for Sustainable Development Requires Resource Governance', *Nature*, 543: 367–72.

Allwood J. M., M. F. Ashby, T. G. Gutowski, and E. Worrell (2013). 'Material Efficiency: Providing Material Services with Less Material Production', *Philosophical Transactions of the Royal Society*, 371: 1–15.

Amnesty International (2016). *This Is What We Die For: Human Rights Abuses in the Democratic Republic of the Congo Power the Global Trade in Cobalt*. London: Amnesty International.

Aurora Energy Research (2017). *The Carbon Price Thaw: Post-Freeze Future of the Great Britain Carbon Price*. Oxford: Aurora Energy Research.

Balcombe, P., K. Anderson, J. Spiers, N. Brandon, and A. Hawkes (2017). 'The Natural Gas Supply Chain: The Importance of Methane and Carbon Dioxide Emissions', *ACS Sustainable Chemistry and Engineering*, 5(1): 3–20.

Banks, D. and G. Taraska (2013). *US Natural-Gas Use Must Peak by 2030*. Washington, DC: Center for American Progress.

BP (2016). *BP Statistical Review of World Energy 2016*. London: BP.

Brandt, A. R., G. A. Heath, E. A. Kort, F. O'Sullivan, G. Pétron, S. M. Jordaan, P. Tans, J. Wilcox, A. M. Gopstein, D. Arent, S. Wofsy, N. J. Brown, R. Bradley, G. D. Stucky, D. Eardley, and R. Harriss (2014). 'Methane Leaks from North American Natural Gas Systems', *Science*, 343: 733–5.

Caney, S. (2016). 'Climate Change, Equity, and Stranded Assets'. Research Backgrounder. Washington, DC: Oxfam America.

Carbon Tracker Initiative (2017). *2 Degrees of Separation: Transition Risk for Oil and Gas in a Low Carbon Economy*. London: Carbon Tracker.

CDP (2017). *Putting a Price on Carbon: Integrating Climate Risk into Business Planning*. London: CDP.

Climate Strategies (2014). *Carbon Control and Competitiveness Post 2020: The Steel Report*. Berlin: German Institute for Economic Research.

Coady, D., I. Parry, L. Sears, and B. Shang (2017). 'How Large Are Global Fossil Fuel Subsidies?' *World Development*, 91: 11–27.

Cust, J., D. Manley, and G. Cecchinato (2017). 'Unburnable Wealth of Nations', *Finance and Development*, 54(1): 46–9.

Dawkins E., M. Chadwick, K. Roelich, and R. Falk (2012). 'Metals in a Low-carbon Economy: Resource Scarcity, Climate Change and Business in a Finite World'. Stockholm: Stockholm Environment Institute.

DECC (2015). 'Industrial Decarbonisation and Energy Efficiency Roadmaps to 2050'. London: UK Government (Department of Energy and Climate Change (DECC)/ Department for Business Innovation and Skills).

de Gouw, J. A., D. D. Parrish, G. J. Frost, and M. Trainer (2014). 'Reduced Emissions of $CO_2$, NOx, and $SO_2$ from US Power Plants Owing to Switch from Coal to Natural Gas with Combined Cycle Technology', *Earth's Future*, 2: 75–82.

Ferreira, F., C. Lakner, and C. Sanchez (2017). 'Against the Tide: A Look at the Countries Where Poverty Has Been on the Rise, Despite Falling Global Numbers'. Washington, DC: World Bank.

Florin, N. and P. Fennell (2010). 'Carbon Capture Technology: Future Fossil Fuel Use and Mitigating Climate Change', Grantham Institute for Climate Change Briefing Paper 3. London: Imperial College.

Frankel, T. C. (2016). 'The Cobalt Pipeline: Tracing the Path from Deadly Hand-Dug Mines in Congo to Consumers' Phones and Laptops', *The Washington Post*, 30 September.

García-Olivares, A., J. Ballabrera-Poy, E. García-Ladona, and A. Turiel (2012). 'A Global Renewable Mix with Proven Technologies and Common Materials', *Energy Policy*, 41: 561–74.

Gençsü, I., M. McLynn, M. Runkel, M. Trilling, L. van der Burg, L. Worrall, S. Whitley, and F. Zerzawy (2017). *Phase-out 2020: Monitoring Europe's Fossil Fuel Subsidies*. London: Overseas Development Institute.

Global Witness (2017). *An Inside Job: Zimbabwe—The State, the Security Forces, and a Decade of Disappearing Diamonds*. London: Global Witness.

Gould, T. and C. McGlade (2017). *The Environmental Case for Natural Gas*. Paris: International Energy Agency.

Howarth, R. (2015). 'Methane Emissions and Climatic Warming Risk from Hydraulic Fracturing and Shale Gas Development: Implications for Policy', *Energy and Emission Control Technologies*, 3: 45–54.

Hughes, N., N. Strachan, and R. Gross (2013). 'The Structure of Uncertainty in Future Low Carbon Pathways', *Energy Policy*, 52: 45–54.

ICMM (2011). *Principles for Climate Change Policy Design*. London: ICMM.

ICMM (2013a). *Adapting to a Changing Climate: Implications for the Mining and Metals Industry*. London: ICMM.

ICMM (2013b). *The Cost of Carbon Pricing: Competitiveness Implications for the Mining and Metals Industry*. London: ICMM.

IEA (2014). *World Energy Investment Outlook 2014: Special Report*. Paris: IEA.

IEA (2015a). *Carbon Capture and Storage: The Solution for Deep Emissions Reductions*. Paris: IEA.

IEA (2015b). *Energy Technology Perspectives 2015: Mobilising Innovation to Accelerate Climate Action*. Paris: IEA.

IEA (2017a). *Renewables 2017: Analysis and Forecasts to 2022*. Paris: IEA.

IEA (2017b). *Coal Falls as Gas Rises: World Energy Balances in 2016*. Paris: IEA.

IEA (2017c). *Gas 2017: Analysis and Forecasts to 2022*. Paris: IEA.

IPCC (2014). *Climate Change 2014: Synthesis Report. Contribution of Working Groups I, II and III to the Fifth Assessment Report of the Intergovernmental Panel on Climate Change*. Geneva: IPCC.

IPCC (2015). *Fifth Assessment Report 2015*. Geneva: IPCC. Available at: https://www.ipcc.ch/report/ar5/.

IPIECA (2013). *Addressing Adaptation in the Oil and Gas Industry*. London: IPIECA.

IPIECA (2015). *The Paris Puzzle: The Pathway to a Low-Emissions Future*. London: IPIECA.

IRP (2017). *Green Technology Choices: The Environmental and Resource Implications of Low-Carbon Technologies—A Report of the International Resource Panel*. Nairobi: United Nations Environment Programme.

Kleijn, R. and E. van der Voet (2010). 'Resource Constraints in a Hydrogen Economy Based on Renewable Energy Sources: An Exploration', *Renewable and Sustainable Energy Reviews*, 14: 2784–95.

Kleijn, R., E. van der Voet, G. J. Kramer, L. van Oers, and C. van der Giesen (2011). 'Metal Requirements of Low-Carbon Power Generation', *Energy* 36(9): 5640–8.

Lazarus, M., K. Tempest, P. Klevnäs, and J. I. Korsbakken (2015). *Natural Gas: Guardrails for a Potential Climate Bridge*. Stockholm: Stockholm Environment Institute.

McCulloch, N. (2017). 'Energy Subsidies, International Aid, and the Politics of Reform', WIDER Working Paper 2017/174. Helsinki: UNU-WIDER.

McGlade, C. and P. Elkins (2015). 'The Geographical Distribution of Fossil Fuels Unused When Limiting Global Warming to 2°C', *Nature*, 517: 187–90.

McJeon, H., J. Edmonds, N. Bauer, L. Clarke, B. Fisher, B. P. Flannery, J. Hilaire, V. Krey, G. Marangoni, R. Mi, K. Riahi, H. Rognor, and M. Tavoni (2014). 'Limited Impact on Decadal-Scale Climate Change from Increased Use of Natural Gas', *Nature*, 514: 482–5.

Müller, D. B., T. Wang, and B. Duval (2011). 'Patterns of Iron Use in Societal Evolution', *Environmental Science and Technology*, 45: 182–8.

Nanivazo, M. and K. Mahrt (2016). 'Growth and Poverty in the Democratic Republic of Congo: 2001 through 2013', in C. Arndt, A. McKay, and F. Tarp (eds), *Growth and Poverty in Sub-Saharan Africa*, 421–46. Oxford: Oxford University Press for UNU-WIDER.

NOAA (2016). 'NOAA's Annual Greenhouse Gas Index'. Boulder, CO: National Oceanic and Atmospheric Administration Earth System Research Laboratory.

OCI (2017). *Dirty Energy Dominance: Dependent on Denial—How the US Fossil Fuels Industry Depends on Subsidies and Climate Denial*. Washington, DC: OCI.

OECD (2013). *Due Diligence Guidance for Responsible Supply Chains of Minerals from Conflict-Affected and High Risk Areas*, 2nd edition. Paris: OECD.

Pauliuk, S., T. Wang, and D. B. Müller (2013). 'Steel All over the World: Estimating In-Use Stocks of Iron for 200 Countries', *Resources, Conservation and Recycling*, 71: 22–30.

Reck, B. and T. E. Graedel (2012). 'Challenges in Metal Recycling', *Science*, 337: 690–5.

RESOLVE (2010). *Tracing a Path Forward: A Study of the Challenges of the Supply Chain for Target Metals Used in Electronics*. New York: RESOLVE.

Rhodes, A., J. Skea, and M. Hannon (2014). 'The Global Surge in Energy Innovation', *Energies*, 7: 5601–23.

Schwietzke, S., O. Sherwood, L. Bruhwiler, J. Miller, G. Etiope, E. Dlugokencky, S. Englund Michel, V. Arling, B. Vaughn, J. White, and P. Tans (2016). 'Upward Revision of Global Fossil Fuel Methane Emissions Based on Isotope Database', *Nature*, 538: 88–92.

Scurfield, T. and D. Manley (2017). 'Negotiating Tanzania's Gas Future: What Matters for Investment and Government Revenues?', briefing. London and New York: NRGI.

Shirai, T. and Z. Adam (2017). *Fossil Fuel Subsidies Are Down, But Not Out.* Paris: IEA.

SOMO (2016). *Cobalt Blues: Environmental Pollution and Human Rights Violations in Katanga's Copper and Cobalt Mines*. Amsterdam: Stichting Onderzoek Multinationale Ondernemingen (SOMO).

Sovacool, B. K. (2017). 'The History and Politics of Energy Transitions: Comparing Contested Views and Finding Common Ground', in D. Arent, C. Arndt, M. Miller, F. Tarp, and O. Zinaman (eds), *The Political Economy of Clean Energy Transitions*, 16–35. Oxford: Oxford University Press for UNU-WIDER.

Sovacool, B. K., M. A. Brown, and S. V. Valentine (2016). *Fact and Fiction in Global Energy Policy: Fifteen Contentious Questions*. Baltimore, MD: Johns Hopkins University Press.

Stern, J. (2017). 'The Future of Gas in Decarbonising European Energy Markets: The Need for a New Approach', OEIS Paper NG 116. Oxford: Oxford Institute for Energy Studies.

Tilley, C. and D. Manley (2017). *In a Low Carbon Future, Better Mineral Governance Could Power Development.* London and New York: NRGI.

UNEP (2011). *Bridging the Emissions Gap: A UNEP Synthesis Report.* Nairobi: UNEP.

USGS (2017). *Mineral Commodity Summaries 2017*. Reston, VA: US Geological Survey, Department of the Interior.

Usher, W. and N. Strachan (2012). 'Critical Mid-Term Uncertainties in Long-Term Decarbonisation Pathways', *Energy Policy*, 41: 433–44.

van Breevoort P., K. Blok, M. Hagemann, H. Fekete, N. Höhne, B. Hare, M. Schaeffer, M. Rocha, and L. Jeffery (2015). 'The Coal Gap: Planned Coal-Fired Power Plants Inconsistent with 2°C and Threaten Achievement of INDCs', Climate Action Tracker Update. Berlin: Climate Action Tracker.

van der Ploeg, R. (2016). 'Fossil Fuel Producers under Threat', *Oxford Review of Economic Policy*, 32(2): 206–22.

Vidal, O., B. Goffé, and N. Arndt (2013). 'Metals for a Low-Carbon Society', *Nature Geoscience*, 6 (October): 894–6.

WMO (2017). 'The State of Greenhouse Gases in the Atmosphere Based on Global Observations through 2016', WMO Greenhouse Bulletin No. 13. Geneva: World Meteorological Organization.

World Bank (2017). *The Growing Role of Minerals and Metals for a Low Carbon Future.* Washington, DC: World Bank.

World Bank/ECOFYS (2017). *Carbon Pricing Watch 2017*. Washington, DC: World Bank.

WWF (2011). *The Energy Report: 100% Renewable Energy by 2050*. Gland (Switzerland): World Wildlife Fund in collaboration with ECOFYS and OMA.

Wykes, S. and A. Scott (2017). *UK Support for Energy in Developing Countries 2010–14.* London: CAFOD.

Yager, T. R. (2014). *The Mineral Industry of Congo (Kinshasa).* Reston, VA: US Geological Survey, Department of the Interior.

Zenghelis, D. and N. Stern (2016). 'The Importance of Looking Forward to Manage Risks: Submission to the Task Force on Climate-Related Financial Disclosures', Policy Paper. London: ESRC Centre for Climate Change Economics and Policy/Grantham Research Institute on Climate Change and the Environment.

ZEP (2013). *$CO_2$ Capture and Storage (CCS) in Energy-Intensive Industries: An Indispensable Route to an EU Low-Carbon Economy*. Brussels: Zero Emissions Platform.

# Part VII

# Leveraging the Direct Impacts of Extractives into Sustainable Development

# 23

# Framework

## The Channels for Indirect Impacts

*Alan Roe and Jeffery Round*

### 23.1 Introduction

The direct economic impact of the mining and oil and gas sectors in general, and of a new project in these sectors in particular, is fairly well established. For a new project this would form part of the project proposal and appraisal. However, the full consequences, macroeconomic or otherwise, are subject to some uncertainty and speculation, not least because any extractives project is costly and non-marginal, and so is likely to impact on a (national or local) economy in many and various ways. In this chapter we consider the possible direct and secondary impacts, and in particular the various channels of effect. Only by identifying these channels in their entirety do we gain an understanding of the range of likely economic and social impacts and thereby find ways of estimating their significance. There is then a basis for formulating sound policy responses to promote the positive effects and to mitigate any negative effects.[1]

We begin in Section 23.2 with a generic overview of the channels of effect, which serves as a framework for later discussion. The social and economic consequences of extractives activities arise from two principal channels: first, the *industry* channel, and secondly the *public spending* channel (AfDB and BMGF 2015a, 2015b). The industry channel traces the income-generating

[1] Thanks are due to the participants at the UNU-WIDER conference in April 2016 who offered valuable comments, and especially to Jim Otto. Our thanks go also to Merl Storr for most helpful editorial suggestions. An extended version of this chapter appeared as a WIDER Working Paper (Roe and Round 2017).

and employment-creating consequences of all the spending that stems from the activity (investment and operations) through the economy. The public spending channel captures the effects of spending the (likely) significant tax and other government revenue generated by the activities (ICMM 2014b; Otto 2017). The industry channel is discussed in more detail in Section 23.3 below, and the public spending channel in Section 23.5.

Primary effects via these two channels potentially lead to secondary effects. These may be substantial and could be at least as large as the primary effects. In the case of the industry channel, some of these effects have been referred to as upstream and downstream effects (including backward and forward linkages), usually in the context of indirect effects on other production activity. But they may also spill over into final demand linkages and other economy-wide effects via employment and income generation. The direction and magnitude of these effects are complex, and are difficult to determine *ex ante*. There have been several attempts to do this empirically; the results are obviously specific to projects, economies, and circumstances. Section 23.4 presents some selective results from applied work in the context of mining activities.

Finally, the second channel—public spending—is elaborated in Section 23.5, where we briefly set out the possible direct and indirect effects on human development via the public spending channel. Unlike the industry channel, the effects here are less dependent on behavioural mechanisms and instead are heavily policy dependent,[2] so it is much more difficult to ascertain outcomes or to derive sound empirical estimates of general applicability. Hence much of our discussion on this matter is of a conceptual nature.

## 23.2 Channels of Impact

The extractives sector (mining and oil and gas) in some low- and middle-income economies is often significant, although there are vast differences in the sector's contribution to the major economic aggregates (gross domestic product (GDP), exports, fiscal revenues, etc.), both in comparison with other sectors in individual countries and in comparisons across countries (ICMM 2014b).

Based on accumulated stylized facts[3] the extractives sector dominates foreign direct investment (FDI) in many lower-income mineral (and oil and gas)

[2] The secondary industry effects are also influenced by policy, of course, but unlike in the public spending channel, behavioural and technological mechanisms play a substantial role.

[3] These stylized facts are based on a series of country case studies undertaken by the International Council for Mining and Metals (ICMM) over the past ten years in a variety of mining countries in Africa, Asia, and Latin America (Roe and Round 2017).

economies (60–90 per cent of FDI); it constitutes a large share of exports (30–60 per cent); mineral taxation is often a significant share of total tax revenue (3–20 per cent); but perhaps surprisingly, the sector typically generates a small share of the country's GDP and GNI (about 3–10 per cent of GNI) relative to other sectors. Mining and oil and gas activities are highly capital intensive, and employment in the mine (or oil field) is typically small relative to the size of its GDP contribution; hence direct employment also tends to be a small share of total national employment (generally only 1–2 per cent of total employment). These broad features are typical of the sector, although presented in this way they mask several channels of wider economic impact which need to be examined in more detail.

Over the life cycle of a typical mining project[4] several phases can usefully be distinguished: exploration, site design and construction, operation, and finally closure and decommissioning. Each phase will create a distinct and potentially quite different direct economic impact. For instance, the exploration phase usually requires a relatively small number of highly skilled and technical personnel, possibly expatriate, with few local inputs. Hence the direct macroeconomic impact of this activity on the economy is probably relatively small. Similarly, in the closing down and decommissioning phase there will be little or no appreciable economic output, engaging a reducing and ultimately low level of employment, although the decommissioning and post-closure costs might be substantial. The phases that have the most significant direct economic impacts are the construction and operating phases, although the nature of the activities involved in each phase may involve quite different direct effects. Clearly, the stage a project has reached within its life cycle will make a huge difference to the initial and overall impacts it will have on the economy (ICMM 2014b).

To illustrate the structure and paths of influence in the wider economy, Figure 23.1 shows in simple schematic form the principal channels of economy-wide impact of a mining or oil and gas activity. Note that here we focus on the *expenditure flows* (in constant prices)[5] and leave possible *price effects* to one side for the moment, even though price changes could affect the size and hence the total influence of the expenditure flows. The direct spending arising out of extractives activities in all phases is shown in the first entry in the first row of Figure 23.1. This direct spending constitutes procurement of goods and services, employment (the payment of wages and salaries), and infrastructure spending. This initiates the 'industry activity' channel (AfDB and BMGF 2015b). The figure shows the ultimate impact on living standards and human well-being. It shows that a primary impact on living standards

[4] For more detail see Roe and Round (2017) and ICMM (2014b: 7).

[5] e.g. the prices in a given year, such as the initial year of the project.

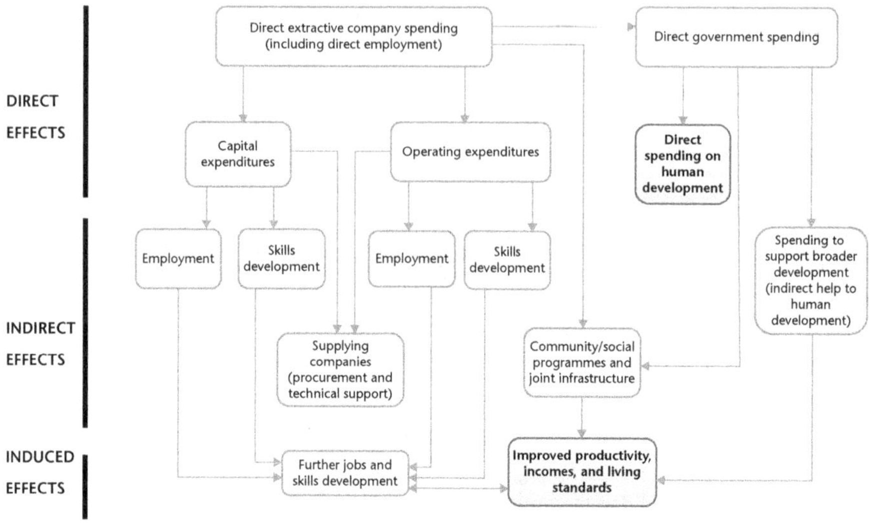


**Figure 23.1.** Extractives activities and development: the main channels
*Source*: authors' illustration based on ICMM (2014b).

derives from the incomes generated by extractives activity thereby directly benefitting households.

The other major direct outcome of extractives activities arises from the revenues accruing to government in the form of taxes, royalties, and other fiscal revenues (Daniels et al. 2010; Otto 2017). This initiates the public spending channel. The revenues generated are likely to be significant, thus bolstering the financial inflows for government spending. However, they are not evenly spread over the life cycle. These tax revenues when they arise can be used to fund a wide range of policy initiatives, both inside and beyond the mining and oil and gas sectors, including those related to economic promotion, to social and human development projects, and to social protection at local and regional levels. It is for governments to design and implement the desired policy initiatives (AfDB and BMGF 2015a, 2015b).

Figure 23.1 also shows expenditures and initiatives directly by extractives companies on community and social programmes, usually made independently (or partly independently) of central or local government. These include spending on schools, health centres, and community buildings in rural areas, where mines and extractive companies are often located.

The resulting impacts of extractives via the channels of impact do not stop at the initial direct spending effects. Secondary effects stem from the direct spending effects, and do so via both the industry activity channel and the public spending channel in potentially different ways. In particular, the direct spending effects through the industry channel can precipitate some indirect effects via backward and forward linkages and can also trigger further induced effects through spending out of the incomes generated. By contrast, the indirect effects via the public spending channel may arise because of induced incomes created by the government's spending to support broader development. Thus, in Figure 23.1 the box showing the ultimate effect on living standards includes the total effect on incomes (direct, indirect, and induced) arising from the increase in economic activity.

Conceptually, it is relatively straightforward to estimate the *short-term* secondary effects of mining (or oil and gas) activity via the industry channel—as we shall demonstrate in Section 23.4. However, it is far less straightforward to estimate the *dynamic and long-term effects*. The best one can do is to identify possible consequences and seek evidence of their existence either directly or from previous studies. The public spending channel will also generate secondary effects that are both short and long term. These effects are *short term* in the sense that they rely on existing capacity to stimulate output, employment, and income. They are *long term* in the sense that public spending today can increase future capacity and so stimulate economic growth, or improve productivity, or make permanent improvements to social or economic institutions and infrastructure. Of course, there is also evidence that the secondary effects may not always be positive. The term 'Dutch disease' emerged in the late 1970s. Corden and Neary (1982) formalized the problem, and Auty (1993) and Sachs and Warner (1995) produced evidence that resource-rich economies tend to grow less rapidly than resource-scarce economies. This evidence, which is reviewed elsewhere in this volume, has been the subject of much subsequent debate, and has prompted a search for policies that have positive secondary outcomes. This is the main hallmark of what Di Boscio (2010) has called the 'post-global' or 'sustainable development' model (see also Lahn and Stevens 2017).

## 23.3 Industry Channel: Downstream and Upstream Multiplier Effects

As a general observation, much of the new mining activity in lower-income economies takes place in more remote areas, away from existing urban areas, where the existing economic activity is most likely based on agriculture (or forestry, etc.). The new mining activity will lead to new communities, and often to new satellite production activities as a consequence. These arise from the creation of production linkage effects, which Hirschman (1958) referred to as 'backward', 'forward', and 'final demand' linkages. Characteristically, in most developing countries the manufacturing and tertiary base is often quite weak. Hence many of the inputs into production (during both construction and operation) have to be imported.[6] However, the supply requirements for mining activity do create opportunities for existing domestic firms to sell to the mine, and also for new firms to enter upstream, in the short to medium term (backward linkages). Likewise, there may be downstream effects (forward linkages). Much mineral processing (e.g. smelting, refining, or material fabrication) usually takes place initially near to the final market rather than near to the mine, which means that many mineral products are exported in their raw (or crude) form. However, in time investment can diversify[7] from mining per se and new profit-making opportunities arise domestically, thereby creating forward linkages within the domestic economy. Auty (2004) cites many examples of this evolutionary process, whereby mining has proved to be the catalyst for a diversified industrial region. These include the West Midlands (UK), Witwatersrand (South Africa), Ciudad Guayana (Eastern Venezuela), and Western Australia, amongst many others. It is therefore important to see the emergence of upstream and downstream effects as part of an evolution, a dynamic process of economic development. In achieving this, investment in transport and power infrastructure is especially critical if the objective is to achieve local processing that is competitive.

Direct final demand linkages from mining activity are less likely. Mining products will not enter into the consumption bundle of households or government directly, nor will it form a part of fixed capital formation. The bulk of output that is not processed domestically will be exported. The principal final demand linkages will arise from spending out of incomes generated: thus they are more likely to be a secondary rather than a primary effect.

As well as considering extractives activity in the context of the whole economy, there is also an important regional dimension to all of this. The impacts of extractives activity are unlikely to be distributed across the whole economy. Many of the direct effects will be heavily localized (due to agglomeration economies, and to the proximity of labour), although some of the secondary effects will be more widely spread throughout the economy.[8] Di Boscio (2010) has described how the pattern of regional development associated with mining activity has evolved and changed over time. He has described how pre-1970 mining activity followed the regional or 'strong local linkage' model, where mineral ventures facing high transport costs were highly labour intensive and generated significant upstream (and often downstream) benefits. This has been supplanted by a more recent global or 'weak local linkage' model, characterized by lower transport costs and high capital intensity (with lower

[6] This is true of goods but less so of services, many of which can be obtained locally.

[7] Östensson (2017) discusses how sometimes this does not happen.

[8] In some (conflict) countries mining often takes place illegally, and this affects the nature and distribution of the activity and of any income generated.

labour requirements), tending to generate lower upstream and downstream benefits. Hence the tendency has been for strong localization of the primary effects to be perpetuated, even with the evolution of mining technology and the changing economy-wide context. Di Boscio (2010) suggests that the global model has been supplanted by the sustainable development model since the 1990s.[9] This reflects the industry's desire to develop an approach that is more sustainable in socio-economic terms. It implies that companies are now more likely to engage with the communities and government, and the effects and benefits are likely to be more widely spread than hitherto. However, there is a limit to how much benefit can be distributed locally, including via job creation, if governments are unwilling or unable to invest enough in supporting local infrastructure and institutions.

All of these primary and secondary effects will be subject to further multiplier responses. Hence, as well as identifying the primary effects of mining activity, there is a question as to how these multipliers might be estimated.

## 23.4 Estimates of Impact Multipliers for Mining Activity

### 23.4.1 *Measuring the Size of Secondary Impacts*

The measurement of secondary impacts of mining activity has proved to be a difficult challenge. Studies (e.g. ICMM 2013, 2014a; Mogilevskii et al. 2015; OPM 2009) have been carried out in the context of either a new mining project or an existing mining enterprise, or as an assessment of the contribution of a mining industry to the economy as a whole. The most obvious method of estimating secondary effects is to apply informed judgement and rules of thumb. Of course, in the absence of reliable economy-wide data this might be the only possible approach, but the major downside is that it is quite difficult to test assumptions, so a more robust formal approach is usually preferred.

Three principal methodologies have been used to assess secondary economic and social impacts. They are: input–output (or multiplier) analysis, computable general equilibrium modelling, and econometric modelling. The first two methods may be described as *ex ante* (simulation) methods, whilst the latter is an *ex post* method that is finding increasing favour in project analysis. However, they ought not to be viewed as alternatives; they have quite different objectives and serve different purposes.

Input–output (I–O) and computable general equilibrium (CGE) models have been the dominant approaches to date for estimating impact multipliers via the industry channel for mining activity. We do not attempt to provide

[9] See Hodge (2017) on the meaning of 'sustainability'.

a comprehensive review of either the methods or the results.[10] Rather, we highlight two studies for illustration, and refer briefly to the results of some other studies. It is not possible to arrive at definitive general conclusions, as the results are bound to be heavily conditioned by economy-specific factors, as well as by the methodology employed.

### 23.4.2 *Input–Output (I–O) Modelling*

Input–output (I–O) modelling is well known and is described in more detail elsewhere (Roe and Round 2017) but we do need to establish two key pieces of terminology to assist present discussion. First, we should distinguish between three commonly used multipliers: output, income, and employment multipliers. The principle is the same for each: a multiplier is defined as the ratio of the total outcome divided by an initial (direct) outcome. For example, the output multiplier is the total gross output (of a particular sector, or of the economy as a whole), taking into account economy-wide interdependences and responses, that results from a unit initial change in gross output of, say, mining.

Second, we have to distinguish between Type I and Type II multipliers. Type I multipliers take account only of the inter-industry independence: these generate the direct and indirect effects (the input–output multipliers). Type II multipliers additionally take account of the Keynesian income–expenditure effects: the direct, indirect, and induced effects. Thus for the purpose of using the results to estimate total economic impacts, only the Type II multipliers are of practical significance.

We now illustrate the use of this methodology in the context of a study of the mining sector in the south-eastern Pará region in *Brazil* (ICMM 2013). The full study was concerned with investigating the social and economic impact of mining at the national and local levels, as represented in a case study of Vale, Brazil's largest mining company in south-eastern Pará.[11] Both Type I and Type II multipliers were calculated. As regards the Type 1 multiplier for mining the results showed that a R$1 increase in the demand for output of mining is likely to lead to R$1.2 of output in the south-eastern Pará region as a whole. This is a small increase, but it is similar to the multipliers for all other sectors. All the multipliers are low because the industry linkage in such a relatively small region is low. The Type II multipliers which include the effects of household spending out of income generated are inevitably larger. The multiplier for mining is 2.3. This means that, taking account of the spending

[10] A discussion can be found in Roe and Round (2017: Sections 4 and 5).

[11] Filgueiras et al. (2017) examine other aspects of the Brazilian/Vale case.

effects of households out of the income created by the increased production activity a further R$1.1 of output is generated by mining in the south-eastern Pará region. Clearly, the multipliers for mining only apply in the operational phase, when there is mining output. But during the construction phase the multiplier effect will be a combination of the output multipliers of sectors stimulated by these local procurements.

The input–output multipliers show equilibrium outcomes with no explicit time dimension. However, the model is often used to estimate the equilibrium outcomes of a series of annual primary effects. While the results do not show the time profile of the incidence of effects, they can be used to show the total effects in relation to the time profile of the primary impacts. This can be illustrated in relation to a study for *Romania* (OPM 2009). It showed both the direct and estimated total contribution to Romania's GDP arising from the Rosia Montana Gold Mine Corporation over a projected eighteen-year period. This covers two years in construction (2011–12) and sixteen years of operation (2013–28). In this case the *direct* contribution to GDP in the two years of the construction phase is lower than that anticipated during the sixteen years of the operation phase. Although there is variation year on year, the estimated *total* effect on GDP from the construction and operational phases of the gold mine is an *additional* 2.0 to 2.5 times the direct contributions (i.e. a multiplier of between 3.0 and 3.5). However, the indirect and induced contributions shown for each year of the projection will not necessarily all materialize in that year; there will be lags as the repercussions work their way through the economic system.

Bocoum (2000) has utilized input–output multipliers to undertake a comparative study of the impacts of mining activities in countries at various stages of development and in different kinds of mining-related sectors. She harmonized input–output tables for *Australia*, the *United States*, and *Chile*, and distinguished sectors for mining, mineral processing, mineral and energy manufacturing, and energy production. The results are reported in terms of Hirschman total linkages,[12] although they are based on (Type I) output multipliers and (Type II) income multipliers (not reported). The main results of her study are:

- The type of mineral activity affects the size and nature of the secondary effects (multipliers).
- Energy (coal, oil, and gas) extraction shows a relatively small direct contribution to GDP, although it is important in terms of total multipliers.

[12] Total linkages (TL) are defined as the sum of backward (BL) and forward (FL) linkages. A sector with a BL or FL greater than 1.0 is considered 'strong', so a sector with a strong BL and FL and a TL greater than 2.0 is considered to be a 'key' sector.

- The overall contribution of non-ferrous mining to output and income is generally low, but it has strong linkages with domestic industries such as energy and other mineral-related activities.
- Non-metallic mining, with the diversity of its output, has the potential for stronger linkages than metallic mining.
- Mining services have strong backward linkages.
- Petroleum and coal transformation generate strong forward linkages.
- The agricultural and industrial chemicals sector is important in terms of output and income and total linkages.

IDC (2013) used the input–output methodology to explore the linkage between the mining and manufacturing sectors in *South Africa*. Based on 2012 data, the Type II output multipliers for the sector were estimated to be about 2.0. IDC (2013) also reports the changing structure of South Africa's mining sector between 1980 and 2012. It is now much more diversified, with gold becoming a diminishing proportion of mining sales. The backward and forward linkages with the manufacturing sectors remain strong.

Rayner and Bishop (2013) applied the input–output methodology to obtain a 'broad measure' of the boom in natural resources in the *Australian* economy from about 2004 to the end of the decade. They used a sectoral context in which they distinguished between 'resource extraction activity' (coal, oil, gas, iron ore, etc.) and 'resource-related activity' (business services, construction, etc.). Based on an input–output table, the total direct contribution of the resource sector to Australian GDP in 2011–12 was 11.5 per cent. On the basis of Type I multipliers, they estimated that a further 6.5 per cent of GDP was attributable to resource-based activity (an implied Type I multiplier of 1.57). If the wage link had been included, then the Type II multiplier would have been well in excess of 2.0.

Finally, a recent large-scale input–output study for the *United States* (NMA 2012) considered the economy-wide contributions of US mining. Based on a large regional input–output table for 2012, the study is an exemplar of its kind. The mining sector includes three subsectors: mine operations, mining support activities, and transport of output from the mine. This underlines one of the problems of comparability across studies—how exactly the 'mining' sector is defined. Using this broad definition of mining activity, the study estimates that the sector contributed 0.64 per cent of US GDP in 2012. But based on a calculation of Type II multipliers, taking account of backward linkages and induced effects, the contribution to GDP rose to 1.4 per cent. The implied Type II multiplier for the sector and for the United States as a whole is 2.2. The variation in the multipliers for different branches of mining across US states is considerable. However, focusing on the results for the mining sector as a whole, the Type II multiplier for California is 4.7, and for New York and New Jersey even higher (9.4 and 11.6 respectively). These values would imply that there is a huge secondary effect of mining activity in these states. But in all cases where the multipliers are very high, the *direct* contribution of mining to GDP, in the state and nationally, is quite modest. Nevertheless, this does suggest that upstream effects might be significant when economic activity is broadly based and well developed.

### 23.4.3 *CGE Modelling*

Relatively few CGE-based studies to assess mining activity per se are available in published work, but two examples will help to illustrate what can be achieved with this approach. A main feature of CGE models is their ability to deal with price shocks and changes in relative prices. For example, unlike I–O models, the real exchange rate effect of Dutch disease can be captured in CGEs. The models are complex and there is often a great deal of specificity in terms of data, assumptions, and model specifications, which makes comparisons difficult to manage.

Downes et al. (2014) attempted to estimate the effects of the mining boom in *Australia* over the decade from the mid-2000s. The 'boom' encapsulated a tripling of the world price of mining exports together with a sharp increase in investment in mining. Their objective was to assess the impact of these shocks on living standards, and to investigate whether the economy experienced negative effects due to Dutch disease. Their overall assessment was that, because of the requirements by mining for manufactured goods, the negative effects due to Dutch disease were not strong. However, their results show that manufacturing output was about 5 per cent below what it would have been without the resource boom. Also, overall living standards were significantly higher than they would have been without the boom, although there were differences across household groups depending on which sectors created their incomes. For example, households relying on income generated in import-competing industries did less well. The study showed that the boom benefitted households as a whole in different ways: higher employment, higher real wages, and lower average tax rates. Unlike the I–O methodology, CGEs show the role of prices, which may precipitate both negative and positive effects of seemingly positive shocks.

An unrelated study by Tourism Research Australia (2013) also carried out a CGE model-based study of the impact of the mining boom on tourism. Many of its key aggregate results were broadly similar to those of Downes et al. (2014), but it found a wide variation of outcomes at a state-wide level. In terms of contributions to GDP, there were losers as well as gainers from the mining boom.

In a classic example of this genre, Mogilevskii et al. (2015) assessed the impact of the Kumtor gold mine on the economic and social development of the *Kyrgyz Republic*. The authors adapted a dynamic CGE model developed by the World Bank, and calibrated it to data for the Kyrgyz economy. Kumtor is the largest enterprise in Kyrgyzstan, so its impact is highly non-marginal. The aim of the study was to estimate not only the direct contributions but also the multiple secondary effects in the context of the country's long-term development. Mogilevskii et al. simulate three scenarios in terms of gold production and then estimate the effects on the main macroeconomic variables. There is little disaggregation within the model to facilitate a detailed impact analysis. Also, the results are heavily conditioned by the model assumptions (i.e. the model 'closure' rules), where in particular the labour market clearing assumption is quite simple. Mogilevskii et al. conclude that there is little scope for increased employment in the mining enterprise, and that a main benefit of Kumtor is its substantial contribution to the government budget and social spending.

CGE simulation models permit several policy-relevant extensions to be added to the basic framework. For example, there are examples where the likely distributional (or gender or environmental) consequences of extractives projects/policies have been examined. Andersen and Faris (2002) examined the possible consequences for the economy of *Bolivia* of an increase in natural gas sales to Brazil. As part of their exercise they demonstrated that, while there would be a likely increase in wages and incomes of all groups in society (and hence the policy would be good for growth and poverty alleviation), there would also be an increase in inequality: smallholders and informal-sector workers would not benefit by as much as those in the formal sector. Likewise, in an interesting series of studies of large-scale mining in *Mongolia*, based in part on CGE models, Kahn et al. (2011) report on the significant gender differences in the benefits (and costs) accruing from the mining boom. In that example, less than 10 per cent of the workforce in extractives was female (and these were employed mainly in service support roles), whereas most of the risks (health, social, and family disruptions) were borne by women.

The assessment of environmental consequences of extractive industries is an ongoing issue—to prevent, mitigate, and control the environmental damages and negative social impacts, improve environmental management, and ensure compliance with environmental standards. There will be primary and secondary environmental effects, which again can be captured via simulation models: both I–O and CGE models can be used to do this. Thiele (1999) used a basic CGE model for environmental policy evaluation of the *South African* mining sector. The results, linking sectoral outputs to hazardous waste (air emissions, waste water, and solid/liquid waste), are indicative of how the models might be developed further.

### 23.4.4 *Econometric Modelling*

The *ex post* econometric approach was originally used by Moretti (2010) in the context of estimating the long-term employment multiplier at the local level in the United States. Specifically, he quantified 'the long-term change in the number of jobs in a city's tradable and non-tradable sectors generated by an exogenous increase in the number of jobs in the tradable sector, allowing for the endogenous reallocation of factors and adjustment of prices' (Moretti 2010: 9). The estimation was based on time series census data for 2001–11. In common with the CGE approach (and unlike I–O), because it uses data on 'initial' employment and 'eventual' employment, effectively the method takes into account endogenous price changes.

Fleming and Measham (2014) applied this approach in order to estimate local job multipliers of mining in *Australia*. The application of the econometric methodology was fairly straightforward: the data were drawn from data on employment changes for two five-year periods, 2001–06 and 2006–11. The application of the Moretti methodology yielded estimates of job elasticities at the sectoral level. Using census data to determine base year employment shares, the authors then calculated local job multipliers. Although it is possible for elasticities to be negative, implying that job increases in mining might lead to job losses in other sectors, virtually all were positive, and seven were found to be statistically significant. These included elasticities for wholesale and retail trade, transport, financial services, real estate, and other services; construction was close to being significant too. Most of these imply job multipliers that are greater than one (i.e. they are greater than 2.0 in concordance with the I–O multiplier definition of multipliers). Elasticity estimates for manufacturing were positive, although agriculture was slightly negative (neither was statistically significant). All of this implied that mining had significantly positive secondary effects in terms of employment across almost all sectors of the economy.

The Moretti approach has also been used by Moritz (2015) on data for a more geographically focused mining region in northern *Sweden*. Due to data limitations, his industrial sector definitions are far more aggregated, and the results are generally not statistically significant. Except for 'government services' all of Moritz's multipliers are positive. However, they are slightly lower than those obtained by Fleming and Measham for Australia; this may be due to the fact that the Swedish regions are smaller and the industrial spread is narrower. Overall, the positive secondary effects of mining activity are confirmed.

## 23.5 Public Spending Channel

### 23.5.1 *Direct Impacts on Human Development*

The second principal channel through which mining activity impacts on human development is the public spending channel. Typically, between 15 per cent and 20 per cent of the production value of mining activity accrues as taxes and other revenue to government (ICMM 2014b), perhaps even more in the cases of some kinds of extractives (e.g. oil and gas). These revenues create the funding which can be used by government to enhance human development in two broad ways:

- Directly, through spending on social sectors: for example, current spending and capital spending on health and education facilities, as well as on social protection programmes including cash transfer schemes.
- Indirectly, through spending to enhance economic growth and development: that is, by boosting capacity in the economy for further income generation and social investment.

'Human development' is open to wide interpretation and definition. It is often defined in terms of three dimensions: health, education, and living standards. Living standards are usually measured in terms of a money-metric criterion, such as real per capita income. Thus 'living standards' are mainly determined by incomes generated in production directly or indirectly, and these are predominantly governed by outcomes through the industry channel discussed previously. However, the living standards dimensions through health and education and other social spending can be strongly influenced through the public spending channel.

As Witter and Jakobsen (2017) argue elsewhere the case for investing natural resources revenues in social sectors is strong.[13] They identify several main channels for potential beneficial impacts. The authors also note that inclusive growth also requires attention to low-income households. A growing body of evidence indicates the potential benefits of cash transfers: these help to tackle poverty in the short run, improve social cohesion, boost local economies, and build human capital in the long term. However, Witter and Jakobsen (2017) also point to a risk of natural resource revenue flows, which have a propensity to distract from results-driven questions ('What do we want to achieve?'), leading instead to a focus on expense-driven questions ('We have funds, what should we spend them on?'). In order to try to address this risk, AfDB and BMGF (2015c) have developed a framework of analysis that identifies the

[13] However, that case does rest on the assumptions that these sectors are given a high priority in the national government's development plans, especially in lower-income countries, and that this spending has the potential to lead to high economic returns.

key questions that can be used to inform investment decisions in the social sectors and so help to produce coherent results.

The *ex ante* decisions for governments in this area are difficult to make. Similarly, the *ex post* effects on outcomes for human development via the public spending channel are difficult to assess. For this purpose it is possible to use indicators such as income per capita or consumption per capita, which are both *outcome* measures. But it is more usual to use *input* measures for health and education—such as school enrolment, hospital visits, or bed occupancy. More broadly, any increase in public spending on education and health that is clearly facilitated by revenues generated by mining activity can itself be seen as a useful indicator of a potential enhancement of human development.[14]

### 23.5.2 *Indirect Effects via the Public Spending Channel*

For many developing countries, the start of a mining- or oil-related activity takes place at an early stage of development (and in many cases it has preceded political independence). Many argue that resource extraction should be seen as a helping hand to achieve the eventual transformation and socio-economic development of the economy, and not as an end in itself (see also Lahn and Stevens (2017), who see transformation and diversification as a key aim of the policy for extractives). Additional public revenues generated from the extractives sector can clearly provide governments with the means to help to facilitate this process of transformation, diversification, and development in various ways (AfDB and BMGF 2015b).

#### 23.5.2.1 INVESTING IN INFRASTRUCTURE

Most LICs, particularly in Africa, are known to have huge infrastructural deficits—a gap between the infrastructure that they need and that which they actually have in place. These deficits in turn constitute a serious brake on growth, irrespective of whether the country concerned has the advantage of natural resources wealth. However, the availability of additional public revenues from extractive resources provides an opportunity to at least make a dent in such deficits. Whilst there is no guarantee that an improvement in capacity and infrastructure will have a positive impact, there are some obvious ways in which such spending can help to generate benefits by enhancing economic growth. Prime examples are improvements in access to the electricity grid, communications, roads, and transport networks. However, Henstridge and Page (2012) caution that the benefits from such investments are not automatic. For example, if construction faces bottlenecks,

[14] Although this does, of course, assume that the money is reasonably well spent and not wasted.

then any new investment could force up costs and prices, thereby reducing output more generally for any increase in nominal investment.[15] Further, the scale of the available resource revenues in most country cases may be small relative to the size of the infrastructure deficit. Hence those revenues need to be used judiciously, and often in partnership with larger sums from private investors, if they are to make any real impact on the infrastructure problem.

#### 23.5.2.2 INVESTING IN STRUCTURAL TRANSFORMATION AND ECONOMIC DIVERSIFICATION

AfDB and BMGF (2015b) cite Chile, Indonesia, and Malaysia as examples of countries that have successfully managed the transformation from economies that were heavily resource-based to those that are now more broadly based. These transformations were, at least in part, the result of a strategic view and subsequent public policy actions taken and associated public spending in each country. The well-known danger here is that of poor selection of the new strategic industries for the future. The availability of large resource revenues can all too easily be used to provide public support to industries whose basic commercial viability is lacking and where eventual financial failure may therefore be the outcome.

Apart from big-ticket strategic actions, other examples include public investments to help strengthen an economy's smaller firms via either technical guidance or financial support; the creation of special economic zones; new training initiatives targeted at new emerging industries; and other improved productivity-enhancing initiatives via research and development and enterprise development. In this regard, ongoing work by Sutton (2009 onwards) and associates on 'enterprise maps' is noteworthy. These are inventories of profiles of leading companies in each major industry in selected countries in sub-Saharan Africa, designed to trace the origins of current industrial capabilities. The aim is to help potential investors, as well as policy makers who are concerned with fostering industrial development, to identify promising targets to link more fully into the supply chains of local extractive companies and their overseas associates. So far five countries have been profiled, spanning industries in agribusiness, manufacturing, and construction.

As presented here, the decisions about the 'best' use of additional public revenues from extractive industries come down to an apparently simple choice: providing direct help to boost local living standards in the short term, or investing public funds in broader development initiatives for the longer term. In practice those decisions—mediated as they are through messy budgetary

[15] They report, for example, that there is some evidence that the marginal costs of construction in Uganda rose in 2010–11 as a result of rapid public investment.

processes that are themselves impacted by political considerations and serious informational deficits (e.g. how much revenue may be available in each future year)—will rarely be taken in any clear and scientific manner. However, wise and well-organized governments will do their best to use such frameworks as are available to guide their thinking and to weigh the choices available to them with the greatest possible care. In particular, there is also a strong case to try to link the benefits of the two public spending routes by embedding decisions about them within a clear set of economic and industrial policies that build explicitly on the stimulating effects that are certain to come from mining and other extractives projects.

## 23.6 Conclusions

The social and economic impacts of the extractive industries in a development context are potentially wide-ranging. In this chapter we have set out the framework and channels of impact, identifying two principal channels, namely the industry channel and the public spending channel. Of these, the public spending channel is perhaps the best known, most discussed, and best documented. The effects on human well-being through this channel are direct and easily understood: they stem from the spending of tax revenues on public projects, especially on health, education, and community projects. This channel is also fundamental to the case that host governments make for extracting more tax revenue from mining activities. But the industry channel—being generally much larger—can also have major impacts on broader economic areas as well as human development and well-being via the generation of income and the increase in domestic economic activity. This chapter has shown that a good deal of this income generation is secondary. It is fundamentally dependent for its magnitude on the existing economic structure of a country and how this structure might develop in future, and this in turn is influenced by economic and industrial policy as well as by public spending. Governments need to address this potential, and their own role in realizing it, with as much seriousness as they lobby for more tax revenue to spend on their own account.

Here we have focused much attention on the industry channel, setting out the various ways in which the secondary effects can be (and have been) ascertained and ultimately measured. Formal methods have relied on economic models which fall into two broad categories: *ex ante* and *ex post* models. *Ex post* models—although currently popular for assessing historical consequences (e.g. non-mining jobs that result from new jobs in mining)—are not so effective when applied *ex ante*. But even the *ex ante* models are only as good as the data and model specifications that are employed. The capacity to do this modelling work—and to collect the data—needs to be embedded in a sustained institutional setting, rather than being generated in one-off efforts.

The overarching conclusion from this chapter is to reinforce the case for a dedicated economic and industrial policy to accompany mining and extractives projects, so that the strongest possible social and economic benefits will flow via both the industry and public spending channels. It is all too easy to overlook especially the induced effects that are more diffuse and so less visible than the other types of mining impacts.

## References

AfDB and BMGF (2015a). 'Delivering on the Promise: Leveraging Natural Resources to Accelerate Human Development in Africa'. Oxford: OPML.

AfDB and BMGF (2015b). 'A Framework: Human Development and the Links to Natural Resources'. Oxford: OPML.

AfDB and BMGF (2015c). 'How to Use Revenues from Extractive Industries to Improve Health and Education in Africa'. Abidjan: Bill and Melinda Gates Foundation and AfDB.

Andersen, L. E. and R. Faris (2002). 'Natural Gas and Income Distribution in Bolivia'. Andean Competitiveness Project Working Paper. Cambridge, MA: Harvard University, Centre for International Development. Available at: http://www.iisec.ucb.edu.bo/assets/publicacion/2002-1.pdf.

Auty, R. (1993). *Sustaining Development in Mineral Economies: The Resource Curse Thesis*. London: Routledge.

Auty, R. (2004). 'Maximising the Positive Socio-economic Impact of Mineral Extraction on Regional Development in Transition Economies: A Review of the Literature', paper prepared for EBRD. Available at: www.ebrd.com/downloads/research/economics/auty.pdf.

Bocoum, B. (2000). 'The Mineral and Energy Sectors and Stages of Economic Development: A Comparative Input-Output Analysis', Economic Research Paper 59. Abidjan: AfDB. Available at: https://www.afdb.org/fileadmin/uploads/afdb/Documents/Publications/00157652-EN-ERP-59.PDF.

Corden, W. P. and P. Neary (1982). 'Booming Sector and Deindustrialisation in a Small Open Economy', *Economic Journal*, 92: 825–48.

Daniels, P., M. Keen, and C. McPherson (2010). *The Taxation of Petroleum and Minerals: Principles, Problems, and Practice*. Washington, DC: IMF.

Di Boscio, N. (2010). 'Mining Enterprises and Regional Economic Development: An Exploratory Analysis of the Sustainable Development Model', PhD thesis, London School of Economics. Available at: etheses.lse.ac.uk/310/1/Di%20Boscio_Mining%20Enterprises%20and%20Regional%20Economic%20Development.pdf.

Downes, P., K. Hanslow, and P. Tulip (2014). 'The Effect of the Mining Boom on the Australian Economy', Research Discussion Paper 2014–08. Sydney: Reserve Bank of Australia. Available at: http://www.rba.gov.au/publications/rdp/2014/pdf/rdp2014-08.pdf.

Filgueiras, L., A. Rabetim, and I. Aché (2017). 'Approaches to Supporting Local and Community Development: Brazil and the Vale SA Model of Corporate Interaction', WIDER Working Paper 2017/80. Helsinki: UNU-WIDER.

Fleming, D. A. and T. G. Measham (2014). 'Local Job Multipliers of Mining', *Resources Policy*, 41: 9–15.

Henstridge, M. and J. Page (2012). 'Managing a Modest Boom: Oil Revenues in Uganda', OxCarre Research Paper 90. Oxford: University of Oxford, Department of Economics.

Hirschman, A. O. (1958). *The Strategy of Economic Development.* New Haven, CT: Yale University Press.

Hodge, A. (2017). 'Towards Contribution Analysis', WIDER Working Paper 2017/19. Helsinki: UNU-WIDER.

ICMM (2013). 'The Mining Sector in Brazil: Building Institutions for Sustainable Development'. Mining: Partnerships for Development. London: ICMM.

ICMM (2014a). *Enhancing Mining's Contribution to the Zambian Economy and Society.* London: ICMM.

ICMM (2014b). *The Role of Mining in National Economies,* 2nd edition. London: ICMM.

IDC (2013). 'The Interface between the Mining and Manufacturing Sectors in South Africa'. Sandton: IDC, Department of Research and Information. Available at: https://www.idc.co.za/images/Content/IDC_research_report_Interface_between_Mining_and_Manufacturing.pdf.

Khan, T. (2011). 'Mongolia: Raising Female Participation in the Large-scale Mining Sector', EASPR Policy Note. Washington, DC: World Bank: Available at: https://openknowledge.worldbank.org/bitstream/handle/10986/16499/775110WP0Box3700scale0mining0sector.pdf?sequence=1&isAllowed=y.

Lahn, G. and P. Stevens (2017). 'The Curse of the One-size-fits-all Fix: Re-evaluating What We Know about Extractives and Economic Development', WIDER Working Paper 2017/21. Helsinki: UNU-WIDER.

Mogilevskii, R., N. Abdrazakova, and S. Chabasova (2015). 'The Impact of the Kumtor Gold Mine on the Economic and Social Development of the Kyrgyz Republic', Working Paper 15. Bishkek: University of Central Asia, Graduate School of Development. Available at: https://www.researchgate.net/publication/315797073_The_Impact_of_Kumtor_Gold_Mine_on_the_Economic_and_Social_Development_of_the_Kyrgyz_Republic.

Moretti, E. (2010). 'Local Multipliers', *American Economic Review, Papers and Proceedings,* 100: 373–7.

Moritz, T. (2015). 'Local Job Multiplier of the Mining Industry in Northern Sweden', MSc thesis, Luleå University of Technology, Economic Unit. Available at: pure.ltu.se/ws/files/103086986/LTU-EX-2015-103069796.pdf.

NMA (2012). 'The Economic Contributions of US Mining'. Washington, DC: National Mining Association. Available at: https://nma.org/pdf/economic_contributions.pdf.

OPM (2009). 'The Economic Impact of the Rosia Montana Gold Project in Romania', Report prepared for Rosia Montana Gold Corporation SA. Oxford: OPM.

Östensson, O. (2017). 'Local Content, Supply Chains and Shared Infrastructure', WIDER Working Paper 2017/96. Helsinki: UNU-WIDER.

Otto, J. M. (2017). 'The Taxation of Extractive Industries: Mining', WIDER Working Paper 2017/75. Helsinki: UNU-WIDER.

Rayner, V. and J. Bishop (2013). 'Industry Dimensions of the Resource Boom: An Input–Output Analysis', Research Discussion Paper 2013–02. Sydney: Reserve Bank of Australia. Available at: www.rba.gov.au/publications/rdp/2013/pdf/rdp2013-02.pdf.

Roe, A. R. and J. I. Round (2017). 'Framework: The Channels for Indirect Impacts', WIDER Working Paper 2017/79. Helsinki: UNU-WIDER. Available at: https://www.wider.unu.edu/sites/default/files/wp2017-79.pdf.

Sachs, J. D. and A. Warner (1995). 'Natural Resource Abundance and Economic Growth', NBER Working Paper 5398. Cambridge, MA: NBER.

Sutton, J. (2009 onwards). 'The Enterprise Map Series'. London: London School of Economics, International Growth Centre. Available at: https://www.theigc.org/project/the-enterprise-map-series/.

Thiele, R. (1999). 'A CGE Model for Environmental Policy Evaluation in the South African Mining Sector', *South African Journal of Economics*, 67(3): 456–74.

Tourism Research Australia (2013). 'The Economic Impact of the Current Mining Boom on the Australian Tourism Industry'. Canberra: Tourism Research Australia. Available at: https://www.tra.gov.au/ArticleDocuments/185/Economic_Impact_of_the_Current_Mining_Boom_on_the_Australian_Tourism_Industry_FINAL.pdf.aspx.

Witter, S. and M. Jakobsen (2017). 'Choices for Spending Government Revenue: The New African Oil, Gas and Mining Economies', WIDER Working Paper 2017/150. Helsinki: UNU-WIDER.

# 24

# Local Content, Supply Chains, and Shared Infrastructure

*Olle Östensson*

## 24.1 Introduction

Extractive industries generate some revenue streams that do not go to the state or the operating enterprise but end up with a multitude of economic subjects in host countries. As seen from Figure 24.1, employment, procurement, and infrastructure typically account for a significantly larger share of spending than government revenue. The recipients of this spending are employees and providers of equipment and inputs. In addition to providing income to the local/national population, extractive industry procurement and employment usually also contribute to building skills, which may raise incomes beyond the life of the project.

The portion of total extractive industry spending that is local or national depends on a number of factors, most of which can be influenced by government policy. The aim of local content policies is to raise the proportion of spending that goes to recipients in the host country and/or in the area immediately surrounding the extractive industry site.

Section 24.2 sets out the scope of the chapter and sketches the background to the policy debate around local content. Section 24.3 discusses corporate behaviour with respect to procurement. Section 24.4 describes and assesses policies for promoting local procurement, using some examples. Section 24.5 reviews practices and cooperation concerning multiple uses of infrastructure, and Section 24.6, finally, attempts to formulate some general conclusions.

## 24.2 Scope and Background

Although the term 'local content' is often used in a wide sense, covering employment of locals and corporate social responsibility activities as well as

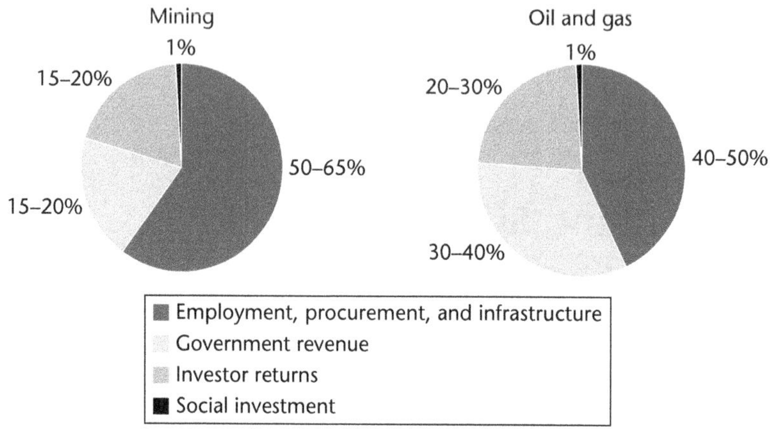


**Figure 24.1.** Average distribution of spending in extractive projects
*Source*: author's estimates.

purchases from national suppliers, in this chapter, the focus is only on procurement of goods and services from national suppliers.

Local content policies in the context of extractive industries have attracted increased interest in recent years. This may be the result of disappointment with the results achieved in more 'traditional' policy areas, particularly maximization of government fiscal revenue. Partly, it is certainly also the result of a realization on the part of policy makers of the potential development effects from local content policies. Thus, recent regulations are moving towards a stronger emphasis on local content and most countries with a significant extractive industry have included local content in their legislation or as a condition in exploitation contracts.

The intellectual roots of the policies, which emphasize the importance of establishing and strengthening various types of linkages, can be found in theories about supply chains and clusters.[1] The concept of linkages is particularly important.

The International Study Group on Africa's Mineral Regimes, whose work laid the foundation for the African Mining Vision, identifies:

> two main groups of linkages. The first includes backward/upstream linkages (to the mine) and forward/downstream linkages (to beneficiators or processors of the mine's output). The second includes sidestream linkages (to industries or

[1] A business cluster is a geographic concentration of interconnected businesses, suppliers, and associated institutions in a particular field. A supply chain is a system of organizations, people, activities, information, and resources involved in moving a product or service from supplier to customer. Supply chain activities involve the transformation of natural resources, raw materials, and components into a finished product that is delivered to the end customer. See, e.g., Porter (1998) and Kaplinsky and Morris (2001).

> organizations providing technological, human resource, and infrastructure inputs) and lateral migration linkages (development of alternative uses of generic technologies used in the industry).
>
> (Economic Commission for Africa and African Union 2009: 102)

Local content policies are concerned with backward linkages. However, where the theories attempt to describe and explain the processes by which industries are established, the policy formulations rest on the conviction that these processes can be influenced by governments, in particular that governments can, by promoting linkages, support the evolution of a more diversified economy, less dependent on exports of natural resource products.

A common reference when discussing linkage policies and industrialization based on natural resource exploitation is the Nordic countries along with Australia, Canada, and the United States. Researchers have argued that the natural resource sector has driven knowledge intensification and industrial growth in these countries (Wright and Czelusta 2004).

However, this inspiration from the older examples is questionable for several reasons that have to do with globalization:[2]

- Transport costs and tariffs are generally lower than they were when the countries mentioned industrialized. Local suppliers are therefore more exposed to international competition, which makes it more difficult to build backward links.
- For the same reasons, processed products are also exposed to international competition, which complicates the establishment of forward links.
- Processes are now standardized and mechanized, which means that the advantage of low local labour costs is relatively less important.
- Skilled and specialized labour is widely available and mobile, so specialists can easily be recruited from other countries, reducing the need to train locals to fill skilled jobs.
- Access to state of the art technology is easy, which reduces the need for local innovation and adaptation of technology to local conditions.

Against this background, it would seem more appropriate for efforts to devise industrial policies based on linkages to build on the experiences of countries that have succeeded more recently in transforming from natural resource dependent to industrialized economies. Section 24.4 will discuss several examples.

[2] Power (2002) disputes both the importance of extractive resources to the economic development of Australia, Canada, and the United States and the validity of comparisons with today's mineral-dependent developing countries.

## 24.3 Local Procurement: The Corporate Perspective

### 24.3.1 *Corporate Procurement*

Before going into details about the factors that guide corporate procurement it is useful to recall some of the differences between oil and gas on the one hand and non-fuel mining on the other, not least because some of these differences determine procurement processes. These differences relate mainly to different structures of costs and to certain differences in how production is organized:

- Typically, in large-scale mining capital costs amount to some 25–50 per cent of the total costs of production. By contrast in oil and gas operations they will normally amount to at least 75 per cent of the total.
- Mining operations typically use unskilled labour for construction and related activities and this can result in such labour becoming a significant part of total costs, especially since local content in service delivery is proportionately quite high.[3] By contrast, in oil and gas operations unskilled labour is rather more concentrated on tasks that are less directly connected to operations, such as catering or security.
- Large mining companies typically use and operate specialized equipment and associated technology obtained from a small number of companies. By contrast, in oil and gas operations it is common for a third-party service company to be both the user and sometimes the owner of the equipment, so large parts of exploration and extraction work are carried out by third-party companies rather than by the oil or gas company holding a concession or contract.
- A final important difference is that the equipment and associated technologies utilized in extracting oil and gas are highly specialized and very high cost, and they also have few possible uses in other industries, meaning that there are high entry barriers to companies aspiring to produce such equipment. By contrast, a lot of the equipment and the technologies used in mining operations have more than one use and are cheaper to supply. This means that entry barriers for potential suppliers are lower and that some suppliers can follow a gradual process, moving from acting as agents for other equipment suppliers to later carrying out some repairs and maintenance, to then producing spare parts, and to finally becoming equipment suppliers in their own right.[4]

[3] e.g. a report by the Chamber of Mines of Zambia and ICMM (2014) concluded that almost all of the services purchased by Zambian mining companies are obtained from local suppliers. This is in sharp contrast to the percentage of goods procured locally.

[4] The experience in Zimbabwe is instructive since it demonstrates that linkages do not necessarily have to be built on a narrow sectoral basis. Small-scale mining companies in that country developed know-how because financial necessity forced them to source spare parts and

A starting point for a discussion of local content is to recognize that most extractive companies have good reasons to form close relationships with local suppliers. Using such suppliers reduces transport costs and waiting times, and it may be easier to build trust and long-term cooperation with a nearby supplier in order to shape the development of products, services, and processes than to deal with an overseas supplier. This being said, large extractive companies with activities in several different countries often find it convenient to use the same supplier for a given type of equipment. This may be the case in particular for equipment that requires training to operate, that has to fit with other pieces of equipment, or that meets specifications included in contracts concerning insurance or other matters where the company's responsibility is engaged.

Local content in the extractive industries is typically low, although much higher for services than for goods. Figure 24.2 shows typical local content shares for mines in developing countries.[5]

A distinction should be made between outsourcing to foreign original equipment manufacturers (OEM) and outsourcing to proximate suppliers that can deliver services such as catering, cleaning, security, and facilities maintenance.[6] Both represent opportunities for local linkages: OEM firms

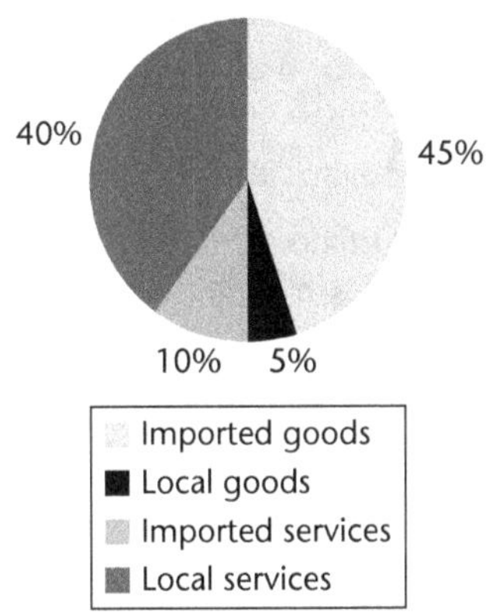


**Figure 24.2.** Shares of total procurement in mines in developing countries, per cent
*Source*: author's estimates.

maintain equipment for their own operations. Eventually, they began selling their services to other companies (Haglund et al. 2014a).

[5] Although percentages may differ, by and large, they are much the same in most developing countries. The reason for this is that the industries producing equipment for extractive industries are often dominated by a few companies with several producing units, either in countries where demand is very important or in their countries of origin. Accordingly, since all of these suppliers are rarely located in the same country, it is rare to see a figure above 30 per cent of local value added even in large and advanced economies.

[6] As a result, between a quarter and half of all staff on a mine site may belong to contractors.

may provide after-sales services requiring the hiring and training of nationals, while locally owned and managed firms may qualify for long-term service contracts.

The general trend towards greater outsourcing has been supported by the emergence of global supply chains. There are three drivers of this trend: (i) falling transport costs, partly due to improvements in logistics; (ii) falling transaction costs due to improved information and communication technologies as well as liberalized financial regulations facilitating international payments; and (iii) reduction of trade barriers, particularly large reductions in tariffs on manufactured goods.

A final trend shaping procurement decisions in the extractive sector concerns broader changes in the international institutional environment. Strengthened legislation concerning corruption may have led companies to be concerned about risks when dealing with local suppliers.

The origins of an extractive industry company's procured goods and services can usefully be split into the following categories according to their local value added (Haglund et al. 2014b):

1. *Direct imports*. No direct linkages beyond benefits to the national economy through customs duties.
2. *Domestic procurement with minimal value addition in the host country*. This includes goods coming through agencies that act as importers for international suppliers. The agencies provide varying but generally low levels of services, often limited to stockholding.
3. *Domestic procurement with moderate-to-high value addition in the host country*. This category includes suppliers of inputs that are at least partly manufactured in the host country. It also includes manufacturers of OEM goods, where the manufacturing takes place abroad, but the product is distributed through branches in the host country. These branches serve to provide a range of services such as installation, operating support, and after-sales service. Lastly, it includes the services (transportation, maintenance) where the value addition takes place largely in-country.

The third category is obviously the one that is of most interest from the point of view of productive linkages.

The company's approach to selecting suppliers will be guided by company policies as well as by local legislation. In environments where public information about suppliers is not easily available, trust and the track record of a firm become particularly important, as do the processes used by companies to actively search for suppliers. When procuring smaller contracts, mining companies often send an inquiry to only a handful of previously successful suppliers, creating barriers to entry for other (potentially more capable) firms

(Haglund 2010). This is partly due to short timeframes facing purchasing managers, who fall back on what they know best and so on existing trusted relationships (Hanlin and Hanlin 2012). Having a track record becomes particularly important in cases concerning complex inputs that are critical to the operation of a mine.

In evaluating suppliers, Kaplinsky and Morris (2001) have argued that firms consider a series of Critical Success Factors (CSFs), representing the particular features of a good or service that is required by the company. If a larger share of activities is outsourced, a correspondingly broader range of factors will be included among the CSFs. The CSFs include price, quality, ability to provide to scale, flexibility, speed of delivery, reliability of delivery, trust, and innovation, and may be reflected in process and product standards (AFDB and BMGF 2015).

For inputs that are critical to the production process, CSFs include timely supply and conforming to adequate standards. These CSFs are likely to be more important than price, which is why such products may command higher margins. High margins are also a characteristic of more complex products, which typically require higher product and process standards. Procurement by mining companies (and by oil and gas companies) can thus be structured along two dimensions as in Figure 24.3: criticality and complexity (Mjimba 2011).

In addition to the challenges of developing specific capabilities that meet the demands of mining companies, would-be suppliers also face a range of more general challenges embedded in their operating environment. These are challenges that are systemic in nature, reflecting the existing policy,

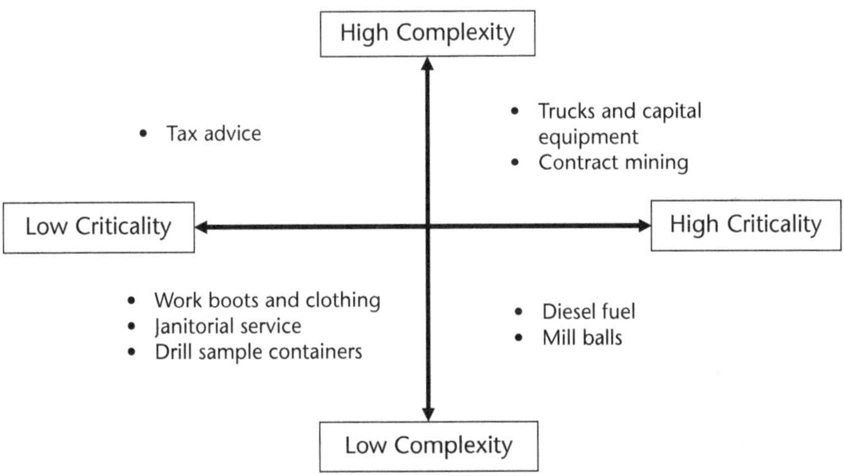


**Figure 24.3.** Supplier matrix and selected supplies to mining companies
*Source*: Mjimba (2011), reproduced with permission.

institutional, and socio-economic framework of the countries in which they operate. Examples of key aspects of the enabling environment include (Haglund et al. 2014b):

- *Access to public skills and small and medium enterprise (SME) support programmes*. As the mining sectors in many resource-rich countries have boomed, shortages in many semi-skilled professions have appeared and skilled individuals are frequently lured away by higher-paying industrial firms. This situation presents a disincentive for smaller companies to invest in building skills, as there is no way of assuring that individuals will remain within the firm.
- *Infrastructure services*. The ability of a domestic SME to be competitive depends on the availability of a range of infrastructure services: for example, reliable supply of utilities (water and electricity) and affordable telecommunications.
- *Availability of local inputs at stable prices*. Where inflation is persistently high, this will undermine the competitiveness of domestic suppliers.
- *Adequate and predictable institutions to support the private sector*. Transparent and simple licensing procedures and independent courts help to provide assurance that property rights will be enforced.
- *Access to finance*. Companies need timely access to finance in order to expand and grow. The ability to access finance depends on the existence of a competitive banking sector as well as having collateral that is acceptable to lenders.

### 24.3.2 *Supplier Development*

In order to overcome the constraints just described, extractive industry companies engage increasingly with suppliers through a process of proactive upgrading and support. In particular where supply is considered critical for the operation of the mine, companies will frequently support the supplier through provision of capital, training, and information. However, supplier development and upgrading typically focuses on firms that are already in the supply chain. Suppliers that gain access to contracts with the lead firm are thus more likely to receive ongoing upgrading support and thus entrench their position further. Lead firms may utilize one or more of the following types of assistance (Haglund et al. 2014b):

*Provision of finance*. Companies may support SMEs through joint financing mechanisms. Other initiatives include partnerships with commercial banks or non-traditional forms of collateral (e.g. by providing invoice factoring, invoice discounting, optimization of payment terms). For example, under the INOVE supplier development programme, which was introduced by the Brazilian mining company Vale in 2008, invoice factoring and working capital loans are provided through Vale in a partnership with commercial banks. This responds to the difficulties faced by suppliers in meeting Vale's 90-day payment terms (ICMM 2013:55).

*Provision of information and supply chain policies*. Support may include provision of data on expected future demand as well as the product and process standards required by the firm, in order to help with pre-qualification. For example, mining companies can optimize payment terms and put in place policies that require primary (so-called 'Tier 1') contractors to pay subcontractors without delay. Other approaches include sharing of supplier databases with Tier 1 contractors so the latter can more easily partner with local firms.

Barrick Lumwana is a large copper mine in Zambia's Solwezi district. Its Local Contractor Development (LCD) programme aims to engage local SMEs and micro-businesses as suppliers in the mine's value chain. After two to three years in this supplier development programme, participant businesses are expected to 'graduate' to the open market. (The Practitioner Hub for Inclusive Business undated).

*Provision of training/capacity-building*. There is some evidence that small firms often benefit more from management training rather than process upgrading—suggesting that an understanding of business management must precede upgrading of technical processes. To encourage capacity-building of domestic suppliers, lead mining firms sometimes require that Tier 1 contractors provide training to local suppliers, including mentoring, health and safety standards, accounting, and finance. Mining companies can support Health, Safety and Environment (HSE) certification, for example, by including suppliers in company HSE training:

> The Diavik Diamond Mine in Canada's Northwest Territories has applied an integrated approach to local training, employment, and procurement, where the mine seeks to expand the skills of Aboriginal employees engaged in entry-level or semi-skilled positions through its own apprenticeship and professional development programmes. In addition, a community and government partnership administers a series of work force development programmes. (ICMM 2010: 48–9)

Although the mine did not meet its Aboriginal employment commitment, reaching 34 per cent instead of the targeted 42 per cent in 2008, this was not an issue, since the absolute number of people was far higher than the original estimate (Östensson and Roe 2013).

*Provide support for product testing and development*. This includes joint product development/testing as well as support whereby the lead mining firm inspects raw material quality or provision of feedback on product quality to suppliers to enable learning over time.

### 24.3.3 *Some Conclusions*

The international extractive industry is highly capital and skill intensive. Its operations have to conform to a large number of regulations concerning financial, environmental, and social matters, and have also to withstand intense scrutiny from media and civil society. This creates an environment where much of its supplies are of a potentially strategic and critical importance and the use of untested local suppliers carries risks. It makes no difference to regulatory authorities or to the media if a faulty installation was due to errors on the part of a supplier or the extractive company itself—and rightly so. The local suppliers may, however, experience the consequent demands placed on them as very challenging.

In response to this situation, and since it is usually in the extractive company's own interest to build close relationships with local suppliers, many companies use supplier development schemes. These can range from providing occasional on-the-job training to highly structured and standardized schemes that are applied in many of the company's areas of operations. Larger companies are clearly better placed to develop and implement such schemes, but cooperation among companies, for instance, under the auspices of a Chamber of Mines, can do much to facilitate the participation of smaller oil and mining companies.

Companies may also go beyond supplier development and accord local suppliers various forms of preferential treatment, including (Esteves et al. 2013a):

- assigning higher preference weightings to local businesses in competitive bidding processes
- sole-sourcing arrangements with local suppliers
- price matching—that is, allowing local suppliers to match the price of other suppliers
- breaking large contracts into smaller ones (unbundling) to create opportunities for smaller local suppliers
- requiring outside suppliers to subcontract locally or enter joint ventures with local suppliers.

## 24.4 Policies for Influencing Supply Chains

### 24.4.1 *Trade-offs and Constraints*

Broadly, governments have two ways of increasing the contribution of the extractive industry to expanding the local industrial base (Esteves et al. 2013a):

- By maximizing revenue from extractive activities and investing this revenue in infrastructure and other policy initiatives that can support the development of the industrial base.
- By maximizing local content, where the host government requires the share of local content to be greater than what would otherwise be the case. The aim of imposing local content measures is to make it easier for local businesses to supply international oil, gas, and mining firms and their large contractors, and benefit from training and transfer of technology.

For companies, however, local content requirements may have an adverse impact on their cost structure and so affect their profitability, thereby reducing the tax base for the government. This would happen if locally available goods and services were not priced competitively when compared with the companies' traditional suppliers. For governments, a balance between the two strategies requires taking into account the fact that while the fiscal instruments typically provide revenues at the national level, local content policies can respond to the interests of local communities, thereby enabling companies to secure a social licence to operate and contribute to economic diversification (Esteves et al. 2013a).

There are numerous definitions of what is a local company. For example, a company may be considered local if it falls within any of the following categories (Paul and Pierre 2010):[7]

- local registration: legal entity is registered under local law
- local ownership: a certain percentage of the company is owned by citizens of the country or by existing locally owned and registered entities
- local workforce: a majority of the company's workforce, whether directly employed or on contract, are citizens of the country
- local value added: a specific percentage of goods/services is produced within the country.

The last definition would seem to be the one most closely aligned with the objectives of local content policies. It is, however, sometimes difficult to apply in practice and governments may settle for a local content definition that is easier to measure.[8] When quantitative targets are used, governments may

[7] In most cases, 'local' actually means national. Nwapi (2015a: 187) criticizes this 'centralist' approach and argues that: 'Given that revenues from extractive resources are managed by national governments in most jurisdictions, LCRs [local content requirements] can provide a mechanism to meet the demands of subnational stakeholders, such as local governments and communities.'

[8] The difference between different measures can be striking. In the case of Zambia, it has been estimated that although the majority of goods are procured from Zambian suppliers, less than 1 per cent of the goods procured by the mining sector are actually manufactured in Zambia (Chamber of Mines of Zambia and ICMM 2014: 66; Kasanga 2012: 6).

also be lobbied by extractive industry companies to apply a measure that makes it easier to meet the target.

Governments may aim to promote backward linkages through regulation. However, countries that are members of the World Trade Organization (WTO) are, in principle, constrained in terms of the requirements they can impose with respect to local content.[9] In particular, as noted by Cordes et al. (2016: 78), 'members of the WTO are bound by the national treatment obligation (NTO) clause under which foreign companies cannot be forced to buy goods from local suppliers or hire suppliers of certain services if a better alternative in terms of price or quality exists abroad'. Many low-income countries (LICs) are or have been subject to exemptions from the WTO rules and several of the relevant agreements provide for some flexibility in the application for lower-income countries, including during transition periods following WTO accession. Currently, 31 LICs are able under the WTO rules to introduce measures that deviate from the NTO clause for a defined period of time on the grounds of their 'individual development, financial, or trade needs, or their administrative and institutional capabilities'. Out of this group, only Angola has introduced explicit local content requirements (Esteves et al. 2013a).

In addition to WTO rules, international investment agreements can also limit the possibility to legislate in favour of local content. Some agreements go further than the WTO provisions and prohibit all types of measures related to local content and performance requirements in general:[10]

> So far only two local content cases related to Trade Related Investment Measures (TRIMs) have been brought under the dispute settlement mechanism. None of them concerned the extractive sector. Possibly, governments are reluctant to challenge tools that they themselves have used or continue to use. Or investors may not find that the WTO dispute mechanism responds to their concerns, particularly as it does not confer monetary compensation and therefore may not repair losses due to contract cancellation or penalties incurred. Most disputes regarding local content requirements in extractives have been brought by investors against host countries under dispute mechanisms of bilateral investment treaties. (Ramdoo 2015: 9)

### 24.4.2 *Government Policies*

Governments use a wide range of instruments and policies to promote local content, from coercive regulations with quantitative targets to efforts aimed at

[9] For an overview of international regulations concerning local content, see Ramdoo (2015).

[10] Performance requirements are measures in law, regulation, or contract that require investors to meet specified goals when entering, operating, or expanding in, or leaving, a host country.

enhancing the enabling environment, including through technology transfer and skills upgrading.

#### 24.4.2.1 QUANTITATIVE TARGETING

Among countries using quantitative targeting, targets may be more or less detailed. In some cases, fairly 'blunt' industry-wide requirements are applied:

*Kazakhstan* requires mining investors to negotiate a binding agreement with the government establishing a certain percentage of local content. The Kazakh Government's Decree No. 367/2010 (Government of Kazakhstan 2010) formalized the measurement of local content in goods, works, and services, using the 'Uniform Method of Kazakh Content Calculation' (the 'Uniform Method'). Subsoil users are required to provide a 20 per cent discount to Kazakh manufacturers (Östensson et al. 2014: 70).[11] Failure to meet the targets can be punished with the loss of the licence. The results of the policy are mixed. During the period from 2010 to 2012, for example, local content in goods was between 13 and 14 per cent, which is not significantly more than in other countries at a similar income level, while in services it rose from 81.5 to 92.1 (Esteves et al. 2013b).

*Ghana* applies different rules for oil and non-fuel mining. The local content policy on mining in Ghana was the product of a dialogue between the industry through the Ghana Chamber of Mines and the regulatory authority. According to Amoako-Tuffour et al. (2015):

> Both mineral title holders and mining support providers must meet the following requirements:
>
> a) a procurement plan, covering five years, has to be submitted by companies
> b) the procurement plan has to specify procurement and use of local products to the extent possible
> c) the procurement plan is guided by a procurement list, developed by the Minerals Commission and to be revised annually
> d) mining businesses with approved procurement and localization plans submit an annual report to the Minerals Commission demonstrating their level of compliance.

The regulations concerning oil are more ambitious. Operators, contractors, and sub-contractors must comply with specified minimum local content levels for goods and services, and companies have to submit plans to increase

[11] Following its joining the WTO in November 2015, Kazakhstan has committed to passing laws and taking measures to review and amend all requirements that are inconsistent with WTO by the end of the transitional period (2021) (World Trade Organization undated).

domestic content over time. All operators must, as far as possible, use goods and services produced or provided in Ghana by Ghanaian companies, even if their prices are higher by up to 10 per cent (AFDB and BMGF 2015).

In *Nigeria*, the Local Content Act (LCA) of 2010 provides for categories of activities to be locally procured, with targets ranging from 45 to 100 per cent for the majority of service categories. For instance, reservoir services as they relate to well and drilling shall have 75 per cent local content level; while 4D seismic data processing services shall have 55 per cent local content level (Nwapi 2015b). 'Where bids are within 1 per cent of each other, the bid containing the highest level of Nigeria content should be selected provided it is at least 5 per cent higher than its closest competitor in terms of its Nigerian content' (AFDB and BMGF 2015: 28). Companies may be subject to penalties for non-compliance, such as cancellation of projects and fines equivalent to 5 per cent of project value. The average local content based on the proportion of the value of contracts awarded to Nigerian companies is estimated to be 70 to 85 per cent. However, the proportion of contract sums spent on Nigerian-made goods is only 12 to 18 per cent (AFDB and BMGF 2015: Annex A). One of the reasons for the relatively disappointing results may be the corruption vulnerabilities that are inbuilt in the LCA (Nwapi 2015b):

> In 2007, the National Petroleum Agency (ANP) of *Brazil* enacted Local Content Certification rules. These rules set up a methodology to calculate the percentage of local content in goods and services acquired in Brazil and provided procedures for the accreditation of independent companies to certify the percentages. A certifying entity is responsible for measuring the local content found in goods and services acquired/contracted by concessionaires in connection with their exploration and production activities in the country. (Galante 2013: 2)

The Brazilian regulations have undoubtedly resulted in Petrobras relying on Brazilian suppliers to a larger extent than would otherwise have been the case. They have also resulted in considerable improvements in the technical capacities of these suppliers. However, the recent revelations of corrupt practices in the awarding of contracts have highlighted the risks associated with strong local content regulations. Where contracts are awarded on other grounds than price and quality, parties to a transaction may be tempted to manipulate the system. It now appears likely that local content rules will be either eliminated or at least given reduced importance when awarding contracts in the future (Reuters 2016).

#### 24.4.2.2 NON-COERCIVE APPROACHES AND THE ENABLING ENVIRONMENT

Several countries pursue policies that focus less on quantitative targets and more on promoting technology transfer and skills improvement. Creating an enabling environment through the removal of constraints can be a part of this strategy. To this end, some countries have adopted policies aimed at directly increasing the participation of local workers and suppliers without establishing legally binding national local content legislation and regulations. Enforcement in these cases relies on mechanisms that range from specific commitments in production-sharing agreements to general agreements on the need to support local content that may not impose restrictions on companies. These types of policies can be initiated not only by governments but also by extractive companies, local and regional organizations, and NGOs, including at the community level.

A report for the World Bank on Central Asia prepared by Ana Mari Esteves and colleagues (World Bank 2014) provided several examples of policies to help improve access to opportunities to local suppliers. These examples included the following:

- ensuring 'full, fair, and reasonable' access to opportunities for local suppliers initiated by national or subnational governments (such as The Australian Industry Participation National Framework)
- compiling a list of 'capable' local suppliers by local government agencies (see for instance the supplier registry system developed by the Industrial Association of Antofagasta and mining companies in Chile)
- harmonizing supplier requirements and encouraging the implementation of certification systems for local suppliers by local government agencies or mining companies (such as in the Atacama mining cluster, Chile). (World Bank 2014: 16)

Such initiatives can be readily implemented because they do not depend on regulations and they can be adapted to the needs of the local economy and the mining companies. However, their effectiveness depends on the existence of competitive local suppliers, and compliance cannot be legally enforced (World Bank 2014: 16).

*Chile's* mining sector framework does not require any formal commitments to increasing local procurement. However, there is a strong culture of public–private collaboration in supporting supplier development and local procurement. Examples include the Mining Skills Council (CCM), in which all mining enterprises provide information on their needs in relation to skilled manpower. Programmes launched in the 1990s included support to large mining companies for spending on training and integration of local suppliers and assisting SMEs in meeting qualification needs, including through technical consultancy sub-contracting exchange, management training, marketing, and export assistance. In 2009 the Enhancing Competitiveness Program (PMC) was launched. Activities include training, tender preparation, and innovation and research capacity support. The Centro de Entrenamiento Industrial y Minero (Industrial and Mining Training Centre, CEIM) has played an important role in improving the capabilities of suppliers. CEIM was established by the Escondida mine in 1999, with the aim of supporting mining clusters within Chile's Region II (Antofagasta), thereby allowing suppliers to benefit from economies of scale and network externalities (Haglund et al. 2014b).

*Norway* has a highly skilled and internationally competitive petroleum-related service and supply industry, developed over more than forty years of petroleum activities. Several regulations were passed in the 1980s and 1990s to support the development of this industry, including requirements for a local content plan to be submitted to the Ministry of Petroleum and Energy for approval before the grant of licence, and provision of training and hiring a certain number of Norwegian personnel. Operators were also required to conclude agreements with the Norwegian Ministry of Petroleum and Energy under which at least 50 per cent of research and development in connection with petroleum activities had to be performed in Norway (Columbia Center on Sustainable Investment 2016). International oil companies are required to announce their tender schedule and the list of companies to be invited. The Ministry of Petroleum and Energy is able to add Norwegian-based firms to the list. General estimates put the level of local content in terms of investment for the exploration and development of the fields at between 50 per cent and 60 per cent (measured by value added) and for maintenance and operations it stands at about 80 per cent (AFDB and BMGF 2015).

## 24.5 Infrastructure

Mining usually requires very substantial investments in infrastructure such as road or rail transport channels—far more typically than is required for oil or gas exploration and production. Offshore oil and gas activities are of course the least demanding in this respect. In particular, most large mines require important investments in infrastructure to enable inputs and machinery to be brought in and finished products to be transported out.

Infrastructure that is built for extractive industries can often be used by local populations and other economic activities and can thereby help to position a region for more rapid economic development and diversification. There are many examples of regions experiencing periods of rapid economic growth due to the infrastructure improvements associated with mining investment (see Box 24.1).

**Box 24.1** EFFECT OF INFRASTRUCTURE IMPROVEMENT: FUNGURUME

In Fungurume, Katanga, Democratic Republic of the Congo, population tripled within a few years to more than 100,000. This was the result mainly of Tenke Fungurume Mining, a mining company that operates a copper mine close to the town, improving the road to the provincial capital, thereby cutting the driving time from two days to four hours, and building a new market. The improved road made it possible for traders from other parts of the province to reach Fungurume, thus increasing local supplies of consumer goods. It also provided local farmers with an outlet for their produce, allowing them to earn cash income. As a result, local incomes improved, the local food price inflation that is often associated with large mining projects has been kept in check, seasonal food price variations declined in amplitude, and the nutritional status of the population improved.

*Source*: Östensson and Roe 2013

Success stories can be replicated elsewhere if mine-related infrastructure is planned and built while taking other economic activities and local opportunities into account. Accordingly, the concept of a 'mineral resource corridor' has attracted increasing attention (World Bank 2012).

The Maputo Development Corridor (MDC), which was officially launched in 1996, is probably the most well-known and successful example of resource corridor development: the MDC represents the shortest road and rail connection between the north-eastern part of South Africa and Botswana and the port in Maputo, Mozambique.

The main infrastructure investments of the MDC include the construction of railway links and a single toll road, rehabilitation of Maputo Port, construction of telecommunication and electricity links, and a one-stop border facility.

A number of industries have been established along the corridor, including:

- the world's third largest aluminium plant, the MOZAL plant developed near Maputo
- the Pande/Temane gas field, with a US$1.4 billion pipeline to South Africa
- the Beluluane Industrial Park (BIP), a 600 hectare industrial free zone next to the MOZAL plant. (World Bank 2012: 21)

Despite the fact that the MDC is widely regarded as a success, there have been setbacks relating to the functioning of the corridor. These include delayed provision of rail services, lack of community engagement, environmental issues, legal matters, governance issues, and investments. Furthermore, the goal of establishing an efficient, one-stop border post between South Africa and Mozambique has not yet been fully realized (World Bank 2012: 22).

The idea of resource corridors has attracted considerable positive attention, not least because of the simplicity and apparent logic of the concept. As often with simple concepts, however, the devil is in the detail and in the practical implementation. Public–private partnerships (PPPs) are central to the resource corridor concept since it requires both the government's capability of establishing an appropriate regulatory framework and its convening power, and the financial resources of the private sector. As pointed out by the International Finance Corporation (IFC 2013), there are, however, very few examples of successful greenfield multi-client/multi-user mining-related infrastructure PPPs in the world—and none of them are in sub-Saharan Africa. This dearth of examples suggests that it is difficult to arrange financing. Given the reliance on project cash flow for repayment, the higher the complexity of the shared-use structure, the less bankable it will be. This means that larger mining companies will most likely have to serve as anchor clients to these projects. It also means that multi-user demands might have to be initially or permanently restricted to secure, first and foremost, the delivery of an efficient mining transport system at the lowest possible cost to its anchor user/client (IFC 2013: 2–3).

Finally, it should be noted that resource corridors may have trade-diverting effects. Mine-related infrastructure typically connects mines directly to the coast. As found by Bonfatti and Poelhekke (2015), such infrastructure may bias a country's transport costs in favour of overseas trade, to the detriment of trade with neighbours and regional integration. Coastal destinations with more mine to coast infrastructure import relatively less from neighbours, and this effect is stronger when the infrastructure overlaps with routes used to import from overseas. The effect is, however, reversed for landlocked destinations, where the mine to coast infrastructure will have to cut through at least one neighbour to reach the coast (Bonfatti and Poelhekke 2015: 1).

## 24.6 Conclusions

In many countries, efforts to leverage greater development from extractive industry investments have in recent years partly focused on raising local content. The industry is usually prepared to participate in such efforts, provided that their impact on costs is limited. In the short term, however, the scope for increasing local content in low-income and mainly agrarian countries may be constrained by the low capacity of potential suppliers, low skill endowments, and a number of other factors constituting the general business environment. A number of extractive industry companies have introduced supplier development programmes that attempt to reduce these constraints and skill gaps. Such measures have had their successes. They are mainly effective in the long term, underlining the need to manage expectations and for local communities to be engaged in the process in order to avoid disappointments.

Governments have adopted varying approaches to local content policies. Some have prescribed ambitious quantitative targets for local content—targets that have often been met. The apparent success of these policies, does not, however, allow one to conclude that they have resulted in the actual accumulation of skills and improved diversification of the host country or local economy.

First, success is often measured in terms of local content provision by firms that are locally owned or registered, rather than in terms of local value added. It is not clear that the policies have actually raised the proportion of locally produced supplies above what would have been achieved in the absence of targets. Second, quantitative targets carry risks of capture by local elites and corruption. The Petrobras scandal in particular should make all governments that consider the introduction of quantitative targets think twice.

Moreover, the international regulatory landscape is evolving in a direction that is not favourable to quantitative local content targeting. While many developing countries benefit from the exemptions from WTO disciplines accorded to LDCs, these exemptions will expire in a few years. Bilateral investment agreements often also prohibit or limit assertive local content policies.

Finally, rigid local content policies carry an economic cost and may reduce other income. For example, they may raise project costs by significant amounts and so undermine project economics, thereby both diminishing the country's standing as an investment destination and reducing government revenue from taxes.

A more constructive non-mandated approach is exemplified by Chile and Norway, where policies have focused on improving skills and raising the capacity of domestic industry to qualify as suppliers to the extractive industry. While these policies have certainly resulted in additional costs for the extractive companies, they have been acceptable to the industry because they have held out a reasonable hope that local suppliers will eventually become fully qualified partners. By focusing on building capacity, much of which can be applied outside the extractive sector, these policies also appear to provide additional resilience to the national industrial fabric, rather than locking local companies into a perpetual dependence on the extractive industry.

Since the shortcomings that constrain the building of backward linkages are usually readily apparent, it would seem reasonable to direct political attention to solving the problems directly and alleviating the shortcomings, thus creating an enabling environment that permits the supplying industry to develop its full potential.

Extractive industries often require significant investment in infrastructure. This infrastructure could be used by local populations and other economic activities and it would seem that this could help to stimulate rapid economic development and diversification. While the potential is certainly significant, the practical implementation has proved more complicated and considerable work needs to be done on PPPs for infrastructure investments since necessary requirements for success include both the government's capability of establishing an appropriate regulatory framework and its convening power, and the financial resources of the private sector.

## References


AFDB and BMGF (African Development Bank and Bill & Melinda Gates Foundation) (2015). 'Paper 6: Creating Local Content for Human Development in Africa's New Natural Resource-rich Countries'. Available at: https://www.afdb.org/fileadmin/uploads/afdb/Documents/Publications/Creating_local_content_for_human_development_in_Africa%E2%80%99s_new_natural_resource-rich_countries.pdf.

Amoako-Tuffour, J., T. Aubynn, and A. Atta-Quayson (2015). *Local Content and Value Addition in Ghana's Mineral, Oil, and Gas Sectors: Is Ghana Getting It Right?* Accra: African Center for Economic Transformation.

Bonfatti, R. and S. Poelhekke (2015). 'From Mine to Coast: Transport Infrastructure and the Direction of Trade in Developing Countries'. CESifo Area Conference on Global Economy. Available at: http://cep.lse.ac.uk/seminarpapers/31-10-14-SP.pdf.

Chamber of Mines of Zambia and ICMM (2014). *Enhancing Mining's Contribution to the Zambian Economy and Society*. London: ICMM.

Columbia Center on Sustainable Investment (2016). 'Local Content: Norway—Petroleum'. Available at: http://ccsi.columbia.edu/files/2014/03/Local-Content-Norway-Petroleum-CCSI-May-2016.pdf.

Cordes, K.Y., O. Östensson, and P. Toledano (2016). *Employment from Mining and Agricultural Investments: How Much Myth, How Much Reality?* New York, NY: Columbia Center on Sustainable Investment.

Economic Commission for Africa and African Union (2009). *Minerals and Africa's Development: The International Study Group Report on Africa's Mineral Regimes*. Addis Ababa: Economic Commission for Africa.

Esteves, A. M., B. Coyne, and A. Moreno (2013a). 'Local Content Initiatives: Enhancing the Subnational Benefits of the Oil, Gas and Mining Sectors', Natural Resource Governance Institute Briefing, July. New York, NY: NRGI.

Esteves, A. M., V. Ogorodnikova, C. Putz, and B. Coyne (2013b). 'Increasing Domestic Procurement by the Mining Industry in Central Asia', report prepared for the World Bank. Washington, DC: World Bank.

Galante, V. (2013). 'Local Content in Brazil', Tauil & Chequer Article. Available at: https://www.mayerbrown.com/pt/publications/detailprint.aspx?publication=9196.

Government of Kazakhstan (2010). Uniform Method of Kazakh Content Calculation. Government Decree No. 367/2010. Astana: Government of Kazakhstan.

Haglund, D. (2010). 'Policy Evolution and Organisational Learning in Zambia's Mining Sector', PhD thesis, University of Bath.

Haglund, D., V. Mjimba, and O. Östensson (2014a). 'Increasing Domestic Procurement in Zimbabwe's Mining Sector', report prepared for the World Bank. Washington, DC: World Bank.

Haglund, D., V. Mjimba, O. Östensson, and E. Sindou (2014b). 'Global Trends in Mining Sector Local Procurement', report prepared under the World Bank project 'Increasing Domestic Procurement by the Mining Sector in Zimbabwe'. Washington, DC: World Bank.

Hanlin, R. and C. Hanlin (2012). 'The View from Below: "Lock-in" and Local Procurement in the African Gold Mining Sector', *Resources Policy*, 37(2012): 468–74.

ICMM (2010). *Mapping In-Country Partnerships*. London: ICMM.

ICMM (2013). 'The Mining Sector in Brazil: Building Institutions for Sustainable Development'. Mining Partnerships for Development—Spotlight Series 17 June. Available at: http://www.opml.co.uk/sites/default/files/The%20mining%20sector%20in%20Brazil_0.pdf.

IFC (International Finance Corporation) (2013). *Fostering the Development of Greenfield Mining Related Transport Infrastructure through Project Financing*. Washington, DC: World Bank.

Kaplinsky, R. and M. Morris (2001). 'A Handbook for Value Chain Research'. Prepared for the IDRC. Available at: https://www.ids.ac.uk/ids/global/pdfs/VchNov01.pdf.

Kasanga, J. (2012). 'Leveraging Zambia's Industrialization with Growth of Copper Mining Investments: Strategy for Expanding Local Manufacturing Capacities to Supply the Zambian Mining Industry', Discussion Paper for the Zambian Mining Local Content Initiative, November. Lusaka.

Mjimba, V. (2011). 'The Nature and Determinants of Linkages in Emerging Minerals Commodity Sectors: A Case Study of Gold Mining in Tanzania'. Making the Most of Commodities Programme (MMCP), Discussion Paper No. 7, March. Available at: http://commodities.open.ac.uk/8025750500453F86/(httpAssets)/F1841F5EFD1C3D388025787E003A7C9E/$file/The%20Nature%20and%20Determinants%20of%20Linkages%20in%20Emerging%20Minerals%20Commodity%20Sectors%20A%20Case%20Study%20of%20Gold%20Mining%20in%20Tanzania.pdf.

Nwapi, C. (2015a). 'Defining the "Local" in Local Content Requirements in the Oil and Gas and Mining Sectors in Developing Countries', *Law and Development Review*, 8(1): 187–216.

Nwapi, C. (2015b). 'Corruption Vulnerabilities in Local Content Policies in the Extractive Sector: An Examination of the Nigerian Oil and Gas Industry Content Development Act, 2010', *Resources Policy*, 46: 92–6.

Östensson, O. and A. Roe (2013). 'Good Practices in the Mining Sector to Contribute to More and Better Jobs', report prepared for the ILO, November. Geneva: International Labour Organization.

Östensson, O., B. Parsons, and S. Dodd (2014). 'Comparative Study of the Mining Tax Regime for Mineral Exploitation in Kazakhstan', report prepared for the World Bank. Washington, DC: World Bank.

Paul, A. and M. Pierre (2010). *Developing Local and Regional Public and Civil Society Capacities to Make Extractive Industries Work for Development and Democracy*. New York, NY: Revenue Watch Institute.

Porter, M. E. (1998). 'Clusters and the New Economics of Competition', *Harvard Business Review*, November–December. Available at: https://hbr.org/1998/11/clusters-and-the-new-economics-of-competition.

Power, T. M. (2002). 'Digging to Development? A Historical Look at Mining and Economic Development'. An Oxfam America Report, September. Boston, MA: Oxfam America.

Ramdoo, I. (2015). 'Unpacking Local Content Requirements in the Extractive Sector: What Implications for the Global Trade and Investment Frameworks?' report prepared for the International Centre for Trade and Sustainable Development and the World Economic Forum. Geneva: ICTSD and WEF.

Reuters (2016). 'Brazil to Ease Local Content Requirements in Oil, Gas Auctions—Report'. 17 October. Available at: http://www.reuters.com/article/brazil-oil-idUSL1N1CN0AI.

The Practitioner Hub for Inclusive Business (undated). 'Barrick Lumwana Mining Company: Lumwana Contractor Development Programme (LCD)'. Available at: http://www.inclusivebusinesshub.org/project/project-profile-barrick-lumwana-mining-company-zambia/.

World Bank (2012). *'Resources Corridors: Experiences, Economics, and Engagement: A Typology of Sub-Saharan African Corridors'*. Extractive Industry Sourcebook. Washington, DC: World Bank.

World Bank (2014). 'Increasing Domestic Procurement by the Mining Industry in Central Asia', Report No. ACS8834. Washington, DC: World Bank.

World Trade Organization (undated). 'Accessions Commitments Database, Article 0897'. Available at: http://acdb.wto.org/tabs.aspx.2016.

Wright, G. and J. Czelusta (2004). 'The Myth of the Resource Curse', *Challenge*, 47(2): 6–38.

# 25

# Downstream Activities

## The Possibilities and the Realities

*Olle Östensson and Anton Löf*

### 25.1 Introduction

The possibilities of downstream integration in extractive industries have traditionally been given close attention by policy makers and many governments have made downstream processing an objective of mineral and energy policy. For instance, according to the African Mining Vision: 'The principal resource endowment opportunities are ... downstream value addition: The use of the locational advantage (CIF-FOB) of producing crude resources to establish resource-processing industries (beneficiation) that could then provide the feedstock for manufacturing and industrialisation' (African Union 2009: 13).[1]

This chapter will discuss the practical possibilities of achieving increased downstream processing and the policies used. It will argue that governments need to be cautious about elevating downstream processing to a priority objective. It will also attempt to show, by way of examples, how attempts to 'correct' market outcomes may carry high costs for the economy.

Section 25.2 will briefly discuss possible alternative explanations for the perceived lack of downstream integration in the extractive industries in lower income countries. Section 25.3 will provide a description of the economics and geography of downstream processing, using examples from three industries: aluminium, copper, and iron ore/steel. Section 25.4 will be devoted to a critical review of policies commonly pursued to promote downstream processing and of practical experiences in a few countries where downstream

[1] A longer version of this chapter, containing more detailed data, has been published as a WIDER Working Paper (Östensson and Löf 2017).

integration has been accorded a high priority. Finally, Section 25.5 will attempt to draw some general conclusions about the kind of policies that could prove successful.

## 25.2 Explanations for the Lack of Downstream Integration in Extractive Industries

If all locations were equally attractive from other points of view, one would expect downstream processing to take place close to the site of the extraction, simply because processing often involves a reduction in volume, which would tend to reduce transport costs. Since this is clearly not always the case, the reasons for the location of processing activities in other sites have to be explained.

### 25.2.1 *Declining Terms of Trade and Unequal Market Power*

Much of the intellectual foundation for the emphasis given to downstream processing and one of the most important possible reasons for the location of processing activities away from primary production was provided more than half a century ago by Hans Singer (1950) and Raúl Prebisch (1950) in the form of the Singer–Prebisch hypothesis. According to this argument, in a world system in which poorer nations specialize in primary products, such as raw minerals and agricultural products that are then shipped to industrialized nations that, in turn, make advanced products to be sold to poorer nations, the major benefits of international trade will go to the wealthy nations, resulting in declining terms of trade for the poorer countries.

The Singer–Prebisch hypothesis attributes the lack of downstream processing in developing countries to the differences in market power between primary producers and producers of processed products. It assumes that developed countries have the power to influence location choices and retain processing capacity in their countries, through trade policy and by other means. Until the 1990s, this view of the world economy did not appear to be in obvious conflict with the observable facts. However, this description no longer appears valid. Price developments for energy and metal commodities since 1960 do not provide unambiguous support for the Singer–Prebisch hypothesis after the year 2000.

There are a number of potential explanations for the absence of downstream processing. Policy measures to support downstream integration would need to be designed on the basis of an understanding of which of these explanations is most relevant to the particular case at hand.

### 25.2.2 *Tariff Escalation*

Tariff escalation (higher import duties on semi-processed products than on raw materials, and higher still on finished products, (WTO n.d.a)) is relevant in situations where the domestic market is not sufficiently large to accommodate the output of downstream processing plants and part or all of the production has to be exported. However, tariffs have declined and tariff escalation is less significant, particularly if concessions within regional or bilateral trade agreements, several of which provide for free access to developed country markets, are taken into account. The large expansion of manufacturing exports from China and other emerging economies could hardly have taken place under the tariffs that were in force in the 1970s or 1980s.

We have compared import tariffs for unprocessed and semi-fabricated metal products, for some selected metals (bauxite/aluminium, copper, and iron ore/steel) and countries and regions (China, the European Union, India, Japan, and the United States).[2] It appears that tariff escalation is significant, although absolute tariffs are relatively low, with averages for semi-fabricated products being well below 10 per cent.

Some qualifications have to be presented, however. First, simple averages of tariffs do not tell the whole story. Products (within a category such as copper) that are deemed to be strategic may be subject to much higher tariffs, thereby effectively reducing imports. However, in the case of metals, the existence of 'tariff peaks' is relatively limited. Table 25.1 shows the range of applied import tariffs for the countries studied. The highest tariffs are still relatively low and would seem unlikely to be prohibitive.

Second, it is important to recall that tariffs are levied on the gross value rather than the value-added. If, say, the tariff on copper concentrates is zero and the tariff on refined copper is 5 per cent while the price of copper concentrates is US$5,000 per metric tonne and that of refined copper US$5,500, the tariff on refined copper is US$275. This corresponds to

**Table 25.1.** Import tariff range, selected metals and countries, %

| | Iron and steel (HS72) | Copper (HS74) | Aluminium (HS76) |
|---|---|---|---|
| China | 1–10 | 0–15 | 0–10 |
| India | 5–10 | 5–10 | 2.5–7.5 |
| European Union | 0–2.2 | 0–5.2 | 0–8.8 |
| Japan | 0–2 | 0–3 | 0–7.5 |
| United States | 0 | 0–3 | 1.3–5.7 |

*Source*: based on tariff data in WTO (n.d.b).

[2] The countries selected are the ones accounting for most of the world imports of the commodities concerned. Detailed results can be found in Östensson and Löf (2017).

55 per cent of the price difference between the two products and in all likelihood to a significantly higher proportion of the value-added, since some inputs will be necessary in the refining. The effective rate of protection provided by the seemingly relatively low tariff would thus in reality be very high.

Third, transportation costs have to be taken into account. These can be considerable and can outweigh the effects of tariff protection. If the difference in transport cost between the refined and the unrefined commodity to a given market is high, it may still make sense to produce the refined commodity and pay the higher tariff.

In conclusion, tariff escalation could be a significant obstacle to downstream processing when exporting to some markets. However, in several cases, the higher tariff on the more processed product may be more than outweighed by the difference in transport costs. Moreover, while access to large developing country markets such as China or India may be hampered by high tariffs on processed products, many developing countries now enjoy tariff-free access to developed country markets (and often to the Chinese market) under various bilateral and regional trade agreements.

### 25.2.3 *Economies of Scale*

Many production processes in the extractive industries have significant economies of scale. For instance, an oil refinery needs a capacity of 200,000 to 250,000 barrels per day to be profitable (Institute for Energy Research 2005). A new alumina plant would not be economic unless it produced at least 1 million tonnes per year. As these examples show, processing the output from a relatively small mine or oil field locally may not always be competitive. Accordingly, downstream processing may have to wait until production has reached required levels or may necessitate the pooling of output from several mines or oil wells, which may be difficult to realize.

### 25.2.4 *Availability of Inputs*

It is not enough to have the basic raw material. Many processes also require specialized inputs. For instance, aluminium smelters need access to low-cost electric power in order to be competitive. Access to low-cost energy is usually the main hurdle. It is maybe illustrative that a legal requirement to process copper concentrate in the Democratic Republic of Congo was delayed after copper producers demonstrated that there was not enough electric power available to smelt and refine the copper mined in the country (Creamer Media 2016a).

### 25.2.5 *Size of Domestic Market*

International markets for energy and minerals are highly competitive and it is difficult to compete without relying on a domestic market where locational advantages in the form of lower transport costs yield higher margins. However, the domestic market for processed energy and mineral products in many developing countries is far too small to provide such a cushion. For instance, while a country may produce enough crude oil to supply a refinery of economic size, local demand for products such as petrol or fuel oil has to be large enough so that most of the production does not need to be exported.

### 25.2.6 *Closeness to Market*

Industrial users of raw materials expect speedy delivery. Many downstream products such as semi-fabricates of steel and nonferrous metals are produced in a wide range of qualities and are used by a variety of industries; it is necessary to be able to supply the needed quality to the customer at short notice, which means that geographic closeness to the market gives a competitive advantage. Accordingly, facilities for producing items such as semi-manufactured copper products tend to be located close to their customers. Producers located far away are at a distinct disadvantage. It is no coincidence that semi-manufacturing of nonferrous products has migrated from Europe and North America to East Asia at the same rate as has the production of manufactured products, as illustrated by statistics of metals use.[3]

### 25.2.7 *Business Environment*

Raw materials are only one ingredient in the production of processed products. Availability of skilled labour, logistics, and financial services is also important. Moreover, the attractiveness of the investment regime needs to be taken into account. While mines have to be located where there are mineral deposits, processing capacity can, in principle, be located anywhere. Therefore, while the quality of a deposit may compensate for unattractive (from the investor's point of view) aspects of the host country's investment regime, these aspects may swing a decision of where to locate processing capacity in favour of a 'safer' site out of the country.

[3] Metals use is usually measured as unwrought metal used in semi-fabricated products. Statistics producers, such as the international metal study groups, use this convention.

### 25.2.8 *Conclusions*

Downstream processing often has an inbuilt advantage of lower transport costs when exporting compared to the export of the unprocessed commodity. However, most other factors such as economies of scale, unavailability of inputs, limited size of the domestic market, distance to export markets, and lack of a favourable business climate may outweigh this advantage. Trade barriers such as tariffs appear to play a limited role, particularly when taking into account the preferences now accorded to developing countries under various trade agreements.

## 25.3 Economics and Geography of Downstream Processing: Some Empirical Data

### 25.3.1 *Extent of Vertical Integration in Selected Mineral Industries*

An indication of the viability of efforts to achieve greater downstream processing should be provided by the extent to which the industry is in fact vertically integrated. Vertical integration would seem to hold advantages if it is high. Accordingly, in order to shed some light on this issue, we look at two aspects of downstream processing. First, we consider whether the degree of vertical integration is higher in developed countries, as predicted by the Singer–Prebisch hypothesis, which would support the case for promotion of downstream processing. Second, we look at the size and variations in processing margins, in order to determine if they are very large compared to the price of the unprocessed commodity and if they are stable enough not to pose unacceptable risks to possible public investment, directly or indirectly, in downstream processing.

The share of output at the unprocessed stage that is processed into the following stage *in the country of mine production* was calculated for three minerals: bauxite/alumina/aluminium, copper, and iron ore/steel, the share being measured both for world totals and for significant producing countries.[4]

Table 25.2 summarizes the results of the analysis and shows trends over the 2000–14 period. It is interesting to note that vertical integration has decreased for all the first steps of the supply chain: bauxite to alumina, copper ore to smelted copper, and iron ore to steel, while the next step shows either increased or stable vertical integration over the period studied. Partly, this is due to the rise of China as a consumer of raw materials and the country's investments in downstream processing. The other main reason is the location

[4] Östensson and Löf (2017) contains detailed data, including on vertical integration measured at the corporate level.

**Table 25.2.** Vertical integration, less processed raw material transformed into more processed material within the same country

| | Bauxite/ alumina | Alumina/ aluminium | Copper ore/ smelted copper | Smelted copper/ refined copper | Iron ore/steel |
|---|---|---|---|---|---|
| | 2000–14 | 2000–14 | 2000–14 | 2000–14 | 2000–14 |
| Average over period | 61.8% | 57.5% | 60.6% | 92.1% | 45.6% |
| High/low | 67.5/54.4% | 69.0/50.4% | 66.3/54.9% | 93.8/90.9% | 52.6/32.5% |
| Trend over period | decreasing | increasing | decreasing | stable | decreasing |
| Standard deviation over period | 4.7% | 7.5% | 2.8% | 0.8% | 5.9% |

*Source*: authors' calculations based on data from Raw Materials Data (n.d.) (bauxite, alumina, and aluminium); USGS (2014) (copper ore, smelted copper, and refined copper); World Steel Association (2015), UNCTAD (2015) (iron ore and steel).

of demand for the processed products, such as steel, which determines where the final step in the processing chain is located. If there are advantages to vertical integration, they may be more important going backward from the final stage than going forward from the first stage.

### 25.3.2 *Processing Margins*

The commercial viability of any processing industry depends on the processing margin: that is, the difference between the prices of the raw material and the processed product. Margins may be inflated by monopoly positions. But in a globalized world with low tariff barriers it is difficult to maintain processing margins containing a large element of monopoly rent. It is difficult to establish the existence of such rents without detailed calculations on the basis of production cost data, which are subject to influences that vary from one plant to another. It is, however, possible to assess if margins are stable or subject to large fluctuations. In that case, it would be reasonable to conclude that investment in downstream processing is associated with large risks. Table 25.3 summarizes data for bauxite/aluminium, copper, and iron ore/steel. For all minerals, the treatment charge, whether explicit or implicit, is seen to be on a decreasing trend as a portion of the price at the following processing stage. Moreover, for both copper and iron ore, the standard deviation is high and increasing, with increased uncertainty and higher risk as a result.

Total processing charges for converting *copper* concentrate to refined copper (TC/RC)[5] have on average been relatively low over an extended period, both

[5] TC stands for 'treatment charges' and refers to the charge for smelting copper concentrate to blister copper. RC stands for 'refining charges' and refers to the charge for refining blister copper to copper cathodes. While blister copper is traded, the bulk of international copper metal trade is in the form of copper cathodes.

**Table 25.3.** Processing margins, cost of turning less processed material into next stage as % of more processed price

| | (Alumina—bauxite)/alumina | (Aluminium—alumina)/aluminium | TC & RC/LME grade copper price | (Rebar (steel)—iron ore)/rebar |
|---|---|---|---|---|
| | Aug 2010–15 | Aug 2010–15 | 2000–15 | 2007–15 |
| Average over period | 80.5% | 82.8% | 9.3% | 70.5% |
| High/low | 85.6/73.2% | 86.5/79.7% | 29.5/0.3% | 83.0/55.7% |
| Trend over period | decreasing | decreasing | decreasing | decreasing |
| Standard deviation over period | 2.9% | 1.6% | 7.6% | 7.9% |

*Sources*: authors' calculations based on data from London Metal Exchange (n.d.) (copper); S&P Global (n.d.a), Metal Bulletin (n.d.) (bauxite to alumina); London Metal Exchange (n.d.), Metal Bulletin (n.d.) (alumna to aluminium); UNCTAD (2015), S&P Global (n.d.b and n.d.c.) (iron ore to rebars).

in absolute dollar terms and also as a proportion of the refined copper price. However, the share of the charges as a proportion of the refined copper price shows violent fluctuations. This implies that the returns from downstream processing of copper are highly uneven.

A comparison of the prices of *bauxite, alumina, and aluminium* over time shows that the bauxite price as a portion of the alumina price fluctuates more than the alumina price as a portion of the aluminium price. Bauxite prices show some larger amplitude fluctuations, but even those are fairly limited compared to copper and iron ore. Alumina and aluminium prices follow each other fairly well. This would seem to imply that the risk associated with additional processing in the bauxite/alumina/aluminium industry is relatively low.

Both *iron ore* and rebar[6] prices showed large amplitude fluctuations. However, unlike in the case of copper, the variations over time in the implicit 'processing charge'—that is, the difference between rebar and iron ore prices—and in the iron ore price itself, are of roughly similar magnitude. Accordingly, the additional processing would not appear to increase the price risk.

## 25.3.3 *Some Conclusions*

The degree of vertical integration varies within the extractive industry and between the three metals studied. In all three cases vertical integration appears to be mainly driven by production economics resulting from technology and transport costs and less by corporate strategies or trade policies. For the

[6] The price of rebars (also known as reinforcing rods, used with concrete in construction) is used since this is the most basic and lowest priced form of commonly traded steel. Import prices for the Middle East were used since they are broadly based and can be taken as roughly representative of the prices paid by most developing countries for imported steel.

bauxite/alumina/aluminium complex, access to low-cost energy is a major determining factor, along with an advantage for aluminium smelters that are located close to markets. For copper, the transport cost for copper concentrate has meant that copper is often smelted and refined close to the mine, except where economies of scale, access to low-cost energy, and the surge in Chinese demand has generated a trade in concentrate and, to some extent, in smelted blister copper. Finally, in the case of iron ore, developments over the past two decades have been completely dominated by the expansion of steel use in China, which has provided Chinese steel mills with a competitive advantage based on location. It should be noted, however, that even discounting the influence of China, a very high proportion of iron ore production has traditionally been exported because the steel markets of most large iron ore producing countries are too small to accommodate processing of more than a share of the iron ore output. It is notable that the degree of downstream processing is relatively low in high-income countries where producers would be expected to be able to exert some influence over markets: for instance, Australia in bauxite and iron ore, Chile in copper, and Sweden in iron ore. It would certainly be technically feasible for these countries to increase downstream processing. The fact that they have not done so would appear to argue that the attraction of downstream integration is not self-evident.

Based on the evidence of price series, processing margins do not appear to have followed any particular trend in the past two decades. The surge in Chinese processing of all three metals could have been expected to change price behaviour. But this has not happened. Longer price series would have allowed more definite conclusions on this point, but they are unfortunately not available. The time series used are probably in any case sufficiently long to reflect accurately the effects of the Chinese expansion of processing. There is considerable variation in the processing margins over time. In cases where the raw material accounts for a large portion of the total production cost and its price is subject to large fluctuations, downstream processing is exposed to considerable risk. This is the case in particular for processing of bauxite, copper concentrate, and iron ore, but less so for the processing of alumina. The risk of periodically seeing margins squeezed dramatically should give pause to any company or indeed government that might consider investing in processing capacity.

## 25.4 Review of Policies

### 25.4.1 *Policy Objectives and Means*

Government policies aimed at increasing downstream processing are rarely introduced with reference to clearly defined and quantifiable objectives. Nevertheless, it is possible to distinguish three broad categories of objectives:

1. *Employment.* It is sometimes argued that increased downstream integration will add a significant number of employment opportunities, although most downstream processing is capital intensive.[7]
2. *Economic diversification and skills development.* While the number of jobs created in downstream industries may be relatively small, incentives to create such industries could be justified if they made the economy more diversified, or if the general skill levels were raised.
3. *Appropriation of rent.* If there are monopoly rents in processing it would be reasonable for the government to try to correct the situation in order both to raise economic efficiency and to increase government revenue.

As already mentioned, specific policy measures to support downstream integration would need to be designed on the basis of an understanding as to which of the three objectives above might be achieved by using any government intervention that is contemplated.

As concerns employment, a WIDER Working Paper on local content (Östensson 2017) argues that measures aimed at strengthening *backward linkages* (e.g. via higher levels of local content) may yield significant results in terms of employment generation and that many of these jobs are likely to be associated with portable skills. It is difficult to see how *downstream processing* could have a comparable impact, since most mineral or energy processing facilities are highly capital intensive and require little of the low-skilled labour in, for example, construction and services that is needed in mining itself or in supplier industries. The higher-skilled jobs in downstream processing are often relatively specialized and skills are thus not easily portable.

The existence, actual or potential, of local markets for extractive industry products would be a positive factor for downstream processing since the development of such markets would reduce dependence on raw material exports and may result in greater economic resilience. However, the mere existence of such markets is not a guarantee that they will benefit fully from the availability of the raw material since the processor may be in a monopoly position with price-setting power.[8]

The existence of large rents that can be appropriated would appear to be a valid reason for promoting downstream processing. However, as discussed in Section 25.3.2, the rents in downstream processing appear to be limited and

[7] Later processing stages may be more labour intensive, but it is difficult to argue that a country needs to have all the intermediate stages in order to arrive at this later, more labour-demanding stage. For an informative analysis of trade patterns for copper contained in semi-fabricated and finished goods, see Tercero et al. (2016).

[8] As the African Mining Vision expresses it: 'TNCs, who often prefer to ... only make the semi-processed resource available to the local market at a monopoly price (import parity price = the alternative imported price of the resource (CIF) in a particular country), if they have a monopoly or oligopoly position in the country concerned' (African Union 2009: 14).

precarious, with large fluctuations over time. Accordingly, policies based on the perceived existence of such rents and aimed at promoting downstream processing could carry large risks: risks to the financial solidity of the resource company and, in the case of positive tax incentives or government subsidies, a risk of wasting taxpayers' money.

Governments typically attempt to influence downstream processing decisions in one or more of three ways:

- consultation with investors aimed at identifying and remedying obstacles to downstream integration
- economic incentives
- bans on the export of unprocessed products.

Government measures may be contained in legislation or may form part of agreements negotiated with investors. While agreements are almost universally used in the oil and gas sector to define rights and obligations of government and companies, they are less common and often more restricted in scope for non-fuel minerals, where conditions are instead usually defined in law. Nevertheless, even where taxation is defined in law, many countries use Mineral Development Agreements to define other obligations of investors. These obligations may cover various aspects of the investment, including employment, local content, and downstream processing. Targets may be defined, or a process for achieving unquantified objectives set out. Consultation processes where plans are worked out and agreed are common in the cases of employment and local content, but less so when it comes to downstream processing. Where the legislation and any agreements are silent on downstream processing, it is of course still possible for the government to try to persuade the investing company to integrate forward. Any government has a number of instruments at its disposal that can be used to influence companies' behaviour in this regard.

Economic incentives to influence companies' decisions with respect to downstream processing are relatively common, particularly in the form of taxes on the export of unprocessed materials.[9]

Positive incentives, for instance, in the form of *tax credits* for downstream processing, have been discussed, but no current examples are known to the authors.

Another positive incentive is *duty exemption* on imported inputs, which is used both at the primary and downstream processing stage by many countries.

*Export bans* on unprocessed products are uncommon, but there are a few examples. The most recent example is the ban that was introduced in Indonesia in 2014, which is discussed in Section 25.4.2.

[9] See Price and Nance (2009) for a number of examples.

### 25.4.2 *Indonesia's Unprocessed Mineral Exports Ban*

The example of downstream processing policies that has attracted most attention in recent years is the Indonesian export ban on unprocessed minerals. This ban took effect in January 2014. It applies to both metal-based and non-metal-based components of unprocessed minerals and specifies certain minimum levels of processing that are required to avoid the ban. The nominal intention of this intervention is to help preserve the country's supplies of the affected resources (Nathan Associates 2013). While mineral ore exports such as nickel and bauxite were banned outright, exports of mineral concentrates—including copper, iron ore, manganese, lead, zinc, and ilmenite—are permitted for a period of three years, during which export taxes are levied at rising rates (Bloomberg 2014).

In order to back up the bans, positive incentives have also been provided for companies that invest in processing facilities.[10]

These measures have had a dramatic impact on Indonesia's mineral exports. The value of unprocessed exports of bauxite, copper, and nickel fell from US$7 billion in 2013 to just under US$2.9 billion in 2014 (UNCTADstat n.d.). The fact that the rules were introduced at a time of falling prices and widespread excess capacity for the minerals concerned has exacerbated these impacts. Some modifications of the rules have already been negotiated with companies. Meanwhile, a number of projects to build processing plants have been announced and a few have started construction.

It is too early to definitively assess the Indonesian experience since 2014. An estimate made for USAid before the rules came into effect found that the economic losses would be very large. In aggregate, accumulated net welfare losses would amount to somewhere between $34 billion and $33 billion in 2020 (Nathan Associates 2013: 2). In the light of the implementation of the legislation and market developments since early 2014, these initial estimates today appear too optimistic. While it is possible that some of the processing capacity now being built will be profitable, this may be due only to the very generous incentives provided.

### 25.4.3 *Export Taxes on Iron Ore in India*

Indian policy with respect to iron ore mining is a good illustration of the difficulties of achieving a balance between the interests of primary producers and processors by using regulations.

In 2011 and 2012, the courts banned mining in the Indian provinces of Goa and Karnataka because of mining outside concession areas and breaches of

[10] See Bellefleur (2014) for details.

environmental regulations. The ban has since been lifted, but there is still a cap on production (Creamer Media 2012b), resulting in reduced domestic supplies.

Iron ore consumption by the Indian steel industry is dominated by lumps and sinter, with most fines that are not suitable for sintering being exported since they cannot be directly used for iron making in the absence of sufficient capacities for agglomeration through pelletizing (Creamer Media 2012a). Most steel producers are equipped to use iron ore pellets from fines as feedstock but do not have pellet-making facilities at their plants (Creamer Media 2012c). As a consequence, the fines have been left in dumps or, when possible, the material has been exported. In order to assure the domestic steel industry of iron ore supplies, railway freight rates are no less than 3.6 times higher for iron ore for exporters than for ore for domestic use (Creamer Media 2015).

A levy on outward shipments of fines was first imposed in December 2009 at 5 per cent. A 15 per cent export tax on lumps followed in April 2010. A uniform rate of 20 per cent was imposed on both lumps and fines in March 2011 and was increased to 30 per cent in December the same year (Creamer Media 2015). The tax was lowered to 10 per cent for lower grade ores in 2016 (Creamer Media 2016b). Royalties on iron ore were raised from 10 to 15 per cent in August 2014 (Creamer Media 2014b). Pelletizing provided a possible way to exploit lower-grade fines since there was no export tax on pellets. However, a tax of 5 per cent was introduced in January 2014 in response to pressure by the steel industry (Creamer Media 2014a, 2014d). Capacity utilization in pelletizing fell to 50 per cent later the same year (Creamer Media 2014c).

The combined result of these various policy interventions can be seen in Figure 25.1. The peak in export and production in around 2008 to 2009 was partly the result of companies exporting from stockpiles of fines. However, once the most easily accessible stockpiles had been exhausted and prices fell following the financial crisis, this activity came to an end. A couple of years later, the court-ordered mine closures strongly influenced the figures. Nevertheless, production and exports would have been expected to recover once mines could be re-opened. This did not happen, due to the policies pursued. Consequently, and remarkably, from being the world's third-largest iron ore exporter, India became a net importer of iron ore in 2015.

### 25.4.4 *Processing of Copper in Zambia*

Zambia levies an export tax on copper concentrate. The tax is intended to promote the smelting and refining of copper concentrate from mines in the country. The tax is 10 per cent of the sales value, which corresponds to about US$470 per tonne of contained copper at mid-2016 prices. The treatment

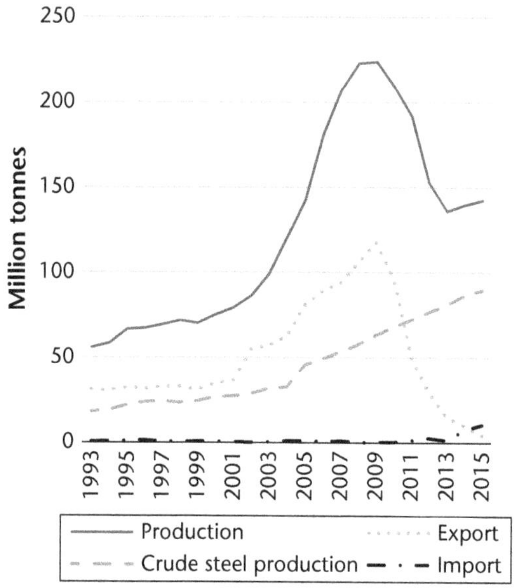


**Figure 25.1.** Iron ore and steel in India 1993–2015, million tonnes
*Source*: authors' calculations based on UNCTAD (2013, 2015), TEX Report (2016).

charge: that is, the cost of smelting concentrate in Zambia to produce blister copper (see footnote 5)—is US$70–75 per tonne of concentrate, while the cost of transporting the concentrate from the mine to an overseas smelter is about US$200 per tonne (corresponding to about US$600 per tonne of contained copper) depending on the route taken (Muller 2013).[11]

Because the cost of transporting concentrate is three times as high as that of transporting blister copper, mining companies would be better off processing the copper even in the absence of the export tax. The smelting charge has to be paid in any case and it does not matter to the miner if it is paid to a smelter in Zambia or overseas. Thus, the export tax does not affect the actions of mining companies. On the other hand, the tax deters investors who for one reason or another are not in a position to smelt concentrates: for example, if a mine produces concentrate that for mineralogical reasons cannot be smelted in existing smelters. Because of the export tax, most such mining projects will not be economically viable in Zambia.

Would this conclusion hold if the smelting charge were to change? As was noted in Section 25.3.2, smelting charges have trended downwards for a long time due to persistent over-capacity at the smelting stage. Several governments

[11] This section is partly based on Chamber of Mines of Zambia and ICMM (2014: Annex I). Copper prices and smelting charges have been updated to, respectively, February 2016 and the average for the 2000–15 period.

have introduced legislation similar to the Zambian export taxes, leading to over-investment in smelting capacity and downward pressure on smelting charges. Moreover, smelters that were originally built to process ore from mines that are now closed often continue operating since the original capital investment has been paid back and running costs can be covered. There are several such smelters in the world.

As for processing the refined copper that is now produced in Zambia, a World Bank study (World Bank 2011) does not hold out much hope. The total market for all copper and copper alloy semis in sub-Saharan Africa (excluding South Africa, which is self-sufficient) can be estimated at around 10,000 tonnes per year, which corresponds to a tiny portion of Zambian refined copper output of over 700,000 tonnes per year. Given the location of Zambia, the principal exit shipment ports, and the main shipping routes, the markets outside Africa that could be best served from Zambia are the Middle East and Southeast Asia. The Asian market is, however, well served by substantial large-scale local capacity.

Accordingly, market prospects do not appear to justify further industrial scale copper and copper alloy semi-manufacturing capability in Zambia in spite of regular political pressures to do just that.

### 25.4.5 *Exports or Local Use of Natural Gas in Tanzania*

Tanzania has been producing natural gas from its Songo Songo fields on the Indian Ocean since 2004. This gas provides about 50 per cent of Tanzania's electricity generation, with the remainder used for industrial purposes (AfDB and BMGF 2015). However, more recent discoveries have radically changed the prospects for gas in Tanzania. In the past decade, there has been very large exploration activity in the concessions so far granted by the Tanzanian Petroleum Development Corporation. These have already resulted in vast commercial finds that are only now beginning to be developed.

The gas can be used in three ways: (i) natural gas can be exported to earn foreign exchange revenue; (ii) it can be used to generate energy (electricity) in the domestic economy; and (iii) it can be used to produce a number of products, including fertilizer, compressed natural gas, and petrochemicals.

The choice between the different uses for the natural gas is partly a technical one that will be dictated by project economics. Mainly, however, the choice is a political one. A large portion of the gas in the largest deep water wells will in any case have to be exported, both because the volumes are too large to realistically be absorbed by the national economy and because producing companies will have to export very large volumes to justify their investment in extraction and liquefied natural gas (LNG) plants. A large share of the export revenues will strengthen the government budget. It has been estimated that the additional income will amount to US$1.4 billion per year on average over the first ten years of production, corresponding to 1.9 per cent of gross domestic product and 9 per cent of government tax revenue from all sources (AfDB and BMGF 2015: 31). There is, however, potential for the contribution to growth to be significantly larger. The Tanzanian economy is capable of delivering at least some of the goods and services needed for gas exploration and extraction and for LNG production. The government has launched an ambitious local content policy which, if successful, should raise the proportion of nationally sourced goods and services significantly (United Republic of Tanzania 2014).

With respect to the possibility of improving access to electrical power, it should be noted that at present only 24 per cent of the Tanzanian population are connected with electricity services. Using the natural gas for this purpose would mean that the national energy supply company will be able to further reduce its dependence on expensive imported feed stocks of diesel, and also reduce further its use of ageing and unreliable hydro plants. In addition, it would be possible to reduce power tariffs, thereby improving living standards and the competitiveness of domestic industry.

A range of processed products can also be produced from the gas and especially from the large deepwater resources if these are developed. Their economic feasibility differs, however. Certain processes for downstream products have considerable economies of scale and for this reason a large part of the production would have to be exported if production was to be on a commercially viable scale. Moreover, several of the uses would require a considerably lower cost of input than, for instance, would LNG production.[12]

A domestic fertilizer facility could make a significant potential contribution to Tanzania's large agricultural sector which at present has a low rate of fertilizer use as well as an expensive subsidy policy to help farmers. A large ammonia/urea plant could be competitive if it could be supplied with gas at US$5–6 per mmBtu: a figure that is more demanding than the US$7–8 supply price needed for a viable LNG export activity. The full capacity output would be significantly higher than the anticipated future demands in East Africa as a whole. So the investment could only be justified if Tanzania had real prospects of also finding large export markets outside the region (Roe 2016).

A competitive methanol plant would consume even more than a fertilizer plant and it would probably need a gas supply price as low as US$4–5. The domestic market is very small, methanol being used mainly as a feedstock into

[12] The following assessment of the feasibility of different types of downstream processes is based on BG Tanzania (2013).

chemical industries. Hence a very large export market would also need to be found to justify such a large capital outlay (Roe 2016).

Converting gas to liquid fuels would be an even more challenging investment for the country. A facility to produce naphtha and high quality diesel would only work competitively if it could acquire gas at a price of around US$3–4 mmBtu. Although this type of project would have some attraction in being able to save Tanzania's large petroleum import bill, the investment needed to achieve this saving would be very large (Roe 2016).

The Tanzanian government has made it clear that it assigns a high priority to downstream processing of natural gas. However, the difficulties of these ambitions are evident. In addition, investors have expressed some concern that the 'domestic market obligation' should not be so large that it threatens the viability of projects.

In conclusion, the case of Tanzania's natural gas illustrates the complex weighing of alternatives that has to take place in order to satisfy several different objectives at the same time.

## 25.5 Conclusions

This chapter has attempted to provide some basic information and analyses that could inform policy choices concerning downstream processing of both metals and oil and gas, and to show both the risks of policy decisions that are oriented towards meeting one single objective as well as the complexities that need to be taken into account in order to arrive at solutions that are both practical and sustainable.

Section 25.2 reviewed the reasons why forward vertical integration is not always an optimal choice for extractive industry companies. While transport costs often argue in favour of such integration, other technical factors, including economies of scale, availability of inputs, size of domestic market, and distance to markets, may outweigh this natural advantage. There appears to be little support for the argument that differences in market power dictate the geography of downstream processing. Relatively high tariffs on processed products appear to play only a limited role.

The review of vertical integration and processing margins in Section 25.3 showed that the degree of vertical integration varies within the extractive industry. It appears to be mainly driven by production economics resulting from technology and less by corporate strategies. It is notable that the degree of downstream processing is relatively low in high-income countries where producers would be expected to be able to exert some influence over markets.

Processing margins tend to fluctuate, in some cases dramatically, which should give pause to all those companies or countries that consider investing in processing capacity.

Section 25.4 discussed government policies for downstream processing. Following a brief description of objectives and instruments, the rest of the section was devoted to a review of four country cases: India, Indonesia, and Zambia on non-fuel minerals, and Tanzania for natural gas.

In Zambia, a simple arithmetic example serves to show that the policy studied—the export tax on copper concentrates—is unnecessary and does not significantly influence processing decisions. It may, however, preclude the exploitation of some mineral deposits, thus needlessly limiting mine production.

In Indonesia, more effort seems to have gone into the formulation of the policy underlying the ban on exports of unprocessed minerals. However, the government has decided to provide very generous incentives to investors in processing capacity, and it is legitimate to ask who pays for the policy: the investor or the Indonesian taxpayer.

The Indian policy concerning iron ore mining and exports illustrates the difficulties of satisfying competing industrial interests who all turn to the government for solutions to their various problems. In trying to help the Indian steel industry to retain international competitiveness, the government succeeded in badly damaging a once successful export industry while leaving the steel industry in an uncertain situation with regard to its raw material supplies.

The Tanzanian example, finally, brings to light an important consideration that is usually absent from the debate about downstream processing, namely the domestic market. In Tanzania the choice may be not between exporting unprocessed or processed products but between satisfying domestic needs—in this case, rural electrification—or maximizing export revenues. This illustrates a basic argument: that downstream processing policies should not be about how to maximize export revenue but about how to meet the needs of the economy.

In conclusion, it is probably reasonable to point out that the presence of raw material resources constitutes only one of the factors that need to be taken into account when taking a decision on the location of a particular type of plant. Incentives or directives to promote downstream processing thus represent only one variation on the familiar theme of picking winners—and one based on a particularly narrow set of criteria. It would probably be more productive in most country cases to instead emphasize industrial policies that focus on removing constraints and bottlenecks that stand in the way of the economy reaching its full potential, including those relating to skills, credit, energy supply, transport infrastructure, and inappropriate regulation.

## References

AfDB (African Development Bank) and BMGF (Bill & Melinda Gates Foundation) (2015). 'Delivering on the Promise: Leveraging Natural Resources to Accelerate Human Development in Africa, Paper 2: Timing and Magnitude of New Natural Resource Revenues in Africa'. Available at: https://www.afdb.org/en/topics-and-sectors/initiatives-partnerships/african-natural-resources-centre/delivering-on-the-promise-leveraging-natural-resources-to-accelerate-human-development-in-africa/.

African Union (2009). *African Mining Vision February 2009*. Available at: http://www.africaminingvision.org/amv_resources/AMV/Africa_Mining_Vision_English.pdf.

Bellefleur, D. (2014). 'Indonesia Export Ban'. *Tractus*, Issue ID2, 17 February.

BG Tanzania (2013). 'Gas Markets Study', unpublished corporate presentation, March.

Bloomberg (2014). 'Constitutional Court in Indonesia Upholds Ore Shipment Ban'. 3 December. Available at: http://www.bloomberg.com/news/articles/2014-12-03/constitutional-court-in-indonesia-upholds-mineral-ore-export-ban.

Chamber of Mines of Zambia and ICMM (2014). *Enhancing Mining's Contribution to the Zambian Economy and Society*. London: ICMM.

Creamer Media (2012a). 'India Invites China to Set up Pelletisation Plants', *Mining Weekly*, 16 May.

Creamer Media (2012b). 'India Lifts Ban on 18 Iron-ore Mines in Karnataka', *Mining Weekly*, 3 September.

Creamer Media (2014a). 'Indian Steel Producers Seek Duty on Overseas Pellet Shipments', *Mining Weekly*, 10 January.

Creamer Media (2014b). 'India to Hike Mineral Royalty Rates to Meet Demands of Provinces', *Mining Weekly*, 13 June.

Creamer Media (2014c). 'Indian Iron Pellet Manufacturers Want to Liquidate Waste Dumps', *Mining Weekly*, 16 June.

Creamer Media (2014d). 'India's Miners Sharply Divided over Iron-ore Export Tax', *Mining Weekly*, 18 June.

Creamer Media (2015). 'Indian Iron-ore Miners' Woes Persist as Government Disregards Pleas to Slash 30% Export Tax', *Mining Weekly*, 10 April.

Creamer Media (2016a). 'Congo Delays Ban on Copper and Cobalt Concentrate Exports', *Mining Weekly*, 13 January.

Creamer Media (2016b). 'Indian Panel to Reconcile Iron-ore Exports Conflict', *Mining Weekly*, 13 April.

Institute for Energy Research (2005). 'US Petroleum Refining: Let the Market Function'. Available at: http://instituteforenergyresearch.org/studies/petroleum-refining-and-the-free-market.

London Metal Exchange (n.d). 'Pricing and Data'. Available at: https://www.lme.com/pricing-and-data/historical-data/.

Metal Bulletin (n.d.) 'Alumina Index'. Available at: https://www.metalbulletin.com/mb-index.html.

Muller, H. (2013). 'Kansanshi Mining plc', unpublished personal communication.

Nathan Associates (2013). 'Economic Effects of Indonesia's Mineral-Processing Requirements for Export', report prepared for USAid. Available at: http://pdf.usaid.gov/pdf_docs/pbaaa139.pdf.

Östensson, O. (2017). 'Local Content, Supply Chains, and Shared Infrastructure', WIDER Working Paper 2017/96. Helsinki: UNU WIDER.

Östensson, O. and A. Löf (2017). 'Downstream Activities: The Possibilities and the Realities', WIDER Working Paper, 2017/113. Helsinki: UNU-WIDER.

Prebisch, R. (1950). *The Economic Development of Latin America and its Principal Problems*. Lake Success, NY: United Nations.

Price, A. H. and S. Nance (2009). 'Export Barriers and Global Trade in Raw Materials: The Steel Industry Experience', report to the Raw Materials Committee of the OECD. Washington, DC: Wiley Rein LLP.

Raw Materials Data (n.d.). https://www.spglobal.com/marketintelligence/en/campaigns/metals-mining.

Roe, A. R. (2016) 'Tanzania: From Mining to Oil and Gas', WIDER Working Paper, 2016/79. Helsinki: UNU-WIDER.

S&P Global (n.d.a). 'SNL Metals & Mining'. Available at: http://marketintelligence.spglobal.com/client-solutions/sectors/metals-mining.

S&P Global (n.d.b). 'Platts Iron Ore Index'. Available at: https://www.platts.com/commodity/metals.

S&P Global (n.d.c). 'Platts Steel Prices'. Available at: https://www.platts.com/commodity/metals.

Singer, H. W. (1950). 'The Distribution of Gains between Investing and Borrowing Countries', *American Economic Review*, 40(2): 473–85.

Tercero, L. E., M. Soulier, and S. Haag (2016). 'An Examination of Copper Contained in International Trade Flows', *Mineral Economics*, 29(2–3): 47–56.

TEX Report (2016). 'India's Iron Ore Imports, Exports in 2015', 6 May. Available at: http://www.texreport.co.jp/xenglish/.

UNCTAD (2013). 'Iron Ore Statistics 2013'. UNCTAD/SUC/2013/3. Geneva and New York: UNCTAD.

UNCTAD (2015). 'The Iron Ore Market 2015'. UNCTAD/SUC/2015/2. Geneva and New York: UNCTAD.

UNCTADstat (n.d.). Available at: http://unctadstat.unctad.org/wds/ReportFolders/reportFolders.aspx?sCS_ChosenLang=en.

United Republic of Tanzania (2014). 'Draft One, Local Content Policy of Tanzania for Oil and Gas Industry'. Dar es Salaam: United Republic of Tanzania.

USGS (United States Geological Survey) (2014). 'Minerals Yearbook'. Available at: https://minerals.usgs.gov/minerals/pubs/myb.html.

World Bank (2011). 'What Is the Potential for More Copper Fabrication in Zambia?' Report No. 62379-ZM. Washington, DC: World Bank.

World Steel Association (2015). *Steel Statistical Yearbook 2015*. Available at: https://www.worldsteel.org/en/dam/jcr:3e501c1b-6bf1-4b31-8503-a2e52431e0bf/Steel+Statistical+Yearbook+2015+r3.pdf.

WTO (n.d.a). 'Glossary'. Available at: https://www.wto.org/english/thewto_e/glossary_e/glossary_e.htm.

WTO (n.d.b). 'Tariffs'. Available at: https://www.wto.org/english/tratop_e/tariffs_e/tariff_data_e.htm.

# 26

# Choices for Spending Government Revenue

## New African Oil, Gas, and Mining Economies

*Sophie Witter and Maja Jakobsen*

## 26.1 Introduction

Extractives projects have the potential to contribute to improved human development, but this is not guaranteed. It is well understood that translating natural resources into individuals living longer and healthier lives, gaining a good education, and having a decent standard of living requires strategic planning and policy choices based around individual country contexts.[1,2]

Public spending of government revenues from extractives projects has the potential to enhance human development either directly or indirectly: a point that is spelled out fully by Roe and Round (2017). Direct routes include spending on health, education, or social protection, while indirect routes include spending on infrastructure or on various measures to enhance economic growth. Both spending channels being important, the balance of expenditure between the two depends on the country context and the expected time profile of natural resource revenues.

This chapter focuses on what it could mean to use natural resource revenues for public spending in social sectors (health, education, and social

[1] We use estimates produced by Haglund et al. (2015) showing the likely timing and magnitude of revenues from new discoveries of oil, gas, or minerals in six African countries, and innovative research by Barca et al. (2015) and Witter et al. (2015a) exploring the potential of new natural resource revenues to improve health and education services and social protection. This chapter is based on research funded by the Bill & Melinda Gates Foundation which was released under Creative Commons Attribution-Share Alike International 4.0 terms (CC BY-SA 4.0). The findings and conclusions contained within are those of the authors and do not necessarily reflect positions or policies of the Bill & Melinda Gates Foundation.

[2] This section draws heavily on Section 2 of Witter et al. (2015a).

protection). The case for investing natural resource revenues in social sectors is strong. It rests on the assumption that these sectors are given a high priority in national government development plans, especially in lower-income countries, and this spending has the potential to lead to high economic returns and the fulfilment of the human rights that are guaranteed by most constitutions.

One of the risks of natural resource revenue flows is their propensity to distract from results-driven questions ('What do we want to achieve?') and rather to focus on expense-driven questions ('We have funds, what should we spend them on?'). Witter et al. (2015a) developed a framework that identifies the key questions that should inform investment decisions in the social sectors; they argue that the same framework can be used by governments to guide policy decisions in this space, irrespective of the price environment—that is, falling or increasing prices.

In general, decisions regarding social-sector expenditure should be based on a comprehensive diagnosis of the needs of the sector, which in turn is based on the development goals of the country. However, decisions regarding the use of natural resource revenues to invest in national priorities require an alignment of such investments against a realistic assessment of the scale, trends, and predictability of revenue flows.

## 26.2 Estimating Revenues from Extractives Projects in Africa

Extractives revenues depend on international commodity prices, which have historically been volatile and notoriously hard to predict. For example, the sharp falls in oil and iron ore prices during the second half of 2014 took most analysts by surprise. While it is generally considered likely that demand for oil, gas, metals, and other commodities will continue to increase over the coming decades due to the growth of emerging markets, there is much greater uncertainty around supply.

As is noted by Roe and Dodd (2017), the past ten to fifteen years have seen many low- and middle-income countries become more economically dependent on extractive industries. This tendency seems likely to continue, despite the recent falls in commodity prices.

A study by Haglund et al. (2015) estimated that in the case of Mozambique, Tanzania, Ghana, Uganda, Sierra Leone, and Liberia, recent discoveries of natural resources are likely to be significant. However, they will not have a transformational impact. Using commodity prices reflecting the 2015 environment, in absolute terms, extractives revenues in these countries are projected to range from an annual average of US$144 million in Liberia to US$2.6 billion in Uganda over the first ten years of production. In relative

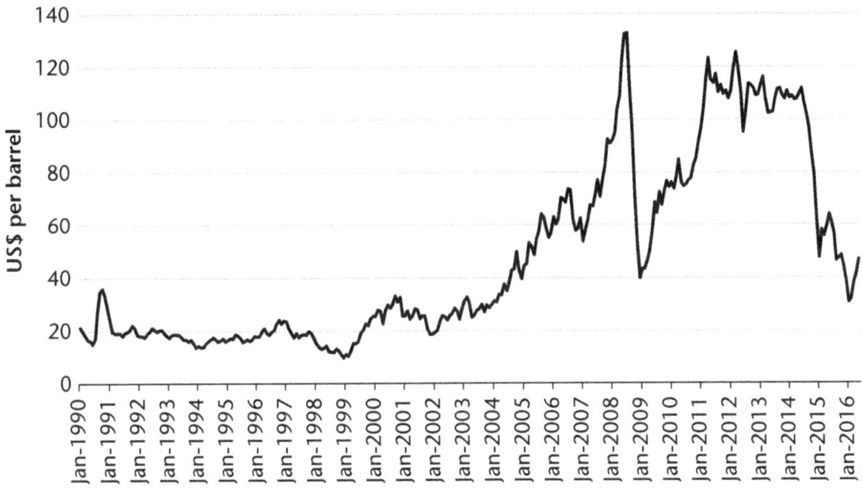


**Figure 26.1.** Europe Brent spot price FOB

*Source*: US Energy Administration (EIA; http://www.eia.gov/dnav/pet/hist/LeafHandler.ashx?n=PET&s=RBRTE&f=M).

terms, they are projected to range from 1.9 per cent of GDP in Tanzania to 5.7 per cent of GDP in Liberia over the same period.

Furthermore, Haglund et al. (2015) showed that revenues will take time to materialize in most countries. It can take up to ten years from discovery for production to start, during which time revenues accruing to the government are minimal, and then a further seven to twelve years for production to reach its peak. In most cases, once revenues hit their peak, they will slowly decline over the next fifty years as resources run out.

Commodity prices have declined further since 2015 (Figure 26.1 shows the example of the oil price), when the Haglund study was completed, and this has almost certainly affected the estimated timing and scale of the figures presented in Haglund et al. (2015). However, the likely timing and broad scale of revenues from extractives in these countries still provide a solid basis for discussions about the policy choices facing governments in these and other natural-resource-rich countries when it comes to public spending on social-sector activities and investments. In fact, price fluctuations are one of the challenges of managing extractives revenues.

## 26.3 Comparison with Spending Gaps and Current Funding for Social Sectors

A comparison of projected revenues from six countries with new extractive discoveries in sub-Saharan Africa (Haglund et al. 2015) with estimated funding

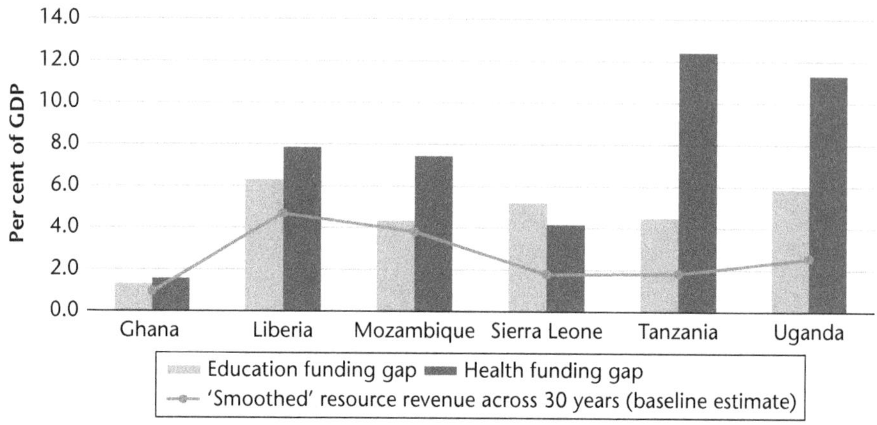


**Figure 26.2.** Health and education funding gaps compared with smoothed natural resource revenues in six African countries, annual average 2016–25

*Note*: It appears that health funding gaps are larger than education funding gaps. This is partly due to the differing methodologies used to calculate gaps, which is related to available data on current spend and estimated needs for each sector and country. 'Smoothed' resource revenues are based on the assumption that government manages revenues from natural resources to create a smooth stream of funds (as a share of GDP) over a period of 30 years.

*Source*: data from Witter et al. (2015a) and Haglund et al. (2015).

gaps for health[3] and education[4] gives an indication of the potential magnitude of what might be contributed by extractives revenues (Witter et al. 2015a). The different scenarios provide evidence that the scale of income from natural resources across selected African countries is in all cases sufficient to cover a significant part of the existing financing gaps in health and education (see Figure 26.2), but shows that the potential in each country is unique as both resource revenues and the size of the funding gaps differ.[5]

Although the revenue estimates have since been downgraded, the overall picture is still illuminating. With a liquefied natural gas (LNG) price of US$11 per million British thermal units (BTU), and with estimated projected

[3] To estimate gaps for health, projected resource revenues were compared with: (1) national health expenditures, based on national health expenditure accounts (NHA) data; (2) funding needs, using international recommendations from McIntyre and Meheus (2014), and estimated health funding gaps; and (3) other potential innovative health-financing sources. On the expenditure side, we constructed a financial programming framework for each country that allowed us to project key economic variables such as growth and domestic revenue, and we used this to estimate resource availability for health.

[4] The estimates of education financing gaps are based on a paper commissioned for the 2010 Education for All Global Monitoring Report (UNESCO 2010). The report calculates education financing needs based on projections of the number of school-age children; teacher salary and classroom construction costs; national targets for pupil–teacher and pupil–classroom ratios; and targets for the proportion of total recurrent costs to be devoted to non-salary spending. The cost of providing adult literacy programmes was also accounted for. Resources available for education were estimated on the basis of GDP projections and the ratio of public education spending to GDP.

[5] This section is adapted from Section 3 of Witter et al. (2015a).

**Table 26.1.** Price variations in smoothed resource revenue projections in relation to the combined health and education financing gaps

| Sample country | New natural resource revenues as a share of total financing gaps in health and education | |
|---|---|---|
| | Low-price scenario | High-price scenario |
| Ghana | 17.6 | 43.7 |
| Liberia | 18.1 | 65.5 |
| Mozambique | 15.6 | 49.4 |
| Sierra Leone | 0.0 | 36.7 |
| Tanzania | 5.4 | 17.4 |
| Uganda | 9.2 | 20.6 |

*Source*: data from Witter et al. (2015a) and Haglund et al. (2015).

revenues smoothed over the next thirty years, Mozambique, for example, could fund most of its education needs, or around a third of the country's need for financing in health, over the next decade. With a crude oil price of US$60 per barrel, Ghana could meet about a third of its combined health and education funding needs over the next decade. In the same time period, Liberia could fill about a third of the combined health and education financing gap (with an iron ore price of US$90 per dry metric ton).

Variations in oil, LNG, and iron ore prices will most likely impact the baseline, or mid-point, estimates presented above. Table 26.1 presents an overview of the variations that could be expected if the baseline price was shifted by +/– 25 per cent (for details see Haglund et al. 2015). In the low-price scenario—which is now closer to the realities post-2015—all sample countries could have natural resource revenues that are sufficient to cover between a tenth and a fifth of their country's combined financing gaps in health and education—assuming that all revenues are allocated to social sectors.

It is important to keep in mind that the needs assessment is high, and is based on the assumption that spending to cover all needs would be possible immediately in both sectors. In reality it would take time to increase spending in both health and education, as absorptive and institutional capacity takes time to develop (not just training staff and building structures but also developing stronger systems to manage them).

### 26.3.1 *Opportunities in the Health Sector*

Analyses of projected natural resource revenues by country for the next decade show different profiles (Figure 26.3);[6] the timing, magnitude, and relative

[6] The decline in Liberia's and Mozambique's estimated health financing gaps as a share of GDP reflects the fact that both countries are expected to experience very rapid GDP growth over the next

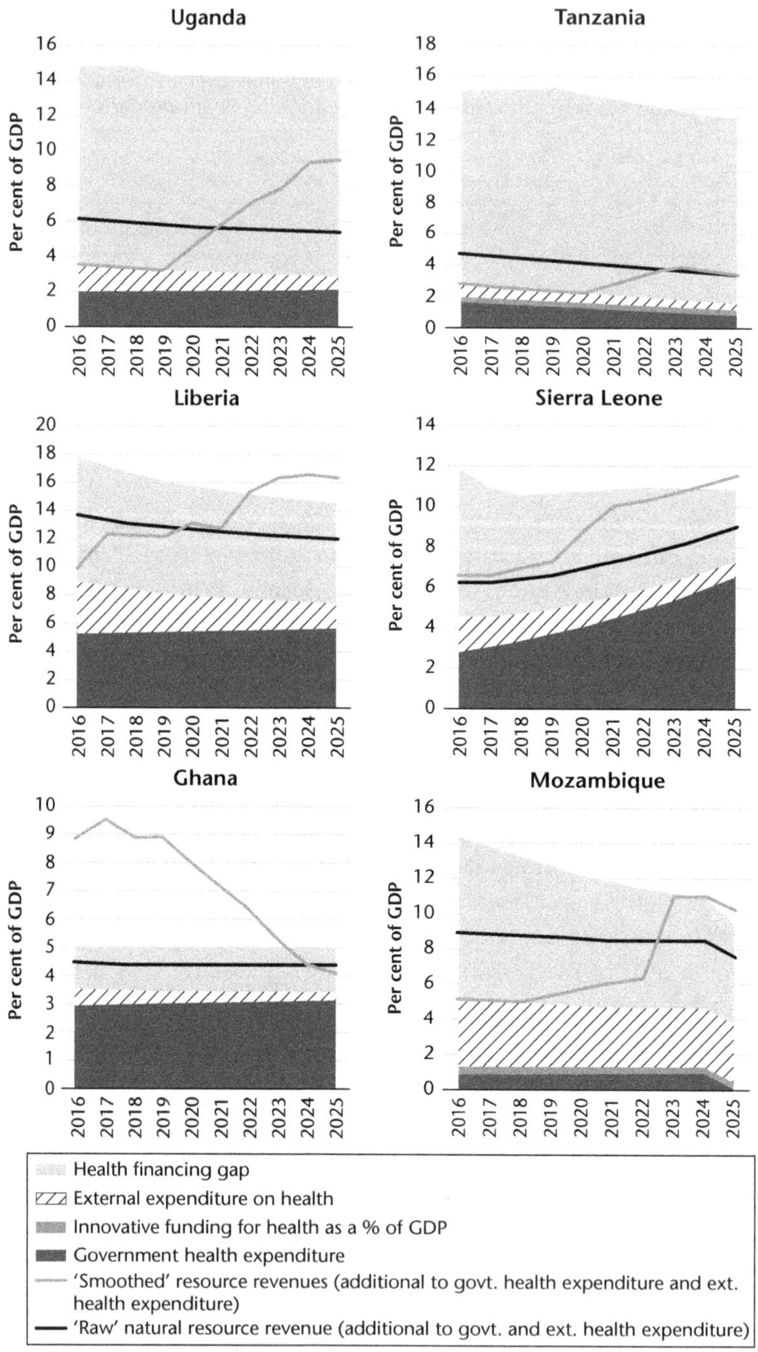


**Figure 26.3.** Natural resources revenues (baseline estimates) compared with health sector financing gaps—profile over time by country

*Note*: These graphs do not show the price sensitivity of the revenue predictions from Haglund et al. (2015). 'Raw' resource revenues are based on the assumption that projected revenues are left unmanaged and allocated directly into the budget available for health and education spending.

*Source*: authors' illustration; data from Witter et al. (2015a) and Haglund et al. (2015).

importance vary across the six countries. In Ghana, Liberia, and Sierra Leone, for example, new resource revenues were already accruing to government in 2015, while in Mozambique, Uganda, and Tanzania revenues are not expected to flow for another four to six years (based on best estimates in 2015).

The graphs shown in Figure 26.3 give a sense of the changing picture in each country and how resource revenues could contribute to improve health services:

- In Ghana and Liberia[7] health financing could be substantially supported by natural resource revenues. The scale of revenues in Ghana is already very high in relation to the various measures of health needs and expenditures; however, if the revenues are smoothed over the next thirty years (as a share of GDP), then just over half of Ghana's health needs can be covered in the next decade.
- In Tanzania and Uganda, the health financing gaps are significant and natural resource revenues are estimated to contribute only partly to those needs and, most likely, not before the early 2020s. However, smoothing revenues over the next thirty years (as a share of GDP) shows that both countries could cover around one-fifth of their health financing needs.
- Mozambique shows a significant dependence on donor funds, which could in fact not be replaced or supplemented by resource revenues for the next five–ten years. However, looking at a scenario where revenues are smoothed over the next thirty years (as a share of GDP) reveals that there is the potential to cover around half of the funding gap in health over the next decade.
- Sierra Leone already has revenues flowing today and, if smoothed over the next thirty years (as a share of GDP), this could close almost half of the country's health financing gap.

Price variations will, however, continue to impact the baseline, or midpoint, estimates presented above and in Figure 26.3. Table 26.2 presents an overview of the variations that are expected if the baseline price is shifted by +/– 25 per cent (for details of this analysis see Haglund et al. 2015). In Tanzania, higher prices result in the possibility that almost a quarter of the country's health gap would be closed. By contrast, a lower price scenario in Sierra Leone could have a devastating impact, leading to cancellation of projects and a zero health contribution.

decade. This means that, although health financing needs in Liberia and Mozambique are expected to rise in absolute terms over this period, they are likely to fall as a share of GDP.

[7] In the projections for natural resource revenues an attempt has been made to take account of the Ebola crisis in both countries.

**Table 26.2.** Price variations in revenue projections in relation to the health financing gap

| Sample country | Projected natural resources revenues as a share of the health financing gap | |
|---|---|---|
| | Low-price scenario | High-price scenario |
| Ghana | 32.2 | 79.9 |
| Liberia | 32.6 | 117.8 |
| Mozambique | 24.6 | 78.2 |
| Sierra Leone | 0.0 | 82.7 |
| Tanzania | 7.3 | 23.6 |
| Uganda | 14.0 | 31.3 |

*Source*: data from Witter et al. (2015a) and Haglund et al. (2015).

Overall, the importance of new natural resource revenues as a possible source of funding for health is potentially large in the sample countries. Other 'innovative' sources of funding for health (such as taxes on remittances, mobile phone levies, and borrowing using domestic bonds for health) would be much smaller than natural resource revenues—even in the low-price scenario—in the two countries where we have estimates of such funding sources (see Figure 26.3).

It is also instructive to compare the projected funding from external donors for health (see Figure 26.3's depiction of external expenditure on health) with the estimated natural resource revenues. Several sample countries cover a substantial part of their health sector spend using external funding; however, this is expected to decline in the medium term, creating a gap that natural resource revenues could close. The magnitude of potential natural resource revenues compared with donor funding is striking for all sample countries—in particular, we observe a cross-over pattern, with falling donor support and rising resources revenues. That said, we need to add a cautionary note and emphasize that both data series are best guesses, with considerable margins of error.

If we compare natural resource revenues with current government expenditure on the health sector, as opposed to the estimated health needs, which are much more ambitious, then their potential contribution to improvement in health services is even clearer.

### 26.3.2 *Opportunities in the Education Sector*

Similarly to the analysis of the health sector, the projected natural resource revenues by country for the next decade show different degrees of potential

**Table 26.3.** Price variations in revenue projections in relation to the education financing gap

| Sample country | Projected natural resources revenues as a share of the education financing gap | |
|---|---|---|
| | Low-price scenario | High-price scenario |
| Ghana | 38.8 | 96.2 |
| Liberia | 40.8 | 147.3 |
| Mozambique | 42.3 | 134.3 |
| Sierra Leone | 0.0 | 65.9 |
| Tanzania | 20.3 | 65.6 |
| Uganda | 27.0 | 60.3 |

*Source*: data from Witter et al. (2015a) and Haglund et al. (2015).

when compared to education financing gaps—in some cases more significant than others (Witter et al. 2015a). Table 26.3 shows a comparison between the projected natural resource revenues from 2016 to 2025 in the six sample countries and the estimated education financing gap needing to be filled in order to achieve Education for All goals (i.e. providing schooling to all pre-primary-, primary- and secondary-aged children).

In Tanzania and Uganda, part of the projected financing gap could be covered by new natural resource revenues, and these revenues could be enough to achieve universal pre-primary and primary education in the high-price scenario. In the low-price scenario (with less available funding), either the financing gap in pre-primary education or about two-thirds of the financing gap needing to be filled to achieve universal primary education could be covered by these revenues. The situation in Sierra Leone is similar, but here a low-price scenario could mean cancellation of the extractives projects and thereby removal of any potential funding available for education.

In Ghana, Liberia, and Mozambique the situation looks different. Our mid-point estimates for the *total* projected new natural resources revenues in these three countries would be sufficient for between two-thirds and three-quarters of the financing gap to be filled, moving close to achieving school attendance for all pre-primary-, primary-, and lower-secondary-aged children. This would mean equipping schools with sufficient classrooms, employing one teacher for every forty-four pupils, and providing the regional standard of learning materials, as well as covering the costs of achieving universal pre-primary education, providing subsidies and supplies to target marginalized pupils, and offering adult literacy programmes.

In the high-price scenario for Ghana, Liberia, and Mozambique, the total estimated financing gap in education could be closed, and in Liberia and

Mozambique there would also be an opportunity to allocate funding to other areas. If all new natural resources revenues were allocated to education in this scenario, there would also be funding available to introduce dramatic improvements to the quality of the education that pupils receive. However, the low-price scenario shows that in that situation less than half of the financing gap could be filled.

The current size of funds from private philanthropic organizations for education further highlights the potential contribution new natural resources revenues could make towards improvements in education. According to the Innovative Finance Foundation (2013), to date there has been only limited financial innovation in the global education arena.

The potential for education sector improvements to be funded by new resources revenues can be further and more specifically illustrated by examining recent reforms in comparable African countries (see Figure 26.4). While the reforms in question were costed for a four- or five-year period, annual figures are presented for ease of comparison. With varying beneficiary populations, these comprehensive system-wide reforms targeted both access and quality outcomes. They included teacher training, textbook development and/or distribution, information and communications technology strengthening, support for school management, and assessment reform.

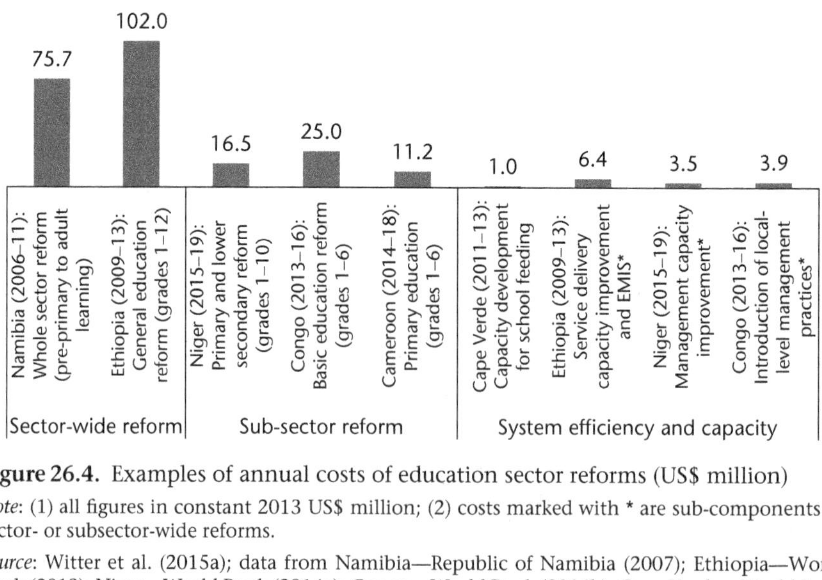


**Figure 26.4.** Examples of annual costs of education sector reforms (US$ million)

*Note*: (1) all figures in constant 2013 US$ million; (2) costs marked with * are sub-components of sector- or subsector-wide reforms.

*Source*: Witter et al. (2015a); data from Namibia—Republic of Namibia (2007); Ethiopia—World Bank (2013); Niger—World Bank (2014a); Congo—World Bank (2014b); Cape Verde—World Food Programme (2013).

The most costly reform shown here—a large-scale programme to improve the quality of primary and secondary education in Ethiopia, with a population almost twice as large as the most populous of the sample countries—was just under US$510 million over five years (World Bank 2013). Costing approximately 7 per cent of the budget for the entire sector plan, this intervention aims to move from a focus on access towards improving quality (Federal Ministry of Education 2008, 2010). Its cost could comfortably be afforded from the projected natural resources revenues for each of the sample countries, assuming they faced similar costs to Ethiopia's.

### 26.3.3 *Opportunities in Social Protection*

Supply-side barriers in relation to health and education spending have been addressed above but outcomes will not improve unless certain demand-side barriers are also addressed. These barriers include poverty, lack of access, and lack of funds for specific social services. Estimates carried out by Barca et al. (2015) showed that, if smoothed over thirty years, new natural resources revenues for the same sample countries were projected to approximate to the cost of a basic social protection package. Rough estimates and existing country data show that a basic national social assistance package (providing cash transfers (CTs) to households in poverty to enable them to access essential social services) costs between 1 per cent and 5 per cent of GDP, exact costs depending on the mix and types of scheme adopted and the demographic profile of the target population (DFID 2011; UNICEF 2009). Only in a handful of southern African countries, where extensive rights-based grant systems and social pensions have been developed, does social assistance expenditure exceed 3 per cent of GDP (Gentilini et al. 2014; World Bank 2012).[8]

It is not suggested that it would be realistic to use all future natural resources revenues for social protection—and governments need to be aware that once social protection schemes are set up they are politically difficult to reverse if money becomes tight. However, the comparisons once again provide a useful order-of-magnitude context. It is not unrealistic to consider spending some part of new natural resources revenues on social protection programmes characterized as recurrent public spending over a thirty-year timeframe.

Even where the additional fiscal space generated by natural resources revenues is not sufficient to cover a full social assistance package, social protection can still be advanced as one possible use of those revenues. Natural resources revenues could, for example, cover CT start-up costs, while the

[8] This sub-section draws on findings from Barca et al. (2015).

government looks for other solutions—perhaps by saving revenues in a sovereign wealth fund, or investing them in strategies to diversify and grow the economy—to finance longer-term recurrent costs. One argument for explicitly linking new natural resources revenues to social protection schemes is that it gives citizens a direct interest in demanding accountability in respect of how the revenues are spent. To maximize their potential, CT programmes have been shown to benefit from close coordination among ministries of finance, health, education, and social welfare.

## 26.4 Policy Choices: What to Consider When Spending Natural Resources in Social Sectors

Witter et al. (2015a) ask the further question of whether the characteristics of extractives revenues result in unique implications for the prioritization of social spending. Table 26.4, which summarizes some of their results, shows that these characteristics may not in fact present significant changes to the normal decision-making required to effectively allocate public resources to, and spend them in, social sectors.[9]

**Table 26.4.** Revenue characteristics and their implications for social sector spending

| | Core and likely characteristics of extractives revenues | Implications for social spending prioritization |
|---|---|---|
| 1 | Of medium-term duration (20–30 years) | Investments should create major additional costs only if projected growth rates suggest they can be maintained over the longer term |
| 2 | Non-renewable | The investments should benefit future generations as well as current generations |
| 3 | Of varying scale and time profile, with rapid scale-up in some cases | They should be capable of rapid introduction, and acknowledge the risk of potential scale-back |
| 4 | Volatile in amount and hard to predict, as extraction and world prices will vary over time and may be affected by shocks | Unless smoothing mechanisms can be found, the investments should focus on discrete interventions, such as systems strengthening and improved value for money, rather than recurrent costs |
| 5 | Associated with weaker accountability to the local or host nation population | Some part should be focused on local communities, who have also had the disadvantages of extraction. Attention should also be paid to creating transparency and accountability mechanisms |

*Source*: data from Witter et al. (2015a).

[9] This section draws heavily on Section 4 of Witter et al. (2015a).

Based on these results, a 'diagnostic framework' was developed which brought together what we know about new natural resources revenues, and also about the needs for funding, the systems constraints, and existing experiences across the social sectors, in order to discuss what approaches are likely to yield good results in low- and middle-income, often fragile, contexts.

A pre-condition for the application of the framework is that there is a willingness to invest in health, education, and social protection. Without this, the framework is redundant. The framework is ideally informed by existing plans for human development, based on the development goals of the country.

Decisions regarding the use of natural resource revenues to invest in national priorities require an alignment of human development investments against a realistic estimate of the scale, trends, and predictability of natural resource revenues. The scale of investments that will be made will depend on the ability to match the costs of investments against the expected revenue flows. Predictability is less straightforward as volatility is a consequence of the uncertainty of future resource prices and demand (Stevens et al. 2013).

However, the diagnostics framework is not only about the revenue flows. It also focuses on three other dimensions: the national context in the social sectors, the financing context for these sectors, and system diagnosis of social sectors. All four dimensions need to be analysed when making decisions regarding the allocation of funds to social sectors. It is important to note that the diagnostics framework does not provide guidance as to how governments should allocate funding *between* sectors—only within each social sector. The full framework is shown in Table 26.5. In Witter et al. (2015a) the framework is applied for illustrative purposes to Ghana and Sierra Leone.

Current public expenditure in the health and education sectors in the sample countries is skewed towards recurrent costs (Witter et al. 2015a). This indicates that natural resource revenues will need to support recurrent expenditure, as this constitutes the bulk of the sectors' needs, even if strategies are used to increase the efficiency of the sectors and to support diverse delivery systems. For education, the range is from 73 per cent in Mozambique to nearly 99 per cent in Sierra Leone, and the bulk of this expenditure is absorbed by salaries (the only exception being Liberia, where non-salary recurrent costs are slightly higher as a percentage). For health, the range is wider, going from 37 per cent in Liberia[10] to 92 per cent in Ghana.

[10] It is possible that the post-conflict setting has generated the relatively higher investment costs in Liberia, but this does not hold true for its education sector, which should be equally affected.

**Table 26.5.** Diagnostics framework—investing natural resource revenues in social sectors

| Core question | Range of answers | Implications |
|---|---|---|
| **Natural resource revenue characteristics** | | |
| Scale of expected resources. | High (more than 60% of public social sector expenditure). | Can fund major new investments and/or restructuring, as well as core inputs. |
| | Medium (20%–60% of public social sector expenditure). | Can fund core inputs and system strengthening. |
| | Low (less than 20% of public social sector expenditure). | Can fund marginal improvements. |
| Volatility of resources. | Government able to smooth out volatility. | Can take on expenditures which are recurrent. |
| | Government not able to remove volatility. | Better suited to funding capital investment and discrete reforms or system strengthening. |
| Duration of resources. | Short-term (5–10 years). | Cannot be used for recurrent expenditures, unless real GDP growth per capita is assumed to be strong at the end of the period. |
| | Medium-term (10–30 years).<br>Long-term (>30 years). | Can commit to longer-term investments. |
| **National context** | | |
| Coverage of health and education systems. | Established, mature health system which can be accessed by all, within WHO norms.<br>Established education system with universal access to early childhood and basic education. | Focus on improving quality, efficiency, and equity, including through CTs, where relevant. |
| | Health and education infrastructure still rudimentary or in recovery from wars and shocks. | Focus on getting full coverage of basic services—infrastructure development, increased staffing, equipment, etc. During recovery focus on reconstruction and peace-building through inclusive health and education systems. |
| Priorities established in a national plan. | There is a medium-term expenditure framework, national health financing strategy, or education sector strategic plan. | Fund priority areas within plan which are not yet supported by existing resources. |
| | There is no medium-term expenditure framework, national health financing strategy, or education sector strategic plan. | Conduct assessment, using tools like OASIS (in health) and Education Sector Review to establish priorities. |
| **Health and education financing context** | | |
| Financial protection. | Financial protection for health care expenditure is adequate (e.g. less than 20% of total health expenditure is out-of-pocket).<br>Mechanisms are in place to offset household expenditure for the poor (e.g. household stipends or school grants). | Check for inequities (are all groups protected or does the average mask substantial differences?).<br>Focus on improving quality of care, efficiency, and equity of access and utilization. CTs may be needed for excluded groups. |
| | Financial protection is currently inadequate. | Fund extension of universal health and education coverage, e.g. through increased public budget, funding to other risk pools, such as social health insurance and more effective public–private partnerships (franchising, social marketing, etc.). |

| | | |
|---|---|---|
| Trends in other funding sources. | The main other financial sources (e.g. government funding and external support) are likely to be maintained and increased over the period. | Natural resource revenues can be seen as supplementary, and can focus on currently neglected areas (e.g. unfunded areas within the sector plan). |
| | The main other funding sources are projected to decline over the period. | Natural resource revenues will need to substitute for current sources and will fund some of the existing commitments. |
| Earmarking. | Within government, there is a willingness to earmark funds. | Natural resource revenues could be deposited within a social fund, to be used for priority areas—especially those which are harder to get external support for. |
| | There is no willingness to earmark funds. | Each year, the sector needs to present its case for additional funding. |
| Bottlenecks in public finance. | Health and education facilities receive adequate funding to cover essential salary and non-salary recurrent costs on a regular, reliable basis. | Funds can be invested in the existing system, which is able to channel resources to front-line providers. |
| | Funds at facility or school level are not adequate, regular, or reliable. | Consider reforming funding of facilities (using capitation or combination of input, output, and quality measures). |
| **System diagnosis of social sectors** | | |
| System preparedness to meet new needs. | The system is resilient to shocks and able to adapt to new needs. | Funds should go to improving and extending the range of services, rather than to systemic investments. |
| | There is evidence of lack of resilience and adaptability. | Systemic investments are needed, including, possibly, re-organization, which can improve local stewardship (e.g. decentralization, increased autonomy), as well as developing supportive national systems. |
| Areas of greatest need in terms of health and education system building blocks. | Adequacy of indicators and performance of:<br>– human resources.<br>– supplies and medicines.<br>– infrastructure and equipment.<br>– governance.<br>– service delivery.<br>– monitoring and evaluation. | Not all pillars can be supported simultaneously, so priorities need to be established using a systems diagnosis, if not already incorporated in sector plans, to establish reform and strengthening sequence. |
| Community engagement and accountability. | There are effective mechanisms for local communities to engage with the health and education systems and to ensure the systems respond to their needs. | These mechanisms do not need additional support but can be used to monitor the use of natural resource revenues. |
| | These mechanisms do not exist or are not effective. | Some part of the natural resource revenues should be earmarked to support local planning and monitoring mechanisms. |
| Non-state engagement. | Private for-profit, not-for-profit, and informal sectors contribute to public goods, including for poorer households. | Natural resource revenues can be focused on funding or extending public provision. |
| | The non-state sector is not well aligned with public health and education policies. | Natural resource revenues can fund pilots relating to changing incentives for the non-state sector and testing new regulatory approaches. |

*Source*: based on Witter et al. (2015a).

### 26.4.1 *Importance of Addressing Demand-side Barriers in Conjunction with Supply*

CTs can help to tackle demand-side barriers to both health and education services—that is, to address the reasons why many people do not access those services (Barca et al. 2015). These barriers include indirect costs associated with accessing those services: for example, travel to hospital, school books and uniforms, and the opportunity cost of time not spent working. CTs can mitigate such costs. A combination of demand-side (CT) and supply-side (health and education) spending can be especially effective in boosting human development.[11]

CTs can be specifically designed to encourage certain behaviours, such as attending health check-ups or sending children to school. This can be achieved through explicit conditionality. However, this can be costly to administer and results have also been achieved through 'nudges', such as: distributing the cash with information suggesting how it could be used; delivering cash at, for example, the start of the school year, when costs of uniforms and books arise; delivering the cash into the hands of a female household member; or even giving the programme a suggestive name, such as 'child support grant'.

There is mounting evidence regarding the most effective ways to design CT programmes that are tailored to each country's needs and fiscal space. Policy decisions include: whether to make CTs universal, or widely or narrowly targeted; the level at which they should be set (often at a meaningful but modest percentage of household budgets); the frequency of payment, with evidence pointing to the benefits of predictability; and whether it is possible to distribute the cash electronically, which has been shown to increase savings rates.

Introducing CT programmes is also in line with the recommendations set out by the African Union, which has been advocating the development of strategies for 'introducing and extending public-financed, non-contributory cash transfers' (African Union 2008). The possibility of supporting existing policy processes in this way is an argument against distributing new natural resource revenue using a 'direct dividend' model: making payments to the population at large. This argument is supported by the size of revenues per capita in many countries (see Box 26.1).

National CT programmes already exist in many African countries, including some that have made recent natural resource discoveries. Kenya, Mozambique, Ghana, and Uganda are among the countries in which programmes have already been developed and are in the process of being scaled up.

[11] This sub-section draws on material in Barca et al. (2015).

**Box 26.1** DIRECT DIVIDEND TRANSFERS

Why are direct dividend transfers less appealing than supporting existing social protection systems for Haglund et al.'s (2015) group of sample countries?

- Overall, it would be a missed opportunity not to support existing government initiatives and build on the know-how and systems developed in recent years by donors and governments alike. The existing 'social contract', social accountability frameworks (including grievance mechanisms and links to community committees), and monitoring and evaluation systems developed for the provision of social assistance would make accountability more effective than the setting-up of a parallel system; it would also contribute to the long-term sustainability of system investments.
- The size of natural resource revenues projected in Haglund et al. (2015) does not predict that any of our sample countries will become the next Angola or Gabon, where distribution of 10 per cent of annual natural resource revenues as direct dividends could eradicate half or more of their average depth in poverty.[12]
- A transfer to all citizens, irrespective of their poverty status or category, in our sample countries would result in very small amounts being distributed per citizen, drastically limiting transformative impacts and undermining the poverty alleviation benefits of CTs. Based on the revenue projections in Haglund et al. (2015), we estimate that between US$13 and US$32 could be distributed to each family per quarter[13] across our sample countries.[14]

These findings are similar to those reported by Giugale and Nguyen (2014). Assuming cost-free, perfectly targeted transfers are made by an all-knowing government in order to close natural resource-rich countries' poverty gaps, they show that most existing resource-rich countries would not be able to eradicate poverty even if they used all their natural resource revenues for this purpose. They conclude that the impact of direct dividends depends as much on the volume of natural resources as it does on demographics and the initial position of the national poverty line.

## 26.5 Conclusion

*There is a strong case for investing natural resource revenues in social sectors.*

- Such investments can create a better educated and healthier workforce that tends to be more productive; they can help to develop domestic human capital to sustain and improve growth, through regional and

[12] As estimated by a recent Centre for Global Development study (Giugale and Nguyen 2014).

[13] It should be noted that monthly transfers of existing CTs are also relatively low but could be significantly increased through natural resource revenues. For example, transfer size in Mozambique is about US$13 month, and in Uganda it is US$9 per month.

[14] Barca et al (2015) calculated a rough estimate of the average direct dividend available to each five-person family every quarter during the first ten years of production in our sample countries. They assume that a quarter of the revenues will have to be spent on administration, and after distribution a third will be collected in tax by the government.

global competition in industries other than natural resources; they can lead to a 'demographic dividend', facilitating a higher input of workers per person and increased GDP per head; they can generate knowledge for the future; they can capture potential temporary macroeconomic externalities; they can build social cohesion and politically stable societies; and they can contribute to sustained and inclusive economic growth.[15]

- The arguments about the macroeconomic effects of health and educational spending are important in providing a response to concerns about the high recurrent cost burden which most social sector programmes imply. Sustainability will be ensured only if the economic benefits of social sector investment are realized through effective and equitable allocation and expenditure that ultimately create and help to sustain broader economic development and progress.

*Part of the financing gap in the social sectors in the African countries studied could be closed by natural resource revenues.*

- The scale of projected resources in the sample countries, allowing for currently unpredictable prices, suggests that a significant contribution could be made by the expected revenues in augmenting existing sources and helping to close some of the considerable financing gaps in the effort to reach universal health care and education. They could also provide a safety buffer if donor resources, which currently play a significant role, dwindle as predicted. In the context of an aspiration to achieve universal coverage for health, education, and social protection, the natural resource revenues could be a critically important source of financing for core services and also for an extension of coverage, quality, and equity—even if not all revenues are allocated to social sectors.

*Many possible interventions are available to start closing the gaps in health and education services; full financing of health and education gaps is not a necessity and may cause scale-up problems.*

- There is potential for education sector improvements to be funded by new resource revenues in all sample countries. For example, a large-scale programme to improve the quality of primary and secondary education in Ethiopia, with a population larger than any of the sample countries, has a cost of just under US$510 million over five years—a cost that could comfortably be afforded from the projected natural resource revenues for each one of the countries considered in this chapter.
- However, closing the financing gaps is not the only challenge to improving health and education. Non-financial constraints to the scale-up of service provision are at least as important. Programmes need to build

[15] This section is based heavily on Section 7 in Witter et al. (2015a).

wider capacity, encourage harmonization between donors and government, bridge institutional divisions in the public sector, and improve collaboration between central and local levels of the health system. Some areas of these weaknesses can be addressed only gradually over time, such as the need to train more doctors. The time taken to build capacity can in some cases dovetail with the development of extractive industries.

- Addressing these structural challenges requires political will, strategic vision, and some room for manoeuvre, which funding can assist. Social sector systems are complex, so the outcome of reforms is inherently largely unpredictable—the scale-up process should therefore be done iteratively, with constant adjustments over time.

*Opportunities and challenges related to the use of natural resource revenues are not fundamentally dissimilar to those associated with other revenue sources, but the political economy risks may well be higher.*

- The lack of natural accountability to citizens, and the high expectations which natural resources tend to generate, increase political economy risks and challenges. In order to manage these risks, it is necessary to taken into account both the positive and negative lessons learned from social funds and other channelling mechanisms used to date, whose documentation also needs to be strengthened.
- There is now quite an extensive body of experience of using social funds. Such funds can increase the transparency of resource use, but they still present important challenges in relation to building capacity and avoiding elite capture. In countries struggling with issues of governance, transparency, and participation, there can be a case for new natural resource revenues being earmarked for specific purposes. However, evidence and opinions on the impact and efficiency of such instruments are mixed. CTs are seen as a route to increasing accountability in the use of natural resource revenues.

*A simple diagnostic framework can help to guide social sector investment decisions in the light of new natural resources revenues.*

- This chapter presents a simple framework that can help to guide investment decisions in health and education in the context of resource revenues via structured questions about natural resource revenues, the socio-economic context, and the sector constraints and priorities. While the framework can be applied to both sectors, as many of the challenges and issues are shared, there are also important differences. In particular, the range of conditions, interventions, and user groups in health collectively present a complex landscape for priority-setting in comparison with education, where services have a more clearly-defined package and target group.

*Further research, analysis, and policy debate are needed.*

- There is a need for more evidence and more systematic documentation of lessons for spending natural resource revenues in the social sectors, especially in low- and middle-income (and fragile) states. In particular, further work is needed to adapt our diagnostic framework to the context of specific countries and to incorporate specific social sector interventions.

## References

African Union (2008). 'Social Policy Framework for Africa.' First Session of the African Union Conference of Ministers in Charge of Social Development, Windhoek, 27–31 October.

Barca, V., L. Pellerano, and M. Jakobsen (2015). 'Paper 5: How to Use Natural Resources Revenues to Enhance Demand for Public Services through Social Protection', Flagship Report Paper Series, Bill & Melinda Gates Foundation and African Development Bank, July.

DFID (UK Department for International Development) (2011). 'Cash Transfers Evidence Paper'. London: DFID.

Federal Ministry of Education (2008). *General Education Quality Improvement Package (GEQIP) Plan*. Addis Ababa: Federal Ministry of Education.

Federal Ministry of Education (2010). *Education Sector Development Program IV (ESDP IV)*. Addis Ababa: Federal Ministry of Education.

Gentilini, U., M. Honorati, and R. Yemtsov (2014). 'The State of Social Safety Nets 2014', World Bank Monitoring Report. Washington, DC: World Bank.

Giugale, M. and N. T. V. Nguyen (2014). 'Money to the People: Estimates of the Potential Scale of Direct Dividends in Africa', Centre for Global Development Policy Paper 043, May. Washington, DC: CGD.

Haglund, D., M. Jakobsen, and C. Hearle (2015). 'Paper 2: Timing and Magnitude of New Natural Resource Revenues in Africa', Flagship Report Paper Series, Bill & Melinda Gates Foundation and African Development Bank, July.

Innovative Finance Foundation (2013). 'Innovative Financing for Global Education', ESP Working Paper Series 2013/58. New York: Open Society Foundations Education Support Program.

McIntyre, D. and F. Meheus (2014). 'Fiscal Space for Domestic Funding of Health and Other Social Services', Working Group on Financing Paper 5. London: Chatham House.

Roe, A. and S. Dodd (2017). 'Dependence on Extractive Industries in Lower-income Countries: The Statistical Tendencies', WIDER Working Paper 2017/98. Helsinki: UNU-WIDER.

Roe, A. and J. Round (2017). 'Framework: The Channels for Indirect Impacts', WIDER Working Paper 2017/79. Helsinki: UNU-WIDER.

Stevens, P., J. Kooroshy, G. Lahn, and B. Lee (2013). *Conflict and Coexistence in the Extractive Industries*. London: Chatham House.

UNESCO (2010). 'Education for All Global Monitoring Report 2010: Reaching the Marginalized'. Paris: UNESCO.

UNICEF (UN Children's Fund) (2009). *Fiscal Space for Strengthened Social Protection*. Dakar, Senegal: UNICEF Regional Office for West and Central Africa and Overseas Development Institute.

Witter, S. and M. Jakobsen (2017). 'Choices for Spending Government Revenue: New African Oil, Gas, and Mining Economies', WIDER Working Paper 2017/150. Helsinki: UNU-WIDER.

Witter, S., R. Outhred, A. Lipcan, and D. Nugroho (2015a). 'Paper 4: How to Use Revenues from Extractive Industries to Improve Health and Education in Africa', Flagship Report Paper Series, Bill & Melinda Gates Foundation and African Development Bank, July.

World Bank (2012). *The Cash Dividend: The Rise of Cash Transfer Programs in Sub-Saharan Africa*. Washington, DC: World Bank.

World Bank (2013). *General Education Quality Improvement Project—APL 1 (GEQIP) (P106855). Project Appraisal Document*. Washington, DC: World Bank.

# 27

# Donor-supported Approaches to Improving Extractives Governance

## Lessons from Nigeria

*Joanna Buckley, Neil McCulloch, and Nicholas Travis*

## 27.1 Introduction: The Rationale for Donor Engagement

Donor agencies are typically focused on poverty reduction.[1] Many poor countries have significant natural resource endowments. A growing number of newly resource-rich countries are low-income and lower-middle-income countries; of the top forty countries with the highest contribution of minerals to exports, fourteen (35 per cent) are classified as low-income and lower-middle-income countries. These include Chad, Democratic Republic of Congo, Guinea, Sudan, and Nigeria (Dietsche et al. 2013; see also Roe and Dodd, Chapter 2, this volume). The resulting logic is that if these resources can be harnessed effectively, they represent an opportunity to reduce poverty faster (Africa Progress Panel 2013). In addition, Africa's natural resource wealth is largely unexplored, so its reserves are likely to be heavily underestimated—the importance of natural resources to Africa's economy is set to increase (Africa Progress Panel 2013). This presents donors with an opportunity both to reduce poverty through encouraging the effective use of the natural resources, and also ultimately to reduce aid.[2]

[1] See DFAT (2014), Global Affairs (n.d.), HM Treasury (2015), and NORAD (2011) for statements of official objectives. In some cases, these objectives also explicitly include contributing towards domestic national interests.

[2] We are grateful for excellent detailed comments and suggestions from Alan Roe and Tony Addison and the participants of the Extractives for Development workshop in Helsinki in May 2016. Helpful comments were also received from participants at the Thinking and Working

However, the 'resource curse' literature has shown that, in many cases, the presence of extractive resources is associated with poor economic performance and poverty (see the review of this evidence in Lahn and Stevens 2017). Resource-rich countries account for nine of the twelve countries at the bottom of the UN Human Development Index.[3] Extractives can also have a wider, pervasive impact—for example, by distorting economic and political structures, and thereby reducing growth (or making it less pro-poor). Corruption in the sector can lead to increased poverty and conflict (DFID 2008: 29). Stevens et al. (2013) find that following an extractive-industry-led development path has often been a factor inhibiting the emergence of independent, strong institutions. This exacerbates public policy failure, with the corruption associated with extractives revenue corroding governance and encouraging high-level state looting (Global Witness 2010).

The Africa Progress Panel (2013: 71) finds compelling evidence that the higher the share of GDP accounted for by resource wealth, the less information is made available to citizens. As a result, several donors have responded to the governance challenges posed by natural resources by implementing a range of projects with business, civil society, and non-governmental organizations (NGOs) focused on greater transparency. Dietsche et al. (2013) suggest that there have been four phases of donor involvement in the extractive industries over the last three decades: (1) attraction of foreign direct investment (FDI) in the late 1980s/early 1990s; (2) rising concerns about the resource curse in the mid-1990s; (3) 'good governance' as the cure in the late 1990s/early 2000s; and (4) cross-sector linkages in the late 2000s. Buur et al. (2013: 56) put forward a similar typology of involvement: (1) linking FDI to the liberalization and structural adjustment agendas of the 1980s and 1990s (reforms which Bourgouin (2011) largely attributes to the World Bank and IMF); (2) an emphasis on strengthening the regulatory and revenue-generating framework for the industry, and improvements in public-sector financial management in the 1990s and 2000s; and (3) a focus on the development effects of natural resources in recent years.

This chapter describes how some donors, such as the UK Department for International Development (DFID), are now experimenting with a new type of engagement. This approach merges the good governance agenda of the late 1990s/early 2000s, and the ongoing focus on how to avoid the resource curse, capture developmental benefits, and build robust regulatory frameworks, with

Politically Community of Practice meeting in London in June 2016. All remaining errors are our own.

[3] The Human Development Index is a summary measure of average achievement in key dimensions of human development: living a long and healthy life, being knowledgeable, and having a decent standard of living. For details, see http://hdr.undp.org/en/content/human-development-index-hdi (accessed 9 January 2017).

an economic development agenda that seeks to mobilize reform through an explicitly political focus.

As Dietsche (2017) argues, there is a need to reframe the political economy of extractive resources away from the negative question 'How can poor outcomes be prevented?' and towards the positive question 'How can positive institutional change for better outcomes be brought about?' This chapter provides a concrete example of how some donors are attempting to do this. Politics has typically been seen as an inhibitor, or a constraint that one has to work around in the implementation of projects. However, the new approach to 'thinking and working politically' represents an attempt to understand, and where possible address, the political challenges associated with reform in a more explicit way.[4]

## 27.2 The Political Economy of Reform: Thinking and Working Politically

There is an increasing acceptance by some donors—including the World Bank, DFID, and the United States Agency for International Development—that in order to bring about large-scale change it may be necessary to take a radically different approach to reform.[5] Instead of focusing first on the provision of technical assistance and then negotiating the complex politics of trying to provide support in an environment that is often not conducive to reform, the new approach starts with the politics. Two simple propositions inform 'thinking and working politically' programming: first, the need to work through domestic political processes to achieve sustainable reform; second, the need for development intervention models that understand these political processes and intervene in iterative, adaptive, and politically informed ways. It draws on the ideas of Andrews (2013) about the difficulties of institutional reform without real political commitment to change, as well as on the methodological insights of Andrews et al. (2012) on solving complex reform problems through problem-driven iterative adaptation.[6]

However, although agencies increasingly *think* politically, many are still trying to understand how to get better at *working* politically. As DFID's chief economist has stated, 'politics is too important for development in general to

[4] For a summary of this approach, see Thinking and Working Politically Community of Practice (n.d.).

[5] Although the approach is, to some degree, contested in all of these agencies, where there are advocates who have more influence in some sectors and countries than others.

[6] In the United States, this approach to engaging with politics in development work typically comes under the heading of 'doing development differently'—see http://bsc.cid.harvard.edu/doing-development-differently (accessed 13 January 2017).

be left to political scientists and governance advisors only—we all need to think about it when we act'.[7] This means continuing to think politically, using political economy analysis (PEA) in order to develop a deep understanding of local context—mapping out the key actors and institutions, and honestly assessing their incentives and appetite for reform. It then means moving beyond the traditional 'transparency agenda' towards a more integrated approach to the politics of reform—creating widespread understanding of the problems, building the capacity of reformers in government, and negating entrenched interests, as well as strengthening the capability of civil society actors to hold institutions and individuals in government to account (Figure 27.1). This approach recognizes that, as popular expectations about the benefits of natural resources grow, so do the stakes for governments, which are under increasing pressure to deliver to their populations. Yet improved transparency is only one pathway to impact, and does not always lead to significant reform.[8] Transparency is necessary but insufficient in itself to drive good governance and more equitable revenue-sharing in the sector—a point repeatedly made by most observers, notably civil society groups (Stevens et al. 2013).

This intimate understanding of context and commitment to reforming the entrenched interests of elites is a potentially uncomfortable area for donors. It requires engagement with local politics, experimentation in order to seek out different strands of commitment for reform within government and

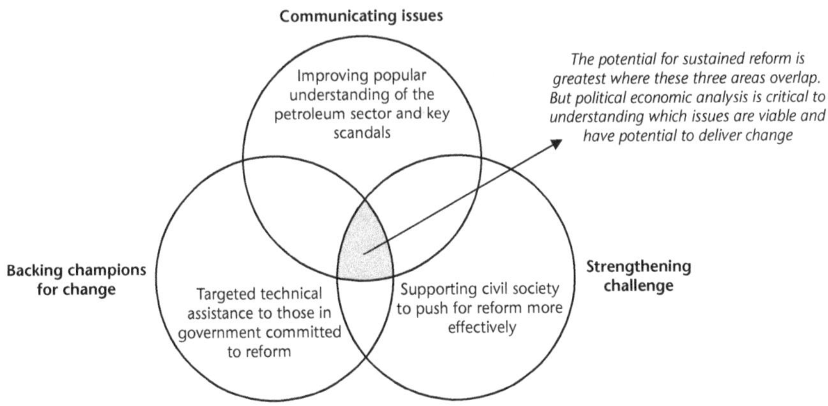


**Figure 27.1.** An integrated approach to supporting reform

*Source*: OPM (2015); reproduced here with permission.

[7] https://twpcommunity.org/what-is-twp/ (accessed 9 January 2017).

[8] See Joshi (2013) for a series of articles on the impact of transparency and accountability initiatives on reform in a variety of sectors.

outside, and sometimes significant risk. But it can have large and measurable results.[9]

Section 27.3 describes a case study of a programme funded by DFID—the Facility for Oil Sector Transparency and Reform (FOSTER) in Nigeria.[10] It demonstrates that DFID sees politics as central to the delivery of effective extractives governance programming.

## 27.3 Facility for Oil Sector Transparency and Reform (FOSTER)

### 27.3.1 *Context*

Nigeria's oil sector is one of the most challenging contexts for governance reform in the extractives sector anywhere in the world. Bhalla et al. (2016: 17) recently described the challenge as follows:

> Oil is the backbone of the Nigerian economy with oil revenues providing over 90% of foreign exchange earnings and at least 70% of government revenues. Predatory elite behaviour aimed at capturing oil rents has defined the character of the Nigerian state in three principal ways: a weak social contract as the government relies on lucrative rents rather than taxes for survival; systematic rigging, intimidation, and violence used to win elections and maintain a grip on the oil sector; and hollow institutions that have little power or capacity to set policy, formulate laws, or effectively regulate the sector.

Repeated attempts to improve governance in the oil sector have met with relatively little success. For example, the Petroleum Industry Bill has still not been passed by parliament, despite being introduced in 2007; repeated high-level government reports have been ignored or remain unimplemented; and the Excess Crude Account[11] has been steadily eroded, from US$20bn in 2008 to a mere US$2bn in 2014.

It is hard to identify a single area of the oil sector in Nigeria which has not been distorted by corruption. Corruption exists in the allocation of licences, which is non-transparent, with licences frequently awarded to politically connected individuals with no capability to operate fields; it exists in extensive crude oil theft (known in Nigeria as 'bunkering'), which amounts to

[9] For a more detailed discussion of the political economy and governance of the extractives sector and the application of a more formal institutional economics model to the sector, see Dietsche (2017).

[10] FOSTER is one of a number of DFID-funded governance programmes focused on the extractives sector employing a PEA driven flexible approach to delivery; others include Ghana Oil and Gas for Inclusive Growth (GOGIG) and the Kenya Extractives Programme (K-EXPRO).

[11] The Excess Crude Account is a Nigerian government account used to save oil revenues above a base amount derived from a defined benchmark price. The Excess Crude Account was established in 2004, and its objective is primarily to protect planned budgets against shortfalls due to volatile crude oil prices.

around 10 per cent of daily oil production (Katsouris and Sayne 2013); it is pervasive in the mechanisms used to sell crude oil, for example through the 'domestic market allocation' of 445,000 barrels per day to the refineries, which in fact rarely operate at more than 20 per cent capacity; it exists in the downstream sector, for example through repeated turnaround maintenance contracts, which never turn around performance; and it exists along the entire import value chain for fuel products, from the shady swap agreements used to purchase product, to the non-transparent allocation of import quotas to marketers, the numerous government institutions involved in approving entry of product, and the corrupt and inefficient subsidy payment mechanism.

Reform in the sector has been difficult because of the strong incentives of a range of powerful actors to oppose reform. For example, until recently, the minister of petroleum was the same individual as the head of the Nigerian National Petroleum Corporation (NNPC).[12] The NNPC therefore was operating as a producer and trader of petroleum products—and also as the de facto regulator of the industry, since the Department of Petroleum Resources reports to and is subordinate to the minister. Successive ruling political parties and elites have benefitted enormously from the rents associated with the capture of Nigeria's oil resources through control over the NNPC and associated institutions. Parliamentarians, it is alleged, have tended to follow the locus of power in order to share some of the rent, and are therefore inclined to block any attempts at reforms that might threaten the flow of resources.

### 27.3.2 *Programme Design*

This complex and highly unconducive environment for reform requires going beyond the provision of technical assistance to specified ministries or government agencies to tackle technical challenges or help build their capacity for effective management. It also requires going beyond supporting reformers to build capacity to demand reform. The challenges to reform in this context are fundamentally political, and many of the actors opposed to reform reside within government institutions. Hence a technocratic approach of merely supplying technical expertise to such organizations would be highly unlikely to be effective. Likewise, a transparency and accountability programme, such as that provided by the Extractive Industries Transparency Initiative (EITI) in several countries including Nigeria, would be unlikely by itself to achieve traction, due to the lack of incentives for reform and the intentional

[12] On 4 July 2016, President Buhari created a separation between the two roles for the first time in decades. However, the minister remains the chair of the board of the NNPC. http://www.premiumtimesng.com/news/headlines/206412-buhari-removes-kachikwu-as-nnpc-boss-appoints-new-board-for-corporation.html (accessed 9 January 2017).

complexity of the rules and regulations governing the industry. To be successful, reform has to bolster both 'demand' for reform through support for civil society (NGOs, think tanks, community-based organizations, and the media) and the capacity for reform on the 'supply' side (government ministries and agencies, including the tripartite Nigerian EITI).[13]

In 2011, DFID selected a bid put forward by Oxford Policy Management for the implementation of FOSTER. Phase 1 of the project ran from 2011 to 2016 with a total value of £14m. The purpose was to reduce the many incentives for misuse of power and capture of oil revenues in Nigeria. The overall intended outcome was the enhanced management of Nigeria's natural resources. This translated into three output areas: (1) an increase in extractive industries revenues identified (and returned to the Federation Account); (2) improved management and accountability of extractive industry resources; and (3) improved policy outcomes for local communities affected by natural resource extraction.

Of the GB£14m, over GB£8.5m was used as part of a managed fund set up with the intention of funding activities on an iterative basis throughout the project life cycle. The rationale was that the specific pathways through which change might occur were not known in advance. With a flexible managed fund, initiatives could be identified as the project went along, and resources allocated to them. Moreover, the project deliberately took a 'portfolio' approach—trying a range of approaches that seemed promising, with the knowledge that not all would necessarily be successful.

The original vision for the project was broad and intentionally open: to support policy reforms, strengthen accountability actors and mechanisms, and increase transparency. The hope was that this would reduce the incentives for the abuse of power and the capture of revenues, which both distort policy and politics in Nigeria and undermine the potential for oil revenues to be used to accelerate economic and social development.

The context in which FOSTER operated changed almost immediately after its launch. President Goodluck Ebele Azikiwe Jonathan was inaugurated in May 2011. His incoming policy priorities made no mention of reforming the oil and gas sector (despite referring to the need for public-sector and power reform). Policy paralysis and uncertainty—and in some cases wilful resistance to fundamental sector reform—characterized the Jonathan administration from February 2011 to May 2015. However, because the managed fund could

[13] In addition, FOSTER attempted to work with the private sector, including the oil majors. However, relatively little was done in this area, in part because the behaviour of the oil companies was determined by the policy environment, making the latter the priority; and also because the incentives of some private-sector actors were not necessarily aligned with good governance, e.g. efforts by companies to minimize fiscal obligations.

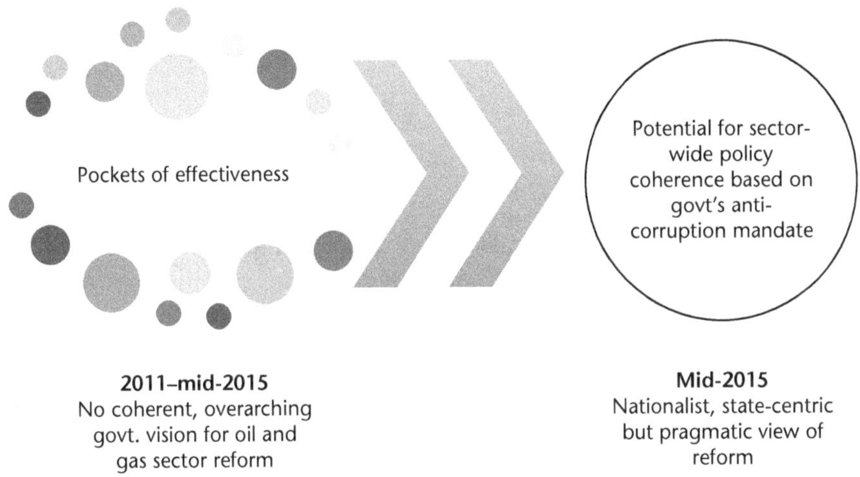


**Figure 27.2.** From pockets of effectiveness to strategic partner
*Source*: OPM (2016); reproduced here with permission.

be allocated flexibly, the change in circumstances could be accommodated by switching activities to focus predominantly on the demand side as a transparency and accountability programme. This enabled the programme to play an important role in building widespread understanding of the failures within the sector. On the supply side, the programme focused on identifying and working with 'pockets of effectiveness' (see Figure 27.2). This approach, located as it was in the largely unsympathetic prevailing political context during the Jonathan administration, mainly secured localized, transactional changes, with little scope for game-changing, sector-wide, transformational reforms. But it also helped to build a reputation for the programme as one that could provide relevant, predominantly local expertise for oil-sector institutions that are genuinely interested in reform (whilst avoiding wasting resources on technical assistance for government ministries or agencies where doing so would not have yielded any results: this would have been the result of a more rigid programme design).

In February 2015, President Buhari was elected. He assumed office in May 2015 on an explicitly anti-corruption platform, leading to the gradual emergence of a more pro-reform environment that has been evident since July 2015. This resulted in FOSTER being reinterpreted to support sector reforms from both the supply and demand sides. In mid-2015, as FOSTER entered the final year of its first project cycle, its patient approach to building relationships with key reformers was quietly vindicated. A project that had begun in the margins was invited to provide assistance to the central institution of management of the oil sector—the NNPC. Over the years FOSTER has evolved subtly and strategically to where it is today, a niche project advancing practical institutional reforms of Nigeria's oil and gas sector (see Figure 27.2).

### 27.3.3 *How FOSTER Worked*

From the onset, the test for the project lay in the very conditions that gave rise to its formulation: making significant progress towards the achievement of ambitious objectives within the challenging political and institutional context that characterized the country's oil sector. FOSTER's response to this was to pioneer and gradually systemize an approach that was built on three principles: (1) undertake deep, regular PEA; (2) use this intelligence to nurture relationships with sympathetic stakeholders, both in government and outside, and craft contextually relevant interventions clearly aligned with the project logframe; (3) work discreetly to minimize risk to DFID, and indeed to the project itself, given the sensitivities around the sector.

Underlying this picture, four forms of deficit were evident in the public sector: (1) a transparency deficit in the centralized, government-controlled oil and gas industry; (2) a monitoring deficit owing to the concentration of power and fiscal control within the executive, and a corresponding lack of institutional checks and balances; (3) a participation deficit because of fundamentally weak state–civil society relations and a deeply embedded culture of corruption; and (4) a data deficit whereby data pertaining to oil were often unreliable, ambiguous, or contradictory.

FOSTER's response to these conditions was to work opportunistically and strategically with pro-reform supply- and demand-side forces within the Nigerian context, seeking out collaboration with 'pockets of effectiveness' (Hickey et al. 2015; Leonard 2008)—testing and confirming the hypotheses and findings from the broader research arena, which suggested that such locales exist, even in weak-governance environments. In particular, FOSTER's theory of change is based on identifying and supporting these pro-reform pockets of effectiveness among two types of organizations:

- Supply-side institutions such as the Petroleum Unit in the Ministry of Finance, the Nigerian EITI, the National Oil Spills Detection and Regulatory Agency, and the National Assembly's committees looking at the Petroleum Industry Bill.
- Demand-side civil society actors from professional bodies, media outlets, think tanks, and research and advocacy organizations.

There were several aspects of FOSTER's approach which were very different from many other donor programmes, and which have been key to FOSTER's success:

#### 27.3.3.1 BEING DRIVEN BY THE PREVAILING POLITICAL ENVIRONMENT

PEA is not a magic bullet. However, FOSTER's use of frequent (quarterly and annual) PEAs enabled the work plan for each quarter to be grounded in a nuanced understanding of the political dynamics and opportunities for reform.[14] Using PEA operationally in a highly politically charged extractives sector demands an implementation team that can undertake sensitive political, technical, and institutional analysis, and can programme accordingly. Thus, as important as the formal PEA was the fact that the local team were embedded in a set of networks, so that they understood the changing circumstances on a daily basis (see Section 27.3.3.2). The daily 'informal' political thinking and the regular formal PEA were complementary—the former shaping the tactics on a day-by-day basis, while the latter provided a mechanism to stand back from the day-to-day changes and look at the likelihood of success of interventions from the broader perspective of the interests and power of the different stakeholders involved.

#### 27.3.3.2 EMPLOYING LOCALLY EMBEDDED 'DEVELOPMENT ENTREPRENEURS'

Success in policy reform programmes depends not just on working with local actors, but also on working with the right local actors: that is, people who support reform and have the capability and inclination to act. Knowing who the right counterparts were required local knowledge, and necessitated appointing staff on the basis of their political networks whilst aiming to ensure that they themselves were politically neutral and sat outside the vested interests/political elites resisting change. It also required a certain type of personality and skill set: individuals who were keen to catalyse, facilitate, and support actions by other individuals and organizations rather than seizing the limelight for themselves, but who also had the ability to repeatedly critically appraise what had worked and what had not, and to adjust accordingly. This type of individual has been called a 'development entrepreneur' (Faustino and Fabella 2011).

The FOSTER team consisted entirely of Nigerian policy entrepreneurs. Collectively, they had a detailed knowledge of the key actors in the sector. However, even with this knowledge, identifying the right partners required a great deal of experimentation. With new partners FOSTER would typically start with a pilot activity to assess the capacity of the partner. If it went well, the activity might be scaled up. In this way, over the five-year life of the first

[14] The quarterly formal PEA was undertaken by a programme advisor who, because they sat outside the core team, could provide a fresh perspective on (and a challenge to) the well-connected and politically savvy team members. Their findings were then checked and discussed with the core team and local partners.

phase of the project, the FOSTER team built up an excellent appreciation of which partners were most effective in pursuing which kinds of activities. This repository of 'network knowledge' was intangible, but essential to the success of the project.

#### 27.3.3.3 USING FLEXIBLE AND ADAPTIVE PROGRAMMING

FOSTER has been successful in large part because it has been able to adapt its programming to the evolving context it has faced. As noted above, it used regular PEA to identify opportunities for reform in the 'pockets of effectiveness' on the supply side. But if there were few or no opportunities for effective programming within government, the design of FOSTER provided it with the flexibility to move programming from the supply to the demand side, to advance public understanding and maintain public pressure for reform. Had its outputs and activities been determined too rigidly at the outset, it would have been much less able to respond to the changing context, and would likely have been much less effective.[15] Many donor projects specify exactly what they want to do up front, and then lay out a set of steps necessary to achieve their goal; they then attempt to work their way through these steps systematically in order to achieve the goal. However, experience in several countries suggests that such 'linear' projects often run into trouble because the circumstances and assumptions that gave rise to the original plan often change (Andrews 2013).

By contrast, the local FOSTER team proactively sought out opportunities for reform—small projects that might push forward reform in one area (see Andrews et al. 2012 for a detailed discussion of such problem-driven, iterative approaches to reform). They were guided in choosing which opportunities to select by an overarching 'logframe': activities had to have some plausible connection with the overall objectives of the programme. But other than this overarching guidance, the project team were at liberty to be 'development entrepreneurs', in the sense of identifying individuals and organizations that were interested in pursuing reform initiatives, and boosting their capacity and knowledge to do so. This flexibility was made possible by ensuring that the project's funds were not pre-allocated to particular activities but held in a flexible £8.5m fund, so that FOSTER staff could act quickly and respond to reform opportunities as they occurred. Programming through the fund relied on an explicitly interactive and dynamic approach to conducting analysis, searching for feasible solutions along the way.

[15] The programme was able to do this due to the early adoption in DFID Nigeria of flexible and adaptive programming. The principles of flexible and adaptive programming have now become embedded in DFID's smart rules for better programme delivery (DFID 2016).

An important element of this way of working was reflecting on failure, failing fast and learning quickly, and using results processes that encouraged adaptation. The interplay of political dynamics is complex, and no matter what local expertise and PEA is conducted, some issues cannot be understood *ex ante*. Having systems (e.g. feedback loops, risk management processes) to 'test the water' and 'learn from doing' is therefore important. Interventions that made sense in Month 1 may no longer make sense in Month 6, and may need to be closed down to avoid waste of resources. Linked to this was the need for results frameworks to be broad and non-prescriptive, whilst keeping long-term outcomes in sight. To ensure accountability (and for the funder to be able to distinguish between an 'adaptive' programme and an underperforming programme), a results framework can instead focus on how well the programme is analysing, learning, and adapting from successes and failures. To that end, forms of ongoing action research may replace traditional mid-/end-programme evaluation.

#### 27.3.3.4 EMBRACING RISK...BUT ALSO MANAGING IT

FOSTER took a very different approach to risk from many other aid projects. Traditional aid projects manage risk by trying to think of all the ways in which an activity might fail, and then putting in place measures to minimize the likelihood of failure and mitigate its impact. Whilst FOSTER did this too for individual activities, it deliberately adopted a 'portfolio approach' to risk management. The attitude was that, as with financial investments, it did not know in advance which activities were going to be successful and which were not. Consequently, FOSTER deliberately selected a wide range of different types of activities, anticipating that the overall portfolio of activities would be successful in pursuing reforms, while accepting in advance that some individual activities would not. However, this approach poses a challenge to traditional forms of aid management. Normally, all individual activities are expected to succeed, and the implementer is criticized if they do not. Hence, although a portfolio approach to risk may be much more sensible in highly uncertain environments, it requires a much higher appetite for risk by funders, and better tools for assessing whether failure is caused by external or internal factors.[16]

#### 27.3.3.5 OPERATING 'UNDER THE RADAR'

External aid was used to facilitate the process of reform through partnerships, not to 'buy' short-term reform. Sometimes donor projects push a particular

[16] It is also important to note that risk can become very personal in sensitive sectors such as the extractives sector. Making fraudulent or inefficient behaviour public poses personal risks for the partners that expose such malfeasance, and potentially also for project staff who support them. Implementers must therefore take their duty of care seriously.

reform agenda. However, in Nigeria, externally promoted reforms are often not appreciated—all the more so when they threaten powerful interests. Rather, the reform actors that have the credibility and access to influence policy are almost entirely local. As a result, the project explicitly chose to identify and support the agendas of local actors—whether in civil society, parliament, or the media—rather than promote its own agenda. In this sense it operated 'under the radar': there was no FOSTER logo, and no external promotion of the project. Rather, it worked with local actors to strengthen their capability to press for change. Sometimes this was done by providing a small grant for a programme of work, at other times it might consist of providing technical assistance or training—but always in support of a work programme with complete ownership by the local partner.

#### 27.3.3.6 TAKING AN 'ARM'S LENGTH' APPROACH TO AID

Working in the way described above necessitates developing deep relationships and networks in a country, as well as the flexibility to exit relationships when reform is not achieved or more effective avenues are identified. This requires trust on the part of the donor. There needs to be sufficient distance from the donor to allow the implementer the flexibility and independence to invest and adapt accordingly, and to create relationships with those who might not be willing to engage directly with a donor (characterized by Booth (2013) as an 'arm's length approach to aid'). On the other hand, the donor needs to be comfortable with the relationships being formed and activities being undertaken, and to be able to articulate this internally. To that end, ongoing dialogue between implementer and funder to develop trust is important, alongside the ability to communicate the results of working in this fashion in a clear manner.

### 27.3.4 *What Has Been Achieved?*

The FOSTER programme has had a significant impact in promoting transparency and accountability in the management of Nigeria's natural resources, especially in the oil sector. In particular, it has made an important contribution in four main areas: the nature of the public debate; the legal framework; revenue savings; and the institutionalization of transparency and accountability in the sector. We discuss each in turn.

#### 27.3.4.1 RAISING THE STATE OF KNOWLEDGE AND CALIBRE OF PUBLIC DEBATE ON OIL AND GAS POLICY

FOSTER's work with a wide range of civil society and media actors has significantly raised the level of understanding about how the oil sector works, and the policies and practices that give rise to corruption and inefficiency. This work has included the creation of online platforms that provide a mechanism for holding government accountable (e.g. an online tool for tracking illegal gas flaring), as well as a deepening of the understanding and capability of the media to report on oil sector issues by training investigative journalists.

#### 27.3.4.2 ACHIEVING SIGNIFICANT AGREEMENT ABOUT WHAT THE LEGAL FRAMEWORK FOR THE SECTOR SHOULD BE

FOSTER invested significant effort in working with legislators and journalists in order to broaden understanding of the strengths and weaknesses of the various drafts of the Petroleum Industry Bill. This has led to a better understanding among all the key stakeholders of the key principles that should underpin legislation in the sector. However, notwithstanding FOSTER's support, the Petroleum Industry Bill did not pass in both the House and the Senate.[17] At face value, therefore, this is an example of a failure. But it also illustrates how measuring success is difficult, because the ideas and associated legal texts discussed during the first phase of FOSTER are forming the basis of renewed attempts to craft framing legislation for the sector under the current administration. The passage of the Petroleum Industry Governance Bill—the first component of a set of smaller packages of legislation—in the Senate in May 2017 demonstrates progress; hence it may be too early to determine whether this support has been successful or not.

#### 27.3.4.3 MAKING LARGE REVENUE SAVINGS FOR THE GOVERNMENT OF NIGERIA

FOSTER has been responsible for the identification and remediation of well over £400m for the government of Nigeria (and possibly much more). This has come from multiple sources including: large savings from support to the Ministry of Finance to ensure the proper audit of fuel subsidy payments; exposure of evidence of the inefficient and corrupt system of oil swaps, leading to the scrapping of the system and its replacement with a more open and transparent mechanism for oil sales; and improvement of the template used to track revenue from oil proceeds, preventing diversion of funds to other uses.

#### 27.3.4.4 INSTITUTIONALIZING TRANSPARENCY AND ACCOUNTABILITY

Through FOSTER's long-term support, the Nigerian EITI has developed to become an effective institution for the promotion of transparency and accountability in the sector (and is widely regarded as one of the best EITI

[17] It passed in the House, but not the Senate, prior to the end of the Jonathan administration.

institutions globally). FOSTER has helped to build the capacity of the National Oil Spill Detection and Response Agency to tackle oil spills and improve the joint investigative visit process. Moreover, FOSTER's recent support for the NNPC has facilitated a sea change in the latter's approach to transparency and accountability, supporting the NNPC to publish its annual report for the first time in over a decade, and strengthening its capability to reform refineries and tackle fuel subsidies.

However, perhaps of most interest is the fact that FOSTER is providing valuable insights about how the international community can engage in efforts to promote reform in the extractive industries sector in challenging contexts (Booth 2016). This is reflected in the design of the second phase of FOSTER, which started in June 2016, and which has explicitly adopted many lessons from the first phase of the project—but also in the adoption of many of the elements of flexible, adaptive, politically savvy programming into the guidance provided by several donors active in the sector (Whaites et al. 2015).

## 27.4 Challenges in Applying the Approach, and Lessons for Donors

### 27.4.1 *Challenges in Applying the Approach*

Models such as FOSTER will not be applicable in every instance of extractive governance programming. Nor are they a quick approach to enabling change. There are also inherent tensions in the approach for development practitioners, donors, and stakeholders engaging in reform. We conclude by outlining some of these tensions and the lessons for future donor engagement.

#### 27.4.1.1 DISRUPTING THE STATUS QUO CAN CREATE CONFLICT

FOSTER overtly aims to disrupt the status quo. This means changing power dynamics, subverting corrupt interests, and aiming to identify reformers who can work against entrenched interests. Working effectively in fragile and conflict-affected situations, such as Nigeria in particular, requires an awareness that interventions shape—and in turn are shaped by—the context within which they take place. Although actors have long been aware of the constraints imposed by conflict on the ability to achieve results, there has been a more limited focus on the wider 'obstacles' of fragility and conflict, and on the need to mainstream such concerns throughout projects. There is increasing institutional recognition within DFID that conflict sensitivity and 'do no harm' approaches are critical in situations where open violence is present, or is a very real threat, and that they can be useful in situations where there are high levels of political tension and/or an unpredictable and constantly changing political context. An awareness that their own interventions impact on conflict dynamics has led donors and NGOs to develop a range of conflict sensitivity tools that facilitate an understanding of these issues. These tools provide a series of structured questions to prompt reflection on the two-way interaction between intervention and conflict. As DFID notes, 'Where international actors fail to invest in good political and conflict analysis, actions can result in more harm than good' (DFID 2010: 8).

#### 27.4.1.2 SPENDING TIME AND MONEY BUILDING RELATIONSHIPS VS SHOWING HOW SUCH AN APPROACH IS EFFECTIVE

In order to be successful, it is essential to be politically informed and skilled in order to negotiate, and to identify reformers and opportunities. However, this entails a large sunk cost in the form of the time that is needed to build alliances, identify or establish possible pockets of effectiveness, and build relationships with key reformers who are willing to trust you with industry information that can be used to press for reform from both within and outside of government. While this is being done, evidence of effectiveness will be minimal. The work is unproven up to the point at which the build-up of small-scale nudges for information and entry points for reform turns into a flash flood of change. This presents a major challenge for showing effectiveness, particularly in the early stages of the project.

#### 27.4.1.3 WORKING ON SUPPLY AND DEMAND AT THE SAME TIME

Part of the success of FOSTER arises because it is able to support reformers in government, while at the same time supporting pressure for reform from outside of government. However, there is an inevitable tension associated with working on supply and demand sides at the same time, since if government counterparts discover that the programme is also supporting advocacy initiatives by elements outside of government, this can erode their trust in the programme (and even spill over into problems for the relationship between the funder and the government). The more the programme is successful in working at the heart of government, the more this becomes an issue, particularly if demand-side partners are saying highly critical things about the government. This tension can be managed by creating a 'Chinese wall' within the programme between the supply- and demand-side work, and by ensuring good communication between the management of the programme and the partner organizations. FOSTER has gained from combining work on the supply and demand sides because this allows information-sharing and coordination between components that is not possible if they are separate. However, in some circumstances it may be necessary to separate such programmes to ensure that those working with government are not 'compromised' by advocacy initiatives funded by the same programme.

#### 27.4.1.4 ACTING POLITICALLY VS MAINTAINING DONOR POLITICAL RELATIONS

Programmes such as FOSTER explicitly attempt to engage in areas that are politically sensitive. They do not do so in a party political sense, but by choosing particular champions and partners, they are implicitly supporting the reform agendas of those actors. While aid is and always has been political,[18] if those opposed to reform become aware that a foreign donor is supporting their opponents, this can create problems for the diplomatic relationship between the two countries. Aid is only one part, and sometimes only a small part, of the diplomatic relationship between two countries, and so in such situations the Foreign Office of the donor country may put pressure on the donor agency to withdraw or modify activities to ensure no damage is done to broader diplomatic relationships. Conversely, there may be certain initiatives which are seen by the donor country as an important part of the relationship with the recipient country and which they wish to see continue, even if PEA or other analysis indicates that they are less directly targeted towards reform. Such tensions can be managed by good communication between the donor—who will be aware of the wider interests of its government—and the programme implementer, as well as substantial trust between the donor and the implementer that activities are not going to create significant problems for the donor.

#### 27.4.1.5 ADHERENCE TO DONOR PROGRAMMING INCENTIVES VS USING PEA AND FLEXIBLE PROGRAMMING

PEA is valuable when linked to forward programming, but there will be times when the PEA indicates that there is no space for reform at the present time. However, donors and implementers alike face pressure to spend programming funds: donors are typically incentivized to fully spend their budgets; implementing organizations may have a milestone-based contract tied to the disbursement of funds. Similarly, implementing organizations are frequently required by donors to meet their forecasted spend with limited or no variance. Not meeting this results in a perception by the donor of poor management on the part of the implementer, even if the mismatch results from a strategic choice not to implement activities which are no longer thought to be valuable. Implementing flexible and adaptive programming therefore requires a change in donor thinking and procedures about what constitutes good planning and financial management.

[18] See the excellent history of the politics of and in aid by Carothers and de Gramont (2013).

### 27.4.2 *Lessons for Donors*

To conclude, we draw out some lessons from FOSTER regarding the ways in which external donors can support improvements in the governance of the extractives sector.

Perhaps the most important lesson is that it is possible. There is an understandable degree of scepticism, both within the development community and outside it, about whether donors can have any significant influence on politically sensitive governance reforms such as those relating to the extractives sector. The experience of FOSTER suggests that it is possible to have an impact, but that doing so requires a rather different approach from that taken in traditional aid programmes. In particular, it requires a programme that 'thinks and acts politically', drawing on local knowledge to navigate the complexities of reform, with the flexibility to adapt its approach accordingly.

However, adopting this approach has risks. Most obviously, not every activity will be successful. The portfolio approach adopted by FOSTER deliberately experiments, trying a wide range of activities with different partners with the intention that they will collectively promote change. But it is almost certain that some of these activities will not be successful. Moreover, it is not even guaranteed that clusters of activities will collectively be successful, as the example of the Petroleum Industry Bill in Nigeria showed. Moreover, this approach requires care to ensure that such activities do not pose reputational risks for the donor.

Making such programmes effective therefore appears to require four things:

- *Local knowledge and leadership.* This entails not only a robust set of analytical frameworks, including PEA, to continuously understand the latest developments, but also a politically savvy local team with the autonomy to propose the key areas and approaches for intervention.
- *Acceptance of risk by the donor.* A key challenge here is to find the right mechanism to ensure accountability for taxpayer funds, whilst providing the necessary flexibility and autonomy for implementation teams on the ground. 'Arm's length' arrangements, in which the implementing organization is accountable to the donor while the local team is held accountable to the implementer, appear to be an effective way of managing these risks.
- *More flexible management processes.* The existence of an un-earmarked managed fund has been important for both programmes, allowing relevant initiatives to be defined as the programme goes along. But better systems are needed to manage the conflict between the donors' desire for complete, timely, and accurate fund disbursement and the variable needs and opportunities for reform.

- *Better evidence about what works, what doesn't, and why.* One of the challenges of an approach that consists of multiple, non-predetermined initiatives implemented by a variety of partners is that it is hard to evaluate impact. New monitoring, evaluation, and learning frameworks are needed for such programmes that focus not only on which types of initiatives are more effective in which contexts, but also on embedding continuous critical reflection into programme management so that lessons can be fed back immediately into practice.

Fundamentally, what is needed is a shift in mindset, from being an implementer of a project to being an enabler of reform. Long-term reform in politically sensitive sectors comes from the collective action of local coalitions, both within and outside of government. External programmes that attempt to provide a technical 'fix' for a sector are likely to fail. Donors therefore need to be more modest and more realistic about what can be influenced, and to become enablers of change—facilitating local actors to take advantage of windows of opportunity. The experience of FOSTER is that, where this approach is taken, donors can play a useful role in supporting significant improvements in the governance of the extractives sector.

## References

Africa Progress Panel (2013). 'Equity in Extractives Africa Progress Report: Stewarding Africa's Natural Resources for All'. Available at: africaprogresspanel.org/wp-content/uploads/2013/08/2013_APR_Equity_in_Extractives_25062013_ENG_LR.pdf.

Andrews, M. (2013). *The Limits of Institutional Reform in Development: Changing Rules for Realistic Solutions, Volume 1*. New York: Cambridge University Press.

Andrews, M., L. Pritchett, and M. Woolcock (2012). 'Escaping Capability Traps through Problem-driven Iterative Adaptation (PDIA)', Working Paper 299. Washington, DC: Center for Global Development. Available at: https://www.cgdev.org/publication/escaping-capability-traps-through-problem-driven-iterative-adaptation-pdia-working-paper.

Bhalla, J., N. Waddell, and R. Ough (2016). 'The Spoils of Oil: Working Politically on Extractives in Nigeria', in D. Booth (ed.), *Politically Smart Support to Economic Development: DFID Experiences*, 17–21. London: Overseas Development Institute.

Booth, D. (2013). 'Facilitating Development: An Arm's Length Approach to Aid'. London: Overseas Development Institute. Available at: https://www.odi.org/publications/7376-facilitating-development-arms-length-approach-aid.

Booth, D. (ed.) (2016). *Politically Smart Support to Economic Development: DFID Experiences*. London: Overseas Development Institute.

Bourgouin, F. (2011). 'The Politics of Large-scale Mining in Africa: Domestic Policy, Donors, and Global Economic Processes', *Journal of the Southern African Institute of Mining and Metallurgy*, 11: 525–9.

Buur, L., O. Therkildsen, M. W. Hansen, and M. Kjær (2013). 'Extractive Natural Resource Development: Governance, Linkages and Aid', Report 2013: 28. Copenhagen: Danish Institute for International Studies. Available at: //www.diis.dk/files/media/publications/import/extra/diis-rp-2013-28_extractive-dev_lbu-mfl_web_1.pdf.

Carothers, T. and D. de Gramont (2013). 'Development Aid Confronts Politics: The Almost Revolution'. Washington, DC: Carnegie Endowment for Peace.

DFAT (2014). 'Australian Aid: Promoting Prosperity, Reducing Poverty, Enhancing Stability'. Barton, Canberra: Department of Foreign Affairs and Trade.

DFID (2008). 'Private Sector Development Strategy: Prosperity for All: Making Markets Work'. London: DFID.

DFID (2010). 'Building Peaceful States and Societies'. Practice Paper. London: DFID. Available at: https://www.gov.uk/government/uploads/system/uploads/attachment_data/file/67694/Building-peaceful-states-and-societies.pdf.

DFID (2016). 'Smart Rules, Better Programme Delivery: Version VI, October 2016–March 2017'. London: DFID. Available at: http://www.gov.uk/government/uploads/system/uploads/attachment_data/file/557366/Smart_Rules-Oct2016.pdf.

Dietsche, E. (2017). 'Political Economy and Governance', WIDER Working Paper 2017/24. Helsinki: UNU-WIDER.

Dietsche, E., S. Dodd, D. Haglund, M. Henstridge, M. Jakobsen, E. Sindou, and C. Slaven (2013). 'Topic Guide: Extractive Industries, Development and the Role of Donors'. Oxford: OPM.

Faustino, J. and R. V. Fabella (2011). 'Development Entrepreneurship', in The Asia Foundation (ed.), *Built on Dreams, Grounded in Reality: Economic Policy Reform in the Philippines*, 253–72. San Francisco: The Asia Foundation.

Global Affairs (n.d.). 'Statement of Priorities'. Available at: http://www.international.gc.ca/development-developpement/priorities-priorites/index.aspx?lang=eng.

Global Witness (2010). 'Donor Engagement in Uganda's Oil and Gas Sector'. Available at: http://www.globalwitness.org/en/archive/briefing-ugandas-donors-must-push-oil-transparency/.

Hickey, S., A.-G. Abdulai, A. Izama, and G. Mohan (2015). 'The Politics of Governing Oil Effectively: A Comparative Study of Two New Oil-rich States in Africa', ESID Working Paper 54. Manchester: University of Manchester.

HM Treasury (2015). 'UK Aid: Tackling Global Challenges in the National Interest', Cm9163. London: HM Treasury and DFID.

Joshi, A. (2013). 'Do They Work? Assessing the Impact of Transparency and Accountability Initiatives in Service Delivery', *Development Policy Review*, 31(1): 29–48.

Katsouris, C. and A. Sayne (2013). 'Nigeria's Criminal Crude: International Options to Combat the Export of Stolen Oil'. London: Chatham House.

Lahn, G. and P. Stevens (2017). 'The Curse of The One-size-fits-all Fix: Re-evaluating What We Know about Extractives and Economic Development', WIDER Working Paper 2017/21. Helsinki: UNU-WIDER.

Leonard, D. (2008). 'Where Are "Pockets" of Effective Agencies Likely in Weak Governance States and Why? A Propositional Inventory', Working Paper 306. Brighton: Institute of Development Studies.

NORAD (2011). 'NORAD's Strategy towards 2015: Results in the Fight against Poverty'. Oslo: Norwegian Agency for Development Cooperation.

Stevens, P., J. Kooroshy, G. Lahn, and B. Lee (2013). 'Conflict and Coexistence in the Extractive Industries: A Chatham House Report'. London: Chatham House. Available at: http://www.chathamhouse.org/sites/default/files/public/Research/Energy,%20Environment%20and%20Development/chr_coc1113.pdf.

Thinking and Working Politically Community of Practice (n.d.). 'The Case for Thinking and Working Politically: The Implications of "Doing Development Differently"'. Birmingham: DLP. Available at: http://www.dlprog.org/research/thinking-and-working-politically-community-of-practice.php.

Whaites, A., E. Gonzalez, S. Fyson, and G. Teskey (2015). *A Governance Practitioner's Notebook: Alternative Ideas and Approaches*. Paris: OECD. Available at: http://www.oecd.org/dac/governance-peace/governance/governance-practitioners-notebook.htm.

# Part VIII

# Capturing Economic and Social Benefits at Community Level

# 28

# The Role of Participation in Sustainable Community Development Programmes in the Extractive Industries

*Catherine Macdonald*

## 28.1 Introduction

### 28.1.1 *Scope*

A central difficulty for almost all extractives activity is that, although the total benefits may be large, these benefits accrue predominantly at the national level and are typically concentrated in the hands of those in power and in capital cities. At the same time, the disruptions in terms of population movements, noise, air and water pollution, and so on are invariably highly localized close to the resource. Local benefits in terms of new employment opportunities frequently disappoint in terms of their magnitude. It has been recognized for many years, at least by those inside the industry, that these imbalances need to be addressed through active local policies and programmes. Increasingly, this corporate response is matched by complementary responses (either specific mandated requirements or discretionary approaches) by host governments.

Over the past decade, there has been an intensification of the efforts of extractives companies to demonstrate their positive contribution to their host communities. This is particularly strong in the remote and underdeveloped areas where many extractives projects are located. Often, an extractives project may be the major, if not the only, economic driver in a region. One action by the industry and by financial institutions has been the development of various international standards and recommended practices (Dodd et al. 2015: 15–17).[1]

[1] The IFC Performance Standards are the most commonly referred to standards for private-sector programmes. For guidance, in the mining sector there is the ICMM *Community Development Toolkit*

At the leading edge of this movement towards universal standards and guidelines has been the role of international finance institutions, such as the IFC and the Equator Principles banks, in insisting upon environmental as well as social responsibility as preconditions to project financing, the varying success levels of which are discussed in Bell (2017). The concepts of corporate social responsibility (CSR) and the social licence to operate (SLO) have gained great currency in recent years and combine to make it almost essential for major projects to manage the social and economic impacts of their developments, ideally leading to an improvement of the standard of living of local communities. In cases where economic and physical displacement or resettlement occur, international standards require that those displaced must regain at least the same standard of living as they had prior to disruption by the project—preferably an improved standard. With the recent addition of human rights and business requirements, a greater focus on the local and regional contributions of extractives projects has emerged.

One area of community engagement that is still deemed controversial by some extractives companies is the issue of free prior informed consent (FPIC) by communities, especially indigenous peoples. Bell (2017) discusses the question of community consent, and the international NGO Oxfam recently prepared a community consent index that analyses corporate commitment to FPIC (Oxfam 2015). This found that no oil and gas companies had made a public commitment to considering the FPIC of indigenous communities, although a growing number of mining companies had made such commitments: fourteen in 2015, compared with five in 2012, when Oxfam prepared the first consent index. Commitments to seeking FPIC from all communities, not just indigenous ones, is still a step too far for the industry, according to Oxfam, although many companies have made commitments to actively seeking community support and agreement before embarking upon operations (Oxfam 2015: 15–20).

The role of host governments in enabling positive local impacts of extractives projects is important. The contribution that can be made by civil-society organizations (CSOs) is also important. The ability of companies to align their efforts with both these key stakeholder groups is essential, as the harnessing of extractives revenues for local development is very much the product of combined and concerted efforts. Such cooperation is discussed in this chapter, and is analysed comprehensively in McPhail (2017), which focuses on multistakeholder processes.

Another trend that may bring benefits to local communities is the recent focus on the negotiation of formal agreements between companies and

(ICMM 2012), and for oil and gas, IPIECA's *Guide to Successful, Sustainable Social Investment* (IPIECA 2008). For further information see Dodd et al. (2015: 15–17).

project-affected communities. Otto's 2017 working paper on model community development agreements discusses this in depth (Otto 2017; see also Otto 2012 and Rio Tinto 2016). The primary aim of this chapter is to identify the optimal approach for companies to take in encouraging sustainable local development among host communities, with a secondary focus on the importance of alignment with other actors.

### 28.1.2 *Structure*

The chapter considers both the conceptual and the practical possibilities for the improved management of extractives resources so as to generate positive and sustainable economic and social outcomes for local populations. The *conceptual* dimensions of these topics are presented via a review of the key recent literature, with a view to determining where gaps have been identified in corporate performance. The *practical* aspects that give life to the scholarship are presented in the form of a long-term case study of the community development programmes of a mining project in Tanzania, with briefer references to other relevant case studies. Additional contextual information was obtained from an unpublished survey of key stakeholders with insights into industry experiences 'at the coal face', which focused on the constraints posed to local development in mining areas (ICMM 2015). This survey was conducted by the author in 2015 for the International Council on Mining and Metals (ICMM) for the information of its members, and is included here by kind permission of ICMM.[2]

The chapter considers the sustainability of local community development programmes: that is, their capacity to endure. There is a focus on community participation—that is, the ability of the intended beneficiaries of programmes to influence their design and implementation—and an assessment of whether this is a major success factor in programmes that have worked. In order to demonstrate what participation means in practice, I draw upon my own intermittent close observation of the community development programmes of a gold mine in Tanzania over a period of fifteen years.

The chapter discusses a number of aspects of community development activities by extractives companies, in both conceptual and practical terms. For example, the degree of autonomy that specific extractives projects have to interact directly with their host communities and local governments in order to be responsive to their needs is considered, as social investment programmes have a tendency to be driven from corporate headquarters, thus minimizing the importance of local content. The degree of inclusiveness of community

[2] Most of the discussion of the survey is contained in Appendix A of Macdonald (2017a).

development programmes in general—that is, whether a wide range of community members are able to participate—is also assessed.

In short, this chapter concentrates on the local and regional (provincial) aspects of the community development programmes of extractives projects in developing economies. It seeks to show how carefully crafted, locally appropriate programmes—especially those led and initiated by the local community members themselves—can make a positive contribution to improving the lives of host communities. The role that can be played by extractives companies is analysed, as well as the best approaches for aligning with other major players.

## 28.2 The Conceptual Dimensions

The literature over the past decade on local and community development associated with extractive industries projects has followed a trajectory from discussions on CSR up to, more recently, a focus on SLO. In most analyses of community development practices at the local level, writers have identified a lack of participation by local communities and their representatives as an obstacle to success. Alignment of development stakeholders, such as companies, local government, and civil society, is also considered, as are matters of resourcing. The most relevant articles are discussed below: that is, those whose primary focus is on 'participation'. This is now receiving renewed attention, having had a long history in the development field, where it predated both the CSR and SLO concepts. Although participation in community development has been the subject of much debate in development analysis for over forty years, it has only emerged in the extractives sector over the last few years: the probable reason for its rise to prominence in the current literature. The new interest of international donor agencies in engaging in programmes in the extractives sector may also be a catalyst for this renewed attention to participation.

As far back as 1969, Arnstein described a 'ladder of citizen participation' in which the eight rungs representing forms of 'participation' range from manipulation at the bottom to citizen control at the top, with consultation sitting in the middle (Arnstein 1969: 217). Somewhat more recent is Pretty's 1990s typology of participation, which ranges from 'passive participation'—in which people are told what will happen by external professionals who have already decided what must be done (not really participation at all, not even consultation)—to 'self-mobilization', in which people take initiatives independent of external institutions (Pretty et al. 1995: 61). Sustainable community development needs to be based upon the greatest possible participation of the intended beneficiaries—actual participation and planning collaboration—and this is where much of the current literature is focused.[3]

### 28.2.1 *Literature Review*

In their introduction to a special volume of the *Community Development Journal* (CDJ) focusing on the extractive industries, community development, and livelihood change in developing countries, Maconachie and Hilson (2013) emphasize the frequent disconnect between local needs and corporate programmes. The CDJ volume contains case studies from many parts of the world, but common themes emerge from the discussion on the deficiencies of many corporate efforts. The editors note that:

> Considerable concerns have been raised by a number of scholars who have highlighted how extractive industries companies often implement community development programmes with little knowledge or understanding of the socio-cultural contexts of the people's lives in which they operate. In extreme cases, an over-emphasis on meeting global performance standards has led to inappropriate and ill-conceived development outcomes at the local level, which have resulted in further fragmentation and inequality. (Maconachie and Hilson 2013: 351)

Thus, although much scholarship has discussed the increase of CSR activities worldwide, and the increase of corporate attention and expenditure being accorded to CSR performance, there is insufficient evidence that this is actually bringing benefits to the communities living near extractives projects. Indeed, Maconachie and Hilson (2013: 353) caution that even when citizens are deliberately invited to participate in 'the community development programmes sponsored by mining or oil and gas companies, the inclusion of local actors in decision-making cannot be assumed'. This observation sends up a flag that mere statements of participation, predicated upon efforts to invite local stakeholders to engage in corporate-sponsored activities, do not guarantee genuine opportunities for community members to steer their own path.

In the CDJ special volume, Banks et al. (2013) explore community development in the localities surrounding four large mines in Papua New Guinea (PNG), and conclude that two principal kinds of local community development emerge from the presence of major mining projects in remote and underdeveloped areas: immanent and intentional. The concept of 'immanent' development is that it takes place neither through the direct intervention of the mine management nor as the result of an explicit development programme,

[3] For a discussion of the development participation debate, see Cornwall (2009) and Cornwall and Pratt (2003).

but builds on the indirect economic growth associated with increased local employment, procurement, and royalties and taxes that accrue to the local area. These can be considerable in PNG, where customary landowners are legally acknowledged as having minerals rights. According to Banks et al. (2013), immanent growth makes up the major part of local and community development. The other form of community development they describe as 'intentional', using the term CCD, or corporate community development. This comprises the programmes and initiatives undertaken by the mining companies in pursuit of their CSR objectives. The authors view these programmes as conservative, often disconnected from the needs and wants of the intended beneficiaries, and hence less successful in terms of human development than actions taken independently by local people. They make a case for improving the outcomes of intentional CCD programmes in the following ways:

> Corporations should also seek higher levels of community participation in project identification, design, monitoring, and evaluation connected with their CCD. Much of the community input into CCD was weak (and seemed particularly so for higher level CSR-type activities), and there was typically little or no monitoring or evaluation of the development or livelihood outcomes of CCD activities. (Banks et al. 2013: 497)

Banks and his colleagues observe that much CSR in PNG was top-down, driven by global corporate imperatives, with little connection to the situation and needs of local communities in the mining areas. They note that it could be harder for community development departments and community members to obtain funding from the companies for locally driven initiatives than it was to gain grants for global corporate programmes such as HIV/AIDS projects. They advise that locally responsive, community-led programmes, especially when integrated with other local activities and developments, were more likely to produce sustainable benefits for the livelihoods of local communities.

In their study of the CSR performance of mining companies in Argentina, Mutti et al. (2012) take a strongly critical stance, accusing companies of undermining local capacity for self-management. They state that:

> CSR is perceived as patronising and paternalistic, when companies undermine knowledge and skills of local communities to identify their own needs and priorities. This leads to a lack of adaptation to local conditions, as companies appear to enforce their own vision of community 'good'. Mining companies are perceived as inflexible to engage with stakeholders, and incorporate community views in design of community development solutions [*sic*]. In terms of performance, the general view is that CSR does not have a substantial impact on poverty reduction or environmental management, and therefore, CSR outcomes have a negligible contribution to a society's welfare. (Mutti et al. 2012: 221)

They draw the conclusion that CSR programmes that are designed to suit the company, at corporate headquarters, rather than utilizing community input, are essentially a failure in terms of human development. Once again, the lack of community participation destroys the good intentions of the programmes.

In their study of the Bolivian context, Campero and Barton (2015) explain that the purpose of SLO is to:

> Generate a binding participation of the community, resulting in the construction of mutual agreements between community, firm, local, and/or central government, and which define the basis on which actions will be taken to promote sustainable local development. (Campero and Barton 2015: 174)

However, the reality they describe is a situation in which firms are voluntarily entering into activities designed to earn them an SLO, but only in a limited manner, as there is no compulsion for them to do so or regulation about how they do it. The limitations are primarily connected to a lack of inclusivity, as civil society actors likely to be antagonistic are often left out, as are many state actors, especially at the regional and local levels, in order to try to minimize costs and the demands that might otherwise be placed upon the company. The result is that there are insufficient levels of 'community participation in decision-making that affects local development', resulting in conflict and 'a persistence of weak local development trajectories within these communities' (Campero and Barton 2015: 168). Several stakeholders interviewed in the study commented that they recognized the need for greater inclusion, which they termed as going 'with God and the Devil', meaning engaging with difficult as well as amenable stakeholders. They also conclude that it was important to include local government representatives and officials, as well as regional ones where possible. One of the notable elements of this study is that it focuses firmly upon inclusion and participation as essential ingredients of the SLO.

Harvey and Bice (2014) advocate an approach to SLO that requires:

> 'collaborative moderation'—defined here as working directly with project-affected stakeholders to achieve accommodation and agreement on issues that are of priority in the local context, as opposed to attempts to respond to an array of deemed universal issues set by regulators and exogenous agencies.
>
> (Harvey and Bice 2014: 328)

They go further in recommending activities that align with the shared value concept made popular in 2011 by Harvard's Porter and Kramer, at least in its development of local economic clusters. The key point Harvey and Bice make for the enhancement of local development is that the emphasis should be on collaboration and mutuality, not on programmes designed externally to solve problems deemed to be universal. Although not explicitly using the term 'participation', that is precisely what they describe.

Maconachie (2014), in discussing youth participation in mining company programmes aimed at livelihood improvement in Sierra Leone, finds it 'useful to make a distinction between autonomous forms of community-led development "from below" and corporate controlled development "from above"' (Maconachie 2014: 281). This distinction echoes that of Banks et al. (2013), noted above, who refer to 'immanent' and 'intentional' community development. Maconachie is very critical of what he describes as decontextualized 'blueprint CSR schemes' and finds that programmes not designed organically with grassroots participation will be unlikely to meet local needs.

Keenan and Kemp (2014), in researching the role of gender in the process of negotiating agreements between companies and communities, discovered that women were more likely to engage in processes that were participatory, open, and transparent. This ties in with recent business scholarship encouraging diversity as beneficial to improved outcomes in most business and community processes. Thus, a participatory approach to community development programme design in the extractives sector is also likely to be more gender-balanced, a goal that is being embraced by an increasing number of extractives companies.[4] The importance of gender balance in participation is also highlighted by Aaron (2012) in his examination of the changing models of community development adopted by Shell and Chevron in the Niger Delta in the form of a Global Memorandum of Understanding (GMoU) signed in 2006. He concludes that, to a large extent, success had been achieved in attaining enhanced participation and ownership of projects. He then points out, however, that it was only men who had participated, in spite of measures in the agreement specifying that a certain number of women had to be involved. In Aaron's view, this suggests that 'while the GMoU model has opened up space for popular participation at the grassroots level, the problem of voicelessness of marginalized groups, such as women, has not been adequately addressed or addressed at all' (Aaron 2012: 270). For a discussion of recent developments in gender and the extractives sector, see Macdonald (2017b).

In an article centred on the local-level development practices of mining companies, Kemp (2010: 199) states that 'many mining companies now use the language of community development and participation as part of sustainable development (SD) and corporate social responsibility (CSR) frameworks'. And yet she found limited evidence of actual community participation on the ground, as companies remained reluctant to cede control over programmes. She also made the important point that 'participatory development requires engagement with individuals, families, households and small groups rather

[4] e.g. see Kemp et al. (2009).

than only engaging at the level of the "community"' (Kemp 2010: 208). This is necessary to avoid exacerbating structural disadvantages within communities by excluding certain sectors, especially the vulnerable and minorities. In her examination of community regeneration projects in the east Kent coalfields of the United Kingdom, Doering (2014) reinforces the importance of not assuming who or what is the 'community' or excluding from the participatory process individuals and groups that do not fit the paradigm of 'community' in the eyes of the company. This links back to the importance of engaging with 'both God and the Devil', as described by Mutti et al. (2012).

Khadiagala (2015: 37) exhorts African governments 'to promote wider participation in natural resource governance', which he sees as part of a movement towards greater transparency and inclusivity in all aspects of governance but as acutely needed in the natural resources sector, where increasing revenues are flowing in but not necessarily being equitably allocated. He advocates the decentralization of mining revenues and the training of local governments to enable them to manage those revenues, or their share of them, more effectively. Although his argument is generally pitched at a higher level than the community, his statement that 'the incorporation of diverse voices in political and economic processes is the next battle in broadening participation' (Khadiagala 2015: 39–40) is valid in the community context.

Suopajärvi et al. (2016) talk about the importance of 'fate control' or the ability for communities affected by mining in northern Europe (northern Finland, Norway, Sweden, and Russia) to guide their own destiny. In their view, the social aspects tend to be left out of discussions of sustainability in favour of economic and environmental aspects; if they are mentioned, it is usually only in terms of impacts or SLO. They state that 'local people need a platform [from which] to be heard during the whole lifecycle of the mine and to be involved in decisions that have impacts on the local communities and different stakeholders' (Suopajärvi et al. 2016: 66) in order for there to be sustainability of local communities. They are talking about not only the right of community members to be heard throughout the life of a mine but also their right to be involved in making decisions that affect them, a good description of participatory development planning. Although they find that modern mining company management recognizes these matters, Suopajärvi et al. (2016: 67) urge that special attention be paid to 'the local people's empowerment when extracting natural resources in the North'.

In an article about a Chinese state-backed mining project in Myanmar—a kind of project structure not often featured in the literature—Tang-Lee (2016: 29) emphasizes the importance of the 'empowerment of communities to participate in development decision-making, given the structured power asymmetry between the mine company and local communities'. She sees public participation as a way of achieving a more equitable distribution of both power and project benefits. Unfortunately, the hierarchical structure of both the Chinese company and the Myanmar ruling establishment has worked against allowing broad-based participation, resulting in tokenistic efforts primarily aimed at placating communities. Nonetheless, she sees that the company is trying to work towards more sustainable community engagement and urges it to continue to do so in line with broader industry practices.

Van Alstine and Afiotis (2013) present a case study of the Kansanshi copper mine in Zambia, in which they discovered an extremely centralized decision-making process. They report that local government councillors were told by the mine that discussions had been held between the company and the government at national level, so there was no space or need for their involvement. This had contributed to antagonism between the mine and the local council. The authors advise that 'factoring in local governments could immensely enhance the understanding of the manner in which the livelihoods of mine-affected communities could be improved' (Van Alstine and Afiotis 2013: 371). They also highlight human resource deficiencies in the community engagement staffing of mining companies, especially among mid-tiers and juniors. These limited resources prevented companies from being able 'to effectively engage with community and sustainable development issues' (Van Alstine and Afiotis 2013: 370).

In the Ugandan oil sector, Van Alstine et al. (2014) describe a situation where local government is left out of the planning process associated with the use of resource revenues, which is highly centralized 'despite the rhetoric of participation'. They observe that a lack of information and investment provided to local officials and communities resulted in their inability 'to take an active role in oil matters' (Van Alstine et al. 2014: 56). In their view, civil society should engage with the industry to improve the alignment of their CSR plans with national and district development plans and enhance their implementation, monitoring, and evaluation. They outline roles for increased participation by donors and civil society, who should 'target research-led engagement that will help overcome these growing information, monitoring, and participation gaps within local government' (Van Alstine et al. 2014: 57).

### 28.2.2 *Conclusion*

Extractive industry companies have moved a long way forward in recent years in accepting that they need to have CSR and must strive to earn and maintain an SLO. In order for the substantial investments in social programmes being disbursed by companies to have a reasonable chance of success and survival, corporates need to understand better the importance of locally designed and managed programmes, with community members at their core. Otherwise, there is a real danger that programmes will not be sustainable, or even appreciated locally, and that company efforts will be seen as corporate window-dressing for the global audience, particularly by the people who are supposed to benefit from the social investment programmes—local communities.

## 28.3 The Practical Aspects

### 28.3.1 *Industry Efforts to Improve Local Community Development Performance*

Although the literature is critical of company performance, the mining and oil and gas industries, particularly through industry bodies such as ICMM and IPIECA, have made considerable efforts to foster improved community development performance among their member companies, and there are examples of projects that have achieved good results. Both these organizations, plus international bodies such as the World Bank and the IFC, have issued guidance publications on a range of community relations topics, including working with indigenous peoples and protecting human rights. ICMM and IPIECA have both produced a large number of toolkits and guides that give practical advice about how to enhance company–community relationships, leading to sustainable development outcomes. These are freely available from their respective websites. IPIECA's *Guide to Successful, Sustainable Social Investment for the Oil and Gas Industry* advises that programmes developed in a participatory way are more likely to be sustainable (IPIECA 2008: 24–5). The ICMM *Community Development Toolkit* (ICMM 2012), first issued in 2005 and revised in 2012, also emphasizes participatory planning as a foundation for successful community development programmes, among many other useful tools and methods. Although there has been no formal monitoring of which companies use which processes in their community programmes, ICMM has found over the years that many companies have incorporated aspects of the Toolkit into their procedures and found the guidance useful.

In a review of the mining industry's contribution to social development internationally, ICMM (2014) found that for local community investments to succeed, there was a need for company activities to align with local governments and civil society organizations and that the capacity of local institutions varied greatly from country to country (ICMM 2014). Thus the challenges of working with local stakeholders on community development programmes could also vary greatly in different contexts and this could affect the potential for successful collaboration. Nonetheless, some of the case studies revealed that companies can and do work in a participatory manner with local communities, with beneficial results. See ICMM (2011) for a good example in the Lao PDR.

Another example of successful participatory planning in the oil and gas sector is that of the BP Tangguh LNG project in West Papua, Indonesia. Community Action Plans, developed through participatory processes, were adopted by project-affected villages and functioned as the basis for community development planning supported by the project for ten years (2004–14), as agreed to in the official Indonesian environmental permit (known as AMDAL). From 2015, BP has reoriented the process in the Tangguh Sustainable Development Program, with a broader focus on indigenous peoples in the wider Bintuni and Berau Bays area, rather than just the nearer villages, while maintaining an emphasis on community participation (BP 2015: 56–7). The new programme is also aimed at supporting the Indonesian government's increased involvement in participatory planning in the project area through several initiatives, working towards greater control by community and government and enhanced sustainability (BP 2015: 50–1).

These examples, although they may not yet represent the majority of community development programmes in the extractives sector, nonetheless demonstrate that when companies work in a participatory manner with local communities, there are positive effects that contribute to programme sustainability.

In an unpublished stakeholder survey in 2015,[5] ICMM discovered that companies needed to dedicate sufficient human resources to building and maintaining broad-based relationships with local communities in order to foster the genuine participation that can lead to sustainable local community development programmes (ICMM 2015). Such an approach builds solid relationships between companies and communities, and enhances the understanding of cultural contexts by corporate community relations personnel and others in the company, thereby leading to greater chances for genuine community participation and effective sustainable community development. Without this understanding, the durability of development programmes is likely to be compromised.

### 28.3.2 *Practical Participation*

There is much talk about participation, but little reflection upon what it means. It does not mean seeking agreement from communities to programmes or projects designed elsewhere, nor does it mean recruiting participants (confusing terminology) for programmes or projects conceived by others. Those may be

[5] See Appendix A of Macdonald (2017a).

appropriate actions to undertake for various activities but, when discussing the participation of local community members in deciding upon their own development priorities and plans, thought should be given to the fact that motivation to become involved must be allowed to emerge from the communities themselves.

Professor Glynn Cochrane, who established Rio Tinto's community relations system in 1995, has commented that 'in order to help the poor, it is necessary to know what they want and the extent to which they can help themselves with their own knowledge and social capital' (Cochrane 2009: 12). He noted further that offering poor people pre-designed projects and programmes to alleviate their poverty 'overlook[s] the fact that poor people have ideas, knowledge, and aspirations; that they wish to be treated with respect and as potential partners in any plans for them' (Cochrane 2009: 3), as he had observed in several decades of involvement in mining sector community development programmes. If they are to succeed in the long term, projects need to be chosen, designed, managed, and implemented by the beneficiaries themselves, with all necessary support from partners and donors.

Of course, if people neighbouring an extractives project are poor and have little infrastructure available, then it may seem easy to see what they apparently need—schools, roads, clinics, economic opportunities, etc. That is not the point. The point is that, in order for them to become economically empowered and self-motivated about their future development, they need to be enabled to plan and chart their own destiny, with the help and support of companies and other development agents. There are numerous examples demonstrating that what people assess as their own needs may be quite different from what 'experts' say they need. A particularly striking example of 'expert ignorance' is discussed in Box 28.1. This results in the failure of projects, which rarely receive community support once any initial grants, funds, or gifts from the project promoters are finished. The way to promote the sustainability of social investments is to empower people by giving them the choice of what projects they want to pursue, and the dignity and autonomy to make their own decisions. This is the essence of participation.

In a widely respected handbook on corporate community engagement, Zandvliet and Anderson (2009) advise the following:

> Company-aided community projects should always connect to and support communities . . . to reach their long-term aspirations. Independent outside NGOs or facilitators . . . can conduct assessments by using a participatory rural appraisal (PRA) to determine the aspirations of the community and to prioritize these based on available resources. . . . communities often prioritize training and strengthening of soft skills if they know a company is prepared to support them

**Box 28.1** THE FAILURE OF NON-PARTICIPATORY PROJECTS

The mismatch of 'expert' good intentions and the needs and priorities of local people has been clearly illustrated by Ernesto Sirolli, originally in his book *Ripples from the Zambezi* (Sirolli 1999) and more recently in a very engaging TED talk (Sirolli 2012). In these works, Sirolli recounts the series of project failures experienced by him and his Italian development colleagues in Zambia. They failed primarily because they did not consult the local people but tried to introduce projects that the community knew were not feasible. Few development practitioners have been so frank in their admission and analysis of their own mistakes, which makes Sirolli's work remarkable and valuable for understanding the need for participation for project success.

> over a longer period of time. Some companies assist local communities... in designing a longer-term development plan [which] can also help the community to solicit support from donors other than the company.
>
> (Zandvliet and Anderson 2009: 148)

This is exactly what is described in the case study below.

### 28.3.3 *Participatory Planning in Action: Tanzanian Examples*

By utilizing participatory planning methods to design social investment programmes in collaboration with local communities, extractives companies and development partners such as governments and donor agencies have a better chance of their investments becoming of sustainable value. This has also been recognized by progressive governments such as Tanzania's, which introduced a participatory planning process for three-year rolling development plans at village level in 2002, called Obstacles and Opportunities for Development (O&OD). See Fjelstad et al. (2010), JICA (2008), and SNV (2008) for research on this programme.

#### 28.3.3.1 CASE STUDY: PARTICIPATORY PLANNING IN NZEGA DISTRICT, TANZANIA

An Australian mining company, Resolute Mining Limited, was the first company to open a modern gold mine in Tanzania.[6] Resolute started producing gold in 1998 at its Golden Pride Mine in Nzega District, Tabora Region, in western Tanzania. In 2000, the company took an innovative decision to try to encourage local communities to empower themselves and take responsibility for their own development plans. Resolute elected to use PRA methods to establish the foundations of its community development programme and

[6] An early version of this case study appeared in ICMM (2012: 170) and an abbreviated version was included in Macdonald et al. (2015).

engaged two experienced Tanzanian facilitators to conduct a participatory planning exercise in the four villages closest to the mine's perimeter. The objective of the programme was to encourage the communities to develop their own Community Action Plans (CAPs), which the company could then work with and support.

This was the first time that Nzega community members had been asked what their development priorities were. Note that the Tanzanian government's O&OD programme had not commenced at this stage and had still not reached Nzega by 2008, so this was a new process for the district and village officials, as well as the community (JICA 2008). Resolute expanded the participatory planning process to cover a wider area and had underwritten community planning processes in twenty-nine villages by the end of 2006 (Mavura 2007). When communities had completed their CAPs, the mine's Community Development team worked with community leaders to select projects to cooperate on, in collaboration and alignment with local government development plans.

The company assessed that over time it had been involved with some activities that were successful, and others that were less so, stating: 'It is difficult to be prescriptive about what you should or should not do' (Sullivan 2014: 27). This is especially the case when enabling communities to decide upon their own priorities, and supporting activities selected by communities. However, in choosing to work with communities in a participatory manner and supporting their priority projects, as long as these are in keeping with the company's parameters for community investment, companies are guaranteed the support and involvement of community members. It is important to remember that throughout the world there is a high level of early failure among small businesses and ventures, so a certain level of project failure is not a reason for abandoning grassroots participatory projects and opting for large-scale 'expert-designed' projects instead. What is essential is that community entrepreneurs and local leaders learn from those mistakes so that they have a greater chance of subsequent success. This is true capacity-building.

The use of PRA in Resolute's community development programmes at its Golden Pride Mine in Tanzania, over a fifteen-year period, demonstrates that this approach can be undertaken in a low-key, cost-effective, and sustainable manner. The PRA approach allows people to build their capacity by making their own choices and learning through their own successes and failures. PRA methods specifically include women and a wide spectrum of society, to ensure that projects are broadly based and widely supported. Although there has not been a published study evaluating the outcomes of the Golden Pride participatory planning programme, the company in 2012 was awarded a Tanzanian Presidential Award for Empowerment Programs for

Communities—recognition by the government of its efforts in community empowerment (Resolute 2013).

The mine ceased production in February 2014 and completed planned closure arrangements and handover to the Tanzanian government in November 2015. This process was guided by a Statutory Mine Closure Plan agreed between the Tanzanian government and Golden Pride Mine management in 2011. It included community development commitments aimed at fostering alternative income generation and specifically mentioned that PRA was the basis of its planning process. Some examples of long-running community development programmes are the use of maize milling machines by small business operators near the mine site. This has improved the income-earning capacity for local farmers, as has an expanding project in pig husbandry that has improved the local livestock supply. The closure process incorporated the handing-over of social infrastructure—including a maternity ward at the local hospital, a medical laboratory at a health centre, and a number of school buildings—to the government, under the auspices of the National Closure Committee. The town water treatment plant was also part of the handover, as was the filling of the mine pit with water suitable for crop irrigation and livestock watering (Resolute 2014). Clearly, the ongoing management and maintenance of this infrastructure will need to be coordinated by government agencies, but the mine's management was hopeful that the years of community and local government capacity-building that they have supported through their participatory planning processes will lead to ongoing success.

#### 28.3.3.2 BUZWAGI GOLD MINE

Another Tanzanian mining sector example is that of African Barrick Gold (now Acacia Mining), which worked with the Kahama District administration in 2010 by supporting the O&OD process in the three villages most affected by its Buzwagi Gold Mine, sponsoring additional training in participatory rural appraisal (PRA) techniques by Professor Francis Lelo, an expert from Kenya. The mine then worked with local government to align the mine's community development programmes with village development priorities.[7] This raises another common theme of good local community development planning—the need to align programmes with existing schemes wherever possible.

[7] Interview with Ms Eliza Bwana, District Executive Director, Kahama, 29 June 2012.

### 28.3.4 *Summary*

Participatory planning, in which community groups determine their own development priorities and design their own community development projects, is an approach to social investment that is more likely to lead to sustainable, community-supported initiatives than projects designed by external experts. Extractives companies can work with their host communities to facilitate the participatory planning process, if needed, and should then cooperate with communities and local governments to work out the intersections between the varying priorities of the different stakeholder groups. The areas where the interests overlap will be the most fruitful ones for development cooperation. Some national governments, such as Tanzania's, have their own participatory planning processes in place (O&OD), so companies can work with local government and communities to discover areas of mutual interest for social and economic investment.

## 28.4 Conclusions

This chapter makes a case for elevating the status of communities and local governments in the development planning processes associated with extractives projects. This study finds that growing levels of corporate expenditure do not usually improve conditions for community members on the ground, largely because most programmes are designed without their involvement, and often do not reflect their needs and priorities.

In addition to identifying a need to increase human resources to enable greater interaction between companies and their local stakeholders, observers advise that expanded efforts to enhance alignment between companies and governments are essential. The decentralization of extractives revenues is also seen as a key step in increasing the capacity of local governments and communities to benefit from the resources sector.

Donor agencies could play a greater role in bringing their expertise in participatory planning, monitoring, and evaluation into the extractives arena. This could contribute much to improved outcomes for the neighbours of resource extraction projects. Although there is no doubt that extractives companies continue to devote resources to the human development of the countries in which their projects are located, a realignment that directs more attention to building relationships and inclusive planning processes at the grassroots level could make these corporate investments more effective. In that way, in the terminology of Banks et al. (2013), 'intentional' development could come closer to 'immanent' development in improving the lives of those affected by extractives projects.

## References


Aaron, K. K. (2012). 'New Corporate Social Responsibility Models for Oil Companies in Nigeria's Delta Region: What Challenges for Sustainability?' *Progress in Development Studies*, 12(4): 259–73.

Arnstein, S. R. (1969). 'A Ladder of Citizen Participation', *Journal of the American Planning Association*, 35(4): 216–24.

Banks, G., D. Kuir-Ayius, D. Kombako, and B. Sagir (2013). 'Conceptualizing Mining Impacts, Livelihoods and Corporate Community Development in Melanesia', *Community Development Journal*, 48(3): 484–500.

Bell, R. G. (2017). 'Protecting the Environment during and after Resource Extraction', WIDER Working Paper 2017/164. Helsinki: UNU-WIDER.

BP Indonesia (2015). 'Tangguh Sustainable Development Plan'. Available at: https://www.bp.com/content/dam/bp-country/en_id/indonesia/Documents/2015%20TSDP%20Book%20-Eng.pdf.

Campero, C. and J. R. Barton (2015). '"You Have to Be with God and the Devil": Linking Bolivia's Extractive Industries and Local Development through Social Licences', *Bulletin of Latin American Research*, 34(2): 167–83.

Cochrane, G. (2009). *Festival Elephants and the Myth of Global Poverty*. Boston, MA: Pearson.

Cornwall, A. (2009). 'Changing Ideals in a Donor Organisation: "Participation" in Sida', IDS Working Paper 317. Brighton: University of Sussex.

Cornwall, A. and G. Pratt (eds) (2003). *Pathways to Participation: Reflections on PRA*. London: Institute of Development Studies.

Dodd, S., M. Jakobsen, E. Dietsche, and C. Macdonald (2015). *Measurement and Reporting of Performance of Social Investment in Oil, Gas, and Mining Companies*. Oxford, London: OPM, DFID.

Doering, H. (2014). 'Competing Visions of Community: Empowerment and Abandonment in the Governance of Coalfield Regeneration', *International Journal of Urban and Regional Research*, 38(3): 1003–18.

Fjeldstad, O.-H., L. Katera, and E. Ngalewa (2010). 'Planning in Local Government Authorities in Tanzania: Bottom-up Meets Top-down', REPOA Brief 18. Dar es Salaam.

Harvey, B. and S. Bice (2014). 'Social Impact Assessment, Social Development Programmes and Social Licence to Operate: Tensions and Contradictions in Intent and Practice in the Extractive Sector', *Impact Assessment and Project Appraisal*, 32(4): 327–35.

ICMM (International Council on Mining and Metals) (2011). 'Utilizing Mining and Mineral Resources to Foster the Sustainable Development of the Lao PDR', Mining: Partnerships for Development, April. London: ICMM.

ICMM (2012). *Community Development Toolkit*. London: ICMM.

ICMM (2014). 'The Role of Mining in National Economies (2nd edition): Mining's Contribution to Sustainable Development'. October. London: ICMM.

ICMM (2015). 'Understanding Local Level Capacity to Participate in the Benefits of Mining', unpublished survey. London: ICMM.

IPIECA (2008). *Guide to Successful, Sustainable Social Investment for the Oil and Gas Industry*. London: IPIECA.

JICA (2008). 'The Study on Improvements of Opportunities and Obstacles to Development (O&OD)'. Planning Process Final Report Summary. Dar es Salaam: United Republic of Tanzania Prime Minister's Office.

Keenan, J. C. and D. L. Kemp (2014). 'Mining and Local-level Development: Examining the Gender Dimensions of Agreements between Companies and Communities'. Brisbane: Centre for Social Responsibility in Mining, University of Queensland.

Kemp, D. (2010). 'Mining and Community Development: Problems and Possibilities of Local-level Practice', *Community Development Journal*, 45(2): 198–218.

Kemp, D., J. Keenan, and J. Davidson (2009). 'Why Gender Matters: A Resource Guide for Integrating Gender Considerations into Communities Work at Rio Tinto'. Brisbane: Rio Tinto and CSRM, University of Queensland. Available at: http://www.riotinto.com/documents/reportspublications/rio_tinto_gender_guide.pdf.

Khadiagala, G. M. (2015). 'Global and Regional Mechanisms for Governing the Resource Curse in Africa', *Politikon*, 42(1): 23–43.

Macdonald, C. (2017a). 'The Role of Participation in Sustainable Community Development Programmes in the Extractives Industries', WIDER Working Paper 2017/28. Helsinki: UNU-WIDER. Available at: https://www.wider.unu.edu/publication/role-participation-sustainable-community-development-programmes-extractives-industries.

Macdonald, C. (2017b). 'The Role of Gender in the Extractives Industries', WIDER Working Paper 2017/52. Helsinki: UNU-WIDER. Available at: https://www.wider.unu.edu/publication/role-gender-extractives-industries.

Macdonald, C., X. Long, M. Jakobsen, S. Dodd, and S. Rai (2015). 'Extractive Industries and Social Investments: Principles for Sustainability and Options for Support', Flagship Report Paper 8. AfDB and BMGF. Available at: http://www.gatesfoundation.org/What-We-Do/Global-Policy/Natural-Resources/Background-Papers.

Maconachie, R. (2014). 'Mining for Change? Youth Livelihoods and Extractive Industry Investment in Sierra Leone', *Applied Geography*, 54: 275–82.

Maconachie, R. and G. Hilson (2013). 'Editorial Introduction: The Extractive Industries, Community Development and Livelihood Change in Developing Countries', *Community Development Journal*, 48(3): 347–59.

Mavura, M. (2007). *Golden Pride Project Community Development 2006 Annual Report*. Nzega: Resolute Tanzania.

McPhail, K. (2017). 'Enhancing Sustainable Development from Oil, Gas and Mining: From an "All of Government" Approach to Partnerships for Development', WIDER Working Paper 2017/120. Helsinki: UNU-WIDER.

Mutti, D., N. Yakovleva, D. Vazquez-Brust, and M. H. Di Marco (2012). 'Corporate Social Responsibility in the Mining Industry: Perspectives from Stakeholder Groups in Argentina', *Resources Policy*, 37: 212–22.

Otto, J. (2012). 'Community Development Agreements: Model Regulations and Example Guidelines', Report prepared for the World Bank (Oil, Gas, and Mining Unit). Available at: http://siteresources.worldbank.org/INTOGMC/Resources/mining_community.pdf.

Otto, J. M. (2017). 'How Do We Legislate for Improved Community Development?' WIDER Working Paper 2017/102. Helsinki: UNU-WIDER.

Oxfam (2015). 'Community Consent Index 2015: Oil, Gas, and Mining Company Public Positions on Free, Prior, and Informed Consent'. Available at: https://www.oxfam.org/en/research/community-consent-index-2015.

Pretty, J. N., I. Guijt, I. Scoones, and J. Thompson (1995). 'A Trainer's Guide for Participatory Learning and Action'. IIED Participatory Methodology Series. London: IIED.

Resolute Mining Limited (2013). *Annual Report 2013*. Perth.

Resolute Mining Limited (2014). *Annual Report 2014*. Perth.

Rio Tinto (2016). 'Why Agreements Matter: A Resource Guide for Integrating Agreements into Communities and Social Performance Work at Rio Tinto'. Available at: http://www.riotinto.com/documents/Rio_Tinto_Why_Agreements_Matter.pdf.

Sirolli, E. (1999). *Ripples from the Zambezi: Passion, Entrepreneurship, and the Rebirth of Local Economies*. Gabriola Island, BC, Canada: New Society Publishers.

Sirolli, E. (2012). 'Want to Help Someone? Shut up and Listen!' TedExEQChCh. Available at: https://www.ted.com/talks/ernesto_sirolli_want_to_help_someone_shut_up_and_listen.

SNV Netherlands Development Organisation (2008). 'Supporting Decentralised Planning in Tanzania', Mbulu Case Study. Dar es Salaam: Netherlands Development Organisation.

Sullivan, P. (2014). 'Bread from Stones: No Single Recipe', in A.Milligan (ed.), *Mining, Agriculture and Development: Bread from Stones?* Proceedings of the Joint Conference of the Crawford Fund and the Africa Australia Research Forum, 24–32. Perth, Western Australia.

Suopajärvi, L., G. A. Poelzer, T. Ejdemo, E. Klyuchnikova, E. Korchak, and V. Nygaard (2016). 'Social Sustainability in Northern Mining Communities: A Study of the European North and Northwest Russia', *Resources Policy*, 47: 61–8.

Tang-Lee, D. (2016). 'Corporate Social Responsibility (CSR) and Public Engagement for a Chinese State-backed Mining Project in Myanmar: Challenges and Prospects', *Resources Policy*, 47: 28–37.

Van Alstine, J. and S. Afiotis (2013). 'Community and Company Capacity: The Challenge of Resource-led Development in Zambia's "New Copperbelt"', *Community Development Journal*, 48(3): 360–76.

Van Alstine, J., J. Manyindo, L. Smith, J. Dixon, and I. Amaniga Ruhanga (2014). 'Resource Governance Dynamics: The Challenge of "New Oil" in Uganda', *Resources Policy*, 40: 48–58.

Zandvliet, L. and M. B. Anderson (2009). *Getting it Right: Making Corporate–Community Relations Work*. Sheffield: Greenleaf.

# 29

# Approaches to Supporting Local and Community Development

## The View from Zambia

*Angel Mondoloka*

### 29.1 Introduction

Mining is the largest sector of the Zambian economy and accounts for 75 per cent of exports and 10 per cent of total private-sector employment. As has been the case elsewhere in the world, however, the relationship between mining companies and their host communities has been fractious. In Zambia, that relationship has evolved at a rapid pace but with no clear consensus on the path towards sustainability.[1]

The end of the so-called super-cycle in 2015 resulted in massive job losses and significant reductions in social investments by the mining companies. In September 2015, for example, the Carlisa Investments-owned Mopani Copper Mines (MCM) informed the government that it intended to lay off more than 3,800 workers because of lower metal prices and high production costs due, in part, to electricity shortages. The challenges to the sustainability of the Zambian mining industry include this notable dependence on foreign investments and markets and the overall national context of poverty.

[1] I wish to acknowledge the contributions of the ILO and WWF to the social and economic development discourse in Zambia. I also express my gratitude to the many individuals and organizations that made this chapter possible through the valuable time, information, and insights they shared with me. This input and support was drawn from the relevant government departments and agencies, mining industry personnel responsible for CSER programming, civil society organizations, and leaders and members of the mining host communities discussed in the chapter. The findings, interpretations, and conclusions expressed in this chapter are mine and do not necessarily reflect the views of any of the parties mentioned above.

In 2011, the World Bank upgraded Zambia to lower-middle-income status but the country's social indicators continue to be unsatisfactory. The country's dire socio-economic situation is evidenced, in part, by a high HIV/AIDS prevalence of 11.6 per cent among adults aged 15 to 49 years,[2] 54.4 per cent of Zambians living below the poverty line (with 40.8 per cent living in extreme poverty),[3] high infant mortality at 45 deaths per 1,000 live births[4] (2015 MDG target: 35.7), an under-five mortality rate of 75 per 1,000 live births (2015 MDG target: 63.6), and a high maternal mortality ratio of 210 per 100,000 live births (2015 MDG target: 162.3). The adverse social and human development context is also underlined by rapid population growth, high rates of urbanization and unemployment, and low and uneven incomes, with high income inequality. Zambia's Gini coefficient of 69 per cent[5] compares with an average of 45 per cent for sub-Saharan Africa.

On the environment front, the country has shortcomings that are exacerbated by pollution arising from mining activity:

> Decades of copper, cobalt, zinc, and lead mining have left many areas of the country contaminated with poisonous substances. Impacts include air pollution from the fumes, gases, and dust; soil contamination from hazardous effluents; water (surface and ground) pollution from effluents, waterways from mines, plants, and dumps; destruction of vegetation and wildlife habitat due to subsistence; deforestation, fumes, and direct health hazards. (Smit 2013: 251)

Zambian copper ore bodies mostly consist of sulphides, consequent to which the copper extraction process produces sulphur dioxide fumes, which cause respiratory and other health problems in people. In addition, the acid rain that results from these toxic fumes causes the acidification of surrounding soils and loss of vegetative cover.

### 29.1.1 *Historical Context of Social Responsibility in Zambian Mining*

Zambia's history of mining dates back to the 1920s, when the country then known as Northern Rhodesia was still a British colony. The mining industry has been the main engine of the country's socio-economic development through a combination of tax contributions and the delivery of social services to its host communities. However, changes in ownership and governance of the industry over the years, together with the cyclicality of commodity prices on the global markets, have had implications for the nature, scale, and scope of the social services delivered by the industry.

[2] ICAP at Columbia University (2016). [3] CSO (2016).
[4] 2015 (CSO, MOH, ICF 2014). [5] CSO (2016).

#### 29.1.1.1 CORPORATE SOCIAL RESPONSIBILITY BEFORE PRIVATIZATION

Prior to independence and the 1969 nationalization of the industry involving the transfer of 51 per cent of all existing mines to the state, the mines in Zambia were owned by the Anglo American Corporation (AAC) and Roan Selection Trust (RST). These two private companies provided housing, education, health, recreation, waste management, and water and sanitation services as part of a comprehensive employment package for their workers. Fraser and Lungu (2007: 8) note that as early as 1929 the private mining companies had become responsible for a wide range of social and medical facilities for both their workers and managers. These included sanitary and orderly housing compounds for employees; food rations such as maize-meal, millet, rice, beans, meat, and fresh vegetables; hospitals, with competent medical personnel at all mining settlements; as well as recreation clubs offering many sporting and entertainment activities.

Post-nationalization, the government continued this social welfare programme and began to expand and leverage it for reasons of political expediency. In 1982, the two nationalized companies were combined to form Zambia Consolidated Copper Mines (ZCCM) which, in effect, became a conduit for this expanded state largesse under the guise of corporate social responsibility (CSR). As Fraser and Lungu (2007) explain in greater detail, ZCCM literally operated 'a cradle-to-grave' welfare policy, even subsidizing burial arrangements for the dead. In this respect ZCCM's actions reflected the development philosophy of the post-colonial Zambia, and were even more generous in scope than those seen during the colonial period—offering things such as free education for miners' children, alongside subsidized housing and food, electricity, water, and transport. Further, as Fraser and Lungu also explain in some detail, the mines extended these social supports to the whole communities in which they had a presence. ZCCM's activities included the maintenance of local roads, refuse collections, and catering facilities. The company also supported shops, farms to supply food to the mine areas, other industrial activities, youth training and development schemes, and women's clubs for home-crafts. Additionally, in mining areas and towns such as Nchanga and Konkola, by the time of privatization, ZCCM had effectively replaced government as the operator of hospital facilities with one or two hospitals at each of its operating divisions; see Fraser and Lungu (2007: 8).

Following the oil crises of 1974 and 1979, Zambia fell into a crippling debt crisis while the country's terms of trade began a declining trend as copper prices collapsed for the next two decades, at the end of which Zambia was ranked among the poorest countries in the world. Predictably, this placed a severe strain on social services provision by ZCCM, which nonetheless continued to be treated as a 'cash cow' by the government.

#### 29.1.1.2 THE ZCCM PRIVATIZATION AND ITS CONSEQUENCES FOR CSR

The coming of multi-party democracy to Zambia in 1991 gave birth to an increasingly active and competitive political arena with a multiplicity of political parties. Economic liberalization, meanwhile, introduced new players and services in industries such as telecommunications, which has achieved one of the highest rates of mobile phone penetration in Africa in recent years. The previous UNIP (United National Independence Party) government had drawn much of its patronage and power from the country's state-owned industrial base, of which the largest and most strategic enterprise was ZCCM. Thus, an aggressive privatization campaign was initiated in 1992 to sell 280 state-owned companies, including the mines. Between 1997 and 2000, ZCCM was split into seven units and sold.

The final and most important stage of privatization was the negotiation and signing of the, now superseded, Development Agreements with each of the companies. 'These secret documents established the terms under which the mines were sold, and the rights and responsibilities of the Zambian state and the new mining companies. The original agreements were negotiated between 1997 and 2000' (Fraser and Lungu 2007: 15). Among the key provisions in the Development Agreements were those on taxation and the environment. To promote greenfield investments and reinvestment during the decline in copper prices at the time, corporate income tax rates were substantially reduced and sales tax was replaced by value added tax (VAT).

On the environment front, during the ZCCM era, the government had set limits, with corresponding penalties, on the amount of pollution the mines could discharge into the rivers and atmosphere. The Development Agreements contained significant exemptions from these provisions as long as the private companies did not exceed the limits that had been set for ZCCM. The exemptions were granted subject to the company's submission, to the Environmental Council of Zambia (ECZ),[6] of an Environmental Management Plan (EMP) whose implementation would be subject to self-reporting, by the company, and monitoring by ECZ.

Although there was a widely held assumption that the Development Agreements did not bind the private mining companies to continuing ZCCM's legacy of corporate social and environmental responsibility, it was found, for example, that Mopani Copper Mines' Development Agreement included the following clause:

> GRZ wishes to ensure that the continued development and exploitation of the commercial deposits of copper and cobalt ore... together with the development

[6] Now the Zambia Environmental Management Agency (ZEMA).

> and operation of the smelter, refinery, concentrators, and cobalt plant will secure the maximum benefit for, and adequately contribute to the advancement and the social and economic welfare of, the people of Zambia, including the people in the vicinity of the Contract Area in a manner consistent with their needs and the protection of the environment and, at the same time, secure an appropriate return on investment for the company. (Fraser and Lungu 2007: 16)

Therefore, it is probable that the Development Agreements did attempt to transfer ZCCM's corporate, social, and environmental responsibility (CSER) legacy to the new owners and that any gaps noted have been due to failures in implementation and regulation.

#### 29.1.1.3 CSER SHORTCOMINGS FOLLOWING PRIVATIZATION

In 2001, Patricia Feeney, then with Oxfam, observed that the 1995

> privatisation of Zambia Consolidated Copper Mines (ZCCM), which was actively encouraged by the donor community as a means of reducing corruption, inefficiency, and waste, was supposed to bring new investment to the Copperbelt, reinvigorating the local economy and restoring its severely degraded and contaminated environment. In reality, despite all the talk about corporate social responsibility, the new mine owners have shown themselves remarkably reluctant to assist cash-strapped local councils [to] improve social service provision. They are even less inclined to shoulder some of the burden of cleaning up ZCCM's environmental legacy. (Feeney 2001)

Just over a decade later, in 2013, in what she termed 'yet another example of the resource curse', Zarina Geloo echoed this observation and commented that despite Zambia's copper industry doing extremely well—and the country now being the third largest copper producer in the world, with record production of over 700,000 tons in 2011—the promise that the industry would provide a huge boost to the government's tax revenues while 'pumping huge amounts into local communities, thanks to the law that five percent of mining revenue must be ploughed back into them', had, in fact, not materialized:

> Outside the fences and fortified walls of the mines, communities continue to suffer from acute poverty and a miserable quality of life—with little access to basic services and even less hope. Firstly, the government has not enforced the five percent rule. Instead, mines have only been made to undertake some kind of corporate social responsibility scheme—usually building a few schools and clinics or digging some boreholes. Seldom is there any genuine consultation with the people in the communities. And very rarely do these gestures foster any long-term change. (Geloo 2013)

Geloo noted further that many workers were retrenched when the mines were privatized. On the ongoing decent work deficit in the industry, she noted that many of the current employees in the industry were low-skilled casual labourers, who were often paid below the minimum wage and were not entitled to accommodation, health care, or education allowances.

Due to the aggressive urbanization associated with mining communities, combined with the lack of technical and financial capacity among the local authorities in these areas, uncontrolled illegal settlements have mushroomed with little or no water, sanitation, or other services. Correspondingly, high unemployment rates in these areas have created soaring rates of crime and various forms of institutional and petty corruption. The reality on the ground in the Copperbelt has belied the general expectation that private investment in the mining sector would translate into job creation for workers and a bigger market for locally produced goods and services. There are grave concerns about the conduct of the mining companies with respect to the labour, health and safety, and environmental laws of the country.

Among the environmental impacts suffered by mining communities are large amounts of toxic sulphur dioxide fumes that are emitted into the air by mine smelters, resulting in serious respiratory and eye problems, which have reportedly caused some deaths. These toxic fumes also cause 'sulphuric' acid rain, which has rendered the soil infertile over vast areas of the country, with adverse consequences for smallholder farmers and other agricultural activity. Another environmental impact has been the contamination of water bodies (such as the Kafue River) and wetlands due to the discharge of toxic waste from the mines' operations. Without external assistance from organizations such as the Southern Africa Resource Watch (SARW), the communities that host these mining operations have generally been powerless to engage the mining companies and other stakeholders in viable debate in order to achieve equitable outcomes. This lack of negotiating power and effective stakeholdership—'abjection'[7] in the terminology of James Ferguson (1999: 236)—on the part of the mining communities has made the resource curse a painful and inescapable reality for them.

There has also been a history of strained relationships with the Zambian government, which is on record as having accused foreign-owned mining businesses of not contributing sufficiently to socio-economic development in Zambia by expatriating the lion's share of their profits, failing to give business to local companies, and flouting tax regimes. For their part, the mining companies have blamed the government for creating volatility in the business environment by unilaterally decreeing minimum wage increases,

[7] 'Abjection can be understood as the process by which mineworkers have been "thrown aside, expelled or discarded" during the economic liberalisation and privatisation of the Copperbelt, leading to a sense of "debasement and humiliation"' (Ferguson 1999: 236).

introducing ad hoc statutory instruments to regulate exports, and, allegedly at one stage, openly encouraging industrial action among mine workers.

Meanwhile, foreign investors and workers have on different occasions been embroiled in a protracted industrial conflict involving illegal strikes and increasingly militant rhetoric from labour and business leaders. There have also been isolated incidents of violence, resulting in deaths of both workers and management representatives, as described in Section 29.1.1.4. Foreign investors widely complain about low labour productivity and the high cost of labour and other operating costs, while workers stress poor working conditions and occupational health and safety concerns. The backdrop of strained industrial relations, mine accidents, incidences of worker abuse, and environmental damage resulting from mine companies' operations has increasingly turned public opinion against foreign direct investors in the industry. The sub-regional debate on resource nationalism is also contributing to the challenges to the sustainability, broadly defined, of mining in Zambia.

#### 29.1.1.4 CHINA'S MINING FOOTPRINT IN ZAMBIA

China entered Zambia's copper mining industry with the 1998 acquisition of the Chambishi copper mine by China Nonferrous Metal Mining Corporation (CNMC) on behalf of its subsidiary Non-Ferrous Company Africa (NFCA). To date, China's mining firms have invested more than US$1 billion in Zambia's copper sector. Politically, China's engagement with Zambia has become an increasingly contentious issue, with claims that Chinese investments are exploitative and that the Chinese mistreat Zambian workers. The following incidents are illustrative:

- The largest health and safety incident since privatization occurred at BGRIMM Explosive (Zambia), a subsidiary of NFCA, when its explosives manufacturing plant in Chambishi was levelled in an explosion that took the lives of around fifty Zambian workers in April 2005 (Fraser and Larmer 2010: 96).
- In 2010, Chinese executives opened fire on workers protesting against poor pay and conditions at the Collum coal mine in Sinazongwe district, in the Southern province. Eleven miners were reportedly hospitalized with wounds.
- In late 2011, Human Rights Watch issued a scathing report on Zambia's copper industry that accused Chinese firms of a number of abuses, including forcing workers to work 12-to-18-hour shifts (Zambian law limits shifts to eight hours) and endangering miners' health and safety by failing to replace damaged equipment and providing inadequate ventilation underground.

- In 2012, a Chinese national was killed during a riot by Zambian workers at Maamba Coal Mine about 250 km from Lusaka. This was consequent to failure by the mine's Chinese owners to pay the, then just implemented, minimum wage of K1,000 for general workers.

In response, the Chinese—who have been lauded by African leaders for their 'no strings attached' approach to aid and investment on the continent—have worked tirelessly to strengthen their relationships with the government regardless of party affiliation.

##### 29.1.1.5 THE AUDITOR GENERAL'S 2014 REPORT ON ENVIRONMENTAL DEGRADATION

In response to the failure of the mining industry to achieve socially and environmentally sustainable mining practices, the Zambian government's Office of the Auditor General (OAG) undertook an audit of the environmental practices of mining companies in Zambia in 2014, together with an assessment of the government's regulatory capacity (OAG 2014). The factors motivating the audit also included public interest concerns raised by several stakeholders, who cited issues such as lack of integrated mining policies, failure to disclose Developmental Agreements, failure by the mines to mitigate the effects of environmental degradation, and unfair resettlements. The main audit findings included the following:

a) Mining companies are not complying with the environmental rules, laws, regulations, and licensing conditions set by government;
b) Measures put in place by government to ensure that the environmental degradation caused by mining activities is adequately managed (or mitigated) are not working effectively; and (consequently),
c) government's national development priority—as set out in the *National Policy on Environment*—of balancing economic growth with environmental protection, to improve the overall quality of life of the people of Zambia, is not being achieved (OAG 2014).

### 29.1.2 *The Zambian Policy Context*

The Zambian government's *Vision 2030* document expresses the national aspiration to become 'A Prosperous Middle-Income Nation by 2030'. The strategic roadmap towards this goal is provided by the government's system of five-year plans, of which the Sixth National Development Plan (SNDP) 2011–15 was superseded by the Seventh National Development Plan (7NDP) 2016–20. In August 2013, the government had issued the Revised Sixth National Development Plan 2013–16 (R-SNDP) with the theme

'People-Centred Development' to focus on inclusive growth, rural development, and job creation in the light of the government's 'realisation that economic growth alone does not inherently contribute to improvements in human development and poverty reduction'.[8] Specific R-SNDP objectives included increased investment in capital projects and programmes in agriculture, infrastructure development, human resource development (focusing on education and skills development), quality health services, water and sanitation, and energy. Through R-SNDP, the government also aimed to accelerate its implementation of the decentralization policy (GRZ 2013).

## 29.2 Mining Impacts and CSER

Mondoloka (2017: Figure 1) illustrates the multidimensional nature of mining's social, economic, environmental, and other impacts at the local, regional, and national levels. CSER is a way for mining businesses to ensure their active compliance with the spirit of laws, ethical standards, and international norms in order to enhance the social and environmental sustainability of their operations. Part of the challenge in Zambia has been the lack of a consensus on CSER and, correspondingly, the lack of a legislative framework for it. In consequence, Zambia navigates CSER through a smorgasbord of legislative and regulatory frameworks that are focused on labour and enterprise development issues. The government's contention has been that actions that are legal obligations cannot be considered to be CSER and has, correspondingly, resisted calls to legislate CSER and make it a legally binding requirement.

Social responsibility, as defined by ISO 26000, is the responsibility of an organization for the impacts of its decisions and activities on society and the environment through transparent and ethical behaviour that: (i) contributes to sustainable development, including the health and welfare of society; (ii) takes into account the expectations of stakeholders; (iii) is in compliance with applicable law and consistent with international norms of behaviour; and (iv) is integrated throughout the organization and practised in its relationships. Sustainable development is about meeting the needs of society while living within the planet's ecological limits and without jeopardizing the ability of future generations to meet their needs (World Commission on Environment and Development 1987). In its *Community Development Toolkit*, the International Council on Mining and Metals (ICMM 2012: 15) defines community development as 'the process of increasing the strength and

[8] Hon. Alexander Chikwanda, Minister of Finance, in a speech at the National Stakeholders Meeting on the Revised Sixth National Development Plan 2013–2016, 22 August 2013.

effectiveness of communities, improving people's quality of life, and enabling people to participate in decision-making to achieve greater long-term control over their lives'.

### 29.2.1 *The Growing Prominence of CSER in Zambia*

As noted earlier, democratization and economic liberalization heightened the public's general awareness of the political, economic, and social issues affecting them. This resulted in a more activist stance by the general public towards the mining companies. Thus, the democratization of information in Zambia, through the political process and the encroachment of technology, has contributed to increased public awareness of the issues and helped create the demand for more sustainable practices and win–win outcomes. In the meantime, there has been a growing recognition among corporate leaders in the industry of the increasing significance of the environmental, social, and economic challenges facing, or being created by, mining operations. This has been partly influenced by the mines' exposure to the international capital markets, where precedents and best practices are set in the context of more advanced legal and regulatory frameworks shaped by heightened sensibilities towards the social and environmental sustainability of economic activity.

In view of the significance of its mining footprint in Zambia, it is instructive to note that China's government has taken steps in recent years to promote the social and environmental sustainability of Chinese investments both at home and abroad. The previously noted 'no strings attached' approach, with its reliance on the acquiescence of political and economic elites, has proved increasingly unsustainable owing to the rise of the grassroots activism noted earlier.[9] In particular, the Chinese government has been encouraging Chinese companies in Africa to adopt more sustainable practices out of concern over the potentially adverse impacts of negative media reports (Russell 2007).

A salient example of this trend is the promulgation, by the Chinese government through the China Chamber of Commerce of Metals Minerals and Chemicals Importers and Exporters (CCCMC 2014), of its *Chinese Guidelines for Social Responsibility in Outbound Mining Investment* (*Chinese Mining Guidelines* for short) in Beijing on 24 October 2014. The guidelines represent the first industry-specific guidance on social responsibility for the Chinese mining industry. CCCMC is a subordinate unit of the Chinese Ministry of Commerce. It represents a membership of more than six thousand companies, the majority of which are Chinese mining companies investing abroad and trading

[9] Chinese mining companies have notably also exhibited better corporate citizenship in jurisdictions, such as South Africa, that have high standards of governance than in other African states where laws and enforcement mechanisms are comparatively weak.

mineral, metal, and hydrocarbon products. The guidelines require Chinese mining companies to integrate social and environmental considerations into their investment decision-making and operations abroad. Their development was informed by a large body of internationally recognized social and environmental responsibility standards.

In the light of the above imperatives, among others, CSER has taken on a new prominence for business owners and managers out of concern for how sustainability issues affect their social licence to operate (SLO) and, hence, the bottom line. In this context, mining companies in Zambia are increasingly looking beyond their traditional business models to integrate social and economic development imperatives into their business strategies and operations.

### 29.2.2 *The Lingering Crisis of Expectations Following the Privatization of ZCCM*

The ZCCM era, which lasted from the 1970s until the commencement of privatization in the mid-1990s, was a golden era for government largesse disguised as CSR. ZCCM paid no taxes but built the infrastructure, and provided practically all the social services, that were, ordinarily, the responsibility of the local authorities in the mining areas. The result has been a lingering confusion in the minds of the public as to the nature of CSER and the respective responsibilities of the government, on the one hand, and the mining companies, on the other. Mining companies in Zambia are, therefore, faced with a massive expectations gap, which is worsened by the government's continuing failure to undertake the necessary development projects and deliver corresponding social services in mining communities.

## 29.3 Case Studies in Zambia's Mining Communities

This section narrates the CSER approaches adopted by two multinational mining corporations and discusses the advantages and disadvantages of their approaches. Kansanshi Mining Plc is a subsidiary of First Quantum Minerals Ltd (FQM). Nonferrous Company Africa Mining Plc (NFCA) is a subsidiary of CNMC.[10]

FQM's sustainability strategy is underlined by a commitment to progress towards sustainable development through economically viable investments,

[10] Mondoloka (2017) discusses these two companies, together with the CSER approaches of an additional three companies: Kalumbila Minerals Ltd, Lumwana Mining Company, and China Luanshya Mine (CLM). These are subsidiaries of FQM, Barrick Gold, and CNMC, in that order.

technically appropriate operations, environmentally sound practices, and socially responsible actions. The company has been relatively successful in mainstreaming its processes for operating its sustainability programmes internally and across its community footprint.

The vision of Barrick Gold, the Canadian mining conglomerate, is 'the generation of wealth through responsible mining—wealth for our owners, our people, and the countries and communities with which we partner' (Barrick Gold Corporation 2016). Its statement on sustainability reads in part:

> As a company and as individuals, we must guide our conduct by the highest standards of honesty, integrity, and ethical behaviour. At Barrick, we are committed to building, operating, and closing our mines in a safe and responsible manner. To do this, we put a priority on developing long-term and mutually beneficial relationships with host governments and communities while working to minimize and mitigate the social and environmental impacts of our activities. (Barrick Gold Corporation 2016)

CNMC issued its first country-level *Social Responsibility Report* for Zambia in 2011, covering its nine subsidiaries in the country. The report asserted its compliance with the *CSR Guidelines for State-owned Enterprises* issued by the Chinese government's State-owned Assets Supervision and Administration Commission (SASAC) of the State Council. The report was also prepared in consultation with ISO 26000 and the *Sustainability Reporting Guidelines* of the Global Reporting Initiative (GRI). As part of the report, the 'CSR Declaration by CNMC Enterprises in Zambia' reads in part:

> We are committed to creating a model of 'China and Zambia Cooperate for Common Development'... We promise to operate in a responsible manner, and fully implement the sustainable development strategy during our strategic planning, decision making, and daily operation. Operating in a transparent way: regularly making public our status quo, planning, and measures in sustainable development; improving communication channels and dialog mechanism; and accepting the supervision by stakeholders on our own initiative. (CNMC 2011: 12)

### 29.3.1 *FQM's Kansanshi Mining*

Kansanshi Mining began its operations 8 km to the north of Solwezi in 2005. It is capable of producing 340,000 tonnes of copper and over 120,000 ounces of gold per year. Kapijimpanga and Kimasala Wards are in proximity to Kansanshi Mine. They have populations of 16,747 people (3,353 households) and 36,287 (7,167 households), respectively (CSO 2010). Solwezi, historically a farming community, is the capital of the North-Western province of Zambia and is a growing mining investment area. The town has never had good roads or other infrastructure and is now suffering from problems caused by ultra-rapid urbanization. The rapid influx of jobseekers, for example, has outstripped the supply of housing.

#### 29.3.1.1 KANSANSHI MINING'S CSER INTERVENTIONS

In addition to investments in social services, such as health and education, Kansanshi Mining has implemented a number of sustainability programmes designed to: (i) mitigate the displacement of indigenous populations; (ii) strengthen employability, entrepreneurship, and agriculture-based livelihoods in the area; (iii) prevent, mitigate, and monitor environmental impacts; and (iv) conserve local wildlife.

To mitigate the geographical and socio-economic displacement that resulted from the creation of its Tailings Storage Facility, the company allocated land within the existing mining licence area to 267 affected households. A Resettlement Working Group (RWG) was formed to agree, through public consultations, on a framework for compensation. The RWG also ensured that information was disseminated back to the other project-affected households.

In the area of skills development for employment and entrepreneurship, the company entered an agreement with the Solwezi Trades Training Institute (SOTTI) in 2011 and invested US$1.4 million in the Kwambula Learnership Programme aimed at providing participants with the opportunity to get a valid craft qualification and, with it, a better chance at getting a good job with the mine or with other employers.

On the agriculture front, the company is running a very successful programme to train local subsistence farmers in conservation farming. The company's target is to train 2,000 farmers from the surrounding rural area, of whom more than 300 have already been trained. Yield increases of up to eight times have been reported by the farmers, and the costs of training are covered in full by the company. This is a promising opportunity for income, food, and nutritional security for the local community through profitable participation in non-core and non-mining value chains.

In the realm of environmental management, the company commissioned environmental impact assessments (EIAs) of various projects, including the New Sulphide Storage Facility, the 12 mtpa[11] Copper Oxide Ore Treatment Facility Upgrade, the No. 5 Sulphuric Acid Plant, the Proposed Copper Smelter, and the Kansanshi Mine Decommissioning and Closure Plan. In addition, the mine's operations are subject to regular environmental monitoring to measure ambient air and dust and noise levels within and near to the mine. The mine has also developed procedures based on the ISO 14001 Environmental Management System (EMS) for the transportation, handling, and storage of acids, mill

[11] million tonnes per annum.

reagents, and other hazardous materials. Being upstream of Solwezi town on the Solwezi river, the main source of water supply in Solwezi district, the mine has been careful to ensure its compliance with Zambian and international effluent discharge standards.

As part of its non-mining value chain community development interventions, the company has implemented an important wildlife conservation programme. The Game Management Area (GMA) within the Kansanshi mining concession has substantial stocks of animals. The company has rolled out a successful anti-poaching programme, with a full-time anti-poaching unit and a community-based wildlife conservation officer, who conducts information sessions with inhabitants of surrounding villages to encourage a culture of nature conservation.

#### 29.3.1.2 PERSPECTIVES AND REACTIONS OF THE SOLWEZI COMMUNITY

While it is clear that the mine has brought new life to Solwezi town, the local community's feelings towards Kansanshi Mining Plc have ranged from ambivalence to outright hostility. Although the mine has created jobs and business opportunities, unemployment remains high, due partly to the influx of jobseekers and vendors from other towns and partly to the high illiteracy levels in the community. Youths bear the brunt of this impact, which has led to delinquent behaviour such as alcohol and drug abuse, and there has not been adequate investment in corresponding social programmes to address these problems. Another source of disquiet for the community is the continuing deficiency in social services owing to the shortage of schools, health facilities, and water and sanitation infrastructure. The community is also unhappy about the dust pollution caused by the mine's activities.

### 29.3.2 *Nonferrous Company Africa Mining*

China's Nonferrous Company Africa Mining Plc (NFCA) took over the Chambishi mine in 1998 and it became operational five years later. NFCA is owned by CNMC and Zambia Consolidated Copper Mines Investment Holdings (ZCCM-IH). NFCA's underground mine produces 6,500 tonnes of copper ore per day, on average. Additional mining activity at the West Ore Body commenced in 2011 and has since been producing some 1 million tonnes per annum. In 2012, NFCA began work on the South-East Ore Body (SEOB), whose capacity is up to 3.3 million tonnes of copper ore per annum.

#### 29.3.2.1 NFCA'S CSER INTERVENTIONS

With the exception of its scholarship programme, NFCA's social responsibility interventions have tended to be philanthropic (i.e. 'give a man a fish'), rather than community empowerment-driven (i.e. 'teach a man to fish'), as illustrated by the examples in this section. In 2008, NFCA donated fourteen butterfly sewing machines and K10,000 for a chicken project to benefit the widows of former miners in Chambishi. This was in addition to anti-malarial indoor spraying activities, on behalf of the Ministry of Health's Kalulushi District Health Management Team (DHMT), at a cost of K18,000. In each of the following three years, the company repeated the spraying exercise, at a cost of K30,000. The company donated K10,000 to the Zambia Table Tennis Association (ZTTA) and also undertook some road rehabilitation work in Chambishi and Kalulushi districts at a cost of US$275,168. In 2010, NFCA donated K500,000 to the Kalulushi Municipal Council for development projects. This social responsibility engagement continued in 2011, when the company made donations to the Nkana Golf Club Tournament, undertook indoor spraying, continued its road rehabilitation work in Kalulushi and Chambishi, procured handheld radios for the provincial headquarters of the Zambia Police in Ndola, and sponsored eighteen students to attend the Copperbelt University (CBU).

In 2012, NFCA purchased an ambulance, a motorbike, and healthcare equipment for K465,000. The company also constructed a classroom block at Chambishi secondary school, at a cost of US$100,000, to increase its seating capacity. NFCA also donated K50,000 to the Zambia Open Golf Tournament Organizing Committee to help it launch the tournament. In the area of infrastructure, the company donated road rehabilitation equipment to the Kalulushi Municipal Council (KMC) worth K696,000. The company's philanthropy continued into 2013, when it donated K12,897 for a bus shelter and a car park in Kitwe. It also sponsored the Zambia Open Golf Tournament at a cost of K270,000. In Kalulushi and Mufulira, the company built a security wall at a total cost of K210,157. It also donated a 140G Motor Grader caterpillar for road rehabilitation to the KMC, at a cost of K730,800.

NFCA is committed to undertaking ecologically safe and environmentally friendly operations. This is partially evidenced by its participation in World Environment Day (5 June) activities each year since 2005. The company promotes awareness of and education in ecological protection and environmental sustainability through various activities.

#### 29.3.2.2 PERSPECTIVES AND REACTIONS OF THE CHAMBISHI COMMUNITY

Despite the company's CSER efforts as noted above, residents of the Chambishi, Musakashi, and Lukoshi townships, which are nearest to the mine's operations, report that their communities do not benefit adequately from the mine's activities. Meanwhile, farmers in the area complain that the company has, on occasion, disrupted their livelihoods. In one instance, sixty-seven farmers from the Lukoshi area alleged that the company had destroyed their farms and crops. Community representatives have also expressed strong reservations about the company's CSER practices and claim that the company's focus is on the politically visible urban areas and that it does nothing to prevent or mitigate the pollution and environmental damage it causes in the immediate surroundings.

## 29.4 Advantages and Disadvantages of the Mining Companies' Approaches

In each of the mining company CSER case studies profiled above, there is a clear divergence between the mining companies' self-assessments and the perceptions of the mines' host communities and other observers. While each of the mining companies rates its CSER engagement favourably, a commonly held view within the communities is that the mining companies' CSER activities are mere window-dressing and that they do not deliver any real, or lasting, benefits on the ground. This asymmetry of perceptions may be rooted in elements of paternalism on the part of the mining companies, and unrealistic expectations resulting from the ZCCM legacy on the part of the host communities. There is also an element of misunderstanding about the role of local government in the area of social services provision in mining areas. Nevertheless, the mismatch in perceptions begs the questions:

- Are the mining companies 'doing the right things'?
- Or are they 'doing things right' and, in effect, merely 'rearranging the deck chairs on the *Titanic*'?

To the extent that they are focused on 'doing things right', mining companies will rate their own CSER interventions very highly. Their well-intentioned, well-crafted, and well-implemented interventions may, however, suffer the serious limitation of lack of congruence with the needs and priorities of the mines' stakeholders (primarily their host communities) due to the failure to consult with them at the design stage, and a corresponding failure to secure the stakeholders' contributions to, and participation in, the implementation and evaluation stages. The mismatch in perceptions is also partly fed by weaknesses in structures and processes for community consultation, on the one hand, and in monitoring and reporting of services delivery, on the other.

As illustrated above, both the Western- and Chinese-owned mining companies in Zambia aspire to the highest international standards of CSER best practice. In particular, the CSER interventions of the Western-owned Kansanshi Mining Plc, Kalumbila Minerals Ltd, and Lumwana Mining Company have the advantage of being driven by clearly articulated, and well-disseminated, sustainability strategies. Each of these mines has a well-staffed sustainability unit that reports to an influential mid-level executive. They are also relatively transparent in their dealings with stakeholders, subject to the limitations of language and literacy in their host communities. In terms of governance, these companies are answerable to the global capital markets, influenced by powerful institutional investors and their socially and environmentally conscious investment mandates. The 'power of the purse' wielded by these institutional investors means that the mining companies are more likely to be responsive to their demands for sustainable mining practices and community engagement.

In contrast, as noted earlier, CNMC and its subsidiaries are answerable to the Chinese government. China has had a history of unwavering commitment to investment in Zambia's mining industry despite the rise and fall of commodity prices on the global markets and the vicissitudes of the political environment in the country. As a result, the Zambian government appears to have applied little pressure on the Chinese copper mining companies to meet national and international labour standards and generally accepted CSER norms. Moreover, Chinese-owned mining companies in Zambia have been relatively free of the fear of divestment by their government. This is added to the fact that it is only within the past five years that the Chinese government has begun to place the spotlight on social and environmental sustainability in its outbound mining investments. Perhaps it is because of the twin factors of assured funding by the Chinese government and lax supervision by the Zambian government that CNMC and other Chinese mining companies have had a relatively weak CSER record, denoted by interventions that fall towards the altruistic, 'window-dressing' end of the spectrum as opposed to the community-led, -owned, and -hosted interventions that characterize substantive social and environmental responsibility. With the possible exception of NFCA and Chambeshi Copper Mines (CCS), another CNMC subsidiary, the corporate culture of Chinese mining companies is generally more opaque to their host communities and other external stakeholders out of the belief that their accountability ends at the highest levels of government. These companies run the risk, therefore, of further alienating their communities through their apparent lack of accountability to their stakeholders.

Another aspect of the agency problem for both the Western- and Chinese-owned mining companies in Zambia relative to their ownership is that of the information asymmetries imposed by the geographical remoteness of their operations from their sources of finance. From an institutional perspective, this is exacerbated by the Zambian government's lack of a CSER regulatory framework, and its capacity limitations as noted in Section 29.1.1 above, which force its regulatory agencies to rely on self-reporting by the mining companies.

### 29.4.1 *Key Shortcomings in CSER Approaches*

At the tactical level, and to varying extents, the key shortcomings in mining companies' CSER approaches in Zambia include:

- Weaknesses in stakeholder coordination and management and the apparent failure to ensure community participation in the design, implementation, and monitoring of interventions.
- Adoption of a paternalistic, top-down, approach to programme design and implementation, which has tended to alienate the host communities while failing to address their needs and priorities.
- Failure to consult the local authority councillors within their host communities about their CSER objectives, strategies, and programmes. Being the gatekeepers to the host communities, the councillors are, potentially, sources of great intelligence about the needs, local politics, and institutional make-up of the host communities.
- Failure to mitigate the barriers imposed by the literacy and capacity limitations within the host communities.
- Failure to develop a sufficient understanding of the power structures and vested interests embedded within their host communities in order to create the appropriate incentive structures
- Non-implementation of mutually acceptable communication channels between the mines and their host communities to help engender transparency, trust, and goodwill.
- Absence of accountability mechanisms involving community-based monitoring and evaluation to promote transparency, community ownership, and buy-in.

Partly due to these deficiencies, mining companies have generally been unable to maximize the involvement of their host communities in their mining and non-mining value chains through employment, entrepreneurship, and participation in social services delivery.

### 29.4.2 *Tripartite Dialogue and Other Success Factors for CSER*

Among the key success factors for CSER engagement and implementation is the existence of a tripartite dialogue involving the mining companies, their host communities, and government. The approach to this tripartite dialogue should be underpinned by the inculcation by the mining companies of a sense of shared responsibility in the counterparties. CSOs have a unique opportunity to act as trusted brokers between the three parties owing to their strategic positioning in the dialogue space between them (as depicted in Mondoloka

2017: Figure 2). Corresponding to the multilevel and multidimensional impacts of mining (Mondoloka 2017: Figure 1), the involvement of a broad cross-section of CSOs is required. In the majority of cases, however, CSOs have failed to achieve the unity of purpose or to establish the common platforms required to enable them to fulfil this role.

An important limitation on the local government side is the tendency to view CSER as a replacement for the government's development programming responsibilities. The mining companies are lobbying strongly and promoting awareness of the local governments' responsibility to use tax revenues received from them for tangible development purposes. In this regard, CSER on the part of the mining companies should be seen as complementary to, and not a substitute for, the government's development responsibility and efforts.

CSER programming should include a requirement for monetary, or in-kind, contributions by host communities to minimize the sense of entitlement and to promote community ownership and stewardship.

## 29.5 Conclusions and Recommendations

An important gap in the Zambian mining context is the government's laissez-faire posture towards CSER on the grounds that it cannot be legislated. This has resulted in the inconsistent CSER practices and virtual lack of accountability that have been noted. This contrasts with international trends that are seeing sub-Saharan countries, such as Ghana, Tanzania, and Mozambique, take measures to introduce national CSER policies aimed at ensuring that organizations promote the welfare of their host communities.

### 29.5.1 *The International Drive to Integrate CSER in Business Strategies*

There is growing international recognition that, to be sustainable, CSER must be part of business strategy and that it must be aligned with the development priorities of its host country or community. An important development noted with respect to Chinese mining investment in Zambia has been the adoption of a CSER-friendly posture buttressed by requirements for compliance by Chinese mining companies through the government's controlling interests or through membership of the CCCMC. As noted in the cases of CLM and NFCA above, there is still a gap between the Chinese government's CSER aspirations and actual conduct on the ground by Chinese mining companies. Enhanced cooperation between the governments of China and Zambia aimed at strengthening the latter's regulatory capacity will help to tie these loose ends.

### 29.5.2 *Towards a Community Self-help Approach*

Paternalism on the part of the mining companies needs to give way to partnership—a transition that is predicated on mining companies repositioning themselves as equal participants with their host communities in the quest for win–win development outcomes. To this end, mining communities need to demonstrate a capacity for self-governance that should better enable them to enter into a partnership with the mining companies. Subject to some notable implementation issues, Senior Chief Mukumbi's Chiefdom in Lumwana provides a useful example of this self-mobilization, self-organization, and self-management (MOM)[12] model of community engagement with mining companies.

Mondoloka (2017: Figure 3) depicts some of the community governance structures that have been established to help manage the interface between the mining company Barrick Lumwana and the community, with the overall objective of enabling a win–win outcome for the mine and its host community. The Lumwana Community Development Forum (LCDF) provides overall development coordination for the community. The board of LCDF represents a broad cross-section of stakeholder interests and includes representation from the Musele and Mumena Royal Establishments, Barrick Lumwana's Sustainability Department, the Zambia Development Agency (ZDA), the area Councillor, Extension Officers from the Ministry of Agriculture and Livestock (MAL), the Ministry of Community Development, Mother and Child Health (MCDMCH), the Ministry of Youth and Sport (MYS), and the Lumwana Community Business Association (LCBA), as well as Community Liaisons.

The structural arrangement is sound, in principle, but there are some issues that have limited its effectiveness in practice. These are summarized in terms of the gaps in economic opportunities and social services for host communities arising from shortcomings in mining companies' CSER interventions (Mondoloka 2017: Figure 4). Figure 29.1 provides a suggested framework for mediating the relationship between mining companies, their host communities, and other key stakeholders.

For the mining community, expected outcomes of the MOM approach include enhanced negotiating power vis-à-vis other stakeholders—principally, the mining companies—and, hence, more equitable participation in the economic fortunes of the community. One aspect of this socio-economic inclusion is that the default consensus among all the stakeholders should be some form of right of first refusal for the community with respect to employment

[12] Author's characterization.

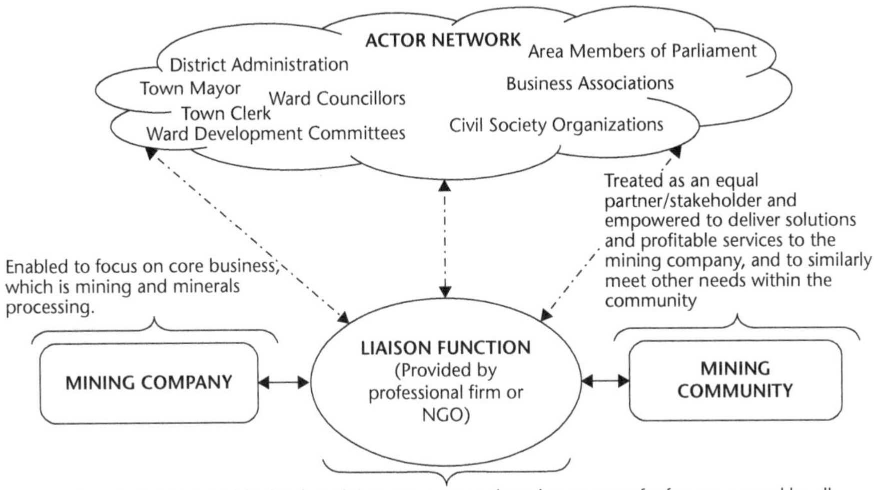


- Appointed and paid for by the mining company under robust terms of reference agreed by all stakeholders;
- Terms of reference to be strong on ensuring and facilitating dialogue between the principal stakeholders;
- Fixed-term performance-based contract of 2 to 3 years' duration to ensure results and quality;
- Clearing house for proposed development interventions to ensure maximum development impact;
- Use of government-allocated constituency development funds (CDFs) to be influenced and reported on by liaison function to promote accountability, economy, and development effectiveness.

**Figure 29.1.** A suggested framework for mediating between the mining company, the community, and other key stakeholders
*Source*: author's construction.

and entrepreneurship opportunities with the mining company and with other large business concerns operating within the community. This is consistent with the spirit of the ILO's Tripartite Declaration of Principles Concerning Multinational Enterprises and Social Policy (ILO 2006).

The underlying principle for all interventions in mining communities must be 'community ownership', which should translate into 'community members solving community problems using community resources'. By extension, this includes the principle that market-based solutions to community problems should, in the first instance, be provided by local entrepreneurs, employing local labour and using other resources sourced, as far as possible, from within the community. This lends itself to a bottom-up approach to mining company engagement, which should be mainstreamed across all mining communities. Mondoloka (2017) cites the case of Youth Alliance for Development (YAD), a NGO operating in the Solwezi district, as a model for this 'bottom-up community engagement'.

## References

Barrick Gold Corporation (2016). *Responsible Mining Governance*. Available at: http://www.barrick.com/responsibility/our-approach/responsible-mining-governance/default.aspx?LanguageId=1.

CCCMC (2014). *Guidelines for Social Responsibility in Outbound Mining Investments*. Beijing: CCCMC.

CNMC (2011). *Zambia Social Responsibility Report*. Kalulushi: China Nonferrous Metal Mining (CNMC).

CSO (2010). *2010 Census of Population and Housing*. Lusaka: Central Statistical Office.

CSO (2016). *2015 Living Conditions Monitoring Survey*. Lusaka: Central Statistical Office.

CSO, MOH, ICF (2014). *Zambia Demographic and Health Survey 2013–14*. Rockville, MD: CSO, Ministry of Health, and ICF International.

Feeney, P. (2001). 'The Limitations of Corporate Social Responsibility on Zambia's Copperbelt'. Available at: http://www.minesandcommunities.org/article.php?a=6655.

Ferguson, J. (1999). 'Abjection and the Aftermath of Modernism', in *Expectations of Modernity: Myths and Meanings of Urban Life on the Zambian Copperbelt*, 234–54. Berkeley, CA: University of California.

Fraser, A. and M. Larmer (2010). 'From Boom to Bust: Diversity and Regulation in Zambia's Privatized Copper Sector', in D. Haglund (ed.), *Zambia, Mining, and Neoliberalism: Boom and Bust on the Globalized Copperbelt*, 1–30. New York: Palgrave Macmillan.

Fraser, A. and J. Lungu (2007). *For Whom the Windfalls? Winners and Losers in the Privatisation of Zambia's Copper Mines*. Lusaka: Civil Society Trade Network of Zambia (CSTNZ).

Geloo, Z. (2013). 'Cursed by Copper: Mining Communities in Zambia'. Available at: http://www.osisa.org/openspace/zambia/cursed-copper-zambia.

GRZ (2013). *Revised—Sixth National Development Plan (R-SNDP) 2013–2016*. Lusaka: Government of Zambia.

ICAP at Columbia University (2016). *Summary Sheet: Preliminary Findings; Zambia Population-Based HIV Impact Assessment*. New York: ICAP.

ICMM (2012). *Community Development Toolkit*. London: ICMM.

ILO (2006). *Tripartite Declaration of Principles Concerning Multinational Enterprises and Social Policy*. Geneva: International Labour Office.

OAG (2014). *Report of the Auditor General on the Management of Environmental Degradation Caused by Mining Activities in Zambia*. Lusaka: Office of the Auditor General.

Russell, A. (2007). 'Jisco Takes Model Approach in S. Africa Project'. Available at: http://www.ft.com/intl/cms/s/0/4f77cfb4-3f83-11dc-b034-0000779fd2ac.html#axzz46FT-o9CVq.

Smit, A. (2013). 'Shaping Corporate Social Responsibility in Sub-Saharan Africa: Guidance Notes from a Mapping Survey'. Bonn: Deutsche Gesellschaft für Internationale Zusammenarbeit (GIZ).

World Commission on Environment and Development (1987). *Our Common Future*. Oxford: Oxford University Press.

# 30

# Approaches to Supporting Local and Community Development

## Brazil and the Vale SA Model of Corporate Interaction

*Liesel Mack Filgueiras, Andreia Rabetim, and Isabel Aché Pillar*

### 30.1 Vale SA: An Overview

According to a Vale SA[1] fact sheet (Vale 2017), Vale is one of the largest mining companies and one of the largest publicly traded companies in the world.[2] Vale has a market capitalization of around US$36 billion, with approximately 270,000 shareholders from all continents. It is the world's largest producer of iron ore and iron ore pellets, and of nickel. Vale also produces manganese ore, ferroalloys, metallurgical and thermal coal, copper, platinum group metals, gold, silver, cobalt, potash, phosphates, and other fertilizer nutrients. The company's global financial position is summarized in Figures 30.1 and 30.2.

Vale's mission is 'to transform natural resources into prosperity and sustainable development', and its corporate vision is 'to be the number one global natural resources company in creating long-term value through excellence and passion for people and the planet' (Vale 2016a).

The company's business strategy is to invest in world-class assets, with long life, low cost, expandability, and high-quality output, capable of creating value through the cycles. Focusing on cost effectiveness, Vale seeks a lean organization,

[1] SA = 'Sociedade Anônima', or 'share company'.

[2] Vale shares are traded on the New York Stock Exchange/NYSE (VALE and VALE.P), the BM&F BOVESPA (Vale3 and Vale5), the Euronext Paris (Vale3 and Vale5), and the Stock Exchange of Hong Kong Limited (codes 6210 and 6230). Average daily trading value was approximately US$320 million in the third quarter of 2015 (3Q15). Vale is an investment-grade company, rated BBB by Standard & Poor's, Baa2 by Moody, BBB+ by Fitch, and BBB (high) by Dominion Bond Rating Service (Vale 2017).

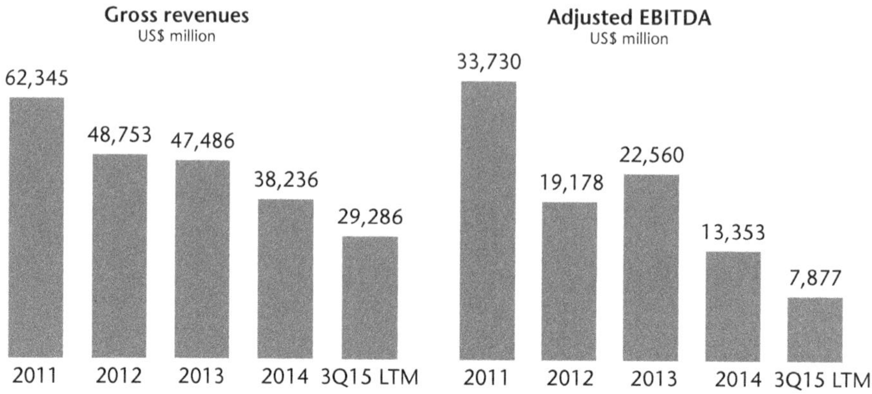


**Figure 30.1.** Vale's financial highlights
*Source*: Vale (2015), reproduced with permission.

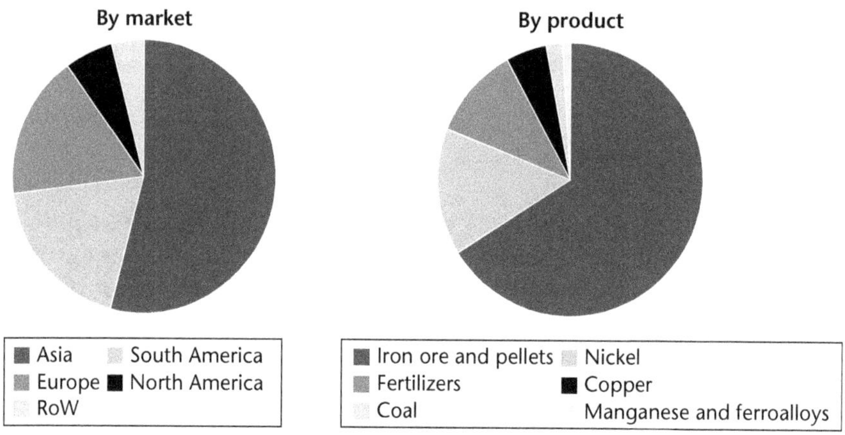


**Figure 30.2.** Vale's gross revenues
*Source*: Vale (2015), reproduced with permission.

with teamwork and accountability, excellence in project execution, and firm commitment to transparency and shareholder value creation. Health and safety, investment in human capital, a positive work environment, and sustainability are also considered critical to Vale's long-term competitiveness.

All of Vale's activities are guided by management policies of transparency, respect for shareholders' rights, protection of the environment, employee development, and improving the quality of life in the communities in which the company operates.

In the last decade, Vale's community development approach has been evolving, especially in terms of contributing to local development in the regions where it has operations. This chapter shares some of the results of this experience and the lessons learned so far.

## 30.2 Vale's Social Investment Evolution

Vale Foundation was established in 1968. When privatized in 1997, Vale focused its Foundation as the main channel for implementing the company's social and community development agendas. Vale Foundation is a non-profit organization, financially maintained by Vale. Like many other companies worldwide, Vale focused the efforts of its Foundation along two main thematic lines of social investment: education and culture. It invested in projects developed by NGOs and specialized companies throughout Brazil. This continued until 2006, after which a process of internationalization of the company began; there were, subsequently, social programmes managed locally in all of the countries in which Vale had operations.

However, following an in-depth strategic analysis of the Foundation's work in 2004, it was clear that the model as used to that date was not the most effective one. There were up to 150 projects managed by the Vale Foundation by 2000 and the results of many of these were unclear. They were not driven by territorial and local development demands, and were isolated and independent efforts. It was also noted that the company was fundamentally dependent on land; that some of its commercial operations had very long-term horizons (up to 100/150 years); that its procedures had the potential to generate both significant positive but also significant negative social, economic, and environmental impacts; and, therefore, that the territorial dimension of sustainability was critical for the company.

Hence, in 2006, Vale Foundation refocused its mission to contribute to the local development of the specific regions where Vale had its operations, by more strategically aligning its work with the nature of the company and also with the relevant specific opportunities and challenges that the mining activity opened up for the various territories of operation.

The first thing that was clear was that the Foundation needed to better understand the socio-economic dynamics of the regions where Vale operated in order to be able to contribute effectively to their local development. So we began to partner actively with specialized institutions and to prepare integrated socio-economic diagnoses of all the regions in Brazil (and later worldwide) where we had operations. That work sought to understand what were the cumulative impacts that Vale's projects and operations had in the mining-affected municipalities[3] and what, in turn, were the possible impacts of the municipalities on Vale's operations. We studied the history of the formation of each of the municipalities, and then identified the main economic drivers and analysed demographic projections, considering both natural

[3] The municipality is the level of public administration in Brazil below the federal and state levels.

growth (births and deaths) and anticipated migration inflows. We also gathered detailed information on public services and infrastructure deficits (current and future), taxes (current and future), land ownership regulation, and so on. We ended up with a very good, comprehensive picture of what were the pre-existing deficits of the municipalities (financial and otherwise); what were the probable pressures that new Vale projects would bring to the region; and, also, the taxes that would be generated and whether the scale would balance itself.

We found that most municipalities—specifically the mining ones that were studied—could have positive balances, in due course, if mining operations reached their planned outputs and the tax resources were adequately applied in infrastructure and public services. Nevertheless, the timing was invariably a challenge. Some impacts would happen before the royalties on production[4] reached the full amount. In other cases, the royalties would be so significant that the execution of all the various affordable projects would be a challenge. We also learned that there were resources available prospectively to municipalities from various federal government programmes, but that the town hall was often not able to access these due to a lack of either technical capacity or proper information. It was clear that capacity-building, planning, and governance (including cross-sector governance) were needed for our host municipalities as well as for some other agencies of government.

Currently, Vale Foundation's mission is to contribute to integrated economic, environmental, and social development in the regions where Vale operates, strengthening human capital in communities and respecting local cultural identities (Vale Foundation n.d.). Figure 30.3 illustrates the model of sustainable development which we follow. The Foundation's investments are addressed mainly to education, health, and income-generation/inclusive businesses initiatives in Brazil.

## 30.3 Public–Private Social Partnership

As a result of these diagnostic insights, we began to develop the concept of the public–private social partnership (PPSP). The starting point for elaborating the concept was Vale Foundation's practical experience, in particular the liaison work it carried out with many federal government ministries and

[4] Compensação Financeira pela Exploração de Recursos Minerais (CFEM: Financial Compensation for the Exploitation of Mineral Resources), the mining royalty in Brazil, is collected by the National Department of Mineral Production (DNPM) and then re-allocated to those states and municipalities where mine production has taken place. Eighty-seven per cent of royalties is captured by the municipal and state governments (OPM, ICMM, and IBRAM 2012).

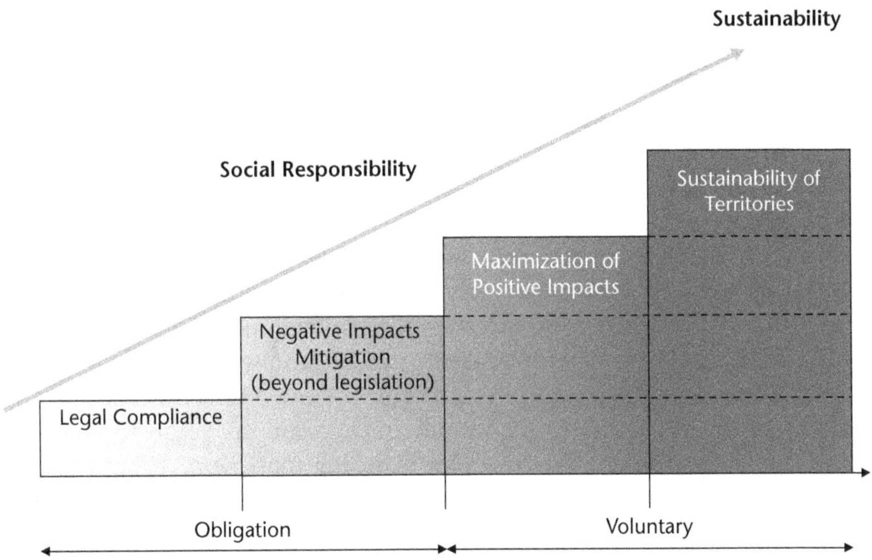


**Figure 30.3.** Sustainable development evolution
*Source*: authors' illustration.

departments with the aim of supporting municipalities to raise funds, especially for deficit reduction in relation to urban infrastructure.

When the Brazilian government launched the Growth Acceleration Program (Programa de Aceleração do Crescimento; Federal Government of Brazil 2007), a national economic policy that invested billions in infrastructural works, in 2007, Vale Foundation saw an opportunity to support the municipalities where Vale operates by elaborating engineering projects in order to allow the municipalities to apply and compete more effectively for federal funds. Since most of these remotes cities have fragile infrastructure, Vale Foundation funds the elaboration of technical projects mainly in connection with housing, sewage, drainage, and pavements. Canaã dos Carajás was one of the municipalities that had its projects approved by the federal government as a result. This has involved fundraising of R$19 million (US$5 million) to finance water system works.

From this experience, Vale Foundation gained a clearer understanding of the public policy framework in Brazil and how to advocate for people to achieve better access to public services. Therefore, the next step was to establish Technical Cooperation Agreements between Vale Foundation and Brazilian ministries, such as the Ministry of Health and the Ministry of Cities, seeking to discuss, plan, and execute programmes of private social investment as aligned as possible with public policies. Concerning these Agreements, there is no transfer of resources between the parties; the main objective is the integration of efforts between the public and private sectors.

One of the consequences of these partnerships was the establishment of the first model of the Public Children's Safe House (Conselho Tutelar) in Canaã in accordance with the Brazilian Human Rights Ministry standards. Initially, the new model was to be set up only in cities with more than five hundred thousand people that were to host large-scale projects such as venues for the Olympics and the World Cup. However, the dynamics of the fast growth of Canaã and its social figures showed the importance, indeed the necessity, of such a social facility for strengthening the local Child Protection Safety Net.

As background to this discussion on PPSP, it is also important to consider the Brazilian historical context. After a military dictatorship lasting for almost two decades (1964–85), the 1989 presidential election was the first since 1960 in which Brazilian citizens were able to vote for the president. Based on the 1988 Constitution, a variety of policies and regulations began to be established. This makes many of the urban planning frameworks a quite recent development in Brazil. The Ministry of Cities—with responsible for housing, land regularization, and sewage—for instance, was created only in 2003. Moreover, until the 1940s, only 30 per cent of the Brazilian population were urban. By the 1990s, about 75 per cent were already living in urban areas. Nevertheless, this population growth in urban areas was not matched in the same proportion by the provision of public services and infrastructure.

In July 2012, based on Vale Foundation's experience after 2006, it was decided to create a working group[5] composed of organizations and experts with vast experience in the areas of sustainable development, urban planning, corporate social investment, and relationships with government, to discuss the concept of PPSP.

Based on the discussions of this working group, a reference text was developed. The PPSP as operated thereafter by Vale Foundation can be summarized as follows. It is a strategy for building a series of inter-sectoral alliances aimed at promoting the sustainable development of territories where the company has large-scale enterprises, through joint efforts, and by mobilizing the resources and knowledge of civil society, governments, and businesses for integrated long-term strategic planning, around a common agenda that includes promoting actions that contribute to:

- the promotion of quality of life and human development
- the strengthening of inter-sectoral and public policy

[5] Members of the PPSP working group: Accenture, MIF (Multilateral Investment Fund of the IDB, the Inter-American Development Bank, IBAM (Brazilian Institute of Municipal Administration), IBRAD (Brazilian Institute of Development Administration), Eupolis Lombardia, Unesco, an expert in monitoring social projects from BNDES (National Bank for Economic and Social Development), and a professor from Dom Cabral Foundation.

- the expansion of democratic participation and the perspective of citizen inclusion
- the effectiveness of private-sector social investments
- the collective construction of the figure of the citizen as simultaneously a beneficiary and co-participant in the process of territorial development.

Vale Foundation's PPSP is required to seek strategies that strengthen the opportunities generated by large mining projects, such as those strategies that can help in: (i) promoting diversification of economic activities; (ii) supporting the qualifications and awareness of civil society, companies, and government regarding the management of territorial development; (iii) strengthening capacity-building and sharing resources and knowledge to achieve the goals defined in the public sphere.

The implementation of the PPSP in the individual territories of Brazil and other countries also requires a political consensus that enables the various initiatives that are proposed to take full advantage of both the management tools (diagnostics, plans, laws, regulations, legal apparatus, websites, etc.) and the institutional arrangements (chambers, trade associations, councils, boards, forums, development agencies) already existing. When gaps are spotted in either of these two areas (management capacity or institutional systems), investment in the capacity-building of local actors and the development of strengthened management tools can be sponsored, but always with a focus on the need to sustain the partnership for sustainable development of the territory, and in a long-term systemic view.

A process of political consensus demands that the parties involved all make commitments and thereby establish their respective roles and responsibilities. From the perspective of a PPSP linked to a large enterprise such as Vale, the opportunities that are to be promoted and the impacts that are to be minimized have to be discussed jointly by the various agents. Thus, we seek to avoid the logic and the potential pitfalls that still prevail in cases where each party assumes a liability in isolation and without the need to interact with others. Specific examples of how particular PPSP projects work in practice are presented later in this chapter.

The work grounded in the PPSP approach may seem like a very laborious way to proceed, as it demands consensus between the parties and the supremacy of the common good over individual vision. In practice, it is often hard and slow work to achieve consensus. However, we have found that it is the best way to develop actions that will enjoy durability and performance characteristics that can contribute to strengthening public policies and the empowerment of local actors. In due course it can, thereby, help to reduce dependence on the mining company—often the primary vector of development for remote municipalities. With this approach, the company itself seeks to promote a qualitative leap, moving from more isolated practice to a joint effort with other actors and players in the territory, integrating networking sites with other spheres, public, private, community, and social.

There is an important and perhaps unique philosophy underlying the approach of PPSP as practised by Vale Foundation. The approach recognizes that the area of 'public' policy extends beyond the government's own actions to involve the engagement of all actors. This helps to generate a different form of social control that enables the continuous improvement of policies and actions developed by communities. PSPP's vision in this sense is based on this concept of the 'public arena', which understands that citizenship is not restricted merely to the scope of the state, and that 'public life' is not just made up of government acts, but also of the acts of groups, institutions, and individuals who all aim to meet social goals.

In other words, the process of building the PPSP system has required a new attitude of governments, enterprises, and society. It is based explicitly on open dialogue to help build a vision of the common good, which is achievable through an integrated and coordinated set of efforts for the generation of local opportunities and the best application of private social investments and governmental resources.

## 30.4 Southeast Pará (Geography, Economy, Vale's Presence)

Vale is present in the State of Pará. Located in the north of Brazil, Pará is the second largest state by area (it is the size of Spain and France combined). However, it is also one of the country's least developed states, ranking twenty-second out of twenty-six states in Brazil in GDP per capita terms.

Comprised of 144 municipalities (elected local governments), it has a population of 8.2 million people (IBGE 2016), making it sparsely distributed but with many inhabitants concentrated in and around the capital, Belém.

A study conducted in 2011 by OPM, the ICMM, and the Brazilian Mining Association, entitled 'The Mining Sector in Brazil: Building Institutions for Sustainable Development' (OPM, ICMM, and IBRAM 2012), provides a good summary of the history of Pará. In terms of development, both economic and social, the north of Brazil has struggled for decades to achieve the same level as the south. Historical and violent conflicts over land between large ranchers, smallholders, small-scale miners, and the indigenous populations of the region have been a major problem since the 1960s. These difficulties have been created mostly by land redistribution and new settlement by farmers, cattle ranchers, and logging companies, and also by the notorious gold rush in

Serra Pelada in the 1980s.[6] Today, the state still faces serious problems of infrastructure deficits, including in water supply, sewage and waste collection, lighting, paving and drainage, road networks, and public transportation. Also, Pará suffers from very large deficiencies in schooling and vocational training.

Also according to OPM, ICM, and IBRAM (2012), the mesoregion of Southeast Pará contains thirty-nine municipalities, separated in the analysis into six 'mining municipalities' with large-scale mining operations (Canaã dos Carajás, Curionópolis, Marabá, Ourilândia do Norte, Parauapebas, and Paragominas) and the other 'non-mining municipalities'.

## 30.5 The PPSP in Southeast Pará: The Canaã dos Carajás Case

In order to illustrate how the PPSP methodology has been implemented, it is helpful to discuss examples from Canaã dos Carajás, a city in Southeast Pará where Vale's largest project, the Carajás S11D Iron Project (Vale 2012), is being implemented.

Around US$16 billion is being invested—$6.8 billion of which is the estimated capital expenditure involved in installing the new mine and processing plant, and the remainder of which is for logistics and infrastructure. Vale is investing heavily to upgrade the Carajás railroad, with some 504 km of track related to expansion and 226 km related to remodelling the existing capacity. This railroad runs all the way from the iron ore mine to the company's export terminal in the Atlantic port in the state of Maranhão. S11D, when fully operational, will supply 90 million metric tons of iron ore per year. It began operating in 2016 (Vale 2016c). Figure 30.4 is a map for the location of Carajás S11D Project and Carajas Railway (expansion and remodelling of existing line).

Canaã dos Carajás separated itself from the city of Marabá in 1994. Before that, it was just a small agricultural settlement, with logging and agricultural activities as its key economic drivers. The first mining cycle happened with the opening of the Sossego mine (for copper extraction) in 2000. In 2001, the city had a total GDP of only R$17 million (US$4.8 m). By 2014, that figure had jumped to R$3 billion (US$857 m), while the city's population had increased more than three times (from 10,000 to 32,000) over the same period.

The fact that Canaã was established as a municipality only recently, in the 1990s, means that it had the daunting challenge of structuring itself as a city at

[6] For a time, the village of Serra Pelada was thought to be the largest gold mine in the world in terms of numbers engaged. Because of the chaotic nature of the operation, estimating the number of miners accurately was difficult, but it was thought to be at least 80,000 people at its peak. Today the Serra Pelada mine is abandoned and the giant open pit that was created by hand has filled with water, creating a lake.

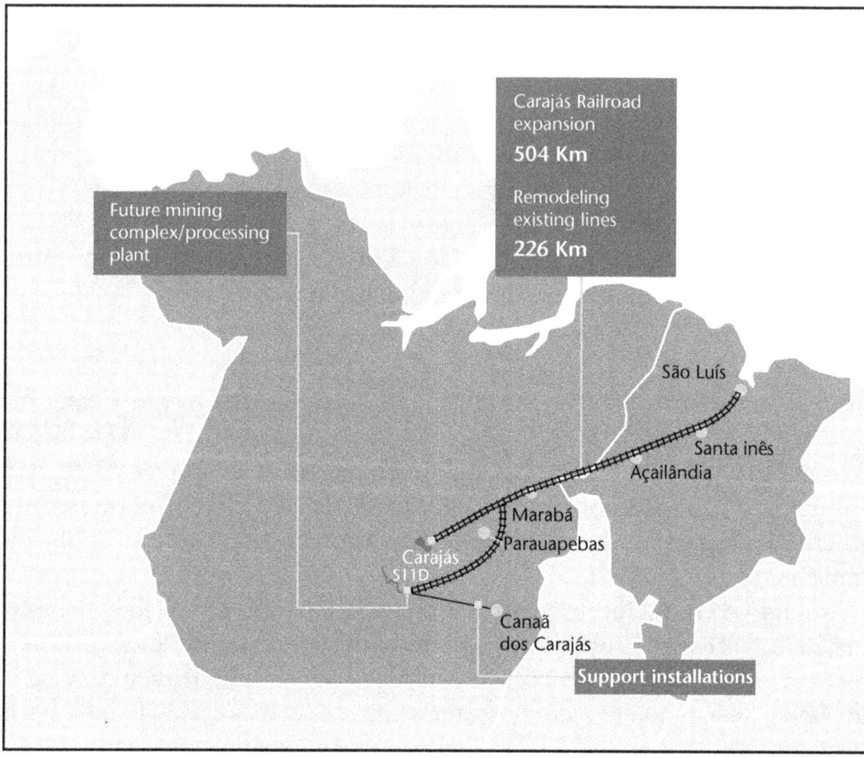


**Figure 30.4.** Carajás mining complex (Pará and Maranhão States)
*Source*: Vale (2012), reproduced here with permission.

the same time that it was starting to receive a very large inflow of mining investments. This was also a challenge for Vale. In addition to meeting all its legal obligations, including the environmental and social requirements needed to obtain and preserve its licence to operate, especially considering the size and dimension of the investment, Vale also recognized the vital need for active collaboration with the various stakeholders involved: territorial development cannot be achieved in isolation. The establishment of a PPSP in Canaã dos Carajás a couple of years ago, through the Vale Foundation, was the result of a learning curve spanning more than a decade in which Vale had participated along with the other parties. This represents an important step towards a governance system for the new municipality wherein government, companies, and civil society have agreed to share responsibilities.

As mentioned in Section 30.2, the robust socio-economic integrated studies and diagnostics have enabled the Vale teams to identify deficits (e.g. in administrative capacities) and associated opportunities for local development of the cities in which the company operates. In the case of Canaã dos Carajás, having those estimated figures has allowed Vale and its Foundation to work in partnership with the town hall, local communities, and the federal government to strengthen a wide range of public policies in order to contribute to the promotion of better-quality education, ill-health prevention and health promotion, inclusive sanitation, and efficient urban mobility; the fostering of local economic activities; and the strengthening of networks for social promotion and protection.

The PPSP meant reshaping the relationship between Vale Foundation and the municipal and federal governments. In the partnerships implemented in Canaã, the town hall began to participate in discussions to build content and results, adopting the attitude of co-participant, and not that of a mere beneficiary of a social project, going much further than making available municipal data and human resources. Moreover, almost all activities under the umbrella of the PPSP are undertaken in partnership with organizations with specialized knowledge, such as NGOs and universities, and also local organizations. Another point that should be highlighted is the fact that, in Canaã, issues are being addressed in an integrated way. For instance, the campaign for traffic safety in the city was possible through the joint effort of the municipal secretariats of transit, education, and sports, in partnership with Vale Foundation, Vale, Vale's suppliers, local businesses, and UNESCO.

### 30.5.1 *A Few of the Main Results of the Cross-sector Partnerships by 2014/15*

A central contribution has been the support to and review of the city's Master Plan,[7] which provides a broad understanding of the city and its growing areas, and specifies guidelines for how the municipality should be developed in future in an organized manner. This work includes an ongoing discussion about the possibility of creating a fund through which to apply and manage the locally received royalties from mining activities.

At the next level, the drafting of Municipal Plans for health, education, sewage, and urban mobility provides crucial inputs to the structural improvement of overall public services provision. Canaã was one the first municipalities among 144 in the state of Pará to have an Urban Mobility Plan.

With the Urban Mobility Plan as a starting point, the following measures were adopted: educational mobility campaigns and improvement of infrastructure services to reduce traffic accidents (a major element of total spending on health in the municipality); construction of bike routes to encourage

[7] This and related plans discussed in this section are unpublished documents developed with the city hall for use by the Mayor and public technicians.

alternative forms of transportation, especially cycling; and the integration of municipal records on health, traffic, and public safety.

This innovative Urban Mobility Plan was intended to benefit the whole population of Canaã dos Carajás (33,632 inhabitants). By 2015, its implementation had included the training of thirty-five educational professionals to enable them to cover traffic and public safety in elementary schools and also some broader educational campaigns for traffic and public safety, with 14,000 people engaged at local schools and churches and in the public sector, as well as 2,000 drivers and pedestrians and forty-five motorcycle riders. In the north of Brazil as a whole there is a precarious situation in terms of sewage and treated water access: almost 70 per cent of households have no sewage system or septic tanks and 45 per cent have no water supply system. Among the activities addressed by the PPSP, ninety technicians and community leaders from Southeast Pará were trained in the policy and management of basic sanitation by Vale Foundation, UNESCO, and the Ministry of Cities, resulting in community engagement and a participatory role in the prioritization of neighbourhoods based on health indicators. Currently, as a result, the municipality has predefined goals and greater social control in the implementation and provision of these services.

A sanitation plan is being implemented based on the Environmental Salubriousness Index (Índice de Salubridade Ambiental). Linking data on sanitation services with characteristics of population density and its relationship to health, the index generates indicators that show the intra-municipal inequalities, considering all components of basic sanitation: water supply, urban cleaning and solid waste management, and urban drainage and management of rainwater. In addition to providing a technical tool to prioritize integrated investment in Canaã's most populous neighbourhoods with the more precarious sanitation services, the index methodology also provides a reference for social participation in public policy and social control. It was created by the Vale Foundation and UNESCO and is available for use in other municipalities.

At the moment, the town hall disposal of waste in landfill sites amounts to about 90.37 tons/day, being 75.37 tons/day of domestic and commercial waste and 15 tons/day of construction waste, which means a per capita value of 1.71 kg/inhabitant/day—more than twice the rate recommended by the WHO of 0.8 kg/inhabitant/day. These data indicate that the useful life cycle of the landfill sites is at risk of saturation within a shorter time than should be expected. Vale supported the transformation of the old dump into a landfill. It was also necessary to improve waste management by seeking to reduce the volume of waste and to stimulate recycling activities. Vale Foundation provided support in drafting the Integrated Management Municipal Plan for Solid Waste, such as specialized technical assistance and training activities that included building the Plan for Selective Collection in a participatory manner with the Cooperative of Waste Collectors of Canaã dos Carajás. Today, the Cooperative has guidelines to improve their own management skills and add value to recycled materials.

Since 2013, the Vale Foundation has also developed health actions in Canaã, based on two major fronts—'Supporting the Public Management' and 'Mobilization and Community Engagement'—emphasizing the strengthening of technical capacity of health services. Concerning the latter, this led to the creation of a group of Young Health Promoters through a process of training young people on issues related to health and citizenship. Over a period of two years, the young people involved developed many educational activities concerning health in their neighbourhoods in a way that was integrated with the Municipal Health Secretariat's calendar, as well as actively participating in forums and promoting the occupation of public spaces. Among the results of the establishment of the Young Health Promoters group have been the participation of two youngsters from Canaã dos Carajás as representatives of the rural areas of Brazil at the 2nd Pan-American Forum of Children and Adolescents (held in 2014 in the Brazilian capital, Brasília) and a digital mapping project that revealed a broader picture of the local context, especially in relation to health, education, environment, social protection, and leisure spaces in three communities of Canaã dos Carajás. This material served as the basis for discussions about the challenges and availability of local resources and also for guiding the drawing up of Vale's Community Action Plan.

One of the developments of the youth mobilization and its partnership with the town hall was the implementation of the national programme Health in Schools (Programa Saúde na Escola—PSE) in all thirteen health units of Canaã. PSE is an inter-sectoral policy for health and education established in 2007, aimed at children, teenagers, and adults enrolled at public schools, to promote health and comprehensive education.

Furthermore, through co-operation between the Vale Foundation and the Health Secretariat of Canaã, actions were established to increase the technical capacities of health services. Among the results achieved was the development of professionals in basic care with 288 hours of training through the years of 2014 and 2015 on issues related to health promotion, healthy food habits, and disease prevention—of Hansen's Disease,[8] among others. With an emphasis on building their capacity to diagnose and treat Hansen's Disease, the professionals examined and accompanied patients on treatment programmes and examined 174 children aged from seven to thirteen, identifying seven new cases of the disease. In Canaã dos Carajás forty-five people were examined and fourteen new cases were identified.

[8] Hansen's Disease is another term for leprosy.

Moreover, all Basic Health Units of Canaã were equipped, reaching a total of around 350 pieces of equipment, in order to improve the ambience/environment of the healthcare facilities, and to stimulate the continuous professional development of those working there. As a result, appointment times were reduced and the execution of collective activities for the community in health education grew. According to the Brazilian Health System, in 2013 the city had primary healthcare coverage of 76 per cent (SUS 2015). In 2015, through this PPSP, that percentage jumped to over 90 per cent, well above the national average of 60 per cent (SUS 2016).

A further challenge faced by the municipal management was that of structuring the public school system. With the commitment of many local actors, in 2013 the Pact for the Improvement of Public Education of Canaã dos Carajás (Pacto pela Melhoria da Educação Pública de Canaã) was established. The results included: reform of twenty schools and training of 400 educational professionals covering 12,000 public school students; greater appreciation of reading through reading groups (*rodas de conversa*), with the participation of five authors of children's and young people's literature, more than 150 teachers, and 500 students; goal-setting by classrooms; establishment of planning and evaluation routines; and increased participation of parents and families in the school environment.

As a result of dialogue with the Ministry of Education, the municipality raised about R$3.6 million (US$1m) from the federal government for construction of day-care facilities for children. Thanks to the efforts of the PPSP, Canaã analysed examination results for 2013, undertook a gap analysis, and established an action plan to address the gaps identified. By 2014, it had already improved its results (compared with 2013) in the National Literacy Assessment (Avaliação Nacional da Alfabetização), a national test that assesses the level and quality of literacy across Brazilian public schools—in nineteen out of twenty aspects, scores were above state averages (Instituto Nacional de Estudos e Pesquisa Educacionais Anísio Teixeira 2015).

According to the results of the Educational Assessment System of the State of Pará released in April 2016, it was possible to observe increases of 30.6 per cent in proficiency in Portuguese and 27 per cent in proficiency in mathematics in the early years of elementary school between 2014 and 2015. For the final years of elementary school, the increase in proficiency was 50.2 per cent in Portuguese and 60.2 per cent in mathematics (Secretaria de Estado de Educação do Pará 2016a, 2016b).

In Canaã the challenge is now to keep continuously improving its results. However, the city's experience was selected to be presented at the 24th International Congress of Educating Cities, which took place in Rosario, Argentina, in June 2016. The event is one of the most important in the world education calendar. Only twenty-three Brazilian municipalities took part in the Congress, with Canaã dos Carajás being the only representative of the North region of the country.

A final example of the support resulting from the PPSP relates to business development. In addition to the employment generated in Canaã dos Carajás (12,000 people employed in the S11D implementation phase and Sossego Copper mine operation; Vale 2016b), Vale Foundation has invested in local vocational and entrepreneurship support, with technical assistance to 110 small businesspeople and support to six local businesses. During the last two years, the average income growth for the six businesses in question was 238 per cent. These results already achieved are part of the Foundation's programme to contribute to increasing non-mining income and generating jobs. Local economic activities such as services, food production, and clothes-making have been supported through training, technical, and management advice; mentoring; direct investment (seed money); and monitoring of projects. Collective businesses or individual entrepreneurs can also be supported. One of the businesses supported that can be highlighted is the Apiculture [bee-keeping] Association of Canaã, which has seen a significant increase in the number of its members since the beginning of the Association's participation in the income-generation and job programme: a jump from twenty-five people to sixty-two.

## 30.6 Challenges Faced and Critical Success Factors for the PPSP Approach

The establishment of Vale's private social investment model and a system of governance that fosters co-responsibility and co-management as a way to achieve long-term territorial development has resulted from three main factors. The first is the trajectory of the Vale Foundation itself and the lessons learned, both successes and failures, from the implementation of its previous projects in several municipalities. The second is the peculiar characteristics of mining activity—and especially its dependence on particular (often less-developed) territories combined with the unusually long-term maturity of its investments. The third is the challenge faced by the Brazilian government in providing quality public services to its citizens in remote areas (such as South-east Pará), in order to comply with the rules established in the Federal Constitution and given the inherent limitations of the more remote municipalities in accessing existing public policies at federal level. The prevailing Brazilian political-administrative division is quite relevant to the context of social investment, to the extent that the Constitution of 1988 took responsibility away from the states and delegated it to municipalities for some powers that are often beyond the municipalities' budget and/or implementation capacity.

During the process of construction and implementation of the PPSP model by Vale, the main challenges were related to the understanding of how to make the engagement between the public and private sectors work in practice. In Brazil, the culture of inter-sectoral integration and collective construction is quite recent. Thus, there is still very active discussion about the appropriate roles and responsibilities of each party, recognizing that the company (Vale) cannot and should not replace the state's role but has to be committed to efforts to implement public policies. Another lesson learned was about the need to balance expectations in relation to the results of the partnership, as the two parties (the state agencies and the company) tend to have different timings and as there was little or no previous experience of this sort of model: it was a pioneering move in Brazil.

The pilot initiative took place in Canaã dos Carajás, as already discussed. It is important to highlight that many important social projects are moving forward well in Canaã mainly as a result of three factors: there is a positive environment in which to establish partnerships;[9] Vale has been willing to develop and integrate its social investment activities in a multidisciplinary way; and there is solid alignment between the company and its Foundation. This alignment is important to allow both institutions to take advantage of the synergies. It makes them more effective in their social investment and in the generation of shared value to local communities.

This alignment between the company and its Foundation is evidenced by a trend derived from the survey 'Benchmarking Corporate Social Investment (BISC)',[10] carried out annually by Comunitas.[11] According to BISC 2015, in 2013 more than half of the funds of institutes and foundations were allocated in a way that was aligned to the core business of the company. From 2013 to 2014, this share increased by 10 per cent (from 53 to 63 per cent) (Comunitas 2016).

In the case of mining, such internal integration is even more necessary since the implementation of mining projects impacts so forcefully on the

[9] This is helped by the relative newness of the municipality and the related absence of any strong vested interests.

[10] Established in 2008, this is an annual study on the current scenario and trends of corporate social investment. Inspired by the methodology of the American-based Committee Encouraging Corporate Philanthropy (CECP), the survey seeks to measure all financial resources, goods, and services applied by private companies in projects and activities of social interest, including: support to social, environmental, and cultural projects; construction of social infrastructure, including that resulting from the installation and operation of enterprises; and permanent or eventual support to formally constituted organizations.

[11] Comunitas is a Brazilian CSO of public interest created in 2000, dedicated to the promotion of social development in Brazil, through the engagement of several sectors of society, stimulating and fostering collective actions with one common purpose: to reduce social inequalities in order to promote sustainable development in the country. Supporting the organization are a number of large corporations and their leaders, who are actively engaged in defining and implementing the organization's programmes.

socio-economic dynamics of the cities in which they take place—especially those, such as Canaã, that start off as very small. Therefore, social investments focusing on contributing to local development and strengthening public services are vital for the sustainability of both the mining operations and the territories.

Finally, there is still plenty of room for improvement, especially in two respects: improved monitoring and impact assessment of the component activities; and the fostering of the establishment of more partnerships. In the case of partnerships, on the one hand, the unfavourable current global economic situation, especially the significant drop in the price of commodities, particularly iron ore, is affecting the amount of resources allocated to social actions. But on the other hand, the logic of partnerships has never been stronger.

Also according to BISC, there has been a drop in the volume of funds invested by a group of 336 organizations, including companies, foundations, and conglomerates, in Brazil: in 2013 the BISC companies invested R$2 billion in the social arena, which represents a decrease of 25 per cent compared to 2012. In 2014, despite investments having increased by 11 per cent compared to 2013, this increase was not enough to recover losses sustained in 2012–13. On the other hand, the logic of partnerships strengthened. In addition, in the BISC 2014, 100 per cent of the organizations surveyed indicated that the identification of public problems required collective action.

Canaã leads the rankings of the Municipal Development Index developed by the Federation of the Industries of the State of Rio de Janeiro (Índice FIRJAN de Desenvolvimento Municipal—IFDM/FIRJAN) in the state of Pará (Canaã dos Carajás—Pará 2016; Jardim 2016). In the overall index ranking, Canaã is also further ahead than other Brazilian capitals such as Aracaju, Porto Velho, Salvador, Maceio, Manaus, and Macapa, and Belém, the state capital. This index can be considered a reference in terms of monitoring Brazilian socio-economic developments since 2005.[12] In 2013, the year on which the research was based, Canaã achieved a 0.7351 rate, a 25 per cent increase since 2005, on a scale from 0 to 1 in which the closer a city is to 1, the greater the development of the city. The municipality also showed an improvement in all three aspects.

Although this chapter has described a specifically Brazilian local experience, this methodology of implementing corporate social investment might be replicated in other country contexts since the basic premises are cross-sector governance and combined efforts, avoiding the overlap of roles and responsibilities. Obviously cultural differences, the level of maturity of organizations, and the institutional environment in which they operate may require

[12] The IFDM takes into account three aspects—employment and income, education, and health—based on official data from the Ministries of Health, Education, and Labour.

adjustments of the partnership framework to particular country circumstances. However, the basic guidelines of PSPP can be followed elsewhere.

## References

Canãa dos Carajás—Pará (2016). 'Diagnóstico Socioeconômico do Município de Canãa dos Carajás, Estado do Pará'. Rio de Janeiro: Canãa dos Carajás—Pará. Available at: http://www.canaadoscarajas.pa.gov.br/arquivos/semdec/DIAGNOSTICO-CANAA-Versao-Final-FEV2016.pdf.

Comunitas (2016). 'BISC: Benchmarking do Investimento Social Corporativo'. Comunitas website. Available at: http://www.comunitas.org/portal/o-bisc/.

Federal Government of Brazil (2007). 'PAC'. PAC website. Available at: http://www.pac.gov.br/.

IBGE (Instituto Brasileiro de Geografia e Estatística) (2016). 'Cidades@'. IBGE website. Available at: http://www.cidades.ibge.gov.br/xtras/home.php.

Instituto Nacional de Estudos e Pesquisa Educacionais Anísio Teixeira (2015). 'Painel Educational Municipal—Canaã dos Carajás, Pará.' Avaliação Nacional da Alfabetização. Brasília: MEC. Available at: http://ana.inep.gov.br/ANA/.

Jardim, L. (2016). 'Cidade no Pará Fica à Frente de Capitais em Índice de Desenvolvimento da Firjan', *O Globo*, 19 February. Available at: http://blogs.oglobo.globo.com/lauro-jardim/post/cidade-paraense-fica-frente-de-sete-capitais-no-indice-de-desenvol vimento-municipal-da-firjan.html.

OPM, ICMM, and IBRAM (2012). 'The Mining Sector in Brazil: Building Institutions for Sustainable Development'. Mining Partnerships for Development—Spotlight Series 17, June. London: ICMM. Available at: https://www.icmm.com/en-gb/publications/mining-partnerships-for-development/the-mining-sector-in-brazil-building-institutions-for-sustainable-development.

Secretaria de Estado de Educação do Pará (2016a). 'Relatório de Resultados 2014'. Sistema Paraense de Avaliação da Educação (SisPAE) Pará website. Available at: https://sispae.vunesp.com.br/reports/RelatorioSISPAE.aspx?c=SEPA1401.

Secretaria de Estado de Educação do Pará (2016b). 'Relatório de Resultados 2015'. Sistema Paraense de Avaliação da Educação (SisPAE) Pará website. Available at: https://sispae.vunesp.com.br/reports/RelatorioSISPAE.aspx?c=SEPA1402.

SUS (Sistema Único de Saúde) (2015). 'Cobertura Populacional da Atenção Básica 2014'. SAGE website. Available at: http://sage.saude.gov.br/.

SUS (Sistema Único de Saúde) (2016). 'Cobertura Populacional da Atenção Básica 2015'. SAGE website. Available at: http://sage.saude.gov.br/.

Vale (2012). 'A New Impetus to Brazil's Sustainable Development'. Carajás S11D Iron Project, June 2012. Canaã dos Carajás: Vale.

Vale (2015). 'Fact Sheet: One of the Largest Mining Companies in the World'. Vale website. Available at: http://www.vale.com/EN/investors/company/fact-sheet/Pages/default.aspx.

Vale (2016a). 'About Vale'. Vale website. Available at: http://www.vale.com/EN/aboutvale/mission/Pages/default.aspx.

Vale (2016b). 'Complexo S11D Eliezer Batista'. Vale website. Available at: http://www.vale.com/brasil/pt/initiatives/innovation/s11d/paginas/default.aspx.

Vale (2016c). 'S11D: First Equipment Testing Starts', press release, 17 February. Canaã dos Carajás: Vale.

Vale (2017). Fact Sheet. Vale website. Available at: http://assets.vale.com/docs/Documents/en/investors/Company/Fact-sheet/factsheeti.pdf.

Vale Foundation (n.d.). 'Sobre a Fundação Vale'. Vale Foundation website. Available at: http://fundacaovale.org/SitePages/quemSomos.aspx.

# 31

# Capturing Economic and Social Benefits at the Community Level

## Opportunities and Obstacles for Civil Society

*Keith Slack*

## 31.1 Introduction

Civil society has a critical role to play in helping local communities capture the benefits produced by extractive industries (EI) and avoiding or mitigating the social and environmental damage these industries can cause. Payment disclosure data is a potentially powerful tool for civil society to hold governments accountable for pro-development expenditure of extractive revenues. Yet analysts have argued that transparency alone does not equate to accountability. In many contexts political dynamics present serious challenges to achieving greater accountability, even with greater public availability of revenue data. In this chapter I will examine these dynamics and the role of civil society in promoting benefit capture from the extractive industries at the local level as well as the promise (and limitations) of increased revenue data disclosure to support that role. I will look at two cases in particular, Ghana and Peru, which offer important lessons on the opportunities and obstacles for civil society engagement with the extractive industries. I will conclude by offering some proposals for strengthening this role.[1]

[1] Views expressed are those of the author and do not necessarily reflect the views of Oxfam America or Oxfam International.

## 31.2 Role of Civil Society in Communities Affected by the Extractive Industries

The term 'civil society' is used in various ways. For the purposes of this chapter I will use the World Bank's definition, which includes not only advocacy- and service-oriented NGOs but also indigenous peoples' federations, trade unions, and religious institutions (World Bank 2013). Civil society organizations (CSOs) have played a variety of roles in extractive-industries-affected communities. They might generally be described as 'watchdogs' that monitor and document the various and often-negative impacts that extractive industries projects have on human rights and the environment in these communities. In a number of contexts, particularly in Latin America, CSOs have helped to organize community opposition to extractive industries projects through local campaigns that are often supported at the national and global levels. CSOs have also helped communities to directly negotiate with companies to resolve human rights and environmental problems (Rees et al. 2012). This has in some cases involved bringing legal action against a company or utilizing global accountability frameworks and bodies, such as the Voluntary Principles on Security and Human Rights, the OECD Guidelines on Multinational Enterprises, and the Compliance Advisor/Ombudsman of the World Bank Group's IFC.

Globally, the normative framework around the extractive industries has seen dramatic changes over the past twenty years, in large part due to pressure from CSOs. To name two examples, payment disclosure and free, prior, and informed consent (FPIC) for indigenous peoples are now part of the policies of extractive corporations and international financial institutions. The EITI is a well-established global norm-setting enterprise that has been supplemented in recent years by the adoption of mandatory extractive industries payment disclosure laws in the United States (through the Dodd–Frank Wall Street Reform and Consumer Protection Act) and European Union (European Accounting Directive). The IFC, which finances extractive industries projects, has adopted both payment disclosure and FPIC in the performance standards its clients are obliged to implement. Likewise, the ICMM, an association of large global mining companies, has also adopted a policy on indigenous peoples that requires respect for FPIC.

There is significant heterogeneity among CSOs involved in the extractive industries space. Some focus primarily on environmental issues, while others are more concerned with human rights impacts. Some espouse a strongly oppositional position in relation to the extractive industries. Others are more interested in promoting reform of the sector through policy dialogue and legislative advocacy work. Corporate-backed NGOs (or 'BONGOs': business-oriented

NGOs) have appeared in some locations to do community development work on behalf of an extractive industries company.

As the extractive industries sector in developing countries has expanded and evolved over the past twenty years, so too have the interests and sophistication of CSOs involved in the space. After focusing initially on human rights and environmental impacts, CSOs have more recently begun to take on difficult technical issues such as the distribution and management of extractive industries revenues, contract analysis, and taxation. Global NGOs such as Oxfam and the NRGI have hired lawyers, tax experts, and other technical specialists to help advance policy reform work in these areas.

Payment disclosure is a critical component for this 'second-order' set of extractive industries accountability issues in which CSOs engage. In Section 31.3 I will examine the potential utility of this information for efforts to promote greater capture of the benefits produced by extractive industries at the local level. I will also frame a discussion of the political obstacles that may undermine the effectiveness of this tool, as well as broader policy reform efforts in this sector.

## 31.3 Origins of the Extractives Revenue Transparency Movement

The global movement for greater revenue transparency in the extractives sector began in 2000 with the launch of British NGO Global Witness's report on Angola's oil sector (Global Witness 2000). The report described in detail the kleptocracy of the Dos Santos regime and its theft of billions of dollars in oil revenue. The report also laid out the complicity of foreign oil companies—notably BP—in this theft. The report included a recommendation that oil companies operating in Angola, including BP, publicly disclose payments as a way for outside actors, including global and Angolan civil society, to hold the government accountable for the use of those revenues. The report brought swift condemnation of Global Witness by the Angolan government. However, the report also captured the attention of billionaire financier George Soros, who had founded the Open Society Institute (OSI). Soros believed that payment disclosure—which was consistent with his 'open society' concept—could be a tool to address problems of corruption and mismanagement in resource sectors globally (Soros 2005).

With support from Soros, Global Witness and other global NGOs established a global campaign to promote extractives payment disclosure. The campaign adopted the name 'Publish What You Pay' (PWYP) and drew participants from a broad sector of civil society across the globe. National and regional chapters of the campaign were set up with funding from Soros and other donors. Building on Global Witness's experience working on BP in Angola, the campaign initially focused on pressuring the UK government to require payment disclosure by British companies. In response to this pressure, the government of UK Prime Minister Tony Blair proposed the creation of the EITI, which was formally announced at the World Summit on Sustainable Development (Rio+10) conference in Johannesburg in September 2002. The EITI was to be a voluntary initiative in which governments, companies, and NGOs would work together to promote greater revenue transparency in the extractive sectors. The initiative was immediately criticized by some NGOs for its voluntary nature, and for promoting only aggregated data disclosure rather than disclosure on the more granular, project-by-project basis that NGOs had called for. Despite the criticism, most NGOs working in the extractive industries space decided to stay engaged with the initiative, and some—including Oxfam, Global Witness, and PWYP itself—accepted seats on the board of the initiative.

To address the weaknesses of EITI, NGOs continued to push for mandatory disclosure regulations. The US coalition of PWYP, led by Oxfam, Revenue Watch Institute (an organization spun off from OSI and now known as the NRGI), and Global Witness, made the furthest progress. Working with a securities lawyer, the coalition developed a proposal for US legislation that would require all companies listed on the New York Stock Exchange to disclose to the Securities and Exchange Commission (SEC) all payments made to governments in countries where they were operating. This information was to be included as part of the standard disclosures required by the SEC. In 2008, at the behest of the coalition, US Senator Charles Schumer and Representative Barney Frank introduced the Extractive Industries Transparency Disclosure Act, which amended the Securities Exchange Act to establish this requirement. The bill made little progress during that year but was reintroduced in 2009 by Senators Ben Cardin and Richard Lugar as the Energy Security Through Transparency Act.

In early 2010, the PWYP-United States coalition saw an opportunity with the introduction of legislation to reform the financial sector following the 2008 financial crisis. The coalition worked closely with Representative Frank, then chairman of the House Financial Services Committee and lead sponsor of the financial reform legislation. Frank agreed to support the extractives transparency legislation as part of the overall financial reform package. Following extensive congressional debate and manoeuvring, the extractives transparency language was added to the bill and approved by the House–Senate conference committee in July 2010. The language appeared as Section 1504 of the 2,300-page bill. The overall financial sector reform bill, including the extractives language, was signed by US President Barack Obama on 21 July 2010.

Following the passage of the legislation, the American Petroleum Institute (API), representing the interests of major oil companies, sued the SEC to block implementation of Section 1504, arguing *inter alia* that the disclosure requirement constituted 'compelled speech' and as such was a violation of its right to free speech under the US Constitution. The US District Court for the District of Columbia, which heard the case, sided with the API and compelled the SEC to remake the rule that would govern the implementation of the 1504 legislation. Various suits and countersuits have followed, with Oxfam America leading the legal effort on behalf of civil society to push for publication and implementation of the rule. The SEC published a new rule in June 2016, which was generally well received by transparency advocates as it maintained most of the key provisions of the original rule.

The passage of Section 1504 had an immediate knock-on effect in Europe, where the European Union began to consider similar revenue disclosure requirements, which were adopted in 2014. Similar disclosure requirements have been adopted in Canada and proposed in South Africa and Australia, all key jurisdictions for the raising of capital for extractives companies. Regrettably, one of the first acts of the Trump administration in February 2017 was to sign legislation vacating the SEC's rule for implementing Section 1504. Although a major setback for extractives transparency proponents, however, the action only requires the SEC to publish a new version of the rule, and does not alter the underlying legislation. Moreover, it does not affect disclosure requirements established in other jurisdictions.

At the heart of the effort to promote extractive industries payment disclosure was a simple idea: that the disclosure of revenue payment information can be a tool for holding governments accountable for where those revenues go. Sunlight, as former US Supreme Court Justice Louis Brandeis famously said, can be the best disinfectant (Brandeis 1913). The 'sunlight' of revenue transparency can 'disinfect' the corruption, mismanagement, and waste that too often characterizes extractive industries revenues. Getting information into the hands of activists, journalists, and ordinary citizens helps to level the playing field with powerful government and corporate actors. It can create popular pressure to address corruption or the misallocation of revenues.

In this sense, the 'theory of change' behind extractive payment disclosure is in line with broader current theory about transparency and data disclosure. We live in an era of 'big data' that holds, according to some proponents, the potential to be a 'game changer' in terms of empowering people to demand better performance from governments, corporations, and other social actors. Data and disclosure can help to rectify asymmetries of information between governing and governed that will ultimately produce broad social benefits. Moreover, resource-rich countries have been empirically found to be less transparent than their non-resource-rich analogues (Williams 2010).

There is some evidence in the extractive sector that greater transparency has indeed begun to produce some of these benefits. In Nigeria, for example, the EITI process revealed US$2.6 billion in missing oil revenues and a failure of oil companies to pay US$9.9 billion in royalties (De Sa 2013). There are some other notable positive examples outside the extractive industries sector. CSOs in Ghana have effectively used information and social accountability to drive improvements in government-provided health and agricultural programmes (Dogbe and Kwabena-Adade 2012). Similarly, in Uganda, a government-sponsored newspaper campaign using data on monthly education funding transferred to local districts led to improved enrolment and test scores (Reinikka and Svensson 2005).

More broadly, however, the literature suggests that information disclosure and transparency have had at best limited or mixed effects on social accountability efforts in developing countries. Some studies have found that voting patterns in developing countries can be influenced by information (Pande 2011), and that transparency can have a generally positive effect in promoting accountability (Kosack and Fung 2014). Sovacool et al. (2016), however, have found that EITI implementation has had no effect on a country's governance performance, although they acknowledge that it may be too soon in the history of EITI to draw definitive conclusions. Mejia-Acosta (2013) points out that there is 'no conclusive evidence' to show that transparency-to-accountability initiatives (such as EITI) have contributed to better development outcomes from extractive industries, and that the causal links between transparency and accountability in the extractive industries sector have not been mapped out.

The effectiveness of transparency is also said to depend significantly on the particular political context in which the transparency data are introduced. Transparency cannot be used, in this analysis, to 'shame the shameless' (Fox 2007) and is unlikely to have much impact in autocratic regimes (Frynas 2010).

There is even some analysis to suggest that transparency and information disclosure can actually have a negative impact on social accountability efforts by *inter alia* creating 'resignation' rather than 'indignation' at the degree of governmental corruption revealed by the disclosures (Bauhr and Grimes 2014). Transparency can also lead to an *increase* in corruption, as it can be used to better identify which officials need to be bribed (Kolstad and Wiig 2009).

As noted at the outset of this discussion, there is a general view among practitioners and academics that transparency alone does not equate to accountability. Simply disclosing information will not produce the desired positive effect for social accountability that one would want to see. Political structures and imperatives can thwart the ability of civil society to make a meaningful impact on how extractive industries revenues are managed

(Collier 2010; Fuhr 2015; Keefer and Khemani 2004; Khemani 2007; Robinson et al. 2006). These political obstacles are in some cases deeply engrained in the political culture and history of particular countries. They also relate to the nature of extractive industries themselves, which require a greater degree of governmental intervention than other industrial sectors. Taken together, these obstacles can defeat even the best-designed policies and the strongest transparency regimes. To make large and powerful sectors such as extractive industries accountable at the local level requires addressing—or at least managing—these issues in some way. I will discuss these obstacles in more detail in Section 31.4.

## 31.4 The Politics of Extractive Industries Revenue Distribution

Ensuring that the benefits of extractive industries are captured at the local level requires an understanding of the inherently political nature of the management of these industries. Because of the direct role that governments must necessarily play in engaging with extractive industries at all stages, opportunities for political manipulation are rife. This begins with the concessioning process, in which extractive industries companies must purchase a permit to do exploration in a particular area. The permit is granted by the relevant government authority, often with no public consultation or other form of oversight.

Some countries require the redistribution of extractive revenues to affected communities to promote local development and compensate for damages caused by extractive industries activities. Examples include Peru, the Dominican Republic, Ghana, DRC, and Burkina Faso. The full amount of revenue owed to local governments is rarely transferred to them, however. This can be due to bureaucratic inertia as well as political bias against redistribution held by central government elites who distrust local government officials. In Ghana, a study by the African Centre for Energy Policy (ACEP) found that political considerations, rather than efficiency or value for money, appear to have influenced the expenditure of oil revenues on infrastructure projects, resulting in only partially funded and incomplete road projects (ACEP 2013).

Politically motivated expenditure is of course not unique to the employment of extractive industries revenues. It undoubtedly affects governance at least to some degree in all countries. Political 'short-termism'—focusing on immediate short-term results rather than longer-term development plans—is also a common problem. Politicians are rewarded not for their long-term vision for development in a community, but for what they can deliver to the population today. This is a particularly concerning problem in the extractive industries context because of the non-renewable nature of the extracted resources. In most cases, communities will only have a limited time to benefit from those resources.

Development planning in rural areas of many developing countries where extractive industries operations take place is often weak. Central governments often demonstrate little interest in promoting such development, because their own political and economic interests are centred in the capital city. Communities in extractive-affected areas are thus doubly marginalized by (i) suffering the impacts of extractive activity and (ii) lacking the political clout to force national elites to address their problems and take action to ensure that benefits are effectively redistributed.

Extractive companies themselves contribute, wittingly or not, to the short-term political time horizons that exist in local communities. Companies' commitments to a community usually exist only as long as their commercial interest. Whether a community actually improves its development prospects as a result of a company's presence can be immaterial to the company's business interests. This is not to suggest that companies are not genuinely interested in communities' welfare. Some extractive industries companies have worked to make positive contributions to local communities through the establishment of community foundations and training programmes, and the construction of infrastructure. But few of these kinds of investment continue after a project closes down, or even necessarily after one company sells the project to another. The company's interests are thus also short-term (even if, as in some cases, they have a presence in the community for fifty years or more).

In light of these challenges of political structures and incentives, it is not easy to envision ways for civil society to overcome them. A key dilemma is for CSOs to engage with these political dynamics without also themselves becoming political (in the sense of politically 'partisan'). The answers may also lie in increasing civic education about what good development and good governance look like. This could help to create demand for better long-term development planning, and for politicians who will articulate a vision for achieving it. Transparency and information disclosure, despite the limitations described above, could be a useful tool for strengthening this education process. In Sections 31.5 and 31.6 I will examine in more detail how these conflicting political imperatives play out in two specific cases: Peru and Ghana.

## 31.5 Ghana

Until recently Ghana enjoyed a reputation as an African 'success' story. Since undergoing structural adjustment in the 1980s, it has had good rates of economic growth and achieved lower-middle income status. It has significantly reduced poverty and improved access to education. The country is generally seen as stable and relatively non-corrupt, certainly in comparison with the 'sorry mess' that is its neighbour Nigeria (Burgis 2015).

Ghana is also a leading minerals producer: the second-largest gold producer in Africa after South Africa, and as of 2011 a mid-level oil producer. Natural resources produced US$1.2 billion in revenues for the government in 2015, according to the EITI. Despite this wealth and Ghana's generally positive reputation for fiscal governance, there is a sense among observers inside and outside the country that it has not benefitted as much as it could have from its resource wealth, particularly from more than a hundred years of industrial mining activity. The social and environmental impacts of large-scale mining have been severe, and poverty remains high in mining-affected areas. Indeed, Ghana's star has dimmed over the last two years in the face of the collapse in oil prices, which forced the country to seek an IMF bailout to cover the government's borrowing against oil futures (Matthews 2016).

If one looks more closely at the Ghana case, one sees in its governance structures and practices a number of the entrenched political dynamics that militate against good resource revenue management at both the national and local levels. These dynamics are not unique to Ghana, but the country offers a particularly stark example, given its resource wealth and otherwise positive governance reputation. These will be discussed in some detail below.

Formally, the Ghanaian parliament has responsibility for approving budgets and, importantly, approving contracts between oil and mining companies and the government. In theory this role presents a potentially strong avenue for accountable management of the extractive sectors. In practice, however, the parliament is dominated by the executive branch, which controls candidate lists and awards loyal members of parliament with ministerial positions and all the financial benefits and trappings those entail (Oxfam America 2016a).

Ghana's annual national budgeting process is largely a paper exercise. There are significant discrepancies between approved budgets and actual expenditures due to 'off-budget' expenditures that the executive branch makes after the formal annual budget has been approved. The role of members of parliament and the ability of civil society to influence the budget development process are limited (Oxfam America 2016a). The dominance of the executive branch over parliament in budgeting and other matters leads Ayee et al. (2011: 15) to state simply that 'in practice, the parliament in Ghana is not an autonomous organ of the state'.

Even more starkly, members of parliament are often appointed to the boards of the local subsidiary of a mining company operating in their district. This has served to undermine the oversight function of parliament on mining activities and has strengthened members' support of the companies (Ayee et al. 2011).

At the local level, Ghana is governed by district assemblies (DAs), institutions that represent the most local-level governance entity in the country. These units were created as part of a decentralization process established with the return to democracy in 1993 following two decades of military rule. In theory, the DAs could be an effective unit of local-level governance and accountability of extractive industries. In reality, however, little progress has been made in giving the DAs real power and authority, and the president himself appoints one third of the DAs' membership, thus ensuring party loyalty to the president (Ayee et al. 2011). Moreover, Standing and Hilson (2013) describe a widespread perception of corruption in the assemblies among Ghanaians and development agencies.

Compounding the difficulties of effective oversight and governance of extractive industries at the local level is the presence of traditional authorities or chiefs. Local governance in Ghana is effectively a hybrid of traditional chieftaincy rule and a modern bureaucratic state system. Under the Ghanaian constitution, chiefs control local landownership and receive a direct disbursement of 20 per cent of mining royalties. Thus the chiefs have a direct financial interest in mining activities, even though ostensibly their role as traditional leaders should require them to look after the interests of their entire community. These funds are used for 'expenditures other than those that benefit the local communities involved' (ICMM 2007: 77). This conflict of interest has led chiefs in some cases to sell land to mining companies without the consent (or even knowledge) of the occupants. This has contributed to conflict and protest in some mining communities (Standing and Hilson 2013).

The accountability problems that Ghana experienced in managing its mining wealth led to a series of efforts to try to avoid such problems with the advent of the country's oil sector, which officially came online in 2011. Most notable among these was the creation of the Public Interest Accountability Committee (PIAC), which was included in the Petroleum Revenue Management Act passed in 2011. The mission of the PIAC is to oversee government management and expenditure of oil revenues. Its membership is comprised of representatives of Ghanaian civil society, trade unions, traditional leaders, and business groups. In principle, the creation of the PIAC was a major step forwards in resource revenue accountability. In practice, however, the PIAC has not been able to do its job effectively, as it has been starved of resources by the parliament. Some analysts have suggested that the Ghanaian executive only accepted the creation of the PIAC due to donor pressure on 'good governance' reforms, and from the beginning was not committed to its success (Oppong 2016).[2]

[2] See also the analysis of the macroeconomic problems that have resulted in Bawumia (2017).

Despite the deeply engrained political and structural impediments, the cause of accountable resource revenue management at the local level in Ghana is not lost. Ghana is possessed of a vociferous and capable civil society, consisting of a broad range of organizations with expertise in community development, environment, human rights, and fiscal governance, among other issues. Civil society played an important role in getting Ghana to join the EITI and in the creation of the PIAC. In 2013, CSOs led a successful popular campaign with the title 'Oil for Agriculture', which convinced the Ghanaian government to commit 15 per cent of oil revenues to small-scale agriculture (Offenheiser 2014).

In the Shama district in Ghana's Western Region, the Ghanaian NGO Friends of the Nation has led innovative work in partnership with the DA to support participatory budgeting and public oversight of expenditure of oil revenues. Friends of the Nation has helped to organize various social accountability mechanisms, including a 'people's forum' and town hall meetings with members of the DA to discuss budget and expenditure issues and development of district development plans. These efforts have resulted in increased expenditure on community development projects, a generally harmonious relationship with and favourable attitude towards the DAs of community members, and, according to one member of the DA, an increased willingness on the part of local citizens to pay their taxes to the DA, given that they believe that they have a meaningful voice in how their taxes are spent (Friends of the Nation 2016).

Whether these bright spots of civil society engagement in Ghana on extractive industries fiscal governance are enough to significantly reshape the broader political dynamics surrounding these issues is an open question. Given its vibrant civil society, relatively subdued ethnic politics, lack of violent conflict, and open political system, Ghana would seem to stand a better chance than most resource-dependent developing countries. Ghanaian civil society may also be better placed than most to take advantage of the revenue disclosure data that will begin to become available. Nevertheless, the challenges even in Ghana remain steep. The experience of another resource-dependent country, Peru, offers another perspective on these challenges—one in which high levels of transparency have not been enough to overcome political obstacles to better deployment of natural resource revenues at the local level.

## 31.6 Peru

Peru is an advanced minerals producer and middle-income country. Like Ghana, it has an active civil society engaging on extractive industries issues from a variety of angles, including human rights, indigenous rights, environment, and fiscal governance. Also like Ghana, the country has struggled to translate its oil and mineral wealth into local-level development. The perceived lack of benefits to local communities from extractive industries projects has driven a cycle of violent conflict around extractive projects that has affected the country over the past fifteen years. The province of Cajamarca, home to Latin America's largest goldmine and the recipient of US$700 million in mining revenues between 2010 and 2015, is now Peru's poorest province (INEI 2015).

The conflicts within Peru's extractive sector have occurred despite a relatively high degree of formal transparency in the sector. The Ministry of Energy and Mines publishes mining project revenue data on its website. The country has been compliant with the EITI since 2012. It scores 75 out of 100 on the Open Budget Index, equal to the United Kingdom and above Germany (Open Budget Index 2015).

Under Peru's *canon minero* legal framework, 50 per cent of mining revenues are to be transferred back to mining production areas. While this is potentially a very significant source of funding for local-level development, expenditure at that level has been problematic due to a lack of local government capacity, poor development planning, and political incentives for short-term spending (Arellano-Yanguas and Mejia-Acosta 2014).

Like Ghana, Peru has a dominant executive and a weak legislative branch that exercises little effective oversight over the executive. The government lacks the capacity and political will to effectively audit extractive industries companies. The tax authority is unable to assess critical data such as production volumes, costs, and transaction prices (Oxfam America 2016b).

Peru's legislative weaknesses stem from the era of former President Alberto Fujimori, who gutted the authority of congress and established a patronage system in which members of congress depend on the executive for their seats. The auditing function of congress has become a 'political game' rather than being designed to hold the executive accountable for the appropriate management of the country's resources. Lack of public faith in political parties has made them largely transient entities that are unable to build the political will or public support to take on corruption and mismanagement. Analysts also describe a 'revolving door' between extractive companies and the government, such that government officials, some of whom have had oversight responsibility for the extractive industries sector, leave to take jobs with extractive industries companies (Oxfam America 2016b).

Peru's media are also heavily concentrated in the hands of a small elite. This is another legacy of Fujimori, who sought control of the media in order to eliminate them as an obstacle to his agenda, which was later revealed to be deeply corrupt and for which he was ultimately jailed (Durand 2016).

Participatory budgeting is widely practised in Peru. Yet the costs of individual participation in budgeting processes can be high. The World Bank (2010) has estimated the costs of participation for a rural Peruvian in an entire year of participatory budgeting at US$195, or 95 per cent of a monthly salary at minimum wage. McNulty (2012) has documented the limitations of the participatory budgeting process, including little engagement by women and women's organizations, complex technical rules, and a lack of formal sanction for politicians who do not carry out the participatory process in the spirit of the law.

The Ministry of Economy and Finance plays a particularly dominant role in the country's fiscal governance, including with regard to EI. This ministry largely controls the national budgeting process. It also makes key decisions about the distribution of mining revenue under the *canon minero*, and can block the transfer of revenue to local governments. The congress is left 'acting like a beggar' vis-à-vis the executive branch, and has no effective control over it on fiscal issues (Mauro 2014).

The breakdown in formal accountability for the use of mining revenues is perhaps seen most starkly in the province of Ancash, home to Peru's largest mine, Antamina. Instead of funding local development priorities, mining revenues have helped to fuel rampant corruption and the creation of what has been called a 'mafia mini-state', characterized by intimidation of the local media and judiciary, and contract murders of political foes of the provincial governor (Bajak 2014a; *The Economist* 2014). The district of San Marcos receives US$50 million per year in mining royalties, but there are no paved roads and no hospital (Bajak 2014b). Ancash is not an isolated case: in 2014, twenty two out of twenty-five regional presidents were under federal investigation for corruption. By the end of 2014, three were in jail, two had resigned, and one was barred from holding public office (Oxfam America 2016b).

As noted above, Peru does get high marks by some transparency standards. Yet the implementation of this transparency is problematic. The country's law on access to information is often not complied with by public institutions (for which there is no effective sanction) and is ignored by private companies. There are also questions about the validity of the information that is disclosed by the government through various portals (Oxfam America 2016b).

Despite the deep challenges presented by Peru's rampant corruption, executive dominance, and weak formal accountability structures, CSOs have made some advances in promoting greater accountability of extractives revenue management. These include identifying discrepancies in the transfer of *canon minero* revenues and highlighting problems with the execution of projects carried out with *canon* funds. The disclosure of additional revenue data through mechanisms such as Dodd–Frank 1504 may further strengthen these efforts.

The Peruvian research organization Propuesta Ciudadana has done an analysis of two cases of local governance that perhaps hold interesting lessons for other countries. In Arequipa in southern Peru, they found that the presence of a regional president who was personally committed to open governance and transparency was an important factor in more effective management of mining revenues in the region. In the northern Peruvian department of San Martin, Propuesta's research concluded that the relatively strong governance performance of the regional government was at least in part connected to investment in governance support by the US Agency for International Development, which was channelled as a result of US interests in fighting coca production in that part of Peru (Grupo Propuesta Ciudadana 2015).

The cases of Ghana and Peru demonstrate the deeply entrenched challenges of political economy to the effective deployment of extractive industries revenues at the local and national levels. These challenges reflect the history and structures of the respective states and their surrounding political cultures. They also reflect internal power dynamics that link to the nature of extractive industries themselves and their geographies (as described above). Addressing these challenges goes beyond the extractive industries, of course, but in countries like Peru and Ghana that are dependent on these industries the path to effective state functioning necessarily runs through them. In Section 31.7 I will address how CSOs (and others) can apply knowledge of these political contexts to promoting greater accountability of the extractive sectors for local development.

## 31.7 Strengthening the Role of Civil Society in Capturing Local Benefits

Ensuring that extractive industries projects produce benefits at the local level for the communities on which these operations impact most directly is one of the most difficult questions in the broader field of extractives' role in development. There are few positive examples globally; success stories in poorly governed developing countries are even scarcer. As noted earlier, this has to do with a number of factors, including the destructive nature of the industries themselves, the mechanisms by which they generate wealth and concentrate it in relatively few hands, governance weakness across the entire production chain, corporate time horizons, and large information asymmetries among key stakeholder groups (governments, companies, and communities.) The deeply political nature of decision-making around these industries, in which poor and politically marginalized communities lose out to elite interests and intra-national power dynamics, also work to limit the benefits that communities obtain.

Transparency and social accountability are in theory a mechanism to push back against these dynamics and empower affected communities to reclaim a fairer share of the benefits derived from resources taken from their lands. However, as described above, transparency does not automatically equate to accountability. The barriers to participation in social accountability initiatives, especially in developing countries, can be high, including time taken away from work and family, and the understandability of the information itself. There is also evidence to suggest that there is a gender bias in such initiatives that impedes the effective participation of women (Bradshaw et al. 2016).

The solutions to these problems will undoubtedly vary significantly according to each case. Getting governments to care about their populations and make decisions in their best interests rather than on parochial political concerns is perhaps the defining problem of political development. It is also key to ensuring that extractive industries provide more benefits to local communities. I propose four broad areas of action that, if focused on in the right contexts, could potentially produce momentum for achieving positive change. These are: (i) strengthening civil society's technical capacity; (ii) providing civic education; (iii) focusing on countries with a minimum level of governance capacity; (iv) identifying and promoting learning from legitimately positive examples. I will take each of these in turn.

### 31.7.1 *Strengthening Technical Capacity*

As we saw earlier, a wide variety of types of organization engaging with extractive industries come under the general rubric of 'civil society'. They vary in their focus and objectives. Standing and Hilson (2013) note that in Ghana urban-based and professionalized NGOs funded by foreign donors do not necessarily represent the interests of affected communities, although they may claim to do so. This is also an issue in other EI-dependent countries such as Peru.

Notwithstanding these caveats, civil society has played a key role in driving change in the extractive industries sector (as seen in the Peru and Ghana cases). However, in many cases they face serious limitations of resources and technical capacity. This is particularly the case with regard to their ability to analyse technical data, including the terms of production contracts, national and local municipal budgets, and environmental impact assessments. In recent years, working with the support of some global NGOs and technical experts, local CSOs have been able to engage on these issues and produce alternative technical analyses that have helped to reframe public debate on extractive industries governance issues (Cordaid et al. 2016; Moran 2001).

The ability of CSOs to prepare these kinds of analysis should be strengthened and made more sustainable. At present, much CSO technical analysis is provided by non-indigenous experts, who do not necessarily build local capacity. The creation of an independent technical support funding mechanism to support the production of technical analyses—and to build local capacity—would be an important innovation that could help to address the asymmetries that so deeply characterize the extractive industries sector. This would include analysing and employing revenue disclosure data. The goal could be the development of a global cadre of independent, national technical extractive industries experts, with some means of remuneration to sustain their work and avoid the temptation of deploying their technical skills for well-paying extractive industries companies.

Technical data analysis skills within local CSOs will also become even more important as more revenue disclosure data become available. Also important are political analysis skills that will enable CSOs to identify opportunities for utilizing the data to drive policy change processes that might ultimately result in more benefits being delivered to local communities. Similarly, engaging with the local media and building their capacity to understand and report on these issues is also essential for generating informed public opinion and mobilizing it to promote change.

### 31.7.2 *Providing Civic Education*

Governance is poor in most developing countries. Governments fail to provide basic public services such as security, infrastructure, healthcare, and education. Likewise, populations may be conditioned to expect poor performance from government, and may not be fully aware of what they have a right to expect from their governments in terms of service delivery. This can create a vicious circle in which poor services lead to lowered expectations, which create more poor services as such services become the norm. Breaking out of this cycle requires education and information about community rights and what good governance performance can and should look like.

Even taking into account the limitations of transparency described above, transparency can be an important complementary tool alongside education. Armed with (i) an awareness of what good governance is and (ii) information to help demand better performance, citizens can press their governments for better delivery. To be effective, data yielded by transparency initiatives must be plugged into a deliberate influencing strategy and approach. Data alone will not be sufficient to produce the desired outcome of greater accountability and delivery of services. CSOs and donors could better coordinate around promising cases that have realistic possibilities of converting into success stories.

### 31.7.3 *Focusing on Appropriate Countries*

In selecting such contexts, it is important to recognize that for some countries, it may simply be too politically difficult to try to fix the problems of the mismanagement of EI. A minimum threshold level of governance is likely required for transparency and other interventions to have positive prospects. This observation is consistent with one of the key recommendations of the World Bank's Extractive Industries Review, which called for 'explicit core and sectoral governance requirements' to be met before the World Bank Group invested in an extractive project (Extractive Industries Review 2003). It is also consistent with observations that transparency is unlikely to have much effect in autocratic regimes.

It may be better for donors and CSOs to focus their efforts in countries with the best chances of progress. This does not mean 'writing off' some countries permanently. It would mean, however, not attempting to address resource mismanagement issues in countries where the current political contexts are such that they likely pose insurmountable obstacles to success. As described above, Peru and Ghana, although facing significant governance challenges, are democratic and possess vibrant civil societies. These conditions would suggest that addressing the problems of persistent rural poverty and poor service delivery should not be beyond the grasp of Peruvian and Ghanaian society. In such contexts, civic education focusing on the 'demand' side of good governance and what citizens have a right to expect (as described earlier) could potentially bear fruit.

### 31.7.4 *Promoting Positive Examples*

If Peru and Ghana were ultimately to become examples of good resource governance, it would be critically important to ensure that those lessons were widely shared and adapted as appropriate to other contexts. In general, the power of positive examples should be better harnessed to demonstrate what is possible in addressing the political aspects of resource governance. Currently there are few positive examples of effective national-level natural resource management in developing countries beyond the oft-cited cases of Chile and Botswana. Relatively little effort has been made to unpack these cases and examine what lessons might (or might not) apply to other countries seeking to implement effective national natural resource revenue management policies. It would be useful to expand the universe of positive examples to find elements of effective resource revenue management at the national level that could be highlighted and studied more thoroughly.

Clearly, the simple existence of a handful of positive examples alone will not be enough to override the political forces that militate against the effective use of resource revenues in developing countries. The learning from these examples needs to be analysed and adapted, where feasible, to other contexts. At present, however, too little is known about what can work in governance-challenged resource states (Ross 2014). Greater coordination among donors could also be beneficial in this realm. Rather than dispersing their efforts across a wide range of countries, it could make sense for the bilateral and multilateral donors referenced above to coordinate their extractive industries and development initiatives more to focus in on a select group of promising cases. Donor countries could also do more to make their own extractive industries companies more transparent and accountable. This is particularly true for countries such as the United States, Canada, Australia, and China, which have significant extractive industries interests in developing countries.

## 31.8 Conclusions

Resource-rich developing countries are at an important 'moment of truth'. Over the past twenty years a massive amount of analysis has been done on the 'resource curse' and the development of policy tools and recommendations to address it. Transparency is one such tool although, as discussed, it is far from being a magic bullet for natural resource management issues. Despite the volume of analysis and the tools developed, resource-rich countries generally did not use the recent commodities price boom (2002–13) to their full advantage to reduce poverty and diversify their economies. Added to this, we are now amidst a commodities price downturn that analysts believe may be long term. This is combined with a renewed push to phase out the use of fossil fuels in order to stave off the worst effects of climate change. While undoubtedly in the best interests of the planet, such a phasing-out will have significant development repercussions for countries that depend on the export of oil, gas, and coal (Caney 2016).

The scenario described above is not necessarily encouraging for resource-dependent countries, or for civil society efforts to try to ensure that local communities capture more of the benefits produced by EI. On a more positive note, however, the years of analysis have identified how governments can 'get it right' with regard to the management of their extractive sectors. This provides the basis for civil society and others to engage with governments and companies to try to ensure that this happens. Whether governments will choose to do the right things (about which they can no longer plead ignorance) comes down now to questions of politics. Civil society will ultimately succeed or fail in driving better development outcomes from extractive industries investments to the extent to which it can effectively engage with, and in some cases overcome, these political dynamics.

## References


ACEP (2013). 'How a Good Law May Not Stop Oil Money from Going Down the Drain'. Accra: African Centre for Energy Policy.

Arellano-Yanguas, J. and A. Mejia-Acosta (2014). 'Extractive Industries, Revenue Allocation, and Local Politics', Working Paper 2014–4. Geneva: United Nations Research Institute for Social Development.

Ayee, J., T. Soreide, G. P. Shukla, and T. M. Le (2011). 'Political Economy of the Mining Sector in Ghana', Policy Research Working Paper. Washington, DC: World Bank.

Bajak, F. (2014a). 'Peru Mining Boom Leaves Highlanders Behind'. New York: Associated Press.

Bajak, F. (2014b). 'Political Murder Betrays Violence, Corruption in Peruvian State's "Mini-Dictatorship"'. New York: Associated Press.

Bauhr, M. and M. Grimes (2014). 'Indignation or Resignation: The Implication of Transparency for Societal Accountability', *Governance: An International Journal of Policy, Administration and Institutions*, 27(2): 291–320.

Bawumia, M. (2017). 'Oil Discovery and Macroeconomic Management: The Recent Ghanaian Experience', WIDER Working Paper 2017/185. Helsinki: UNU-WIDER.

Bradshaw, S., B. Linnker, and S. Overton (2016). 'Gender and Social Accountability: Ensuring Women's Inclusion in Citizen-led Accountability Programming Relating to Extractive Industries', Oxfam America Research Backgrounder. Washington, DC: Oxfam America.

Brandeis, L. (1913). 'What Publicity Can Do', *Harper's Weekly*, 20 December.

Burgis, T. (2015). *The Looting Machine: Warlords, Oligarchs, Corporations, Smugglers, and the Theft of Africa's Wealth*. New York: Public Affairs.

Caney, S. (2016). 'Climate Change, Equity, and Stranded Assets', Oxfam America Research Backgrounder. Washington, DC: Oxfam America.

Collier, P. (2010). 'The Political Economy of Natural Resources', *Social Research*, 77(4): 1105–32.

Cordaid, T., Kenya Civil Society Platform on Oil and Gas, and Resources for Development Consulting (2016). 'Potential Government Revenues from Turkana Oil'. Nairobi: Kenya Civil Society Platform on Oil and Gas.

De Sa, P. (2013). 'From Transparency to Accountability'. Washington, DC: World Bank. Available at: http://www.worldbank.org/en/news/feature/2013/12/19/from-transparency-to-accountability.

Dogbe, T. and J. Kwabena-Adade (2012). 'Ghana: Budget Monitoring by SEND-Ghana and its Partners Helps Improve Nutrition for Children and Support Local Farmers'. Washington, DC: International Budget Partnership.

Durand, F. (2016). 'Cuando el Poder Extractivo Captura el Estado: Lobbies, Puertas Giratorias y Paquetazo Ambiental en Peru'. Lima: Oxfam America.

*Economist, The* (2014). 'Peru's Italian Job', *The Economist*, 12 April.

Extractive Industries Review (2003). 'Striking a Better Balance: The World Bank Group and Extractive Industries: The Final Report of the Extractive Industries Review'. Jakarta: Extractive Industries Review.

Fox, J. (2007). 'The Uncertain Relationship between Transparency and Accountability', *Development in Practice*, 17(4–5): 663–71.

Friends of the Nation (2016). 'Awakening Active Citizenship through Social Accountability'. Accra: Friends of the Nation.

Frynas, J. G. (2010). 'Corporate Social Responsibility and Societal Governance: Lessons from the Oil and Gas Sector', *Journal of Business Ethics*, 93: 163–79.

Fuhr, H. (2015). 'Avoiding the Resource Curse: Political, Economic, Domestic, and International Factors to Foster Competitive Arrangements', Proceedings of the International Conference on Perspectives on the Development of Energy and Minerals Resources. Honolulu: University of Hawaii at Manoa.

Global Witness (2000). 'A Crude Awakening: The Role of the Oil and Banking Industries in Angola's Civil War and the Plunder of State Assets'. London: Global Witness.

Grupo Propuesta Ciudadana (2015). 'Buenas Practicas de Gestion Decentralizada'. Lima: Grupo Propuesta Ciudadana.

ICMM (2007). 'Ghana Country Case Study. The Challenge of Mineral Wealth: Using Resource Endowments to Foster Sustainable Development'. London: ICMM.

INEI (2015). 'Sintesis Estadistica 2015'. Lima: Instituto Nacional Estadistica e Informatica.

Keefer, P. and S. Khemani (2004). 'Democracy, Public Expenditures, and the Poor'. Washington, DC: World Bank.

Khemani, S. (2007). 'Can Information Campaigns Overcome Political Obstacles to Serving the Poor?' Washington, DC: World Bank.

Kolstad, I. and A. Wiig (2009). 'Is Transparency the Key to Reducing Corruption in Resource-rich Countries?' *World Development*, 37(3): 521–32.

Kosack, S. and A. Fung (2014). 'Does Transparency Improve Governance?' *Annual Review of Political Science*, 17: 65–87.

Matthews, C. (2016). 'Ghana's Success Story Built on Gold, Oil and Cocoa Is Foundering', *The Guardian*, 26 January.

Mauro, R. (2014). 'La Construccion Politica del Presupuesto del Republica', in E. Toche (ed.), *Peru Hoy: Mas a la Derecha Comandate*, 213–32. Lima: Centro de Estudios y Promocion del Desarrollo.

McNulty, S. (2012). 'An Unlikely Success: Peru's Top-down Participatory Budgeting Experience', *Journal of Public Deliberation*, 8(2): article 4.

Mejia-Acosta, A. (2013). 'The Impact and Effectiveness of Accountability and Transparency Initiatives: The Governance of Natural Resources', *Development Policy Review*, 31(1): S89–105.

Moran, R. (2001). 'An Alternative Look at a Proposed Mine in Tambogrande, Peru'. Washington, DC: Oxfam America.

Offenheiser, R. (2014). 'What Do Oil Revenues Have to Do with Smallholder Farmers?' Oxfam America. Available at: http://politicsofpoverty.oxfamamerica.org/2014/01/oil-revenues-smallholder-farmers/.

Open Budget Index (2015). 'Open Budget Survey 2015: Peru'. Available at: http://www.internationalbudget.org/wp-content/uploads/OBS2015-CS-Peru-English.pdf.

Oppong, N. (2016). 'Ghana's Public Interest and Accountability Committee: An Elusive Quest for "Home-Grown" Transformation in the Oil Industry', *Journal of Energy & Natural Resources Law*, 34(3): 313–36.

Oxfam America (2016a). 'The Weak Link: The Role of Local Institutions in Accountable Natural Resource Management: Ghana'. Washington, DC: Oxfam America.

Oxfam America (2016b). 'The Weak Link: The Role of Local Institutions in Accountable Natural Resource Management: Peru'. Washington, DC: Oxfam America.

Pande, R. (2011). 'Can Informed Voters Enforce Better Governance? Experiments in Low Income Democracies'. Available at: http://scholar.harvard.edu/files/rpande/files/caninformedvotersenforcebettergovernance.pdf.

Rees, C., D. Kemp, and R. Davis (2012). 'Conflict Management and Corporate Culture in the Extractive Industries: A Study in Peru', Corporate Social Responsibility Initiative Report 50. Cambridge, MA: Harvard University, John F. Kennedy School of Government.

Reinikka, R. and J. Svensson (2005). 'Fighting Corruption to Improve Schooling: Evidence from a Newspaper Campaign in Uganda'. Washington, DC: World Bank.

Robinson, J. A., R. Torvik, and T. Verdier (2006). 'Political Foundations of the Resource Curse'. *Journal of Development Economics*, 79: 447–68.

Ross, M. (2014). 'What Have We Learned about the Resource Curse?' *Annual Review of Political Science*, 18: 239–59.

Soros, G. (2005). 'Transparency Can Alleviate Poverty', *Financial Times*, 17 March.

Sovacool, B., G. Walter, T. Van De Graaf, and N. Andrews (2016). 'Energy Governance, Transnational Rules, and the Resource Curse: Exploring the Effectiveness of the Extractive Industries Transparency Initiative (EITI)', *World Development*, 83: 179–92.

Standing, A. and G. Hilson (2013). 'Distributing Mining Wealth to Communities in Ghana: Addressing Problems of Elite Capture and Political Corruption'. Bergen: Anti-Corruption Resource Centre.

Williams, A. (2010). 'Shining a Light on the Resource Curse: An Empirical Analysis of the Relationship between Natural Resources, Transparency, and Economic Growth'. *World Development*, 39(4): 490–505.

World Bank (2010). 'Peru: Evaluacion del Presupuesto Participativo y su Realcion con el Presupuesto por Resultados'. Washington, DC: World Bank.

World Bank (2013). 'Defining Civil Society'. Available at: http://web.worldbank.org/WBSITE/EXTERNAL/TOPICS/CSO/0,,contentMDK:20101499~menuPK:244752~pagePK:220503~piPK:220476~theSitePK:228717,00.html.

# 32

# How Do We Legislate for Improved Community Development?

*James M. Otto*

## 32.1 Challenges of Non-renewable Resources-led Development

It is the nature of non-renewable natural resources that an extractive minerals project has a finite lifespan and that once the minerals have been depleted, the project will cease operation. Governments have choices in deciding what they want to achieve from the extraction of minerals during a project's life and have a variety of regulatory and other tools by which to achieve their objectives. Historically, the primary objective of many governments was to obtain fiscal revenues from an extractive project and to use larger projects as a means to build infrastructure that will benefit society. More recently, governments are additionally looking to achieve benefits for local communities both in the near term while operations are ongoing but also on a sustainable basis. According to MMSD:

> At the local level, sustainable development is about meeting locally defined social, environmental, and economic goals over the long term. Interactions between the mine and community should add to the physical, financial, human, and information resources available—not detract from them. The challenge is to ensure that the effects of interactions are regarded as positive by those affected locally as well as by the promoters of the project, and that communities develop in ways that are consistent with their own vision. This may be realized through, for example, the provision of social services, income, or skills development. (MMSD 2002: 198)

Many large projects implement corporate social responsibility (CSR) programmes which often include benefits for nearby communities. In some cases, such programmes can include binding contracts with communities or more informal written approaches, such as a non-enforceable memorandum of understanding. Increasingly, governments are mandating in legislation that extractive industry projects must have programmes for community development rather than relying on voluntary efforts.

One means that regulators have to achieve benefits at the community level is to require extractive companies to enter into community development agreements (CDAs). Not all governments that require community development programmes require CDAs; however, CDAs can be a useful approach that provides both the project and the community with a mutually agreed means to define the attributes of project-assisted development in a way that can help to manage expectations. This chapter looks at one aspect of CDAs—the legislation that can be used to regulate CDA requirements. The concept of a CDA, a contract between an economic enterprise and a community, can be applied to both the mining and oil and gas sectors, but at the present time, the use of CDAs for mining projects is becoming commonplace although they are rarely used in the petroleum sector. Here, the emphasis is on mining-sector CDA legislation.

The term 'community development agreement' is sometimes used in a broad sense but for the purposes of this chapter, a CDA is narrowly defined as 'a legally binding contract between the holder of an authorization granting the rights to extract minerals, and a community (or communities) that will be affected by the exercise of those rights, that addresses matters concerning community development'.

## 32.2 Respective Roles of Government and Investors in Community Development

The respective roles of government and extractive companies with regard to community development are complex and vary considerably from nation to nation and from project to project. There is a history of mines contributing to community development in ways ranging from the building of company towns, to guaranteeing infrastructure bonds, to paying locally imposed property taxes, to building schools and health clinics, to providing financing and training for local enterprises, and to supporting local sourcing of goods, services, and employees. There has long been a dialogue within the political science field about whether community development should be government- or private-sector-led.[1] At the heart of the argument is the issue

[1] For an introduction to mining and its growing role in local and sustainable development, and its potential negative impacts, the author recommends Evans et al. (2001), Richards (2009),

of sustainability: the miner will eventually depart when the mine closes but the government will remain and thus their respective planning horizons and long-term objectives may be quite different. Also, miners are expert at mining, while governments are experienced in developing health, education, and community infrastructure. Historically, miners have tended to regard local communities from the viewpoint of how the community can serve the mine—for example, as a source of employees and housing—rather than how the mine can serve the community, sometimes resulting in non-sustainable enclave-type development. Regardless of the academic argument about whether community development should be government- or miner-led, in practice, most communities do not care, as long as beneficial development takes place:

> Few areas present a greater challenge than the relationship between mining companies and local communities.... Widespread community demands for relevant, direct, and sustained benefits from mineral wealth are a relatively recent phenomenon, so frequently neither government institutions nor companies or communities themselves have been properly equipped to respond to them. In areas of weak governance, communities often turn to the operating companies, which have found themselves providing development services to obtain or to maintain their social licence to operate. A new relationship is beginning to emerge, based on recognition of the rights of communities and the need for community participation in decision-making. Moreover, new initiatives seek to avoid the company assuming the role and responsibilities of government, but rather focus on improving the capacity of local government and other local institutions to deliver mine-derived benefits over the long term. (MMSD 2002: xix)

The legal system can be used to define the respective roles of extractive companies in the community development process. Requiring CDAs in fragile states may have particular significance. If the central government is unable or unwilling to fund development at the community level, a CDA may provide development opportunities otherwise unobtainable. A challenge in drafting CDA legislation is how to balance the respective development roles of government and the miner. Ideally, development efforts under a CDA will act to complement government-led development, not displace it.

## 32.3 Voluntary vs Mandatory Approaches

If a government desires that mining companies play a role in community development, it may decide to allow that role to be defined voluntarily by the

Crowson (2003, 2007), and the various publications of the Mining, Minerals and Sustainable Development (MMSD) project (IIED n.d.).

miner or may impose mandatory requirements. In the last two decades there have been many mining projects worldwide that have been unable to move from discovery to development because of community opposition. Oftentimes, even if a project is built, it may attract the future ire of communities if their current and evolving expectations are not met. This has led many of the world's mining companies to place an increased emphasis on CSR programmes including community development. Additionally, mining-sector organizations such as the ICMM have done work to help define community development good practice. In 2003, the council committed its members to a sustainable development framework of ten principles including '9. Contribute to the social, economic, and institutional development of the communities in which we operate'.[2]

In support of this commitment the organization has since prepared a number of supporting 'tools' for optional use by its members including: Community Development Toolkit (ICMM 2005, 2012) and Partnerships for Development Toolkit (ICMM 2011). The first of these is discussed and assessed in detail in Macdonald (2017).

Relying on companies to voluntarily assist in community development is risky—not all firms are good corporate citizens and not all are competent to know how to offer such assistance. For this and other reasons, an increasing number of governments are now mandating the use of CDAs or other community development tools in their mining legislation rather than relying simply on voluntary approaches. While many companies baulk at the concept of increased regulation rather than being able to achieve desired outcomes according to their own means, most would agree that CDAs, whether required by legislation or not, can be a useful tool to manage community expectations.

The use of community development statutory requirements does not preclude voluntary programmes—they are not mutually exclusive. Hybrid approaches combining mandated elements and voluntary activities are not uncommon.[3] In nations where the mining or environmental law requires community development, that requirement can form the core of or complement a CSR programme. Likewise, law-drafters can utilize concepts originating in voluntary initiatives. For example, when drafting the model CDA regulations published by the World Bank (Otto 2010), the author took into account the ICMM Community Development Toolkit and examined many existing 'voluntary' CDAs. An attribute of a good law is being able to achieve a degree of consensus between policy makers, regulators, beneficiaries, and those that will be regulated on issues such as objectives, processes, and good practice.

[2] See ICMM (n.d.) for the full set of sustainable development principles.

[3] A good example is Brazil with the situation there being described in some detail in ICMM (2013).

### 32.3.1 *Legal Mandates: Acts of Law and State Mining Agreements*

Governments have two main ways in which to legally impose community development requirements on the private sector—state agreements and statutes. A state agreement is a contractual agreement between the owner of the rights of extraction and the government and sets out the mutually agreed obligations and rights of the parties. During the post-colonial era up through the 1990s, such agreements for individual large projects were commonplace throughout the developing world, but in recent times, their use has been reduced as laws relating to mining (mining law, environmental law, labour law, income tax law, etc.) generally have improved, thereby negating the need for project-specific agreements. Provisions in agreements can find their way into statutes over time. For example, Guinea developed a model mining agreement in 2006 that contained extensive CDA provisions. It introduced a new mining law in 2011 containing similar CDA requirements, obviating the need to address that subject in future agreements (AMLA n.d.a). Like in Guinea, other nations have now addressed community development requirements in their mining laws. According to Schott et al. (2015):[4] 'Since the mid-1980s thirty-two countries have adopted community development provisions in mining codes, with nine countries currently in the process' (Schott et al. 2015).

Table 32.1 lists countries whose mining law (or model agreement) requires some sort of community development action. The list includes a wide variety of community development related requirements, not all of which include a CDA as defined for the purposes of this chapter. For readers interested in a short description of the specific community development requirement in most of the jurisdictions listed in the table, the author recommends Penagos et al. (2014). For the African nations listed in Table 32.1, their mining laws, including the community development requirements, are accessible through the African Mining Legislation Atlas (AMLA n.d.b).

### 32.3.2 *World Bank Model CDA Regulations and Later Improvements*

Given the growing interest in CDAs and their application to extractive projects, in 2010 the World Bank launched a multi-year specialized research project intended to investigate the nature and usage of CDAs and to provide associated information and guidance to governments, industry, communities, and other concerned stakeholders. The work was conducted in several stages. In the initial stage, a review was done of existing CDA agreements and community development extractive industries initiatives worldwide. A conceptual framework was developed to identify key building blocks to

[4] Based largely on work by Dupuy (2014).

**Table 32.1.** Countries where community development is required by law or agreement

| Country | Country |
|---|---|
| Afghanistan | Liberia |
| Autonomous Region of Bougainville | Mali |
| Burkina Faso (pending) | Malawi (pending) |
| Canada (several subnational jurisdictions) | Mongolia |
| Central African Republic | Mozambique |
| China | Myanmar (pending) |
| Colombia | Namibia |
| Democratic Republic of Congo | Nicaragua |
| Ecuador | Niger |
| Equatorial Guinea | Nigeria |
| Eritrea | Oman |
| Ethiopia | Papua New Guinea |
| Fiji (pending) | Peru |
| Ghana | Philippines |
| Guatemala | Sierra Leone |
| Guinea | South Africa |
| India | South Sudan |
| Indonesia | Togo |
| Kazakhstan | Vietnam |
| Kenya | Yemen |
| Kyrgyzstan | Zimbabwe |
| Laos | |

*Source*: derived from Dupuy (2014), Penagos et al. (2014), and various CDA-related projects executed by the author.

enable extractive-led community development and that framework was then used to draft preliminary model CDA regulations and guidelines suitable for use with a mining law. The second phase of the project involved extensive consultations with a broad spectrum of stakeholders including reviews of the draft regulations and guidelines. Following this preliminary work, field research was commissioned to assess the community development practice in selected countries and to further evaluate the results of the preliminary phases. Finally, research was done on the processes and frameworks used to negotiate, structure, and implement CDAs. The project culminated with the publication of a major multi-volume sourcebook on CDAs (World Bank 2012).[5] For readers who are interested in CDAs, the sourcebook remains one of the most comprehensive sources of information and case studies on the subject.

After publication of the World Bank's CDA sourcebook, the author of this chapter undertook a number of assignments to assist nations to draft or amend their mining laws. Based on the author's subsequent experience in drafting mining laws and regulations for nations in the African, Asian, and Pacific regions, including provisions mandating CDAs, he updated and

[5] The author of this chapter was lead author of the World Bank model CDA regulations (Otto 2010).

improved the model CDA mining regulations, and they have been published by the United Nations.[6] The model provisions include both articles for use in the primary mining law as well as detailed regulations. While the model provisions provide a 'good practice' basis for legislating CDAs, they are intended only as a starting point for policy makers and legislators. The development of actual mining law provisions and regulations must be crafted to meet unique jurisdictional needs and requirements.

## 32.4 Legislated Requirements

### 32.4.1 *What Is a CDA?*

There are many forms of agreements that aim to provide a formal or semi-formal linkage between an extractive industry project and nearby communities. Such agreements go by many names such as: impact and benefit agreements; access and benefits agreements; indigenous land use agreements; partnering agreements; contracts with the community; landowner agreements; shared responsibilities agreements; community joint venture agreements; empowerment agreements; benefits sharing agreements, and so forth. Some of these agreements are intended as an informal, non-binding means by which the signatories mutually express their views on certain topics, such as in a memorandum of understanding. Others take the form of a legally binding contract or even a treaty. Today, the use of various forms of agreements between extractive companies and communities is becoming widespread, but CDA use is not yet considered standard practice in many nations. However, in some nations, such as Canada, their use is nearly universal for large extractive projects.[7] Some agreements are held confidentially, but many are available on publicly accessible databases.[8] As was indicated at the beginning, for the purposes of this chapter regarding CDA legislation, a CDA means a legally binding contract between the holder of an authorization granting the rights to extract minerals and a community (or communities) that will be affected by the exercise of those rights that addresses matters concerning community development.

[6] The model CDA mining law articles and regulations may be accessed in the annex section of Otto (2017).

[7] In Canada, the use of 'impacts and benefits agreements' between aboriginal groups and companies that extract minerals from their lands is widespread, and although such agreements are not required by statute, they are common, with over 180 in use in 2012. Miningfacts.org (2012) defines an impact and benefits agreement as 'a formal contract outlining the impacts of the project, the commitment and responsibilities of both parties, and how the associated Aboriginal community will share in benefits of the operation through employment and economic development'. Sosa and Keenan (2001) provide a good overview of impact and benefits agreements in Canada.

[8] See for example the Agreements, Treaties and Negotiated Settlements Project collection (ATNS n.d.).

One of the advantages of a legislated approach to CDA requirements, versus just allowing a miner to determine whether one is required and what it should contain, is that it avoids problems that arise when every mine is handled on an ad hoc basis. For example, over the past decade in the Solomon Islands, companies exploring for gold, nickel, and other minerals began entering into ad hoc land access agreements, which differed greatly. When it became known to communities that some agreements were substantially more beneficial than others, the viability of some agreements became questionable and pressure was brought on politicians and local leaders to ensure that all communities got a fair deal. As a result, the government requested assistance to draft a standardized model land access agreement that would be suitable for introduction as a requirement under its mining regulations. O'Faircheallaigh has observed:

> Agreements between commercial developers and local communities are becoming more common in virtually all parts of the world, from inner city America to remote mining regions. The need for such 'community development agreements' (CDAs) is especially acute in mining, where environmental and social costs are often borne by communities while project benefits accrue in national capitals and global financial centres, leading to conflict between local people and miners. (O'Faircheallaigh 2012: 222)

It is important not to confuse community development requirements, CDA or otherwise, with impact compensation—these are two separate concepts. Most mining laws require that if a mine causes damage to personal or community property or property rights, the owner of that property or right must be compensated by the miner. Compensation payments are different from an investment by the miner in community development. A compensation payment is usually a one-time payment for a real property loss, while CDA funding requirements can be ongoing and seek to achieve development objectives (for example, human capital enhancement, microbusiness creation, etc.).

### 32.4.2 *Parties to a CDA*

If a CDA is required by law, who should be a party to the agreement? Obviously, the community and the company holding the extraction rights should be parties, but what is the role of district, provincial, and national government? In many situations, it makes sense for the community's local government (for example, an elected city council, tribal elders, or mayor's office) to be a party to the agreement, but where such a government is absent or not respected or accepted by the community, it can be a challenge to identify who is best positioned to represent the community. This is a challenge for policy makers and law-drafters particularly in nations that have a combination of elected and traditional leaders. One approach is to require in the law that a CDA be ratified by the community according to such process and in such manner as is customary for such community to make decisions on matters affecting the community as a whole. This approach allows flexibility from one locale to another. In a legislated approach, it can be made clear who the CDA parties are, avoiding the uncertainty that may arise where a purely miner-led voluntary approach is allowed.

The role of higher levels of government can be to act as the regulator of the CDA mandate rather than be a party to it. For example, a ministry of mines can be given the role of: ensuring that all miners who are required to have CDAs do indeed have them; ensuring that a CDA addresses all mandatory subject matter; monitoring a CDA to make sure it is being implemented; taking appropriate actions (such as levying fines or suspension or cancellation of rights) if implementation is not proceeding as required, and so forth. The responsible ministry can also act as a repository of CDAs, provide information about CDA requirements tailored for use by communities (such as maintaining a website), and make compliance reports available to the public.

### 32.4.3 *Participatory Processes: Meeting the Needs of Disparate Communities*

The process by which CDAs are negotiated and agreed can be complicated. One of the benefits of a legislated CDA approach is that a clear but flexible roadmap can be provided to guide the parties with regard to the CDA negotiation process. This can be particularly helpful when there are a number of communities involved that may have distinct cultural differences and varying expectations. The use of sociologists, mediators, and others who specialize in local-level negotiations can be an important part of the CDA process.[9] An agreement that constitutes a great fit for one community may be inappropriate for another.

There is a real possibility that a community may not want a mine nearby, especially if mine development requires a resettlement process. Its future vision may emphasize preservation of the status quo or slow improvements that will not endanger social value systems. Evans has summarized the differing objectives that sometimes occur between well-intentioned miners and local communities:

> Today, mining companies are clamoring to take the lead in defining sustainability, offering schools, and hospitals and jobs in return for the mineral wealth, supposedly extracted with minimal long-term environmental harm. However, for

[9] Gibson and O'Faircheallaigh (2010) offer CDA negotiations advice, including whom to include in negotiation processes, in the form of a detailed negotiations toolkit.

> communities facing the bulldozer, the concept of and the reality of 'sustainable mining' is not necessarily the same. (Evans et al. 2001: 250)

In most nations, minerals belong to the state and it is often in the state's interest to see its mineral resources developed for the good of all its citizens, even if a local community opposes mining. Participatory processes where a community is hostile to a mine can thwart the dialogue necessary to create a CDA. A mining law that requires CDAs needs to address this situation, either by embracing the concept of prior informed consent or by providing an exemption from a CDA requirement in certain situations.

### 32.4.4 *Community Contract Negotiating Capacity*

As defined in this chapter, a CDA is a legally binding contract. A basic principle underlying contract law is that a contract is a legally binding agreement between two or more competent and consenting parties. Gibson and O'Faircheallaigh have discerned:

> Once a decision to negotiate is made, a community and its leaders need to undertake a hard-headed assessment of their position in relation to the company, the government authorities that will approve or reject the project, and the wider economic and political context. (Gibson and O'Faircheallaigh 2010: 11)

However, many communities lack the capacity and competency effectively to negotiate or understand the ramifications of terms in a community development agreement. Lacking such capacity, they are vulnerable to agreeing to terms that may not be in their long-term best interest. Over time, a lop-sided agreement risks becoming obsolete when the community realizes that the agreement is suboptimal. For this reason, some CDA legislation requires that CDAs address key mandatory issues and that the community either be provided with experienced counsel or be assisted to develop its own negotiating capacity. For example Section 123 (10) of the mining law of the Autonomous Region of Bougainville requires:

> If in the opinion of the Secretary a qualified community that is entitled to a community development agreement lacks the capacity to effectively negotiate a community development agreement, the holder of a large-scale mining lease shall assist to build that capacity including the provision of such funds to the qualified community for capacity-building and preparation as are reasonable in the circumstances.[10]

If CDAs are created in an unregulated environment, there can be a greater likelihood that a community may not have the capacity and competence protection that the Bougainville legislation provides.

[10] Bougainville Mining Act 2015, Section 123(10).

In some cases, NGOs may be able to play an important advisory role where such legal protection is unavailable. For example, in Ghana, NGOs were made part of the group that negotiated the CDAs for the Ahafo gold project.[11] In a case-study-based analysis of CDAs, Brereton et al. (2011: 15) note that 'the success of a CDA relies heavily on all parties having the capacity to participate constructively in the agreement making process, support the agreement over time and deliver on their respective commitments'.

## 32.5 Mining Act and Mining Regulations Provisions

In this section, the core issues that underlie a CDA requirement in a mining act and the associated regulations are introduced. Such core issues also can apply to a petroleum law but few nations have moved to apply CDA requirements to oil and gas projects. It is typical in mining acts that CDA-related provisions are few, short, and to the point with details instead provided in the mining act regulations. In addition to the general examples provided in this section, detailed and comprehensive CDA model mining act provisions and regulations recommended by this author may be accessed in the annex of Otto (2017).

### 32.5.1 *Which Operations Require CDAs?*

The mining industry is comprised of many types and sizes of projects and not all operations are amenable to CDA requirements. For example, imposing CDA requirements on an artisanal miner or a small quarry would not be practical. This is also true for operations that will have a short duration or that will generate only minimal revenues. It is therefore necessary for a mining law to define the types and scale of operations that will be subject to CDA requirements. Generally, CDAs make sense where the scale and duration of a project have the potential to make a substantial contribution to local sustainable development without imperilling the economic viability of the project. One approach is to determine whether CDA requirements make sense on a case-by-case basis. However, an approach that relies on administrative discretion to make such a determination poses risks for the project investor, government, and the affected communities. Other approaches can provide greater certainty.

If the mining law makes provision for several types of mining authorizations, the CDA requirement can be applied to specified licence types, such as a large-scale mining licence, or alternatively, some licence types can be exempted (artisanal mining licence, quarry licence, etc.). The following

[11] For case studies of CDA processes and participants see Sarkar et al. (2010).

example illustrating a linkage between licence type and a CDA requirement is from South Sudan:

> 68. Community Development Agreements
>
> (1) A Large-Scale Mining Licence Title Holder shall
>
> ...
>
> (b) enter into Community Development Agreements with such communities in cooperation with relevant government authorities.[12]

Another similar approach is to require CDAs if certain scale-of-operations criteria are exceeded. An example of this approach is found in the Sierra Leone mining act:

> 139. Where community development agreement is required.
>
> (1) The holder of a small-scale or large-scale mining licence is required to have and implement a community development agreement with the primary host community if its approved mining operation will or does exceed any of the following limits:
>
> (a) in the case of extraction of minerals from primarily alluvial deposits, where annual throughput is more than one million cubic metres per year;
>
> (b) underground mining operations, where annual combined run-of mine ore and waste production is more than one hundred thousand tonnes per year (waste material not exiting mine mouth to be excluded);
>
> (c) in the case of open-cast mining operations extracting minerals from primarily non-alluvial deposits, where annual combined run-of mine ore, rock, waste and overburden production is more than two hundred and fifty thousand tonnes per year; or
>
> (d) where the licence holder employs or contracts more than one hundred employees or workers at the mine site on a typical working day (including all shifts).[13]

### 32.5.2 *Which Communities Qualify for a CDA?*

If a mining law requires that a miner enter into community development agreements it is important to define what constitutes a community for the purposes of such a requirement. For the purposes of CDAs, 'community' can generally be considered 'a particular area or place considered together with its inhabitants'.[14]

[12] Section 68, Mining Act, 2012 (South Sudan).

[13] Section 139, The Mines and Minerals Act, No.12 of 2009 (Sierra Leone).

[14] This is one of several definitions for the word 'community' offered by Oxford Dictionaries, at: http://www.oxforddictionaries.com/us/definition/american_english/community (accessed 15 April 2016).

However, in order to be practical, this broad definition needs to be refined. For example, the area around a mine may be populated by numerous clans or small family units each of which may consider themselves a community. To require a separate agreement with every clan or family would place an undue burden on both the miner and the regulator. Additionally, since being party to a CDA implies the receipt of benefits, every community would like to qualify for an agreement, even if located far from the project. If the number of communities is too large, the benefits may be too diluted to achieve meaningful sustainable development. This then leads to the concept of 'qualified community' where the law defines the term using parameters that limit the sphere of beneficiaries.

Some governments may desire to focus the community development agreement effort on a single qualified community that can act as a regional facilities hub for other neighbouring communities. By concentrating expenditure on a single 'host' community, it may be possible to build infrastructure, such as a hospital, that would not be possible if the available CDA resources were disbursed among several or many communities. This single host community approach was embodied in the Sierra Leone mining act:

> 139. Where community development agreement is required.
>
> (1) The holder of a small-scale or large-scale mining licence is required to have and implement a community development agreement with the primary host community...
>
> (2) The primary host community is the single community of persons mutually agreed by the holder of the small-scale or large-scale mining licence and the local council, but if there is no community of persons residing within thirty kilometres of any boundary defining the large-scale mining licence area, the primary host community shall be the local council.
>
> (3) If the holder of the small-scale or large-scale mining licence and local council cannot agree on which community is the primary host community, the licence-holder may notify the Minister requesting clarification, and the Minister shall notify the licence holder and local council within sixty calendar days from the date of such notice, specifying which community is the primary host community.[15]

Similar single host community provisions are provided in the 2011 Mozambique model mining agreement,[16] the Nigerian mining regulations,[17] and the mining

[15] Section 139, The Mines and Minerals Act, No.12 of 2009 (Sierra Leone).
[16] Section 19.2, unpublished 2011 Mozambique Model Mining Agreement.
[17] Section 193(5), Nigerian Minerals and Mining Regulations, S.I. 47 of 2011.

regulations of Yemen.[18] One of the challenges of the primary host community approach is that other communities nearby may feel slighted, resulting in possible hostile attitudes. To alleviate this, broader poverty reduction initiatives that encompass communities that do not have a CDA may be useful. A possible problem with a mandated primary host community requirement is that once a company has met its statutory CDA requirement, it may have little incentive to work further afield with other communities.

While some governments focus CDA resources on a single qualified community, others take a more egalitarian approach and a project may be required to enter into CDAs with all communities that meet prescribed criteria. This approach has merit in that it can avoid situations where one community is perceived to receive benefits to the detriment of other communities, perhaps setting the stage for a hostile situation. The following example is from the mining law of the Autonomous Region of Bougainville in Papua New Guinea:

> 2. INTERPRETATION.
>
> (1) In this Act, unless the contrary intention appears:
> 'Qualified Community' means any community of more than 1,000 persons who by tradition or by circumstances constitute a social community that usually reside within:
>
> (a) a large-scale mining lease area and any associated lease for mining purposes area; or
>
> (b) fifteen (15) kilometres of any boundary defining a large-scale mining lease; or
>
> (c) an area-of-influence, identified in an environmental impact assessment prepared as a requirement under the PNG Environment Act 2000, that will be affected in a major way by large-scale mining lease operations; or
>
> (d) a village or township that will house more than ten (10) per cent of the workers employed or contracted by a large-scale mining lease holder, and which is thus eligible to enter into a community development agreement with the holder of a large-scale mining lease;
>
> ...
>
> 123. COMMUNITY DEVELOPMENT AGREEMENTS
>
> (5) Subject to Subsection (6), the holder of a large-scale mining lease is required to have and implement community development agreements with all qualified communities that are willing to enter into a community development agreement.[19]

[18] Article 73, Executive Regulation of Law No. (22) for the year 2010 regarding Mines and Quarries (Yemen).

[19] Sections 2 and 123, Bougainville Mining Act, 2014 (Autonomous Region of Bougainville, Papua New Guinea).

Where a mining law seeks to provide an egalitarian approach, it may strive to provide CDA benefits to a wide number of small disparate groups by requiring such 'sub-communities' to aggregate into a more manageable 'qualified community' that is then the party to the CDA. Since the benefits that derive from a CDA can be substantial, there is a strong incentive for smaller communities to join together to form a community that qualifies for a CDA. The following example is also from the Autonomous Region of Bougainville in Papua New Guinea:

> 91. INTERPRETATION FOR THIS PART.
>
> In this Part, words and expressions shall have the meanings ascribed to them in the Act and as follows:
>
> ...
>
> 'Sub-Community' means any social community of less than 1,000 persons that otherwise meets the definition of a qualified community.
>
> 92. IDENTIFICATION OF QUALIFIED COMMUNITIES.
>
> ...
>
> (3) Sub-communities may join to form a Qualified Community and any such joining shall be recorded in a registry of sub-communities comprising the Qualified Community.
>
> (4) Sub-communities do not qualify for a community development agreement.[20]

### 32.5.3 *Mandatory CDA Obligations under the Mining Act*

In this author's opinion, when drafting CDA requirements for a mining law, at least four core obligations should be addressed for projects that are required to have CDAs:

- the project has a development obligation with regard to certain communities
- the project's development activities must be agreed in its CDAs
- the agreed CDA activities must be implemented
- the activities must be periodically reported so that the regulator can verify that the CDA is being implemented.

Below are sample clauses extracted from the model mining act CDA article recommended by the author in Otto (2017) that address these obligations:

> The holder of a large-scale mining licence must assist in the development of qualified communities affected by its operations to promote sustainable

[20] Sections 91 and 92, Mining Regulations, 2015 (Autonomous Region of Bougainville, Papua New Guinea).

development, enhance the general welfare and the quality of life of the inhabitants and must recognize and respect the rights, customs, and traditions of local communities.

The holder of a large-scale mining licence is required to have and implement community development agreements with all communities that meet the definition of a qualified community that are willing to enter into a community development agreement.

The holder of a large-scale mining licence is required to reasonably comply with its approved community development agreements.

The holder of a large-scale mining licence must:

(a) expend on community development no less than [x per cent] of its annual gross sales revenues, in such manner, at such time and on such activities as are prescribed; and

(b) submit annually, at such time and in such form and manner as are prescribed, a community development expenditure report;

(c) submit semi-annually, at such time and in such form and manner as are prescribed, a community development agreement report for each community development agreement associated with its mining licence; and

(d) periodically, as is prescribed, update its community development agreement(s).

The requirement in the last example clause also requires the licence holder to annually expend a minimum amount on community development (including but not limited to expenditure on CDA activities) based on a specified percentage of its gross revenue. In this author's experience, this is an important option for policy makers to consider. By providing a sure and unambiguous income stream, there is a much higher prospect that CDA activities will be implemented, and also that the beneficiary communities will have a better understanding of funding levels, thus acting as a way to manage expectations. Such a provision also aids CDA negotiations because the funding requirement is set out in the law and is not open to negotiation. In discussions with industry, the author learned that many mines voluntarily expend around 2 per cent of gross revenues on CSR programmes, but the amount is highly variable from project to project. Thus, a statutory requirement for CDA-related expenditure of around 2 per cent would probably not be considered as unreasonable by many mining companies.

As a further aid to CDA negotiations, the author also recommends that mining law regulations be adopted to statutorily set out topics that a CDA must address. In practice, such regulations provide a blueprint for negotiations and protect the interests of communities by ensuring that key subject matter is covered. The model regulations provided in Otto (2017) contain a list of mandatory topics that a CDA must cover. One of the key topics addressed in the regulations is a requirement for a CDA to contain a 'development plan' jointly prepared by the miner and the community which forms the core of the CDA. The model regulations articles that related to a development plan were inspired by the findings of a major study of mining and its role in sustainable development undertaken by the Mining, Minerals, and Sustainable Development Project (MMSD 2002: xxviii):

> The CSDP *[community sustainable development plan]* should be based on the community's concept of how the mine can best contribute to achieving its social, environmental, and economic goals. The plan should provide the fundamental framework for relationships among the company, the community, and the government (and any other parties) through the project life and into post-closure. It should identify the specific actions needed and the respective roles and responsibilities to achieve the agreed-upon vision. It could also create some obligations, on all sides, for taking those steps. Independent mechanisms for monitoring and evaluation, including clear and agreed indicators of performance, need to be included. The plan will need to evolve and be amended over the life of the project to reflect changing priorities and capacities. (MMSD 2002: xxviii)

While the model regulations in Otto (2017) set out a list of mandatory topics that must be addressed in a CDA, they do not specify how each of the topics is to be resolved. For example, the regulations require that a CDA must contain grievance and dispute resolution provisions but do not say what they are—that is for the miner and the community to negotiate and agree. In this way, each agreement is different, recognizing that every community and project will have their own unique needs and attributes. It is important to distinguish between obligations imposed under the mining law and those that arise under the terms of a negotiated CDA: the former are statutory obligations and the latter, contractual obligations.

### 32.5.4 *Enforcement, Offences, and Penalties*

A mining law that requires the use of CDAs and that imposes related obligations often contains provisions to enforce compliance. Failure to comply with a statutory obligation is deemed an offence and a substantial fine may result for non-compliance. Additionally, linking non-compliance to an administrative action, such as suspension or cancellation of the extractive right, is a valuable enforcement tool for regulators. For obligations arising under the CDA, in contrast to those arising under the mining law, the parties to the agreement have recourse to the appropriate court or other dispute resolution methods set out in the agreement.

### 32.5.5 *Applicability to Pre-existing Mining Rights*

One of the challenges in drafting a mining law that requires CDAs is how to handle pre-existing projects and arrangements. Should all such projects be exempt? How should pre-existing arrangements between a mine and community be addressed? One approach is to allow such projects a defined time period in which to obtain required CDAs that meet new statutory requirements. The author has drafted the following example of such an approach: 'the holder of an authorization to conduct large-scale mining operations that currently is in force but that was granted prior to the adoption of this Act is required to comply with this article and must be in compliance with this article no later than two years from the effective date of this Act'.

Another approach, which the author of this chapter does not advocate, is for a new mining law to remain silent on the transition issue. In this latter approach (taken by Nigeria and Sierra Leone in their most recent mining laws) all miners of a certain scale must have CDAs, but no time period for effecting the requirement by pre-existing operations is stated; in effect, upon the day that the law becomes effective they are liable to have CDAs. It is then left to the regulator to determine when to commence an enforcement action. This approach allows a great deal of flexibility, but some regulators may be hesitant to commence enforcement.

It is a fundamental nature of a new law that the 'rules of the game' change, otherwise why not just keep the existing law? When a country imposes CDA requirements in a new law, the policy principle driving those requirements is that the nation has certain expectations with regard to community development that must be met according to the new system, not the old one. Thus, if a miner has a pre-existing approach to community development it must ensure that its approach conforms to the new law.

### 32.5.6 *Legal Effect of a CDA*

As has been mentioned earlier in this chapter, there are many forms of agreements between extractive firms and communities. A law requiring CDAs should clearly define the legal nature of required CDAs. A CDA, as defined in this chapter, is a legally binding contract and is enforceable by recourse to the appropriate court.

### 32.5.7 *Effect of Transfer of the Underlying Exploration/Mining Right*

Extractive industry projects typically go through several phases including exploration, development, extraction, rehabilitation, and post-closure. For many projects, ownership of the respective rights granted during the various phases may change. This is particularly true of the transition from the exploration phase to the development phase. A key part of a law addressing CDAs is to define at what phase such agreements are required and to require that a CDA entered into by a company will be honoured by a successor company. Many laws that require CDAs impose the requirement at the development stage. Requiring a CDA at the exploration stage is impractical because the nature and location of the resource will at that stage be unknown.

## 32.6 Conclusions

The use of CDA legislation to achieve community-level benefits is just one option for governments. Some governments may place no direct emphasis on community development with the expectation that a community that is located in close proximity to a mine will benefit from project-related employment and other linked opportunities. Many extractive companies have CSR programmes that may be directed, at least in part, toward community development. Other nations prefer that, instead of benefits flowing to a community via a direct link between the project and the community, the community will be a recipient instead of a portion of the fiscal revenues generated by the project.[21] The widespread use of CDA legislation is a relatively new phenomenon and the efficacy of CDAs as a means to achieve community-level benefits is still to be determined. However, when compared to approaches that rely entirely on voluntary actions by companies/projects, statutorily mandated CDAs greatly reduce the risk that sustainable community development will not take place. If the CDA legislation is robust, such as the model legislation appearing in Otto (2010) and Otto (2017), it can provide a clear roadmap for mandatory processes, approvals, monitoring, and enforcement, all of which are lacking in an unregulated approach to mine-assisted community development. Some of the advantages of a regulated CDA approach versus a mining law that is silent about community development include:

- the different roles and responsibilities of the miner, communities, and government can be made clear
- communities that are qualified for development assistance are identifiable
- qualified communities will have a written and enforceable contract with the miner that identifies their rights and obligations regarding development

[21] Examples of where a portion of a project's fiscal revenues are allocated by law, rather than through the budgeting process, back to local communities: property taxes in the United States; a statutory portion of income tax in Peru; a statutory portion of royalty in Brazil.

- the expectations of the community can be aligned to a practical level of funding
- miners will know the minimum level of annual development funding that they must provide
- CDA minimum content guidance can ensure that key issues are addressed
- formal grievance and dispute resolution approaches can be established
- the probability of development implementation is high because non-compliance can result in fines, penalties, or possible cancellation of the right to mine in addition to civil action arising through the application of contract law.

One of the downsides of a regulated CDA requirement is that it can impose an administrative burden on the regulatory agency. It may be necessary to hire officers and provide them with appropriate training and resources.

While a CDA—either one required by law or one entered into voluntarily—is no guarantee that the boom and bust cycle that communities experience when a mine closes will be avoided; if the agreement includes objectives that address sustainable development, it can be hoped that mine closure will have a lesser impact than had the CDA not been in place. Miners have always had the option to assist in community development. The question for government is whether it is satisfied with allowing miners to offer development assistance on an ad hoc basis or whether that assistance should be required by law and regulated.

## References

AMLA (African Mining Legislation Atlas) (n.d.a). 'Article 130, Code Minier 2011, Loi L/2011/006/CNT (Guinea)'. Article 130, Code Minier 2011, Loi L/2011/006/CNT (Guinea). Available at: http://www.a-mla.org.

AMLA (African Mining Legislation Atlas) (n.d.b). African Mining Legislation Atlas. Available at: http://www.a-mla.org.

ATNS (Agreements, Treaties and Negotiated Settlements Project) (n.d.). 'ATNS Database'. Available at: http://www.atns.net.au.

Brereton, D., J. Owen, and J. Kim (2011). 'Good Practice Note—Community Development Agreements'. St Lucia: Centre for Social Responsibility in Mining, University of Queensland. Available at: http://www.eisourcebook.org/cms/files/csrm_good_practice_notes_on_cdas_document_final_260911.pdf.

Crowson, P. (2003). *Astride Mining: Issues and Policies for the Minerals Industry*. London: Mining Journal Books.

Crowson, P. (2007). 'Adding Public Value: The Limits of Corporate Responsibility'. Oxford: Oxford Policy Institute.

Dupuy, K. (2014). 'Community Development Requirements in Mining Laws', *The Extractive Industries and Society*, 1(2): 200–15.

Evans, G., J. Goodman, and N. Lansbury (2001). *Moving Mountains: Communities Confront Mining and Globalisation*. Otford, New South Wales: Mineral Policy Institute and Otford Press.

Evans, G., G. Russell, and R. Sullivan (2001). 'An International Regulatory Framework?' in G. Evans, J. Goodman, and N. Lansbury (eds), *Moving Mountains: Communities Confront Mining and Globalisation*, 207–22. Otford, New South Wales: Mineral Policy Institute and Otford Press.

Gibson, G. and C. O'Faircheallaigh (2010). 'IBA Community Toolkit: Negotiation and Implementation of Impact and Benefit Agreements'. Toronto: Gordon Foundation. Available at: https://www.afoa.ca/afoadocs/L3/L3a%20-%20IBA_toolkit_March_2010_low_resolution.pdf .

ICMM (2005). *Community Development Toolkit*. London: ICMM.

ICMM (2011). *Mining: Partnerships for Development Toolkit*. London: ICMM. Available at: http://www.icmm.com/mpd.

ICMM (2012). *Community Development Toolkit—Revised*. London: ICMM. Available at: https://www.icmm.com/en-gb/publications/mining-and-communities/community-development-toolkit.

ICMM (2013). *The Mining Sector in Brazil: Building Institutions for Sustainable Development*. London: ICMM.

ICMM (n.d.). 'ICMM 10 Principles: Principle 9'. Available at: https://www.icmm.com/en-gb/about-us/member-commitments/icmm-10-principles/icmm-principle-9.

Macdonald, C. (2017). 'The Role of Participation in Sustainable Community Development Programmes in the Extractives Industries', WIDER Working Paper 2017/28. Helsinki: UNU-WIDER.

Miningfacts.org (2012). 'What Are Impact and Benefit Agreements (IBAs)?' Vancouver: Fraser Institute. Available at: http://www.miningfacts.org/Communities/What-are-Impact-and-Benefit-Agreements-(IBAs)/.

MMSD (2002). *Breaking New Ground*. London: Mining, Minerals and Sustainable Development Project/IIED. Available at: http://pubs.iied.org/pdfs/G00910.pdf.

O'Faircheallaigh, C. (2012). 'Community Development Agreements in the Mining Industry: An Emerging Global Phenomenon', *Community Development*, 44(2): 222–38. Available at: http://www.tandfonline.com/doi/abs/10.1080/15575330.2012.705872.

Otto, J. M. (2010). 'Mining Community Development Agreements: Source Book (Vol. 4): Community Development Agreement Model Regulations and Example Guidelines'. Washington, DC: World Bank. Available at: http://documents.worldbank.org/curated/en/2010/06/16798660/mining-community-development-agreements-source-book-vol-4-4-community-development-agreement-model-regulations-example-guidelines.

Otto, J. M. (2017). 'How Do We Legislate for Improved Community Development?' WIDER Working Paper 2017/102. Helsinki: UNU-WIDER. Available at: https://www.wider.unu.edu/publication/how-do-we-legislate-improved-community-development.

Penagos, M. D., J. Mandelbaum, and P. Toledano (2014). 'Requirements for Community Development Agreements in Mining Laws'. Available at: http://ccsi.columbia.

edu/files/2015/01/Community-Development-Requirements-in-Mining-Laws-Matrix-November-2014-.pdf.

Richards, J. (ed.) (2009). *Mining, Society and a Sustainable World.* Heidelberg: Springer.

Sarkar, S., A. Gow-Smith, T. Morakinyo, R. Frau, and M. Kuniholm (2010). 'Mining Community Development Agreements—Practical Experiences and Field Studies'. Washington, DC: Environmental Resources Management for the World Bank. Available at: http://www.sdsg.org/wp-content/uploads/2011/06/CDA-Report-FINAL.pdf.

Schott, J., F. M. Conteh, N. King, D. Desai, and N. Menzies (2015). 'Development from the Ground up? Mining Community Development Agreements in Sierra Leone.' World Bank 'Governance for Development' blogsite. Available at: http://blogs.worldbank.org/governance/development-ground-mining-community-development-agreements-sierra-leone.

Sosa, I. and K. Keenan (2001). *Impact Benefit Agreements between Aboriginal Communities and Mining Companies: Their Use in Canada.* Toronto: Canadian Environmental Law Association & Environmental Mining Council of British Columbia. Available at: http://citeseerx.ist.psu.edu/viewdoc/download?doi=10.1.1.473.2025&rep=rep1&type=pdf.

World Bank (2012). *Mining Development Agreements Sourcebook.* Washington, DC: World Bank. Available at: http://siteresources.worldbank.org/INTOGMC/Resources/mining_community.pdf.

# 33

# Conclusions

*Tony Addison and Alan Roe*

## 33.1 Challenges

The extractive industries have invariably occupied a somewhat uncomfortable position in development debate and practice: they might be likened to a rich but somewhat bossy relative who must be invited to the family parties but whom no one really wants to talk to. The very word 'extraction' conjures up images of forceful (and painful) removal. The media image is frequently one of despoiling nature, sometimes abusing and destroying the environment, including the resources (water, forests, soils etc.) essential to human life. Nor does mining infrastructure offer the same attractive photo opportunities for local politicians or for visiting ministers from aid-donor countries as do schools and clinics. In some cases there can be an imbalance of power between large extractives companies and host governments (as well as local communities)—and corruption and non-transparency are still to be found.

Donor assessments of private-sector development opportunities have often given only grudging acknowledgement (and sometimes none) to the role of extractives sectors in attracting investment, creating economic growth, and generating export earnings: even in some country cases where mining is demonstrably the largest private-sector activity. Within donor agencies and development banks, there has been a tendency to keep departments dealing with the extractives sector (e.g. providing finance to encourage investments) at arm's length from the more 'mainstream' work in economic policy and poverty reduction. Some agencies find it uncomfortable to liaise with large companies in the design and implementation of community projects in some poorer economies even when they are prepared to spend large amounts on these.

Yet human history, together with the growth in living standards down the centuries, are associated with the creation of new and better tools, derived ultimately from the planet's renewable and non-renewable resources: this has included metals and fossil fuels. From the Stone Age through the Bronze and Iron ages and up to today's world of advanced material sciences, human civilizations have extracted and refined metals, materials, and various fuels. In the past two decades, this has continued to be the case but now with a significant shift of emphasis. Specifically, most of those involved in some way with the extractive sectors—either as producers, regulators, or commentators—are far more concerned than ever before to reduce the environmental footprint. The task has much more clearly become one of using resource wealth well, for purposes of sustainable and inclusive development. That is an easy objective to articulate but a far more difficult one to accomplish. The world faces an immense challenge: how to sustain the expansion of living standards and reduce poverty—which will continue to involve a huge use of natural resources—while simultaneously avoiding calamitous climate change.

The demands on human ingenuity to create new solutions are unprecedented. The new technologies and ways of organizing societies that make up the low-carbon future require another leap forward in knowledge. But they also require metals, materials, and fuels—in large amounts. The extractives sector is central to creating an environmentally sustainable future of continuing and rising prosperity. Within the extractives sector, fossil fuels will eventually retire—but only after several more decades. Yet as this happens, the need for the metals and materials to construct renewable energy systems, zero-carbon buildings and new forms of transportation will grow. More recycling can meet some of the extra demand, but the challenge for the extractives sector is to ramp up production while reducing its own emissions and local environmental impact. The sector will be further encouraged to green itself by the demands of manufacturers to ensure environmentally and socially sound supply chains, in turn encouraged by increased consumer knowledge and concern, as well as socially responsible investors. Producing countries will need to adjust to the economic impact of the resulting shifts in markets implied by the transition to a low-carbon future.

This book seeks to demonstrate that *no serious development scholar or practitioner can today ignore the role of extractive industries in the big debates about sustainable development* (see Message 1 of Chapter 1). These industries have become far more significant to the economies of low-income countries (LICs) and many middle-income countries (MICs) over the past two decades (Roe and Dodd, Chapter 2, this volume). For reasons discussed in the book, this is only partially a consequence of the super-cycle in commodity prices during that period. The economic significance of the extractives sector is likely to increase substantially further for countries with resources vital to the technologies needed for the low-carbon future (Addison, Chapter 22, this volume). Those countries with oil and gas will see continued demand in the immediate and medium-term future, as natural gas will most likely constitute an 'energy bridge'—as it replaces coal—to a renewable energy future (and the transition from the internal combustion engine to electrically powered vehicles is still in its infancy). But the slippage in oil and gas prices in recent years from their historic highs of a decade ago, and the resulting pressures on the public finances of some oil and gas economies, are an early-warning signal to national leaders of the need to adjust and create new economic activities. Time is not on the side of fossil-fuel exporters.

Another theme of this book is the changing nature of the extractives industry itself and attitudes and policies towards it—especially the large improvements in practice seen since the turn of the millennium. Despite this, amongst development specialists, both practitioners and researchers, attitudes towards extractives still easily slip into a 'them and us' dichotomy. Mining and oil and gas companies (the 'them') are often characterized as exploitative and wholly disrespectful of local tax and other laws, environmental and other regulations—thereby causing harm to the communities, governments, and nations (the 'us').[1] This is one strong manifestation of the general ambivalence towards the private sector, and especially a distrust of large businesses, that has permeated much development debate down the years.

Consequently, development debates have arguably devoted too little attention to the role of large private companies, both domestic and foreign, and their potential to generate growth and jobs. Yet, extractive sectors that work for development almost always require very large-scale private investment, and the kinds of technical and commercial knowledge provided only by large-scale mining and oil and gas companies. The resulting revenues generated by their activities will often far exceed inflows of foreign aid, while the stimulus of the extractives sector to the economy can (if managed well) deliver thousands of new jobs.[2]

But what might replace the 'them and us' mentality when we do decide to engage more actively with the extractives sectors and bring them more centrally into the development debate? The answer given in this book is that we should recognize more explicitly the *wide variation* in behaviours across both host governments and extractive companies (a variation discussed as well as illustrated schematically in Figure 1.2 in the introductory chapter). Some governments do have strong institutions favourable to delivering national benefits

[1] The 'us' in this dichotomy is invariably characterized as having positive motives, contrary to Message 6 in the introductory chapter, which reminds us that some governments can be ineffective and divisive, with some local elites showing little concern for the interests of the nation or those of communities.

[2] Foreign aid has a vital role to play in development, and our argument here does not imply that foreign direct investment (FDI) is a substitute for aid.

from extractives development, while others are run for the benefit of wealthy elites (with little concern for the broader local society), just as there are differentiated companies: some good and socially responsible and some rogues.

We therefore need to be more aware of differentiation across host governments as well as across extractives companies. Many top-down policy initiatives, from both national and international programmes, are likely to fail if they do not first assess, diagnose, and factor in the large differences that do exist in the real world! This point has been underplayed in research on extractives. Econometric methods have helped to identify some broad patterns in the relationship between resource wealth and economic performance, but given the number of individual country outliers from most regression results, it is unhelpful to generalize—and worse still derive policy—from econometric studies alone (concluding, for instance, that resource wealth is inevitably disastrous and advising unhelpfully that it should invariably be 'kept in the ground'). Equally, care must be exercised in generalizing from country experiences by, for example, majoring on failing states and downplaying the situation in better-performing countries. International policy prescriptions can easily fall into the trap of 'one-size-fits-all' recommendations. It is unhelpful, for example, to suggest that all resource-rich LICs and MICs should establish a sovereign wealth fund: some LICs and MICs may have plentiful high-return domestic opportunities that can productively absorb the available funds.

It is also important to acknowledge that the numerous national and international initiatives of the past two decades—one of many examples discussed in the book being the Extractives Industries Transparency Initiative (EITI)—have already improved the general performance quality of many host governments and many extractives companies (in terms of the positioning represented stylistically in Figure 1.2 in the introductory chapter), notwithstanding some regression in a few cases. We now see more reasonably enlightened companies working with reasonably effective and inclusive governments (i.e. at or near Zone B in Figure 1.2) than was evident twenty years ago. Future policy reforms and international initiatives must continue that process; more countries and companies can then achieve compliance with the high standards proposed by, for example, the NRGI benchmarks and the IFC and the ICMM corporate performance standards.

This book has explored in detail what good policy and practice looks like, in its many dimensions (macroeconomic, environmental, taxation, local content, downstream planning, community development, gender equality, etc.). It could be argued that in many of these areas we know what 'good policy' looks like. However, while much improvement is evident it remains crucial to: (i) recognize the continuing failings of actual practice in specific cases (both governments and companies); (ii) advance a learning process based in part on the experiences of better-performing countries and companies, in order to help others move into compliance with best practice; and (iii) continue to look for innovative ways of facilitating that process, especially in the most difficult situations where there is some hope of progress but where standard prescriptions have so far proven less than effective (especially Zones A and D, in Figure 1.2 in the introductory chapter). But these tasks require policy analysts more explicitly to recognize difference: that is, to assess the very substantial variations in behaviours between different host governments and different extractive companies, and to tailor their recommendations accordingly.

## 33.2 Articulating and Implementing a Vision

A key theme of this book is that the many dimensions of extractives policy analysed in the individual specialized chapters should not be seen—as they all too often are—as separate distinct areas of policy to be contained within their own technical silos. On the contrary, for real progress in making extractives an effective development driver, these policies need to be part of an *integrated and comprehensive package.* By welding the component policies together around a strongly maintained and coordinated vision, a country is far more likely to turn the promise of extractives into a genuinely better, more sustainable, and more inclusive economic future for its citizens. That is the reason why this book, in addition to its specialist policy chapters, devotes significant space to four main overarching and linking themes.

First, the book makes a strong case for what can be called an *'all of government approach'* (especially in Chapter 17 by Kathryn McPhail). Many LICs and MICs have recently articulated 'vision statements' for their long-term futures. Yet in practice we rarely see a strongly coordinated approach to the management of extractive resources that gives credible substance to such statements. Indeed, individual ministries and agencies frequently find it extremely difficult to cooperate and co-ordinate over extractives activity, and often take their own individual paths. Opportunities are then missed, especially in maximizing economic benefits (such as local economic development linked to extractive operations).

Second (especially in Chapter 6 by Evelyn Dietsche), it tries to unbundle a great many ideas around the concept of 'institutions'. While there is now an almost universal consensus that the building of 'good institutions' is essential if extractives are to play an effective role in converting resource wealth into sustainable and effective development, the complex nature of this process, and the true underlying meaning of the term 'institutions' (and its component elements), are far less commonly spelled out.

Third, the book seeks to broaden significantly the traditional set of ideas about the impact of extractive industries on their host economies. Specifically, Chapter 18 by Anthony Hodge articulates the case for developing systematic methods to define, evaluate, and monitor the sector's contribution to human, economic, and ecosystem well-being over the entire project and product life cycle: the author adopts the label '*contribution analysis*' for this. Such analysis involves a more demanding test than the one used in most current practices of monitoring and evaluation (M&E) because it explicitly combines a *comprehensive* picture of the various impacts (thereby fitting well with the all-of-government approach); combines economic and non-economic impacts; provides a greater opportunity for the perspectives of all interest groups to be properly heard; and takes a long-term (life-cycle) perspective on extractive activity. The practical application of such an approach might be demanding relative to the capacities available to many host countries. However, the examples provided by Hodge do indicate that elements of the approach are certainly practical in at least some contexts (and capacities to deliver more should be built). Above all, because the contributions of extractives to both human and ecosystem well-being are so significant and controversial, these contributions need to be both identified, and more actively tracked over time.

Fourth, the book also seeks to fill something of a gap in the available literature by showing how extractive industries can fit within the frameworks of some at least of the *new industrial economics* debates (in Chapter 7 on this topic by Evelyn Dietsche).[3] In this area of debate our earlier comment about the uncomfortable positioning of extractives in the development storybook is particularly pertinent. Many who write on the topic do so without engaging significantly (if at all) with the role of mining and oil and gas. In some manifestations of the literature these extractive sectors are definitely seen as inimical to the longer-term productivity gains needed to drive economic growth: manufacturing by contrast, with its potential for hosting large productivity improvements, is seen as the *sine qua non* of the necessary policy direction. This book recognizes explicitly that diversification must be at the centre of any long-term strategy for the use of a non-renewable resource (Message 4 from Chapter 1). However, the time horizon that defines 'long term' can often be very long indeed, providing innumerable opportunities to use a large extractive resource boom to help catalyse a whole range of non-extractive activity for the future.

Chapter 7 notes several of the key characteristics of manufacturing, notably in the East Asian model that has made this sector the central focus of many of the recommendations from the new industrial economics literature. But most

[3] Or the new 'structural' economics as it is referred to in some parts of the literature.

of these characteristics are arguably also present in the extractive industries, including: (i) they are export-led and so do not face significant demand constraints from small domestic markets; (ii) they address the foreign-exchange problem of countries that might otherwise have been limited by that constraint; (iii) they provide a convenient basis for learning-by-doing and the absorption of new technologies from abroad, which are critical elements for discovery and learning and productivity gains; (iv) they provide relatively easily taxable revenues and so help to boost public spending capacity; and (v) being centred on a discrete number of enterprise units, they provide for a relatively natural system of accountability for those revenues, as compared to the problem of taxing a huge and dispersed set of, say, agriculture-based businesses.

So, the challenge we present to the development community is to consider these *positive* characteristics of extractives alongside the various *negative* characteristics (that are much more familiar from the huge resource-curse literature) before leaping to any final judgement about the extractive industries and their role in development. This is an agnostic position on our part. We have no illusions that in some extractive-dependent LICs and MICs (especially those in Zones C and D of our Figure 1.2 in Chapter 1) there is any early prospect of a pro-active and successful industrial policy based on the extractive resources of those countries. But in many other countries—especially those with a very long time horizon until resource depletion—there is an evident *potential* for a pro-active policy approach. This can encourage, for example, commercially logical local content, viable downstream activities, and more effective programmes for small- and medium- sized enterprises (SMEs)—all acting as major catalysts for successful economic diversification. The main and truly demanding challenge for host governments in this context is to somehow create a governance framework for a *long-term* industrial strategy and the appropriate policy interventions to support that strategy over periods of time spanning far more than one electoral cycle. But to be provocative one might even argue that to forgo this future potential—because of inadequate analysis and the failure to adopt the necessary promotional policies (including those that have succeeded in other countries)—could be as costly to the host economy as would the squandering of the government revenues from extractives through white elephant investment projects and widespread rent-seeking!

## 33.3 Final Words

Many of the issues covered by this book will be familiar to specialists who major on the development dimensions of the extractive industries. Still, it is impossible to be a specialist in every aspect of what is a large and fast-moving field, and so we trust that specialists will find much that is thought-provoking in the book.

The book should also be of considerable interest to the larger number of parties (in government, academia, NGOs, and donor agencies) who have some contact with, or a degree of interest in, specific narrow issues around the topic but who might benefit from the analysis, insights, and suggestions on a broad range of component issues that extend well beyond those in their immediate fields of vision. For that audience, we hope that our aspiration to provide a well-informed and comprehensive coverage of the complex and extensive range of component issues that connect the extractive sectors with the broader development agenda will enhance their understanding of the subject and possibly provoke some marginal adjustments of their own maintained positions.

A book such as this, written by some thirty or more experienced and independent specialists, could easily have just covered a multitude of disparate and disconnected topics and still perhaps have been classed as 'comprehensive'. But from the start of this UNU-WIDER initiative, the objective has always been to achieve 'comprehensiveness' in a different sense.[4] Specifically, the aim has been to deliver a structured flow of the materials across the book's eight parts—the many different angles, ideas, and policy positions of the contributors—then link them together coherently. But that structuring process did not start out, and does not finish, with any 'maintained hypothesis' about the benefits or failings of extractives as a contributor to the development process. The book certainly does not seek to present a rose-tinted view of the development benefits of extractives, and nor does it subscribe to the most negative manifestations of the resource-curse thesis. Rather, the editors and the contributors hope that the book will encourage its readers to explore the complexity of the real world of extractives and development, to take away an increased awareness of the huge potential for good management of the extractives sector to improve the lives of millions of people, and to be more fully apprised of the policies and actions, national and global, necessary to achieve this. It is a demanding task and we in the development community can do it much better than we have in the past!

[4] This book is part of the UNU-WIDER project *Extractives for Development*. Further papers together with blogs and links to video interviews with many of the authors can be found at: https://www.wider.unu.edu.

# Index

The manufacturer's authorised representative in the EU for product safety is Oxford University Press España S.A. of el Parque Empresarial San Fernando de Henares, Avenida de Castilla, 2 – 28830 Madrid (www.oup.es/en or product.safety@oup.com). OUP España S.A. also acts as importer into Spain of products made by the manufacturer.

www.ingramcontent.com/pod-product-compliance
Lightning Source LLC
LaVergne TN
LVHW011853200926
848484LV00002B/447

* 9 7 8 0 1 9 8 8 1 7 3 6 9 *